Power Systems

Electrical power has been the technological foundation of industrial societies for many years. Although the systems designed to provide and apply electrical energy have reached a high degree of maturity, unforeseen problems are constantly encountered, necessitating the design of more efficient and reliable systems based on novel technologies. The book series Power Systems is aimed at providing detailed, accurate and sound technical information about these new developments in electrical power engineering. It includes topics on power generation, storage and transmission as well as electrical machines. The monographs and advanced textbooks in this series address researchers, lecturers, industrial engineers and senior students in electrical engineering.

Power Systems is indexed in Scopus

Narayanaswamy P. R. Iyer

Inverters and AC Drives

Control, Modeling, and Simulation Using Simulink®

Springer

Narayanaswamy P. R. Iyer
Myna Electrical and Electronics Consultancy
North Kellyville, NSW, Australia

ISSN 1612-1287 ISSN 1860-4676 (electronic)
Power Systems
ISBN 978-3-031-62786-6 ISBN 978-3-031-62784-2 (eBook)
https://doi.org/10.1007/978-3-031-62784-2

This Springer imprint is published by the registered company Springer Nature Switzerland AG
The registered company address is: Gewerbestrasse 11, 6330 Cham, Switzerland

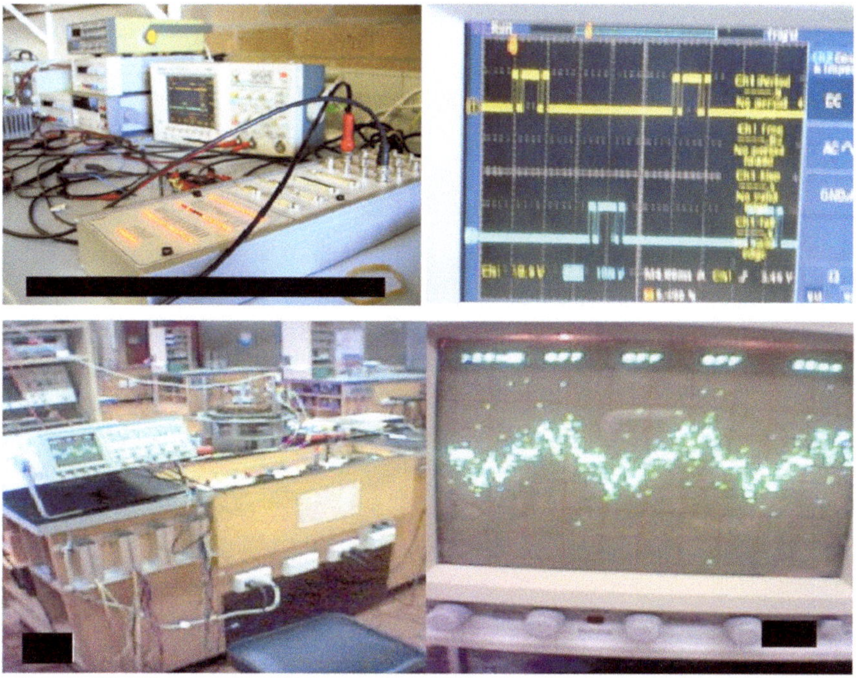

Dedicated to Power Electronics and Drives research community who have strived hard to develop power electronic converters and electric drives to produce clean sustainable energy.

Preface

Modelling of power electronic converters and converter fed electric drives is a growing field. There are several software packages available for this purpose. This field is only around two and a half decades old and has become a course of study and research in Universities/Institutions and also a tool for designing new power electronic converters and converter fed drive topologies in industries and research and development organizations. This book provides models for single, multi-triangle carrier, selective harmonic elimination (SHE) and space vector (SV) pulse width modulation (PWM) techniques for three-phase two-level, multilevel and modular multilevel converters (MMC), various boost control techniques for Z-source (ZS), quasi Z-source (QZS), switched inductor (SL), active switched capacitor (ASC) and diode assisted extended boost (DAEB) inverters. In addition, models for six-step inverter fed permanent magnet synchronous motor (PMSM), brushless DC motor (BLDCM) and induction motor (IM) drives, vector-controlled PMSM and IM drives, space vector PWM (SVPWM) inverter fed direct torque controlled (DTC) IM drives, fuzzy logic controlled (FLC) converter fed PMSM and IM drives are presented.

As mentioned, modelling of power electronic converters and converter fed electric drives is a growing field with great emphasis given in the university course curriculum, industries and research and development organizations. Now it has become a common industrial practice to model power electronic converters and converter fed electric drives, study the model response for machine speed, torque and current and make relevant design changes before hardware implementation. Sample references in support of this statement are given below.

- H. Burzanowska et al.: "Redundant Drive with Direct Torque Control (DTC) and Dual-Star Synchronous Machine, Simulations and Verification", IEEE-DOI: https://doi.org/10.1109/EPE.2007.4417422.
- ABB Automation Products GmbH, Germany: "AC500 with Simulink PLC Coder", Application Note, Copyright© 2020 ABB.

This book thus gains importance. Software SIMULINK® developed by Mathworks Inc., USA, is used to develop the models presented in this book. The term model in this book refers to Simulink model. Brief content of each chapter is given below.

Chapter 1 provides the introduction and the model with simulation result for the traditional three-phase sine PWM two-level inverter, book novelty and outline. Models for sine, harmonic injection (HI), third harmonic injection (THI), clipped sine (CS), dead-band sine (DBS), modified sine (MS) PWM techniques for three-phase two-level inverters, multicarrier sine level shift (MSLS) and multicarrier sine phase shift (MSPS) PWM techniques for three-phase diode clamped three level inverter (DCTLI), five level cascade H-bridge (CHB) inverter (FLCHBI), respectively, and models for single- and three-phase PWM MMC are presented in Chap. 2. SHE-PWM techniques for three-phase two-level unipolar and bipolar switched inverters and dSPACE implementation of gate drive for the former topology, three-phase DCTLI, flying capacitor three level inverter (FCTLI), CHB inverter (CHBI), single-phase SHE-PWM five-level DCTLI and FCTLI are presented in Chap. 3. In Chap. 4, models for space vector modulation (SVM) of three-phase two-level inverter and that for eliminating even order harmonics from line-to-ground voltage, and SVM of DCTLI and FCTLI are presented. Chapter 5 provides models for three-phase ZS and QZS inverters when controlled by simple boost (SB), maximum boost (MB), third harmonic injection maximum boost (THIMB), third harmonic injection maximum constant boost (THIMCB) and SVM techniques. Models for SL-ZS inverter (SL-ZSI), SL QZS inverter (SL-QZSI), ASC-QZSI and DAEB-QZSI using SB, THIMCB and SVM control techniques are presented in Chap. 6. Interactive models for six-step continuous current and discontinuous current mode inverter fed PMSM drive, the laboratory experimental verification of the latter type, model for vector-controlled three-phase PMSM drive and that fed by three-phase SVPWM inverter and models for three-phase inverter fed BLDCM drive in open loop, with hysteresis current controller in closed loop and with phase angle advance control, are presented in Chap. 7. Models for three-phase IM using dq0-axis flux linkage equations in state space and the performance of this IM when fed by sine PWM, THIPWM, CSPWM, SVPWM inverter, the multi carrier sine level shift PWM (MCSLSPWM) DCTLI, vector control, DTC of three-phase IM drive fed by SVPWM inverter using classical and modified switching table, voltage to frequency ratio (V/f) control of IM when fed by three-phase inverter, sine PWM inverter and thyristor controller are presented in Chap. 8. Models for three-phase inverter, sine PWM inverter and thyristor controller fed IM drive with voltage to frequency ratio (V/f) control and also its DTC, vector control of three-phase IM and that of three-phase SVPWM inverter fed PMSM drive all using FLC are presented in Chap. 9. All chapters are presented with examples and case studies. Source codes of Embedded MATLAB Function from the models in various chapters, model projects and answers to selected model projects from all chapters are presented in Appendices I, II and III, respectively.

Knowledge of fundamental semiconductor electronics, basic power electronics and AC machines theory are a prerequisite for this course.

I am thankful to Mathworks Inc. USA for providing sponsored licence to use MATLAB R2023a along with the required Simulink toolbox for this book project.

This book is suitable for senior level undergraduates and postgraduates doing an advanced course on power electronic inverters, AC drives and their modelling. This book is equally suitable for researchers and practicing professional engineers in the industry working in the area of power electronic inverters and AC drives.

Models for SVPWM of three-phase two-level inverter have been developed by two methods, one by a rigorous approach presented in Chap. 4 and another by a simple fundamental approach presented in Chaps. 5 and 8. The former method is suitable for postgraduate degree course level and the latter is suitable for senior undergraduate degree course level.

The models presented in Sects. 2.5, 2.6, 2.8, 2.9, 2.11, 2.13, 2.14, 2.16, 2.18, 2.19 and 2.20, Chaps. 3, 4, 5, and 6, Sects. 7.4 to 7.7, Sects. 8.6 to 8.14 and Chap. 9 were developed by me here as an electronics consultant. The rest of the models are based on my Master of Engineering by Research degree thesis.

I am highly grateful to the late Dr. Venkat Ramaswamy, formerly senior lecturer, School of Electrical Engineering, The University of Technology Sydney, Broadway, NSW, Australia, for introducing me to the research area of modelling power electronic converters and electric drives using Simulink and for providing me the teaching fellowship under the Research Training Scheme. I am also thankful to Dr. Jianguo Zhu, Head, School of Electrical Engineering, The University of Technology Sydney, Broadway, NSW, for his encouragement and suggestion to take up the modelling and laboratory work on Lybotec six-step inverter fed PMSM drive. The author wishes to thank all the technical staff in the laboratories of the Center for Electrical Machines and Power Electronics (CEMPE), Faculty of Engineering, The University of Technology Sydney, Broadway, NSW, for their support in doing the experimental work.

Model files of examples and case studies in this book are available on the publisher's website https://link.springer.com. Students doing a course on power electronic inverters and AC drives, faculty instructors engaged in teaching and research in this subject area based on this book and professional engineers from industry who are using this book can download these model files of examples and case studies. In addition an "Instructor Manual" is prepared giving detailed solution of selected model projects along with model files. Instructors engaged in teaching this course based on this book can download a copy of this "Instructor Manual" from https://sites.google.com/springernature.com/extramaterial/lecturer-material.

I hope students, researchers, faculty members and practicing professional engineers in the industry alike will find this book useful.

I also would like to thank Michael McCabe, Senior Editor, Applied Sciences, Springer Nature, and A. Thiyagarajan, Production Editor, Springer Nature, for the keen interest shown in the development of this book.

Finally I would like to thank my wife Mythili Iyer for her patience, understanding and support in spite of her busy office and domestic schedule. Without her support, this work would not have been possible.

North Kellyville, NSW, Australia
December 2023

Narayanaswamy P. R. Iyer

Contents

About the Author

Narayanaswamy P. R. Iyer received his M.E. degree by Research and Ph.D. degree both in the area of Power Electronics and Drives from the University of Technology Sydney, NSW, and Curtin University of Technology, Perth, WA, Australia, respectively. He received his B.Sc. (Engg) (Electrical) and M.Sc. (Engg) (Power Systems) degrees from the University of Kerala and the University of Madras, respectively. He worked as a part time faculty with the Department of Electrical Engineering, University of New South Wales (UNSW), Kensington, NSW, Australia, and University of Technology Sydney (UTS), NSW, Australia. Earlier he had worked as a full time faculty in the Department of Electrical Engineering, Government Engineering College (GEC), Thrissur, Kerala; Vellore Engineering College (VEC), Vellore; and Raja Rajeswari Engineering College (RREC), Chennai. He was a research fellow in the Faculty of Engineering, Power Electronics Machines and Control (PEMC) group, the University of Nottingham, England, during July 2012 to March 2014. Presently he is an Electronics Consultant managing his own consultancy organization. He has more than two decades of experience in the modelling of electrical and electronic circuits, power electronic converters and electric drives using various software packages like MATLAB/SIMULINK, PSIM, PSCAD, PSPICE, MICROCAP, etc. and has published several papers in this area in leading conferences and journals. He has to his credit several discoveries such as "Three Phase Clipped Sinusoid PWM Inverter", "Single

Programmable and Dual Programmable Rectifier using Three Phase Matrix Converter Topology", "A Novel AC to AC Converter Using a DC Link" and a submitted patent on "A Single Phase AC to PWM Single Phase AC and DC Converter" with alternative title "Swamy Converter". He has published two books: "Power Electronic Converters Interactive Modelling Using Simulink", CRC Press, USA, 2018, and "AC to AC Converters Modelling, Simulation and Real-Time Implementation Using SIMULINK", CRC Press, USA, 2019. He is a chartered professional engineer of Australia.

Chapter 1
Introduction

1.1 Introduction

The category of the power electronic converter which converts DC voltage and current to variable frequency variable magnitude AC voltage and current is known as DC to AC converter or inverter. The input to the inverter is a fixed DC voltage source, and the output is a variable frequency and variable or fixed magnitude AC voltage either single phase or three phase with two levels for the line-to-ground voltage. These inverters are also known as two-level inverters. Voltage-source inverters (VSIs) are built using semiconductor switches which can be any one of bipolar junction transistor (BJT), metal oxide semiconductor field effect transistor (MOSFET), insulated-gate bipolar transistor (IGBT) or gate-turn-off thyristor (GTO) [1]. Basic topology of these two-level inverters is shown in Fig. 1.1a, b [1, 2].

Multilevel inverters were proposed to obtain higher output voltages without the use of step-up transformer closely following a sinusoidal waveform. Additionally, the harmonic content in the output voltage waveform is greatly reduced in comparison with conventional two-level inverters [3–10, 28]. Multilevel inverters are classified as (a) diode clamped, (b) capacitor clamped or flying capacitor and (c) cascade H-bridge, as shown in Fig. 1.1c–e. The line to ground voltage level of multilevel inverter can be three or even more. The output phase voltage of these two-level and multilevel inverters are less than the DC link voltage. Modular multilevel converters (MMC) were proposed which are either cascade half H-bridge or full H-bridge topologies with the individual DC voltage source of each of the cascade cell replaced by individual capacitor for each submodule (SM) and charged by a single DC link voltage source. This is shown in Fig. 1.1f. The MMCs find wide application in the high-voltage DC (HVDC) transmission and medium-voltage drives (MVD) [71, 72]. To obtain inverter DC link voltage and

---

Supplementary Information The online version contains supplementary material available at https://doi.org/10.1007/978-3-031-62784-2_1.

N. P. R. Iyer, *Inverters and AC Drives*, Power Systems,
https://doi.org/10.1007/978-3-031-62784-2_1

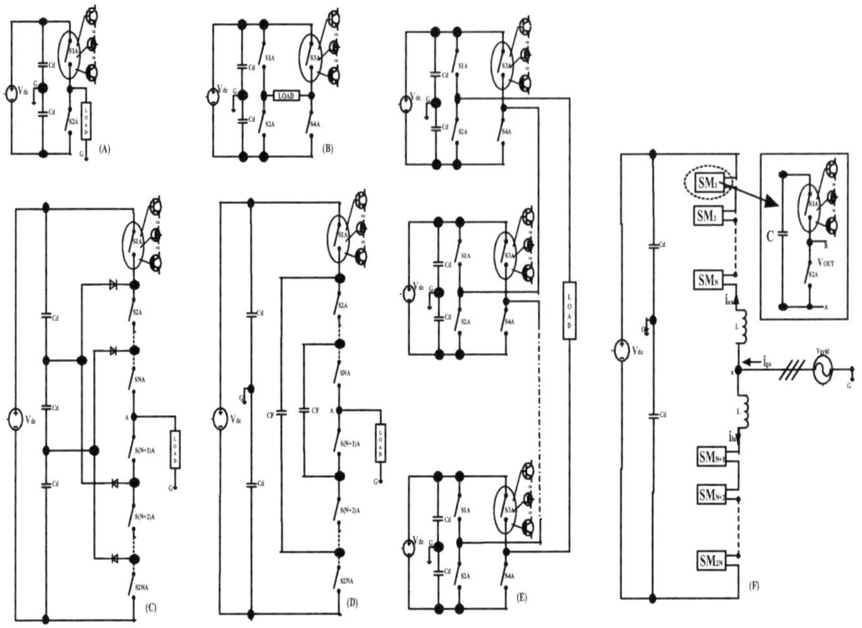

Fig. 1.1 Single-phase inverter topology: (**a**) half H-bridge, (**b**) full H-bridge (**c**), diode clamped multilevel, (**d**) flying capacitor multilevel, (**e**) cascade H-bridge multilevel and (**f**) modular multilevel

output phase voltage higher than the DC source voltage, Z-source inverter (ZSI) and quasi-Z-source inverter (QZSI) were proposed [11–16, 73, 74]. These are shown in Fig. 1.2a–d. These types of Z-source and quasi-Z-source inverters permit shoot through state during which the upper and lower switches in the same leg of any one or more phase of the inverter are turned on simultaneously. The higher the shoot through time duration, the greater is the inverter output phase voltage and DC link voltage compared to DC source voltage.

Inverters find wide applications such as in adjustable speed drives (ASDs), uninterrupted power supplies (UPS), static var. compensators, active filters, voltage compensators, flexible AC transmission systems (FACTS), photovoltaic power generation, renewable energy and grid integration [17–21].

Output voltage control of conventional inverters is achieved by pulse-width modulation (PWM) which are classified as sine pulse width modulation (SPWM), harmonic injection pulse width modulation (HIPWM), third-harmonic injection pulse-width modulation (THIPWM) [22, 23], selective harmonic elimination pulse-width modulation (SHE-PWM) and space vector pulse-width modulation (SVPWM) [29–42]. A new type of PWM for two-level inverters known as clipped sine pulse-width modulation (CSPWM) was proposed in the literature [24–27]. In all these PWM techniques, excluding SHE-PWM and SVPWM, modulating SINE/HISINE/THISINE/CSINE wave is compared with a triangle carrier using comparator, and the resulting gate pulse is used to turn on the inverter switches. For

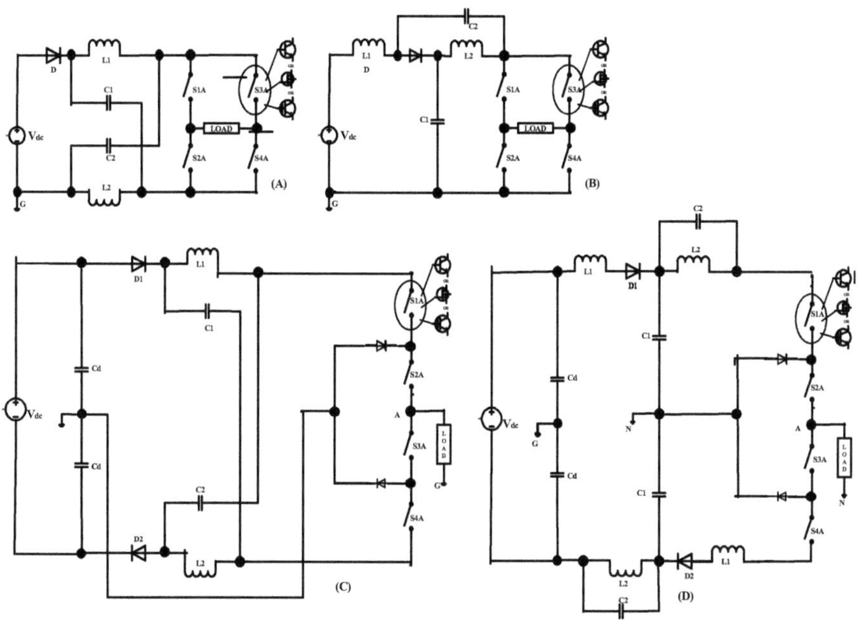

Fig. 1.2 Z-source and quasi Z-source inverter topologies: (**a**) Z-source two level, (**b**) quasi Z-source two level, (**c**) Z-source neutral point clamped three level and (**d**) quasi Z-source neutral point clamped three level

SHE-PWM inverter, the given line to ground voltage is expressed as a Fourier series, and the angles $\alpha 1, \alpha 2, \alpha 3 \ldots .\alpha n$ are calculated to eliminate a given set of $n + 1$ odd or even harmonics, and the inverter switch turn on and off is carried out corresponding to these angles. In SVPWM, the abc-axis line to neutral voltage is transformed to dq-axis voltages forming six sectors. The dwell timing T_i and $T_{(i\ +\ 1)}$ for any two neighbouring vectors in a given sector and the dwell time To for the zero vector are calculated using given sample time T_S, and the timings for the upper and lower switches of each phase are calculated, and the switches turned on according to this timing. For multilevel inverters depending on the number of levels, triangle carriers with in-phase, phase opposition, or alternate phase opposition configuration are compared with the modulating SINE/HISINE/THISINE wave in comparators, and the resulting gate pulse are used to turn on the switches. For Z-source and quasi Z-source inverters, shoot through state is generated by using sine modulating wave and triangle carrier wave using techniques known as simple boost, maximum boost, maximum constant boost, third harmonic injection maximum boost and third harmonic injection maximum constant boost [11–16]. Switched inductor, switched capacitor, diode-assisted extended boost Z-source and quasi Z-source inverters are another classification of conventional Z-source and quasi Z-source inverters which have higher boost factor and voltage gain compared to the later conventional ZSI and QZSI topologies [43–48]. These topologies are shown in Fig. 1.3a–d.

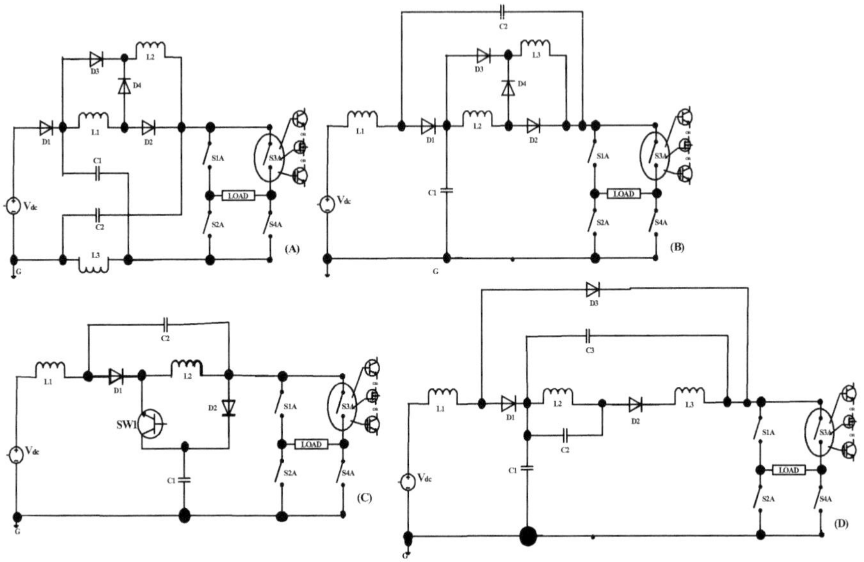

Fig. 1.3 Z-source and quasi Z-source inverter topology: (**a**) Z-source switched inductor, (**b**) quasi Z-source switched inductor, (**c**) quasi Z-source active switched capacitor and (**d**) quasi Z-source diode-assisted extended boost

Permanent magnet synchronous motor (PMSM) and brushless DC motor (BLDCM) drives find wide application as industrial drives and in electric two wheelers [21, 49–58]. The former has a sinusoidal back e.m.f and the later has a trapezoidal back e.m.f. These motors invariably are inverter driven and require sensing of rotor position information to generate gate pulse for the inverter to rotate the rotor in the forward direction [49–58]. The sensing of rotor position can be using sensors which work on Hall effect, phototransistors and disc encoders. Sensorless control by directly sensing the back e.m.f. of motor, input DC voltage and current are also possible. Interactive models for six-step continuous and discontinuous conduction mode inverter fed PMSM drive are reported [26, 58–62]. Models for vector-controlled PMSM drives fed by three phase sine wave AC voltage source and also by SVPWM inverter with hardware-in-the-loop (HIL) simulation are reported [64–67]. Models for three-phase inverter fed BLDCM drive with hysteresis current control, PWM control [68, 69] and also for extension of speed above rated base speed using phase angle advance control are reported [68–70].

Three-phase induction motor (IM) drives find wide applications in the industry, in electric vehicles and in electric traction where large horsepower output is required. To study the variations in stator and rotor currents, rotor torque and speed when the externally applied mechanical load suddenly changes or when the input frequency deviates from the original value when fed by power electronic converter, dq0-axis voltage current and flux linkage equations are used to develop the dynamic model [52, 53, 58, 63]. Interactive model for three-phase IM in synchronous reference frame using dq0-axis flux linkage equations in state space is presented, and this model is studied when the IM is fed by sine PWM, THI PWM, CSPWM, SVPWM inverters and also by multi-carrier sine level shift PWM (MCSLSPWM) diode-clamped three-level inverter (DCTLI). Also models for three-phase vector-controlled and direct torque-controlled (DTC) IM and this IM drive performance when fed by SVPWM inverter are presented [64, 67, 75–78].

For the intelligent control of AC drives, to reduce amplitude of overshoot/ undershoot and the number of oscillations in the motor speed and torque response encountered with proportional-integral (PI) controller and to reach steady-state values quickly, fuzzy logic controller (FLC) is used. Models of three-phase converter fed IM and PMSM drives with FLC from fuzzy logic toolbox in Simulink are presented [79–85].

1.2 Three-Phase Sine PWM Inverter

In this method, the modulating three-phase sine wave AC signal at the inverter switching frequency fm Hz is compared with triangular carrier wave having frequency fc Hz using op.amp. Comparators in each of the three phases. The resulting PWM gate pulse and their respective inverted gate pulse drives the respective upper and lower semiconductor switches in each of the three phases. The peak value of the sine wave vm is measured at the centre of the half cycle, i.e. at the interval Tm/4 where Tm is the period 1/fm seconds.

The three-phase inverter is shown in in Fig. 1.4, and its sine PWM gate drive model is shown in Fig. 1.5 (Model file: EXAMPLE 1_1). Three-phase sine wave AC

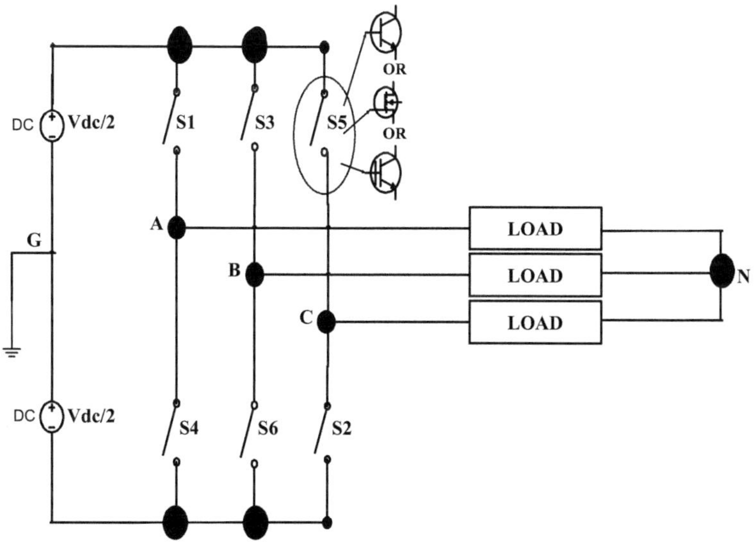

Fig. 1.4 Three-phase inverter

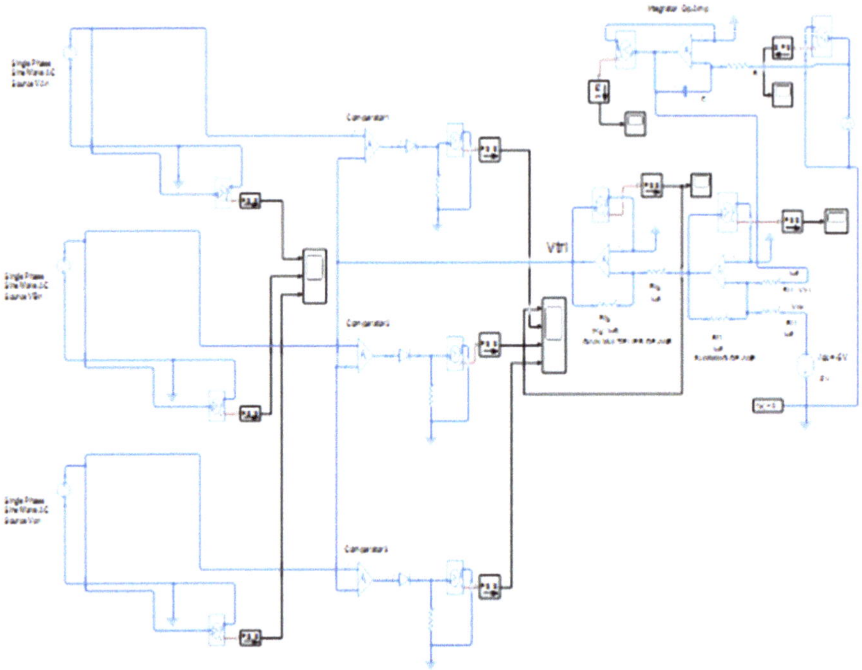

Fig. 1.5 Three-phase sine PWM inverter gate drive

signal of suitable amplitude Vm and frequency fm Hz in each phase is compared with a triangle carrier wave of peak value +Vc/-Vc and frequency fc Hz using op. amp comparators. The a.m. index M_a is Vm/Vc and f.m. index M_f is fc/fm. In Fig. 1.5, the triangle carrier is generated using op. amp.integrator, summer, gain multiplier and DC voltage source. The component level gate drive model simulation results are shown in Fig. 1.6a–c and in Fig. 1.7a–d.

In the under modulation region for $0 < Ma <= 1$, the maximum value of the fundamental—component of the line-to-line output voltage Vab1 can be expressed as follows:

$$v_{ab1} = \frac{M * \sqrt{3} * V_{dc}}{2} \text{ for } M_a \leq 1 \tag{1.1}$$

Similarly in the overmodulation region, the maximum value of the nth harmonic component of line-to-line output voltage is given below:

$$v_{abn} = \frac{4 * \sqrt{3} * V_{dc}}{2 * n * \pi} \text{ for } M_a > 1 \tag{1.2}$$

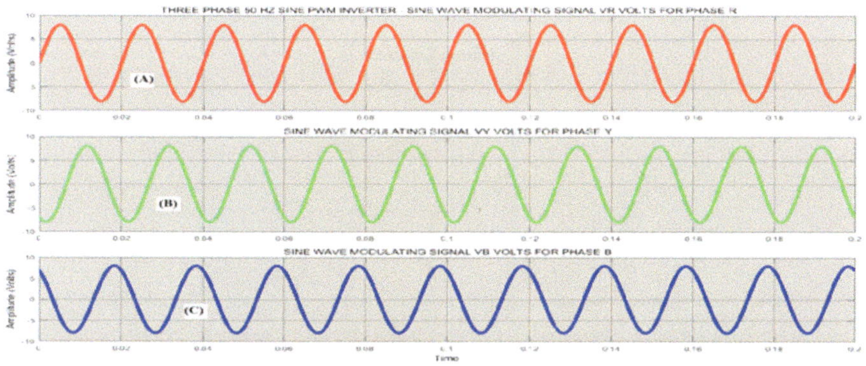

Fig. 1.6 Three-phase sine PWM inverter gate drive simulation results: sine wave modulating signals for phases A, B and C (top to bottom)

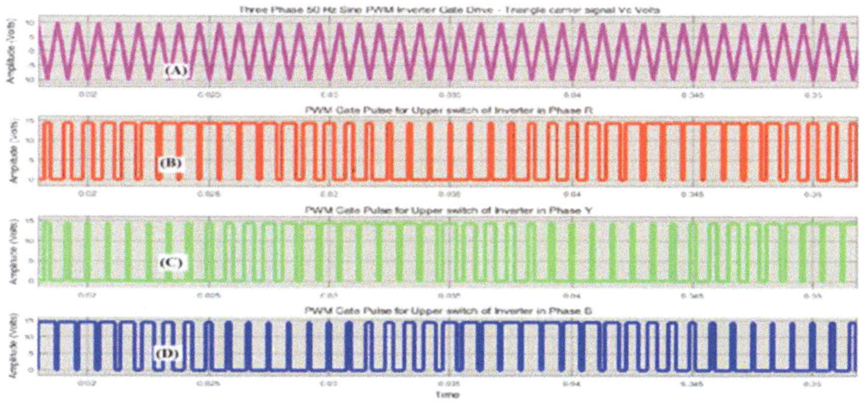

Fig. 1.7 Three-phase 50 Hz sine PWM inverter gate drive simulation results. (**a**) Triangle carrier, (**b**) (**c**), (**d**) PWM gate pulse for upper switches in phases A, B, and C

1.2.1 Modelling of Three-Phase Sine PWM Inverter

The model of the three-phase sine PWM inverter is shown in Fig. 1.8 (Model file: EXAMPLE 1_2). The modulating signal which is a three-phase sine wave AC voltage having frequency fm Hz and amplitude Vm Volts is generated using Embedded MATLAB function. This is shown in Program segment 1.1 in the model file EXAMPLE 1_2.

Here fm is 50 Hz and Vm is 8 V, respectively. The triangle generator along with function block Fcn1 generates a triangle carrier with amplitude +/-vc (+/−10 V) and frequency fc Hz (1200 Hz). The output of embedded MATLAB function Vcon_a, Vcon_b and Vcon_c is compared with the triangle carrier using three relational operator comparator blocks. The output of the three relational operator blocks in

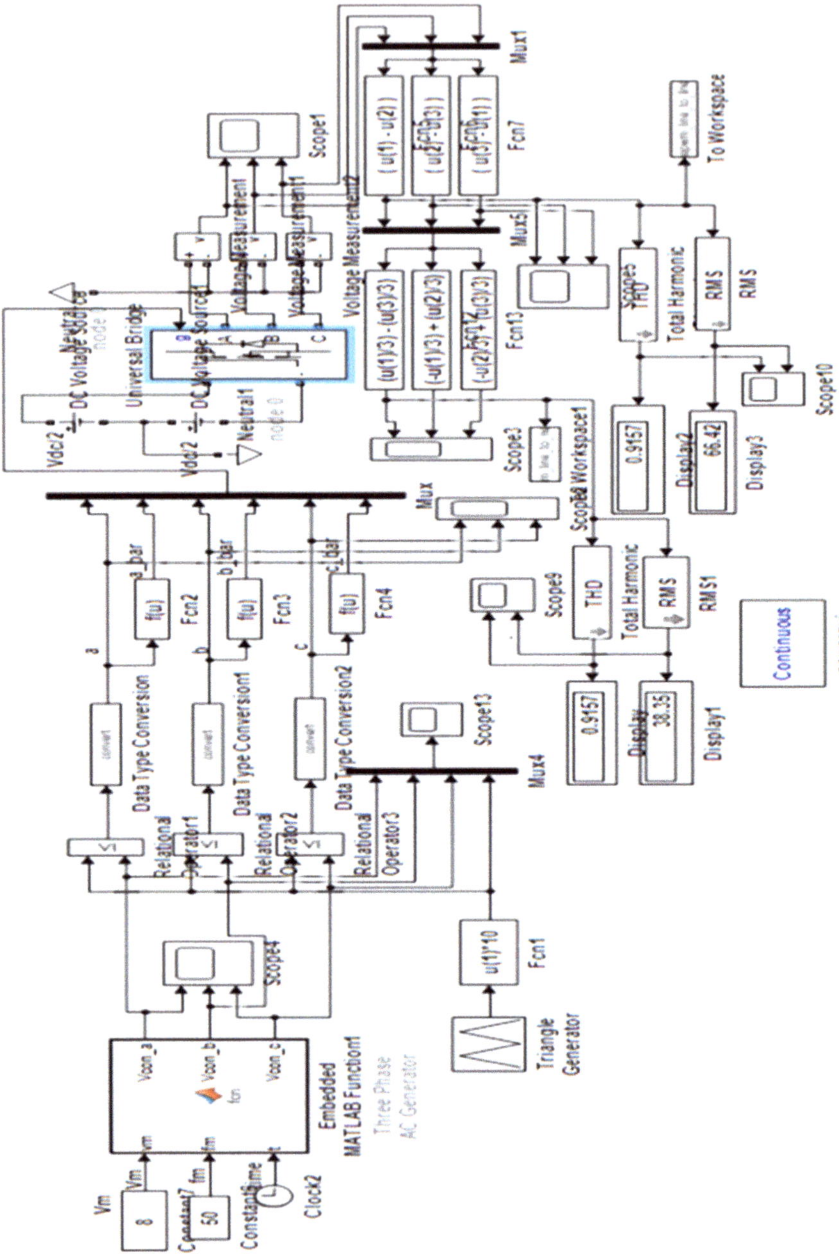

Fig. 1.8 Three-phase sine PWM inverter model

each phase and the respective inverted outputs using Fcn blocks Fcn2, Fcn3 and Fcn4 form the PWM gate drive for the three-phase inverter. Function blocks Fcn2, Fcn3 and Fcn4 invert the gate pulse input using the relation (1 – $u(1)$). In Fig. 1.8, universal bridge subsystem forms the three-phase inverter. The two DC voltage sources with midpoint grounded form the DC link. The three-phase line-to-line output voltage is formed by subtracting the respective line to ground output voltage using Fcn blocks, and the line-to-neutral output voltage is formed by matrix transformation of line-to-line output voltage using Fcn blocks, as given in Eqs.1.3 and 1.4.

$$\begin{bmatrix} V_{RY} \\ V_{YB} \\ V_{BR} \end{bmatrix} = \begin{bmatrix} 1 & -1 & 0 \\ 0 & 1 & -1 \\ -1 & 0 & 1 \end{bmatrix} * \begin{bmatrix} V_{RG} \\ V_{YG} \\ V_{BG} \end{bmatrix} \tag{1.3}$$

$$\begin{bmatrix} V_{RN} \\ V_{YN} \\ V_{BN} \end{bmatrix} = \frac{1}{3} * \begin{bmatrix} 1 & 0 & -1 \\ -1 & 1 & 0 \\ 0 & -1 & 1 \end{bmatrix} * \begin{bmatrix} V_{RY} \\ V_{YB} \\ V_{BR} \end{bmatrix} \tag{1.4}$$

1.2.2 Simulation Results

The simulation of the three-phase sine PWM inverter was carried out using SIMULINK [80]. The ode23tb (stiff/TR-BDF2)) solver was used. The model parameters are shown in Table 1.1. The model is shown in Fig. 1.8. In Table 1.1, A.M. Index Ma is the ratio of vm/vc and F.M.

Index Mf is the ratio of fc/fm. Simulation results are tabulated in Table 1.2. The simulation results of line-to-line and line-to-neutral voltage for Ma of 0.8 and Mf of 24 p.u. are also shown in Fig. 1.9a–j. The harmonic spectrum of VRY and VRN is shown in Fig. 1.10 and Fig. 1.11. The three-phase sine-wave modulating signal and triangle carrier are shown in Fig. 1.12. Other advanced pulse-width modulation techniques for inverters are presented in the respective chapters. The book novelty is presented in the next section.

Table 1.1 Model parameters

Parameter	Value	Unit
DC link voltage	100	Volts
Modulating sine wave signal frequency	50	Hz
Triangle carrier frequency	1200	Hz
Amplitude modulation index ma	0.2, 0.4, 0.6, 0.8, 1.2, 1.4, 1.6	–
Frequency modulation index mf	24	–

Table 1.2 Three-phase sine PWM inverter simulation results

Sl. No	Ma	VLL (rms) V	VLL1 (rms) V	VLN (rms) V	VLN1 (rms) V	THD of VLL	THD of VLN	Remarks
1	0.2	29.98	10.4	17.38	6.076	2.878	2.871	Under modulation
2	0.4	42.54	22.6	24.61	13.16	1.908	1.901	–
3	0.6	55.65	35.81	32.13	20.73	1.285	1.28	–
4	0.8	66.42	49.00	38.35	28.28	0.916	0.916	–
5	1.2	78.21	68.23	45.09	39.49	0.5528	0.5628	Over modulation
6	1.4	80.36	71.48	46.19	41.42	0.5113	0.5181	–
7	1.6	80.94	72.73	46.73	42.32	0.4862	0.4883	–

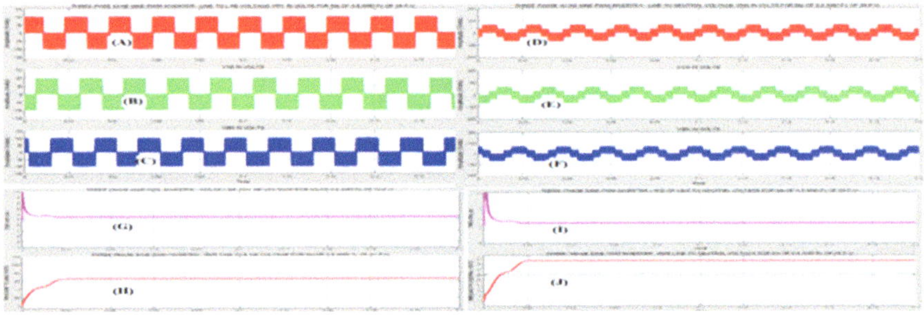

Fig. 1.9 Three-phase sine PWM inverter simulation results. (**a**), (**b**), (**c**) Line-to-line voltage VRY, VYB and VBR. (**d**), (**e**), (**f**) Line-to-neutral voltage VRN, VYN, VBN. (**g**), (**h**) THD and RMS value of VRY. (**i**), (**j**) THD and RMS value of VRN

1.3 Book Novelty

The following are some of the features associated with this textbook.

- Clipped sine (CS) PWM technique is a new finding. This technique is found to be better than sine, harmonic injection (HI) and third harmonic injection (THI) PWM techniques from the point of view of total harmonic distortion (THD) of line-to-neutral and line-to-line voltages of three-phase two-level and NPC three-level inverter.
- Models for dead-band sine (DBS) and modified sine (MS) PWM technique for three-phase two-level inverter are new.
- Models for multi-carrier sine level shift (MCSLS) PWM technique for three-phase diode-clamped three-level inverter (DCTLI) are new.
- Models for three-phase clipped sine level shift PWM (CSLSPWM) and third harmonic injection sine level shift PWM (THISLSPWM) DCTLI are presented which are new.

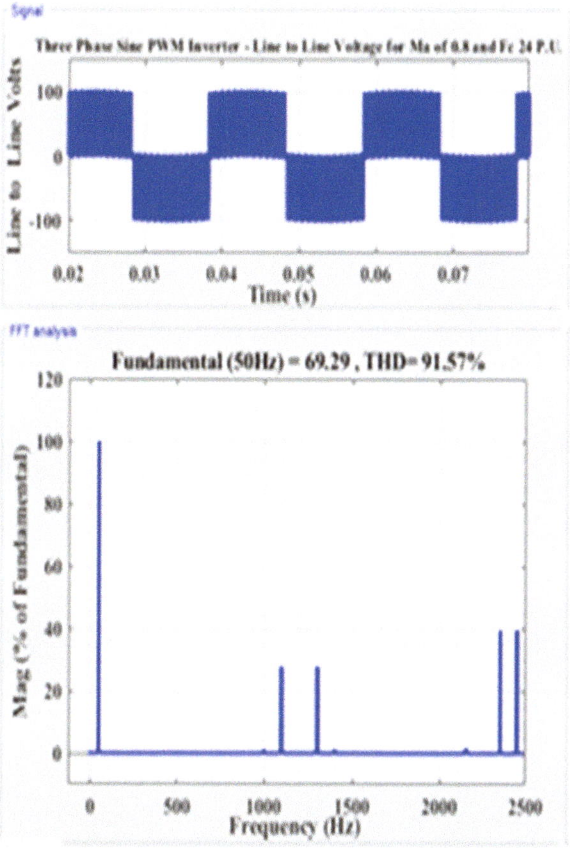

Fig. 1.10 Three-phase sine PWM inverter line-to-line voltage (top) and harmonic spectrum (bottom)

- Models for three-phase five level cascade H-bridge inverter (FLCHBI) by multi-carrier clipped sine phase shift PWM (MCCSPSPWM) and by third harmonic injection sine phase shift PWM (MCTHISPSPWM) are presented which are new.
- Models for single-phase and three-phase MMC are presented. A modified analogue technique for submodule capacitor voltage balance is presented for a single-phase MMC with simulation results.
- Models are presented to eliminate selected fifth and seventh harmonics from the line-to-ground voltage of three-phase cascade H-bridge inverter (CHBI), DCTLI and flying capacitor three-level inverter (FCTLI) which are new.
- Models for single-phase selective harmonic elimination PWM (SHE-PWM) diode clamped five-level inverter (DCFLI) and flying capacitor five-level inverter (FCFLI) with desired wave shape selection using logic gate arrays are presented which are new.
- Models for space vector modulation (SVM) of three-phase two-level inverter, DCTLI and FCTLI are presented in a unique way.

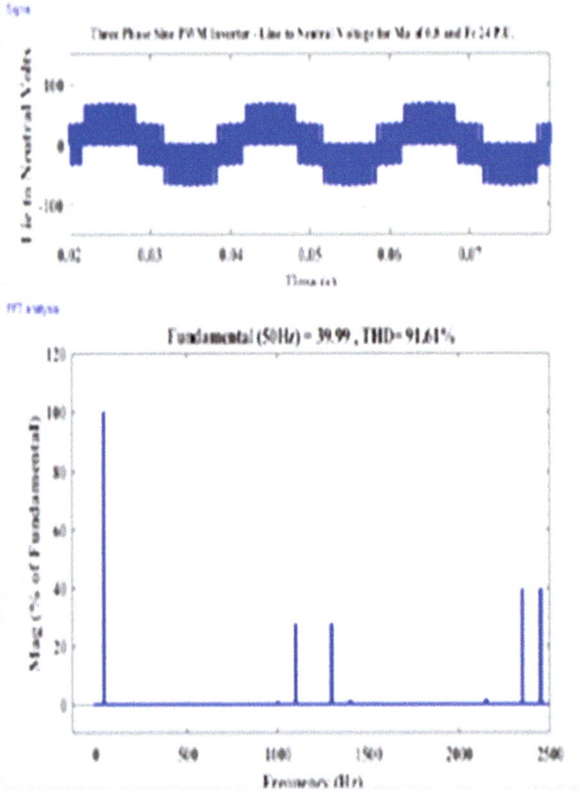

Fig. 1.11 Three-phase sine PWM inverter line-to-neutral voltage (top) and harmonic spectrum (bottom)

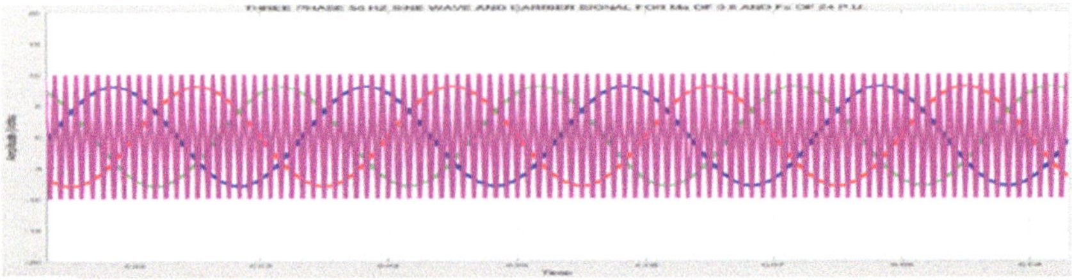

Fig. 1.12 Three-phase sine PWM inverter—three-phase 50 Hz sine wave modulating signal and triangle carrier

- For the models of SVM of three-phase DCTLI and FCTLI, dwell time calculation for reference voltage space vector is shown in detail. Similarly, region identification of reference voltage space vector is carried out in a new different way.
- Models for SVM of a three-phase flying capacitor three-level inverter (FCTLI) and two-level inverter with even order harmonics elimination switching are presented.

- The expressions for the shoot through ratio, boost factor and voltage gain for Z-source (ZS) and quasi Z-source (QZS) inverter when controlled using maximum boost, third harmonic injection maximum boost, third harmonic injection maximum constant boost are derived from fundamentals which are new features.
- Models for SVM of a three-phase Z-source and quasi Z-source two-level inverter are presented which are new.
- Models for Z-source and quasi Z-source neutral-point-clamped (NPC) three-level inverter are presented.
- The derivation for the boost factor and the model of three-phase switched inductor Z-source inverter (SL-ZSI) using simple boost and maximum constant boost control are new.
- Model of SVM of a three-phase switched inductor quasi Z-source (SL-QZS), active switched capacitor quasi Z-source (ASC-QZS), diode-assisted extended boost quasi Z-source (DA-QZS) and SL-ZS inverter are presented which are new.
- Interactive models for six-step continuous and discontinuous current mode inverter fed PMSM drive are presented.
- The criteria for inverter switching angle advance φ_{MT} for maximum electromagnetic torque of six-step continuous current mode inverter fed PMSM drive is studied by modelling technique.
- Experimental verification of six-step discontinuous conduction mode Lybotec inverter fed PMSM drive with sensorless control is presented.
- Detailed derivation for calculating the proportional-integral (PI) controller constants for the speed and current control loop along with the models for three-phase vector-controlled PMSM fed by sine wave voltage source and by SVPWM inverter are presented.
- Models for three-phase inverter fed BLDCM drive in open loop and in closed loop with hysteresis current control without and with phase angle advance facility are presented. The advantage of phase angle advance control for extending speed range is established.
- A gate drive model using op.amp comparators and logic gates for a six-step discontinuous current mode inverter operation along with a new method for trapezoidal back e.m.f. generation for a three-phase BLDCM drive is presented.
- Model for three-phase IM using qd0-axis flux linkage equations in state space is presented.
- Models of three-phase IM fed by CSPWM, SVPWM and MCSLSPWM inverter are new.
- Detailed derivation for calculating the PI controller constants for the speed and current control loop along with the models for three-phase vector-controlled IM fed by sine wave AC voltage source and by SVPWM inverter are presented.
- Models for DTC of three-phase IM fed by two-level SVPWMM inverter are presented with and without the use of PI controller using both classical and modified switch tables.
- Two methods for developing the models for SVPWM of three-phase two-level inverter are presented, one by a rigorous approach which is new and another by a simplified fundamental approach.

- Models for constant voltage to frequency (V/f) control of three-phase IM fed by six-step 180 degree mode inverter and that by sine PWM inverter using PI controller are presented.
- Models for three-phase converter fed IM and PMSM drives using FLC are presented to illustrate the advantage of FLC over PI controller to control their rotor speed and torque response for changes in mechanical load.

Book outline is presented in the next section.

1.4 Book Outline

The book covers nine chapters. Chapter 1 provides the introduction, model with simulation results for the three-phase two-level sine PWM inverter, the book novelty and outline.

Chapter 2 covers other advanced PWM techniques for three-phase two-level and diode-clamped three-level inverter (DCTLI). This includes harmonic injection (HI), third harmonic injection (THI), clipped sinusoid (CS), dead-band sine (DBS) and modified sine (MS) PWM techniques for three-phase two-level inverter and multi-carrier sine level shift (MCSLS) PWM for three-phase DCTLI. Case studies are presented for three-phase clipped sine level shift PWM (CSLSPWM) and third harmonic injection sine level shift PWM (THISLSPWM) DCTLI. Also model for three-phase multi-carrier sine phase shift PWM (MCSPSPWM) FLCHBI is presented along with case studies for three-phase MCCSPSPWM and MCTHISPSPWM methods. Also model for three-phase MMC is presented along with a case study for a single-phase MMC model using an analogue technique for submodule capacitor voltage balancing.

Models to eliminate 5th and 7th harmonics from the line-to-ground voltage of three-phase unipolar, bipolar, cascade H-bridge, DCTLI and FCTLI are presented in Chap. 3. Case studies are presented for single-phase selective harmonic elimination PWM (SHE-PWM) diode-clamped five-level inverter (DCFLI) and flying capacitor five-level inverter (FCFLI).

Models for SVPWM of three-phase two-level inverter, DCTLI and FCTLI, are presented in Chap. 4. Case studies are presented for SVM of a three-phase flying capacitor three-level inverter (FCTLI) and two-level inverter with even order harmonics elimination switching.

Models for Z-source (ZS) and quasi Z-source (QZS) inverter using simple boost control (SBC), maximum boost control (MBC), third harmonic injection maximum boost control (THIMBC), third harmonic injection maximum constant boost control (THIMCBC) are presented in Chap. 5. Case studies are presented for SVM of a three-phase Z-source and quasi Z-source inverter. Also presented are the models for three-phase Z-source and quasi-Z-source FLCHBI and NPC three-level inverter.

Models for three-phase SL-ZSI, switched inductor quasi-Z-source inverter (SL-QZSI), active switched capacitor quasi-Z-source inverter (ASC-QZSI), diode-assisted extended boost quasi-Z-source inverter (DA-QZSI) using simple boost control (SBC) and third harmonic injection maximum constant boost control (THIMCBC) are presented in Chap. 6. Case studies are presented for SVM of a three-phase SL-QZSI, ASC-QZSI, DA-QZSI and SL-ZSI.

Interactive models for six-step continuous current mode and discontinuous current mode inverter fed PMSM drive along with laboratory experimental results for the later type inverter fed PMSM drive are presented in Chap. 7. Models for three-phase inverter fed BLDCM drive in open loop, hysteresis current controlled inverter without and with phase angle advance facility in closed loop are presented. Also model for three-phase vector-controlled PMSM fed by sine wave voltage source is presented. Case studies are presented for a three-phase space vector PWM (SVPWM) inverter fed vector-controlled PMSM drive and also for a three-phase hysteresis current controlled inverter fed BLDCM drive with phase angle advance facility.

Interactive model for three-phase IM using flux linkage equations in state space is presented in Chap. 8. This is followed by models for three-phase IM fed by (a) sine PWM, (b) THIPWM, (c) CSPWM and (d) MCSLSPWM inverter. Also case studies are presented where models for three-phase IM fed by SVPWM inverter and vector-controlled three-phase IM fed by SVPWM inverter are presented. Models for DTC of IM fed by three-phase SVPWM two-level inverter with and without using PI controller are presented using classical switching and as a case study using modified switching table. Models for voltage to frequency ratio (V/f) control of IM fed by six-step 180-degree mode inverter and that by sine PWM inverter using PI controller are presented. A case study is presented for the speed control of IM fed by three-phase AC to AC thyristor controller using PI controller.

Fuzzy logic control (FLC) of three-phase converter fed IM and PMSM drives are presented in Chap. 9 to illustrate the advantage of FLC over PI controller in controlling their speed and torque response for changes in external load. A case study is presented for the FLC-based speed control of IM fed by three-phase SCR controller.

References

1. Issa Batarseh: "Power Electronic Circuits", John Wiley, 2004, pp. 36–67.
2. J.R. Espinoza: "Inverters" in "Power Electronics handbook" Ch.15, Editor: M.H. Rashid, Elsevier, 2011, pp. 357–374.
3. Clark Hochgraf, Robert Lasseter, Deepak Divan and T.A. Lipo: "Comparison of Multilevel Inverters for Static VAR Compensation"; IEEE; 1994; pp. 921–928.
4. J.S. Lai and F.Z. Peng: "Multilevel convertors—A New Breed of Power Convertors"; IEEE Transactions on Industry Applications; Vol.32; No.3; May/June 1996; pp. 509–517.
5. F.Z. Peng, J.S. Lai and J.W. Mckeever and J. VanCoevering "A Multilevel Voltage Source Inverter with Separate DC Sources for Static Var Generation"; IEEE Transactions Industry Applications; Vol. 32; No. 5; Sept/Oct. 1996; pp. 1130–1133.

6. Jose Rodriguez, J.S. Lai and F.Z. Peng: "Multilevel Inverters: A Survey of Topologies Control and Applications", IEEE Transactions on Industrial Electronics Vol.49, No.4, August 2002, pp. 724–738.
7. Pradeep M. Bhagwat and V.R. Stefanovic: "Generalized Structure of a Multilevel PWM Inverter"; IEEE Transactions on Industry Applications; Vol.IA-19; No.6; November/December 1983; pp. 1057–1069.
8. Akira Nabae, Isao takahashi and Hirofumi Akagi: "A Neutral-Point Clamped PWM Inverter", IEEE Transactions on Industry Applications, Vol.IA-17, No.5, September/October 1981, pp. 518–523.
9. Surin Khomfoi and L.M. Tolbert: "Multilevel Power Converters" in "Power Electronics handbook", Editor: M.H. Rashid, Ch. 17, Elsevier, 2011, pp. 455–473.
10. J.W. Dixon. "Multilevel Converters" in Power Electronic Converters and Systems Frontiers and Applications, Ed. Andrzej M. Trzynadlowski, pp. 43–72, The Institution of Engineering and Technology, 2016.
11. F.Z. Peng: "Z-Source Inverter", IEEE Transactions on Industry Applications, Vol.39, No.2, March/April 2003, pp. 504–510.
12. F.Z. Peng, M. Shen and Z. Qian: "Maximum Boost Control of the Z-Source Inverter", IEEE Transactions on Power Electronics, Vol. 20, No. 4, July 2005, pp. 833–838.
13. M. Shen, J. Wang, Alan Joseph, F.Z. Peng, L.M. Tolbert and D.J. Adams: "Constant Boost Control of the Z-Source Inverter to minimize current ripple and voltage stress", IEEE Transactions on Industry Applications, Vol.42, No.3, May/June 2006, pp. 770–778.
14. M. Shen, J. Wang, Alan Joseph, F.Z. Peng, L.M. Tolbert and D.J. Adams: "Maximum Constant Boost Control of the Z-Source Inverter", IEEE Industry Applications Society Conference, 2004, pp. 142–147.
15. F.Z. Peng, Alan Joseph, Jin Wang, M. Shen, Lithua Chen, Zhiguo Pan, E.O. Riviera and Yi Huang: "Z-Source Inverter for Motor Drives", IEEE Transactions on Power Electronics, Vol. 20, NO. 4, JULY 2005, pp. 857–863.
16. Y. Li, J. Anderson, F.Z. Peng and D. Lu: "Quasi-Z-Source Inverter for Photovoltaic Power Generation Systems', IEEE APEC, February 2009, Washington DC, USA, pp. 918–924.
17. Adel Nasiri: "Uninterruptible Power Supplies" in "Power Electronics handbook", Editor: M.H. Rashid, Ch. 24, Elsevier, 2011, pp. 627–638,
18. Lana El Chaar: "Photovoltaic System Conversion" in "Power Electronics handbook", Editor: M.H. Rashid, Ch. 27, Elsevier, 2011, pp. 711–721,
19. C.V. Nayar, S.M. Islam, H. Dehbonei, K. Tan and H. Sharma: "Power Electronics for Renewable Energy Sources" in "Power Electronics handbook", Editor: M.H. Rashid, Ch. 28, Elsevier, 2011, pp. 723–764,
20. E.H. Watanabe et al. "Flexible AC Transmission Systems" in "Power Electronics handbook", Editor: M.H. Rashid, Ch. 32, Elsevier, 2011, pp. 861–876.
21. M.F. Rahman et al.: "Motor Drives" in "Power Electronics handbook", Editor: M.H. Rashid, Ch. 34, Elsevier, 2011, pp. 923–959.
22. J.A. Houldsworth and D.A. Grant: "The Use of Harmonic Distortion to Increase the Output Voltage of a Three Phase PWM Inverter"; IEEE Transactions on Industry Applications; Vol.IA-20; NO.5; pp. 1224–1227, 1984.
23. M.A. Boost and P.D. Ziogas: "State of the Art Carrier PWM Techniques: A Critical Evaluation"; IEEE Transactions on Industry Applications; Vol.24; No.2; pp. 271–280; March / April 1988.
24. Narayanaswamy P.R. Iyer, Venkat Ramaswamy and Jianguo Zhu: "SIMULINK Model for Three Phase SINE PWM Inverter fed Induction Motor Drive, 40th International Universities Power Engineering Conference, Cork, Ireland September 2005; pp. 1180–1184.
25. Narayanaswamy P. R. Iyer, Venkat Ramaswamy and Jianguo Zhu: "Three Phase Clipped Sinusoid PWM Inverter–A New technique for Pulse Width Modulation", 40th International Universities Power Engineering Conference (UPEC 2005), Cork, Ireland September 2005, pp. 1185–1189.

26. Narayanaswamy P. R. Iyer: "MATLAB/SIMULINK Modules for Modelling and Simulation of Power Electronic Converters and Electric Drives"; M.E. (Research) Thesis; University of Technology Sydney, NSW, 2006.
27. Narayanaswamy P R Iyer: "Models for three phase pulse width modulated inverters", Research seminar, School of Electrical and Information Engineering, The University of Sydney, Redfern, NSW, Australia, 4th November 2008.
28. N P R Iyer: "Power Electronic Converters: Interactive Modeling using SIMULINK", CRC Press, USA, Chapter 9 and 10, pp. 243–249 and pp. 315–319., 2018.
29. H.S. Patel and R.G. Hoft: "Generalized Techniques of Harmonic Elimination and Voltage Control in Thyristor Inverters: Part I-Harmonic Elimination", IEEE Transactions on Industry Applications, Vol. IA-9, No.3, pp. 310–317. May/June 1973.
30. H.S. Patel and R.G. Hoft: ""Generalized Techniques of Harmonic Elimination and Voltage Control in Thyristor Inverters: Part II—Voltage Control Technique", IEEE Transactions on Industry Applications, Vol.IA-10, No.5, pp. 666–673, September/October 1974.
31. P.N. Enjeti, P.D. Ziogas and J.F. Lindsay: "Programmed PWM Techniques to Eliminate Harmonics: A Critical Evaluation" IEEE Transactions on Industry Applications", vol. 26, No.2, pp. 302–316, March/April 1990.
32. J. Pontt, J. Rodriguez, R. Huerta and J. Pavez: "A Mitigation Method for Non-Eliminated Harmonics of SHEPWM Three-Level Multipulse Three-Phase Active Front End Converter", IEEE ISIE, Vol.1, pp. 258–263, 2003.
33. J. Pontt, J. Rodriguez, R. Huerta and J. Pavez: "Mitigation of Noneliminated Harmonics of SHEPWM Three-Level Multipulse Three-Phase Active Front End Converters with Low Switching Frequency for Meeting Standard IEEE-519-92", IEEE PESC, pp. 531–536, 2003.
34. J. Pontt, J. Rodriguez and R. Huerta: "Mitigation of Noneliminated Harmonics of SHEPWM Three-Level Multipulse Three-Phase Active Front End Converters with Low Switching Frequency for Meeting Standard IEEE-519-92", IEEE Transactions on Power Electronics, Vol. 19, No. 6, November 2004, pp. 1594–1600, November 2004.
35. V.G. Agelidis, A.I. Balouktsis and C. Cossar: "On Attaining the Multiple Solutions of Selective Harmonic Elimination PWM Three-Level Waveforms Through Function Minimization", IEEE Transactions on Industrial Electronics, Vol. 55, No. 3, pp. 996–1004, March 2008.
36. J.N. Chiasson, L.M. Tolbert, K.J. McKenzie and Z. Du: "A Complete Solution to the Harmonic Elimination Problem", IEEE Transactions on Power Electronics, Vol. 19, No.2, pp. 491–499, March 2004.
37. J. Chiasson, L. Tolbert, K. Mckenzie and Z. Du: "Eliminating Harmonics in a Multilevel Converter using Resultant Theory", IEEE PESC, Cairns, Australia, June 2002, pp. 503–508.
38. P. Enjeti And J.F. Lindsay: "Solving Nonlinear Equations of Harmonic Elimination PWM In Power Control", Electronics Letters, Vol. 23, No.12, June 1987. Pp. 656–657.
39. V.G. Agelidis, A. Balouktsis I. Balouktsis and C. Cossar: "Multiple Sets of Solutions for Harmonic Elimination PWM Bipolar Waveforms:Analysis and Experimental Verification", IEEE Transactions on Power Electronics, Vol.21, No.2, March 2006, pp. 415–421.
40. John N. Chiasson, Leon M. Tolbert, Keith J. McKenzie and Zhong Du: "A Unified Approach to Solving the Harmonic Elimination Equations in Multilevel Converters', IEEE Transactions on Power Electronics, Vol. 19, No. 2, March 2004, pp. 478–490.
41. Bin Wu: "High-Power Converters and AC Drives", IEEE Press, pp. 101–111 and pp. 143–162, 2006.
42. Amit Kumar Gupta and Ashwin M. Khambadkone; 'A Space Vector PWM Scheme for Multilevel Inverters Based on Two-Level Space Vector PWM", IEEE Transactions On Industrial Electronics, Vol. 53, No. 5, October 2006, pp. 1631–1639.
43. M. Zhu, K. Yu, and F. L. Luo, "Switched inductor Z-source inverter," IEEE Trans. Power Electronics., Vol. 25, No. 8, pp. 2150–2158, August 2010.
44. Hailong Liu, Yuyao He, Mingyang Zhang Liu: "Quasi-Switched-Inductor Z-Source Inverter", IEEE 8th International Power Electronics and Motion Control Conference (IPEMC-ECCE Asia), 2016.

45. M. K. Nguyen, Y. C. Lim, and G. B. Cho, "Switched-inductor quasi-Z-source inverter," IEEE Trans. Power Electronics., vol. 26, no. 11, pp. 3183–3191, November 2011.
46. Anh-Vu Ho and Tae-Won Chun: "Topologies of Active-Switched Quasi-Z-source Inverters with High-Boost Capability"; Journal of Power Electronics, Vol. 16, No. 5, pp. 1716–1724, September 2016.
47. C. J. Gajanayake, F. L. Luo, H. B. Gooi, P. L. So, and L. K. Siow, "Extended boost Z-source inverters," IEEE Trans. Power Electronics., Vol. 25, No. 10, pp. 2642–2652, Oct. 2010.
48. F.L. Luo and H. Ye: "Power Electronics: Advanced Conversion Technologies", CRC Press, 2010, pp. 460–477.
49. J.F. Gieras and M. Wing: "Permanent Magnet Motor Technology"; Marcell Dekker Inc.; Newyork; 2002; pp. 211–212.
50. T. Kenjo and S. Nagamori: "Permanent-Magnet and Brushless DC Motors"; Oxford Press; 1985; pp. 58–64.
51. T.J.E. Miller: "Brushless Permanent and Reluctance Motor Drives"; 1993; pp. 80–83.
52. Paul C. Krause: "Analysis of Electric Machinery"; McGraw Hill Inc.;1986; pp. 178–197.
53. R. Krishnan: "Electric Motor Drives Modeling Analysis and Control"; Prentice Hall Inc.; 2001.
54. P.C. Krause, R.R. Nucera R.J. Krefta and O. Wasynczuk: "Analysis of a Permanent magnet Synchronous Machine Supplied From A 180o Inverter With Phase Control"; IEEE Transactions on Energy Conversion; Vol.EC-2; No.3; September 1987; pp. 423–431.
55. P. Pillay and R. Krishnan: "Modeling, Simulation and Analysis of Permanent – Magnet Motor Drives, Part I: The Permanent magnet Synchronous Motor Drive"; IEEE Transactions on Industry Applications, Vol.25, No.2, March/April 1989; pp. 265–273.
56. R.R. Nucera, S.D. Sudhoff and P.C. Krause: "Computation of Steady-State Performance of an Electronically Commutated Motor", IEEE Transactions on Industry Applications, Vol.25, No.6, Nov./Dec. 1989, pp. 1110–1117.
57. P. Pillay and R. Krishnan: "Modeling of Permanent magnet Motor Drives"; IEEE Transactions on Industrial Electronics; Vol.35; No.4; November 1988; pp. 537–541.
58. Chee-Mun-Ong: "Dynamic Simulation of Electric machinery Using MATLAB/SIMULINK"; Prentice Hall Inc.; 1998.
59. N.P.R. Iyer and J.G. Zhu: Interactive Modeling and Simulation of A Six Step Continuous Current Mode Inverter fed PMSM Drive Using SIMULINK", 17th International Conference on Electrical Machines, Greece, OTM2-4, September 2–5, 2006.
60. N.P.R. Iyer and J.G. Zhu: "Modeling and Simulation of A Six Step Discontinuous Current Mode Inverter Fed Permanent Magnet Synchronous Motor Drive Using SIMULINK" , Sixth IEEE International Conference on Power Electronics and Drive Systems (PEDS05), Kualalumpur, Malaysia, Nov.-Dec. 2005, pp. 1056–1061.
61. N.P.R. Iyer and J.G. Zhu: "Interactive Modelling of a Six Step Discontinuous Current Mode Inverter fed Permanent Magnet Synchronous Motor Drives", 37th IEEE Power Electronics Specialists Conference, June 18–22, 2006, pp. 574–579, Jeju, Korea, PS 1–94, IEEE 1-4244-9717-7.
62. N.P.R. Iyer and J.G. Zhu: "Modelling and Simulation of a Six Step Discontinuous Conduction Mode MOSFET Inverter Fed Permanent Magnet Synchronous Motor Drive"; Australasian Universities Power Engineering Conference; Melbourne, December 10–13, 2006.
63. Narayanaswamy. P.R. Iyer, Venkat Ramaswamy and Jianguo Zhu: "Interactive Modeling of A Three Phase Induction Motor in All Reference Frames Using dq0-Axis Flux Linkage Equations in State Space", Australasian Universities Power Engineering Conference (AUPEC 2006), Melbourne, December 10–13, 2006, ISBN: 978 1 86272 669 7.
64. Ned Mohan: "Advanced Electric Drives Analysis, Control and Modeling Using SIMULINK", MNPERE, 2001, pp. 3–1 to 3–30 and pp. 5–1 to 5–16.
65. A Consoli, A. Raciti and A. Testa: "Sensorless Vector and Speed Control of Brushless Motor Drives", IEEE Transactions on Industrial Electronics, Vol. 41, No. 1, February 1994, pp. 91–96.

66. J. Mat Lazi, Z. Ibrahim, S.N. Mat Isa, A. Anugerah, F.A. Patakor and R. Mustafa: "Sensorless Speed Control of PMSM Drives using dSPACE DS1103 Board", IEEE International Conference on Power and Energy (PECon), 2–5, December 2012, Malaysia, pp. 922–927.
67. D.W. Novotny and T.A. Lipo: "Vector Control and Dynamics of AC Drives", Oxford University Press, New York, 1996, pp. 240–243. and pp. 257–275.
68. P. Pillay and R. Krishnan: "Modeling, Simulation, and Analysis of Permanent-Magnet Motor Drives, Part II: The Brushless DC Motor Drive", IEEE Transactions On Industry Applications, Vol. 25, No. 2, March/April 1989, pp. 274–279.
69. Chau, K. T.: "Electric Vehicle Machines and Drives: Design, Analysis and Application", John Wiley & Sons, Inc., 2015. pp. 69–94.
70. B. Nguyen and C. Ta, "Phase advance approach to expand the speed range of brushless DC motor," International Conference on Power Electronics and Drive Systems, 2007, pp. 1255–1262.
71. J. Dragan and K. Ahmad: "High Voltage Direct Current Transmission: Converters, Systems and DC Grids", John Wiley and Sons, 2015.
72. A. Lesnicar and R. Marquardt, "An innovative modular multilevel converter topology suitable for a wide power range," in Proc. IEEE Power Tech Conf., Bologna, Italy, 2003, vol. 3, pp. 6–11.
73. Liu, Yushan, et al.: "Impedance Source Power Electronic Converters, John Wiley & Sons, Incorporated, 2016. Chapter 12, pp. 196–198.
74. Oleksandr Husev, Carlos Roncero-Clemente, Enrique Romero-Cadaval, Dmitri Vinnikov, Tanel Jalakas: "Three-level three-phase quasi-Z-source neutral-point-clamped inverter with novel modulation technique for photovoltaic application", Electric Power Systems Research, Elsevier, 2016, pp. 10–21.
75. Takahashi, I. and Noguchi, T.: "A new quick-response and high efficiency control strategy of an induction motor". IEEE Trans. Ind. Appl, IA-22(5), 1986, pp. 820–827.
76. Depenbrock, M: "Direct self-control (DSC) of inverter-fed induction motors", IEEE Trans. Power Electronics, 3(4), 1988, pp. 420–429.
77. Haitham Abu-Rub et al.: "High Performance Control of AC Drives with Matlab/Simulink Models", John Wiley and Sons, 2012, Chapter 5, pp. 171–199.
78. P. Vas and P. Tiitinen: "Sensorless Vector and Direct Torque Controlled Drives" in M.H. Rashid [Editor]: "Power Electronics Handbook', Chapter 28, pp. 743–766, Academic Press, 2001.
79. Zadeh, L. A., "Fuzzy sets," Information and Control, Vol. 8, pp. 338–353, 1965.
80. The Mathworks Inc., www.mathworks.com: "MATLAB/SIMULINK", version 9.14/10.7 and "Fuzzy Logic Toolbox" Version 3.1, R2023a, May 25, 2023.
81. Bimal K. Bose: "Expert System, Fuzzy Logic, and Neural Networks in Power Electronics and Drives" in B.K. Bose [Editor]: "Power Electronics and Variable Frequency Drives: Technology and Applications", Chapter 11, IEEE Press, 1997.
82. Bimal K. Bose: "Modern Power Electronics and AC Drives", Chapter 11, Prentice Hall PTR, 2002,
83. Bimal K. Bose: "Expert System, Fuzzy Logic, and Neural Network Applications in Power Electronics and Motion Control", Proceedings of the IEEE. Vol. 82, No. 8, August 1994, pp. 1308–1314.
84. Ahmed Rubaai et al.: "Fuzzy Logic Applications in Electrical Drives and Power Electronics" in M.H. Rashid [Editor]: "Power Electronics Handbook: Devices, Circuits and Applications", Elsevier, 2011, Chapter 37, pp. 1115–1120 and pp. 1126–1137.
85. Ahmed Rubaai: "Fuzzy Logic in Electric Drives" in M.H. Rashid [Editor]: "Power Electronics Handbook", Academic Press, 2001, pp. 779–790.

Chapter 2
Advanced Carrier-Based Pulse-Width Modulation of Three-Phase Two-Level and Multilevel Inverters

2.1 Introduction

Many industrial applications require voltage control of inverters to maintain constant output voltage irrespective of variations in the input dc voltage and for constant voltage and frequency control requirements. This is achieved by carrier-based pulse-width modulation (PWM) technique. For three-phase inverters, the commonly used PWM techniques are (1) sine PWM (SPWM), (2) third harmonic injection PWM (THIPWM), (3) harmonic injection PWM (HIPWM), (4) modified sine PWM (MSPWM), (5) dead-band sine PWM (DBSPWM) and so on. The SPWM is the fundamental or basic PWM technique, and its model is presented in the first chapter. This chapter provides the modelling of the HI, THI, MS and DBS PWM techniques and in addition presents the model of a newly discovered PWM known as clipped sinusoid or clipped sine PWM (CSPWM) technique. Also model for multi-carrier sine level shift PWM (MCSLSPWM) three-phase diode-clamped three-level inverter (DCTLI) is presented. Case studies for multi-carrier clipped sine (MCCS) and multi-carrier third harmonic injection sine (MCTHIS) level shift PWM (LSPWM) three-phase DCTLI and MCCS and MCTHIS phase shift PWM (PSPWM) three-phase cascade H-bridge inverter (CHBI) are also presented. Modular multilevel converters (MMCs) using strings of submodules (SM) built using either half H-bridge or full H-bridge semiconductor switches are used in high-voltage DC (HVDC) transmission and in medium-voltage drives (MVD). Detailed analysis of modulation technique and model for a three-phase MMC is presented. Case study of a single-phase MMC with submodule capacitor voltage balance is presented. In both the MMC models, half H-bridge topology is used.

Supplementary Information The online version contains supplementary material available at https://doi.org/10.1007/978-3-031-62784-2_2.

N. P. R. Iyer, *Inverters and AC Drives*, Power Systems,
https://doi.org/10.1007/978-3-031-62784-2_2

2.2 Three-Phase Third Harmonic Injection PWM Inverter

To increase the output line-to-line voltage and to reduce its third harmonic components in a three-phase inverter, in comparison with sine PWM inverter, THIPWM technique was proposed [1, 2]. In THIPWM, third harmonic component having appropriate amplitude is added to the fundamental sine reference waveform. The reference modulating signal waveform is given below [1, 2, 4–7]:

$$y = 1.15^{*} \sin \ (\omega.t) + 0.19^{*} \sin \ (3.\omega.t) \tag{2.1}$$

Using the above PWM techniques, it is possible to achieve up to 15% increase in the line-to-line voltage of the inverter without over modulation, pulse dropping and without any material effect on the harmonic content of the line-to-line waveform, as compared to a conventional sine PWM inverter [1, 2].

2.2.1 Modelling of Three-Phase Third Harmonic Injection PWM Inverter

The model of the three-phase THIPWM inverter is shown in Fig. 2.1 (Model file: EXAMPLE 2_1). The modulating signal y defined in Eq. 2.1 is generated using Embedded MATLAB function. This is shown in program segment 2.1 in the model file EXAMPLE 2_1. In program segment 2.1, k1, k2, f and vm are, respectively, 1.15, 0.19, 50 Hz and 8 V, respectively, where f and vm represent the frequency and amplitude of the third harmonic sine wave modulating signal. The triangle generator along with function block Fcn2 generates a triangle carrier with amplitude vc (+/− 10 V) and frequency fc Hz. The output of embedded MATLAB function vthr, vthy and vthb is compared with the triangle carrier using three relational operator comparator blocks. The output of respective relational operator blocks and the inverted outputs using Fcn blocks form the PWM gate drive for the three-phase inverter. In Fig. 2.1, Universal Bridge subsystem forms the three-phase inverter. The two DC sources with midpoint grounded form the DC link. The line-to-neutral output voltage is formed by matrix transformation of line-to-line output voltage using Fcn blocks.

2.2.2 Simulation Results

The simulation of the three-phase THIPWM inverter was carried out using SIMULINK [3]. The ode15s (stiff/NDF) solver was used. The model parameters are shown in Table 2.1. In Table 2.1, Amplitude Modulation Index (A.M. Index) Ma is the ratio of vm/vc, and Frequency Modulation Index (F.M. Index) Mf is the ratio

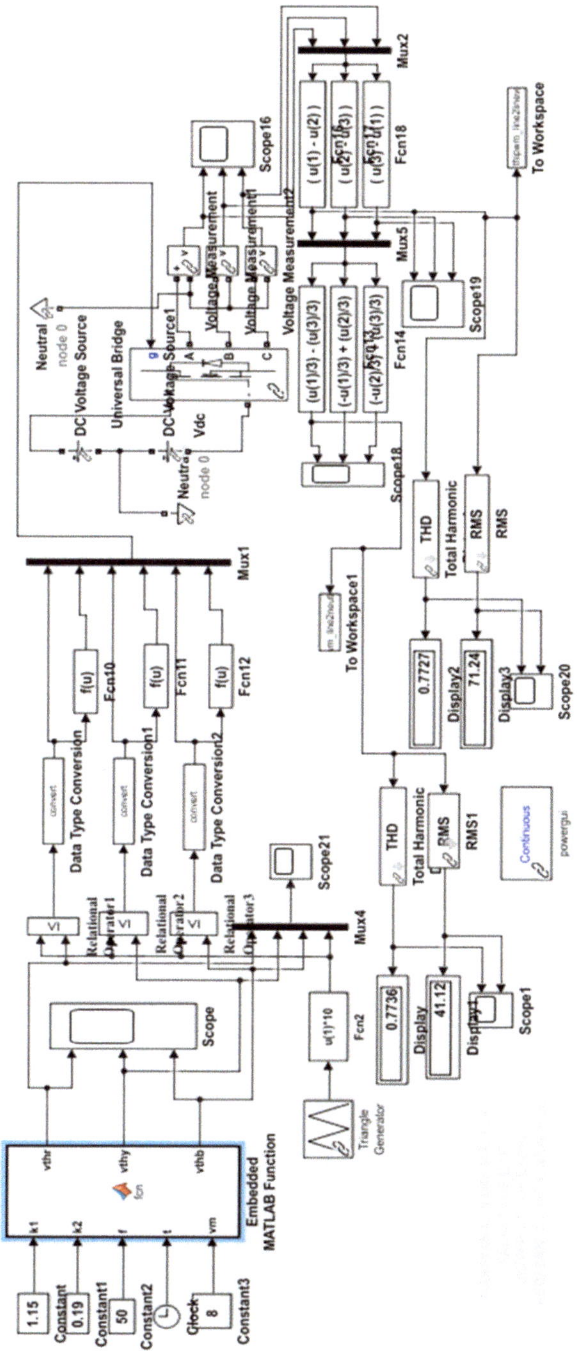

Fig. 2.1 Three-phase third harmonic injection PWM inverter

Table 2.1 Model parameters

Parameter	Value	Unit
DC link voltage	100	Volts
Modulating sine wave signal frequency	50	Hz
Triangle carrier frequency	1200	Hz
Amplitude Modulation Index Ma	0.2, 0.4, 0.6, 0.8, 1.2, 1.4, 1.6	–
Frequency Modulation Index Mf	24	–

Table 2.2 Three-phase THIPWM inverter simulation results

Sl. No	Ma	VLL (rms) V	VLL1 (rms) V	VLN (rms) V	VLN1 (rms) V	THD of VLL	THD of VLN	Remarks
1	0.2	31.74	15.38	18.33	7.931	2.715	2.714	Under modulation
2	0.4	45.12	24.53	26.05	14.19	1.773	1.773	–
3	0.6	62.18	42.93	35.89	24.82	1.047	1.047	–
4	0.8	71.41	56.98	41.22	32.78	0.7459	0.7484	–
5	1.2	81.45	73.45	47.00	42.32	0.4556	0.4565	Over modulation
6	1.4	81.63	75.15	47.15	43.36	0.4185	0.4194	
7	1.6	81.62	75.51	47.13	43.74	0.3858	0.3863	

of fc/fm. Simulation results are tabulated in Table 2.2. The simulation results of line-to-line and line-to-neutral voltage for Ma of 0.8 and Mf of 24 p.u. are also shown in Fig. 2.2a–j. The harmonic spectrum of VRY and VRN is shown in Figs. 2.3 and 2.4. The three-phase third harmonic sine wave modulating signal and triangle carrier are shown in Fig. 2.5. Simulation results are tabulated in Table 2.2.

2.3 Three-Phase Harmonic Injection PWM Inverter

The HIPWM is a variation of the THIPWM technique discussed in Sect. 2.2. In HIPWM, additional harmonics apart from third harmonics are added to the sine reference waveform. The modulating reference waveform is given below [2, 5]:

$$y = 1.15^{*}\ \sin(\omega.t) + 0.27^{*} \sin\ (3.\omega.t) - 0.029^{*} \sin\ (9.\omega.t) \tag{2.2}$$

The resulting flat-topped waveform allows over modulation while improving the frequency spectra of ac terms and dc terms waveforms [2]. The modulating three-phase sine wave signals with frequency component fm, 3*fm and 9*fm are multiplied by appropriate constants and added with proper signs to generate the modulating reference waveform given by Eq. 2.2 for the three phases and are compared with a triangular carrier wave in a comparator and the resulting pulse-width-

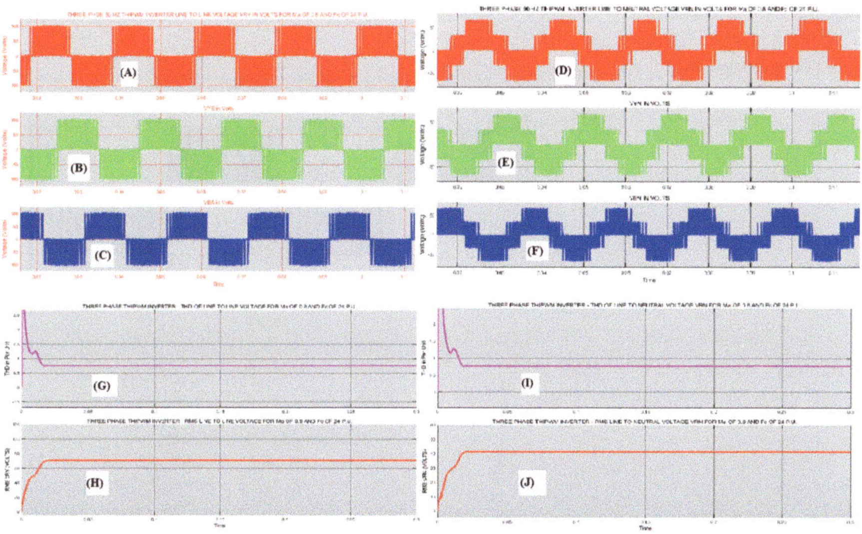

Fig. 2.2 Three-phase THIPWM inverter simulation results. (**a**), (**b**), (**c**) Line-to-line voltages VRY, VYB and VBR. (**d**), (**e**), (**f**) Line-to-neutral voltages VRN, VYN, VBN. (**g**), (**h**) THD and RMS value of VRY. (**i**), (**j**) THD and RMS value of VRN

modulated gate drive for each phase, and their respective inverted gate pulses are used to drive the upper and lower arms of semiconductor switches in the three arms of the inverter.

2.3.1 *Modelling of Three-Phase Harmonic Injection PWM Inverter*

The model of the three-phase HIPWM inverter is shown in Fig. 2.6 (Model file: EXAMPLE 2_2). The modulating signal y defined in Eq. 2.2 is generated using Embedded MATLAB function. This is shown in Program segment 2.2 in the model file EXAMPLE 2_2. In program segment 2.2, $k1$, $k2$, $k3$, f and vm are, respectively, 1.15, 0.27, 0.029, 50 Hz and 8 V, respectively, where f and vm represent the frequency and amplitude of the third harmonic sine wave modulating signal. The triangle generator along with function block Fcn2 generates a triangle carrier with amplitude +vc/-vc (+/−10 V) and frequency fc Hz. The remaining part of the model is the same as explained in Sect. 2.2.1.

2.3.2 *Simulation Results*

The simulation of the three-phase HIPWM inverter was carried out using SIMULINK [3]. The ode15s (stiff/NDF) solver was used. The model parameters are shown in Table 2.1. Simulation results are tabulated in Table 2.3. The simulation

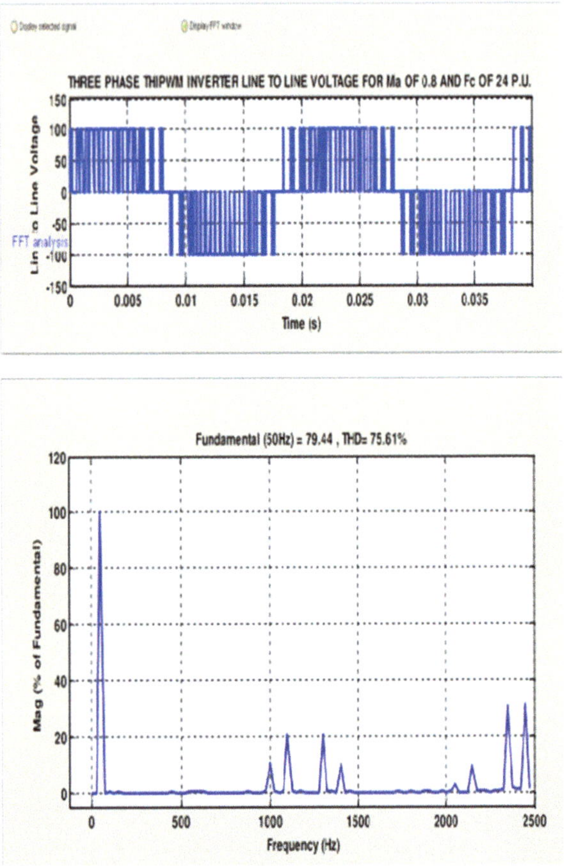

Fig. 2.3 Three-phase THIPWM inverter line-to-line voltage (top) and harmonic spectrum (bottom)

results of line-to-line and line-to-neutral voltage for Ma of 0.8 and Fc 24 p.u. are also shown in Fig. 2.7a–j. The harmonic spectrum of VRY and VRN is shown in Figs. 2.8 and 2.9. The three-phase harmonic sine wave modulating signal and triangle carrier are shown in Fig. 2.10.

2.4 Three-Phase Clipped Sine PWM Inverter

This section describes the discovery of the new PWM technique known as three-phase CSPWM technique [4–6]. This is explained below:

In this method, the modulating three-phase sine wave AC signal at the inverter switching frequency fm Hertz is clipped at desired voltage levels using zener diode clippers with breakdown voltages +Vclip and –Vclip placed back to back in series. The resulting three-phase AC clipped sine wave at the frequency fm Hertz is compared with triangular carrier wave using op.amp. comparators in each of the

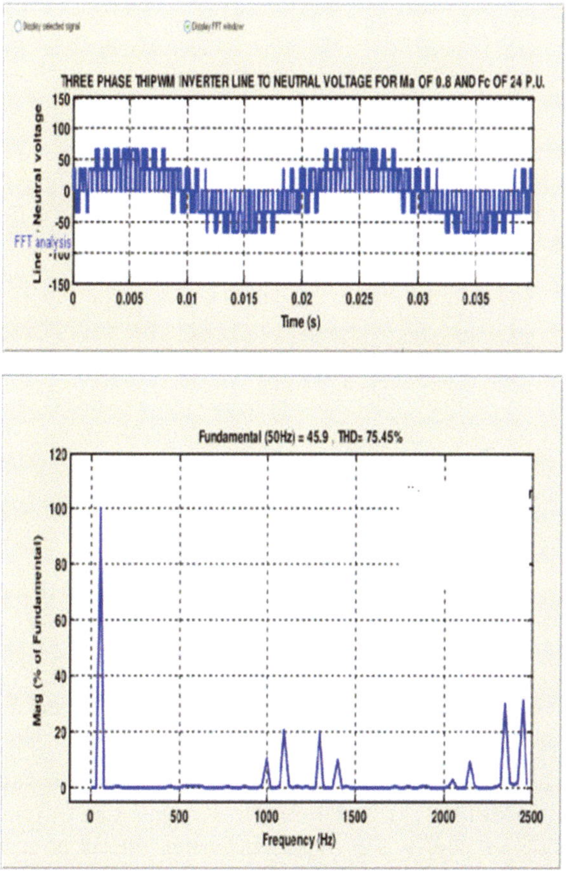

Fig. 2.4 Three-phase THIPWM inverter line-to-neutral output voltage (top) and harmonic spectrum (bottom)

three phases. The resulting PWM gate pulse for each of the three phases and their respective inverted gate pulse drive the respective upper and lower semiconductor switches in each of the three phases. The peak value of the clipped sine wave Vclip is measured at the centre of the half cycle, i.e. at the interval Tm/4 where Tm is the period 1/fm s. The three-phase CSPWM inverter gate drive model is shown in Fig. 2.11 (Model file: CSPWM_INV_GATE_DRIVE). Three-phase sine wave AC signal of suitable amplitude and frequency fm Hz in each phase is clipped using back-to-back Zener diodes to a value +Vclip and –Vclip. This is then compared with a triangle carrier wave of peak value +Vc/-Vc and frequency fc Hz using op. amp comparators. The a.m. index Ma is Vclip/Vc and f.m. index Fc is fc/fm. In Fig. 2.11, the triangle carrier is generated using op. amp.integrator, summer, gain multiplier and DC voltage source. The component level gate drive model simulation results are shown in Fig. 2.12a–f and in Fig. 2.13a–d.

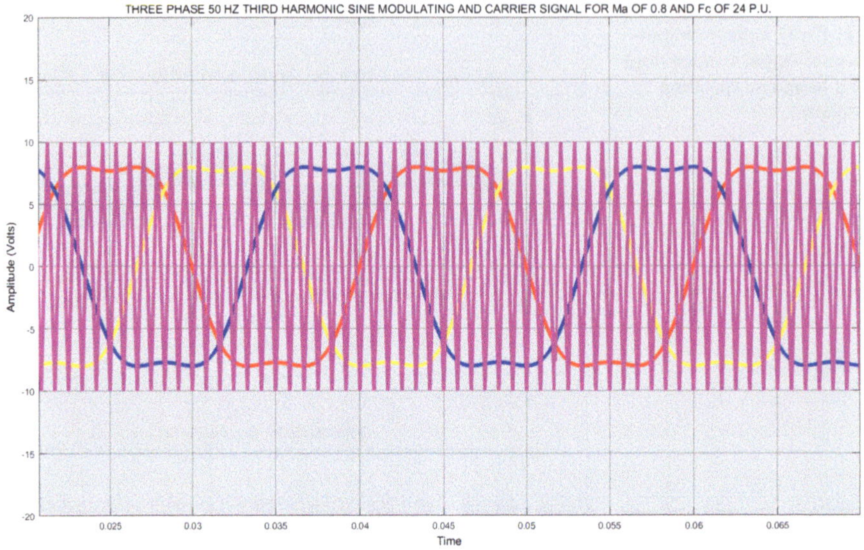

Fig. 2.5 Three-phase THIPWM inverter—three-phase 50 Hz third harmonic sine wave modulating signal and triangle carrier

2.4.1 Modelling of Three-Phase Clipped Sine PWM Inverter

The model of the three-phase CSPWM inverter is shown in Fig. 2.14 (Model file: EXAMPLE 2_3). The three-phase clipped sine wave modulating signal with amplitude +v_clip/-v_clip V and frequency fm Hz is generated using two Embedded MATLAB functions. These are, respectively, marked "Three Phase Sine Wave AC Generator" and "Three Phase Clipped Sine wave AC Generator". The three-phase sine wave used to generate clipped sine wave has an amplitude vm V and frequency fm Hz. The source codes for generating the three-phase sine wave and clipped sine wave AC generator are given in Program segment 2.3 and 2.4 in the model file EXAMPLE 2_3. The triangle generator along with function block Fcn generates a triangle carrier with amplitude +vc/-vc (+/-10 V) and frequency fc Hz. In this case, Ma is v_clip/vc, and Fc is fc/fm. The remaining part of the model is the same as explained in Sect. 2.2.1.

2.4.2 Simulation Results

The simulation of the three-phase CSPWM inverter was carried out using SIMULINK [3]. The ode15s (stiff/NDF) solver was used. The model parameters are shown in Table 2.1. Simulation results are tabulated in Table 2.4. The simulation results of line-to-line and line-to-neutral voltage for Ma of 0.8 and Fc 24 p.u. are also

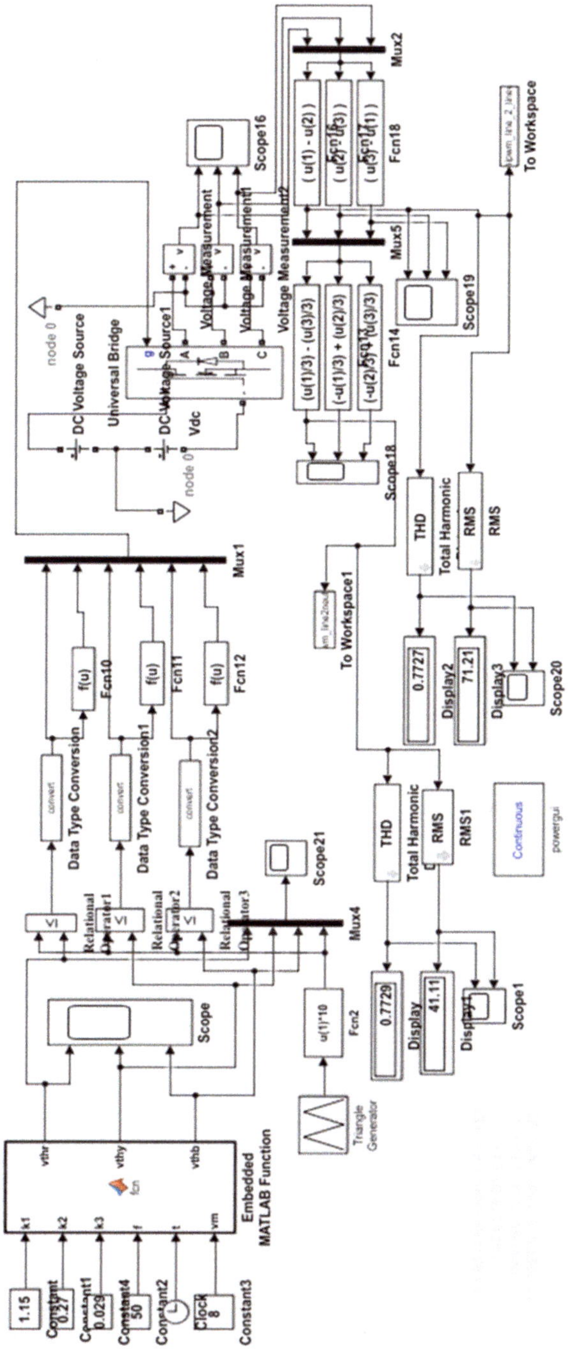

Fig. 2.6 Three-phase harmonic injection PWM inverter

Table 2.3 Three-phase HIPWM inverter simulation results

Sl. No	Ma	VLL (rms) V	VLL1 (rms) V	VLN (rms) V	VLN1 (rms) V	THD of VLL	THD of VLN	Remarks
1	0.2	35.62	14.07	20.57	8.11	2.319	2.323	Under modulation
2	0.4	50.37	28.24	29.08	16.20	1.481	1.483	–
3	0.6	61.68	42.45	35.61	24.52	1.063	1.063	–
4	0.8	71.21	56.31	41.11	32.52	0.7727	0.7729	–
5	1.2	81.65	75.11	47.14	43.39	0.4252	0.4256	Over modulation
6	1.4	81.65	76.02	47.14	43.90	0.3926	0.3933	–
7	1.6	81.65	76.38	47.14	44.07	0.3784	0.3789	–

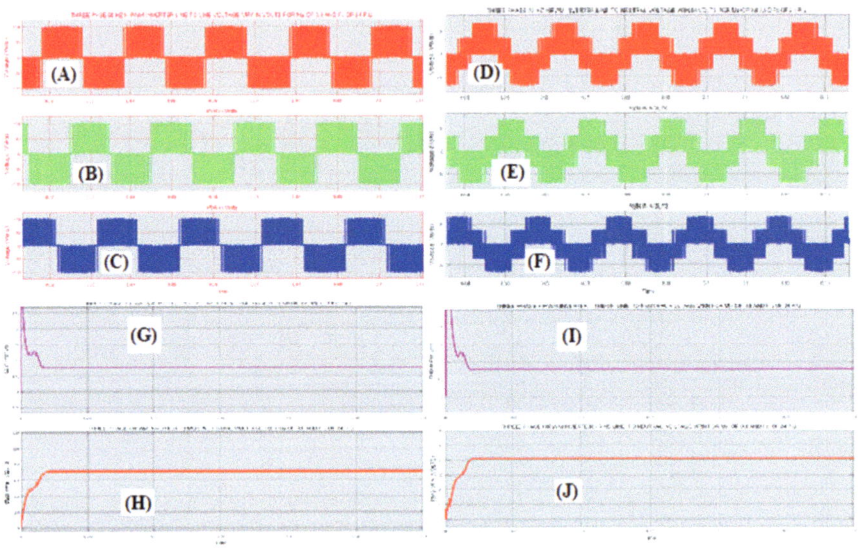

Fig. 2.7 Three-phase HIPWM inverter simulation results—(**a**), (**b**), (**c**) Three-phase line-to-line voltage. (**d**), (**e**), (**f**) Three-phase line-to-neutral voltage. (**g**), (**h**) THD and RMS value of line-to-line voltage. (**i**), (**j**) THD and RMS value of line-to-neutral voltage

shown in Fig. 2.15a–j. The harmonic spectrum of VRY and VRN are shown in Figs. 2.16 and 2.17. The three-phase clipped sine wave modulating signal and triangle carrier are shown in Fig. 2.18.

2.5 Three-Phase Dead-Band Sine PWM Inverter

Reference modulating waveform continuity is not a required condition for the implementation of switching pattern for three-phase PWM inverters if the load is star connected. Dead-band refers to the discontinuities introduced in the reference

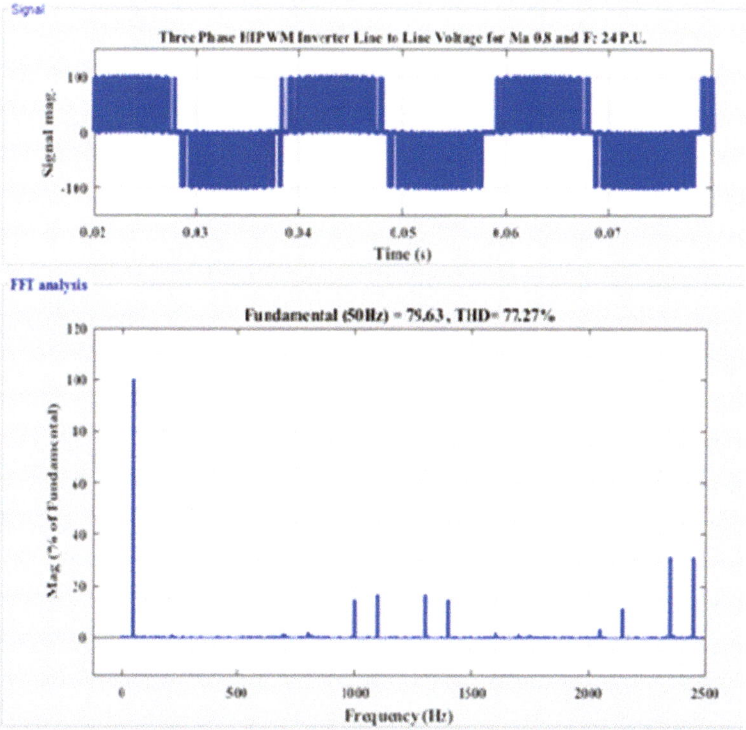

Fig. 2.8 Three-phase HI PWM inverter line-to-line voltage (top) and harmonic spectrum (bottom)

modulating waveform which do not degrade the quality of the output voltage waveform. Dead-band PWM improves output voltage waveforms and provides better performance for inverters compared to that with non-dead-band or continuous PWM technique. Switching frequency can be reduced by avoiding intersection of modulating reference and carrier waveforms for an interval of time. This minimises switching losses and reduces lower order harmonics in the output voltage compared to continuous PWM technique [6, 7]. For the dead-band sine PWM inverter, the modulating reference signal is defined below:

$$f_{1a}(\omega.t) = \left.\begin{array}{c} M_a * \sin\left(\omega.t + \dfrac{\pi}{3}\right) \; for \; 0 \le \omega.t \le \dfrac{\pi}{3} \\ 1 \; for \; \dfrac{\pi}{3} \le \omega.t \le \dfrac{2\pi}{3} \\ M_a * \sin(\omega.t) \; for \; \dfrac{2\pi}{3} \le \omega.t \le \pi \end{array}\right] \tag{2.3}$$

In Eq. 2.3, Ma is the ratio vm/vc where vm and vc are the amplitude of modulating sine and triangle carrier wave, respectively, and ω is (2π*fm). Amplitude of vc is +/−1 V.

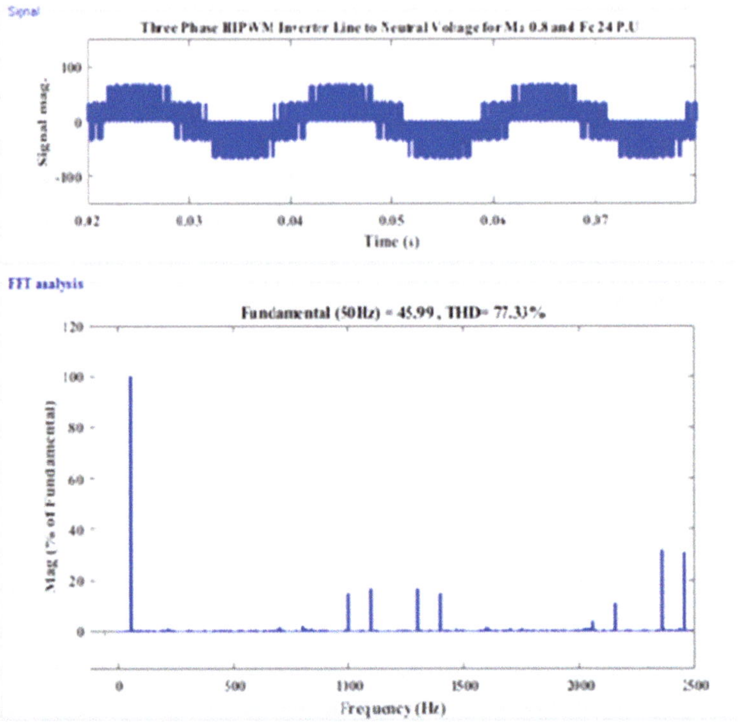

Fig. 2.9 Three-phase HIPWM inverter line-to-neutral voltage (top) and harmonic spectrum (bottom)

2.5.1 *Modelling of Three-Phase Dead-Band Sine PWM Inverter*

The model of the three-phase dead-band sine PWM inverter is shown in Fig. 2.19 (Model file: EXAMPLE 2_4). The model gate drive defined in Eq. 2.3 is developed using Embedded MATLAB function. The source code to develop the gate drive is shown in Program segments 2.5, 2.6 and 2.7 in the model file EXAMPLE 2_4. In Program segment 2.5, input k is one and fm is 50 Hz. Output y1 takes values 1–6 corresponding to every 60-degree interval of the modulating sine wave. The y2 and y3 are sine and cosine functions, and sector switch functions ssf1 to ssf6 are defined in Program segment 2.6. For example, ssf1 is 1 if y1 or sector takes value 1 and is 0 otherwise. Similar definition holds good for ssf2 to ssf6. Modulation sine functions defined in Eq. 2.3 are developed using Program segment 2.7. Here the inputs are Ma, fm, ssf1 to ssf6 and time t. Outputs are the modulation sine functions for phase a, b and c, marked f1a, f1b and f1c. If sector y1 takes vale 1 or 4, ssf1 or ssf4 is 1 and f1a takes value ma*sin(2*pi*fm*t + pi/(3)). When sector is 2 or 5, ssf2 or ssf5 is 1, and f1a takes value +1 or − 1. When sector is 3 or 6, ssf3 or ssf6 is 1 and f1a value is

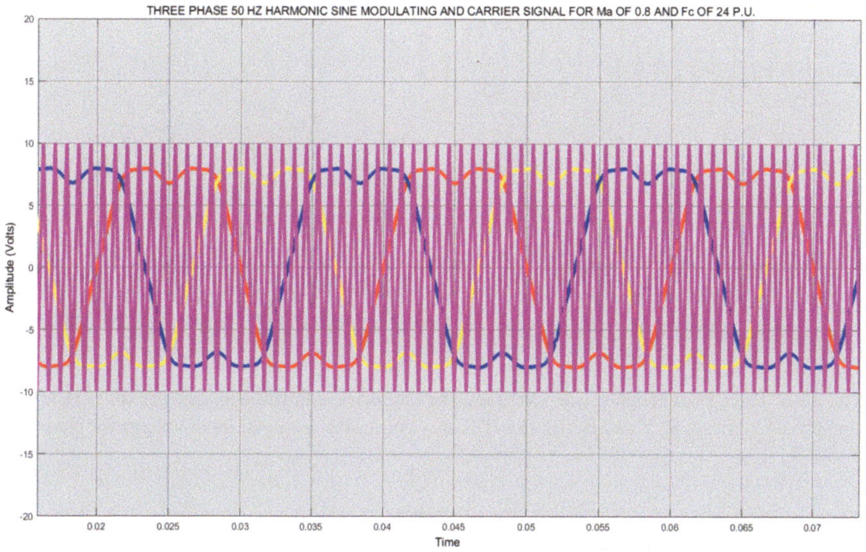

Fig. 2.10 Three-phase HIPWM inverter: three-phase 50 Hz harmonic sine wave modulating signal and triangle carrier

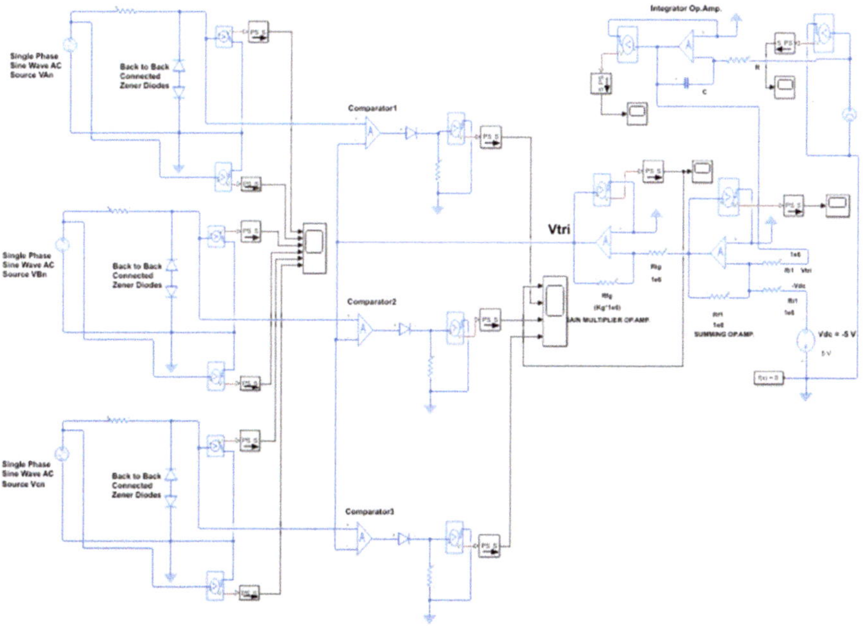

Fig. 2.11 Three-phase CSPWM inverter gate pulse generator

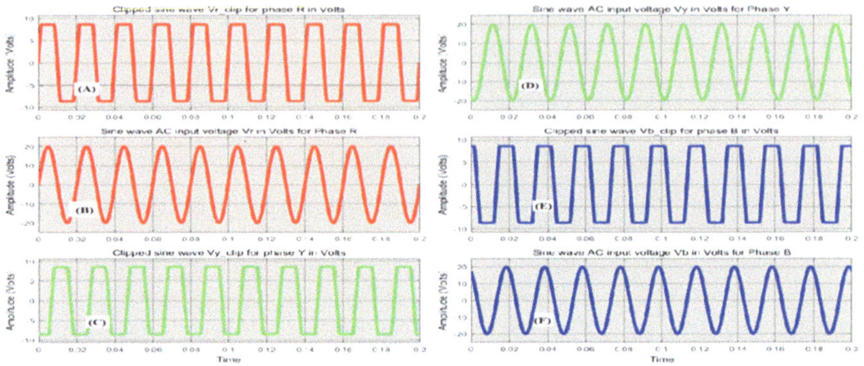

Fig. 2.12 Three-phase clipped sine PWM inverter gate drive simulation results. (**a**), (**c**), (**e**) Clipped sine wave for phases R, Y and B. (**b**), (**d**), (**f**) Three-phase sine wave AC input voltage for phases R, Y and B

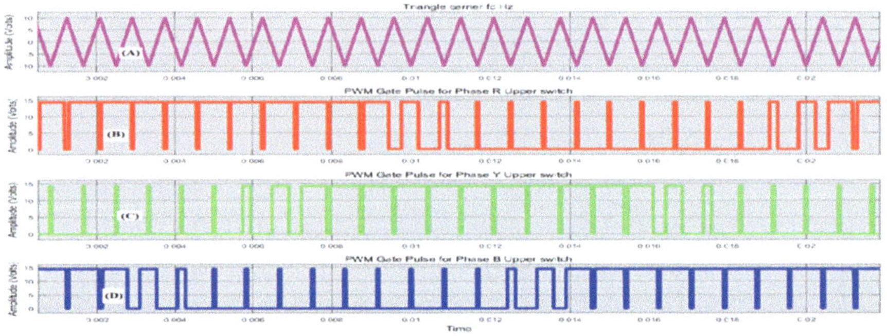

Fig. 2.13 Three-phase clipped sinusoid PWM inverter: triangle carrier wave and gate pulse for upper switch in phases R, Y and B (top to bottom)

ma*sin(2*pi*fm*t). Similar arguments hold good for f1b and f1c. Here a saw-tooth carrier having a frequency fc (1200) Hz having a peak value +1 V and minimum value zero for the period 1/(2*fm) of the modulating sine wave positive half and − 1 V and zero during the remainder period 1/(2*fm) of the modulating sine wave negative half is used. This carrier wave is generated using sawtooth generator block with frequency fc (1200) Hz and amplitude +/−1 V along with Fcn block having multiplication constant 0.5*(1-u(1)), pulse generator blocks 1 and 2, multiplier product blocks 1, 2 and an adder block. The pulse generator 1 and 2 blocks have amplitude +1 and − 1 V, frequency fm (50) Hz and phase of zero and 180 degrees (10e-3 Sec), respectively. This carrier wave is compared with f1a, f1b and fic using relational operator comparator blocks. The respective gate pulse output for each phase is then inverted using Fcn blocks using the relation (1-u(1)). This forms the gate drive for the Universal Bridge block which forms the inverter. The remaining part is the same as explained in Sect. 2.2.1.

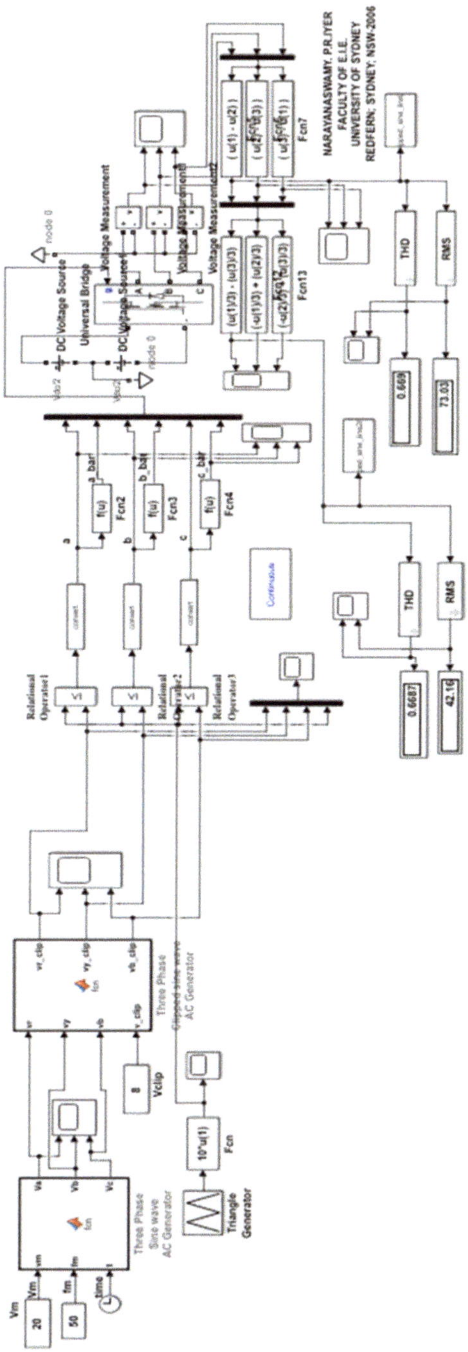

Fig. 2.14 Three-phase clipped sinusoid PWM inverter

Table 2.4 Three-phase CSPWM inverter simulation results

Sl. No	Ma	VLL (rms) V	VLL1 (rms) V	VLN (rms) V	VLN1 (rms) V	THD of VLL	THD of VLN	Remarks
1	0.2	36.51	15.56	21.08	8.97	2.115	2.124	Under modulation
2	0.4	51.64	31.00	29.81	17.9	1.33	1.335	–
3	0.6	63.25	46.04	36.51	26.58	0.9392	0.939	–
4	0.8	73.03	60.69	42.16	35.05	0.669	0.6687	–
5	1.2	81.65	74.67	47.14	43.12	0.4379	0.4426	Over modulation
6	1.4	81.65	74.67	47.14	43.12	0.4379	0.4426	–
7	1.6	81.65	74.67	47.14	43.12	0.4379	0.4426	–

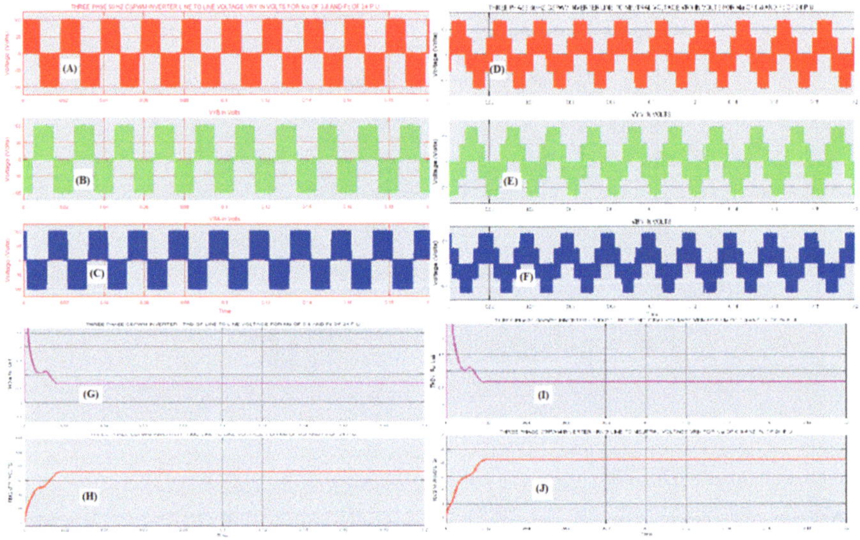

Fig. 2.15 Three-phase CSPWM inverter simulation results: (**a**), (**b**), (**c**) line-to-line voltage. (**d**), (**e**), (**f**) line-to-neutral voltage. (**g**), (**h**) THD and RMS value of line-to-line voltage. (**i**), (**j**) THD and RMS value of line-to-neutral voltage

2.5.2 *Simulation Results*

The simulation of the three-phase dead-band sine PWM inverter was carried out using ode23tb (stiff/TR-BDF2) solver in SIMULINK [3]. The simulation parameters are the same as in Table 2.1. Simulation results are tabulated in Table 2.5. The simulation results of line-to-line and line-to-neutral voltage for Ma of 0.8 and Fc 24 p.u. are also shown in Fig. 2.20a–j. The harmonic spectrum of VRY and VRN is shown in Figs. 2.21 and 2.22. The three-phase dead-band sine PWM modulating signal and triangle carrier are shown in Fig. 2.23.

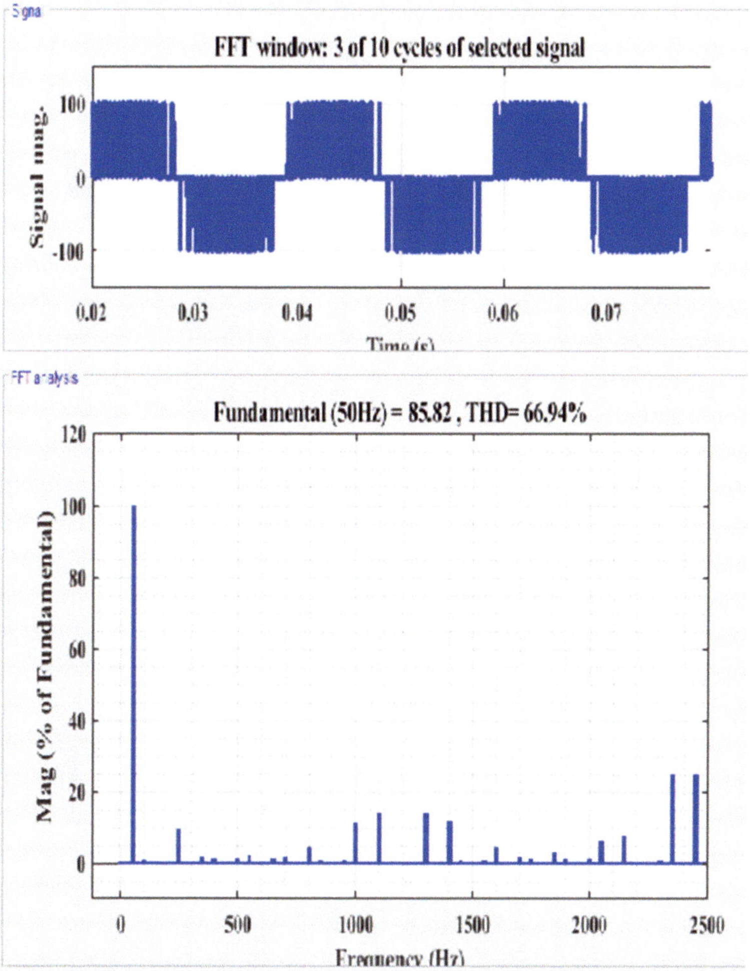

Fig. 2.16 Three-phase CSPWM inverter line-to-line voltage (top) and harmonic spectrum (bottom)

2.6 Three-Phase Modified Sine PWM Inverter

For the modified sine PWM inverter, the modulating reference signal waveform is defined below [6, 7]:

$$f_{1a}(\omega.t) = \begin{bmatrix} M_a * \sin(\omega.t)\ for\ 0 \le \omega.t \le \dfrac{\pi}{3} \\ 1\ for\ \dfrac{\pi}{3} \le \omega.t \le \dfrac{2\pi}{3} \\ M_a * \sin(\omega.t)\ for\ \dfrac{2\pi}{3} \le \omega.t \le \pi \end{bmatrix} \quad (2.4)$$

In Eq. 2.4, Ma and ω are as defined in Sect. 2.5. Amplitude of vc is +/−1 V.

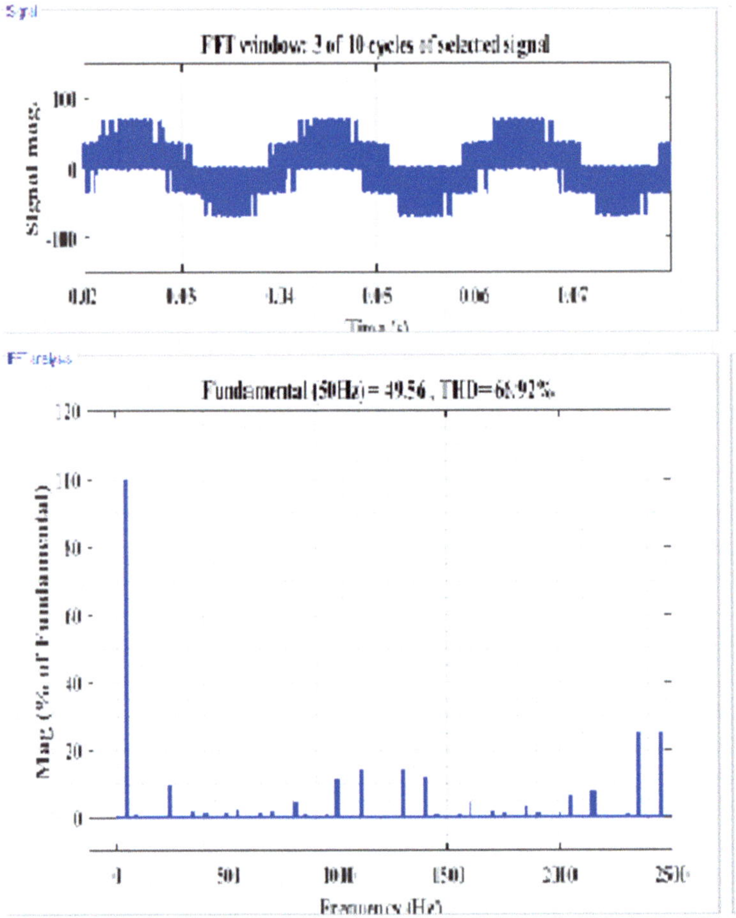

Fig. 2.17 Three-phase CSPWM inverter line-to-neutral voltage (top) and harmonic spectrum (bottom)

2.6.1 Modelling of Three-Phase Modified Sine PWM Inverter

The model of the three-phase modified sine PWM inverter is shown in Fig. 2.24 (Model file: EXAMPLE 2_5). The model gate drive defined in Eq. 2.4 is developed using Embedded MATLAB function. The source code to develop the gate drive is shown in program segments 2.5, 2.6 and 2.8 in the model file EXAMPLE 2_4. Program segments 2.5 and 2.6 are the same as presented in Sect. 2.5.1. In program segment 2.7, the modulation function f1a for the (ω.t) interval from 0 to $\pi/3$ is replaced with Ma*sin(2*pi*fm*t). Similar changes apply to f1b and f1c. The remaining part of the model is the same as explained in Sect. 2.5.1.

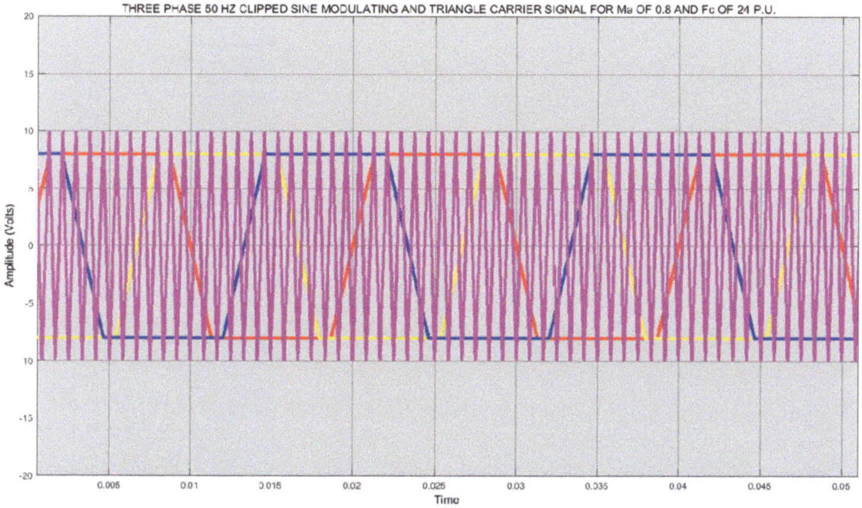

Fig. 2.18 Three-phase CSPWM inverter—three-phase 50 Hz clipped sine modulating signal and triangle carrier

2.6.2 Simulation Results

The simulation of the three-phase modified sine PWM inverter was carried out using ode23tb (stiff/TR-BDF2) solver in SIMULINK [3]. The simulation parameters are the same as in Table 2.1. Simulation results are tabulated in Table 2.6. The simulation results of line-to-line and line-to-neutral voltage for Ma of 0.8 and Fc 24 p.u. are also shown in Fig. 2.25a–j. The harmonic spectrum of VRY and VRN is shown in Figs. 2.26 and 2.27. The three-phase modified sine PWM modulating signal and triangle carrier are shown in Fig. 2.28.

2.7 Discussion of Results

Models for advanced PWM technique for three-phase inverter are presented in this chapter. Clipped sine PWM technique was newly proposed [4–6]. Graphs showing THD of line-to-line and line-to-neutral voltage of three-phase inverter by the various PWM techniques such as sine, THI, HI, clipped sine, dead-band sine and modified sine are shown in Figs. 2.29 and 2.30, respectively. Results for sine PWM are obtained from Table 1.2 of Chap. 1. Results indicate that clipped sine PWM has a lower THD for line-to-line and line-to-neutral voltages as compared to that for sine, THI and HI PWM throughout the under modulation region and in the over modulation region up to an A.M. Index of 1.2. Dead-band sine PWM has lower THD for

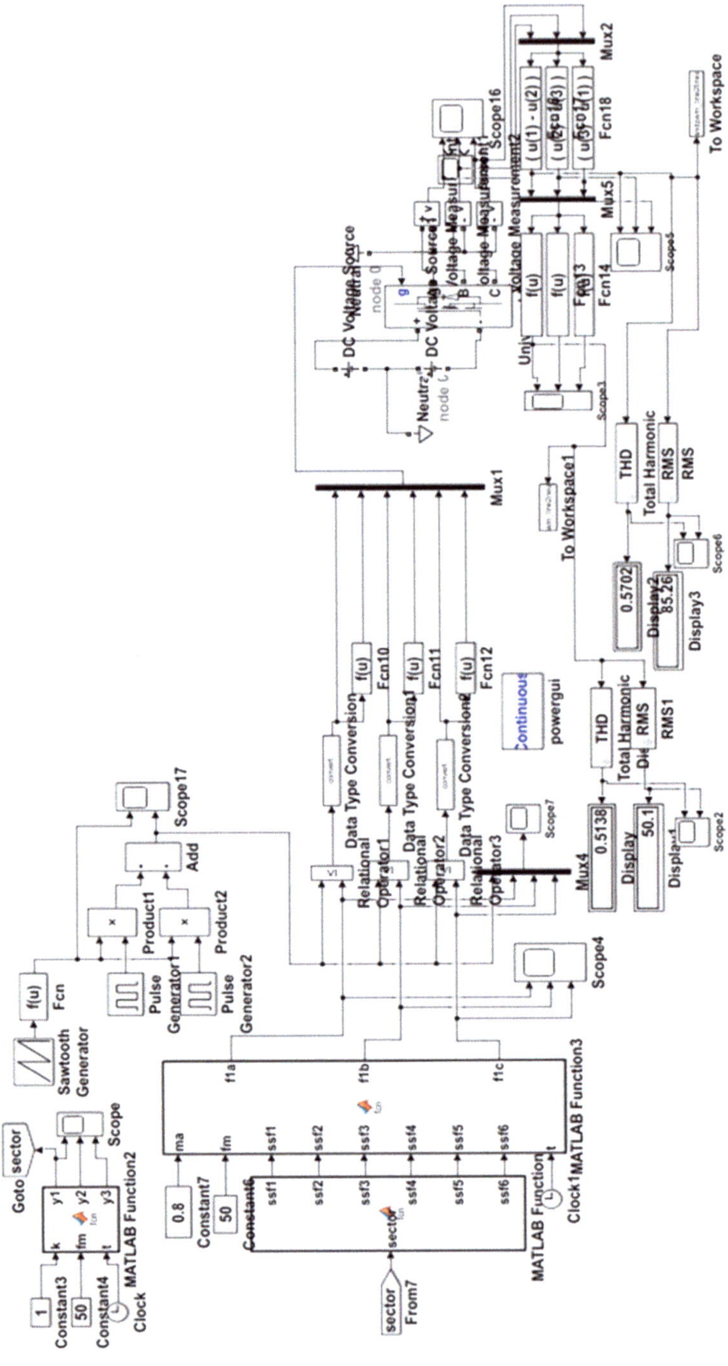

Fig. 2.19 Model of three-phase dead-band sine PWM inverter

Table 2.5 Three-phase DBSPWM inverter simulation results

Sl. No	Ma	VLL (rms) V	VLL1 (rms) V	VLN (rms) V	VLN1 (rms) V	THD of VLL	THD of VLN	Remarks
1	0.2	65.96	50.00	41.83	33.33	0.861	0.76	Under modulation
2	0.4	73.16	56.96	44.86	36.95	0.806	0.69	–
3	0.6	79.55	65.18	47.61	40.74	0.7	0.605	–
4	0.8	85.26	74.07	50.1	44.56	0.57	0.514	–
5	1.2	90.15	84.71	52.05	48.92	0.364	0.364	Over modulation
6	1.4	88.8	84.31	51.26	48.75	0.33	0.325	–
7	1.6	88.12	83.90	50.87	48.50	0.321	0.3165	–

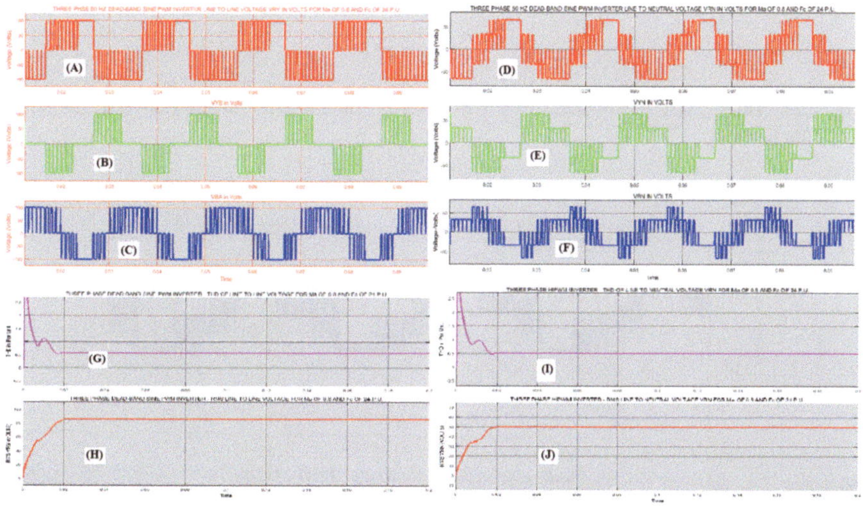

Fig. 2.20 (**a**) to (**j**): Three-phase dead-band sine PWM inverter simulation results. (**a**), (**b**), (**c**) Three-phase line-to-line voltage. (**d**), (**e**), (**f**) Three-phase line–to-neutral voltage. (**g**), (**h**) THD and RMS value of line-to-line voltage. (**i**), (**j**) THD and RMS value of Line-to-Neutral Voltage

both line-to-line and line-to-neutral voltages throughout the under and over modulation regions, whereas for modified sine PWM up to an A.M. index of 0.8 compared to clipped sine PWM.

2.8 Carrier-Based PWM Techniques for Three-Phase Diode-Clamped Three-Level Inverter

The methods used for pulse-width modulation (PWM) of conventional two-level inverters can be extended for multilevel converters [8–10]. The schematic of a three-phase diode-clamped three-level inverter (DCTLI) is shown in Fig. 2.31. The three

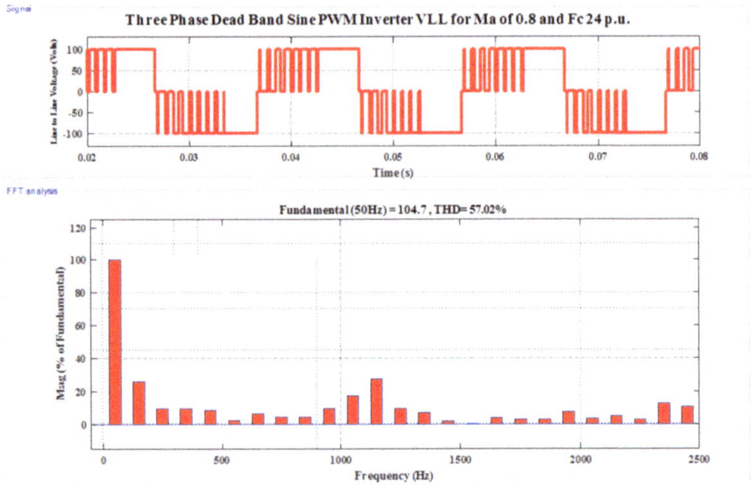

Fig. 2.21 Three-phase dead-band sine PWM inverter—line-to-line voltage (top) and harmonic spectrum (bottom)

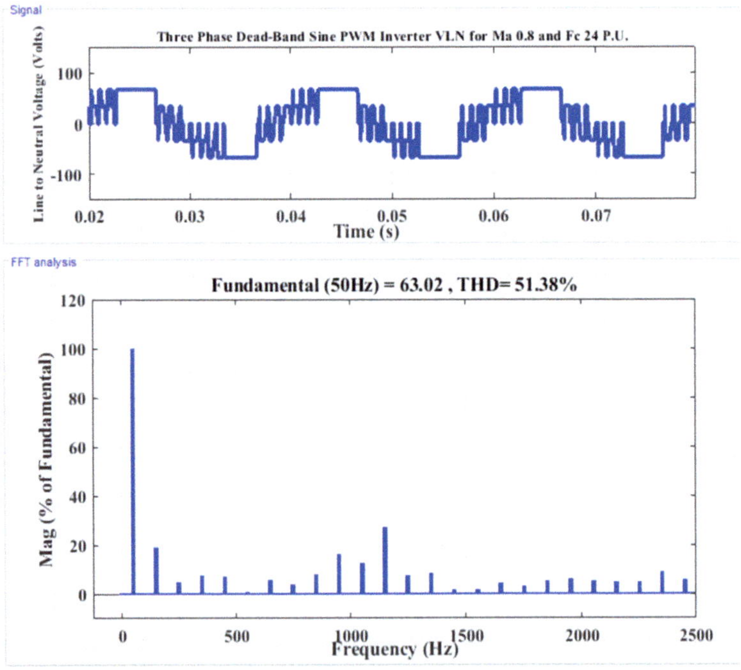

Fig. 2.22 Three-Phase Dead-Band Sine PWM Inverter—Line-to-Neutral Voltage (Top) and Harmonic Spectrum (Bottom)

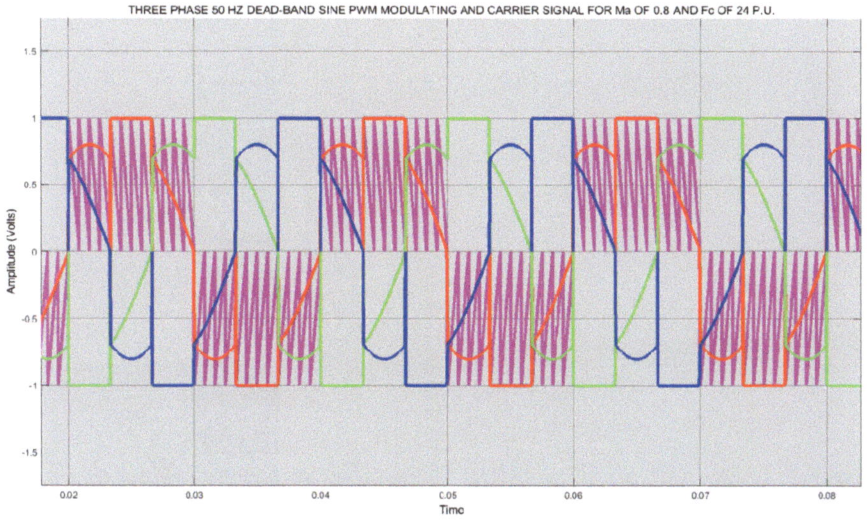

Fig. 2.23 Three-phase dead-band sine PWM inverter: three-phase 50 Hz dead-band sine modulating signal and triangle carrier

PWM methods used for multilevel converter can be classified as (1) multilevel carrier-based sine PWM or multicarrier sine PWM, (2) selective harmonic elimination PWM and (3) multilevel space vector PWM.

Multicarrier-based sine PWM method can be classified into two groups: (1) multicarrier sine phase shift PWM (MSPSPWM) and (2) multicarrier sine level shift PWM (MSLSPWM). In this method, depending on the voltage level of the multilevel converter, triangle carriers having same peak value Ac and carrier frequency fc Hz are either shifted in phase or shifted in voltage level and then compared with a sine wave reference signal or modulating signal whose peak value is Am and frequency fm Hz and its zero centered at the middle of the triangle carrier set. The reference sine wave signal is compared with each of the triangle carrier signal. If the triangle carrier signal is less than sine reference signal, then the switch corresponding to that carrier is turned ON and if the triangle carrier signal is greater than sine reference signal, the relevant switch is turned OFF. In multilevel converters having line-to-ground output voltage levels m, the amplitude modulation and frequency modulation indices m_a and m_f are defined below [8, 9]:

$$M_a = \frac{A_m}{(m-1) * A_c} \tag{2.5}$$

$$M_f = \frac{f_c}{f_m} \tag{2.6}$$

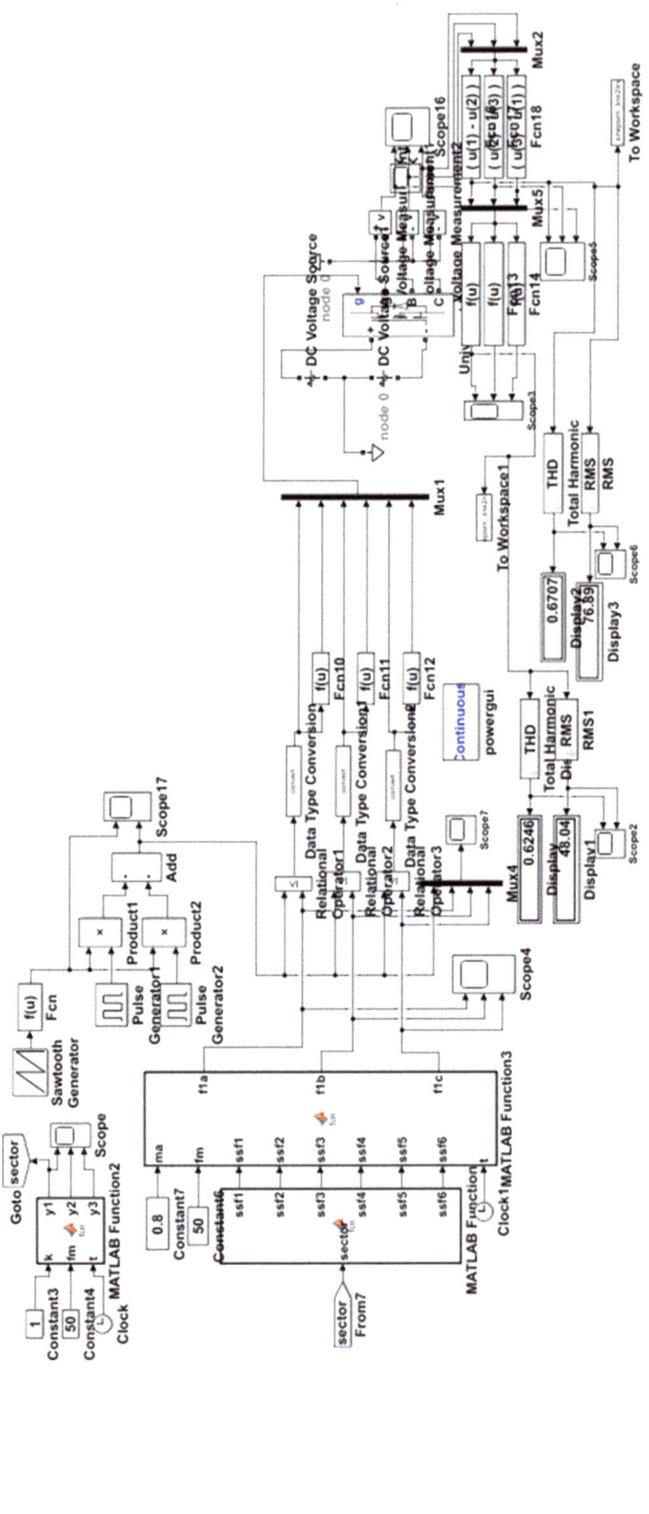

Fig. 2.24 Model of three-phase modified sine PWM inverter

Table 2.6 Three phase MSPWM inverter simulation results

Sl. No	Ma	VLL (rms) V	VLL1 (rms) V	VLN (rms) V	VLN1 (rms) V	THD of VLL	THD of VLN	Remarks
1	0.2	63.29	49.44	41.03	32.76	0.7994	0.7545	Under modulation
2	0.4	54.17	56.96	43.46	35.43	0.7669	0.71	–
3	0.6	72.76	59.01	45.79	38.11	0.721	0.6676	–
4	0.8	76.89	63.89	48.04	40.77	0.671	0.6246	–
5	1.2	84.06	73.27	52.15	46.09	0.5627	0.5295	Over modulation
6	1.4	86.4	76.57	53.2	47.71	0.5225	0.4989	–
7	1.6	86.71	77.78	53.22	48.12	0.4929	0.4728	–

where A_m and f_m are the peak-to-peak amplitude and frequency of the modulating sine wave and A_c and f_c are the amplitude and frequency of the triangle carrier. System level models for MSLSPWM generator are reported in the literature [11]. In this chapter, model for MSLSPWM is presented using triangle carrier generator, operational amplifier (Op.Amp) summers, sign changers, comparators and digital logic gates for a three-phase diode-clamped three-level inverter (DCTLI) [3]. Simulation results are presented.

2.8.1 Model of Three-Phase Multi-carrier Sine Level Shift PWM Diode-Clamped Three-Level Inverter

The model of the MSLSPWM three-phase DCTLI is shown in Fig. 2.32 (Model file: EXAMPLE 2_6). The various subsystems are shown in Fig. 2.33a–c. The parameters shown in Table 2.1 are used to develop the model. Here triangle carrier generator with peak value +/−1 volt with frequency ftri (1200 Hz) from control and measurements/pulse and signal generators library from specialised technology block set is used along with operational amplifier (Op.Amp.) summers, sign changers and comparators to develop four-level shifted triangle carriers, each level shifted by one volt with peak value Ac (1 volt and frequency ftri 1200 Hz). These triangle carriers are compared using Op.Amp. comparators with a three-phase modulating sine wave voltage with frequency fm (50) Hz and peak value Am Volt to generate PWM gate pulse for the switches of three-phase DCTLI. The value of Am lies within the range 0 to 2 volts for under modulation.

Figure 2.33a shows the model of the triangle carrier generator, op.amp summers and sign changers to create eight different patterns of triangle carriers each having a peak value of 1 volt and frequency ftri (1200 Hz), each level shifted and phase shifted by 1 volt and 180 degrees, respectively, with respect to one another. Referring to Fig. 2.33a, the triangle generator generates Vg_tri with peak value +/−1 volt and frequency ftri (1200 Hz).

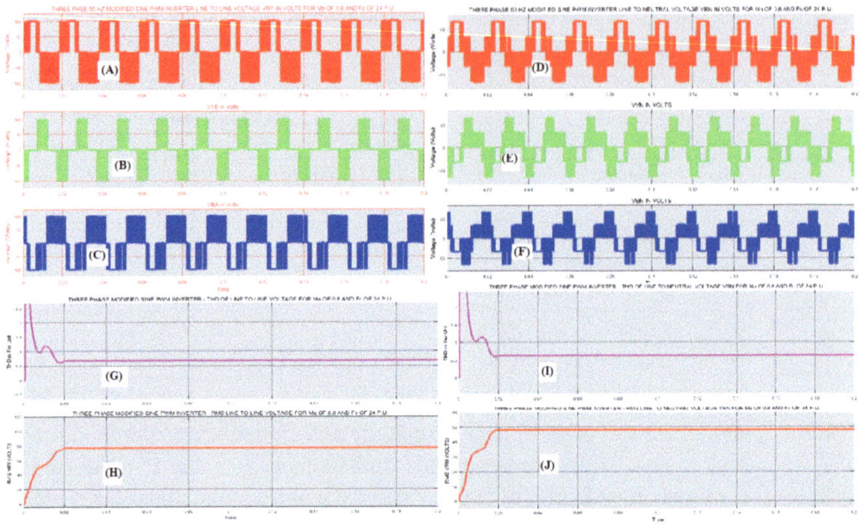

Fig. 2.25 (**a**) to (**j**) Three-phase modified sine PWM inverter simulation results. (**a**), (**b**), (**c**) Line-to-line voltage. (**d**), (**e**), (**f**) Line-to-neutral voltage. (**g**), (**h**) THD and RMS value of line-to-line voltage. (**i**), (**j**) THD and RMS value of line-to-neutral voltage

The output Vtri2 of summing amplifier1 can be expressed as follows:

$$V_{\text{tri2}} = V_{\text{ref1}} * \left(\frac{R_7}{R_6 + R_7}\right) * \left[1 + \frac{R_2}{R_1}\right] - V_{g\_\text{tri}} * \left(\frac{R_2}{R_1}\right) \tag{2.7}$$

Summing amplifier 2, 3 and 4 output can be, respectively, expressed as follows:

$$V_{\text{tri1}} = V_{\text{ref2}} * \left(\frac{R_{14}}{R_{13} + R_{14}}\right) * \left[1 + \frac{R_4}{R_3}\right] - V_{\text{tri2}} * \left(\frac{R_4}{R_3}\right) \tag{2.8}$$

$$V_{\text{tri7}} = -V_{\text{ref3}} * \left[1 + \frac{R_{20}}{R_{21}}\right] - V_{\text{tri1}} * \left(\frac{R_{20}}{R_{21}}\right) \tag{2.9}$$

$$V_{\text{tri8}} = V_{\text{ref4}} * \left[1 + \frac{R_{22}}{R_{23}}\right] - V_{\text{tri2}} * \left(\frac{R_{22}}{R_{23}}\right) \tag{2.10}$$

Sign changer 1, summing amplifier 5 and sign changer 6 output, respectively, can be expressed as follows:

$$V_{\text{tri3}} = -V_{\text{tri2}} * \left(\frac{R_8}{R_5}\right) \tag{2.11}$$

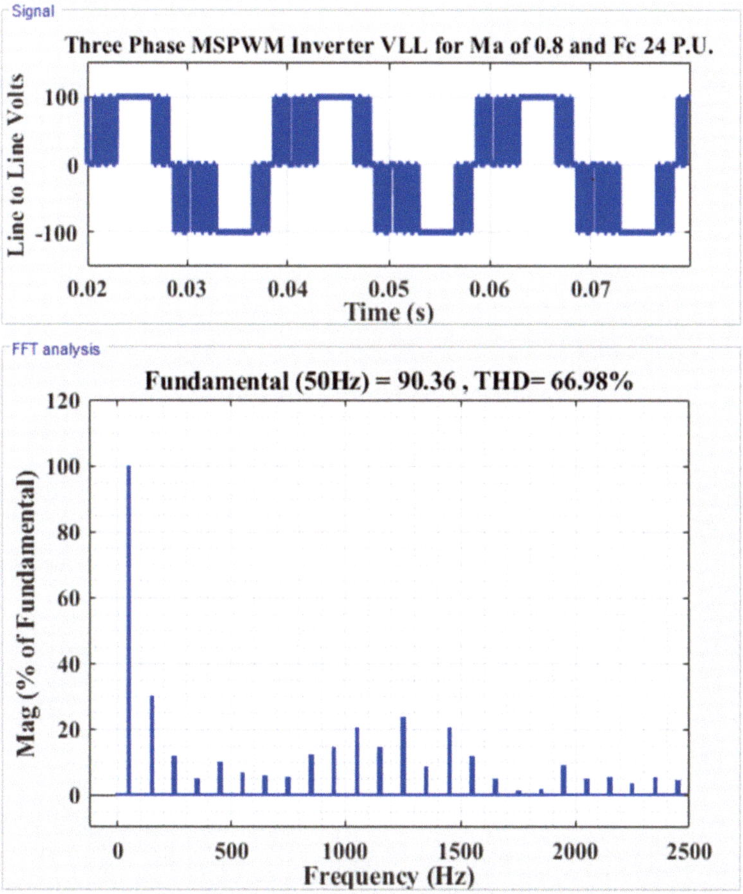

Fig. 2.26 Three-phase modified sine PWM inverter—line-to-line voltage (top) and harmonic spectrum (bottom)

$$V_{\text{tri4}} = V_{\text{ref5}} * \left[1 + \frac{R_{10}}{R_9}\right] - V_{\text{tri3}} * \left(\frac{R_{10}}{R_9}\right) \tag{2.12}$$

$$V_{\text{tri6}} = -V_{\text{tri4}} * \left(\frac{R_{11}}{R_{12}}\right) \tag{2.13}$$

Sign changer 3 output which is $-V_{\text{ref5}}*[R_{17}/R_{18}]$ is given to the non-inverting pin of summing amplifier 6 whose output can be expressed as follows:

$$V_{\text{tri5}} = -V_{\text{ref5}} * \left(\frac{R_{17}}{R_{18}}\right) * \left[1 + \frac{R_{15}}{R_{16}}\right] - V_{\text{tri3}} * \left(\frac{R_{15}}{R_{16}}\right) \tag{2.14}$$

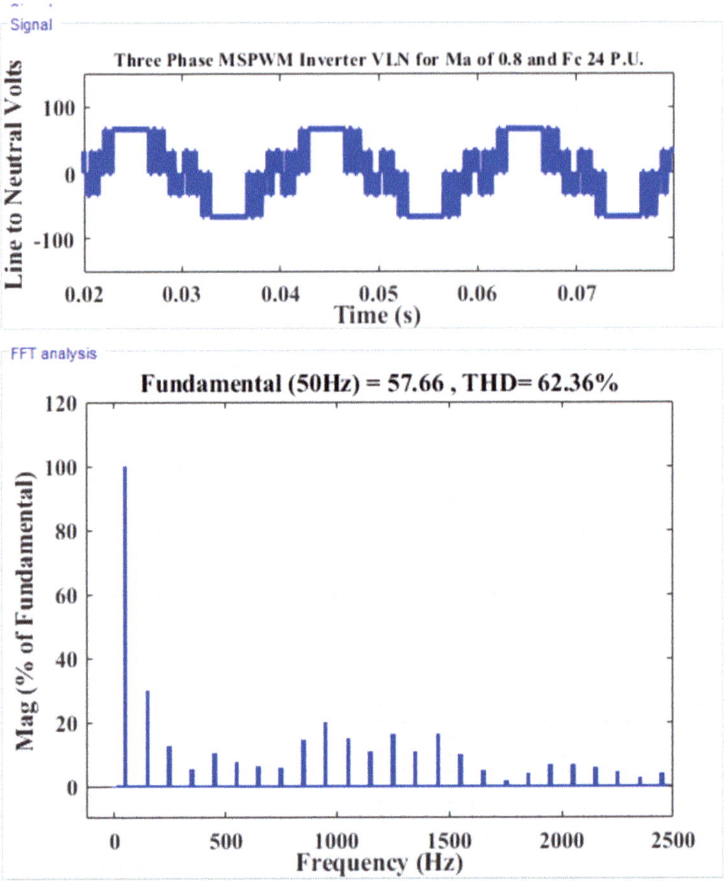

Fig. 2.27 Three-phase modified sine PWM inverter—line-to-neutral voltage (top) and harmonic spectrum (bottom)

The simulation results for the eight triangle carrier are shown in Fig. 2.34. The results are tabulated in Table 2.7. Figure 2.33b shows the three-phase sine wave modulating signal, eight triangle carriers, the four selected triangle carrier Vg_pwm1 to Vg_pwm4 as per Table 2.7, Op.Amp. comparators and NOT gates to generate PWM gate drives for the switches of three-phase DCTLI.

For the three-phase DCTLI shown in Fig. 2.31, when the gate pulse vg_s1A and vg_s2A for switches S1A and S2A are HIGH (logic 1) and the gate pulse vg_s3A and vg_s4A for switches S3A and S4A are LOW (logic 0), output voltage VAG is +Vdc/2 volts. Similarly when vg_s2A and vg_s3A are HIGH and vg_s1A and vg_s4A are LOW, VAG is zero volts. When vg_s3A and vg_s4A are HIGH and vg_s1A and vg_s2A are LOW, then VAG is –Vdc/2 volts. Thus, it can be seen that vg_s3A is (~vg_s1A) and vg_s4A is (~vg_s2A) where ~sign represents the logical NOT operation. This principle is used here to develop the PWM gate pulse for the switches of three-phase DCTLI.

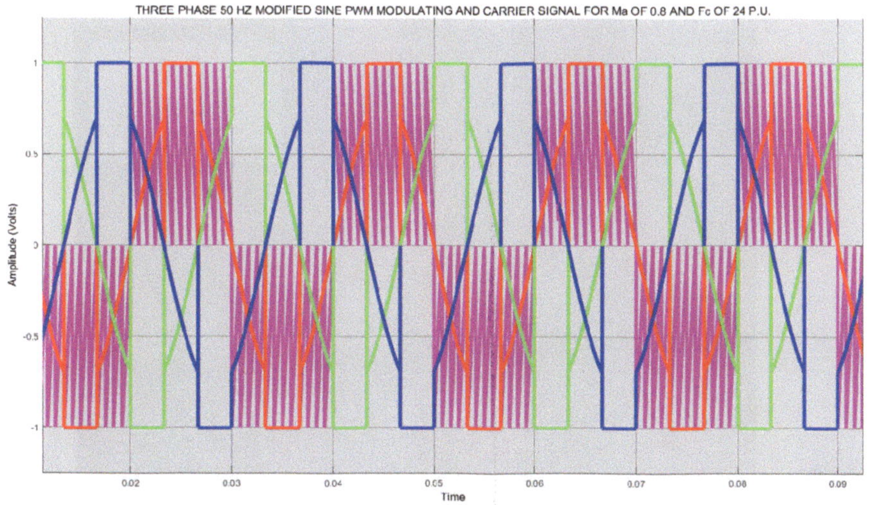

Fig. 2.28 Three-phase modified sine PWM inverter—Three-phase 50 Hz modified sine modulating signal and triangle carrier

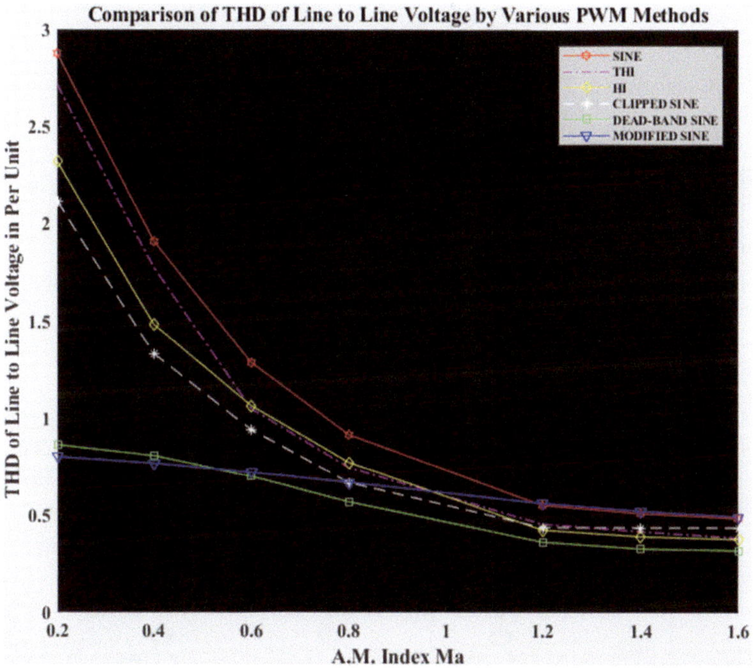

Fig. 2.29 Comparison of THD of line-to-line voltage of three-phase inverter by various PWM methods

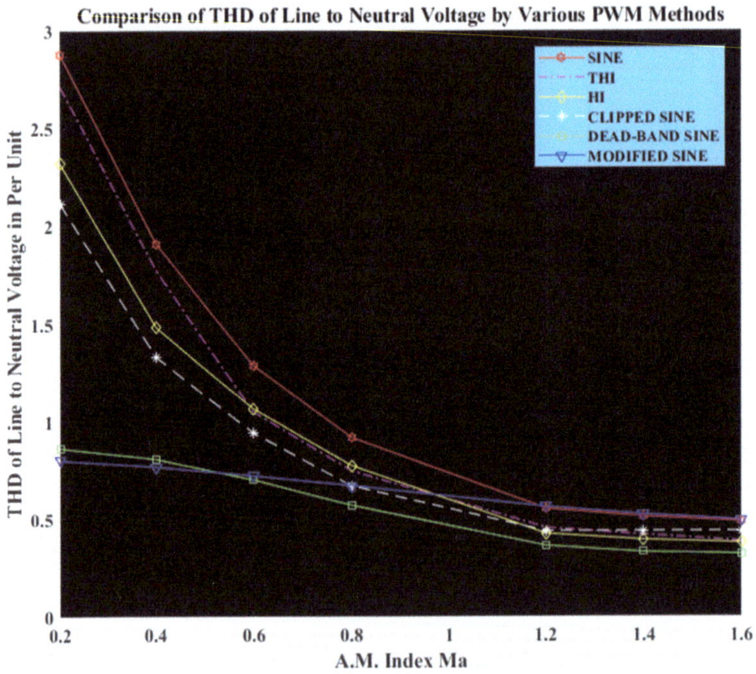

Fig. 2.30 Comparison of THD of line-to-neutral voltage of three-phase inverter by various PWM methods

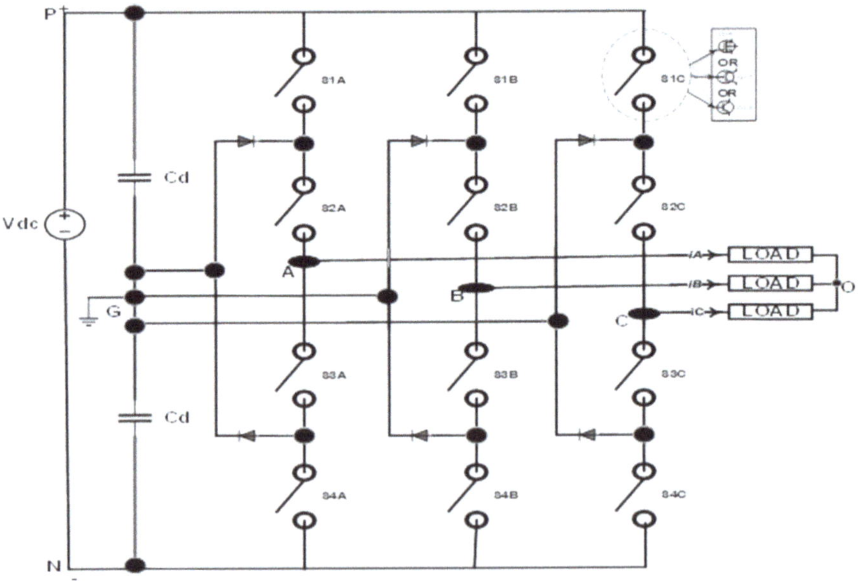

Fig. 2.31 Three-phase diode clamped three-level inverter

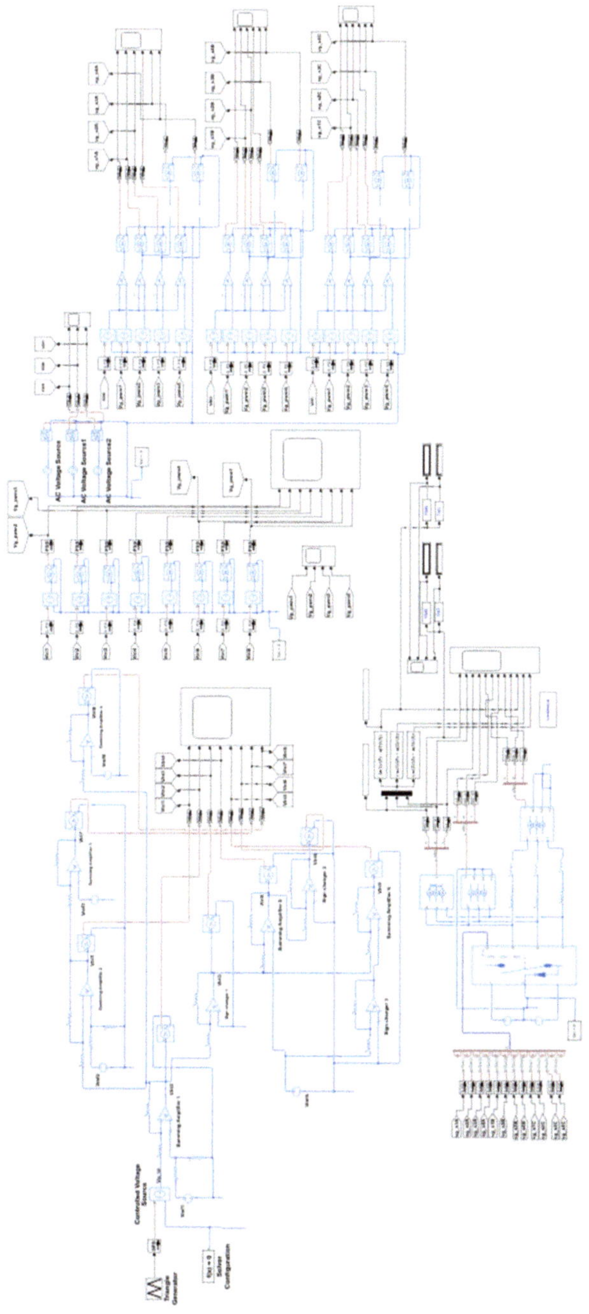

Fig. 2.32 Model of MSLSPWM three-phase diode clamped three-level inverter

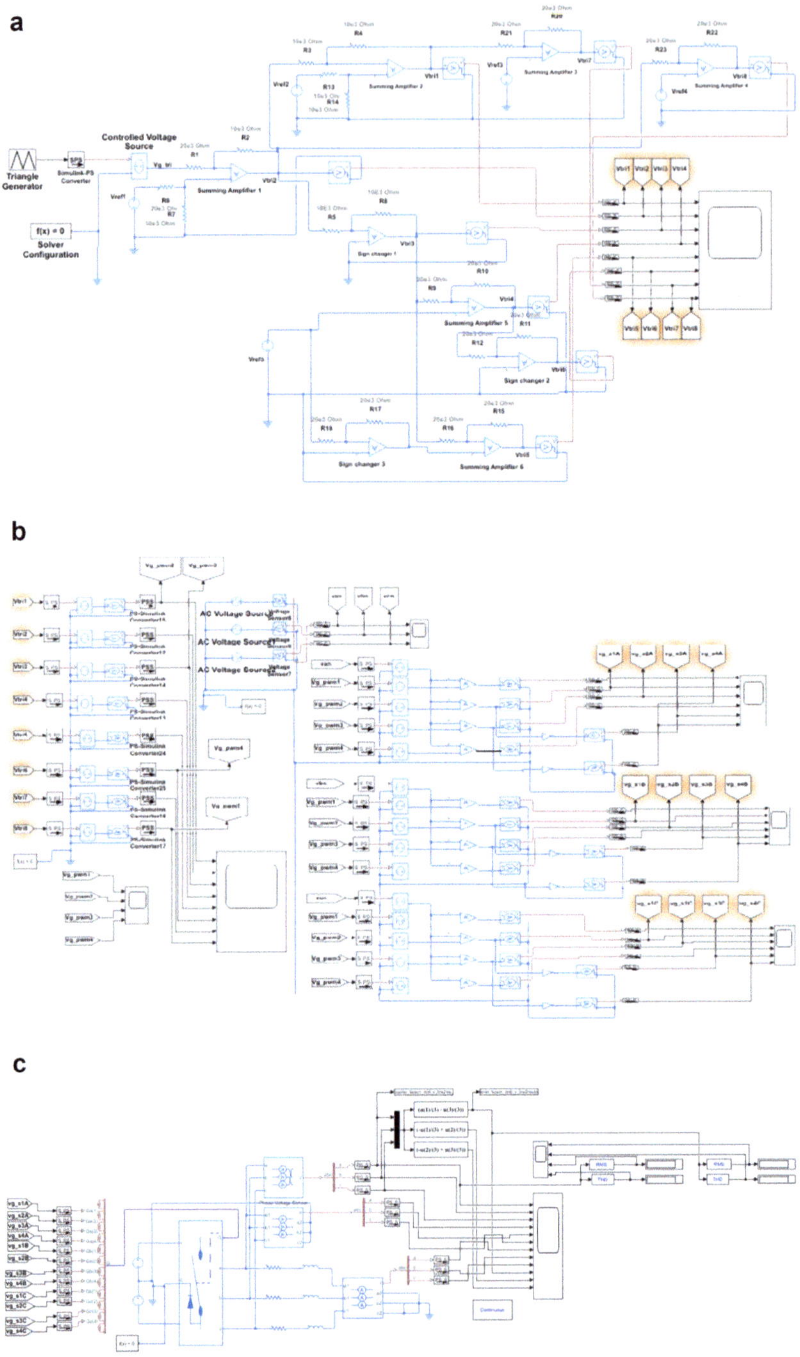

Fig. 2.33 (**a**) Triangle carrier generator, operational amplifier summers and sign changers. (**b**) Three-phase sine wave modulating signal, triangle carrier and operational amplifier comparators for gate pulse Generation. (**c**) Three-phase diode-clamped three-level inverter

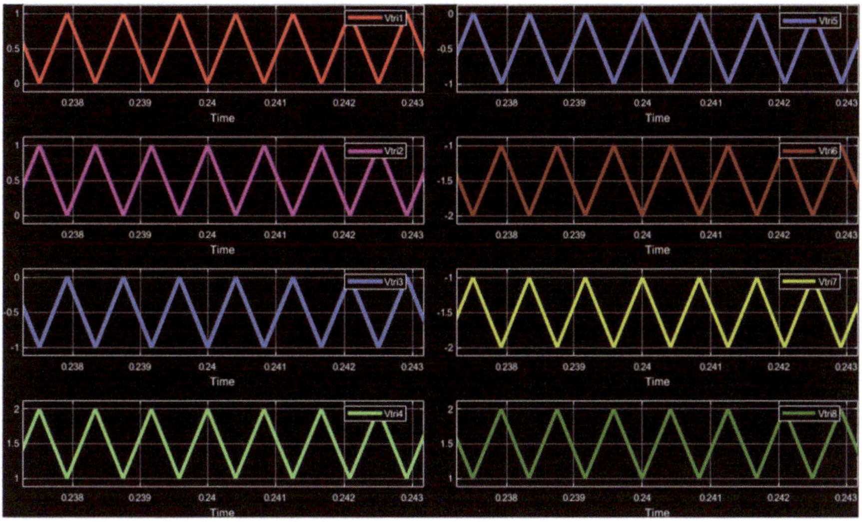

Fig. 2.34 Three-phase MCSLSPWM DCTLI—eight triangle carriers

Table 2.7 Triangle carrier pattern summary

Sl. No	Vref volts	Vref2 volts	Vref3 volts	Vref4 volts	Vref5 volts	Triangle carrier selected Vg_pwm1, Vg_pwm2, Vg_pwm3, Vg_pwm4	Triangle carrier pattern
1	−+1	+1	+0.5	+1	+0.5	*Vtri8, Vtri1, Vtri3, Vtri6*	In phase disposition (IPD)
2	−+1	+1	+0.5	+1	+0.5	*Vtri8, Vtri1, Vtri5, Vtri7*	Alternate phase opposition disposition (APOD)
3	−+1	+1	+0.5	+1	+0.5	*Vtri4, Vtri2, Vtri3, Vtri6*	Alternate phase opposition disposition (APOD)
4	−+1	+1	+0.5	+1	+0.5	*Vtri4, Vtri1, Vtri5, Vtri6*	Phase opposition disposition (POD)
5	−+1	+1	+0.5	+1	+0.5	*Vtri8, Vtri2, Vtri3, Vtri7*	Phase opposition disposition (POD)

The three-phase sine wave AC voltage source is developed using three single-phase sine wave AC voltage source from electrical source library. Each sine wave AC voltage source van, vbn and vcn has an amplitude Am (1.6) volts, frequency fm (50) Hz and having phase 0, $-2\pi/3$ and $-4\pi/3$ radians, respectively. The four selected triangle carriers vg_pwm1 to vg_pwm4 correspond to that shown against serial number 1 in Table 2.7 for in-phase disposition. Referring to Phase A of modulating signal van, the triangle carrier vg_pwm2 and vg_pwm3 are compared with van using two Op.Amp. comparators and the resulting output of the comparators form PWM gate drive vg_s1A and vg_s2A for the switches S1A and S2A in Fig. 2.31. The PWM gate pulse vg_s3A and vg_s4A for switches S3A and S4A in Fig. 2.31 are obtained by inverting vg_s1A and vg_s2A using NOT gates. The same principle applies to Phase B and C switches where the modulating sine wave signals are vbn and vcn, respectively.

Figure 2.33c shows the three-phase DCTLI model. The three-phase DCTLI and 12 input gate multiplexer are from semiconductors library, and the two 50 volts DC source is from electrical sources library in Simscape. The 12 gate pulses corresponding to Phase A, B and C switches from vg_s1A to vg_s4C in Fig. 2.31 are connected to the 12 input gate multiplexer as shown in Fig. 2.33c. Line-to-neutral voltage across load VAO, VBO and VCO are derived from line-to-line voltage by matrix transformation using Fcn blocks. A three-phase series R-L load of 16.4 Ω and 14.2 m.H. is used.

2.8.2 Simulation Results

The simulation of the three-phase MCSLSPWM DCTLI is carried out using fixed step ode 14*x* (extrapolation) solver in SIMULINK [3]. The simulation parameters are the same as in Table 2.1. Simulation results for IPD of triangle carriers are tabulated in Table 2.8. The simulation results of line-to-line, line-to-ground, load current, line-to-neutral voltage, RMS value and THD of line-to-neutral and line-to-line voltage for Ma of 0.8 and Mf of 24 p.u. are also shown in Fig. 2.35a–p. The harmonic spectrum of VAB and VAO is shown in Fig. 2.36 and 2.37. The three-phase MCSLSPWM modulating signal and triangle carrier are shown in Fig. 2.38, and the four PWM gate pulses for the switches in phase A of DCTLI are shown in Fig. 2.39.

Now for the phase opposition disposition (POD) corresponding to serial number 4 in Table 2.7, simulation is carried out for various A.M. Index. Simulation results for POD of triangle carriers are tabulated in Table 2.9. The simulation results of line-

Table 2.8 Three-phase MCSLSPWM DCTLI simulation results

Sl. No	Ma	VLL (rms) V	VLL1 (rms) V	VLN (rms) V	VLN1 (rms) V	THD of VLL	THD of VLN	Remarks
1	0.2	25.46	15.28	14.7	8.822	1.33	1.33	IPD
2	0.4	44.42	40.6	25.65	23.44	0.4443	0.4443	IPD
3	0.6	66.19	63.49	38.18	36.62	0.2946	0.2944	IPD
4	0.8	70.88	68.81	40.92	39.73	0.2471	0.2471	IPD
5	1.2	74.62	72.89	43.08	42.08	0.2193	0.2193	IPD
6	1.4	75.2	73.6	43.42	42.49	0.2097	0.2097	IPD
7	1.6	75.74	74.07	43.73	42.76	0.2137	0.2137	IPD

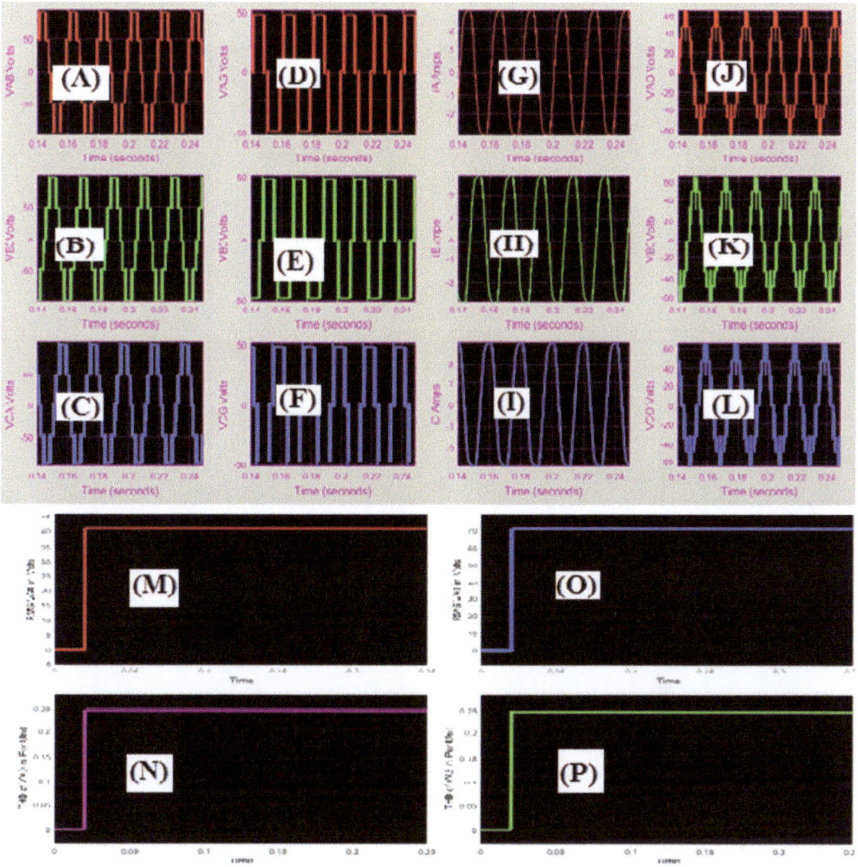

Fig. 2.35 MCSLSPWM three-phase DCTLI simulation results: (**a**), (**b**), (**c**) Line-to-line voltage. (**d**), (**e**), (**f**) Line-to-ground voltage. (**g**), (**h**), (**i**) Load current. (**j**), (**k**) (**l**) Line-to-neutral voltage across load. (**m**), (**n**) RMS value and THD of line-to-neutral voltage. (**o**), (**p**) RMS value and THD of line-to-line voltage

to-line, line-to-ground, load current, line-to-neutral voltage, RMS value and THD of line-to-neutral and line-to-line voltage for Ma of 0.8 and Mf of 24 p.u. are also shown in Fig. 2.40a–p. The harmonic spectrum of VAB and VAO is shown in Figs. 2.41 and 2.42. The three-phase MCSLSPWM modulating signal and with POD of triangle carriers are shown in Fig. 2.43, and the four PWM gate pulses for the switches in phase A of DCTLI are shown in Fig. 2.44.

Now for the alternate phase opposition disposition (APOD) corresponding to serial number 2 in Table 2.7, simulation is carried out for various A.M. Index. Simulation results for APOD of triangle carriers are tabulated in Table 2.10. The simulation results of line-to-line, line-to-ground, load current, line-to-neutral voltage, RMS value and THD of line-to-neutral and line-to-line voltage for Ma of 0.8 and Mf of 24 p.u. are also shown in Fig. 2.45a–p. The harmonic spectrum of VAB

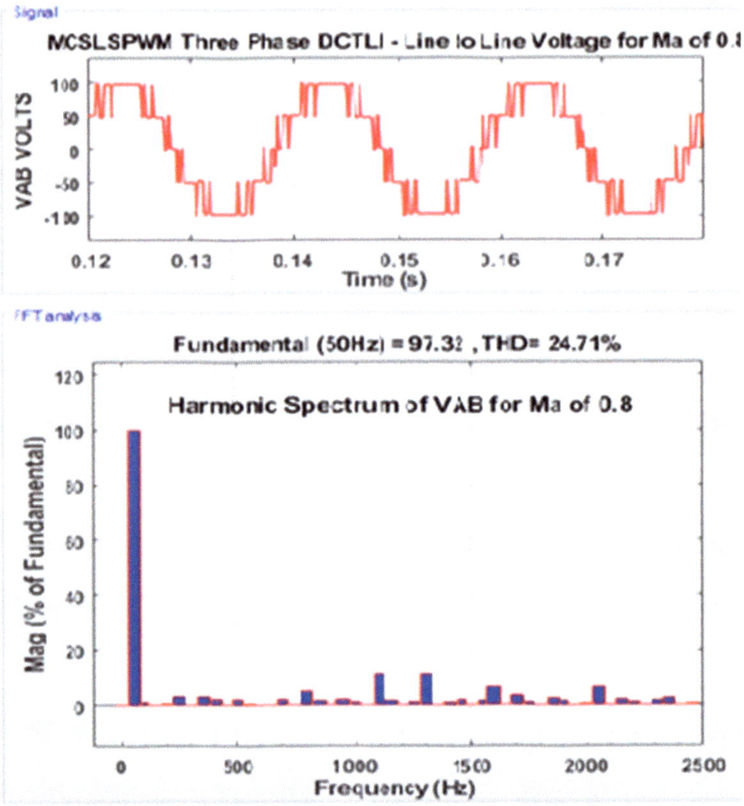

Fig. 2.36 MCSLSPWM Three-Phase DCTLI—Line-of-Line Voltage (Top) and Harmonic Spectrum (Bottom)

and VAO is shown in Fig. 2.46 and 2.47. The three-phase MCSLSPWM modulating signal and with APOD of triangle carriers are shown in Fig. 2.48, and the four PWM gate pulses for the switches in phase A of DCTLI are shown in Fig. 2.49.

POD and APOD of triangle carriers, the triangle carriers corresponding to Vg_pwm2 and Vg_pwm3 are out of phase by 180 degrees. The A.M.index Ma below one is under modulation and that above one is over modulation.

2.9 Case Study: Three-Phase Clipped Sine Level Shift PWM Diode-Clamped Three-Level Inverter

The model of the CSLSPWM three-phase DCTLI is shown in Fig. 2.50 (Model file: CASE_STUDY_EX2_1). The various subsystems are shown in Fig. 2.33a–c. In Fig. 2.33b, three-phase sine wave modulating signal is replaced by three-phase

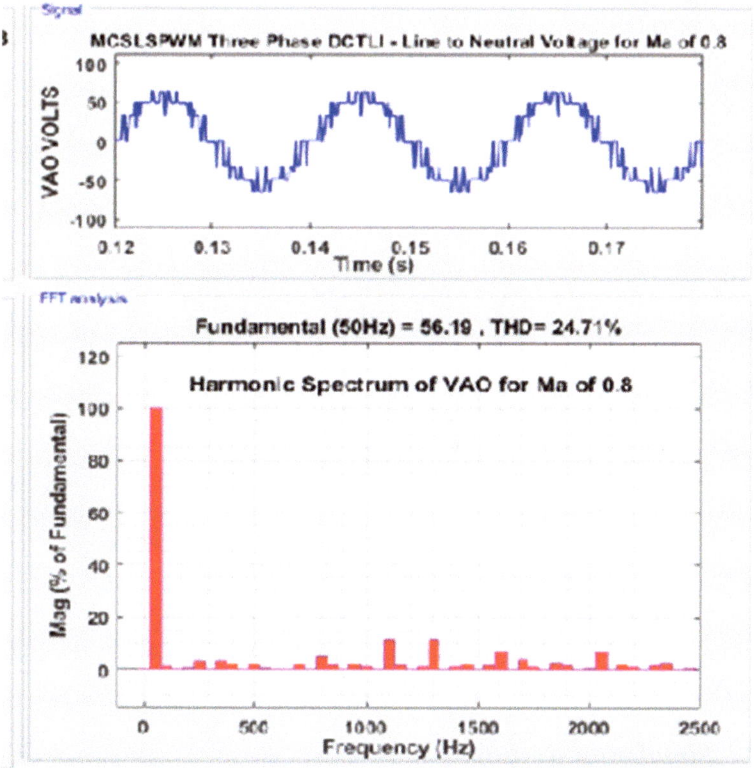

Fig. 2.37 MCSLSPWM Three-Phase DCTLI—Line-of-Neutral Voltage (TOP) and Harmonic Spectrum (Bottom)

clipped sine PWM generator whose subsystem is shown in Fig. 2.51. The parameters shown in Table 2.1 are used to develop the model. Here triangle carrier generator with peak value +/−1 volt with frequency ftri (1200 Hz) from control and measurements/pulse and signal generators library from specialised technology block set is used along with operational amplifier (Op.Amp.) summers, sign changers and comparators to develop four-level shifted triangle carriers, each level shifted by one volt with peak value Ac (1 volt and frequency ftri 1200 Hz). These triangle carriers are compared using Op.Amp. comparators with a three-phase clipped sine modulating sine wave voltage with frequency fm (50) Hz and peak value Am volt to generate PWM gate pulse for the switches of three-phase DCTLI. The value of Am lies within the range 0 to 2 volts for under modulation.

The various model subsystems are already explained in Sect. 2.8.1. The clipped sine PWM generator shown in Fig. 2.51 is already explained in Sect. 2.4.1 [4, 5]. In the model file CASE_STUDY_EX2_1, Program segment 2.3 and 2.4 refer to clipped sine PWM generation. In Fig. 2.51, sine wave modulating signal amplitude is 4 volts, and clipper value is 1.6 volts which corresponds to an A.M. index Ma of 0.8.

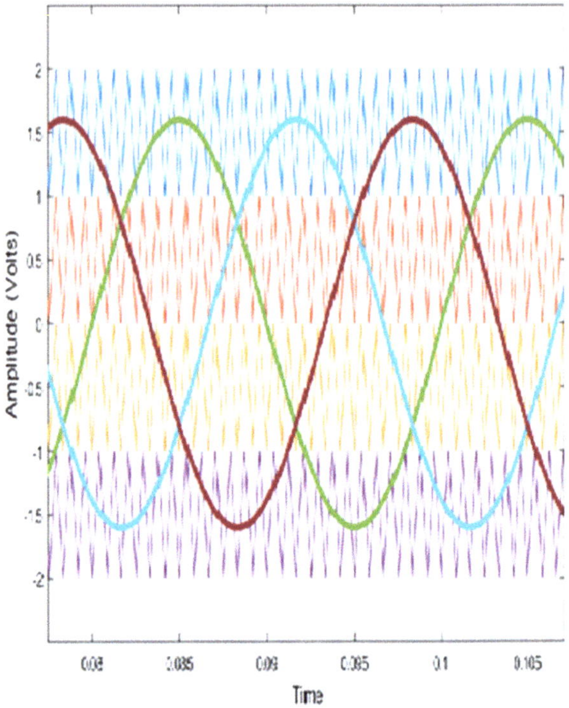

Fig. 2.38 MCSLSPWM three-phase DCTLI—four triangle carrier with in-phase disposition and three-phase sine wave modulating signals

2.9.1 Simulation Results

Simulation of the three-phase clipped sine level shift PWM (CSLSPWM) DCTLI is carried out using ode 15 s solver in Simulink [3]. The parameters shown in Table 2.1 are used. An A.M. index Ma of 0.8 is used. For the IPD of carriers, simulation results are shown in Figs. 2.52a–p, 2.53 and in Table 2.11. For the POD of carriers, simulation results are shown in Figs. 2.54a–p, 2.55 and in Table 2.12. For the APOD of carriers, simulation results are shown in Figs. 2.56a–p, 2.57 and in Table 2.13. Comparison graphs of THD of line-to-line and line-to-neutral voltage of three-phase DCTLI by sine and clipped sine PWM techniques for IPD, POD and APOD of carriers are shown in Figs. 2.58a–c, respectively. Data from Table 2.8 to Table 2.13 are used for plotting the graphs in Figs. 2.58a–c.

2.10 Discussion of Results

The model for three-phase MCSLSPWM and CSLSPWM DCTLI is compared. Simulation results are shown for the four triangle carriers having in-phase disposition (IPD), phase opposition disposition (POD) and alternate phase opposition (APOD). For any given A.M. index, the THD of line-to-line and line-to-neutral voltage for three-phase MCSLSPWM DCTLI with IPD, POD and APOD of triangle

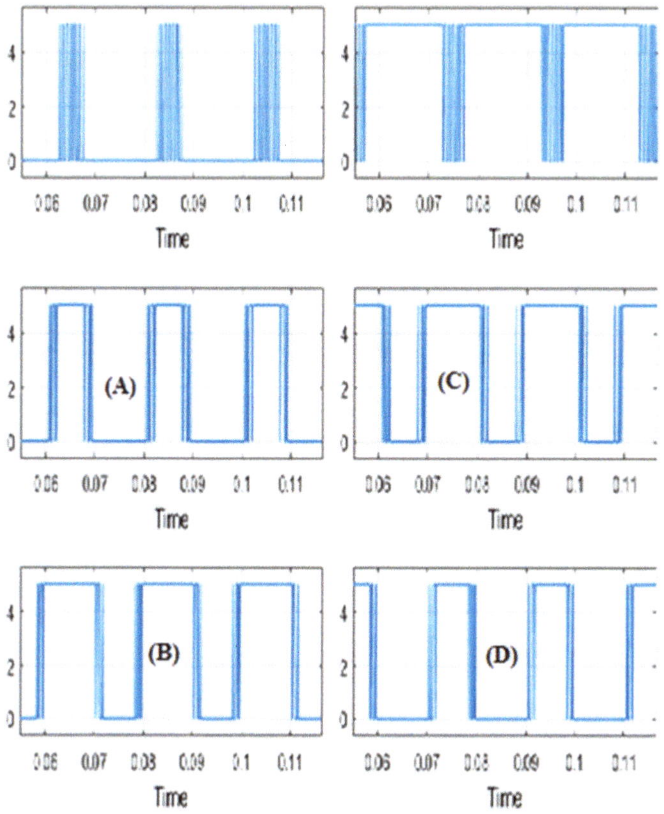

Fig. 2.39 MCSLSPWM Three-Phase DCTLI—(**a**), (**b**) PWM gate pulse for switches S1A and S2A. (**c**), (**d**) PWM gate pulse for switches S3A and S4A. (Note: The top two PWM Gate Pulse on the Left and Right side are not used)

Table 2.9 Three-phase MCSLSPWM DCTLI simulation results

Sl. No	Ma	VLL (rms) V	VLL1 (rms) V	VLN (rms) V	VLN1 (rms) V	THD of VLL	THD of VLN	Remarks
1	0.2	32.67	15.27	18.82	8.77	1.9244	1.9157	POD
2	0.4	53.65	41.66	30.78	23.89	0.8118	0.8121	POD
3	0.6	65.41	62.36	37.59	35.83	0.3166	0.3173	POD
4	0.8	70.78	68.74	40.71	39.51	0.2471	0.249	POD
5	1.2	74.73	73.2	42.99	42.09	0.2055	0.2084	POD
6	1.4	75.63	74.37	43.52	42.75	0.1837	0.1905	POD
7	1.6	75.89	74.53	43.67	42.86	0.192	0.1953	POD

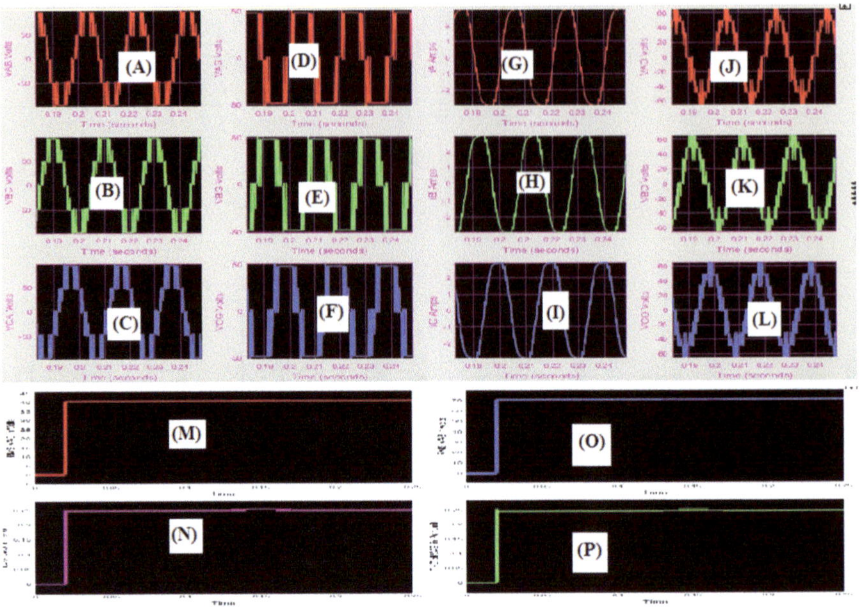

Fig. 2.40 MCSLSPWM three-phase DCTLI with triangle carriers in POD simulation results. (**a**), (**b**), (**c**) Line-to-line voltage. (**d**), (**e**), (**f**) Line-to-ground voltage. (**g**), (**h**), (**i**) Load current. (**j**), (**k**), (**l**) Line-to-neutral voltage across load. (**m**), (**n**) RMS value and THD of line-to-neutral voltage. (**o**), (**p**) RMS value and THD of Line-to-line voltage

carriers is much less compared to that for two-level inverter having sine, HI, THI and CS PWM techniques. Also it is seen from Table 2.9 and Table 2.10 that the THD of line-to-line and line-to-neutral voltage for various A.M. index are almost the same. A case study of the three-phase multi-carrier clipped sine level shift PWM DCTLI for IPD, POD and APOD of carriers is also presented. A.M. Index Ma below one is under modulation and above one is over modulation. Graphs are presented comparing the THD of line-to-line and line-to-neutral voltage of three-phase DCTLI by sine and clipped sine PWM techniques. It is seen that the THD values of both line-to-line and line-to-neutral voltages by clipped sine PWM are much lower than that for multi-carrier sine PWM in the under modulation region and almost equal in the over modulation region for IPD, POD and APOD of carriers.

2.11 Case Study: Three-Phase Third Harmonic Injection Sine Level Shift PWM Diode-Clamped Three-Level Inverter

The model of the third harmonic injection sine level shift PWM (THISLSPWM) three-phase DCTLI is shown in Fig. 2.59 (Model file: CASE_STUDY_EX2_2). The various subsystems are shown in Fig. 2.33a–c. In Fig. 2.33b, three-phase sine wave

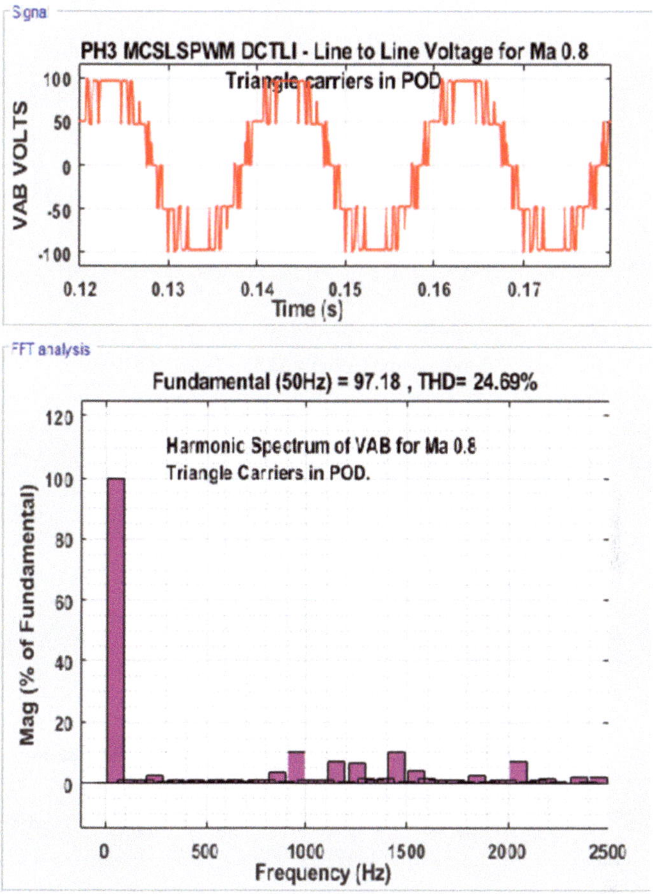

Fig. 2.41 MCSLSPWM three-phase DCTLI—line-to-line voltage VAB (top) and harmonic spectrum (bottom). Triangle carriers in POD

modulating signal is replaced by three-phase third harmonic injection sine PWM generator whose subsystem is shown in Fig. 2.60. The parameters shown in Table 2.1 are used to develop the model. Here triangle carrier generator with peak value +/−1 volt with frequency ftri (1200 Hz) from control and measurements/pulse and signal generators library from specialised technology block set is used along with operational amplifier (Op.Amp.) summers, sign changers and comparators to develop four-level shifted triangle carriers, each level shifted by one volt with peak value Ac (1 volt and frequency ftri 1200 Hz). These triangle carriers are compared using Op.Amp. comparators with a three-phase third harmonic injection sine modulating wave with frequency fm (50) Hz and peak value Am Volt to generate PWM gate pulse for the switches of three-phase DCTLI. The value of Am lies within the range 0–2 volts for under modulation.

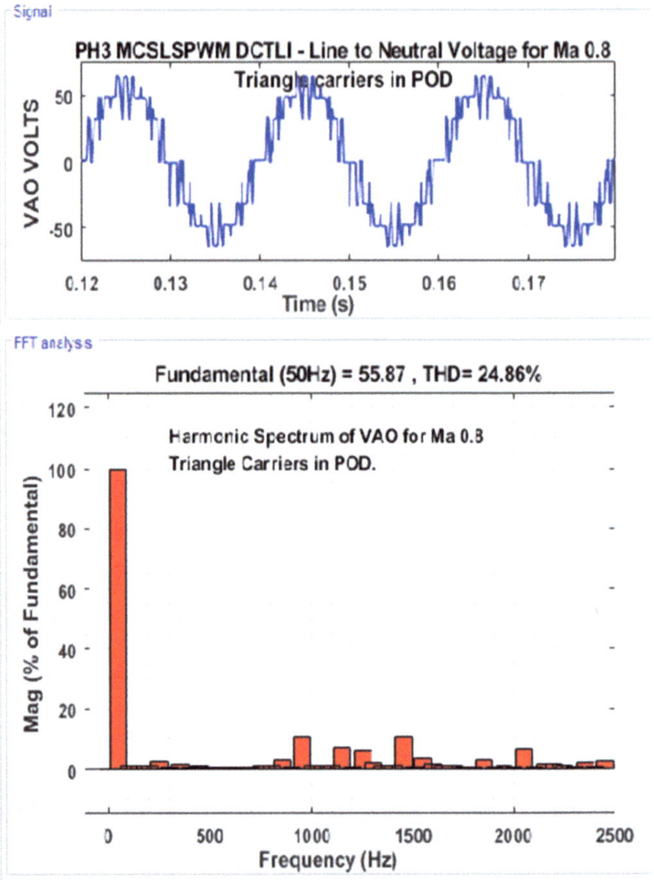

Fig. 2.42 MCSLSPWM Three-Phase DCTLI—Line-to-Neutral Voltage VAO (Top) and Harmonic Spectrum (Bottom). Triangle carriers in POD

The various model subsystems are already explained in Sect. 2.8.1. The third harmonic injection sine PWM generator shown in Fig. 2.60 is already explained in Sect. 2.2.1 [1, 2, 4, 5]. In the model file CASE_STUDY_EX2_2, Program segment 2.1 refers to third harmonic injection sine PWM generation. In Fig. 2.60, sine wave modulating signal amplitude is 1.6 volts which corresponds to an A.M. index Ma of 0.8.

2.11.1 Simulation Results

Simulation of the three-phase THISLSPWM DCTLI is carried out using ode 15 s solver in Simulink [3]. The parameters shown in Table 2.1 are used. An A.M. index Ma of 0.8 is used. For the IPD of carriers, simulation results are shown in

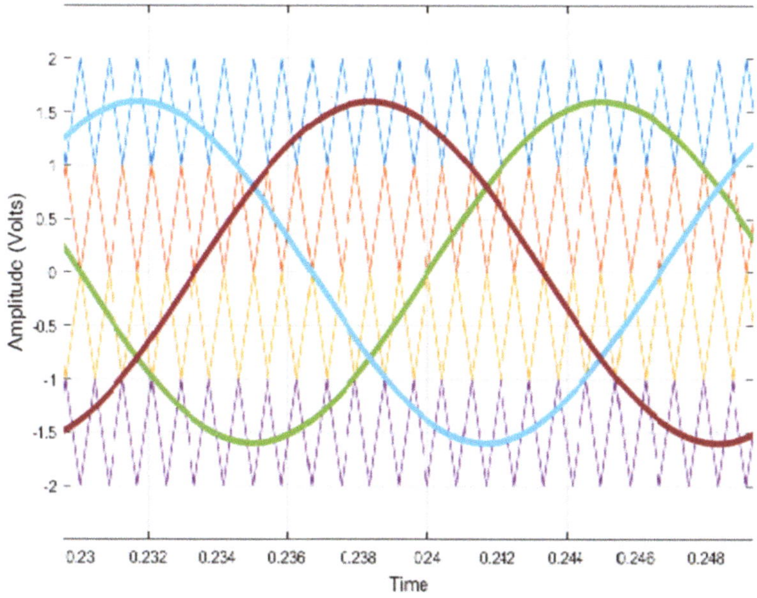

Fig. 2.43 MCSLSPWM three-phase DCTLI—our triangle carriers in phase opposition disposition and three-phase sine wave modulating signals

Figs. 2.61a–p and 2.62 and in Table 2.14. For the POD of carriers, simulation results are shown in Fig. 2.63a–p, 2.64 and in Table 2.15. For the APOD of carriers, simulation results are shown in Figs. 2.65a–p, 2.66 and in Table 2.16. Comparison graphs of THD of line-to-line and line-to-neutral voltage of three-phase DCTLI by sine and third harmonic injection sine PWM techniques for IPD, POD and APOD of carriers are shown in Figs. 2.67a–c, respectively. Data from Table 2.8 to 2.10 and Table 2.14 to 2.16 are used for plotting the graphs in Fig. 2.67a–c.

2.12 Discussion of Results

The model for three-phase MCSLSPWM and CSLSPWM DCTLI is compared. Simulation results are shown for the four triangle carriers having in-phase disposition (IPD), phase opposition disposition (POD) and alternate phase opposition (APOD). For any given A.M. index, the THD of line-to-line and line-to-neutral voltage for three-phase MCSLSPWM DCTLI with IPD, POD and APOD of triangle carriers are much less compared to that for two-level inverter having sine, HI, THI and CS PWM techniques. Also it is seen from Tables 2.9 and Table 2.10 that the THD of line-to-line and line-to-neutral voltage for various A.M. indices are almost the same. A case study of the three-phase multi-carrier third harmonic injection sine level shift PWM (THISLSPWM) DCTLI for IPD, POD and APOD of carriers is also

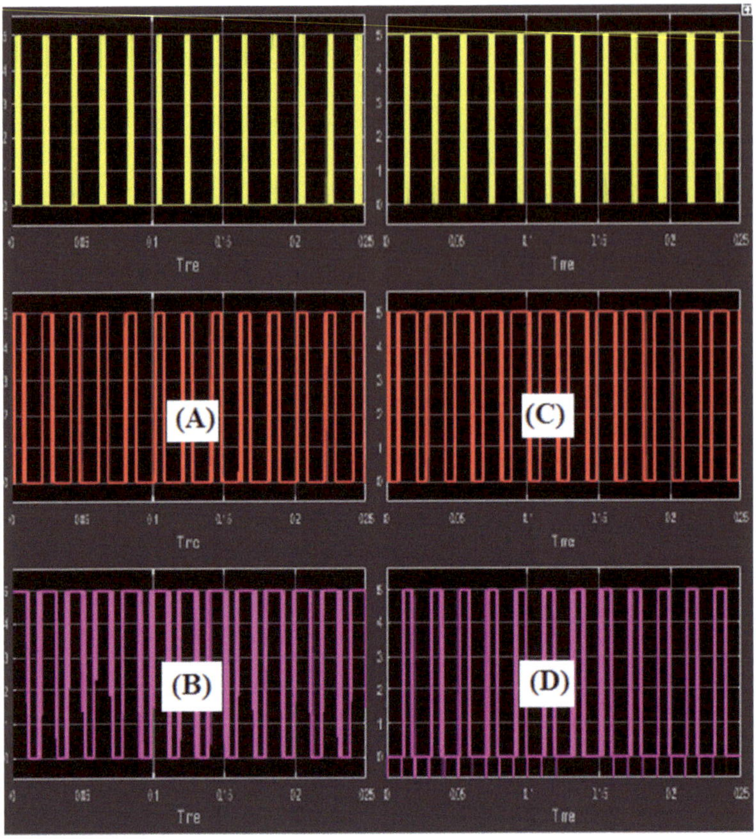

Fig. 2.44 MCSLSPWM three-phase DCTLI—(**a**), (**b**) PWM gate pulse for switches S1A and S2A. (**c**), (**d**) PWM Gate pulse for switches S3A and S4A

Table 2.10 Three-phase MCSLSPWM DCTLI simulation results

Sl. No	Ma	VLL (rms) V	VLL1 (rms) V	VLN (rms) V	VLN1 (rms) V	THD of VLL	THD of VLN	Remarks
1	0.2	32.67	15.27	18.82	8.77	1.9244	1.9157	APOD
2	0.4	53.65	41.66	30.78	23.89	0.8118	0.8121	APOD
3	0.6	65.41	62.36	37.59	35.83	0.3166	0.3173	APOD
4	0.8	70.78	68.74	40.71	39.51	0.2471	0.249	APOD
5	1.2	74.73	73.2	42.99	42.09	0.2055	0.2084	APOD
6	1.4	75.63	74.37	43.52	42.75	0.1837	0.1905	APOD
7	1.6	75.89	74.53	43.67	42.86	0.192	0.1953	APOD

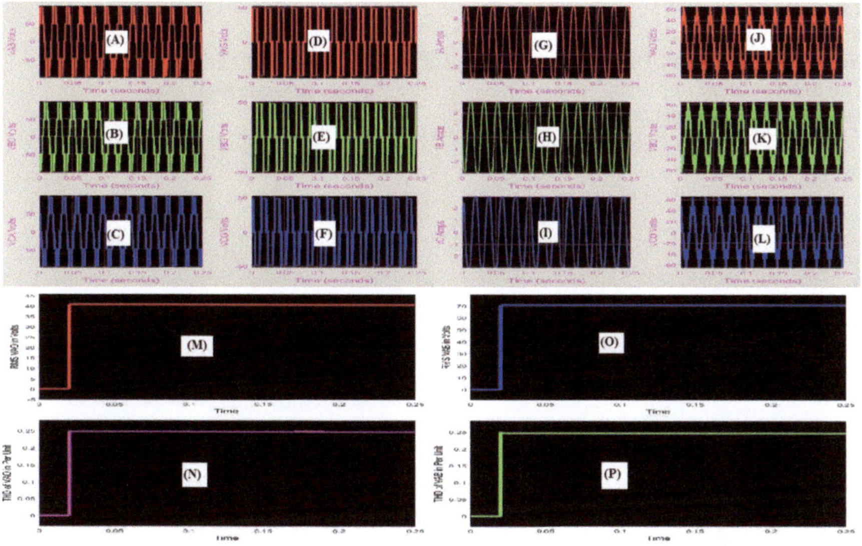

Fig. 2.45 MCSLSPWM three-phase DCTLI with triangle carriers in APOD simulation results. (**a**), (**b**), (**c**) Line-to-Line Voltage. (**d**), (**e**), (**f**) Line-to-ground voltage. (**g**), (**h**), (**i**) Load current. (**j**), (**k**), (**l**) Line-to-neutral voltage across load. (**m**), (**n**) RMS value and THD of line-to-neutral voltage. (**o**), (**p**) RMS value and THD of line-to-line voltage

presented. A.M. Index Ma below one is under modulation and above one is over modulation. Graphs are presented comparing the THD of line-to-line and line-to-neutral voltage of three-phase DCTLI by sine and third harmonic injection sine PWM techniques. It is seen that the THD values of both line-to-line and line-to-neutral voltages by third harmonic injection sine PWM are much lower than that for multi-carrier sine PWM in the under modulation region and almost equal in the over modulation region for IPD, POD and APOD of carriers.

2.13 Single-Phase Cascade H-Bridge Inverter

The conventional three-phase five-level cascade H-bridge inverter (TPFLCHBI) topology, its modelling and the model for phase shift PWM technique are reported [9, 11]. For clarity, a single-phase five-level cascade H-bridge inverter (SPFLCHBI) topology is shown in Fig. 2.68. Here, the upper and lower arm bridges have a DC link voltage each Vdc volts. The mode of operation of this SPFLCHBI is given below:

- When switches S2A, S3A, S6A, S7A are only ON, output voltage Vout across load is +2Vdc volts.
- When switches S1A, S4A, S5A, S8A are only ON, Vout is -2Vdc volts.

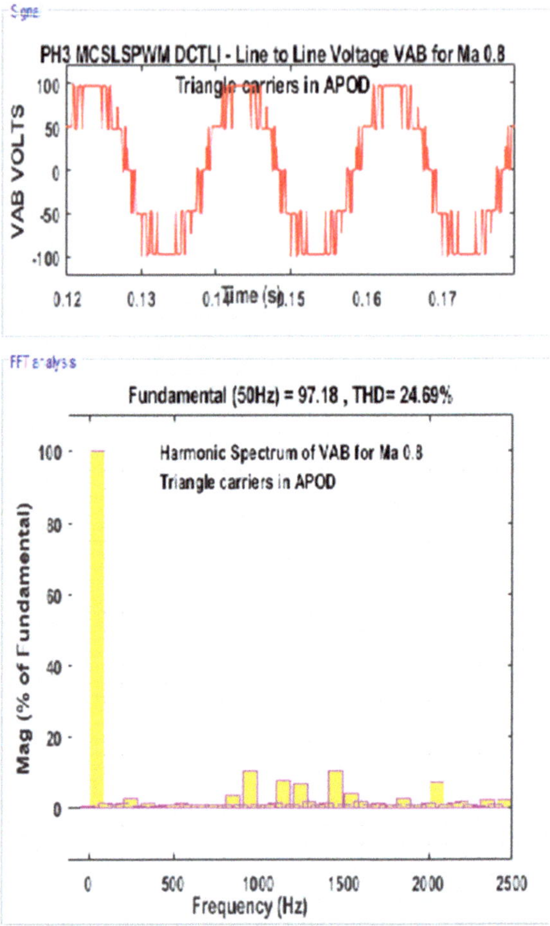

Fig. 2.46 MCSLSPWM three-phase DCTLI with triangle carriers in APOD: line-to-line voltage (top) and harmonic spectrum (bottom)

- When switches S2A, S3A with either of S5A, S7A or S6A, S8A are only ON, Vout is +Vdc.
- When switches S6A, S7A with either of S1A, S3A or S2A, S4A are only ON, Vout is +Vdc.
- When switches S1A, S4A with either of S5A, S7A or S6A, S8A are only ON, Vout is -Vdc.
- When switches S5A, S8A with either of S1A, S3A or S2A, S4A are only ON, Vout is -Vdc.
- When switches S1A, S3A or else S2A, S4A along with either S5A, S7A or else S6A, S8A are only ON, Vout is zero.
- Above arguments hold good for TPFLCHBI.

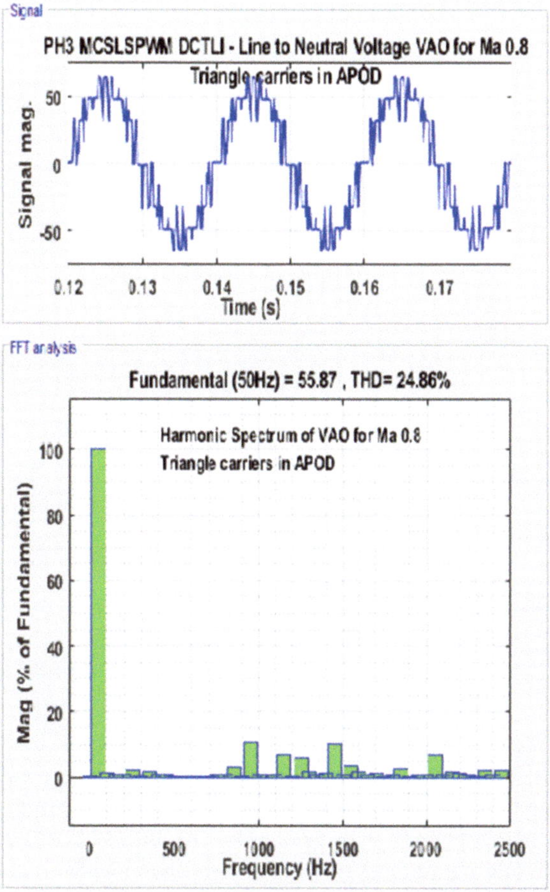

Fig. 2.47 MCSLSPWM three-phase DCTLI with triangle carriers in APOD: line-to-neutral voltage (top) and harmonic spectrum (bottom)

This truth table for the operation and the model of the TPFLCHBI are reported [11]. Thus the phase-to-ground voltage Vout has five levels: +2Vdc, +Vdc, 0, -Vdc and -2Vdc.

2.13.1 *Model of Three-Phase Multi-carrier Sine Phase Shift PWM Five-Level Cascade H-Bridge Inverter*

The model of the three-phase multi-carrier sine phase shift PWM five-level cascade H-bridge inverter (MCSPSPWMFLCHBI) is shown in Fig. 2.69 (Model file: EXAMPLE 2_7). The model subsystems are also shown in Fig. 2.70a–c. The DC link voltage source for the upper and lower H-bridge each has a value of 50 volts. The inverter output frequency is 50 Hz, and the triangle carrier switching frequency

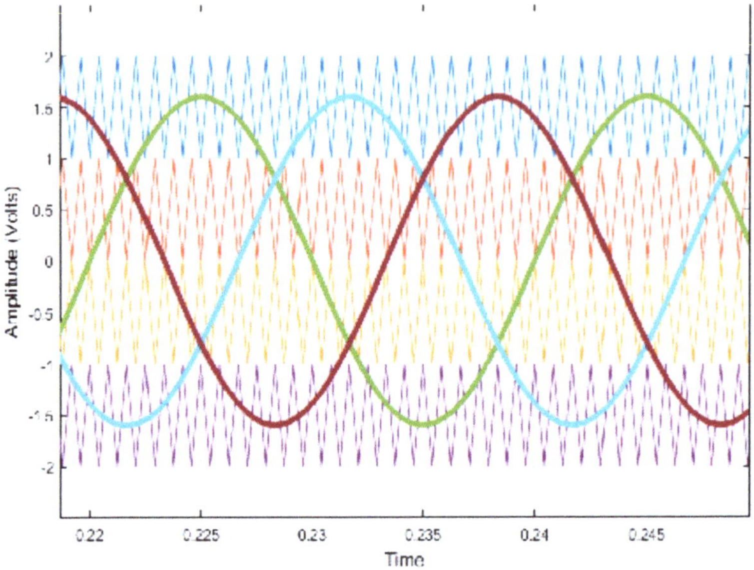

Fig. 2.48 MICSLSPWM three-phase DCTLI: four triangle carriers in APOD and three-phase sine wave modulating signals

is 1200 Hz. Two triangle carriers are used as there are two H-bridge inverters in cascade. The phase shift between the two triangle carriers is (π/N) radians where N is the number of H-bridge inverters in cascade. The value of N here is two. The data shown in Table 2.1 are used to develop the model. Here two triangle carrier generators each with peak value Ac of +/−1 Volt with frequency ftri (1200 Hz) having phase delay of zero and π/2 radians, from control and measurements/pulse and signal generators library in the specialised technology block set are used. The three sine wave generators each has an amplitude Am(0.8) volts, frequency fm (50 Hz) and phase delay zero, 2π/3 and 4π/3 radians, respectively. These three sine wave generator outputs are also inverted by 180 degrees by multiplying with a gain constant of −1 using three gain blocks.

Referring to gate drive for switches in Phase A shown in Fig. 2.70a, the triangle carrier with zero phase delay is compared with sine wave generator output for phase A in a relational operator comparator block. The relational operator output is HIGH (logic 1) if triangle carrier signal is less than or equal to sine wave modulating signal, else its output is LOW (logic 0). This relational operator comparator output gives the gate pulse VGS1A for switch S1A which is inverted using a NOT gate to obtain the gate pulse VGS2A for switch S2A. The same triangle carrier output is now compared with inverted sine wave modulating signal from gain block using another relational operator comparator block and the output VGS3A for switch S3A is derived. This VGS3A output is inverted using a NOT gate to obtain gate pulse VGS4A for the switch S4A. These form the gate pulse for upper H-bridge switches.

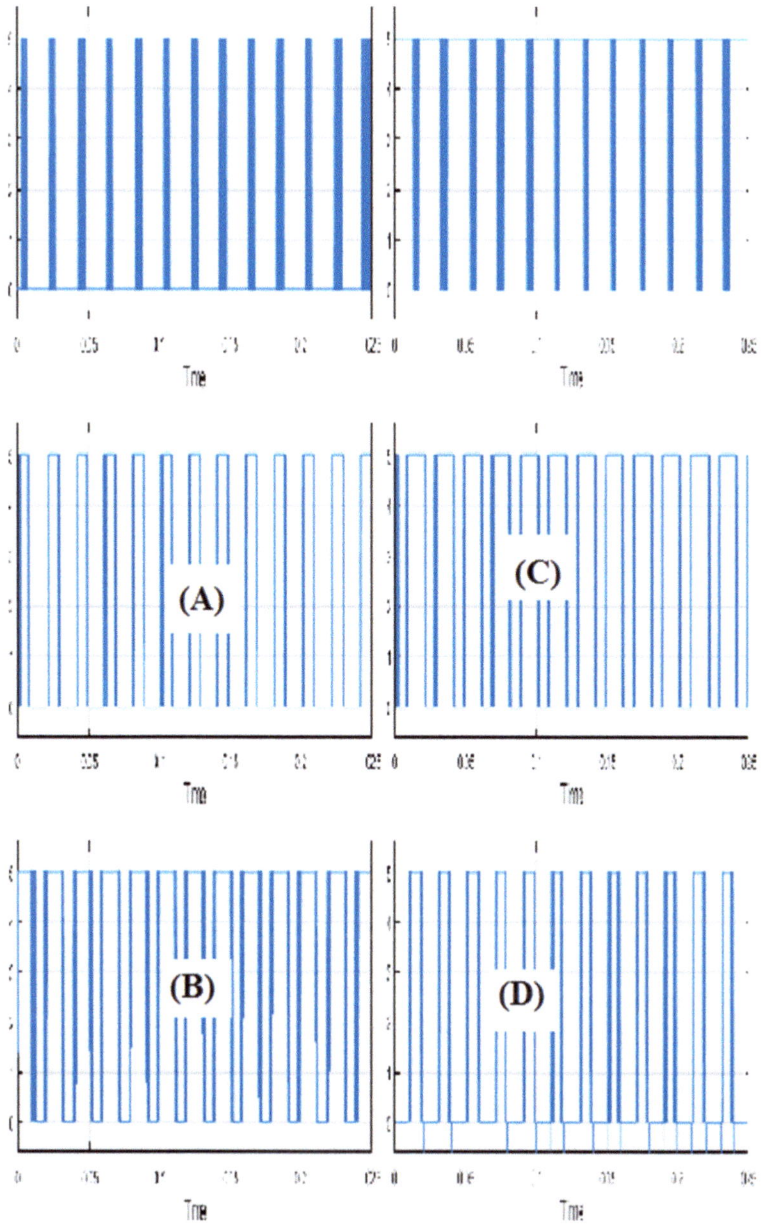

Fig. 2.49 MCSLSPWM three-phase DCTLI with four triangle carriers in APOD. (**a**), (**b**) PWM gate pulse for switches S1A and S2A. (**c**), (**d**) PWM gate pulse for switches S3A and S4A

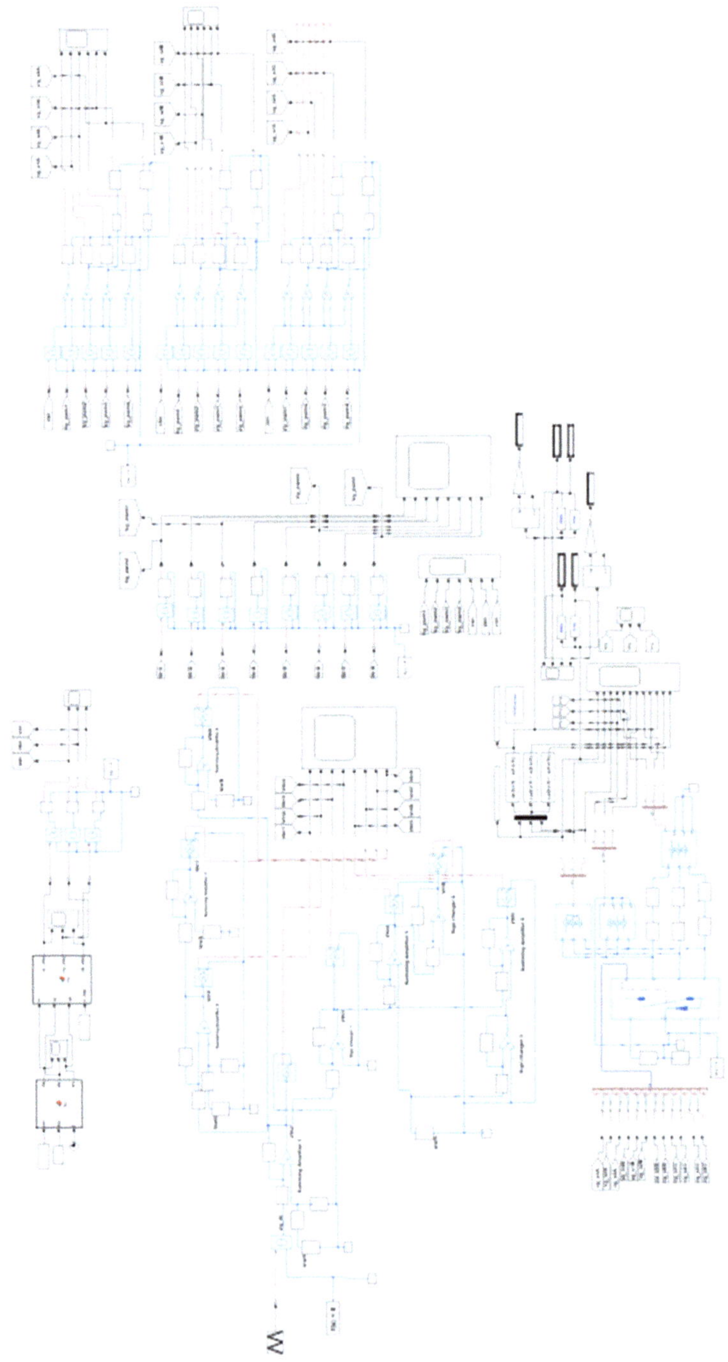

Fig. 2.50 Model of three-phase clipped sine level shift PWM Diode clamped three-level inverter

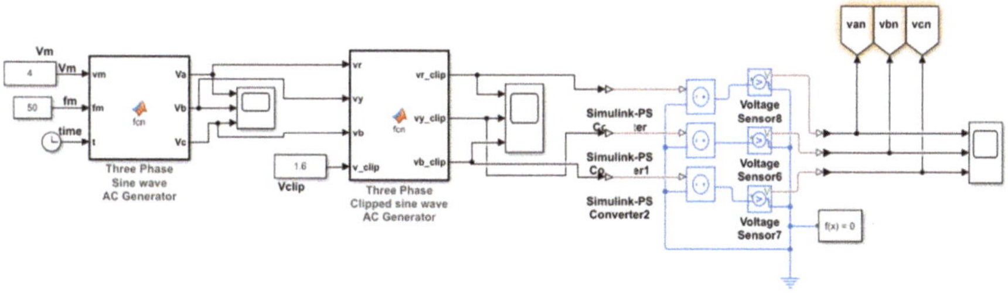

Fig. 2.51 Three-phase clipped Sine PWM generator

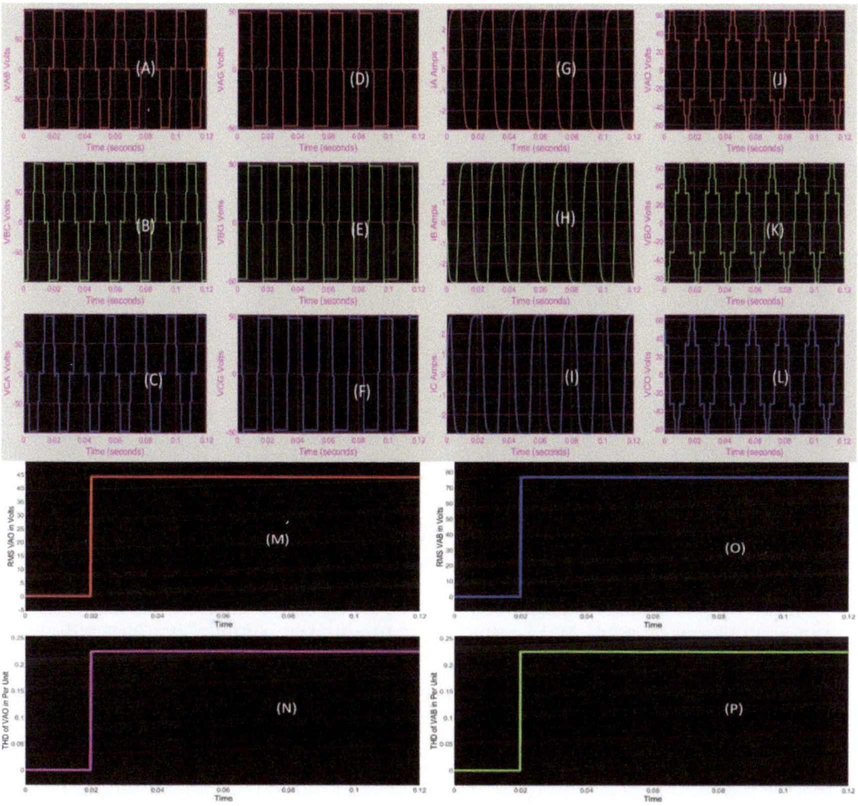

Fig. 2.52 Three-phase CSLSPWM DCTLI simulation results for IPD of carriers. (**a**), (**b**), (**c**) Line-to-line voltage, (**d**), (**e**), (**f**) Line-to ground voltage, (**g**), (**h**), (**i**) Load current and (**j**), (**k**) and (**l**) Line-to-neutral voltage across load. Three-Phase CSLSPWM DCTLI simulation results for IPD of Carriers: (**m**), (**n**) RMS value and THD of line-to-neutral voltage. (**o**), (**p**) RMS value and THD of line-to-line voltage

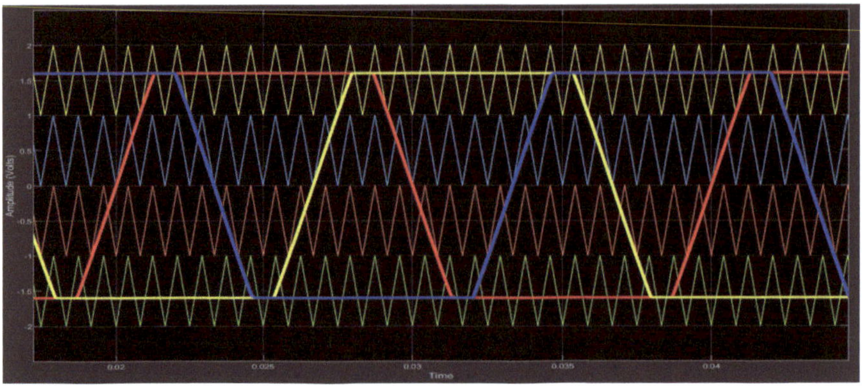

Fig. 2.53 Three-phase CSLSPWM DCTLI simulation result: four IPD of triangle carriers and three-phase clipped sine modulating signal

Table 2.11 Three-phase CSLSPWM DCTLI simulation results

Sl. No	Ma	VLL (rms) V	VLL1 (rms) V	VLN (rms) V	VLN1 (rms) V	THD of VLL	THD of VLN	Remarks
1	0.2	34.41	28.65	19.86	16.54	0.6646	0.6646	IPD
2	0.4	64.18	59.48	37.06	34.34	0.405	0.4051	IPD
3	0.6	76.66	74.78	44.26	43.17	0.225	0.2256	IPD
4	0.8	76.66	74.7900	44.26	43.18	0.2248	0.2246	IPD
5	1.2	76.66	74.77	44.26	43.17	0.2257	0.2255	IPD
6	1.4	76.66	74.79	44.26	43.18	0.2248	0.2247	IPD
7	1.6	76.66	74.78	44.26	43.18	0.2253	0.225	IPD

For the lower H-bridge switches, the same procedure mentioned above for upper H-bridge switches is repeated with triangle carrier generator replaced with the one having ($\pi/2$) radians phase delay.

For switches in phase B and C, sine wave modulating signal generator with zero phase delay is replaced with the one having delay of -$2\pi/3$ and + $2\pi/3$ radians, respectively. The triangle carrier and phase shifted triangle carrier with zero and $\pi/2$ radians delay and method of gate pulse generation remain the same as above.

Figure 2.70b shows the switch arrangement of CHBI for phase A, and Fig. 2.70c shows the measurement circuit for RMS, THD and Fourier component of line-to-line and line-to-neutral output voltage. In Fig. 2.70b, a gain block with a gain constant of ten is used for the gate drive circuit, so that the gate pulse amplitude is greater than the gate threshold voltage of the semiconductor switches.

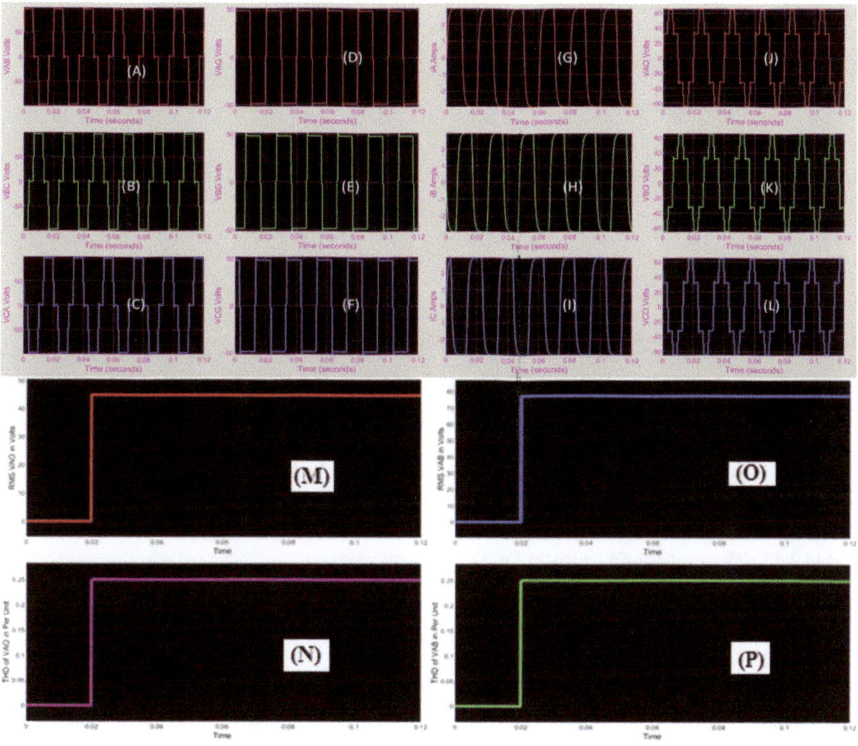

Fig. 2.54 Three-phase CSLSPWM DCTLI simulation results for POD of carriers—(**a**), (**b**), (**c**) Line-to-line voltage. (**d**), (**e**), (**f**) Line-to-ground voltage. (**g**), (**h**), (**i**) Load current. (**j**), (**k**) and (**l**) Line-to-neutral voltage across load. Three-phase CSLSPWM DCTLI simulation results for POD of carriers—(**m**), (**n**) RMS value and THD of line-to-neutral voltage. (**o**), (**p**) RMS value and THD of line-to-line voltage

2.13.2 Simulation Results

The simulation of the three-phase MCSPSPWMFLCHBI is carried out using ode 23tb(stiff/TR-BDF2) solver in Simulink [3]. The data shown in Table 2.1 are used for simulation. The simulation results for an A.M. index Ma of 0.8 are shown in Fig. 2.71a–d. The triangle carrier and sine wave modulating signal for phase A are shown in Fig. 2.72. The simulation results for all A.M. index are tabulated in Table 2.17. In Fig. 2.71c, d, the fundamental and all the harmonic component values displayed are peak voltage in volts.

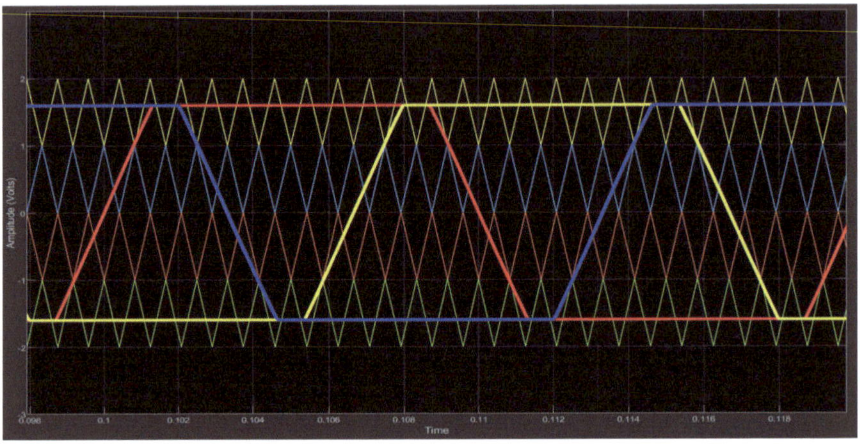

Fig. 2.55 Three-phase CSLSPWM DCTLI simulation result: four POD of triangle carriers and three-phase clipped sine modulating signal

Table 2.12 Three-phase CSLSPWM DCTLI simulation results

Sl. No	Ma	VLL (rms) V	VLL1 (rms) V	VLN (rms) V	VLN1 (rms) V	THD of VLL	THD of VLN	Remarks
1	0.2	49.92	28.74	28.82	16.59	1.42	1.42	POD
2	0.4	69.56	59.7	40.16	34.47	0.5974	0.5976	POD
3	0.6	77.53	75.22	44.76	43.42	0.2491	0.2496	POD
4	0.8	77.53	75.2	44.76	43.41	0.2504	0.2507	POD
5	1.2	77.53	75.21	44.76	43.42	0.25	0.25	POD
6	1.4	77.53	75.18	44.76	43.41	0.2513	0.2514	POD
7	1.6	77.53	75.2	44.76	43.42	0.2504	0.2503	POD

2.14 Case Study: Model of Three-Phase Multi-carrier Clipped Sine Phase Shift PWM Five-Level Cascade H-Bridge Inverter

The model of the three-phase multi-carrier clipped sine phase shift PWM five-level cascade H-bridge inverter (MCCSPSPWMFLCHBI) is shown in Fig. 2.73 (Model file: CASE_STUDY_EX2_3). The various subsystems are shown in Fig. 2.70a–c. In Figs. 2.69 and 2.70a, three-phase sine wave modulating signal is replaced by three-phase clipped sine PWM generator with amplitude Am(0.8) and frequency fm(50 Hz) whose subsystem is shown in Fig. 2.74. The parameters shown in Table 2.1 are used to develop the model. Here two triangle carrier generators with peak value +/−1 volt with frequency ftri (1200 Hz) and phase shifted by π/2 radians from control and measurements/pulse and signal generators library from specialised technology block set are used along with three-phase clipped sine generator and relational operator comparators to develop gate pulse for the switches of FLCHBI, as already explained in Sect. 2.13.1.

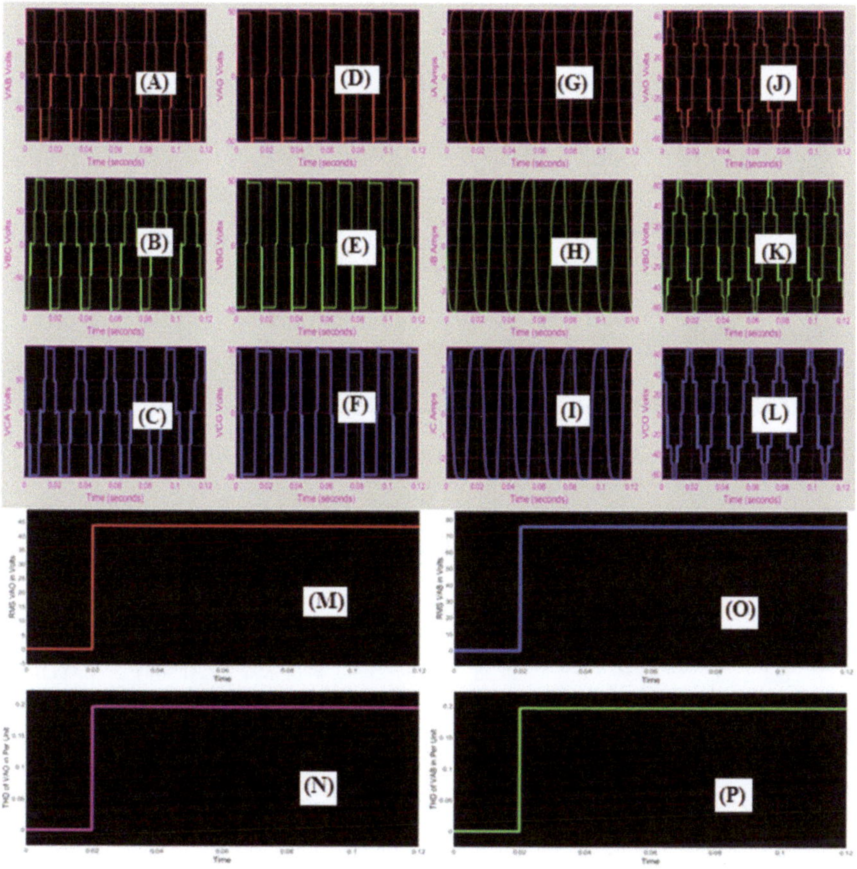

Fig. 2.56 Three-phase CSLSPWM DCTLI simulation results for APOD of carriers—(**a**), (**b**), (**c**) Line-to-Line Voltage. (**d**), (**e**), (**f**) Line-to-ground voltage. (**g**), (**h**), (**i**) Load current. (**j**), (**k**) and (**l**) Line-to-neutral voltage across load. Three-phase CSLSPWM DCTLI simulation results for APOD of carriers—(**m**), (**n**) RMS value and THD of line-to-neutral voltage. (**o**), (**p**) RMS value and THD of line-to-line voltage

The clipped sine PWM generator shown in Fig. 2.74 is already explained in Sect. 2.4.1 [4, 5]. In the model file CASE_STUDY_EX2_3, Program segment 2.3 and 2.4 refers to clipped sine PWM generation. In Fig. 2.74, sine wave modulating signal amplitude is 2 volts, and clipper value is 0.8 volt which corresponds to an A.M. index Ma of 0.8.

2.14.1 Simulation Results

The simulation of the three-phase MCCSPSPWMFLCHBI is carried out using Simulink [3]. The data shown in Table 2.1 are used for simulation. The simulation

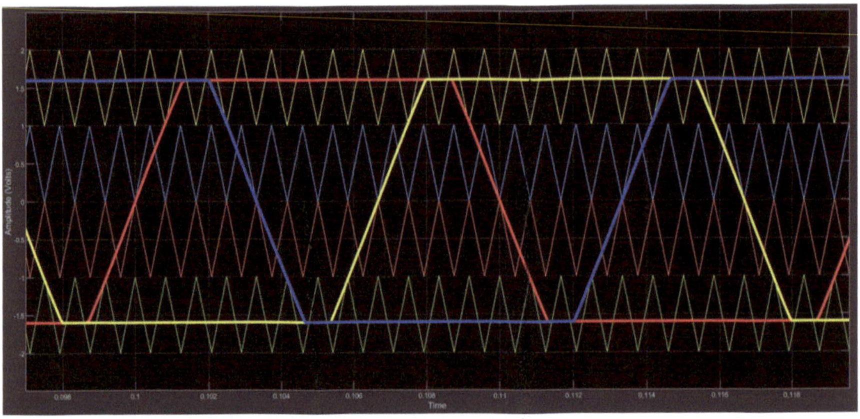

Fig. 2.57 Three-phase CSLSPWM DCTLI simulation result: four APOD of triangle carriers and three-phase clipped sine modulating signal

Table 2.13 Three-phase CSLSPWM DCTLI simulation results

Sl. No	Ma	VLL (rms) V	VLL1 (rms) V	VLN (rms) V	VLN1 (rms) V	THD of VLL	THD of VLN	Remarks
1	0.2	48.48	28.56	27.99	16.49	1.372	1.371	APOD
2	0.4	68.2	59.25	39.38	34.21	0.5697	0.5694	APOD
3	0.6	75.79	74.37	43.76	42.93	0.1961	0.1965	APOD
4	0.8	75.79	74.35	43.76	42.94	0.197	0.1959	APOD
5	1.2	75.79	74.34	43.76	42.93	0.1978	0.1969	APOD
6	1.4	75.79	74.34	43.76	42.92	0.198	0.1979	APOD
7	1.6	75.79	74.35	43.76	42.93	0.1976	0.1972	APOD

results for an A.M. index Ma of 0.8 are shown in Fig. 2.75a–d. The triangle carriers and clipped sine wave modulating signal for phase A are shown in Fig. 2.76. The simulation results for all A.M. index are tabulated in Table 2.18. In Fig. 2.75c, d, the fundamental and all the harmonic component values displayed are peak voltage in volts. Comparison of total harmonic distortion (THD) of line-to-line and line-to-neutral voltage and that of RMS values of their fundamental components by multi-carrier sine and clipped sine PWM techniques is shown in Figs. 2.77 and 2.78, respectively.

2.15 Discussion of Results

Referring to Fig. 2.77, it is seen that the THD of line-to-line voltage by MCCSPSPWM technique is low compared to MCSPSPWM technique up to a a.m. index of 0.5 and the former value is high compared to later value for a.m. index in the range 0.5–0.9, and thereafter, the former value becomes low

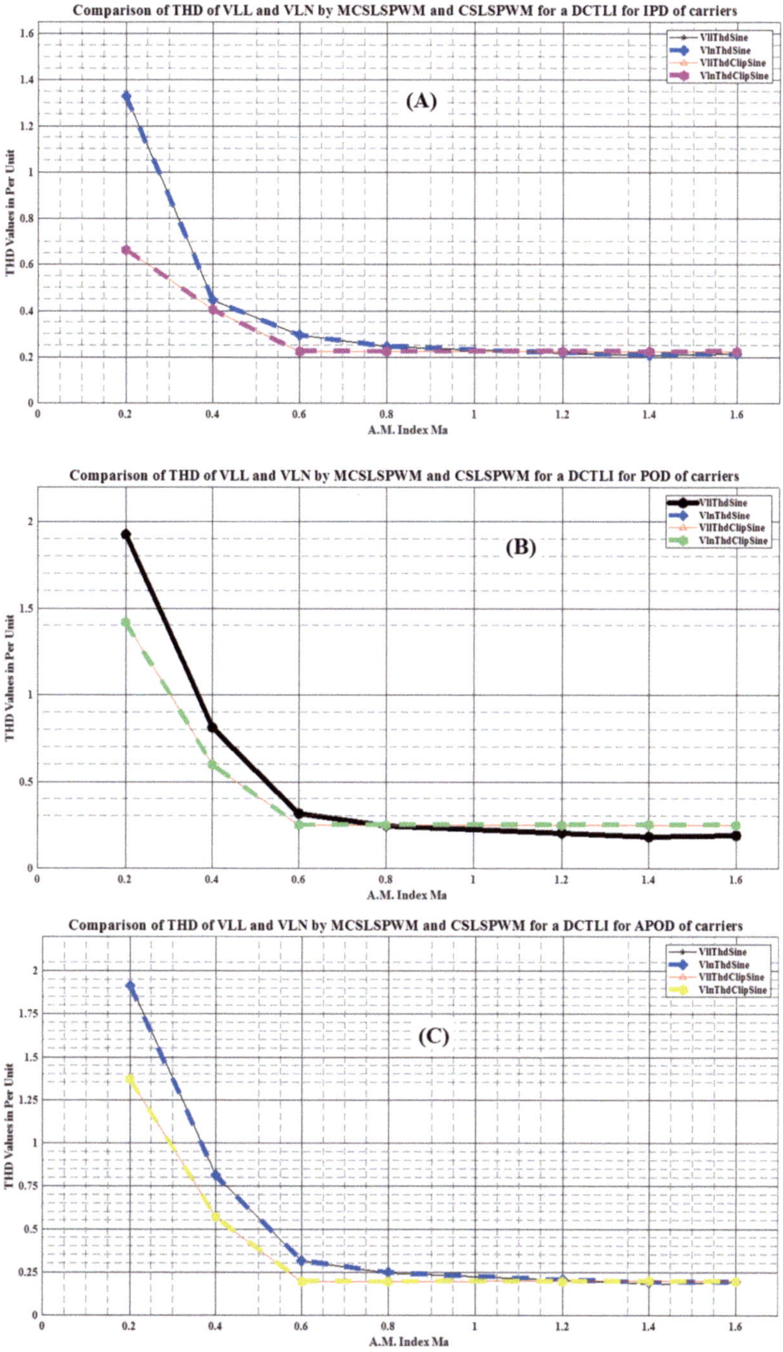

Fig. 2.58 Comparison of THD of line-to-line and line-to-neutral voltage of three-phase DCTLI by sine and clipped sine PWM techniques: (**a**) IPD of carriers, (**b**) POD of Carriers and (**c**) APOD of carriers

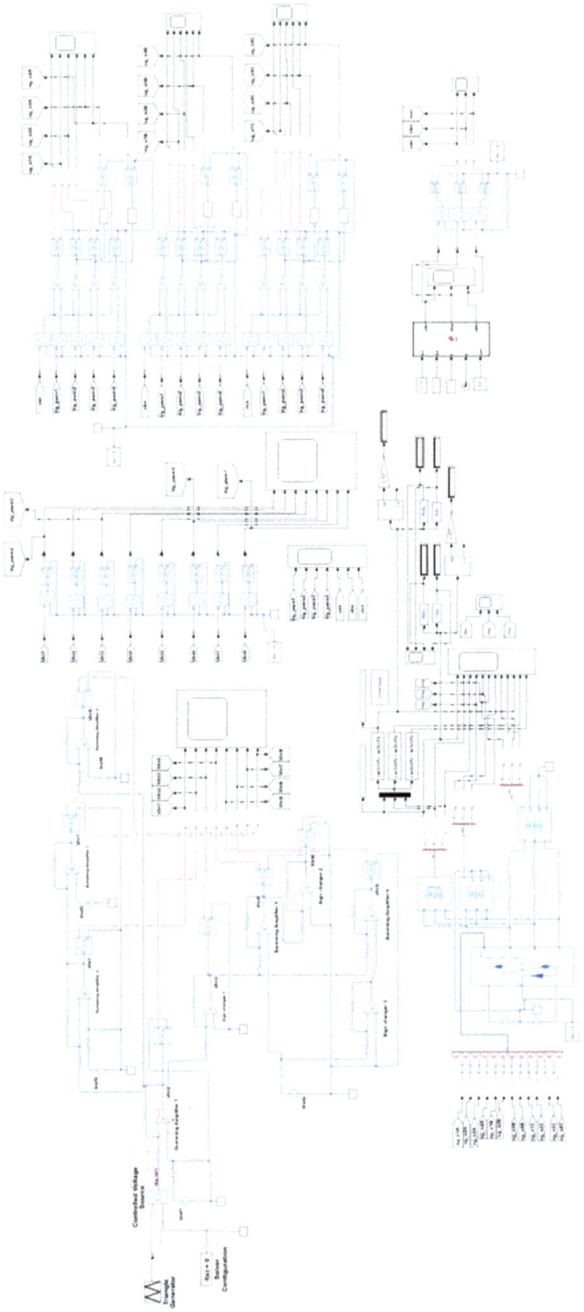

Fig. 2.59 Model of three-phase third harmonic injection sine level Shift PWM DCTLI

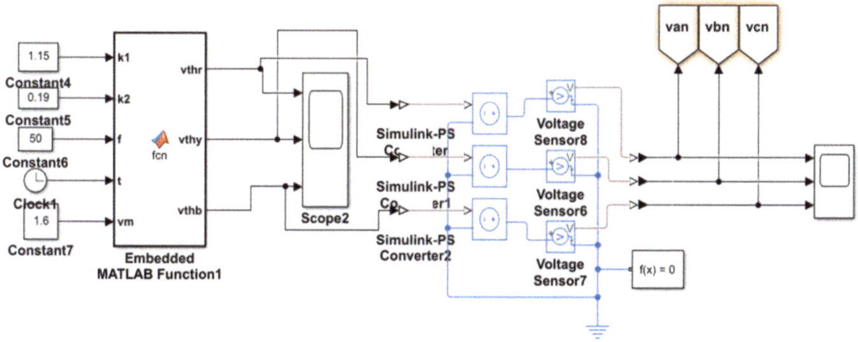

Fig. 2.60 Three-phase third harmonic injection sine PWM generator

compared to that of the later. Referring to Fig. 2.78, comparing the RMS value of fundamental component of line-to-line voltage and line-to-neutral voltage by MCCSPSPWM technique with that obtained by MCSPSPWM technique, it is seen that the former method gives a higher value compared to the later method for all a.m. indices.

2.16 Case Study: Model of Three-Phase Multi-carrier Third Harmonic Injection Sine Phase Shift PWM Five-Level Cascade H-Bridge Inverter

The model of the three-phase multi-carrier third harmonic injection sine phase shift PWM five level cascade H-bridge inverter (MCTHISPSPWMFLCHBI) is shown in Fig. 2.79 (Model file: CASE_STUDY_EX2_4). The various subsystems are shown in Fig. 2.70a–c. In Figs. 2.69 and 2.70a, three-phase sine wave modulating signal is replaced by three-phase third harmonic injection sine PWM generator with amplitude Am(0.8) and frequency fm(50 Hz) whose subsystem is shown in Fig. 2.80. The parameters shown in Table 2.1 are used to develop the model. Here two triangle carrier generators with peak value +/−1 volt with frequency ftri (1200 Hz) and phase shifted by $\pi/2$ radians from control and measurements/pulse and signal generators library from specialised technology block set are used along with three-phase third harmonic injection sine generator and relational operator comparators to develop gate pulse for the switches of FLCHBI, as already explained in Sect. 2.13.1.

The third harmonic injection sine (THIS) PWM generator shown in Fig. 2.80 is already explained in Section. 2.2 [1, 2]. In the model file CASE_STUDY_EX2_4, Program segment 2.1 refers to THIS PWM generation. In Fig. 2.80, sine wave modulating signal amplitude Am is 0.8 volts and frequency 50 Hz. This corresponds to an a.m. index of 0.8.

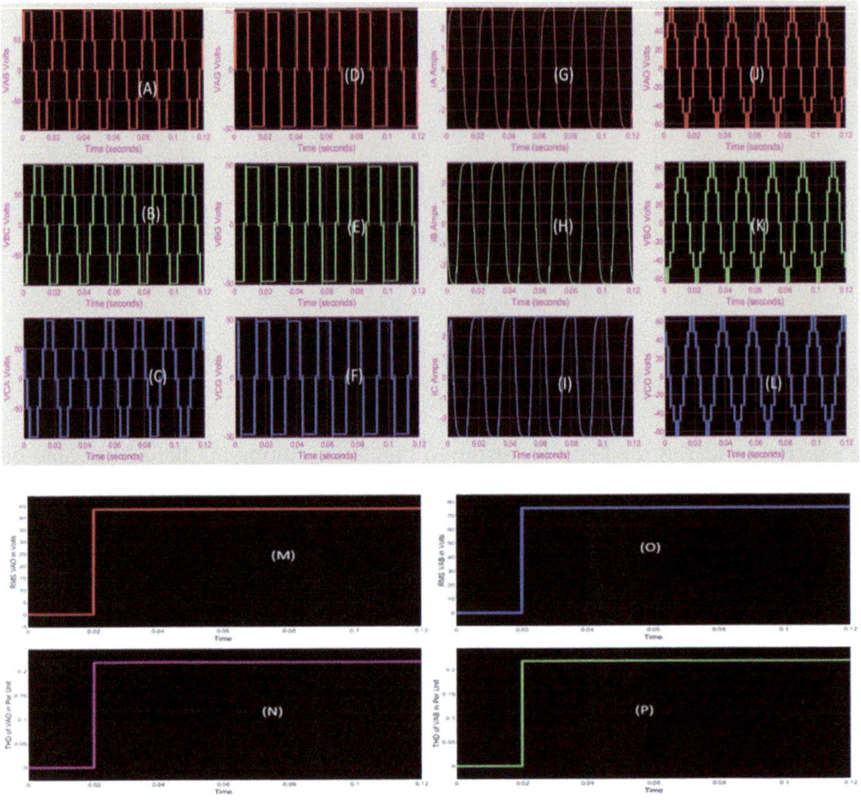

Fig. 2.61 Three-phase THISLSPWH DCTLI simulation results for IPD of carriers—(**a**), (**b**), (**c**) Line-to-line voltage, (**d**), (**e**), (**f**) Line-to-ground voltage, (**g**), (**h**), (**i**) Load current and (**j**), (**k**) and (**l**) Line-to-neutral voltage across load. Three-phase THISLSPWM DCTLI simulation results for IPD of carriers—(**m**), (**n**) RMS value and THD of line-to-neutral voltage, (**o**), (**p**) RMS value and THD of line-to-line voltage

2.16.1 Simulation Results

The simulation of the three-phase MCTHISPSPWMFLCHBI is carried out using Simulink [3]. The ode15s(stiff/NDF) solver is used. The data shown in Table 2.1 are used for simulation. The simulation results for an A.M. index Ma of 0.8 are shown in Fig. 2.81a–d. The triangle carriers and third harmonic sine wave modulating signal for phase A are shown in Fig. 2.82. The simulation results for all A.M. index are tabulated in Table 2.19. In Fig. 2.81c, d, the fundamental and all the harmonic component values displayed are peak voltage in volts. Comparison of total harmonic distortion (THD) of line-to-line and line-to-neutral voltage and that of RMS values of their fundamental components by multi-carrier sine and third harmonic injection sine PWM techniques is shown in Figs. 2.83 and 2.84, respectively.

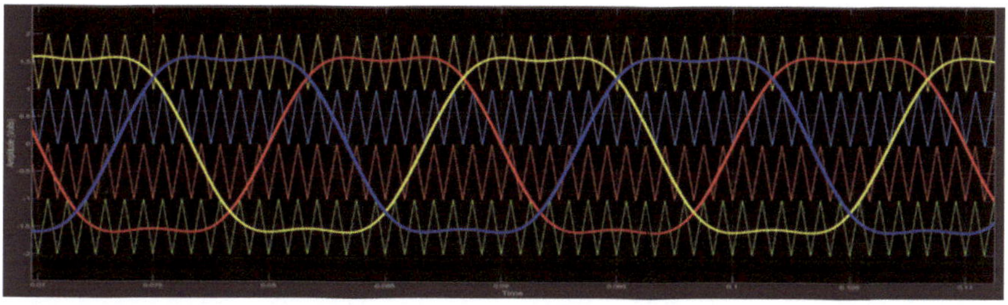

Fig. 2.62 Three-phase THISLSPWM DCTLI simulation result: four IPD of triangle carriers and three-Phase Third harmonic Injection. Sine Modulating Signal

Table 2.14 Three phase THISLSPWM DCTLI simulation results

S1. No	Ma	VLL (rms) V	VLL1 (rms) V	VLN (rms) V	VLN1 (rms) V	THD of VLL	THD of VLN	Remarks
1	0.2	33.72	25.73	19.47	14.85	0.8471	0.8471	IPD
2	0.4	58.1	53.88	33.55	31.11	0.4033	0.4032	IPD
3	0.6	73.74	71.7	42.58	41.4	0.2396	0.2397	IPD
4	0.8	75.4	73.67	43.53	42.53	0.2174	0.2174	IPD
5	1.2	76.68	74.79	44.27	43.18	0.2258	0.2259	IPD
6	1.4	76.83	74.86	44.36	43.22	0.2301	0.2299	IPD
7	1.6	76.96	74.92	44.43	43.26	0.2337	0.2331	IPD

2.17 Discussion of Results

Referring to Fig. 2.83, it is seen that the THD of line-to-line voltage by MCTHISPSPWM technique is low compared to MCSPSPWM technique up to an a.m. index of 0.5, and the former value is high compared to later value for a.m. index in the range 0.5 to 0.9, and thereafter, the former value becomes low compared to that of the later. Referring to Fig. 2.78, comparing the RMS value of fundamental component of line-to-line voltage and line-to-neutral voltage by MCTHISPSPWM technique with that obtained by MCSPSPWM technique, it is seen that the former method gives a higher value compared to the later method for all a.m. indices.

2.18 Three-Phase Modular Multilevel Converter

The modular multilevel converter (MMC) is a viable alternative to diode-clamped multilevel converters for medium- and high-voltage applications [12–15]. The MMC is an ideal choice for high-voltage DC (HVDC) transmission, medium-voltage drives (MVD), converters for renewable energy and offshore wind farms. One of the notable features of MMC is that it is modular and scalable. Modularity refers to the technique of developing larger systems by combining smaller

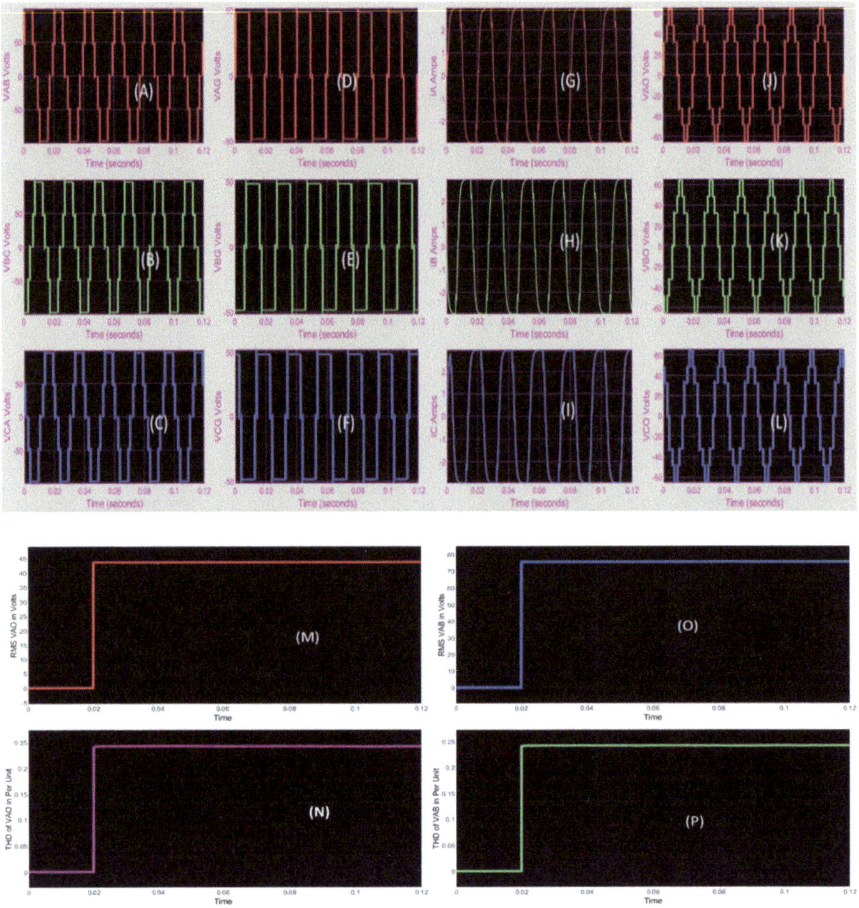

Fig. 2.63 Three-phase THISLSPWM DCTLI simulation results for POD of carriers. (**a**), (**b**), (**c**) Line-to-line voltage, (**d**), (**e**), (**f**) Line-to-ground voltage, (**g**), (**h**), (**i**) Load current and (**j**), (**k**) and (**l**) Line-to-neutral voltage across load. Three-phase THISLSPWM DCTLI simulation results for POD of carriers—(**m**), (**n**) RMS value and THD of line-to-neutral voltage, (**o**), (**p**) RMS value and THD of Line-to-line voltage

subsystems. These smaller subsystems are known as submodules (SM) or cells which are built using semiconductor switches and floating capacitors. A cascade connection of these converter cells or SM forms a chain link which is an ideal solution to reach high-voltage and high-quality near sinusoidal waveform. In order to transfer active power, isolated dc sources are required for cascade-connected converters. This problem is overcome by the use of one DC source along with floating capacitors in each cell which replace separate DC sources [13]. Depending on the output voltage requirement, SM can be added or removed which makes MMC

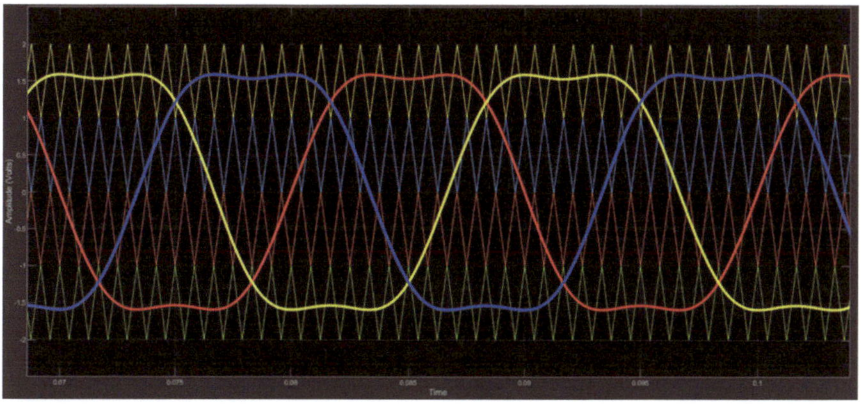

Fig. 2.64 Three-phase THISLSPWM DCTLI simulation result: four POD of triangle carriers and Three-phase third harmonic injection sine modulating signal

Table 2.15 Three phase THISLSPWM DCTLI simulation results

Sl. No	Ma	VLL (rms) V	VLLl (rms) V	VLN (rms) V	VLN1 (rms) V	THD of VLL	THD of VLN	Remarks
1	0.2	45.16	25.73	26.07	14.86	1.442	1.442	POD
2	0.4	63.78	53.88	36.82	31.11	0.6329	0.6329	POD
3	0.6	73.86	71.66	42.64	41.37	0.2485	0.2489	POD
4	0.8	75.53	73.4	43.61	42.38	0.2419	0.2418	POD
5	1.2	77.55	75.21	44.77	43.42	0.251	0.2513	POD
6	1.4	77.64	75.23	44.83	43.44	0.2545	0.2543	POD
7	1.6	77.73	75.25	44.88	43.45	0.2583	0.258	POD

flexible and scalable. The MMC semiconductor switch topology is either half bridge or full bridge. Semiconductor multilevel switch topologies using diode-clamped and flying capacitors are also in use.

The configuration of a three-phase MMC using half-bridge SM is shown in Fig. 2.85. Here each SM is connected in series, and there are N cells each in the upper and lower arms. The upper and lower arms together constitute one leg. Each arm is connected to an inductor L to limit the circulating current harmonics caused by switching the cell. Each cell in this half-bridge topology consists of an upper switch S1, lower switch S2 and a capacitor C connected as shown in Fig. 2.85b. The output phase to ground voltage has N + 1 levels. Although there are 2 N switches in the upper and lower arm put together, only N switches in total from the upper and lower arm put together will be ON any time. Referring to Fig. 2.85b, if switch S1 is ON and S2 is OFF, the output cell voltage Vout is V_c, and if S1 is OFF and S2 is ON, cell voltage Vout is zero. These two SM states are defined as ON and OFF states, respectively. The DC link voltage is V_{dc}. The SM has three modes of operation given below:

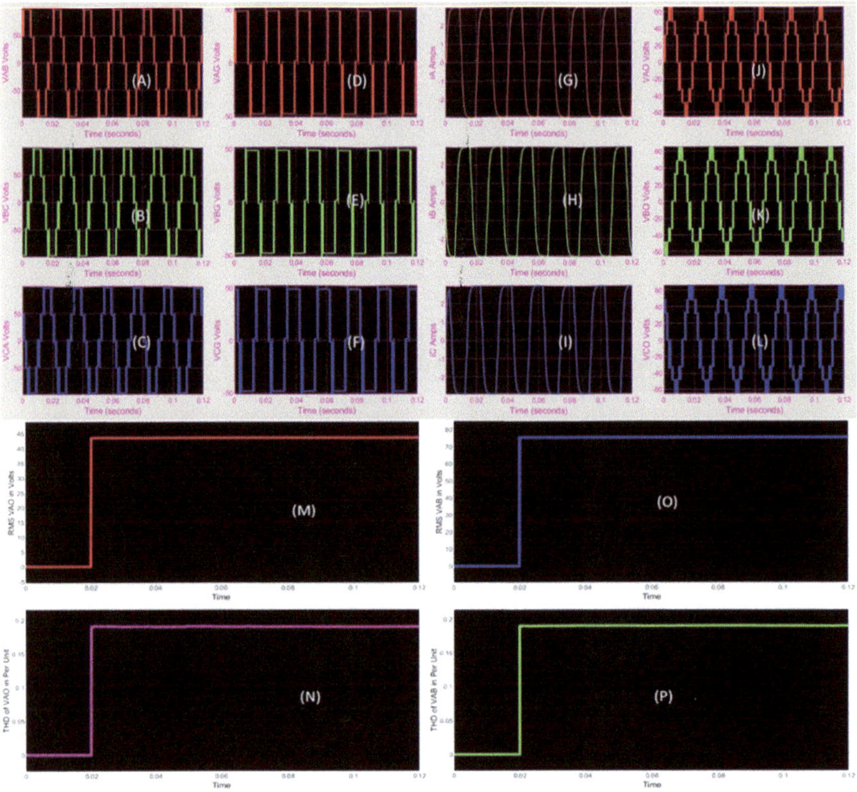

Fig. 2.65 Three-phase THISLSPWM DCTLI simulation results for APOD of carriers—(**a**), (**b**), (**c**) line-to-line voltage, (**d**), (**e**), (**f**) Line-to-ground voltage, (**g**), (**h**), (**i**) Load current and (**j**), (**k**) and (**l**) Line-to-neutral voltage across load. Three-phase THISLSPWM DCTLI simulation results for APOD of carriers—(**m**), (**n**) RMS value and THD of line-to-neutral voltage (**o**), (**p**) RMS value and THD of Line-to-Line Voltage

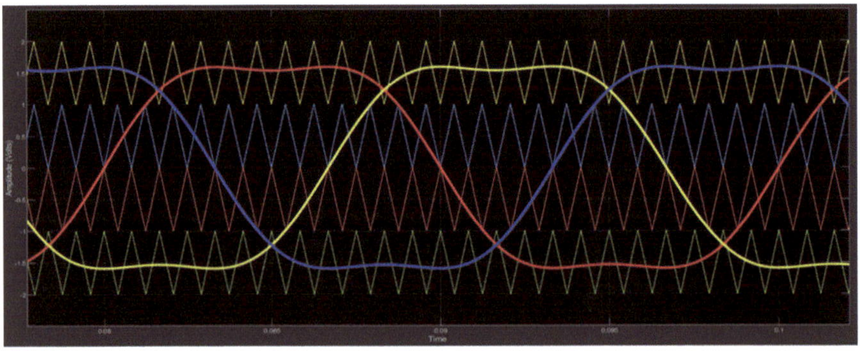

Fig. 2.66 Three-phase THISLSPWM DCTLI simulation result: four APOD of triangle carriers and three-phase third harmonic injection. Sine modulation signal

Table 2.16 Three phase THISLSPWM DCTLI simulation results

S1. No	Ma	VLL (rms) V	VLL1 (rms) V	VLN (rms) V	VLN1 (rms) V	THD of VLL	THD of VLN	Remarks
1	0.2	44.89	25.75	25.92	14.87	1.428	1.428	APOD
2	0.4	63.5	53.9	36.66	31.12	0.6228	0.6226	APOD
3	0.6	73.64	71.74	42.51	41.42	0.2304	0.2303	APOD
4	0.8	75.27	73.94	43.46	42.69	0.19	0.1901	APOD
5	1.2	75.82	74.37	43.77	42.93	0.1972	0.1978	APOD
6	1.4	76.01	74.47	43.88	43	0.2036	0.2024	APOD
7	1.6	76.18	74.58	43.98	43.07	0.2071	0.2063	APOD

1. Insert Mode: Switch S1 ON and S2 OFF. During this mode capacitor either charges or discharges.
2. Bypass Mode: Switch S1 is OFF and S2 is ON. No change in capacitor voltage.
3. Block Mode: Switch S1 and S2 are both OFF. During this mode capacitor only charges through freewheeling diodes of switches but never discharge.

Capacitor voltage in each SM is related to switch state and direction of arm current. This is shown in Table 2.20. In Table 2.20, i_{ux} and i_{lx} are the upper arm and lower arm currents and $x \in a, b, c$, respectively.

2.18.1 MMC Mathematical Model

The equivalent circuit for one phase of the MMC is shown in Fig. 2.86. In Fig. 2.86, v_{ox} and i_{ox} are the output phase voltage and load current where $x \in a, b, c$. Also v_{ux}, v_{lx}, i_{ux} and i_{lx} are the upper and lower arm voltages and currents, respectively. Referring to Fig. 2.86, the voltage expressions for the upper and lower arms can be expressed as follows:

$$\frac{V_{dc}}{2} - V_{ux} + R * i_{ux} + L * \frac{di_{ux}}{dt} - V_{ox} = 0 \tag{2.15}$$

$$\frac{V_{dc}}{2} - V_{lx} + R * i_{lx} + L * \frac{di_{lx}}{dt} + V_{ox} = 0 \tag{2.16}$$

Now define the following currents:

$$i_{ox} = (i_{ux} - i_{lx}) \tag{2.17}$$

$$i_{cx} = \frac{(i_{ux} + i_{lx})}{2} \tag{2.18}$$

Adding and then subtracting Eqs. 2.15 and 2.16 and using Eqs. 2.17 and 2.18, we have the following:

$$V_{dc} - (V_{ux} + V_{lx}) + 2 * R * i_{cx} + 2 * L * \frac{di_{cx}}{dt} = 0 \tag{2.19}$$

$$(V_{lx} - V_{ux}) + R * i_{ox} + L * \frac{di_{ox}}{dt} - 2 * V_{ox} = 0 \tag{2.20}$$

Now define the following:

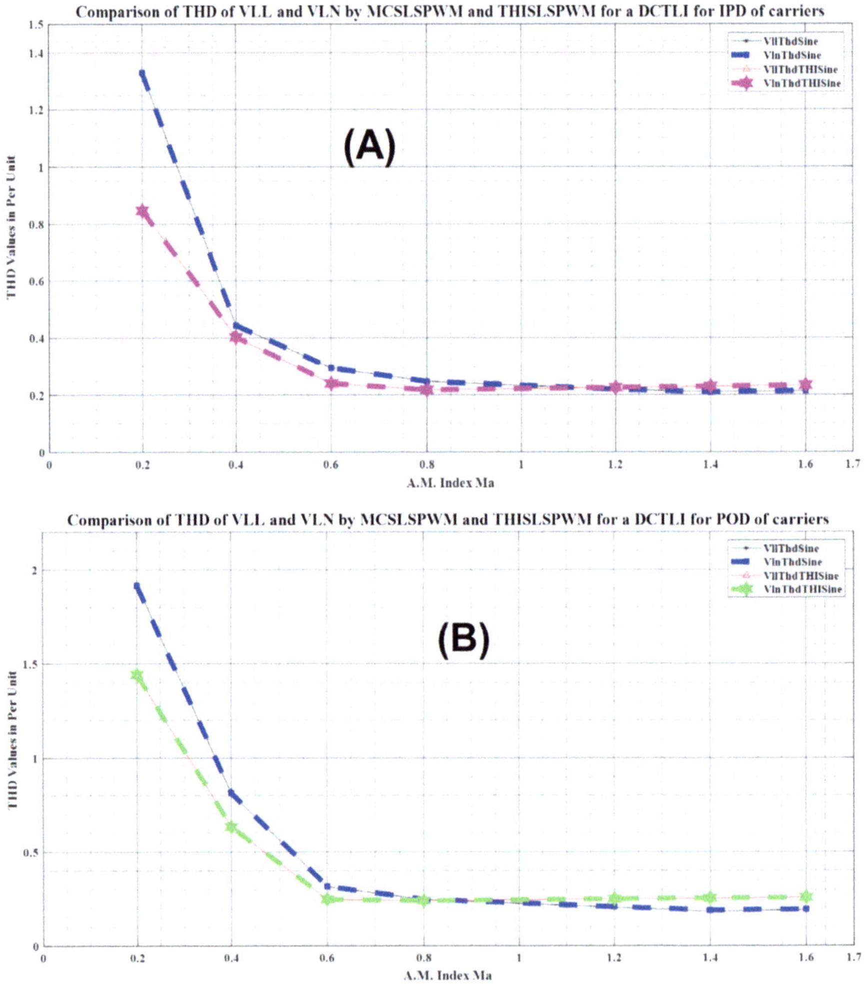

Fig. 2.67 Comparison of THD of line-to-line and line-to-neutral voltages of DCTLI by MCSLSPWM and THISLSPWM techniques—(**a**) IPD, (**b**) POD and (**c**) APOD of triangle carriers

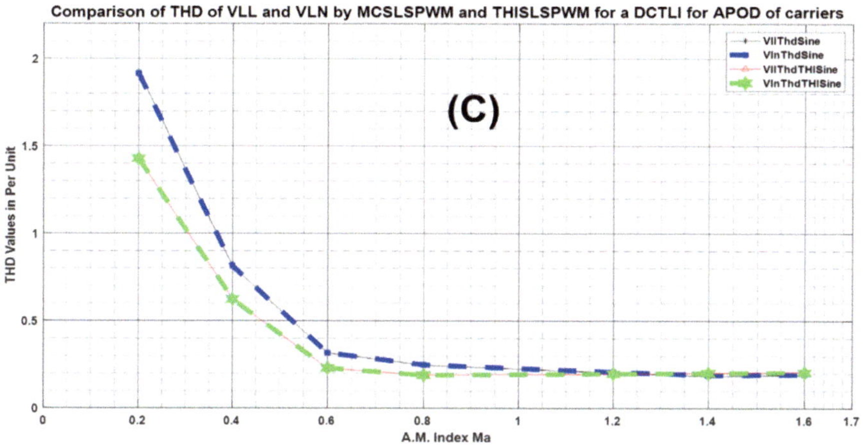

Fig. 2.67 (continued)

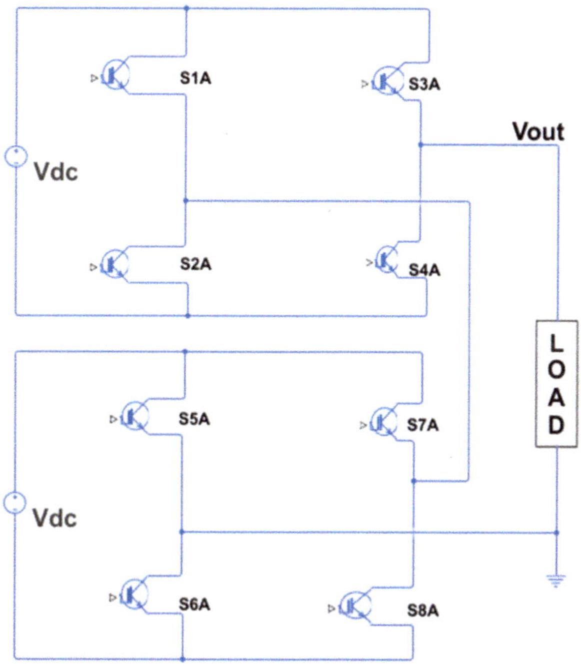

Fig. 2.68 Single-phase five-level cascade H-bridge inverter

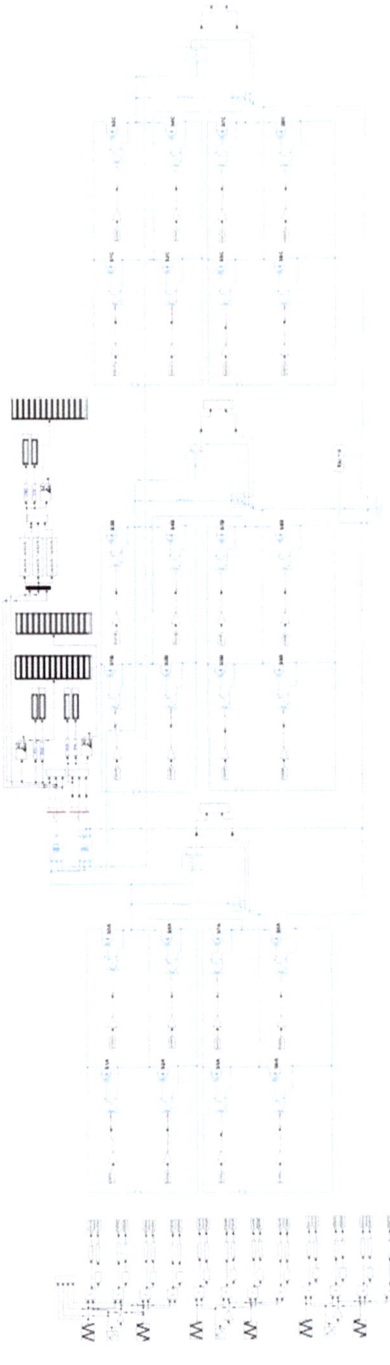

Fig. 2.69 Model of three-phase sine phase shift PWM five-level cascade H-bridge inverter

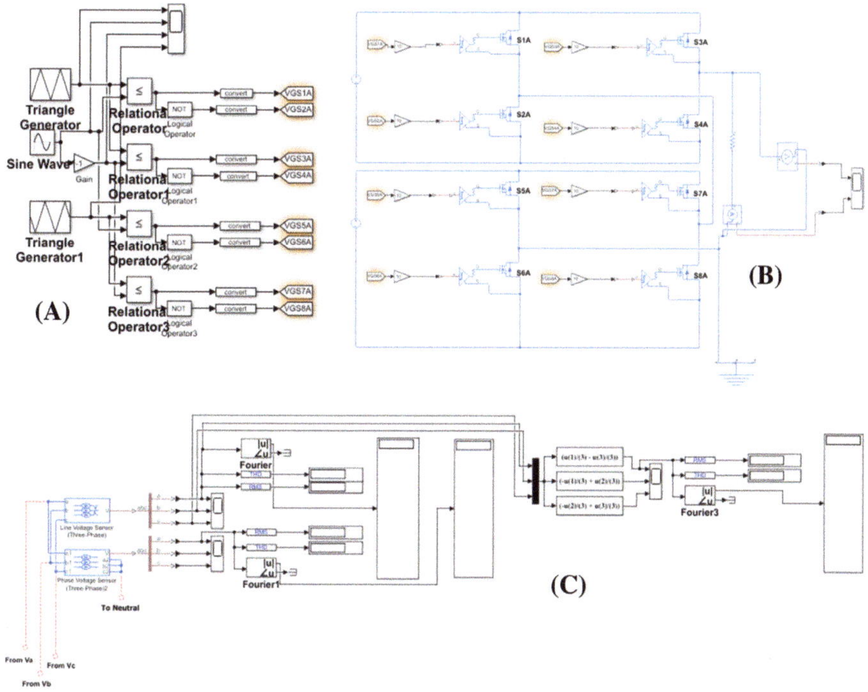

Fig. 2.70 MSPSPWM FLCHBI: (**a**) Gate drive for phase A and (**b**) Switch arrangement for phase A MSPSPWM FLCHBI: (**c**) Measurements unit

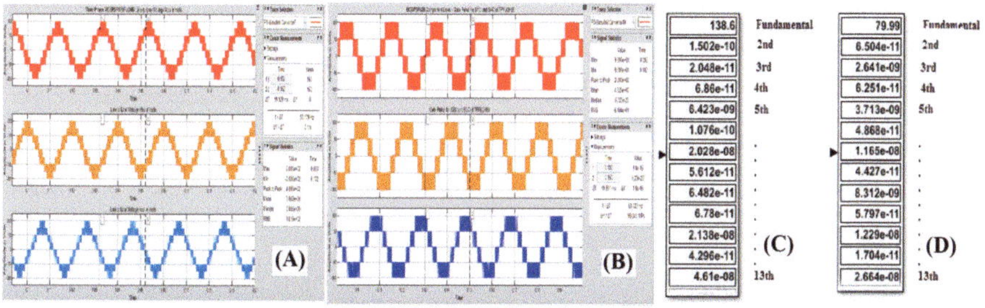

Fig. 2.71 Three-phase MCSPSPWMFLCHBI simulation results: (**a**) Line-to-line voltage (top to bottom), (**b**) Line-to-ground voltage (top to bottom), (**c**) Harmonic spectrum of line-to-line voltage and (**d**) Line-to-ground voltage

$$\frac{(V_{lx} - V_{ux})}{2} = V_{sx} \tag{2.21}$$

$$\frac{(V_{lx} + V_{ux})}{2} = V_{cx} \tag{2.22}$$

Using Eqs. 2.19, 2.20, 2.21 and 2.22, the following expressions:

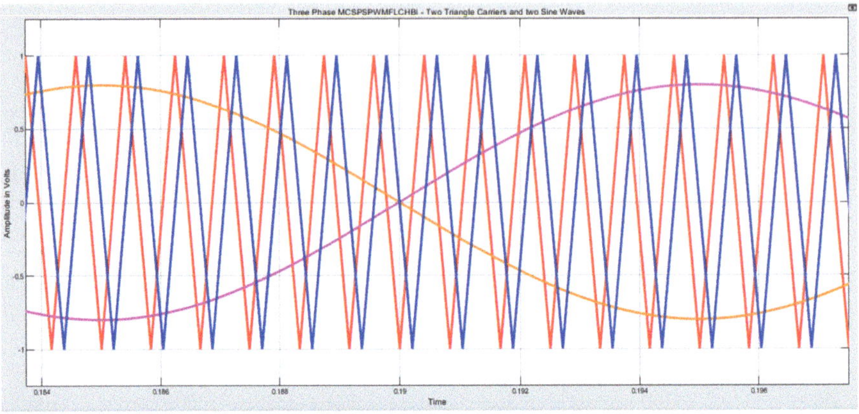

Fig. 2.72 Three-phase MCSPSPWMFLCHBI simulation results: two triangle carriers and two sine wave modulating signals. Sine wave peak 0.8 volts

Table 2.17 Three phase MCSPSFWM FLCHB inverter simulation results

S1. No	Ma	VLL (rms) V	VLL1 (rms) V	VLN (rms) V	VLN1 (rms) V	THD of VLL	THD of VLN	Remarks
1	0.2	41.69	24.498	25.22	14.144	1.378	1.477	Under modulation
2	0.4	58.97	49.00	35.67	28.29	0.67	0.7688	–
3	0.6	75.87	73.48	46.44	42.43	0.2573	0.4451	–
4	0.8	102.2	98.02	60.58	56.57	0.2965	0.3833	–
5	1.2	137.5	135.29	79.93	73.147	0.181	0.2174	Over modulation
6	1.4	142.8	141.44	83.58	81.683	0.14	0.2193	———
7	1.6	146.2	145.12	86.04	83.73	0.1258	0.2354	———

$$R * i_{cx} + L * \frac{di_{cx}}{dt} = V_{cx} - \frac{V_{dc}}{2} \tag{2.23}$$

$$\frac{R}{2} * i_{ox} + \frac{L}{2} * \frac{di_{ox}}{dt} = V_{ox} - V_{sx} \tag{2.24}$$

In Eqs. 2.17 and 2.18, i_{ox} and i_{cx} are the output current and circulating current, respectively.

Also referring to Fig. 2.85, in any phase if $k1$ SMs are ON in the upper arm and $k2$ SMs are ON in the lower arm, then the relation connecting $k1$, $k2$, total number of cells per arm N and the voltage across cell capacitors can be expressed as follows:

$$N = (k_1 + k_2) \tag{2.25}$$

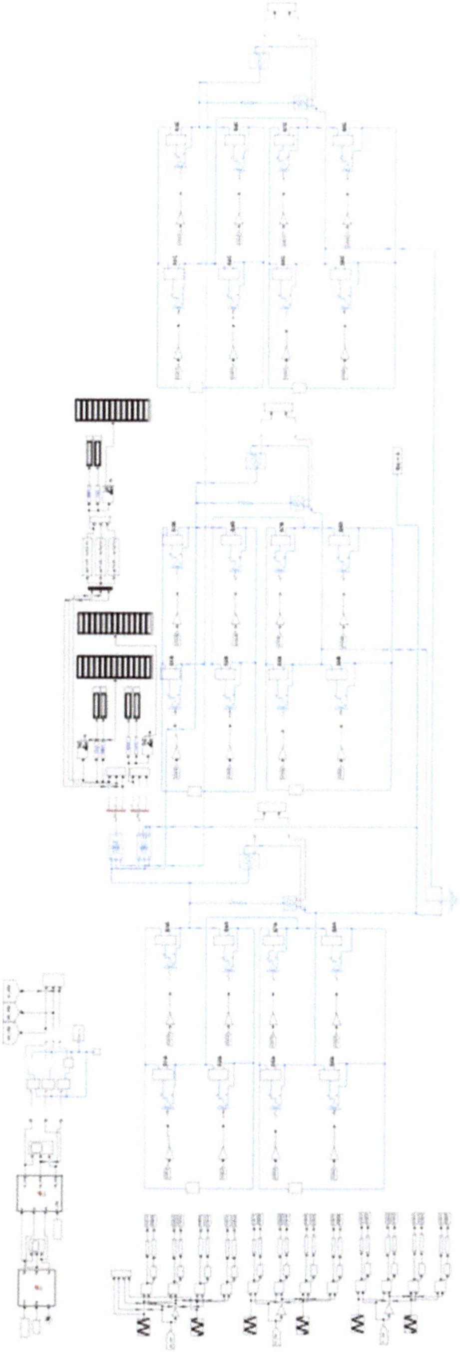

Fig. 2.73 Model of three-phase multicarrier clipped sine phase shift PWM five-level cascade H-bridge inverter

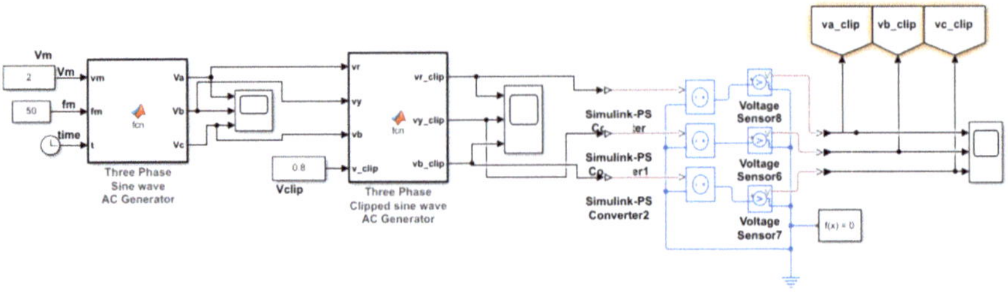

Fig. 2.74 Model of three-phase clipped sinusoid PWM generator

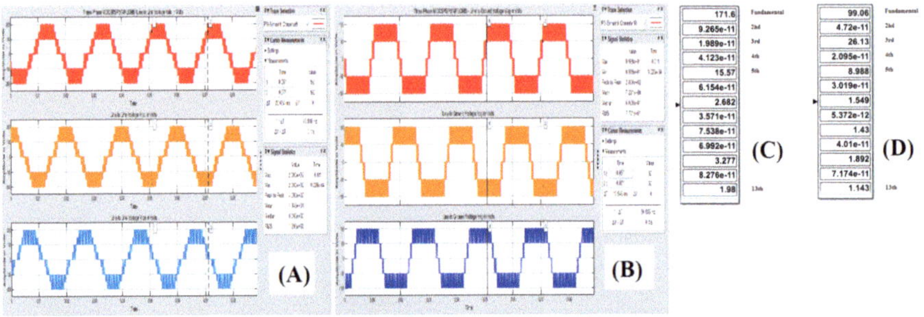

Fig. 2.75 Three-phase MCCSPSPWMFLCHBI simulation results: (**a**) Line-to-line voltage (top to bottom), (**b**) Line-to-neutral voltage (top to bottom), (**c**) Harmonic spectrum of line-to-line voltage and (**d**) Line-to-neutral voltage

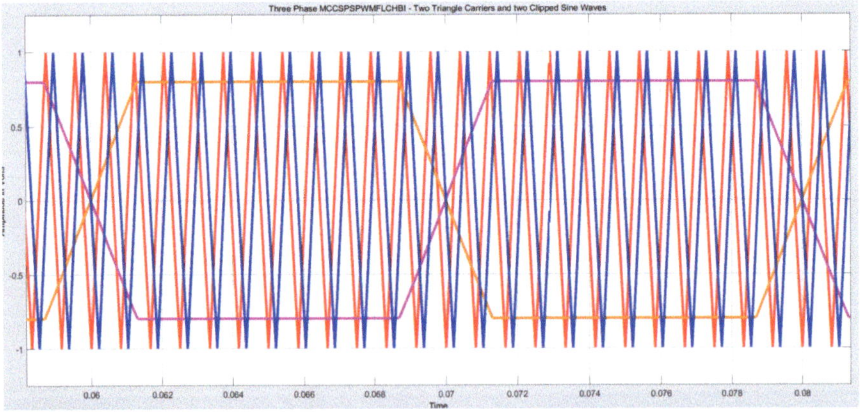

Fig. 2.76 Three-phase MCCSPSPWMFLCHBI simulation results: two triangle carriers and two clipped sine waveforms. Clipped sine wave peak 0.8 volts

Table 2.18 Three phase MCCSPSPWMFL CHB inverter simulation results

S1. No	Ma	VLL (rms) V	VLL1 (rms) V	VLN (rms) V	VLN1 (rms) V	THD of VLL	THD of VLN	Remarks
1	0.2	50.96	31.13	31.07	17.977	1.296	1.41	Under modulation
2	0.4	71.21	61.93	43.23	35.77	0.567	0.679	–
3	0.6	98.67	92.15	59.46	53.20	0.3831	0.5	–
4	0.8	127.9	121.36	76.48	70.06	0.3349	0.4382	–
5	1.2	150.3	149.2	89.23	86.14	0.1213	027	Over modulation
6	1.4	150.3	149.22	89.23	86.14	0.1213	027	
7	1.6	150.3	149.22	89.23	86.14	0.1213	0.27	

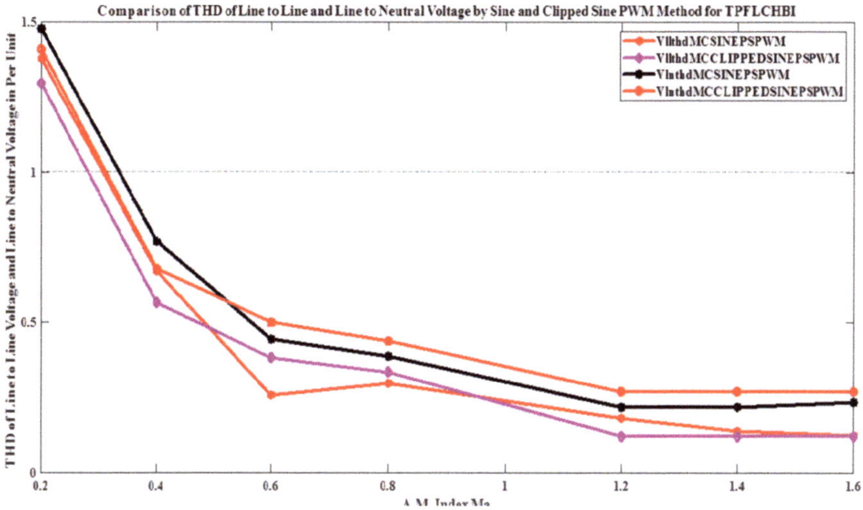

Fig. 2.77 TPFLCHBI THD comparison for line-to-line and line to neutral voltage by sine and clipped sine phase shift PWM methods

$$V_{ci} = \frac{V_{dc}}{N}, \mathrm{i} = 1, 2, 3, \ldots..N \tag{2.26}$$

2.18.2 MMC Modulation

The number of inserted cells in the ON state in any one leg depends on the control signal and can vary between 0 and N. The control signal m is a sine function which is synchronised with reference coordinate frame using phase-locked loop (PLL), as defined below:

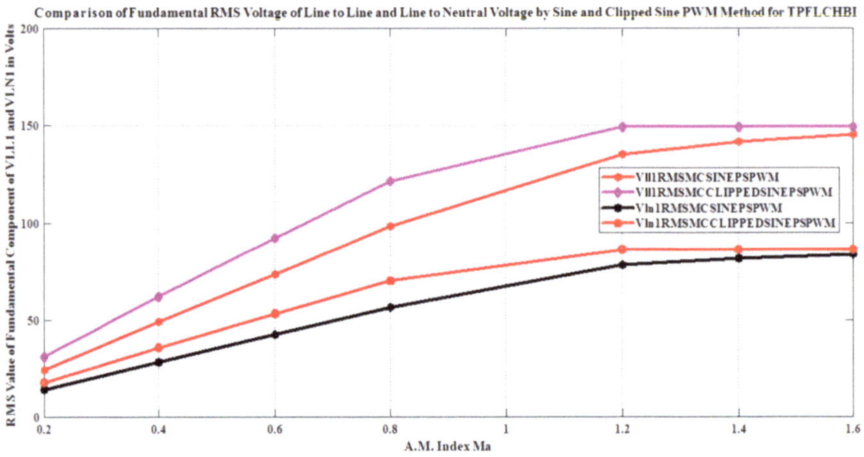

Fig. 2.78 TPFLCHBI comparison of fundamental RMS line-to-line and line-to-neutral voltage by sine and clipped sine phase shift PWM methods

$$m_x = M * \cos(\omega.t + \theta_x) \text{ for } 0 < M < 1 \text{ and } -1 < m < +1 \tag{2.27}$$

where $\theta_x = 0, -2\pi/3, -4\pi/3$ for $x = a, b, c$, respectively, and ω is the angular frequency of the inverter output voltage.

The control signals for each arm are derived to satisfy the DC voltage balance in Eq. 2.23. Thus, control signals for the upper and lower arms, m_u and m_l, can be expressed as follows:

$$m_{ux} = \frac{1}{2} - \frac{M}{2} * \cos(\omega.t + \theta_x) \text{ for } 0 < m_u < 1 \tag{2.28}$$

$$m_{lx} = \frac{1}{2} + \frac{M}{2} * \cos(\omega.t + \theta_x) \text{ for } 0 < m_l < 1 \tag{2.29}$$

Considering inserted capacitors in each arm, the dynamic equations for the upper and lower arm voltage in any phase can be expressed as follows:

$$\frac{dv_{ux\_\max}}{dt} = \frac{m_u * i_{ux}}{C_{u_{\text{arm}}}} \tag{2.30}$$

$$\frac{dv_{lx\_\max}}{dt} = \frac{-m_l * i_{lx}}{C_{l\_\text{arm}}} \tag{2.31}$$

where C_{u_arm} and C_{l_arm} are the total value of inserted capacitors in the upper and lower arms, respectively. If $k1$ and $k2$ are the number of inserted capacitors in the

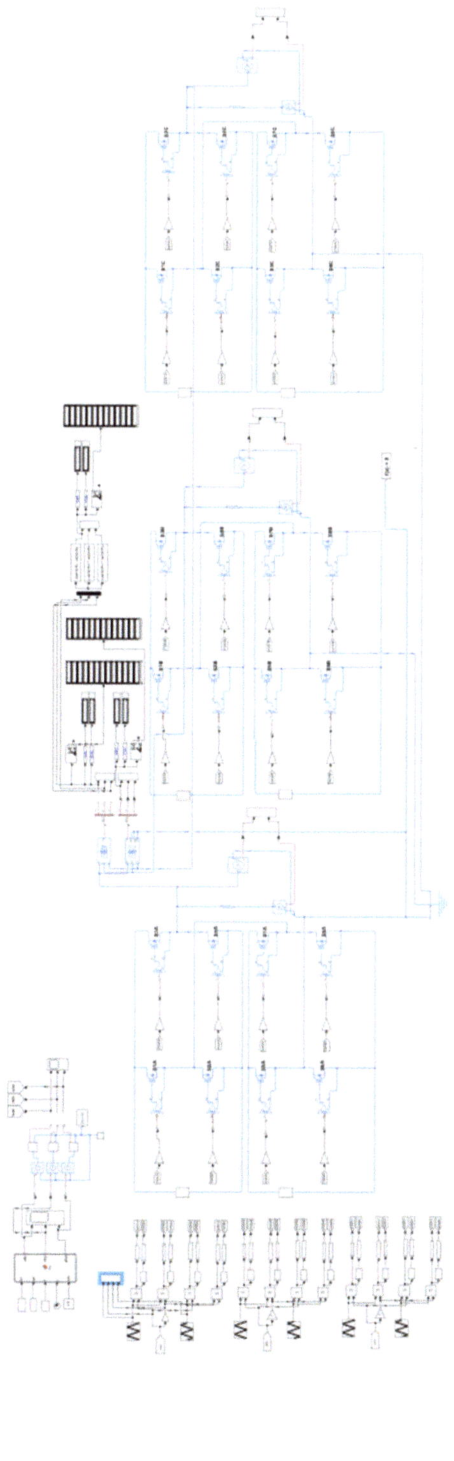

Fig. 2.79 Model of three-phase multicarrier third harmonic injection sine phase shift PWM five-level cascade H-bridge inverter

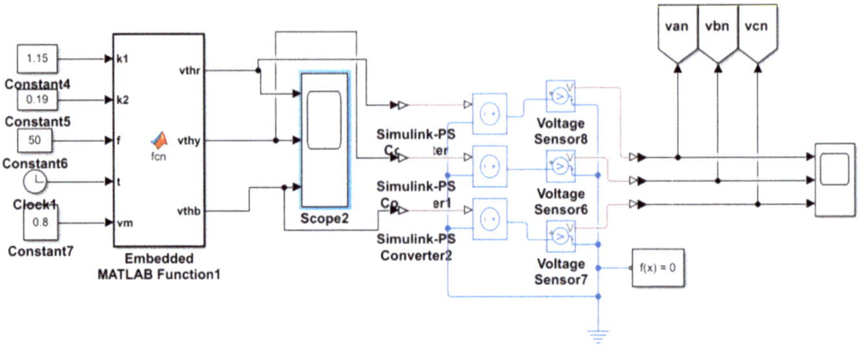

Fig. 2.80 Three-phase third harmonic injection sine wave generator

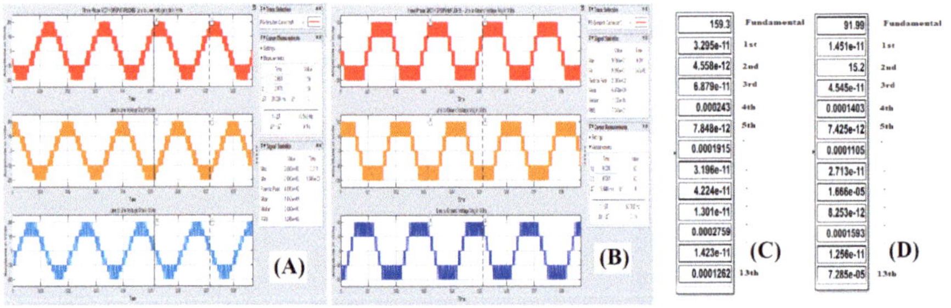

Fig. 2.81 Three-phase MCTHISPSPWMFLCHBI simulation results: (**a**) Line-to-line voltage (top to bottom), (**b**) Line-to-neutral voltage (top to bottom), (**c**) Harmonic spectrum line-to-line voltage and (**d**) Line-to-neutral voltage

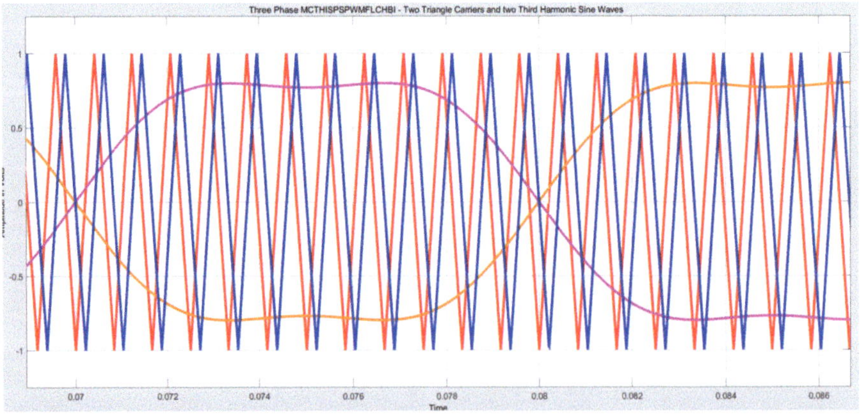

Fig. 2.82 Three-phase MCTHISPSPWMFLCHBI simulation results: two triangle carriers and two third harmonic injection sine wave

Table 2.19 Three phase MCTHISPSPWMFLCHB inverter simulation results

Sl. No	Ma	VLL (rms) V	VLL1 (rms) V	VLN (rms) V	VLN1 (rms) V	THD of VLL	THD of VLN	Remarks
1	0.2	46.48	28.17	27,78	16.266	1.313	1.385	Under Modulation
2	0.4	65.73	56.34	39.29	32.53	0.60	0.6778	–
3	0.6	89	84.5	52.88	48.79	0.33	042	–
4	0.8	118.7	112.66	69.81	65.056	0.3315	0.3894	–
5	1.2	149.5	148.444	88.62	85.714	0.123	0.2646	Over Modulation
6	1.4	152	150.778	90.67	87.058	0.13	0.29	
7	1.6	153.7	152.19	92.07	87.836	0.143	0.3141	

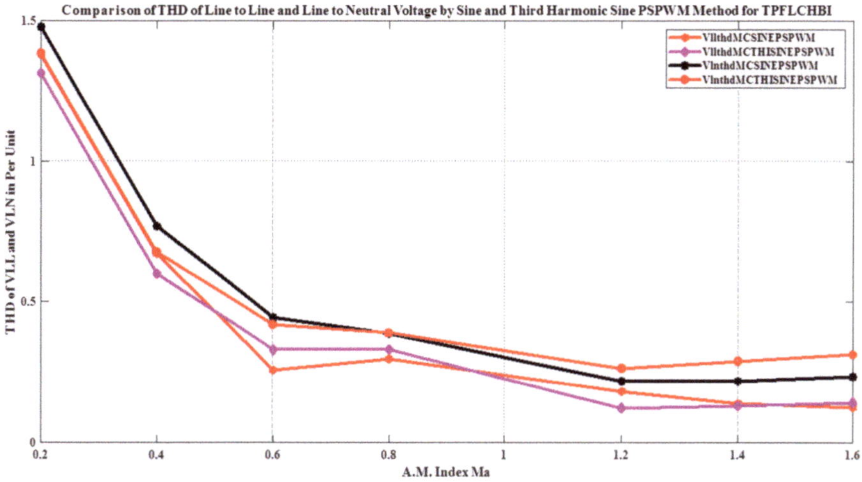

Fig. 2.83 TPFLCHBI comparison of THD of line-to-line and line-to-neutral voltage by sine and third harmonic injection sine phase shift PWM methods

upper and lower arms, respectively, and C is the capacitor in a single SM, then using Eqs. 2.17 and 2.18, Eqs. 2.30 and 2.31 can be expressed as follows:

$$\frac{dv_{ux\_\max}}{dt} = \frac{m_u}{C/k_1} * \left(i_{cx} + \frac{i_{ox}}{2}\right) \tag{2.32}$$

$$\frac{dv_{lx\_\max}}{dt} = \frac{m_l}{C/k_2} * \left(i_{cx} - \frac{i_{ox}}{2}\right) \tag{2.33}$$

The product of sine waves m_u, m_l with i_{ox}, produces a DC and second harmonic voltage which forces circulating current through the arm inductor. The arm inductor value is so chosen to limit this circulating current. The phase-shifted carrier (PSC) modulation is presented below:

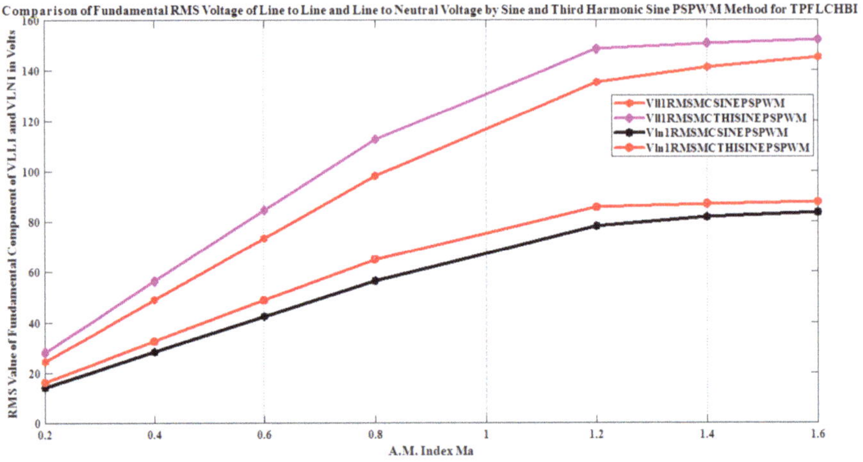

Fig. 2.84 TPFLCHBI comparison of fundamental RMS value of line-to-line and line-to-neutral voltage by sine and third harmonic injection sine phase shift PWM methods

Here a total of 2 N triangle carriers with the same frequency fc Hz is used, N triangle carriers each for the upper and lower arm phase shifted by $2\pi/N$ radians between neighbouring triangle carriers. Thus the ith triangle carrier in the upper and lower arm has a phase shift θ_{ui} and θ_{li} defined below:

$$\theta_{ui} = \frac{2\pi}{N} * (i - 1) \tag{2.34}$$

$$\theta_{li} = \theta + \frac{2\pi}{N} * (i - 1) \tag{2.35}$$

In Eq. 2.35, θ is the displacement angle between upper and lower arm carriers and varies from 0 to π/N radians. This value of θ is chosen to minimise the output phase voltage harmonics and to eliminate the circulating current harmonics [14]. The upper and lower arm triangle carriers along with sine wave modulation functions m_u and m_l are shown in Fig. 2.87. The gate switching pulse for each SM is generated by comparing the reference modulating sine wave signal with the respective triangle carrier.

2.18.3 MMC Capacitor Voltage Balancing

The capacitor voltages of the individual SMs must be monitored and kept equal to the value in Eq. 2.26. The voltages of the capacitors are periodically measured and sorted according to their value in the descending order by software. The modulation

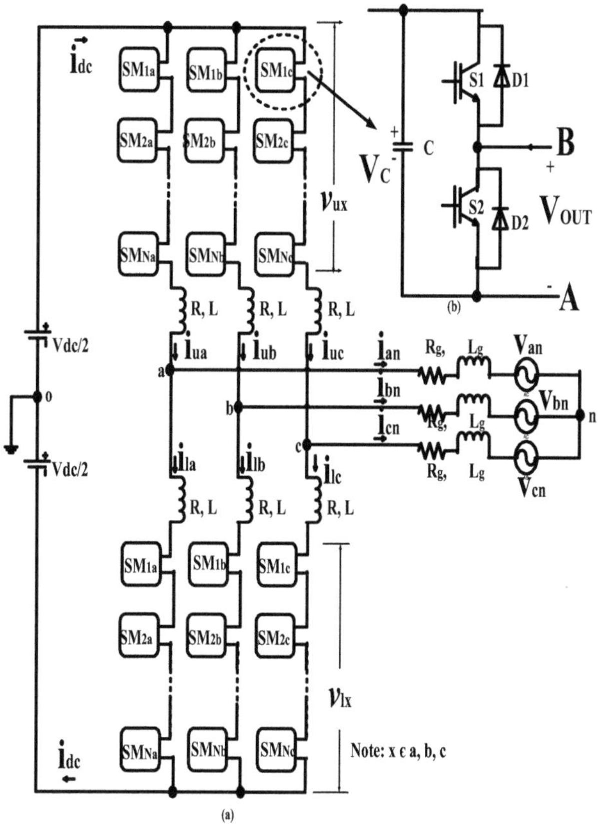

Fig. 2.85 (**a**) Three-phase modular multilevel converter and (**b**) Half bridge submodule

Table 2.20 Submodule capacitor state

Arm current (i_{ux} or i_{lx})	Switch state	SM state	Capacitor state	Capacitor voltage V_C	SM voltage Vout	SM mode
Positive (>0)	S1=ON; S2=OFF	ON	Charging	Increases	Vc	Insert
Positive (>0)	S1=OFF; S2=ON	OFF	Bypass	NO change	0	Bypass
Negative(<0)	S1=ON; S2=OFF	ON	Discharging	Decreases	Vc	Insert
Negative(<0)	S1=OFF; S2=ON	OFF	Bypass	NO change	0	Bypass

strategy determines the number of SMs that should be ON in the upper and lower arms of the MMC. Capacitor voltage values and also the direction of the arm currents are used to select switch S1 of the SMs in the upper and lower arm and to determine which SMs should be switched ON. When the current in the upper or lower arm is

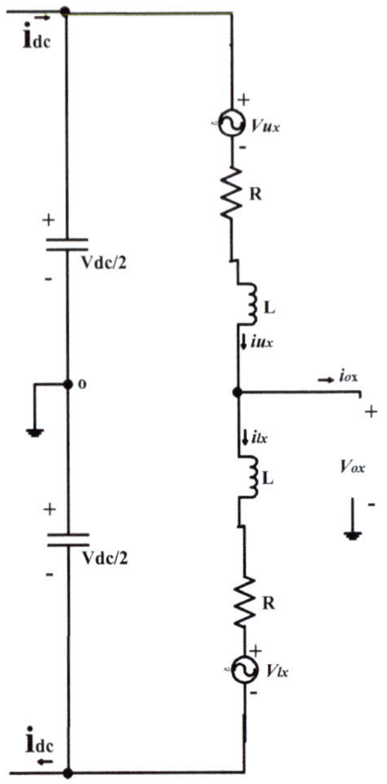

Fig. 2.86 Modular multilevel converter equivalent circuit

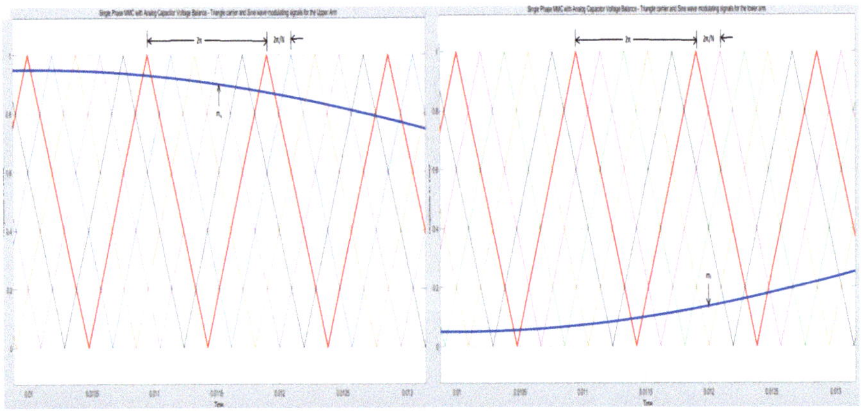

Fig. 2.87 Triangle carriers and sine wave modulating signal for upper arm (left) and lower arm (right)

positive indicating charging, to apply the desired arm voltage, the required SMs with the lowest voltages are switched ON. When the current in the upper or lower arm is negative indicating discharging, to apply the desired arm voltage, the SMs with the highest voltages are turned ON. Regardless of the direction of the upper or lower arm current, if S1 of the SM is OFF, the corresponding capacitor will be bypassed, and its voltage remains unchanged. By this method, continuous balancing of the capacitor voltages is achieved.

Analogue technique using operational amplifier comparators to achieve capacitor voltage balancing is reported [14]. This analogue technique requires no sorting of cell capacitor voltage in the ascending or descending order. The reference modulating sine wave signal is adjusted by adding a small voltage proportional to the difference between reference capacitor voltage and actual voltage of the capacitor in each SM which either increases or decreases the ON time thereby controlling the charge/discharge duration of the SM capacitor [14]. Here a modified form of this analogue technique is given below:

Referring to Eqs. 2.28 and 2.29, the new modulating sine wave functions for the upper and lower arms, m_{ux_new} and m_{lx_new}, can be expressed as follows:

$$m_{ux_{new}}(i) = m_{ux} + \delta m_{ux}(i) \tag{2.36}$$

$$m_{lx_{new}}(i) = m_{lx} + \delta m_{lx}(i) \tag{2.37}$$

The incremental modulation functions $\delta m_{ux}(i)$ and $\delta m_{lx}(i)$ for the uuper and lower arms, respectively, are defined below:

$$\delta m_{ux}(i) = K_P * (V_{\mathrm{cux}}(i) - V_{ci}) * i_{ux}/I_{dc_{\mathrm{ref}}} \tag{2.38}$$

$$\delta m_{lx}(i) = K_P * (V_{\mathrm{clx}}(i) - V_{ci}) * i_{lx}/I_{dc_{\mathrm{ref}}} \tag{2.39}$$

where $i = 1, 2, 3, \ldots .N$ and $x \in a, b, c$. K_P is a proportional constant. If P is the power output of the MMC, then neglecting switching losses, I_{dc_ref} is defined below:

$$I_{dc\_ref} = \frac{P}{V_{dc}} \tag{2.40}$$

Equations 2.28, 2.29, 2.36, 2.37, 2.38, 2.39 and 2.40 can be implemented using function generators, error amplifiers, multipliers and analogue op.amp comparators for each of the cells in the upper and lower arms. As the measured value of cell capacitor voltage comparison with capacitor reference voltage and the action to increase or decrease the charge/discharge time duration in the insert mode or to implement bypass mode takes place at the same instant of time for all cell capacitors in the upper and lower arms, no ascending or descending order sorting based on cell capacitor voltage magnitude is required.

2.19 Model of a Three-Phase Modular Multilevel Converter

The model of a three-phase MMC without implementing the capacitor voltage balance is shown in Fig. 2.88 (Model file: EXAMPLE 2_8). The submodule and gate pulse generator are shown in Fig. 2.89a, b. The model of a single-phase MMC with capacitor voltage balance implemented is presented in the next section.

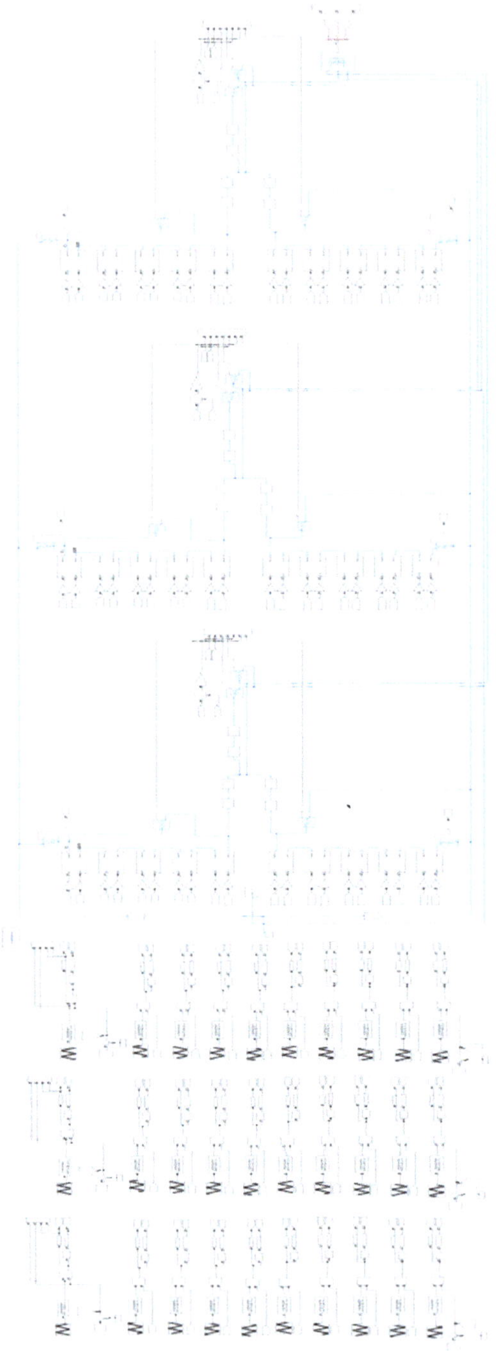

Fig. 2.88 Model of a three-phase modular multilevel converter

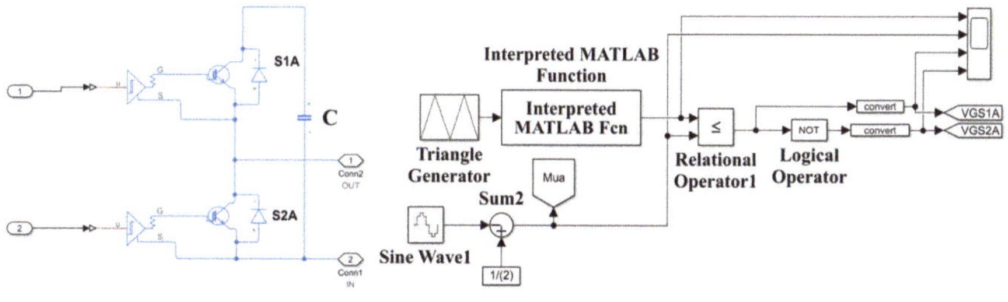

Fig. 2.89 Three-phase MMC model subsystems: (**a**): Submodule or cell (left) and (**b**): Gate pulse generator (right)

The SM1 in phase a having two IGBT switches with freewheeling diodes is connected to the cell capacitor C as shown in Fig. 2.89a. The half-bridge submodule truth table is shown in Table 2.20. There are five such submodules per arm per phase. The gate pulse generator is shown in Fig. 2.89b. In Fig. 2.89b, the triangle generator generates triangle carrier having frequency fc Hz, amplitude +1/−1 volt. This is given to the interpreted MATLAB Fcn block with value [(1 + u(1))*0.5], giving an output triangle carrier having frequency fc Hz, amplitude +1 volt and minimum value zero. The triangle carriers for cells 2–5 in the upper arm successively differ in phase by 2π/N radians. The same principle holds good for generating the triangle carrier for the cells in the lower arm. The phase displacement θ defined in Eq. 2.35 between the upper and lower arm carriers is optional. Here θ value is zero. The triangle carriers for upper and lower arm are shown in Fig. 2.87. In Fig. 2.89b, sine wave1 generator generates the sine wave with peak value $M/2$ ($0 < M < 1$), angular frequency ω rad/Sec, phase shift $[(\pi/2) + \theta\_x]$, where $\theta x = 0, -2\pi/3, -4\pi/3$ for $x = a, b, c$, respectively. This sine wave output is subtracted from ½ using Sum2 block to generate modulation function mu as defined in Eq. 2.28 for the upper arm. Similarly, the sine wave output of sine wave function generator is added with ½ using sum block to get the modulation function ml as defined in Eq. 2.29 for the lower arm. The triangle carrier from interpreted MATLAB Fcn block output is compared with modulation function mu in the relational operator1 block which forms the comparator. The comparator output is HIGH if the triangle carrier is less than or equal to modulation function mu or else its output is LOW which forms the gate drive for the switch S1A. The comparator output is inverted using logical operator block performing NOT logic operation which forms the gate drive for the switch S2A. The gate drive for all other switches in the upper and lower arms is developed in the same way. In this model, there is no SM capacitor voltage compensation.

2.19.1 Simulation Results

The simulation of the three-phase MMC is performed using fixed step ode1be (Backward Euler) solver in Simulink [3]. Three-phase MMC model data are tabulated in Table 2.21 [15]. The model simulation results are shown in Figs. 2.90 and 2.91.

Table 2.21 Three-phase MMC model parameters

Sl. no.	Parameter	Value	Unit	Sl. no.	Parameter	Value	Unit
1.	Number of phase	3	–	7.	Submodule capacitor C	3.3E-3	Farads
2.	Power output P	10	kVA	8.	Arm inductor L	3.1E-3	Henries
3.	DC link voltage V_{dc}	500	Volts	9.	Arm resistor R	0.8	Ohms
4.	MMC output frequency ω	2*π*50	Rad/sec.	10.	Load inductance Lg	1.5E-3	Henries
5.	Triangle carrier frequency fc	1050	Hz	11.	Load resistance Rg	50	Ohms
6.	Submodules per arm N	5	–	12.	Modulation index M	0.9	–

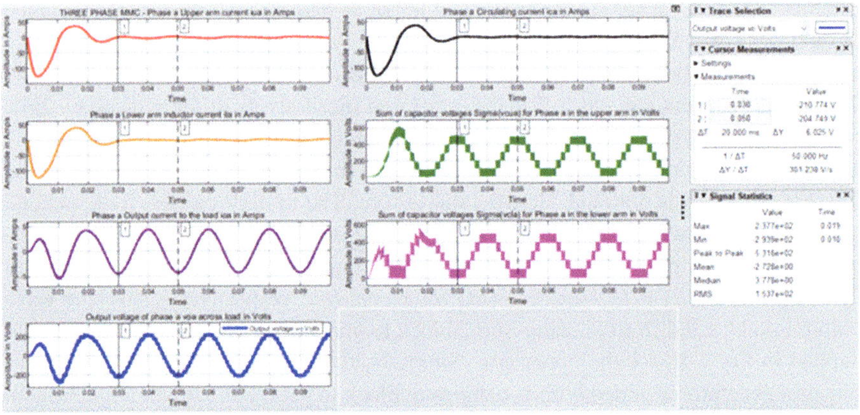

Fig. 2.90 Three phase MMC model simulation results for phase a: upper arm current iua, lower arm current ila, Load current ioa, load voltage voa (left column top to bottom), circulating current ica, sum of upper arm capacitor voltage vua, sum of lower arm capacitor voltage vla (right column top to bottom)

2.19.2 *Discussion of Results*

The simulation results in Fig. 2.90 show that the sum of capacitor voltage for the upper and lower arm is initially distorted for the first one cycle and thereafter displays balanced six distinct (N + 1) voltage levels. The three-phase line-to-line voltages in Fig. 2.91 are well balanced. Output phase-to-ground and line-to-line voltages are sinusoidal even without using an output filter. Capacitor voltage balance is not incorporated in the model due to number of blocks exceeding the maximum allowed limit for the Simulink version used here [3].

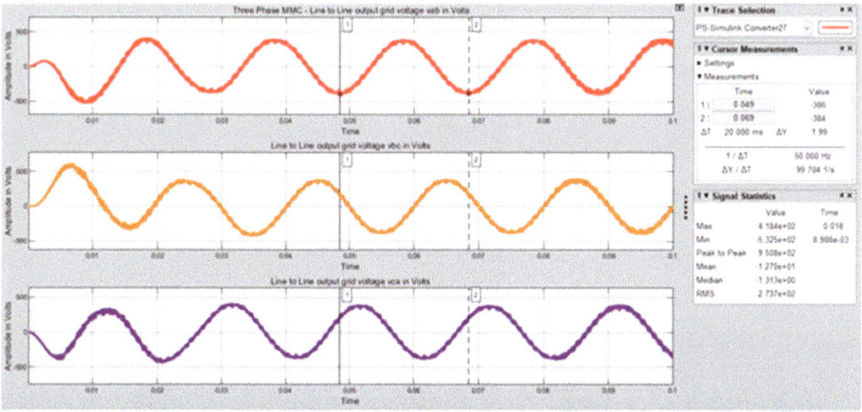

Fig. 2.91 Three-phase MMC model simulation results: line-to-line voltage vab, vbc and vca (top to bottom)

2.20 Case Study: Model of a Single Phase Modular Multilevel Converter with Capacitor Voltage Balance

The model of the single-phase MMC with capacitor voltage balancing is shown in Fig. 2.92 (Model file: CASE_STUDY_EX2_5). The model subsystem for the gate drive is shown in Fig. 2.93. The model subsystem for submodule is the same as shown in Fig. 2.89a except that here a cell capacitor voltage measurement unit, i.e. voltmeter block is added. The model parameters are given in Table 2.21. The model subsystem for generating the gate drive is described below:

The method of generating the modulation functions m_{ux} and m_{lx} defined in Eqs. 2.28 and 2.29 is the same as explained in Sect. 2.19. Here the modulation functions m_{ux_new} and m_{lx_new} defined in Eqs. 2.36 and 2.37 have to be generated for SM capacitor voltage balancing. The incremental modulation functions δm_{ux} and δm_{lx} defined in Eqs. 2.28 and 2.29 for the upper and lower arms are generated using two separate Embedded MATLAB functions as shown in Fig. 2.92. In Fig. 2.93, δm_{ux} for the cell 1 of upper arm dvc1ua thus generated is added with m_{ux} in the Sum12 block. The output of Sum12 block gives m_{ux_new} which is compared with the triangle carrier from interpreted MATLAB Fcn block using relational operator1 comparator block, as explained in Sect. 2.19. The output of this comparator block and its inverted output using NOT gate form the gate drive for switches S1A and S2A, respectively. The gate pulse for the other switches in the upper and lower arm is generated in the same way.

The source code for generating δm_{ux} for all the five cells of upper arm using Embedded MATLAB Function MMC_UPPER_ARM in Fig. 2.92 is shown in Program segment 2.9 in the model file CASE_STUDY_EX2_5. For the Embedded MATLAB Function MMC_UPPER_ARM, the inputs are the capacitor voltages of each cell Vc1ua to Vc5ua measured from the model, Vc_ref defined in Eq. 2.26,

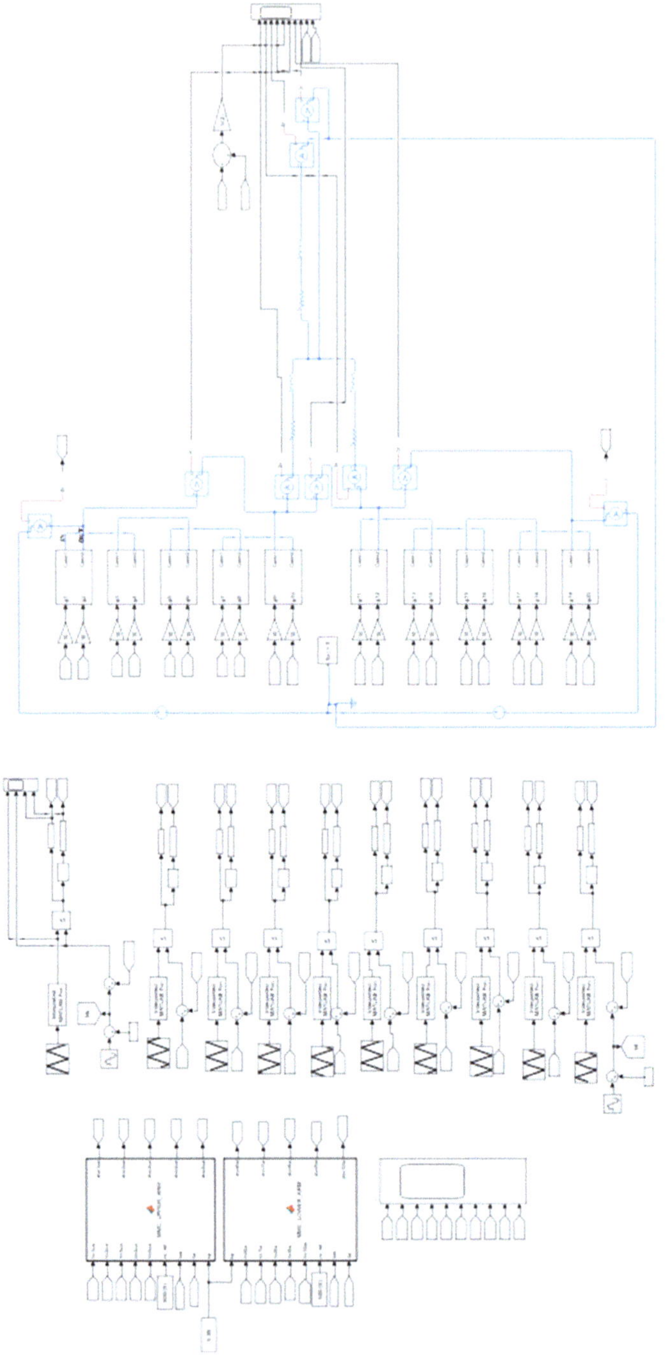

Fig. 2.92 Model of single-phase MMC with capacitor voltage balance

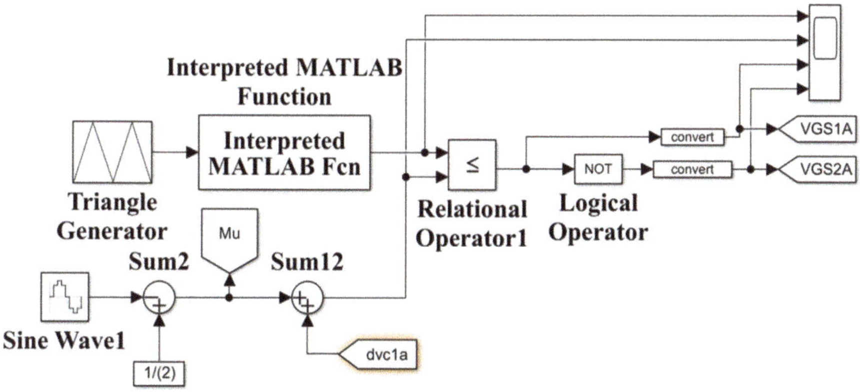

Fig. 2.93 Single-phase MMC with capacitor voltage balance—gate pulse generator

upper and lower arm currents iua, ila measured from the model, the proportionality constant Kp and the outputs are dvc1ua to dvc5ua defined in Eq. 2.38. Embedded MATLAB Function MMC_LOWER_ARM generates δm_{lx} for the five cells of lower arm using source code shown in Program segment 2.10 in the model file CASE_STUDY_EX2_5. For the Embedded MATLAB Function MMC_LOWER_ARM, the inputs are the capacitor voltages of each cell Vc1la to Vc5la measured from the model, Vc_ref defined in Eq. 2.26, upper and lower arm currents iua, ila measured from the model, the proportionality constant Kp and the outputs are dvc1la to dvc5la defined in Eq. 2.39.

2.20.1 *Simulation Results*

The simulation of the single-phase MMC is performed using fixed step ode1be (Backward Euler) solver in Simulink [3]. Single-phase MMC model data are tabulated in Table 2.21 [15]. The model simulation results are shown in Figs. 2.94 and 2.95 for a *Kp* value of zero and in Figs. 2.96 and 2.97 for a *Kp* value of 0.05.

2.20.2 *Discussion of Results*

The simulation results in Figs. 2.94 and 2.95 with *Kp* set to zero correspond to no voltage balance implemented for cell capacitors. Referring to Figs. 2.94, simulation result for output phase voltage voa shows initial transient disturbances, and the sum of lower arm capacitor voltages shows unbalanced distortions during the first cycle. In Fig. 2.95, all the cell capacitor voltages in the upper and lower arm show an overshoot as high as 123.26 volts and finally settle close to 100 volts. The simulation

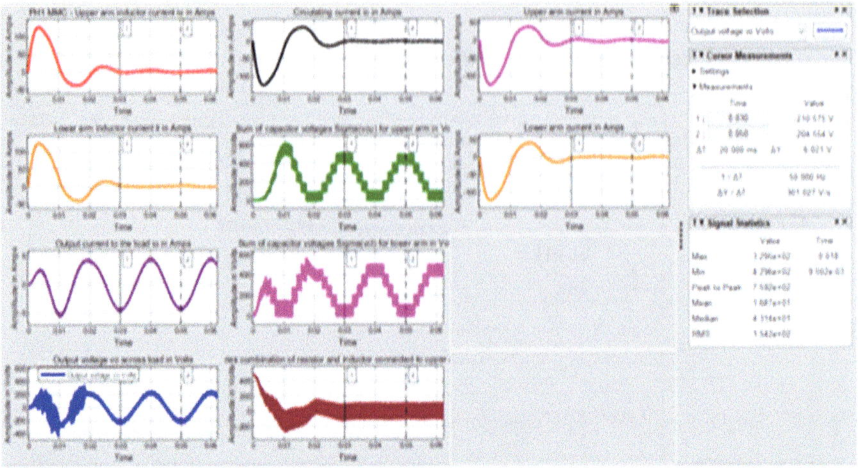

Fig. 2.94 Single-phase MMC model simulation results with Kp set to zero—Upper arm inductor current, lower arm inductor current, output current ioa, output phase voltage von (First column top to bottom), circulating current ica, sum of capacitor voltages of upper arm vcua, sum of capacitor voltages of lower arm vela, voltage across upper am and lower arm resistor inductor combination (Second column top to bottom), upper arm and lower arm currents (Third column top to bottom)

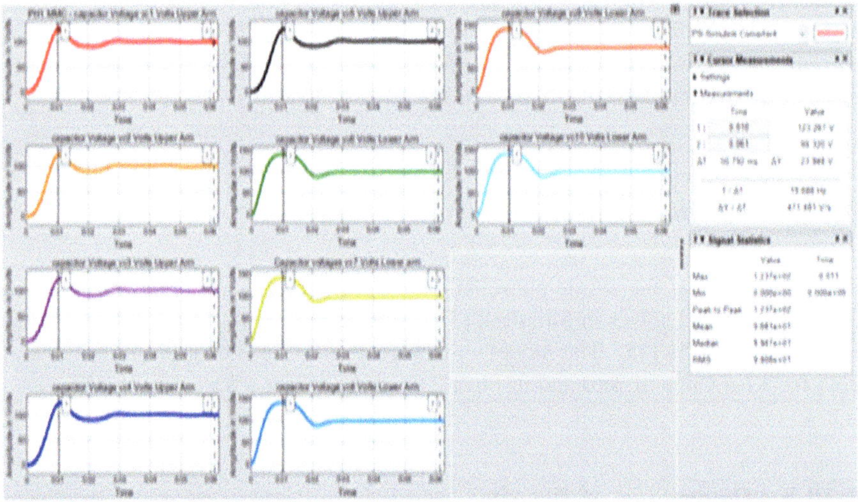

Fig. 2.95 Single-phase MMC model simulation results with Kp set to zero—Capacitor voltages of cell 1 to cell 4 in the upper arm (First column top to bottom), capacitor voltage of cell 5 in upper arm, cell 1 to cell 3 in the lower arm (Second column top to bottom), capacitor voltage of cell 4 and cell 5 in the lower arm (Third column top to bottom)

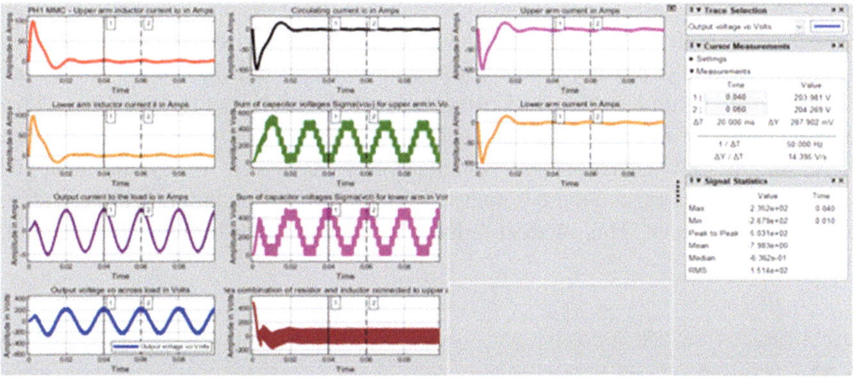

Fig. 2.96 Single-Phase MMC Model simulation results with Kp set to 0.05—Upper arm inductor current, lower arm inductor current, output current ioa, output phase voltage voa (First column top to bottom), circulating current ica, sum of capacitor voltages of upper arm vcua, sum of capacitor voltages of lower arm vela, voltage across upper am and lower arm resistor inductor combination (Second column top to bottom), upper arm and lower arm currents (Third column top to bottom)

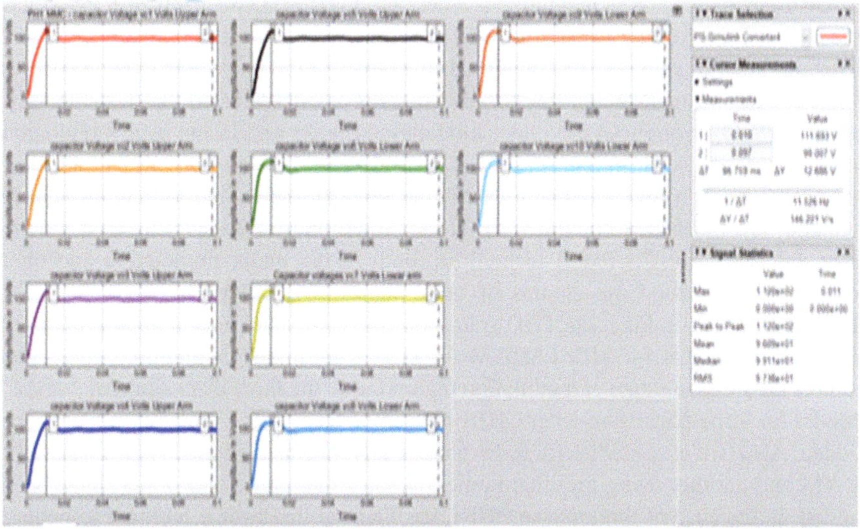

Fig. 2.97 Single-phase MMC model simulation results with Kp set to 0.05—Capacitor voltages of cell 1 to cell 4 in the upper arm (First column top to bottom), Capacitor voltage of cell 5 in upper arm, cell 1 to cell 3 in the lower arm (Second column top to bottom), capacitor voltage of cell 4 and cell 5 in the lower arm (Third column top to bottom)

results in Figs. 2.96 and 2.97 with Kp set to 0.05 correspond to voltage balance implemented for all cell capacitors. Referring to Figs. 2.96, simulation result for output phase voltage voa shows NO initial transient disturbances, and the sum of lower arm capacitor voltages is well balanced without any distortion during the first cycle. In Fig. 2.97, all the cell capacitor voltages in the upper and lower arm show an overshoot as high as 111.69 volts and finally settle close to 100 volts. The cell capacitor voltage overshoot is much less, and its initial unbalanced distortions are well eliminated with Kp set to 0.05 compared to that with Kp set to zero. Thus with a suitable choice of *Kp*, it is possible to achieve perfect cell capacitor voltage balance. Here, all the cell capacitors are simultaneously taken care of by the analogue feedback mechanism, and there is no necessity for prioritising them in the ascending or descending order. This method reduces the software/hardware implementation burden.

2.21 Conclusions

In this chapter, models for all the advanced carrier-based PWM techniques for three-phase two level inverter are presented. Also models for three-phase MCSLSPWM, CSLSPWM and THISLSPWM DCTLI with four triangle carriers having IPD, POD and APOD are presented for under modulation and over modulation region. Three-phase clipped sine PWM technique is a new proposition. Dead-band sine PWM has the lowest THD for line-to-neutral and line-to-line voltages throughout the under and over modulation regions. For two-level inverters, clipped sine PWM shows lower THD for line-to-neutral and line-to-line voltages as compared to that for sine, THI and HI PWM throughout the under modulation region and in the over modulation region up to an A.M. index Ma of 1.2. Also for the three-phase DCTLI with four triangle carriers in IPD, POD and APOD, the THD of line-to-line and line-to-neutral voltages are much less compared to that for three-phase two-level inverter having sine, HI, THI and CS PWM techniques, both in the under modulation and over modulation region. Case studies of three-phase CSLSPWM and THISLSPWM DCTLI model reveal that the THD values of line-to-line and line-to-neutral voltages are lower than that for MCSLSPWM DCTLI in the under modulation region and almost the same in the over modulation region for all the three disposition of carriers. Model for three-phase five-level CHBI with sine phase shift PWM carriers has been added. Also two case studies for three-phase CHBI one using clipped sine phase shift PWM and another using third harmonic injection sine phase shift PWM have been added. In the case of three-phase CHBI, the THD of line-to-line and line-to-neutral voltage by clipped sine and third harmonic injection sine is lower than that of sine phase shift PWM only for a limited range; however, the magnitude of fundamental component of line-to-line and line-to-neutral voltages for the former two methods is higher compared to that for sine phase shift PWM method. Also model for three-phase MMC without cell capacitor voltage balance is presented. A case study for a single-phase MMC with a modified analogue method for SM capacitor voltage balancing is presented. The modified analogue method indicates the advantage of SM capacitor voltage balancing.

References

1. J.A. Houldsworth and D.A. Grant: "The Use of Harmonic Distortion to Increase the Output Voltage of a Three Phase PWM Inverter"; IEEE Transactions on Industry Applications; Vol.IA-20; NO.5; pp. 1224–1227, 1984.
2. M.A. Boost and P.D. Ziogas: "State of the Art Carrier PWM Techniques: A Critical Evaluation"; IEEE Transactions on Industry Applications; Vol.24; No.2; pp. 271–280; March / April 1988.
3. The Mathworks Inc.: www.mathworks.com: "MATLAB/SIMULINK Version 9", R2021a, 2021, USA.
4. Narayanaswamy P.R. Iyer, Venkat Ramaswamy and Jianguo Zhu: "Three Phase Clipped Sinusoid PWM Inverter—A New technique for Pulse Width Modulation", 40th International Universities Power Engineering Conference (UPEC 2005), Cork, Ireland September 2005, pp. 1185–1189.
5. Narayanaswamy P.R. Iyer: "MATLAB/SIMULINK Modules for Modelling and Simulation of Power Electronic Converters and Electric Drives"; M.E. (Research) Thesis; University of Technology Sydney, NSW, 2006.
6. Narayanaswamy P R Iyer: "Models for three phase pulse width modulated inverters", Research seminar, School of Electrical and Information Engineering, The University of Sydney, Redfern, NSW, Australia, 4th November 2008.
7. V.G. Agelidis, P.D. Ziogas and G. Joos: ""Dead-Band" PWM Switching Patterns"; IEEE Transactions on Power Electronics; Vol. 11, No.4; July 1996; pp. 522–529.
8. J.W. Dixon. 2016. "Multilevel Converters" in Power Electronic Converters and Systems Frontiers and Applications, Ed. Andrzej M. Trzynadlowski, 43–72, The Institution of Engineering and Technology.
9. Surin Khomfoi and Leon M. Tolbert. 2011. "Multilevel Poweronverters" in Power Electronics Handbook, Ed. M.H. Rashid, 455–484, Elsevier, Butterworth-Heinemann.
10. Bin Wu: "High Power Converters and AC Drives"; IEEE Press, 2006, Ch.8, pp. 170–174.
11. N P R Iyer: "Power Electronic Converters: Interactive Modeling using SIMULINK", CRC Press, USA, Chapter 9 and 10, pp. 243–249 and pp. 315–319., 2018.
12. J. Dragan and K. Ahmad: "High Voltage Direct Current Transmission: Converters, Systems and DC Grids", John Wiley and Sons, 2015.
13. A. Lesnicar and R. Marquardt, "An innovative modular multilevel converter topology suitable for a wide power range," in Proc. IEEE Power Tech Conf., Bologna, Italy, 2003, vol. 3, pp. 6–11.
14. B. Lee, R. Yang, D. Xu, G. Wang, W. Wang and D. Xu: "Analysis of the Phase-Shifted Carrier Modulation for Modular Multilevel Converters", IEEE Transactions on Power Electronics, Vol. 30, No. 1, January 2015, pp. 297–310.
15. Lennart Harnefors, Antonios Antonopoulos, Staffan Norrga, Lennart Ängquist and Hans-Peter-Nee: "Dynamic Analysis of Modular Multilevel Converters", IEEE Transactions on Industrial Electronics, Vol. 60, No. 7, July 2013, pp. 2526–2534.

Chapter 3
Selective Harmonic Elimination Pulse-Width Modulation of Three-Phase Two-Level and Multilevel Inverter

3.1 Introduction

Selective harmonic elimination pulse-width modulation (SHE-PWM) technique is one of the methods used to control the fundamental component of the output voltage of inverters and to eliminate specifically chosen harmonics. Many algorithms to calculate switching angles to eliminate specified voltage harmonics with analytical and hardware experimental verification are reported. This chapter presents a model to verify the performance of (1) three-phase unipolar SHE-PWM inverter, (2) three-phase bipolar SHE-PWM inverter, (3) three-phase cascade H-bridge inverter, (4) three-phase multilevel inverter using diode-clamped and flying capacitor topology. Also two case studies are presented one for single-phase diode-clamped five-level inverter (DCFLI) and another for flying capacitor five-level inverter (FCFLI) to eliminate the 11th, 13th and 15th harmonics from the line-to-ground voltage with two different wave shapes and different total harmonic distortion (THD) and the method to select the desired wave shape. The switching angles to eliminate the desired harmonics from the line-to-ground voltage of the inverter are calculated using Symbolic Math Toolbox in MATLAB. Model simulation results are presented.

3.2 Three-Phase Unipolar SHE-PWM Inverter

The performance of a voltage source inverter depends on the choice of the pulse-width modulation technique used. Where high switching efficiency is desired, one method is to choose the switching angles such that the desired fundamental is achieved and specified harmonics from the waveform are eliminated. This is widely

---

Supplementary Information The online version contains supplementary material available at https://doi.org/10.1007/978-3-031-62784-2_3.

N. P. R. Iyer, *Inverters and AC Drives*, Power Systems,
https://doi.org/10.1007/978-3-031-62784-2_3

known as selective harmonic elimination [1–8]. There are many papers which deal with the verification of the performance of SHE-PWM inverters using analytical and hardware approach [7, 8]. A model to verify three-phase bipolar SHE-PWM inverter is reported [9, 10]. Recently, a number of software packages are available which can be used to verify the performance of power electronic converters [11–13]. This section provides a model to verify the performance of three-phase unipolar SHE-PWM inverter. The switching angles are calculated using "solve" command in the Symbolic Math Tool Box in MATLAB [11].

3.2.1 Review of Harmonic Elimination in Unipolar Switched Inverters

The harmonic elimination technique makes use of Fourier transform. For a symmetrically defined unipolar voltage waveform of inverter shown in Fig. 3.1, the DC link voltage is 2 p.u. Fourier series expansion is given in Eq. (3.1).

$$V(\theta) = \frac{4}{\pi} \cdot \left[\sum_{k=1,3,5,\ldots}^{\infty} \frac{\sin(k.\theta)}{k} \cdot \left(\sum_{i=1,2,3,\ldots}^{N+1} (-1)^i . \cos(k.\alpha_i) \right) \right]. \quad (3.1)$$

The waveform shown in Fig. 3.1 has a quarter-wave symmetry, and the number of angles to be found is $N + 1$, where N is the number of non-triplen odd harmonics to be eliminated from the PWM waveform. The following set of Eqs. (3.2), (3.3), (3.4) and (3.5) has to be solved.

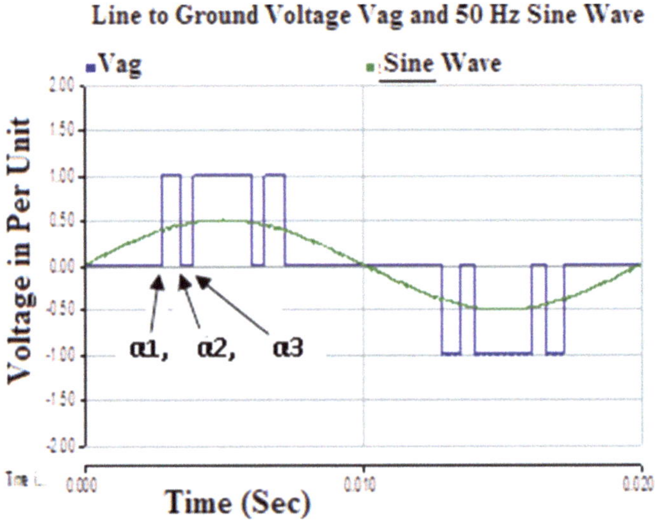

Fig. 3.1 Unipolar voltage waveform

$$\begin{bmatrix} \sum_{i=1}^{N+1} (-1)^{i+1}.\cos(\alpha_i) - M = 0 \\ \sum_{i=1}^{N+1} (-1)^{i+1}.\cos(5.\alpha_i) = 0 \\ \sum_{i=1}^{N+1} (-1)^{i+1}.\cos(n.\alpha_i) = 0 \end{bmatrix} \tag{3.2}$$

where

$$n = 3N + 1, \quad \text{for} \quad N = \text{even} \tag{3.3}$$

$$n = 3N + 2, \quad \text{for} \quad N = \text{odd} \tag{3.4}$$

$$0 \leq M \leq 1 \tag{3.5}$$

In general, if 2*Vdc is the DC link voltage of the inverter, then the Fourier expression for the unipolar waveform shown in Fig. 3.1 can be expressed as follows:

$$V(\omega t) = \sum_{n=1,3,5,\ldots}^{\infty} \frac{4.V_{dc}}{n.\pi} * [\cos(n.\alpha_1) - \cos(n.\alpha_2) + \cos(n.\alpha_3)] * \sin(n.\omega t) \tag{3.6}$$

where modulation index M is defined below:

$$M = \frac{V_1}{\left(\frac{4.V_{dc}}{\pi}\right)} \tag{3.7}$$

V_1 is the amplitude of the fundamental component of the inverter unipolar line to ground output voltage. For a DC link voltage 2*Vdc of 2 p.u., Eq. 3.7 simplifies to the following:

$$M = \frac{V_1}{(4/\pi)} \ p.u. \tag{3.8}$$

Now consider that the non-triplen odd harmonics 5th and 7th have to be eliminated from the line-to-ground unipolar output voltage of inverter. Here N is 2, and the number of switching angles is three. These angles are $\alpha 1$, $\alpha 2$ and $\alpha 3$, and from Eq.3.6, the following expression can be obtained:

Table 3.1 Three-phase unipolar SHE-PWM inverter switching angles

Sl.No.	M	α1 degrees	α2 degrees	α3 degrees
1	0.5	50. 06528	62. 26,686	71. 12,892
2	0.8	23. 63,032	38. 06067	47. 83,966
3	0.8	13. 30,409	72. 43,925	82. 61,393

$$\left.\begin{aligned} \cos(\alpha_1) - \cos(\alpha_2) + \cos(\alpha_3) = M \\ \cos(5.\ \alpha_1) - \cos(5.\ \alpha_2) + \cos(5.\ \alpha_3) = 0 \\ \cos(7.\ \alpha_1) - \cos(7.\ \alpha_2) + \cos(7.\ \alpha_3) = 0 \end{aligned}\right\} \tag{3.9}$$

For a given modulation index in the range 0 to 1, set of Eq. 3.9 can be solved by Newton-Raphson method [1, 2], Nelder–Mead [7] or theory of Resultant algorithm [8, 14]. Here, Eq. 3.9 is solved using Symbolic Math Toolbox in MATLAB [11]. This is shown in Program segment 3.1 for M value of 0.8. Here "solve" command is used to get values for $\alpha 1$, $\alpha 2$ and $\alpha 3$.

```
%% Program segment 3.1
%% Three Phase Unipolar SHE-PWM Inverter
%% Dr. Narayanaswamy P R Iyer
s1 = solve('cos(alfa1) - cos(alfa2) + cos(alfa3)-0.8 = 0','cos
(5*alfa1) - cos(5*alfa2) + cos(5*alfa3)= 0','cos(7*alfa1) - cos
(7*alfa2) + cos(7*alfa3) = 0');
s2 = solve('cos(alfa1*0.0175) - cos(alfa2*0.0175) + cos(alfa3*0.0175)
- 0.8 = 0','cos(5*alfa1*0.0175) - cos(5*alfa2*0.0175) + cos
(5*alfa3*0.0175)= 0','cos(7*alfa1*0.0175) - cos(7*alfa2*0.0175) +
cos(7*alfa3*0.0175) = 0');
```

From the printed values for $\alpha 1$, $\alpha 2$ and $\alpha 3$ in radians (degrees), only those values that satisfy the relation $\alpha 1 < \alpha 2 < \alpha 3 < \pi/2$ are selected. These values are tabulated in Table 3.1 for M value 0.5 and 0.8.

3.2.2 Model for Three-Phase Unipolar SHE-PWM Inverter

The gate drive model block diagram for the three-phase unipolar SHE-PWM inverter is shown in Fig. 3.2 [15]. The switching angles to eliminate any two non-triplen harmonics from the line-to-ground voltage are alfa1, alfa2 and alfa3. The phase A reference sine wave voltage 1.sin($\omega.t$) during positive half cycle is compared with sine of alfa1, alfa2 and alfa3 in separate comparators marked Comparator X1, X2 and X3. When phase A sine wave reference voltage is greater than or equal to Sin ($\alpha 1$), Comparator 1 output is HIGH, or else its output is LOW and similarly when the former reference sine voltage is only less than Sin($\alpha 2$), and Comparator 2 output is HIGH. Comparators X1 and X2 output are given to AND gate U1. Similarly,

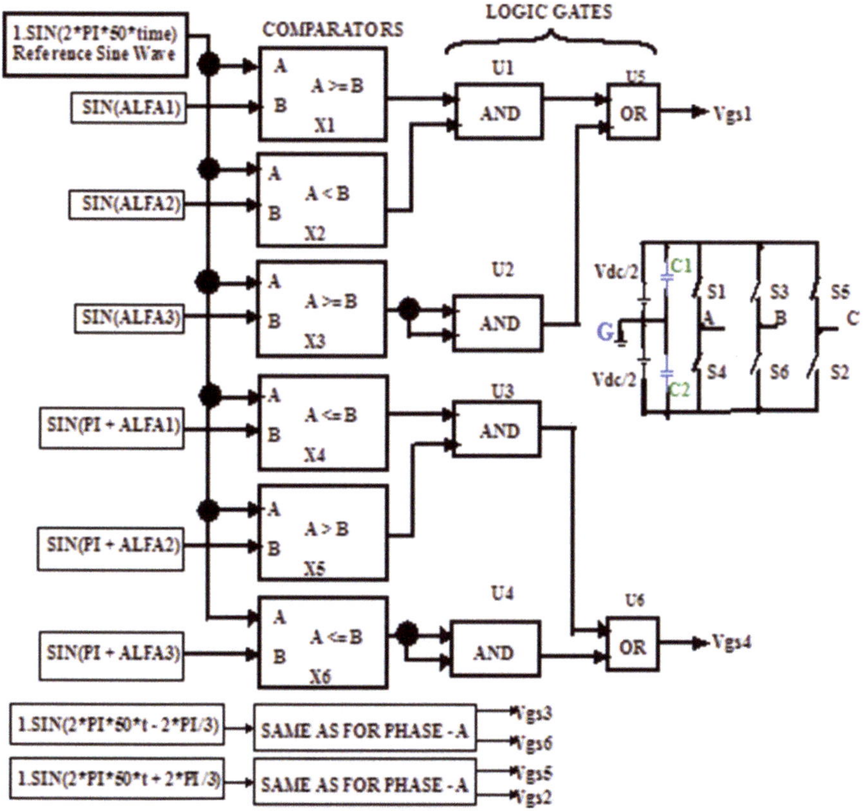

Fig. 3.2 Three-phase unipolar SHE-PWM inverter gate drive model block diagram

reference sine voltage is connected to Comparator X3 whose output is HIGH only when sine wave voltage is equal to or greater than Sin(α3). Comparator X3 output through AND gate marked U2 and the U1 output are connected to OR gate U5. The U5 output Vgs1 forms the gate drive for the inverter upper switch in Phase A marked S1. Similar procedure is adopted to compare the reference sine wave output for Phase A with sine of (pi+α1), (pi+α2) and (pi+α3) in three separate Comparators marked X4, X5 and X6. The resulting output of these three Comparators X4 to X6 through AND gates U3 and U4 are connected to OR gate U6. The output Vgs4 of U6 forms the gate drive for the inverter lower switch in Phase A marked S4. For Phase B and C, the reference sine wave voltages are (1.sin($\omega.t$ – 2*pi/(3)) and (1.sin($\omega.t$ + 2*pi/(3)), respectively. The remaining procedure is the same as for Phase A given above. Here $\omega = 2\pi*50$ is used. The following action takes place for various intervals of the Phase A reference sine wave voltage:

Case 1: $0 < \omega.t < \alpha1$ radians. During this time interval of the reference sine wave voltage for Phase A, the following operation takes place:

1. Comparators X1, X2, X3 outputs are LOW, HIGH, LOW and that of X4, X5 and X6 are LOW, HIGH and LOW, respectively.
2. Output of logic gates U1 to U6, respectively, are all LOW.
3. Output Vgs1 and Vgs4 are both LOW.
4. Switches S1 and S4 are both OFF, and the output voltage VAG is 0 volts.

Case 2: $\alpha 1 < \omega.t < \alpha 2$ radians. During this time interval of the reference sine wave voltage for Phase A, the following operation takes place:

1. Comparators X1, X2, X3 outputs are HIGH, HIGH, LOW and that of X4, X5 and X6 are LOW, HIGH and LOW, respectively.
2. Output of logic gates U1 and U5 are HIGH and that of U2 to U4 and U6 are LOW.
3. Output Vgs1 and Vgs4 are HIGH and LOW, respectively.
4. Switch S1 is ON and S4 is OFF, and the output voltage VAG is +Vdc/2 volts.

Case 3: $\alpha 2 < \omega.t < \alpha 3$ radians. During this time interval of the reference sine wave voltage for Phase A, the following operation takes place:

1. Comparators X1, X2, X3 outputs are HIGH, LOW and LOW and that of X4, X5 and X6 are LOW, HIGH and LOW, respectively.
2. Outputs of logic gates U1 to U6 are all LOW.
3. Output Vgs1 and Vgs4 are both LOW.
4. Switches S1 and S4 are both OFF, and the output voltage VAG is 0 volts.

Case 4: $\alpha 3 < = \omega.t < \pi/2$ radians. During this time interval of the reference sine wave voltage for Phase A, the following operation takes place:

1. Comparators X1, X2, X3 outputs are HIGH, LOW and HIGH and that of X4, X5 and X6 are LOW, HIGH and LOW, respectively.
2. Outputs of logic gates U2 and U5 are HIGH and that of U1, U3, U4 and U6 are all LOW.
3. Output Vgs1 is HIGH and that of Vgs4 is LOW.
4. Switch S1 is ON and S4 is OFF, and the output voltage VAG is +Vdc/2 volts.

Case 5: $\pi < = \omega.t < (\pi + \alpha 1)$ radians. During this time interval of the reference sine wave voltage for Phase A, the following operation takes place:

1. Comparators X1, X2, X3 outputs are LOW, HIGH and LOW and that of X4, X5 and X6 are LOW, HIGH and LOW, respectively.
2. Outputs of logic gates U1 to U6, respectively, are all LOW.
3. Outputs Vgs1 and Vgs4 are both LOW.
4. Switches S1 and S4 are both OFF, and the output voltage VAG is 0 volts.

Case 6: $(\pi + \alpha 1) < = \omega.t < (\pi + \alpha 2)$ radians. During this time interval of the reference sine wave voltage for Phase A, the following operation takes place:

1. Comparators X1, X2, X3 outputs are LOW, HIGH and LOW and that of X4, X5 and X6 are HIGH, HIGH and LOW, respectively.

2. Outputs of logic gates U3 and U6 are HIGH and that of U1, U2, U4 and U5 are LOW.
3. Outputs Vgs1 and Vgs4 are LOW and HIGH, respectively.
4. Switch S1 is OFF and S4 is ON, and the output voltage VAG is -Vdc/2 volts.

Case 7: $(\pi + \alpha 2) <= \omega.t < (\pi + \alpha 3)$ radians. During this time interval of the reference sine wave voltage for Phase A, the following operation takes place:

1. Comparators X1, X2, X3 outputs are LOW, HIGH and LOW and that of X4, X5 and X6 are HIGH, LOW and LOW, respectively.
2. Outputs of logic gates U1 to U6 are all LOW.
3. Outputs Vgs1 and Vgs4 are both LOW.
4. Switches S1 and S4 are both OFF, and the output voltage VAG is 0 volts.

Case 8: $(\pi + \alpha 3) <= \omega.t < 3\pi/2$ radians. During this time interval of the reference sine wave voltage for Phase A, the following operation takes place:

1. Comparators X1, X2, X3 outputs are LOW, HIGH and LOW and that of X4, X5 and X6 are HIGH, LOW and HIGH, respectively.
2. Outputs of logic gates U4 and U6 are HIGH and that of U1, U2, U3 and U5 are all LOW.
3. Output Vgs1 is LOW and that of Vgs4 is HIGH.
4. Switch S1 is OFF and S4 is ON, and the output voltage VAG is -Vdc/2 volts.

Block diagram model shown in Fig. 3.2 is developed using SIMULINK [11]. This model is shown in Fig. 3.3 (Model file: EXAMPLE 3_1), and the gate pulse subsystem is shown in Fig. 3.4. Here a three-phase 50 Hz sine wave AC voltage is used as a reference. The value of $\alpha 1$, $\alpha 2$ and $\alpha 3$ for M = 0.5 shown in Table 3.1 is used in the model. The inverter switching frequency is 50 Hz. The DC link voltage of inverter is 100 volts. The three-phase inverter is the Universal Bridge subsystem from Power Systems block set. The reference sine waves for the three phase are generated using Embedded MATLAB function and the sine of switching angles for $\alpha 1$, $\alpha 2$, $\alpha 3$ and that for $(\pi + \alpha 1)$, $(\pi + \alpha 2)$ and $(\pi + \alpha 3)$ are generated using another Embedded MATLAB function. The source codes are given in Program segment 3.2 and 3.3 in the model file EXAMPLE 3_1. In Fig. 3.3, the switching angles $\alpha 1$, $\alpha 2$ and $\alpha 3$ in degrees are entered in order in the constant block separated by a space. The selector block selects or directs the first output $\alpha 1$, second output $\alpha 2$ and third output $\alpha 3$. The model subsystem to generate gate pulse for phase A switches of inverter is shown in Fig. 3.4. Here reference sine wave for Phase A (1. Sin(2*π*50*t)) is compared with sine of $\alpha 1$, $\alpha 2$, $\alpha 3$, $(\pi + \alpha 1)$, $(\pi + \alpha 2)$ and $(\pi + \alpha 3)$ in six relational operator blocks which form the comparators X1 to X6. The resulting output of each comparator is connected to AND and OR logic gates U1 to U6 as shown in Fig. 3.4 to generate gate pulse Vgs1 and Vgs4 for the upper and lower of inverter in Phase A. Similar comparators and logic gates are used to compare Phase B and Phase C reference sine waves to generate gate pulse for the upper and lower switches in Phase B and C of the inverter.

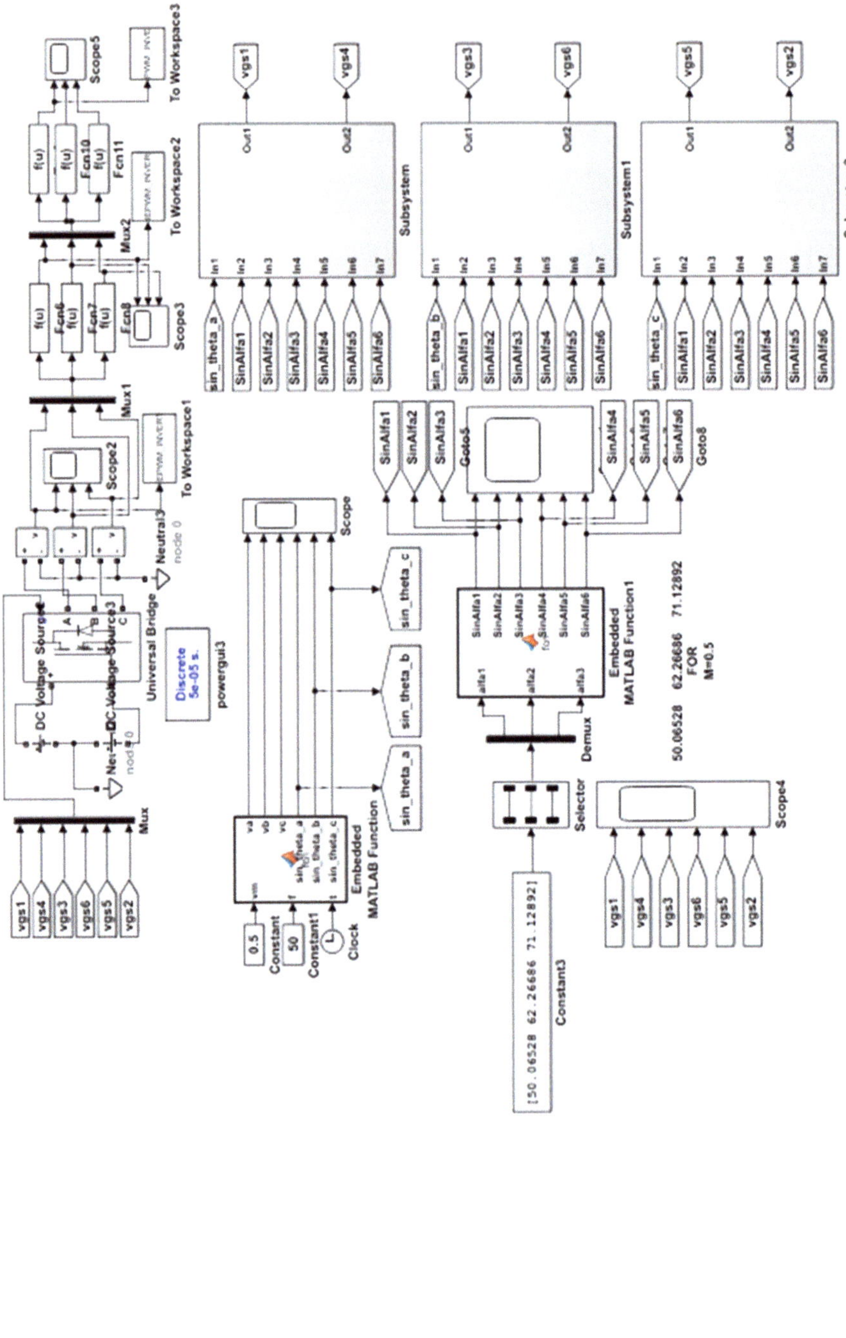

Fig. 3.3 Model of three-phase unipolar SHE-PWM inverter with three switching angles

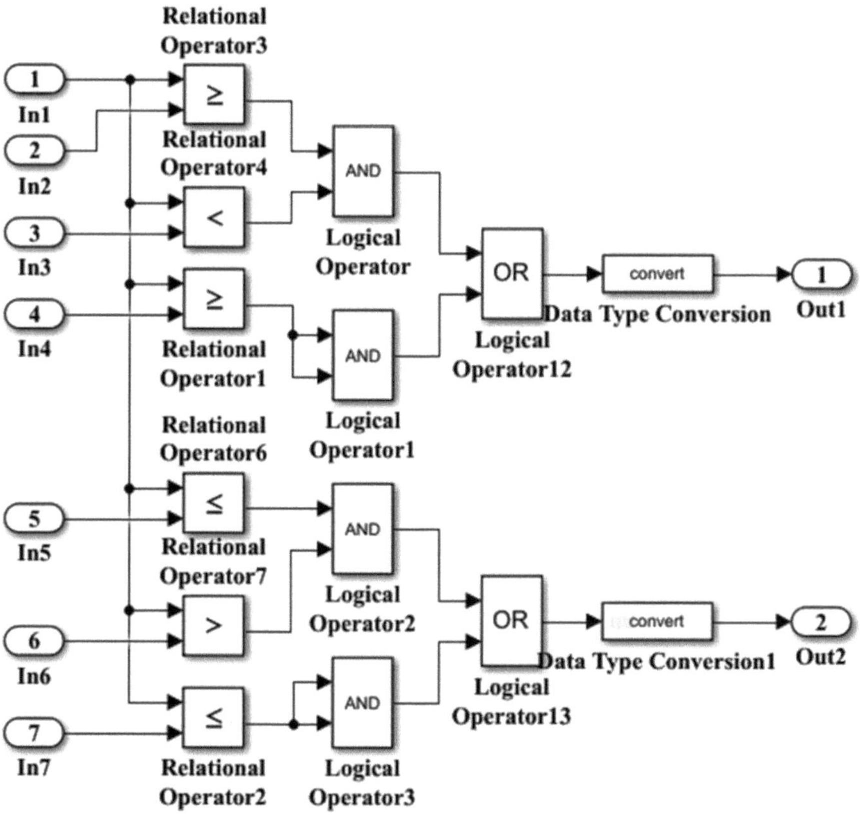

Fig. 3.4 Three-phase unipolar SHE-PWM inverter gate pulse generator

3.2.3 *Simulation Results*

The model simulation of the three-phase unipolar SHE-PWM inverter was carried out using ode23tb (stiff/TR-BDF2) solver. Harmonic spectrum for a three-phase 50 Hz inverter for M value of 0.5 to eliminate 5th and 7th harmonics from the line-to-ground voltage is shown in Fig. 3.5. The three-phase line-to-ground voltage is shown in Fig. 3.6. Harmonic spectrum for the line-to-ground voltage and the three-phase line-to-ground voltage for M of 0.8 and alfa start group 23.63032 degrees are shown in Figs. 3.7 and 3.8 and that for alfa start group of 13.30409 degrees are shown in Figs. 3.9 and 3.10, respectively.

3.2.4 *dSPACE Implementation of Single-Phase Unipolar SHE-PWM Inverter Gate Drive*

The single-phase unipolar SHE-PWM inverter gate drive model connected to DAC C3 and DAC C4 blocks from library rtilib 1104 is shown in Fig. 3.11a, and DS1104

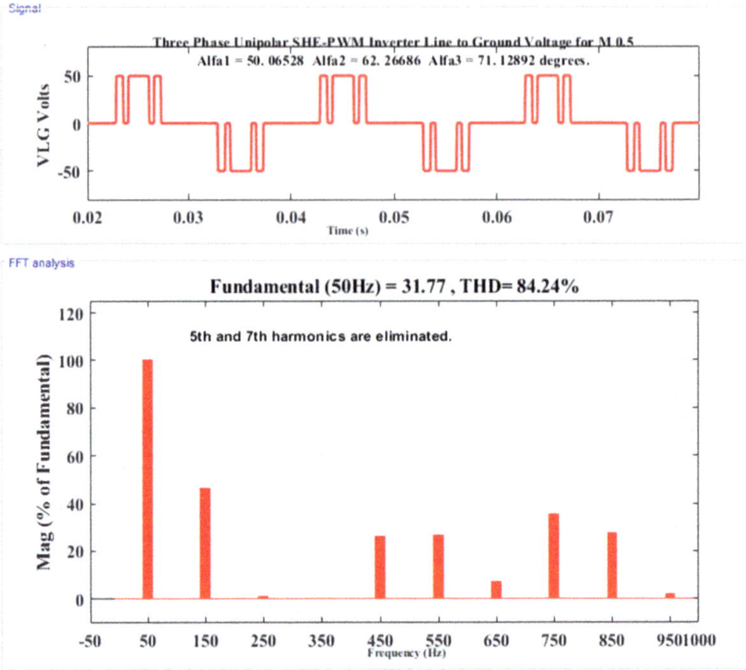

Fig. 3.5 Three-phase SHE-PWM inverter line-to-ground voltage (top) and harmonic spectrum with 5th and 7th harmonic eliminated (bottom)

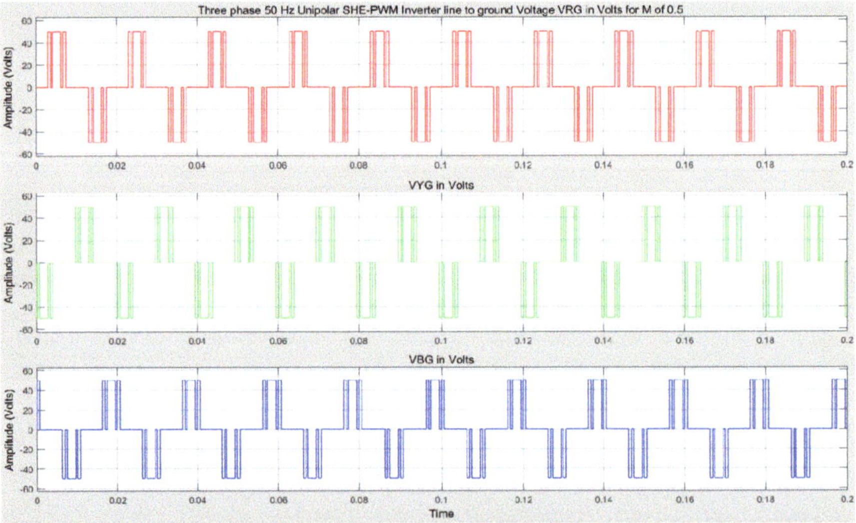

Fig. 3.6 Three-phase unipolar SHE-PWM inverter line-to-ground voltage for *M* of 0.5 and alfa start group of 50.06528 degrees

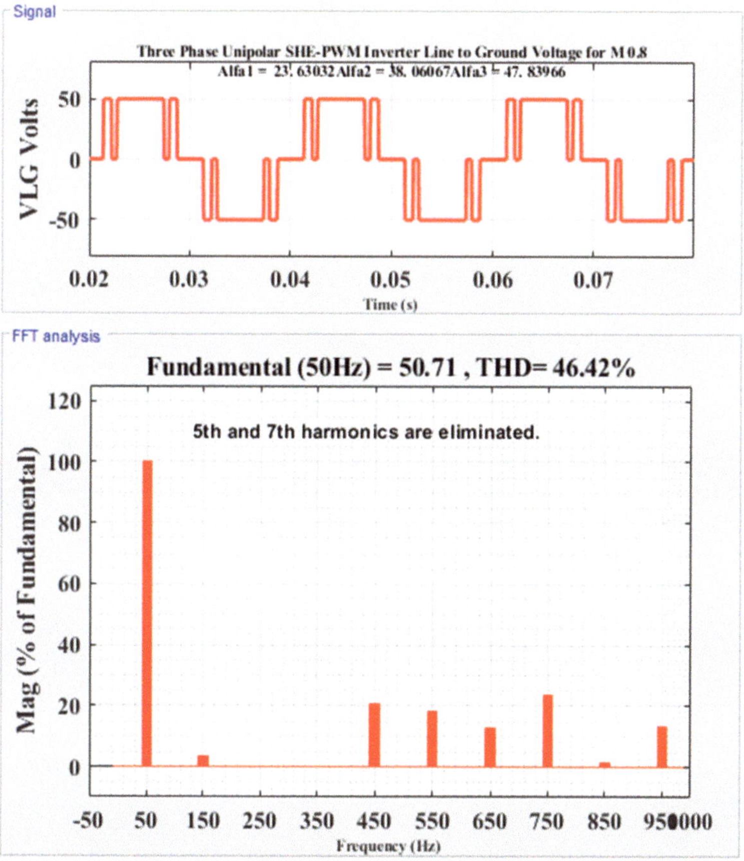

Fig. 3.7 Three-phase SHE-PWM inverter line-to-ground voltage (top) and harmonic spectrum with 5th and 7th harmonics eliminated (bottom)

hardware controller board connection is shown in Fig. 3.11b. In Fig. 3.11a blocks marked Fcn, Fcn1 and Fcn2 calculate the sine of angles $\alpha 1$, $\alpha 2$ and $\alpha 3$ and those marked Fcn3, Fcn4 and Fcn5 calculate the sine of angles $(\pi + \alpha 1)$, $(\pi + \alpha 2)$ and $(\pi + \alpha 3)$, respectively. The remaining part is the same as presented in Sect. 3.2.2.

Photograph of the dSPACE DS1104 hardware implementation scheme and the waveform of gate drive for the single-phase unipolar SHE-PWM inverter are shown in Figs. 3.12 and 3.13, respectively.

3.2.5 Discussion of Results

The simulation results for the harmonic spectrum of line-to-ground voltage shown in Figs. 3.5, 3.7 and 3.9 reveal that the 5th and 7th harmonics are well eliminated. Also for M value of 0.5 and 0.8 for a DC link voltage of 100 V, using Eq.3.7, the peak

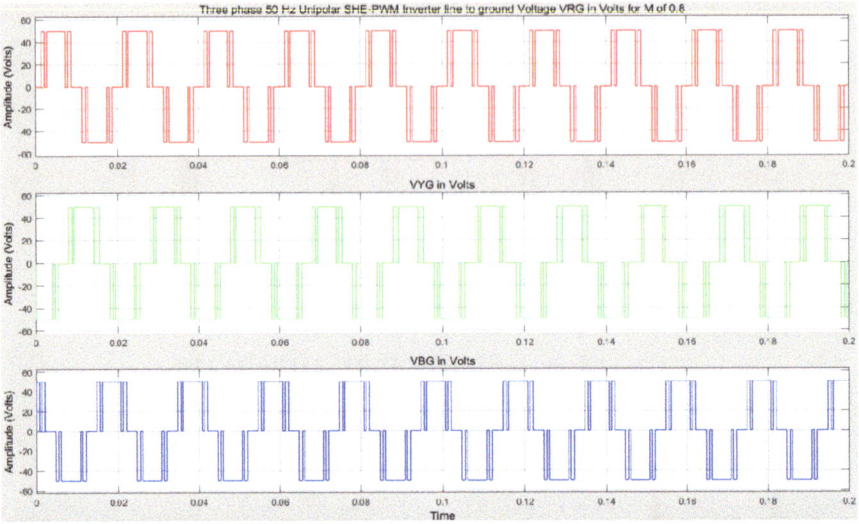

Fig. 3.8 Three-phase unipolar SHE-PWM inverter line-to-ground voltage for *M* of 0.8 and alfa start group of 23.63032 degrees

fundamental voltage V_1 is found to be 31.831 V and 50.93 V, respectively, whereas that by simulation of model is found to be 31.77 V for the former, 50.71 V and 51.3 V for the later value of M. Also for M value of 0.8, the THD of line-to-ground voltage is minimum for values of switching angles corresponding to serial number 2 of Table 3.1.

3.3 Three-Phase Bipolar SHE-PWM Inverter

Selective harmonic elimination PWM (SHE-PWM) technique provides selected lower-order harmonic elimination from the output voltage of inverters [9, 10]. The method involves finding.

the switching angles to eliminate the lower order harmonics by solving a set of non-linear equations relating to the output voltage of inverter. These non-linear output voltage equations are derived using Fourier transform analysis. It is also possible to control the amplitude of the fundamental component of output voltage by suitable choice of modulation index. The general solution algorithm proposed generates two sets of straight lines with positive and negative slopes that closely approximate the exact solution pattern of the non-linear equations obtained by Fourier transform [16]. The total number of odd non-triplen harmonics to be eliminated added with one forming *N* harmonics is plotted as a graph against switching angles $\alpha 1$, $\alpha 2$, $\alpha 4$, etc. with average positive slope mp and against switching angles $\alpha 3$, $\alpha 5$, $\alpha 7$, etc. with average negative slope mn from which an exponential curve fit is made for positive and negative slope of switching angles mp

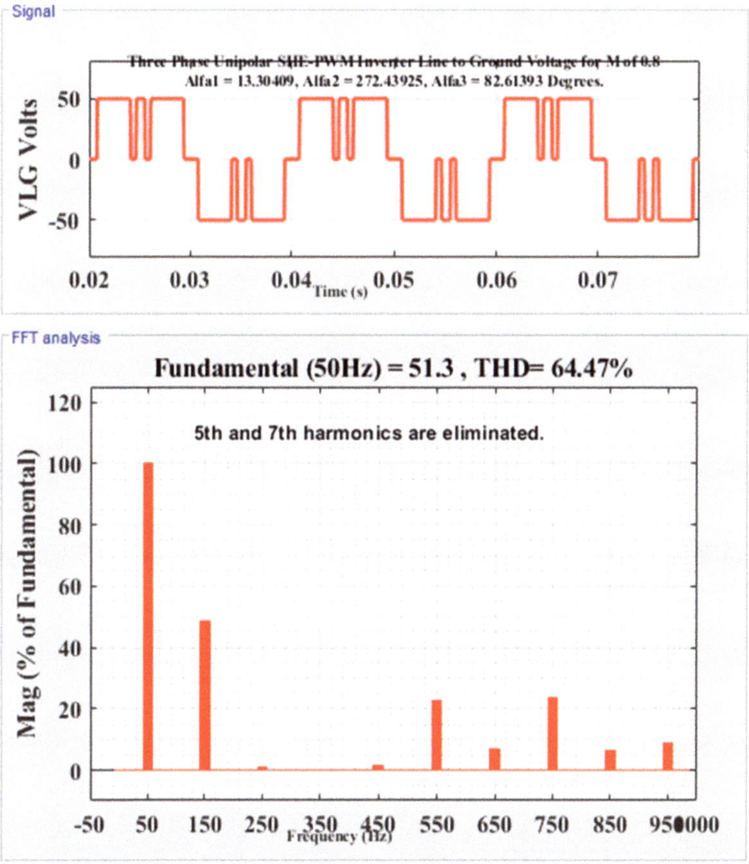

Fig. 3.9 Three-phase unipolar SHE-PWM inverter line-to-ground voltage (top) and harmonic spectrum with 5th and 7th harmonics eliminated (bottom)

and mn [16]. The switching angles αk can be expressed in the form $\alpha k = m^*M + Ck$ where M is the modulation index, $m =$ mp for $k = 1,2,4$, etc. and mn for $k = 3,5,7$, etc. and Ck is the constant determined using exponential curve fit equation [16]. A method of finding all switching angles to eliminate selected harmonics from output voltage using sequential homotopy is reported [18]. This homotopy method finds multiple solutions for degree of freedom $M = m$ from those for $M = \mathrm{m}^{-1}$ sequentially by mathematical induction by varying a fundamental component value as the homotopy parameter [18]. Another method is to make an initial guess of the switching angle and solve the non-linear harmonic expression for the voltage by Newton-Raphson iteration Method [1, 2]. The method of solution to find switching angles by resultant theory is reported [8, 14]. In the method by resultant theory, integer multiples cosine of switching angles are expressed as a power series using sine and cosine of switching angles and the resulting polynomial equation is expressed as a determinant which is solved to find the switching angles [8, 14]. In

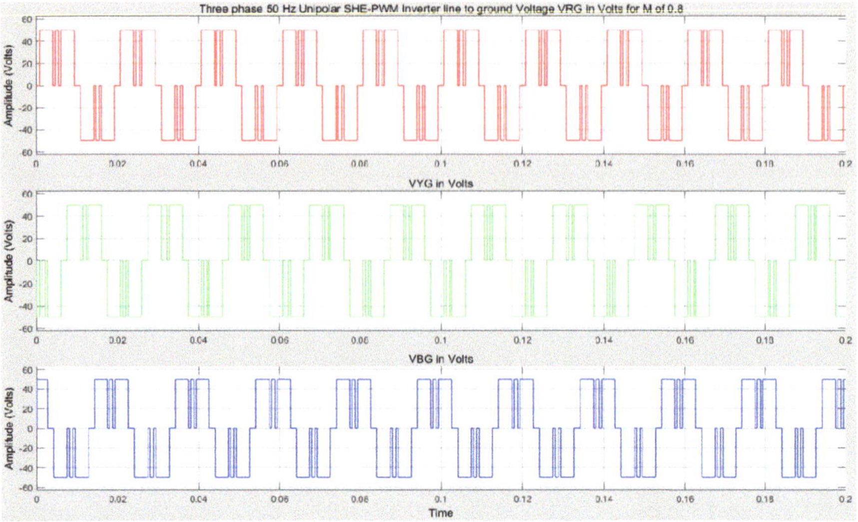

Fig. 3.10 Three-phase unipolar SHE-PWM inverter line-to-ground voltage for *M* of 0.8 and alfa start group of 13.30409 degrees

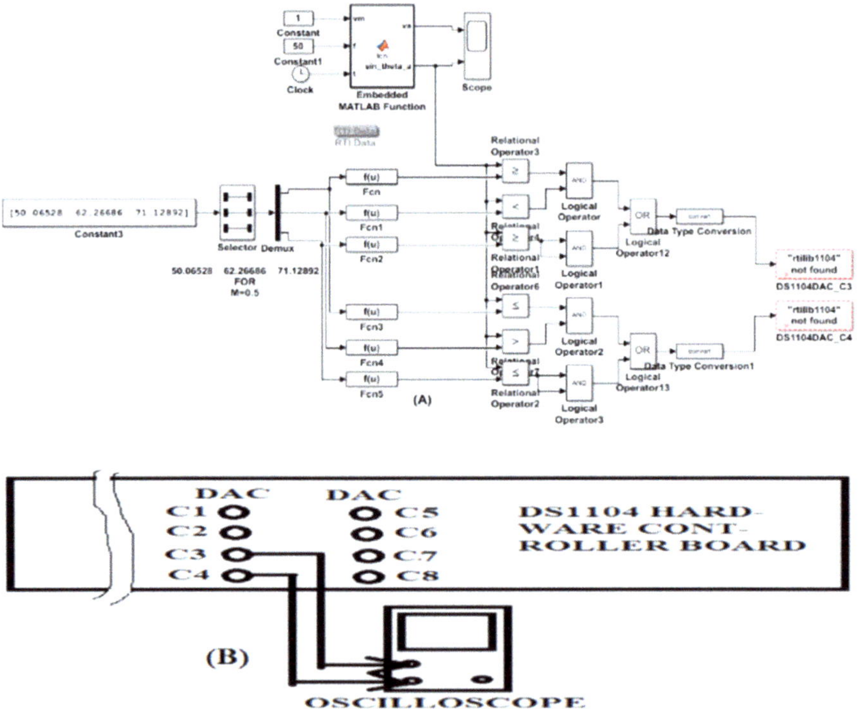

Fig. 3.11 Single-phase unipolar SHE-PWM inverter gate drive implementation (**a**) Model and (**b**) DS1104 hardware controller board connection

Fig. 3.12 Single-phase unipolar SHE-PWM inverter implementation using dSPACE DS1104 hardware controller board

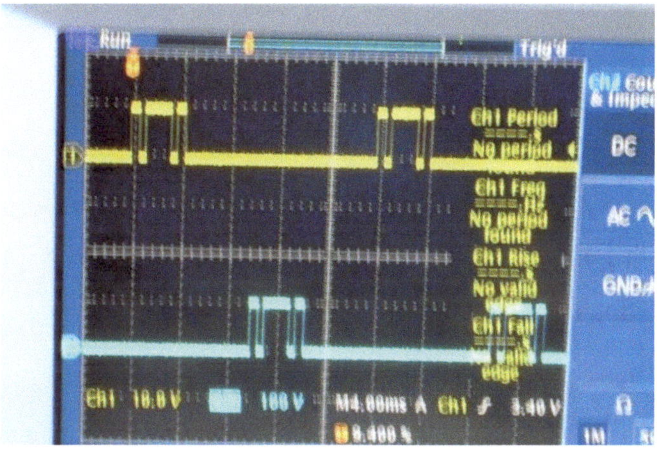

Fig. 3.13 Single-phase unipolar SHE-PWM inverter gate drive waveforms

the method presented in this section, a model is developed for the three-phase bipolar SHE-PWM inverter. The switching angles are calculated using “solve” command in the Symbolic Math Tool Box in MATLAB [11].

3.3.1 Review of Harmonic Elimination in Bipolar Switched Inverters

Here a standard H-bridge inverter is considered where choosing three switching angles for the bipolar switching pattern results in an output line-to-ground voltage

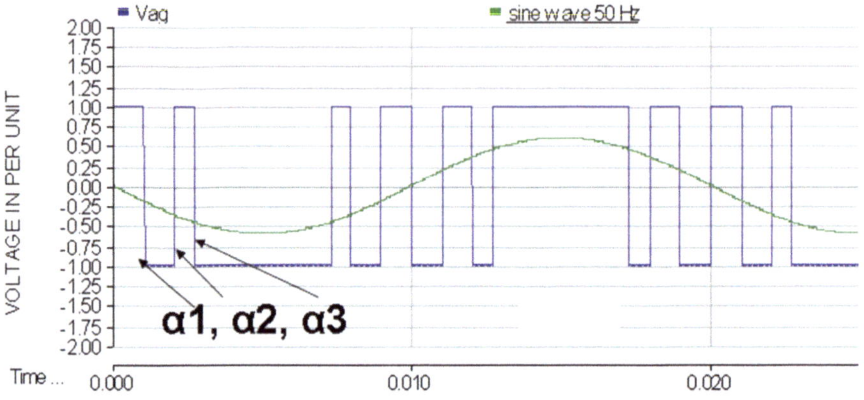

Fig. 3.14 Bipolar waveform

waveform of the form shown in Fig. 3.14. The bipolar SHE-PWM technique offers solution to both single-phase and three-phase inverters. The generalised case is to find appropriate angles $\alpha 1$, $\alpha 2$, $\alpha 3$,... αi where $i = N + 1$ so that N non-triplen odd harmonics are eliminated and control of the fundamental is also achieved. The DC link voltage of the inverter is assumed to be 2 p.u. The Fourier series expansion of the waveform in Fig. 3.14 is given by Eq. 3.10 [8, 17]:

$$V(\theta) = \frac{4}{\pi} * . \left\{ \sum_{k=1,3,5,\ldots}^{\infty} \frac{\sin\ (k.\theta)}{k} * \left(1 + 2. \sum_{i=1,2,3,\ldots\ldots}^{N+1} (-\ 1)^i . \cos(k.\alpha_i) \right) \right\} \tag{3.10}$$

The waveform shown in Fig. 3.14 has a quarter-wave symmetry. To eliminate N non-triplen odd harmonics, $N + 1$ angles have to be found. Eqs. 3.11 and 3.12 must be solved.

$$\begin{bmatrix} 1 + 2. \sum_{i=1}^{N+1} (-1)^i . \cos(\alpha_i) - M = 0 \\ 1 + 2. \sum_{i=1}^{N+1} (-1)^i . \cos(5.\alpha_i) = 0 \\ 1 + 2. \sum_{i=1}^{N+1} (-1)^i . \cos(n.\alpha_i) = 0 \end{bmatrix} \tag{3.11}$$

where

$$0 \leq |M| \leq 1 \tag{3.12}$$

In general, if 2*Vdc is the DC link voltage, then the Fourier series expansion of Fig. 3.14 is as given in Eq. 3.13.

$n = 3, 5, 7\ldots.2.N + 1$ for single-phase systems
$n = 5, 7, 11,\ldots.3.N + 1$ when N is even for three-phase systems
$n = 5, 7, 11,\ldots.3.N + 2$ when N is odd for three-phase systems

$$V(\omega.t) = \frac{-4.V_{dc}}{\pi} * \left\{ \sum_{k=1.3.5,7.....}^{\infty} \frac{\sin(k.\omega.t)}{k} * (1 - 2.\cos(k.\alpha_1) + 2.\cos(k.\alpha_2) - 2.\cos(k.\alpha_3)) \right\} \tag{3.13}$$

where $\theta = \omega.t$ is the angular frequency of the output voltage. The modulation index M is defined below:

$$M = \frac{V_1}{\left(\frac{4.V_{dc}}{\pi}\right)} \tag{3.14}$$

V_1 is the amplitude of the fundamental component of the bipolar inverter line-to-ground output voltage. For a DC link voltage 2*Vdc of 2 p.u., Eq. 3.14 simplifies to the following:

$$M = \frac{V_1}{(4/\pi)}\ p.u. \tag{3.15}$$

Given the fundamental voltage amplitude V_1, to eliminate 5th and 7th harmonics from the line-to-ground voltage of the inverter, Eq. 3.16 must be solved.

$$\left.\begin{aligned} 1 - 2.\cos(\alpha_1) + 2.\cos(\alpha_2) - 2.\cos(\alpha_3) &= -M \\ 1 - 2.\cos(5.\alpha_1) + 2.\cos(5.\alpha_2) - 2.\cos(5.\alpha_3) &= 0 \\ 1 - 2.\cos(7.\alpha_1) + 2.\cos(7.\alpha_2) - 2.\cos(7.\alpha_3) &= 0 \end{aligned}\right\} \tag{3.16}$$

For a given modulation index in the range 0 to 1, set of Eq. 3.16 can be solved by Newton-Raphson method [1, 2], Nelder–Mead [7] or theory of Resultant algorithm [8, 14]. Here Eq. 3.16 is solved using Symbolic Math Toolbox in MATLAB [11]. This is shown in Program segment 3.1 using "solve" command to get values for $\alpha 1$, $\alpha 2$ and $\alpha 3$. In this "solve" command, Eq. 3.9 is replaced with Eq. 3.16. From the printed values for $\alpha 1$, $\alpha 2$ and $\alpha 3$ in radians (degrees), only those values that satisfy the relation $\alpha 1 < \alpha 2 < \alpha 3 < \pi/2$ are selected. These values are tabulated in Table 3.2 for M value 0.5 and 0.8.

Table 3.2 Three-phase bipolar SHE-PWM inverter switching angles

Sl. No.	M	$\alpha 1$ degrees	$\alpha 2$ degrees	$\alpha 3$ degrees
1	0.5	5. 7056	68. 4650	82. 9911
2	0.5	20. 9355	35. 7758	51. 1467
3	0.8	8. 9320	75. 0757	80. 2314
4	0.8	14. 4942	37. 4962	43. 5127

3.3.2 Model for Three-Phase Bipolar SHE-PWM Inverter

The block diagram model of the three phase bipolar SHE-PWM inverter is shown in Fig. 3.15. The reference sine wave is marked (1.sin(2*π*f*t)). Sine wave zero crossing comparator (ZCC) X7 output is +5 V when the input sine wave crosses zero and goes positive and – 5 V when the input sine wave crosses zero and goes negative. Comparators X1, X3, X8 marked A > =B have their output HIGH if input A is greater than or equal to B or else their output is LOW. The B input for X8 is zero, and this comparator functions as a diode rectifier. Thus X8 output is HIGH during the positive half cycle and LOW during the negative half cycle of the reference sine wave voltage. Similar argument holds good for comparators marked $A > B$, $A < =B$ and $A < B$, where the symbols >, <= and < correspond to greater than, less than or equal to and less than, respectively. The reference sine wave voltage 1*Sin($\omega.t$) is connected to the A input of comparators X1 to X6 and to the ZCC X7. The sine of α1, α2 and α3 is connected to the B input of X1, X2 and X3. Similarly sine of (π + α1), (π + α2), (π + α3) is connected to the B input of X4, X5 and X6. The outputs of X1 and X2 are given to AND gate U1, that of X3 is given to AND gate U2, that of X4 and X5 are given to AND gate U3 and that of X6 is given to AND gate U4. The output of all AND gates U1 to U4 is given to OR gate U5. The outputs of diode X8 and OR gate U5 are given to an EXCLUSIVE-OR gate U6. The output of U6 is inverted using a NOT gate U7. The output of U6 and U7 forms the gate pulse for the switches S1 and S4 in Phase A of inverter. For the gate pulse of switches in Phase B and C, comparison is made with reference sine wave voltages 1*Sin($\omega.t$ – 2π/3) and 1*Sin($\omega.t$ + 2π/3), respectively. The following action takes place for various intervals of the Phase A reference sine wave voltage:

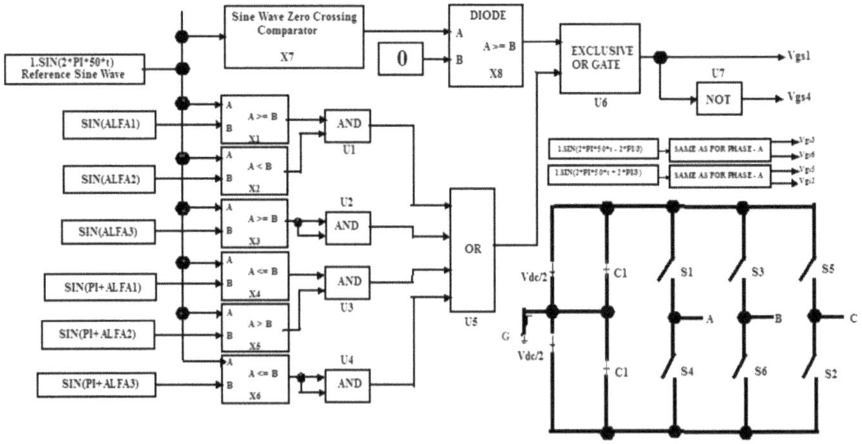

FIG. 3.15 Three-phase BIPOLAR SHE-PWM inverter model

Case 1: $0 <= \omega.t < \alpha 1$ radians. During this time interval of the reference sine wave voltage for Phase A, the following operation takes place:

1. Comparators X1, X2, X3 outputs are LOW, HIGH and LOW and that of X4, X5 and X6 are LOW, HIGH and LOW, respectively. Outputs of X7 and X8 are HIGH.
2. Outputs of logic gates U1 to U5 and U7 are all LOW and that of U6 is HIGH.
3. Output Vgs1 is HIGH and that of Vgs4 is LOW.
4. Switch S1 is ON and S4 is OFF, and the output voltage VAG is +Vdc/2 volts.

Case 2: $\alpha 1 <= \omega.t < \alpha 2$ radians. During this time interval of the reference sine wave voltage for Phase A, the following operation takes place:

1. Comparators X1, X2, X3 outputs are HIGH, HIGH and LOW and that of X4, X5 and X6 are LOW, HIGH and LOW, respectively. Outputs of X7 and X8 are HIGH.
2. Outputs of logic gates U1, U5 and U7 are HIGH and that of U2 to U4 and U6 are LOW.
3. Outputs Vgs1 and Vgs4 are LOW and HIGH, respectively.
4. Switch S1 is OFF and S4 is ON, and the output voltage VAG is -Vdc/2 volts.

Case 3: $\alpha 2 < \omega.t < \alpha 3$ radians. During this time interval of the reference sine wave voltage for Phase A, the following operation takes place:

1. Comparators X1, X2, X3 outputs are HIGH, LOW and LOW and that of X4, X5 and X6 are LOW, HIGH and LOW, respectively. Outputs of X7 and X8 are HIGH.
2. Outputs of logic gates U1 to U5 and U7 are all LOW and that of U6 is HIGH.
3. Output Vgs1 is HIGH and that of Vgs4 is LOW.
4. Switch S1 is ON, that of S4 is OFF, and the output voltage VAG is +Vdc/2 volts.

Case 4: $\alpha 3 <= \omega.t < \pi/2$ radians. During this time interval of the reference sine wave voltage for Phase A, the following operation takes place:

1. Comparators X1, X2, X3 outputs are HIGH, LOW and HIGH and that of X4, X5 and X6 are LOW, HIGH and LOW, respectively. Outputs of X7 and X8 are HIGH.
2. Outputs of logic gates U2, U5 and U7 are HIGH and that of U1, U3, U4 and U6 are all LOW.
3. Output Vgs1 is LOW and that of Vgs4 is HIGH.
4. Switch S1 is OFF and S4 is ON, and the output voltage VAG is -Vdc/2 volts.

Case 5: $\pi <= \omega.t < (\pi + \alpha 1)$ radians. During this time interval of the reference sine wave voltage for Phase A, the following operation takes place:

1. Comparators X1, X2, X3 are outputs LOW, HIGH and LOW and that of X4, X5 and X6 are LOW, HIGH and LOW, respectively. Outputs of X7 and X8 are LOW.

2. Outputs of logic gates U1 to U6 are all LOW and that of U7 is HIGH.
3. Output Vgs1 is LOW and that of Vgs4 is HIGH.
4. Switch S1 is OFF and S4 is ON, and the output voltage VAG is –Vdc/2 volts.

Case 6: $(\pi + \alpha 1) <= \omega .t < (\pi + \alpha 2)$ radians. During this time interval of the reference sine wave voltage for Phase A, the following operation takes place:

1. Comparators X1, X2, X3 outputs are LOW, HIGH, LOW and that of X4, X5 and X6 are HIGH, HIGH and LOW, respectively. Outputs of X7 and X8 are LOW.
2. Output of logic gates U3, U5 and U6 are HIGH and that of U1, U2, U4 and U7 are LOW.
3. Output Vgs1 is HIGH, and Vgs4 is LOW.
4. Switch S1 is ON and S4 is OFF, and the output voltage VAG is +Vdc/2 volts.

Case 7: $(\pi + \alpha 2) <= \omega .t < (\pi + \alpha 3)$ radians. During this time interval of the reference sine wave voltage for Phase A, the following operation takes place:

1. Comparators X1, X2, X3 outputs are LOW, HIGH and LOW and that of X4, X5 and X6 are HIGH, LOW and LOW, respectively. Outputs of X7 and X8 are LOW.
2. Outputs of logic gates U1 to U6 are all LOW and that of U7 is HIGH.
3. Output Vgs1 is LOW and that of Vgs4 is HIGH.
4. Switch S1 is OFF and S4 is ON, and the output voltage VAG is –Vdc/2 volts.

Case 8: $(\pi + \alpha 3) <= \omega .t < 3\pi/2$ radians. During this time interval of the reference sine wave voltage for Phase A, the following operation takes place:

1. Comparators X1, X2, X3 outputs are LOW, HIGH and LOW and that of X4, X5 and X6 are HIGH, LOW and HIGH, respectively. Outputs of X7 and X8 are LOW.
2. Outputs of logic gates U4, U5 and U6 are HIGH and that of U1, U2, U3 and U7 are LOW.
3. Output Vgs1 is HIGH and that of Vgs4 is LOW.
4. Switch S1 is ON and S4 is OFF, and the output voltage VAG is +Vdc/2 Volts.

Block diagram model shown in Fig. 3.15 was developed using SIMULINK [11]. This model is shown in Fig. 3.16 (Model file: EXAMPLE 3_2), and the gate drive subsystem is shown in Fig. 3.17a, b. Three-phase 50 Hz sine wave AC voltage is used as a reference. The value of $\alpha 1$, $\alpha 2$ and $\alpha 3$ for $M = 0.8$ shown in serial number 4 of Table 3.2 is used in the model. The inverter is from Universal Bridge subsystem from Power Systems block set. The reference sine wave voltages for the three phase are generated using Embedded MATLAB function, and the sine of switching angles for $\alpha 1$, $\alpha 2$ and $\alpha 3$ and that of $(\pi + \alpha 1)$, $(\pi + \alpha 2)$ and $(\pi + \alpha 3)$ are generated using another Embedded MATLAB function. The source codes are given in Program segment 3.2 and 3.3 in the model file EXAMPLE 3_2. In Fig. 3.16, the switching angles $\alpha 1$, $\alpha 2$ and $\alpha 3$ in degrees are entered in order in the constant block separated by a space. The selector block directs the output $\alpha 1$, $\alpha 2$ and $\alpha 3$ in sequence.

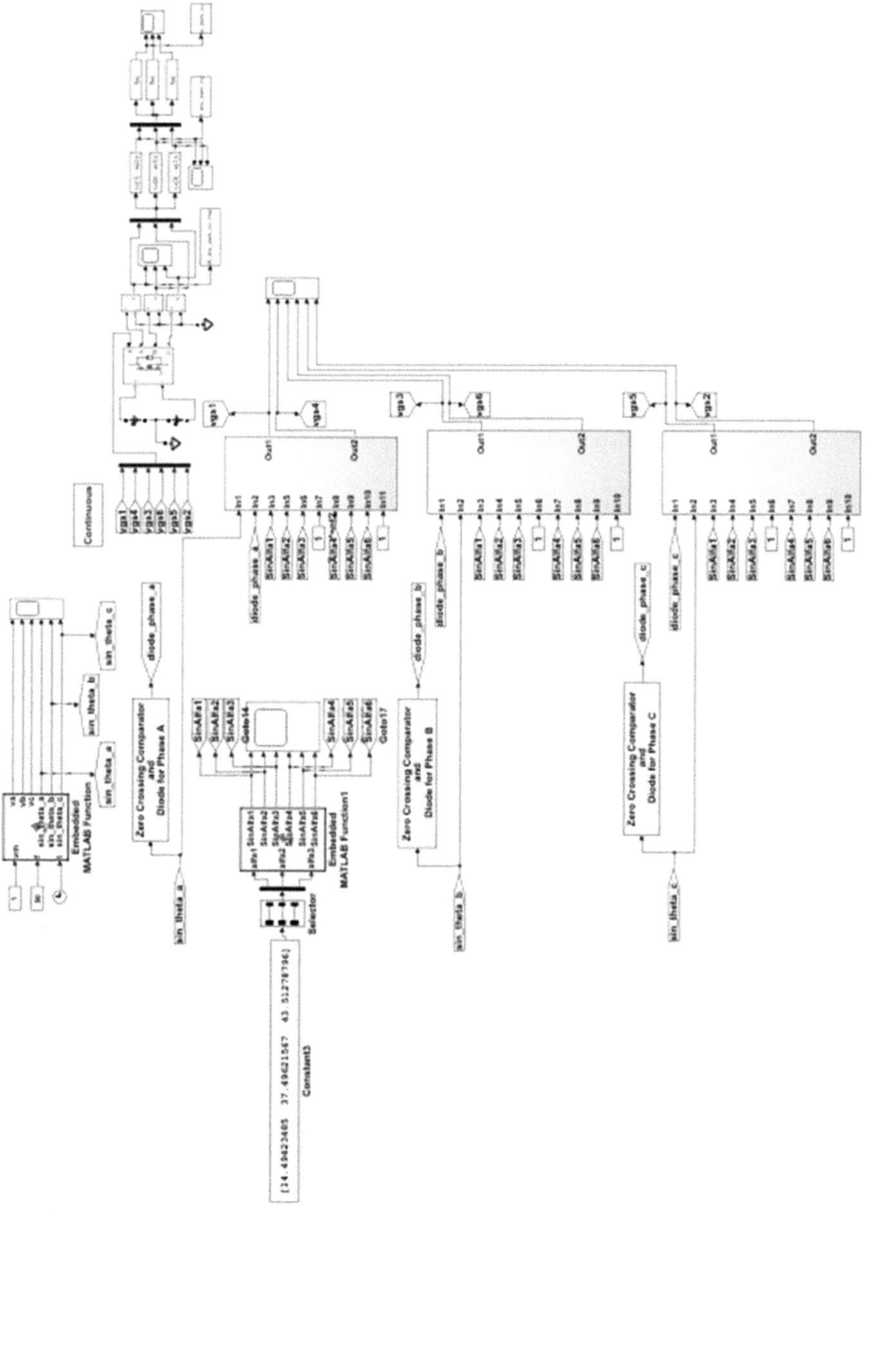

Fig. 3.16 Model of three-phase bipolar SHE-PWM inverter with three switching angles

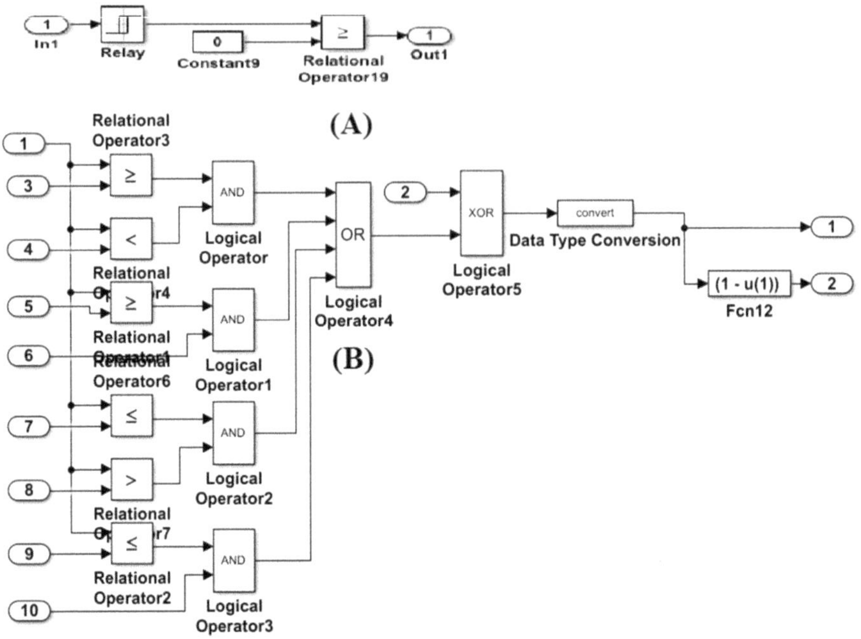

Fig. 3.17 Three-phase bipolar SHE-PWM inverter model—(**a**) Zero crossing comparator and diode for phase A and (**b**) Gate pulse generator for phase A

The model subsystem to generate gate pulse for phase A switches of inverter is shown in Fig. 3.17a, b. Here as shown in Fig. 3.17a, reference sine wave voltage for Phase A is given to relay block which forms zero-crossing comparator (ZCC) giving square wave output with positive and negative half cycles when the input sine wave voltage crosses zero and goes positive and negative, respectively. The ZCC output is compared with zero using a relational operator block which forms the comparator whose output is HIGH when the input is positive or greater than or equal to zero and LOW when the input is negative. The ZCC and comparator together operate as a diode rectifier. Reference sine wave voltage for phase A ($1.\mathrm{Sin}(2*\pi*50*t)$) is compared with sine of $\alpha1$, $\alpha2$, $\alpha3$, $(\pi + \alpha1)$, $(\pi + \alpha2)$ and $(\pi + \alpha3)$ in six relational operator blocks which form the comparators, as shown in Fig. 3.17b. The resulting output of each comparator is connected to four AND gates. The four AND gate outputs are given to an OR logic gate. The OR gate output and the diode rectifier output are given to EXCLUSIVE-OR gate as shown in Fig. 3.17b to generate gate pulse Vgs1 and the inverted pulse using NOT gate to generate Vgs4 for the upper and lower switches of inverter in phase A. Similar method is used along with reference sine wave voltages for phase B and C to generate gate pulse for the upper and lower inverter switches in the respective phase.

3.3.3 Simulation Results

The model simulation of the three-phase bipolar SHE-PWM inverter was carried out using ode23tb (stiff/TR-BDF2) solver. The inverter switching frequency is 50 Hz, and DC link voltage is 100 V. The switching angles in Table 3.2 are used to eliminate 5th and 7th harmonics from the line-to-ground voltage. The harmonic spectrum of line-to-ground voltage and the three-phase line-to-ground voltage for M values of 0.5 and 0.8 in the respective order as in Table 3.2 are shown in Figs. 3.18 to 3.25, respectively.

3.3.4 Discussion of Results

The simulation results for the harmonic spectrum of line-to-ground voltage shown in Figs. 3.18, 3.20, 3.22 and 3.24 reveal that the 5th and 7th harmonics are well eliminated. Also for a DC link voltage of 100 V, using Eq.3.14, the peak fundamental voltage V_1 is found to be 31.831 and 50.93 V for M of 0.5 and 0.8, respectively, whereas that by simulation of model is found to be 31.72 V and

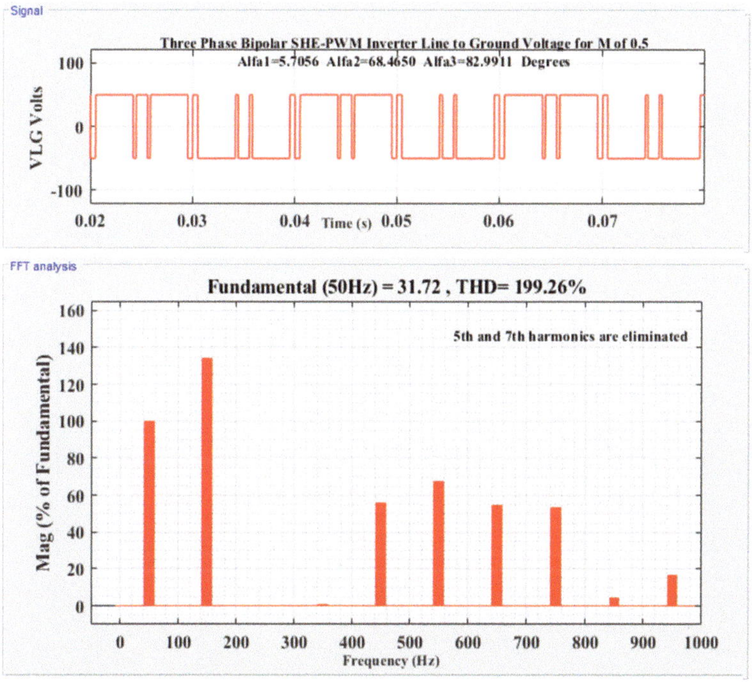

Fig. 3.18 Three-phase bipolar SHE-PWM inverter line-to-ground voltage (top) and harmonic spectrum with 5th and 7th harmonic eliminated

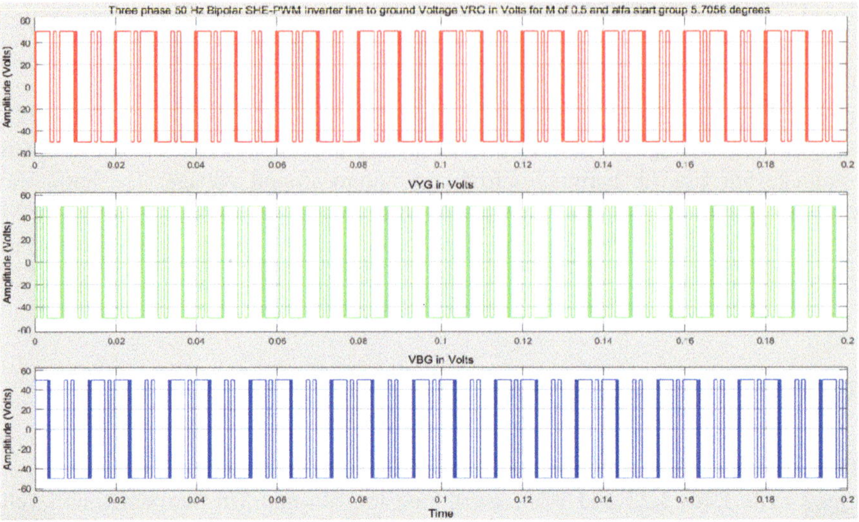

Fig. 3.19 Three-phase bipolar SHE-PWM inverter line-to-ground voltage for *M* of 0.5 and alfa start group of 5.7056 degrees

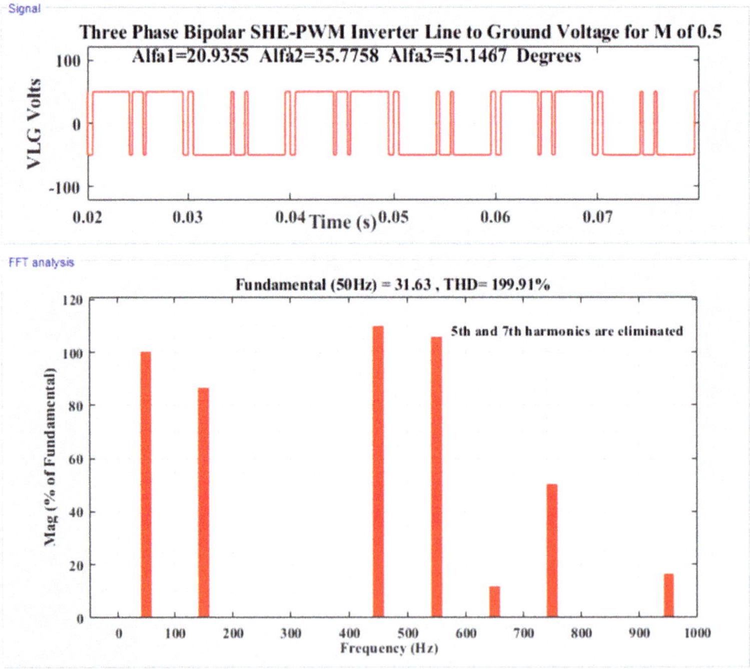

Fig. 3.20 Three-phase bipolar SHE-PWM inverter line-to-ground voltage (Top) and harmonic spectrum with 5th and 7th harmonic eliminated (bottom)

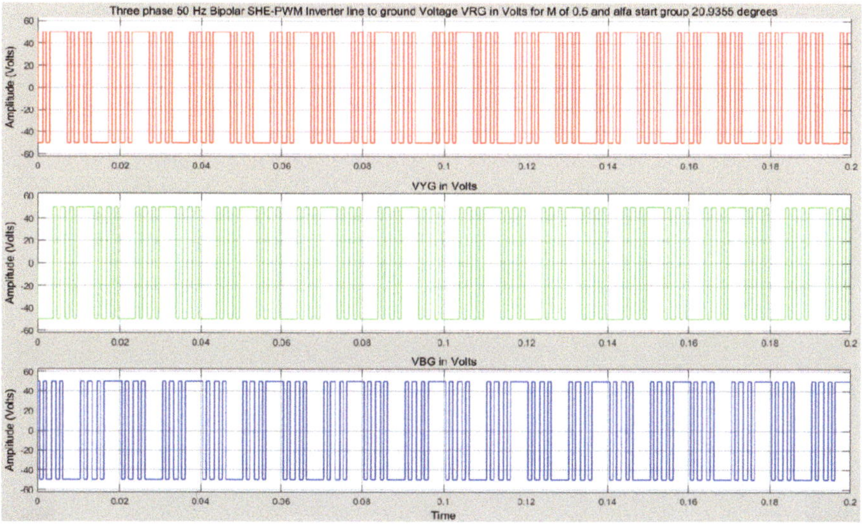

Fig. 3.21 Three-phase bipolar SHE-PWM inverter line-to-ground voltage for M of 0.5 and Alfa start group of 20.9355 degrees

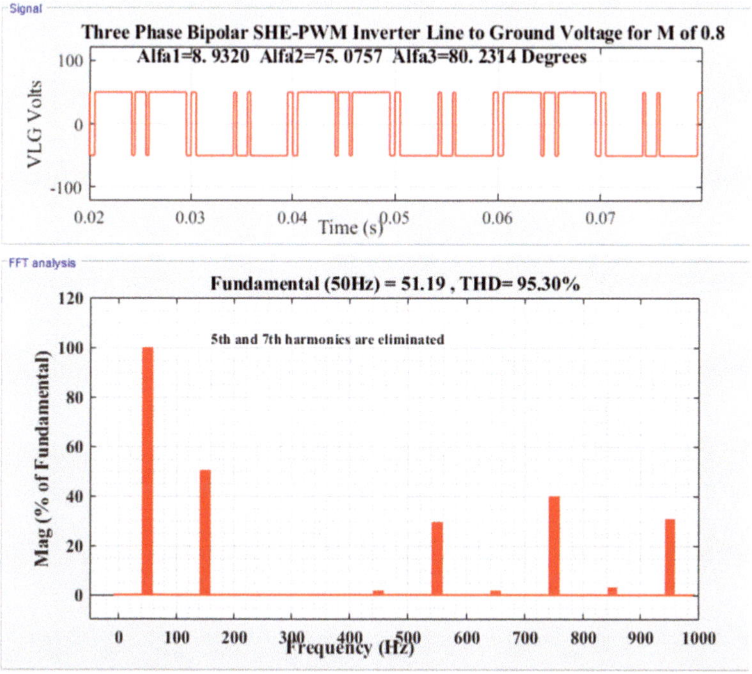

Fig. 3.22 Three-phase bipolar SHE-PWM inverter line-to-ground voltage (top) and harmonic spectrum with 5th and 7th harmonic eliminated (bottom)

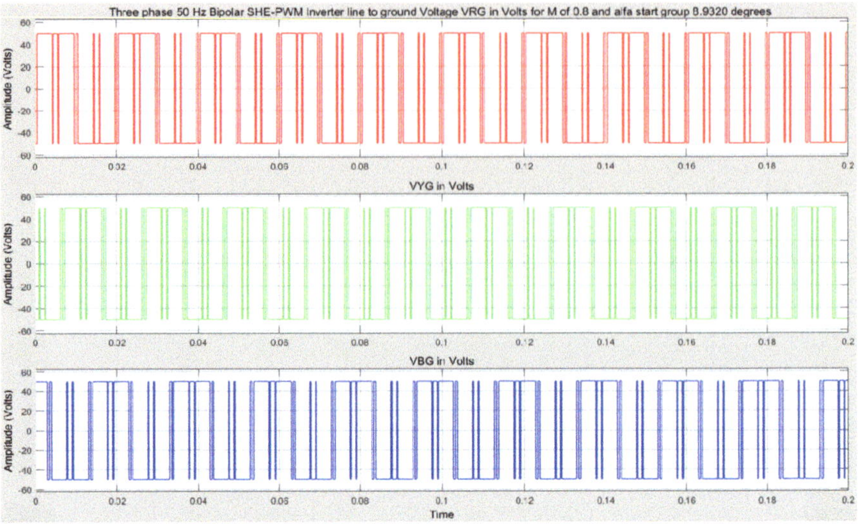

Fig. 3.23 Three-phase bipolar SHE-PWM inverter line-to-ground voltage for *M* of 0.8 and alfa start group of 8.9320 degrees

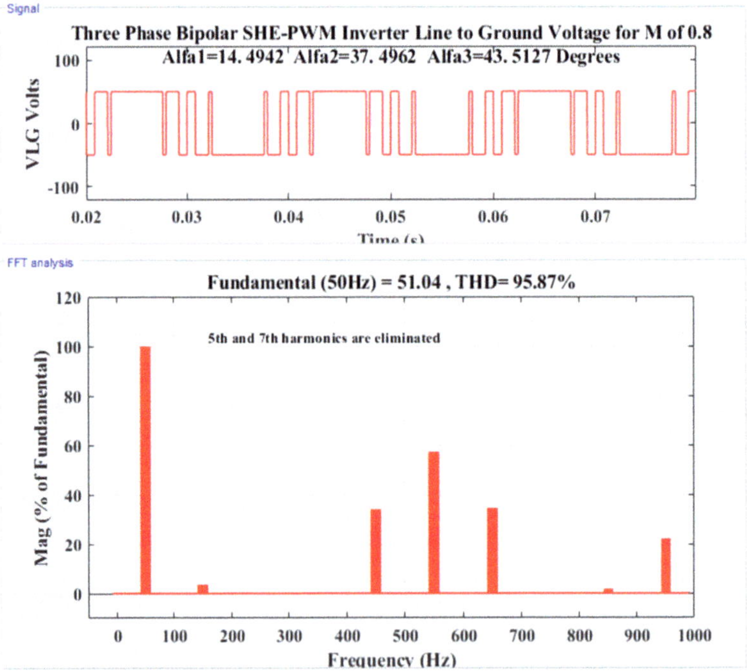

Fig. 3.24 Three-phase bipolar SHE-PWM inverter line-to-ground voltage (top) and harmonic spectrum with 5th and 7th harmonic eliminated (bottom)

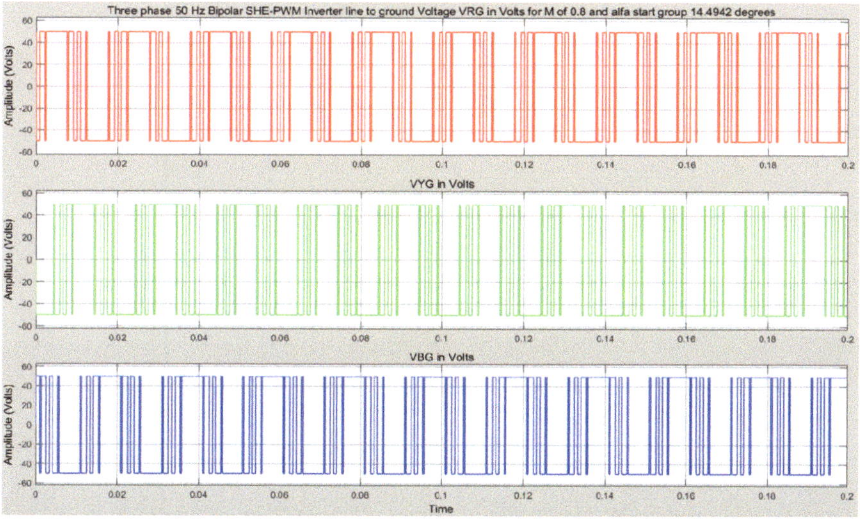

Fig. 3.25 Three-phase bipolar SHE-PWM inverter line-to-ground voltage for *M* of 0.8 and alfa start group of 14.1542 degrees

31.63 V for *M* of 0.5 and 51.19 V and 51.04 V for *M* of 0.8. Also for M value of 0.8, the THD of line-to-ground voltage is found to be much less than that for *M* value of 0.5.

3.4 Three-Phase Cascade Multi-cell SHE-PWM Inverter

The schematic of a single-phase cascade multicell inverter also known as cascade H-bridge inverter (CHBI) with two cells is shown in Fig. 3.26 where H1 is the lower H-bridge and H2 is the upper H-bridge [19, 20]. The DC link voltage source for the lower and upper cell is Vdc1 and Vdc2, respectively. The two voltage sources can be equal or different. The output voltages of lower and upper cells are marked V_1 and V_2 and the resultant output voltage of the multicell inverter VRN$(t) = V_1(t) + V2(t)$. If Vdc1 = Vdc2 = Vdc, then V_1 and *V2* each can take the value -Vdc, 0 and + Vdc by proper switching combination in the respective cells. Output voltage *VRN* has five levels: −2.Vdc, −Vdc, 0, +Vdc and + 2.Vdc, respectively. These waveforms are shown in Fig. 3.27.

In Fig. 3.27, the Fourier transform f$(\omega.t)$ of *VRN(t)* is the sum of the Fourier transform f1$(\omega.t)$ of $V_1(t)$ and f2$(\omega.t)$ that of $V_2(t)$.

$$f(\omega.t) = f_1(\omega.t) + f_2(\omega.t) \tag{3.17}$$

$f_1(\omega.t)$ is odd and half-wave symmetric and the Fourier coefficient bn is given below:

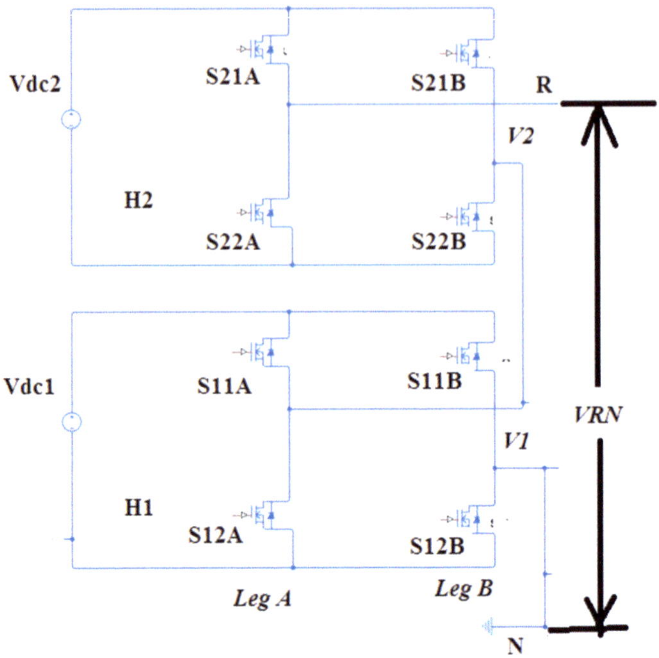

Fig. 3.26 Single-phase multicell inverter

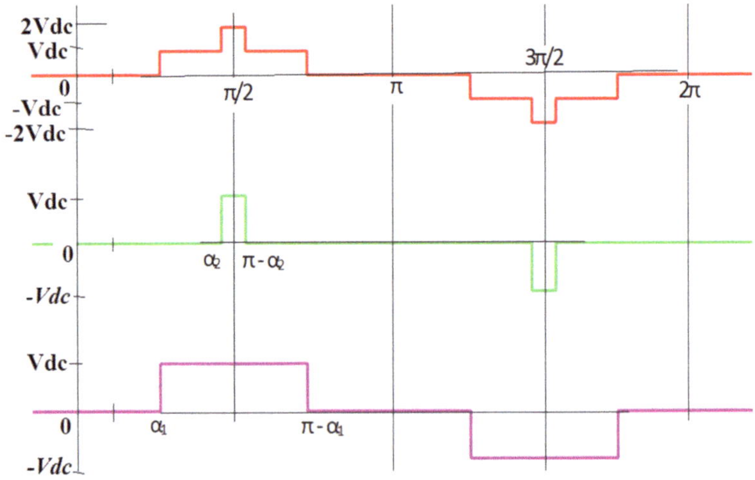

Fig. 3.27 Staircase waveform produced by a two-cell cascade H-bridge inverter: VRN (top), V2 (middle) and V1 (bottom)

$$b_n = \frac{2^*}{\pi}\left[\int_{\alpha_1}^{\pi-\alpha_1} V_{dc}{}^* \sin(n.\omega t) * d(\omega t)\right]$$
$$b_n = \frac{2.V_{dc}{}^*}{\pi.n}\left[-\cos n.\omega t\right]_{\alpha_1}^{\pi-\alpha_1} \tag{3.18}$$

$$b_n = \sum_{n=1,3,5,7,11,13....}^{\infty} \frac{4^* V_{dc}{}^*}{\pi.n} \cos(n.\alpha_1) \tag{3.19}$$

$$f_1(\omega.t) = \sum_{n=1,3,5,7,9,11,13....}^{\infty} \frac{4^* V_{dc}{}^*}{\pi.n} \cos(n.\alpha_1) * \sin(n.\omega t) \tag{3.20}$$

By a similar argument as given above for $f_1(\omega t)$, $f_2(\omega t)$ for $V_2(t)$ can be expressed as follows:

$$f_2(\omega.t) = \sum_{n=1,3,5,7,9,11,13....}^{\infty} \frac{4 * V_{dc}}{\pi.n} * \cos(n.\alpha_2) * \sin(n.\omega t) \tag{3.21}$$

Thus f(ωt) for *VRN(t)* which is also the line-to-ground voltage V_1 can be expressed as follows:

$$f(\omega t) = V(\omega t) = \sum_{n=1,3,5,7,9,11,13....}^{\infty} \frac{4 * V_{dc}}{\pi.n} * [\cos(n.\alpha_1) + \cos(n.\alpha_2)] * \sin(n.\omega t) \tag{3.22}$$

Thus, to eliminate 5th harmonics from *VRN(t)*, the following set of Eq. 3.23 has to be solved:

$$\left.\begin{aligned} \cos(\alpha_1) + \cos(\alpha_2) &= M \\ \cos(5.\alpha_1) + \cos(5.\alpha_2) &= 0 \end{aligned}\right\} \tag{3.23}$$

In Eq. 3.23, modulation index $M = V_1/(4^*V_{dc}/\pi)$.

Now consider the case when the DC link voltage source for each cell is different. In Fig. 3.26, let the bottom cell H1 has the DC link voltage source Vdc and that of the top cell H2 has Vdc/2 volts. Thus bottom cell H1 can have –Vdc, 0 and +Vdc voltage levels for V_1 and the top cell H2 can have –Vdc/2, 0 and +Vdc/2 voltage levels for V_2. Thus, the output of the multicell *VRN(t)* can have –3Vdc/2, –Vdc, –Vdc/2, 0, +Vdc/2, +Vdc and + 3Vdc/2 voltage levels. Thus, in this case, Fig. 3.26 is a seven-level cascade H-bridge inverter, and the output voltage waveform is shown in Fig. 3.28. The Fourier transform of *VRN(t)* in Fig. 3.28 is derived below.

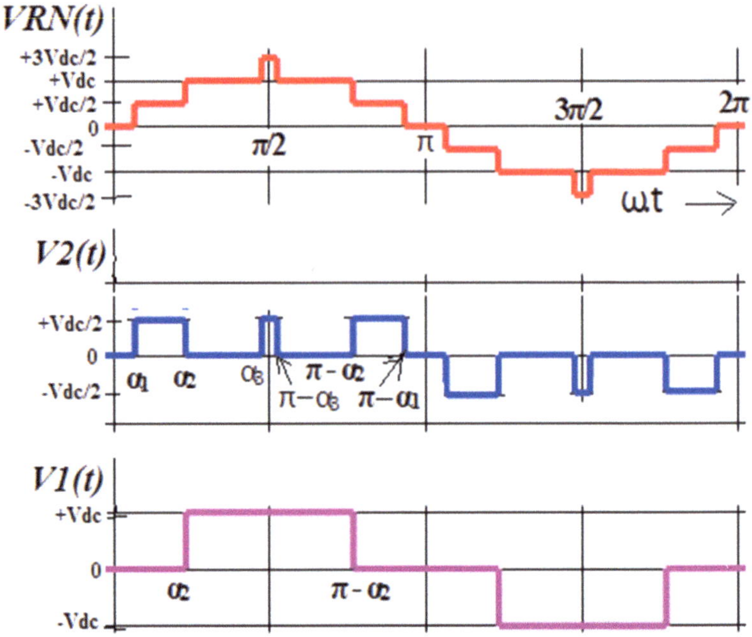

Fig. 3.28 Seven-level output voltage waveform of a two-cell cascade H. Bridge Inverter: VRN (t) (top), V2(t) (middle) and V1(t) (bottom)

Following the same method shown for Fig. 3.27, noting the limits $\alpha2$ and π-$\alpha2$, the Fourier transform $f_1(\omega t)$ of $V_1(t)$, using Eq. 3.18, can be expressed as in Eq. 3.24:

$$f_1(\omega.t) = \sum_{n=1,3,5,7,9,11,13....}^{\infty} \frac{4 * V_{dc}}{\pi.n} * \cos(n.\alpha_2) * \sin(n.\omega t) \tag{3.24}$$

The Fourier transform $f_2(\omega t)$ of $V_2(t)$ in Fig. 3.28 is derived below:

$V_2(t)$ is odd and half-wave symmetric, and the Fourier coefficient bn can be expressed as in Eq. 3.25.

$$\frac{2^*}{\pi}\left[\int_{\alpha_1}^{\alpha_2} \frac{V_{dc}{}^*}{2} \sin(n.\omega t) * d(\omega t) \quad b_n = \right.$$
$$\left. + \int_{\alpha_3}^{\pi-\alpha_3} \frac{V_{dc}{}^*}{2} \sin(n.\omega t) * d(\omega t) + \int_{\pi-\alpha_2}^{\pi-\alpha_1} \frac{V_{dc}{}^*}{2} \sin(n.\omega t) * d(\omega t)\right] \tag{3.25}$$

Simplifying Eq. 3.25,

$$b_n = \frac{V_{dc}^*}{\pi.n}\{[-\cos n.\omega t]\}\Bigg]_{\alpha_1}^{\alpha_2} + \frac{V_{dc}^*}{\pi.n}\{[-\cos n.\omega t]\}\Bigg]_{\alpha_3}^{\pi-\alpha_3} + \frac{V_{dc}^*}{\pi.n}\{[-\cos n.\omega t]\}\Bigg]_{\pi-\alpha_2}^{\pi-\alpha_1} \quad (3.26)$$

Simplifying Eq. 3.26, $f_2(\omega t)$ can be expressed as follows:

$$f_2(\omega t) = \sum_{n=1,3,5,7,9,11,13....}^{\infty} \frac{2 * V_{dc}}{\pi.n} * [\cos(n.\alpha_1) - \cos(n.\alpha_2) + \cos(n.\alpha_3)] * \sin(n.\omega t) \quad (3.27)$$

Adding Eq. 3.24 and 3.27, the Fourier transform $f(\omega t) = V(\omega t)$ for $VRN(t)$ in Fig. 3.28 can be expressed as follows:

$$f(\omega t) = V(\omega t) = \sum_{n=1,3,5,7,9,11,13....}^{\infty} \frac{2 * V_{dc}}{\pi.n} * [\cos(n.\alpha_1) + \cos(n.\alpha_2) + \cos(n.\alpha_3)] * \sin(n.\omega t) \quad (3.28)$$

To eliminate 5th and 7th harmonics from the line-to-ground voltage $VRN(t)$ in Fig. 3.28, the following set of equations must be solved:

$$\left.\begin{aligned} \cos(\alpha_1) + \cos(\alpha_2) + \cos(\alpha_3) = M \\ \cos(5.\alpha_1) + \cos(5.\alpha_2) + \cos(5.\alpha_3) = 0 \\ \cos(7.\alpha_1) + \cos(7.\alpha_2) + \cos(7.\alpha_3) = 0 \end{aligned}\right\} \quad (3.29)$$

In Eq. 3.29, M is the ratio of $V_1/(2*\text{Vdc}/\pi)$ where V_1 is the amplitude of the fundamental component of the output voltage $VRN(t)$. Modulation index ma = M/k where k is the number of DC link voltage sources used.

3.4.1 Modelling of Three-Phase Two Cell Cascade H-Bridge SHE-PWM Inverter with Unequal DC Voltage Sources

The model of three-phase two-cell H-bridge SHE-PWM inverter shown in Fig. 3.26 for one phase is presented in this section. Here Vdc1 for lower cell is Vdc, and Vdc2 for upper cell is Vdc/2. The switching angles to eliminate 5th and 7th harmonics from the line-to-ground voltage are calculated using Eq. 3.29. The source code using "solve" command is shown in Program segment 3.1 where Eq. 3.9 is replaced with Eq. 3.29. From the printed values for $\alpha 1$, $\alpha 2$ and $\alpha 3$ in radians (degrees), only those values that satisfy the relation $\alpha 1 < \alpha 2 < \alpha 3 < \pi/2$ are selected. These values are tabulated in Table 3.3 for M/ma values of 1.6/0.8 and 1.8/0.9.

The model block diagram for the three-phase two-cell cascade H-bridge SHE-PWM inverter is shown in Fig. 3.29 [15]. The switching angles are alfa1,

Table 3.3 Three-phase two-cell cascade H-bridge SHE-PWM inverter switching angles

Sl.No.	M/ma	$\alpha 1$ degrees	$\alpha 2$ degrees	$\alpha 3$ degrees
1	1.6/0.8	18.9554	52.30388	87.1887
2	1.8/0.9	11.794	41.599	85.486

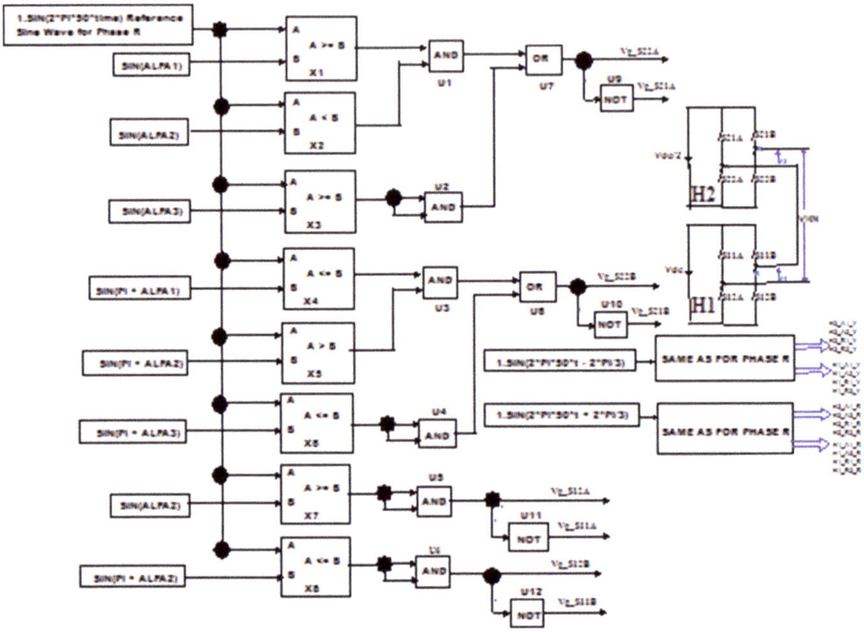

FIG. 3.29 Three-phase two-cell cascade H-bridge SHE-PWM inverter gate drive model block diagram

alfa2 and alfa3. The sine wave reference voltage is greater than or equal to Sin(α1), comparator X1 output is HIGH or else its output is LOW and similarly when the former reference sine voltage is only less than Sin(α2), and Comparator X2 output is HIGH. Comparators X1 and X2 output are given to AND gate U1. Similarly reference sine voltage connected to comparator X3 output is HIGH only when sine wave voltage is equal to or greater than Sin(α3). Comparator X3 output through AND gate marked U2 and the U1 output are connected to OR gate U7. The U7 output Vg_S22A forms the gate drive for the switch in the leg A of upper cell H2 marked S22A. The output of U7 is inverted using NOT gate U11 whose output Vg_S21A forms the gate drive for switch S21A in the upper H2 cell. Similar procedure is adopted to compare the reference sine wave output for phase R with sine of (pi+α1), (pi+α2) and (pi+α3) in three separate Comparators marked X4, X5 and X6. The resulting output of these three comparators through AND gates U3 and U4 is connected to OR gate U8. The output of U8 and its inverted output from NOT gate U10 form the gate drive Vg_S22B and Vg_S21B for the inverter switches marked S22B and S21B in leg B of the upper cell H2.

For the switches in the lower cell H1, reference sine wave voltage for phase R is compared with Sin(α2) in comparator X7 whose output is HIGH only when the former input is greater than or equal to the later value. This output through AND gate U5 forms the gate pulse Vg_S11A for switch S11A, and the inverted output through NOT gate U11 Vg_S12A forms the gate pulse for switch S12A in the lower H1 cell. Similar procedure is used to compare reference sine wave voltage with Sin(pi+α2) using comparator X8 whose output is HIGH only when the former input is less than or equal to the later input else its output is LOW. The resulting output of X8 through AND gate U6 and the inverted output through NOT gate U12 form the gate pulse Vg_S11B and Vg_S12B for switches S11B and S12B, respectively, in the lower H1 cell. For phase Y and B switches, similar to phase R, comparison is made with reference sine waves (1.sin(ω.t—2*pi/(3)) and (1.sin(ω.t + 2*pi/(3)), respectively. Here ω vlue is taken as 2*π*50 rad/second. The following action takes place for various intervals of the phase A reference sine wave voltage:

Case 1: $0 <= \omega.t < \alpha 1$ radians. During this time interval of the reference sine wave voltage for phase A, the following operation takes place:

1. Comparators X1, X2, X3 outputs are LOW, HIGH and LOW, that of X4, X5 and X6 are LOW, HIGH and LOW, and that of X7, X8 are LOW and LOW, respectively.
2. Outputs of logic gates U1 to U8 are all LOW and that of U9, U10, U11 and U12 are HIGH.
3. Output Vg_S22A, Vg_S22B, Vg_S12A and Vg_S12B are all LOW and that of Vg_S21A, Vg_S21B, Vg_S11A and Vg_S11B are all HIGH.
4. Switches S22A, S22B, S12A and S12B are all OFF and that of S21A, S21B, S11A and S11B are all ON, and the output voltage V_1, V_2 and VRN are, respectively, 0 volts.

Case 2: $\alpha 1 <= \omega.t < \alpha 2$ radians. During this time interval of the reference sine wave voltage for phase A, the following operation takes place:

1. Comparators X1, X2, X3 outputs are HIGH, HIGH and LOW; that of X4, X5 and X6 are LOW, HIGH and LOW; and that of X7 and X8 are LOW and LOW, respectively.
2. Outputs of logic gates U1, U7, U10, U11 and U12 are all HIGH and that of U2 to U6, U8 and U9 are LOW.
3. Outputs Vg_S22A, Vg_S22B, Vg_S12A and Vg_S12B are all LOW and that of Vg_S22A, Vg_S21B, Vg_S11A, Vg_S11B are all HIGH.
4. Switches S21A, S22B, S12A and S12B are all OFF; that of S22A, S21B, S11A and S11B are all ON; and the output voltage V_1, V_2 and VRN are zero, +Vdc/2 and + Vdc/2 volts, respectively.

Case 3: $\alpha 2 <= \omega.t < \alpha 3$ radians. During this time interval of the reference sine wave voltage for phase A, the following operation takes place:

1. Comparators X1, X2, X3 outputs are HIGH, LOW and LOW; that of X4, X5 and X6 are LOW, HIGH and LOW; and that of X7 and X8 are HIGH and LOW, respectively.

2. Outputs of logic gates U5, U9, U10 and U12 are all HIGH and that of U1 to U4, U6, U7, U8 and U11 are LOW.
3. Outputs Vg_S22A, Vg_S22B, Vg_S11A and Vg_S12B are all LOW and that of Vg_S21A, Vg_S21B, Vg_S12A and Vg_S11B are all HIGH.
4. Switches S22A, S22B, S11A and S12B are all OFF; that of S21A, S21B, S12A and S11B are all ON; and the output voltage V_1, V_2 and VRN are +Vdc, zero and + Vdc volts, respectively.

Case 4: $\alpha 3 <= \omega.t < \pi/2$ radians. During this time interval of the reference sine wave voltage for phase A, the following operation takes place:

1. Comparators X1, X2 and X3 outputs are HIGH, LOW and HIGH; that of X4, X5 and X6 are LOW, HIGH and LOW; and that of X7 and X8 are HIGH and LOW, respectively.
2. Outputs of logic gates U2, U5, U7, U10 and U12 are HIGH and that of U1, U3, U4, U6, U8, U9 and U11 are all LOW.
3. Outputs Vg_S21A, Vg_S22B, Vg_S11A and Vg_S12B are all LOW and that of Vg_S22A, Vg_S21B, Vg_S12A and Vg_S11B are all HIGH.
4. Switches S21A, S22B, S11A and S12B are all OFF; that of S22A, S21B, S12A and S11B are all ON; and the output voltage V_1, V_2 and VRN are +Vdc, +Vdc/2 and + 3Vdc/2 volts, respectively.

Case 5: $\pi <= \omega.t < (\pi + \alpha 1)$ radians. During this time interval of the reference sine wave voltage for phase A, the following operation takes place:

1. Comparators X1, X2 and X3 outputs are LOW, HIGH and LOW; that of X4, X5 and X6 are LOW, HIGH and LOW; and that of X7 and X8 are LOW and LOW, respectively.
2. Outputs of logic gates U9 to U12 are HIGH and that of U1 to U8 are LOW.
3. Outputs Vg_S22A, Vg_S22B, Vg_S12A and Vg_S12B are all LOW and that of Vg_S21A, Vg_S21B, Vg_S11A and Vg_S11B are all HIGH.
4. Switches S22A, S22B, S12A and S12B are all OFF; that of S21A, S21B, S11A and S11B are all ON; and the output voltage V_1, V_2 and VRN are zero, zero and zero volts, respectively.

Case 6: $(\pi + \alpha 1) <= \omega.t < (\pi + \alpha 2)$ radians. During this time interval of the reference sine wave voltage for phase A, the following operation takes place:

1. Comparators X1, X2 and X3 outputs are LOW, HIGH and LOW; that of X4, X5 and X6 are HIGH, HIGH and LOW; and that of X7 and X8 are LOW and LOW, respectively.
2. Outputs of logic gates U3, U8, U9, U11 and U12 are HIGH and that of U1, U2, U4 to U7 and U10 are LOW.
3. Outputs Vg_S22A, Vg_S21B, Vg_S12A and Vg_S12B are all LOW and that of Vg_S21A, Vg_S22B, Vg_S11A and Vg_S11B are all HIGH.
4. Switches S22A, S21B, S12A and S12B are all OFF; that of S21A, S22B, S11A and S11B are all ON; and the output voltage V_1, V_2 and VRN are zero, –Vdc/2 and –Vdc/2 volts, respectively.

Case 7: $(\pi + \alpha 2) <= \omega .t < (\pi + \alpha 3)$ radians. During this time interval of the reference sine wave voltage for phase A, the following operation takes place:

1. Comparators X1, X2 and X3 outputs are LOW, HIGH and LOW; that of X4, X5 and X6 are HIGH, LOW and LOW; and that of X7 and X8 are LOW and HIGH, respectively.
2. Outputs of logic gates U6, U9 to U11 are all HIGH and that of U1 to U5, U7, U8 and U12 are LOW.
3. Outputs Vg_S22A, Vg_S22B, Vg_S12A and Vg_S11B are all LOW and that of Vg_S21A, Vg_S21B, Vg_S11A and Vg_S12B are all HIGH.
4. Switches S22A, S22B, S12A and S11B are all OFF; that of S21A, S21B, S11A and S12B are all ON; and the output voltage V_1, V_2 and VRN are −Vdc, zero and –Vdc volts, respectively.

Case 8: $(\pi + \alpha 3) <= \omega .t < 3\pi/2$ radians. During this time interval of the reference sine wave voltage for phase A, the following operation takes place:

1. Comparators X1, X2 and X3 outputs are LOW, HIGH and LOW; that of X4, X5 and X6 are HIGH, LOW and HIGH; and that of X7 and X8 are LOW and HIGH, respectively.
2. Outputs of logic gates U4, U6, U8, U9 and U11 are HIGH and that of U1, U2, U3, U5, U7, U10 and U12 are all LOW.
3. Outputs Vg_S22A, Vg_S21B, Vg_S12A and Vg_S11B are all LOW and that of Vg_S21A, Vg_S22B, Vg_S11A and Vg_S12B are all HIGH.
4. Switches S22A, S21B, S12A and S11B are all OFF; that of S21A, S22B, S11A and S12B are all ON; and the output voltage V_1, V_2 and VRN are −Vdc, −Vdc/2 and –3Vdc/2 volts, respectively.

The model of the three-phase two-cell cascade H-bridge inverter is developed using SIMULINK [11]. This is shown in Fig. 3.30 (Model file: EXAMPLE 3_3). The gate drive subsystem for phase R is shown in Fig. 3.31. Here a three-phase sine wave AC voltage is used as a reference. The value of $\alpha 1$, $\alpha 2$ and $\alpha 3$ for M/ma value 1.8/0.9 shown in Table 3.3 is used in the model. The inverter switching frequency is 50 Hz. The DC link voltage of inverter is 100 volts for the lower cell and 50 volts for the upper cell. The three-phase cascade H-bridge inverter (CHBI) cell is the Universal Bridge subsystem from Power Systems block set. The reference sine waves for the three phase are generated using Embedded MATLAB function and the sine of switching angles for $\alpha 1$, $\alpha 2$ and $\alpha 3$ and that of $(\pi + \alpha 1)$, $(\pi + \alpha 2)$ and $(\pi + \alpha 3)$ are generated using another Embedded MATLAB function. The source codes are given in Program segment 3.2 and 3.3 in the model file EXAMPLE 3_3. In Fig. 3.30, the switching angles $\alpha 1$, $\alpha 2$ and $\alpha 3$ in degrees are entered in order in the constant block separated by a space. The selector block selects or directs the outputs $\alpha 1$, $\alpha 2$ and $\alpha 3$ in successive order. The model subsystem to generate gate pulse for phase R switches is shown in Fig. 3.31. Here reference sine wave voltage for phase R (1. sin(2*π*50*t)) is compared with sine of $\alpha 1$, $\alpha 2$, $\alpha 3$, $(\pi + \alpha 1)$, $(\pi + \alpha 2)$ and $(\pi + \alpha 3)$ in six relational operator blocks which form the comparators. The resulting output of each comparator is connected to AND and OR logic gates as shown in Fig. 3.31.

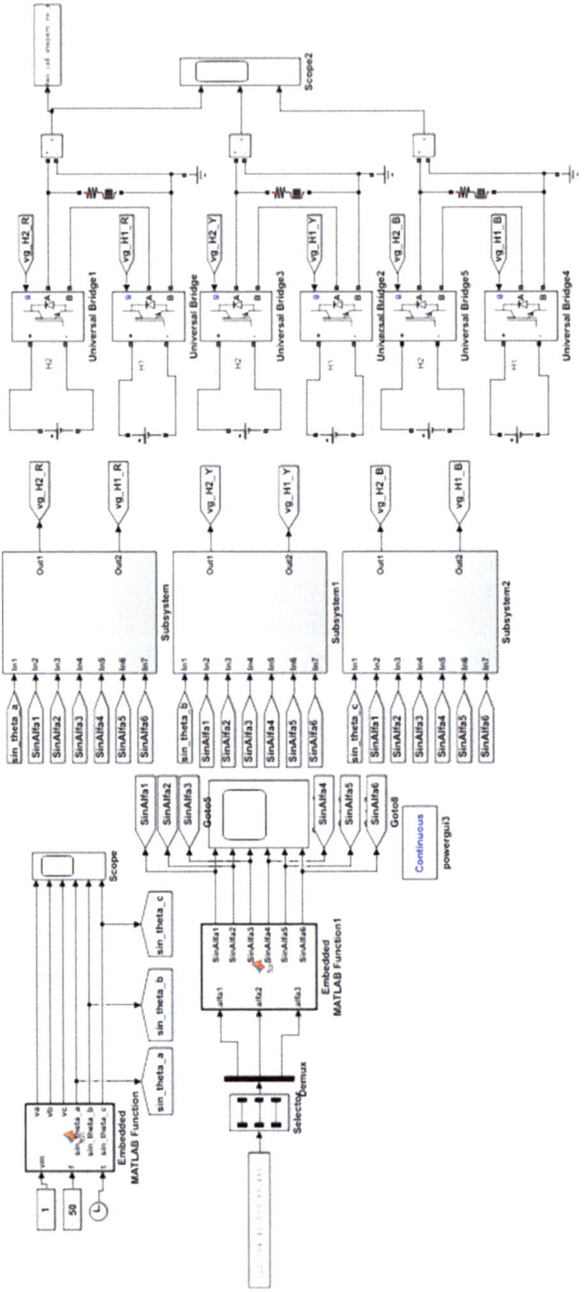

Fig. 3.30 Model of three-phase two-cell cascade H-Bridge SHE-PWM inverter

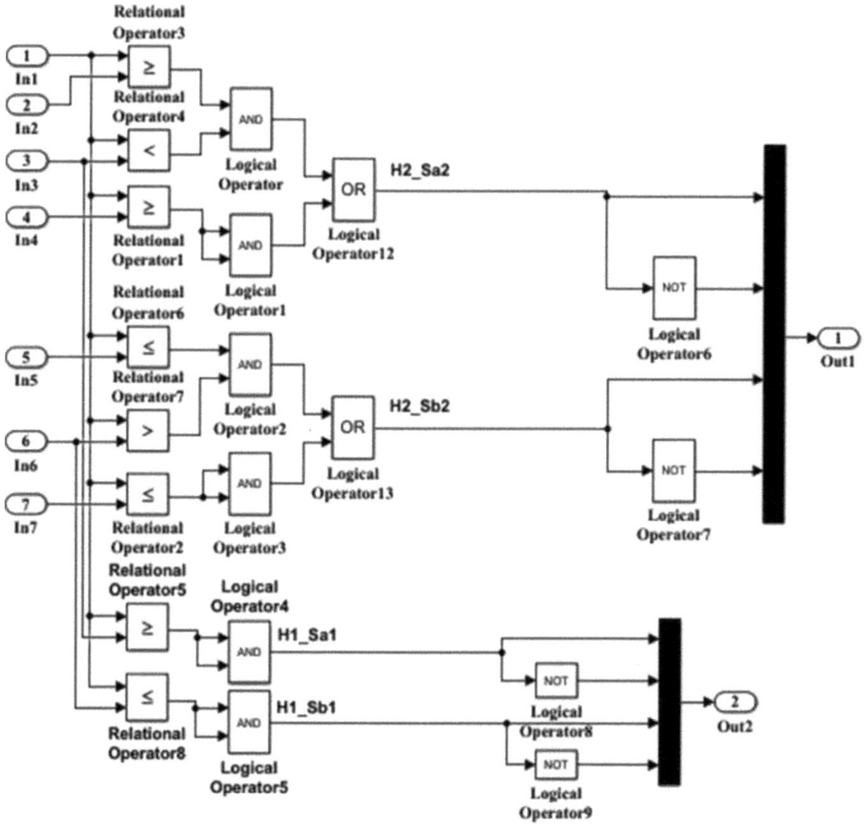

Fig. 3.31 Three-phase two-cell cascade H-Bridge SHE-PWM inverter gate pulse generator

These gate pulse outputs from OR gates and their inverted pulse using NOT gate form gate drives for the switches in the upper cells of CHBI. Similarly, the above reference sine wave voltage for phase R is compared with sine of $\alpha 2$ and $(\pi + \alpha 2)$ in two relational operator blocks 5 and 8. The output of these comparators through AND gates and their inverted output using NOT gate generate gate pulse for the switches in lower cells of CHBI. Similar comparators and logic gates are used to compare phase Y and phase B reference sine wave voltages to generate gate pulse for the upper and lower switches in phase Y and B of CHBI.

3.4.2 Simulation Results

The model simulation of the three-phase SHE-PWM two-cell CHB inverter was carried out using ode23tb (stiff/TR-BDF2) solver. The inverter switching frequency is 50 Hz, and DC link voltage is 100 V for lower cell and 50 volts for upper cell. The switching angles in Table 3.3 are used to eliminate 5th and 7th harmonics from the line-to-ground voltage. The harmonic spectrum of line-to-ground voltage and the three-phase line-to-ground voltage for M/ma values of 1.6/0.8 and 1.8/0.9 in the respective order are shown in Fig. 3.32.

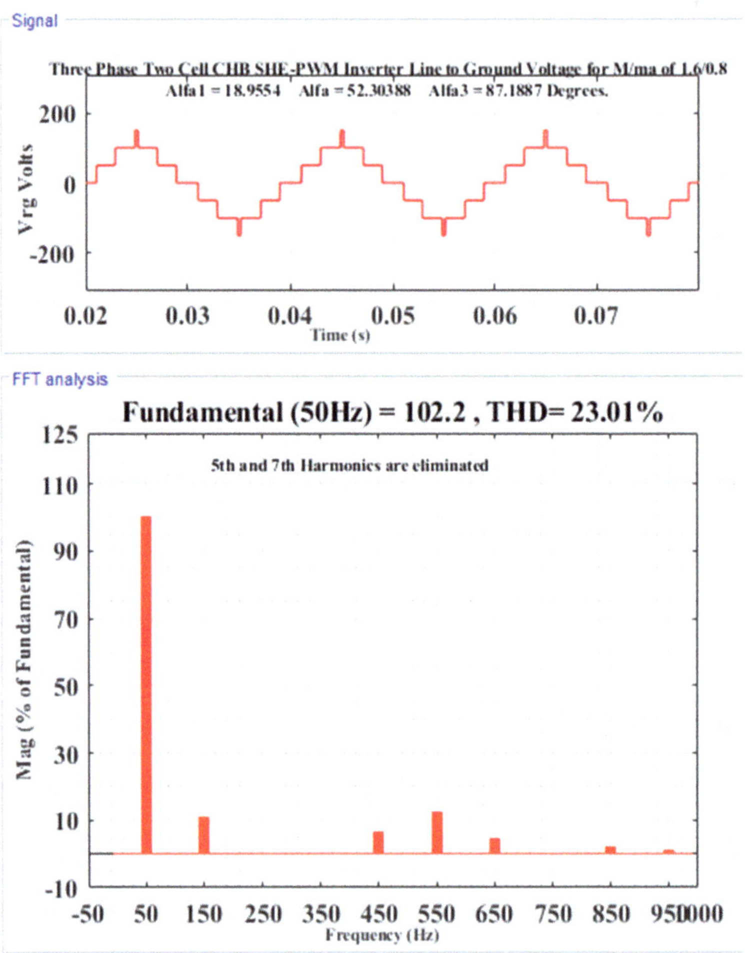

Fig. 3.32 Three-phase two-cell CHB inverter simulation results: line-to-ground voltage and harmonic Spectrum for M/ma value 1.6/0.8 (top left), three-phase line-to-ground voltage for M/ma value of 1.6/0.8 (top right), line-to-ground voltage and harmonic spectrum for M/ma value 1.8/0.9 (bottom left), three-phase line-to-ground voltage for M/ma value of 1.8/0.9 (bottom right)

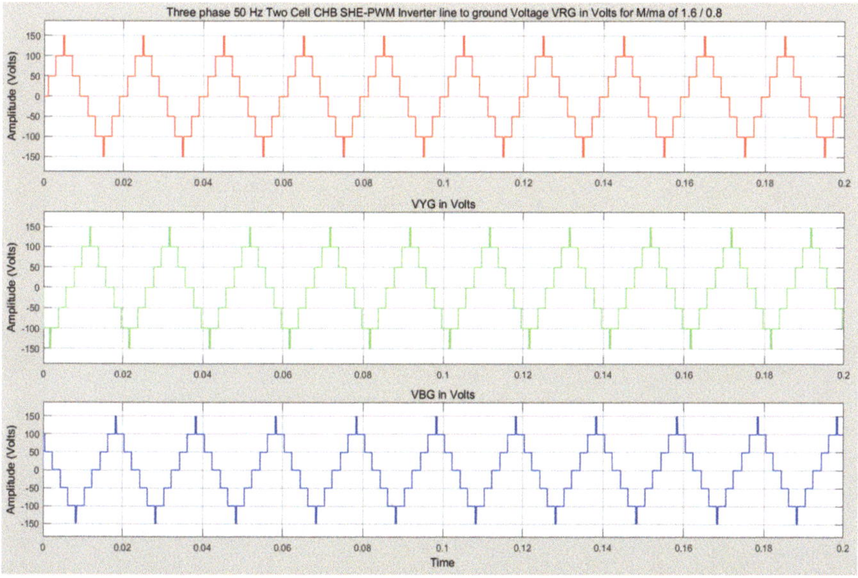

Fig. 3.32 (continued)

3.4.3 Discussion of Results

The simulation results for the harmonic spectrum of line-to-ground voltage shown in Fig. 3.32 reveal that the 5th and 7th harmonics are well eliminated for both the M/ma values in Table 3.3.

Also for a DC link voltage of 100 V for the lower cell, using the equation for M given in Sect. 3.4, the peak fundamental voltage V_1 is found to be 101.91 and 114.65 V for M of 1.6 and 1.8, respectively, whereas that by simulation of model these values is found to be 102.2 and 114.9 V, respectively. Also for M/ma value of 1.8/0.9, the THD of line-to-ground voltage is found to be much less than that for 1.6/ 0.8.

3.5 Three-Phase Diode-Clamped and Flying Capacitor Three-Level SHE-PWM Inverter

The circuit schematic for one phase of the diode-clamped three-level inverter (DCTLI) and that of the flying capacitor three-level inverter (FCTLI) are shown in Fig. 3.33a, b, respectively [21–23]. Multilevel converters having three, four, five, six, etc. for the line-to-ground output voltage were proposed to obtain higher voltages without the need for a step-up transformer. These converters find

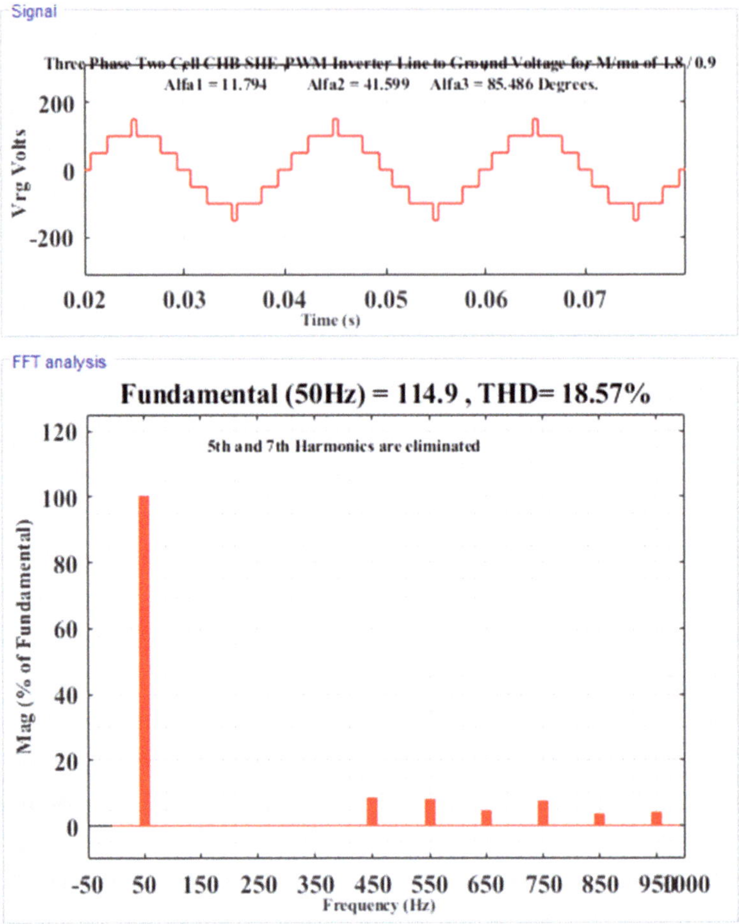

Fig. 3.32 (continued)

applications in STATCOM, variable speed AC motor drives and high voltage system interconnection [23–29]. Models for a three-phase DCTLI and FCTLI are reported [30, 31].

In this chapter, models for three-phase SHE-PWM DCTLI and FCTLI are presented. Neglecting PWM, the truth tables for the operation of the DCTLI and FCTLI derived from Fig. 3.33 are shown in Tables 3.4 and 3.5, respectively. There are three levels +Vdc/2, 0 and –Vdc/2 for the line-to-ground output voltage. Each voltage level has a phase (time duration) of $\pi/2$ radians (T/4 s) where T is the period of switching of the inverter. By properly switching on and off the inverter at calculated switching angles, it is possible to eliminate the desired harmonics from the line-to-ground output voltage. This is shown in Fig. 3.34 where this inverter is alternatively turned on and off at intervals $\alpha 1$, $\alpha 2$ and $\alpha 3$ degrees to eliminate any

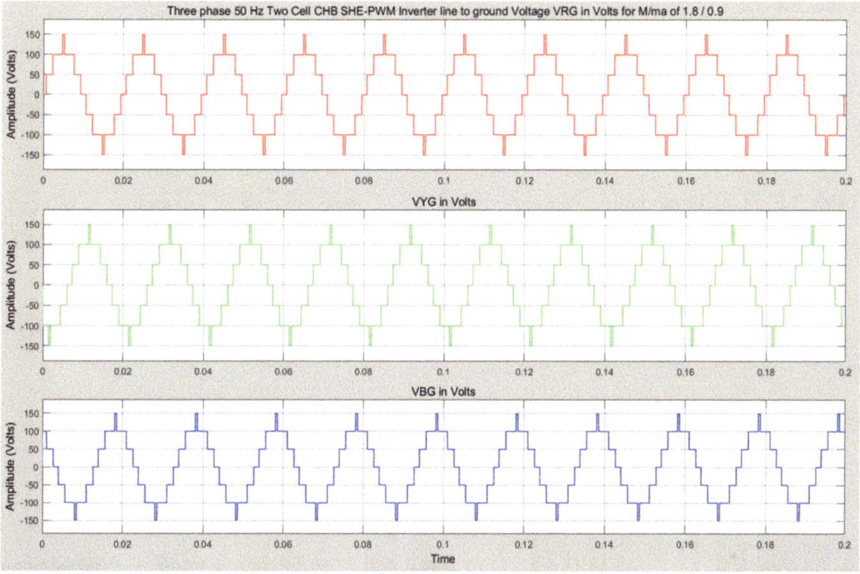

Fig. 3.32 (continued)

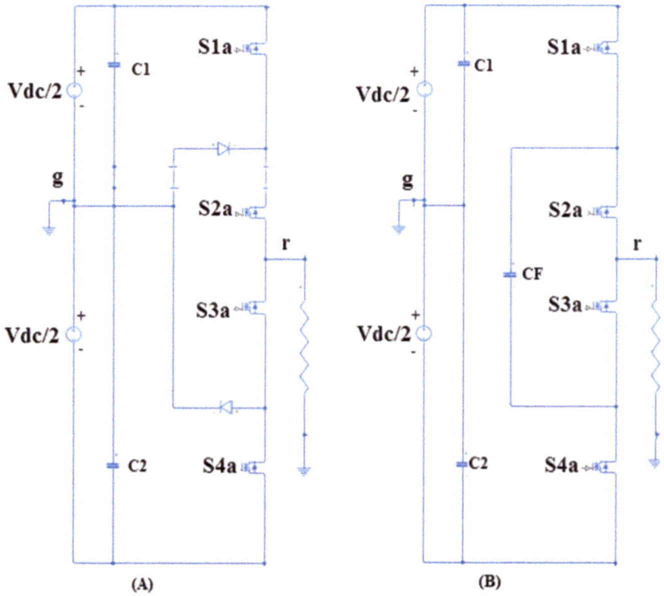

Fig. 3.33 Three-level inverter circuit schematic for one phase—(**a**) Diode clamped and (**b**) Flying capacitor

Table 3.4 DCTLI truth table

Sl.No.	S1a	S2a	S3a	S4a	Vrg (V)
1	1	1	0	0	+Vdc/2
2	0	1	1	0	0
3	0	0	1	1	-Vdc/2

Table 3.5 FCTLI truth table

Sl.No.	S1a	S2a	S3a	S4a	Vrg (V)
1	1	1	0	0	+Vdc/2
2	1	0	1	0	0
	0	1	0	1	0
3	0	0	1	1	-Vdc/2

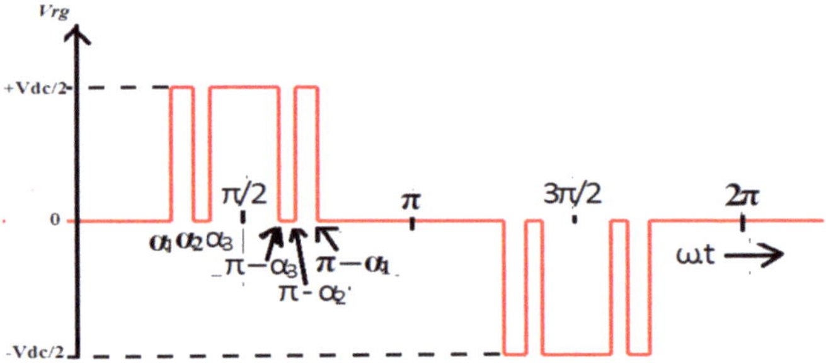

Fig. 3.34 Line-to-ground voltage of a three-phase DCTLI and FCTLI

two non-triplen harmonics from the line-to-ground output voltage. Here angles $\alpha 1$, $\alpha 2$ and $\alpha 3$ are calculated to eliminate 5th and 7th harmonics from the line-to-ground voltage Vrg for both DCTLI and FCTLI. The Fourier transform of Fig. 3.34 is given by Eq. 3.27 derived in Sect. 3.4. To eliminate 5th and 7th harmonics from the line-to-ground voltage of DCTLI and FCTLI, the following equations must be solved:

$$\left.\begin{aligned}\cos(\alpha_1)-\cos(\alpha_2)+\cos(\alpha_3)&=M\\ \cos(5.\alpha_1)-\cos(5.\alpha_2)+\cos(5.\alpha_3)&=0\\ \cos(7.\alpha_1)-\cos(7.\alpha_2)+\cos(7.\alpha_3)&=0\end{aligned}\right\}\tag{3.30}$$

In Eq. 3.30, M is the ratio of $V_1/(2*\mathrm{Vdc}/\pi)$ where V_1 is the amplitude of the fundamental component of the output voltage *Vrg*.

3.5.1 Modelling of Three-Phase SHE-PWM Diode-Clamped Three-Level Inverter

The model block diagram for the gate drive of SHE-PWM DCTLI topology is shown in Fig. 3.35. The three switching angles to eliminate any two non-triplen harmonics

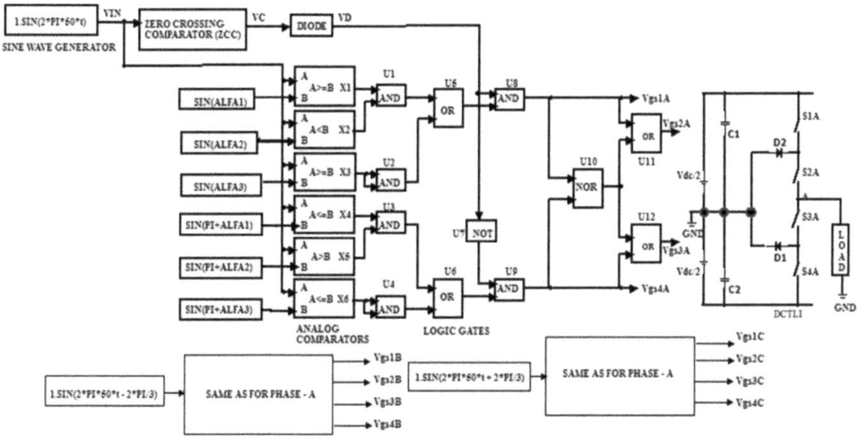

Fig. 3.35 Three-phase DCTL SHE-PWM inverter gate drive model block diagram

from line-to-ground output voltage are alfa1, alfa2 and alfa3. The phase A reference sine wave voltage 1.sin(ω.*t*) during positive half cycle is compared with sine of alfa1, alfa2 and alfa3 in separate comparators marked X1, X2 and X3. When phase A sine wave reference voltage is greater than or equal to Sin(α1), Comparator X1 output is HIGH or else its output is LOW, and similarly when the former reference sine voltage is only less than Sin(α2), Comparator X2 output is HIGH. Comparators X1 and X2 output are given to AND gate U1. Similarly reference sine voltage is connected to Comparator X3 whose output is HIGH only when reference sine wave voltage is equal to or greater than Sin(α3). Comparator X3 output through AND gate marked U2 and the U1 output are connected to OR gate U5. Similar procedure is adopted to compare the reference sine wave output for phase A with sine of (pi+α1), (pi+α2) and (pi+α3) in three separate comparators marked X4, X5 and X6. The resulting output of these three Comparators through AND gates U3 and U4 are connected to OR gate U6. The output of U5 and the reference sine wave for phase A through zero-crossing comparator and diode are connected to AND gate U8, and the inverted output of diode from NOT gate U7 and the output of U6 are connected to AND gate U9. The outputs of U8 and U9 are connected to NOR gate U10. Outputs of U8 and U10 are given to OR gate U11 and that of U9 and U10 are given to OR gate U12. Output of U8, U11, U12 and U9 forms the gate drive for the switches S1a, s2a, s3a and S4a of DCTLI. For phase B and C, the reference sine waves are (1.sin (ω.*t* – 2*pi/(3)) and (1.sin(ω.*t* + 2*pi/(3)), respectively, are used for comparison as per the method for phase A given above.

Referring to Fig. 3.34, the principle of operation of the gate drive in Fig. 3.35 can be explained as follows:

Case 1: $0 <= \omega.t < \alpha 1$ radians. During this time interval of the reference sine wave for phase A, the following operation takes place:

1. Comparators X1, X2 and X3 are LOW, HIGH and LOW and that of X4, X5 and X6 are LOW, HIGH and LOW, respectively.
2. Outputs of logic gates U1 to U9, respectively, are all LOW.
3. Outputs of logic gates U10, U11 and U12 are all HIGH.
4. Vgs1a, Vgs2a, Vgs3a and vgs4a are, respectively, LOW, HIGH, HIGH and LOW.
5. Switches S2a and S3a are only ON, and the output voltage is 0 volts.

Case 2: $\alpha 1 <= \omega.t < \alpha 2$ radians. During this time interval of the reference sine wave for phase A, the following operation takes place:

1. Comparators X1, X2 and X3 are HIGH, HIGH and LOW and that of X4, X5 and X6 are LOW, HIGH and LOW, respectively.
2. Outputs of logic gates U1, U5 and U8 are HIGH and U2, U3, U4, U6, U7 and U9 are LOW.
3. Outputs of logic gates U10, U11 and U12 are LOW, HIGH and LOW, respectively.
4. Vgs1a, Vgs2a, Vgs3a and vgs4a are, respectively, HIGH, HIGH, LOW and LOW.
5. Switches S1a and S2a are only ON, and the output voltage is +Vdc/2 volts.

Case 3: $\alpha 2 <= \omega.t < \alpha 3$ radians. During this time interval of the reference sine wave for phase A, the following operation takes place:

1. Comparators X1, X2 and X3 are HIGH, LOW and LOW and that of X4, X5 and X6 are LOW, HIGH and LOW, respectively.
2. Outputs of logic gates U1 to U9 are all LOW.
3. Outputs of logic gates U10, U11 and U12 are all HIGH.
4. Vgs1a, Vgs2a, Vgs3a and vgs4a are, respectively, LOW, HIGH, HIGH and LOW.
5. Switches S2a and S3a are only ON, and the output voltage is 0 volts.

Case 4: $\alpha 3 <= \omega.t < \pi/2$ radians. During this time interval of the reference sine wave for phase A, the following operation takes place:

1. Comparators X1, X2 and X3 are HIGH, LOW and HIGH and that of X4, X5 and X6 are LOW, HIGH and LOW, respectively.
2. Outputs of logic gates U1, U3, U4, U6, U7 and U9 are LOW and that of U2, U5 and U8 are HIGH.
3. Outputs of logic gates U10, U11 and U12 are LOW, HIGH and LOW, respectively.
4. Vgs1a, Vgs2a, Vgs3a and vgs4a are, respectively, HIGH, HIGH, LOW and LOW.
5. Switches S1a and S2a are only ON, and the output voltage is +Vdc/2 volts.

Case 5: $\pi < \omega.t < (\pi + \alpha 1)$ radians. During this time interval of the reference sine wave for phase A, the following operation takes place:

1. Comparators X1, X2 and X3 are LOW, HIGH and LOW and that of X4, X5 and X6 are LOW, HIGH and LOW, respectively.
2. Outputs of logic gates U1 to U6, U8 and U9 are LOW and that of U7 is HIGH.
3. Outputs of logic gates U10, U11 and U12 are all HIGH.
4. Vgs1a, Vgs2a, Vgs3a and vgs4a are, respectively, LOW, HIGH, HIGH and LOW.
5. Switches S2a and S3a are only ON, and the output voltage is 0 volts.

Case 6: $(\pi + \alpha 1) < = \omega.t < (\pi + \alpha 2)$ radians. During this time interval of the reference sine wave for phase A, the following operation takes place:

1. Comparators X1, X2 and X3 are LOW, HIGH and LOW and that of X4, X5 and X6 are HIGH, HIGH and LOW, respectively.
2. Outputs of logic gates U1, U2, U4, U5 and U8 are LOW and that of U3, U6, U7 and U9 are HIGH.
3. Outputs of logic gates U10, U11 and U12 are LOW, LOW and HIGH, respectively.
4. Vgs1a, Vgs2a, Vgs3a and vgs4a are, respectively, LOW, LOW, HIGH and HIGH.
5. Switches S3a and S4a are only ON, and the output voltage is –Vdc/2 volts.

Case 7: $(\pi + \alpha 2) < = \omega.t < (\pi + \alpha 3)$ radians. During this time interval of the reference sine wave for phase A, the following operation takes place:

1. Comparators X1, X2 and X3 are LOW, HIGH and LOW and that of X4, X5 and X6 are HIGH, LOW and LOW, respectively.
2. Outputs of logic gates U1 to U6, U8 and U9 are LOW and that of U7 is HIGH.
3. Outputs of logic gates U10, U11 and U12 are all HIGH.
4. Vgs1a, Vgs2a, Vgs3a and vgs4a are, respectively, LOW, HIGH, HIGH and LOW.
5. Switches S2a and S3a are only ON, and the output voltage is 0 volts.

Case 8: $(\pi + \alpha 3) < = \omega.t < 3\pi/2$ radians. During this time interval of the reference sine wave for phase A, the following operation takes place:

1. Comparators X1, X2 and X3 are LOW, HIGH and LOW and that of X4, X5 and X6 are HIGH, LOW and HIGH, respectively.
2. Outputs of logic gates U1, U2, U3, U5 and U8 are LOW and that of U4, U6, U7 and U9 are HIGH.
3. Outputs of logic gates U10, U11 and U12 are LOW, LOW and HIGH, respectively.
4. Vgs1a, Vgs2a, Vgs3a and vgs4a are, respectively, LOW, LOW, HIGH and HIGH.
5. Switches S3a and S4a are only ON, and the output voltage is –Vdc/2 volts.

A model of the three-phase SHE-PWM DCTLI is developed in SIMULINK [11]. This model is shown in Fig. 3.36 (Model file: EXAMPLE 3_4), and the model subsystems are shown in Fig. 3.37a, b. Three-phase 50 Hz sine ware AC voltage is used as a reference. The values of $\alpha 1$, $\alpha 2$ and $\alpha 3$ for M value of 0.5 and 0.8

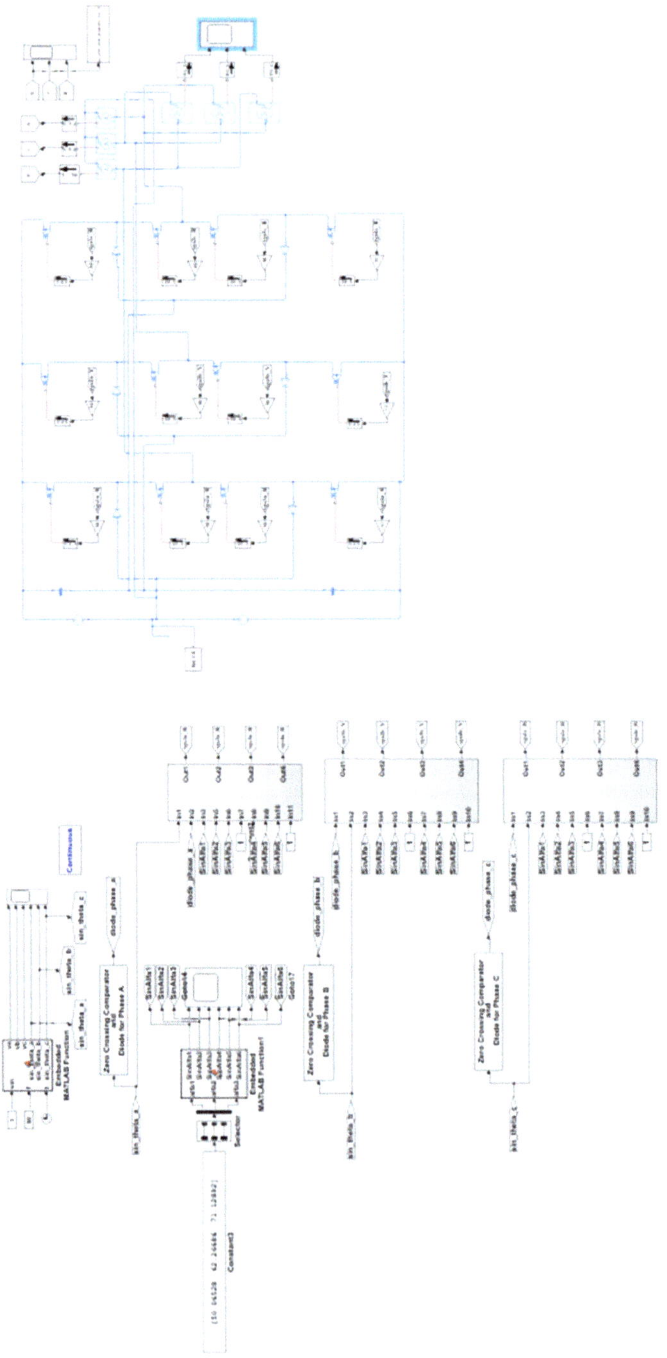

Fig. 3.36 Model of three-phase SHE-PWM DCTLI with three switching angles

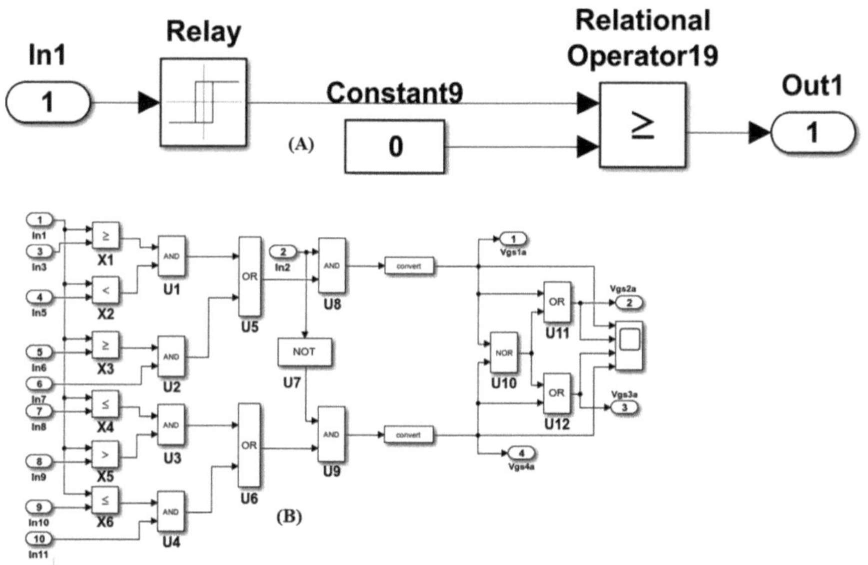

Fig. 3.37 Three-phase SHE PWM DCTLI model subsystems: (**a**) Zero crossing comparator and diode. (**b**) Gate drive for DCTLI switches in Phase A

to eliminate 5th and 7th harmonics from the line-to-ground voltages of DCTLI are shown in Table 3.1. The values of switching angles corresponding to M value of 0.5 are used in the model. The inverter switching frequency is 50 Hz. The DC link voltage of inverter is 100 volts. The three-phase DCTLI NMOSFET switches are from Semiconductors/Fundamental components library in the Simscape components block set. The reference sine waves for the three phase are generated using Embedded MATLAB function and the sine of switching angles for $\alpha 1$, $\alpha 2$, $\alpha 3$ and that for $(\pi + \alpha 1)$, $(\pi + \alpha 2)$, $(\pi + \alpha 3)$ are generated using another Embedded MATLAB function. The source codes are given in Program segment 3.2 and 3.3 in the model file EXAMPLE 3_4. In Fig. 3.36, the switching angles $\alpha 1$, $\alpha 2$ and $\alpha 3$ in degrees are entered in order in the constant block separated by a space. The selector block selects or directs the first output $\alpha 1$, second output $\alpha 2$ and third output $\alpha 3$.

The model subsystem to generate gate pulse for phase A switches of inverter is shown in Fig. 3.37a, b. Here as shown in Fig. 3.37a, reference sine wave for phase A is given to relay block which forms zero-crossing comparator (ZCC) giving square wave output when the input sine wave crosses zero and goes positive and negative, respectively. The ZCC output is compared with zero using a relational operator block which forms the comparator. The ZCC and comparator together operate as a diode. This comparator output is HIGH when the ZCC input is greater than or equal to zero or else its output is LOW. Reference sine wave for phase A ($1.\sin(2*\pi*50*t)$) is compared with sine of $\alpha 1$, $\alpha 2$, $\alpha 3$, $(\pi + \alpha 1)$, $(\pi + \alpha 2)$ and $(\pi + \alpha 3)$ in six relational operator blocks which form the comparators X1 to X6, as shown in Fig. 3.37b. The resulting output of each comparator X1 to X6 and that of ZCC and diode is connected to logic gates U1 to U12 in the same way shown in Fig. 3.35. The output of U8, U11, U12 and U9 forms the gate pulse for switches S1a, S2a, S3a and S4a in phase A. Similar method is used along with reference sine waves for phase B and C to generate gate pulse for the four switches in the respective phase. In Fig. 3.36, the gate pulse is multiplied by ten using gain blocks, so that the amplitude of the gate pulse exceeds the threshold voltage of the semiconductor switch.

3.5.2 Simulation Results

The model simulation of the three-phase SHE-PWM DCTLI was carried out using ode45 (Dormand-Prince) variable step solver. The inverter switching frequency is 50 Hz, and DC link voltage is 100 V. The switching angles in Table 3.1 are used to eliminate 5th and 7th harmonics from the line-to-ground voltage. The harmonic spectrum of line-to-ground voltage and the three-phase line-to-ground voltage for M values of 0.5 and 0.8 in the respective order are shown in Figs. 3.38 to 3.43 respectively.

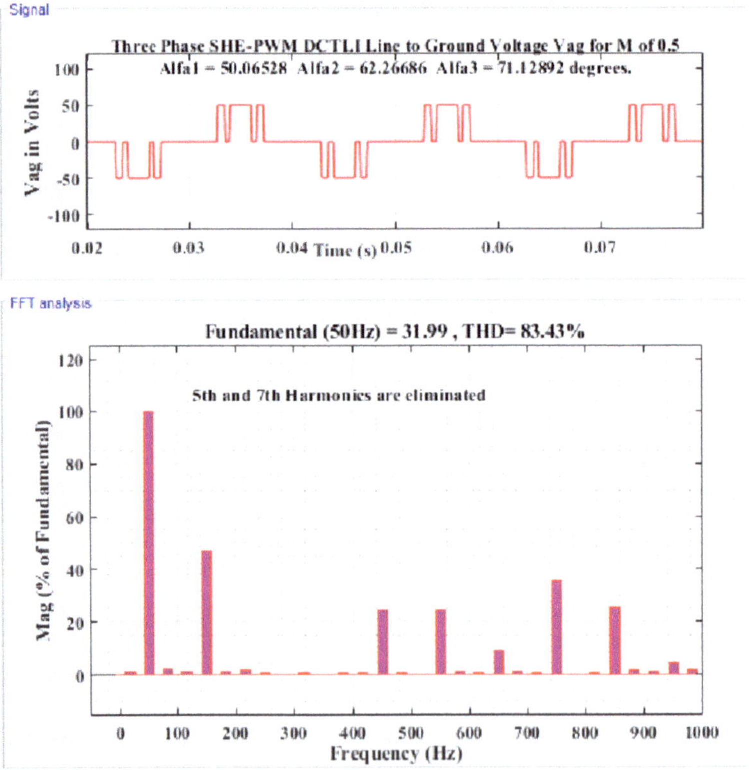

Fig. 3.38 Three-phase SHE-PWM DCTLI line-to-ground voltage (top) and harmonic Spectrum with 5th and 7th harmonics eliminated (bottom)

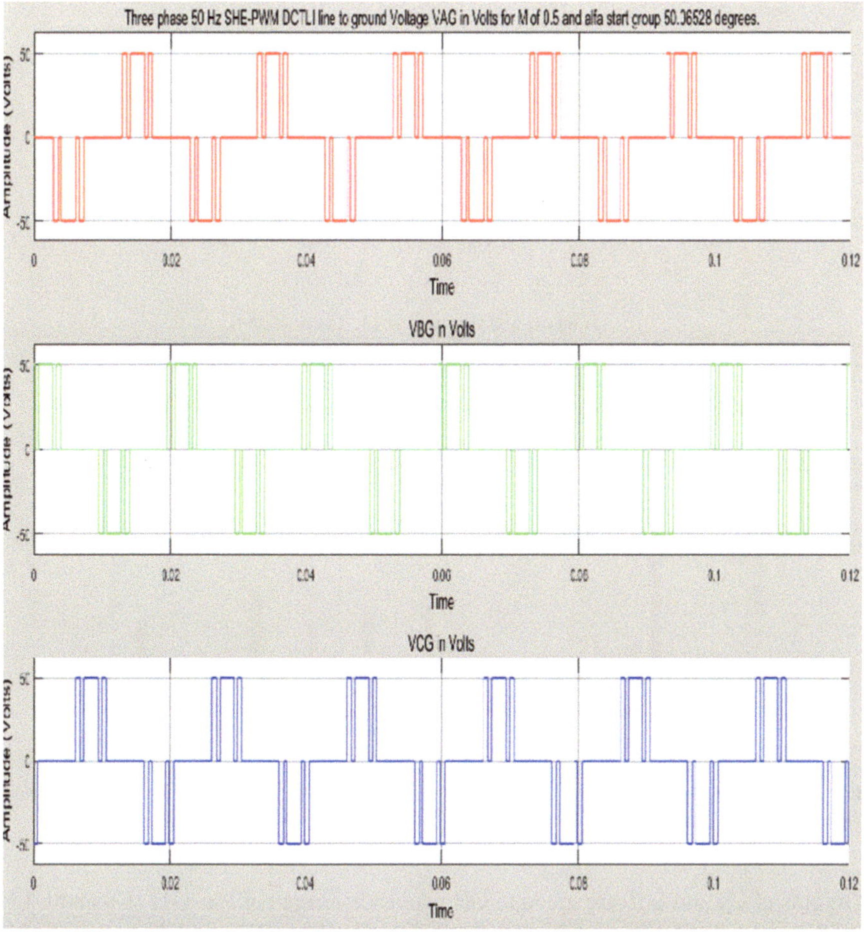

Fig. 3.39 Three-phase SHE-PWM inverter line-to-ground voltage for M of 0.5 and alfa start group of 50.006528 degrees

3.5.3 Modelling of Three-Phase SHE-PWM Flying Capacitor Three-Level Inverter

The model block diagram for the gate drive of three-phase SHE-PWM FCTLI is shown in Fig. 3.44. Here the reference sine wave for phase A is compared with sine of $\alpha1$, $\alpha2$, $\alpha3$, $(\pi + \alpha1)$, $(\pi + \alpha2)$ and $(\pi + \alpha3)$ in six comparators X1 to X6 in the same way as shown in Fig. 3.35. Outputs of X1 to X6 and that of zero-crossing comparator and diode are connected to logic gates U1 to U9 which are identically the same as shown in Fig. 3.35. The difference is in the logic gates U10, U11 and U12. Outputs of AND gates U8 and U9 are given to NOT gates U10 and U11, respectively.

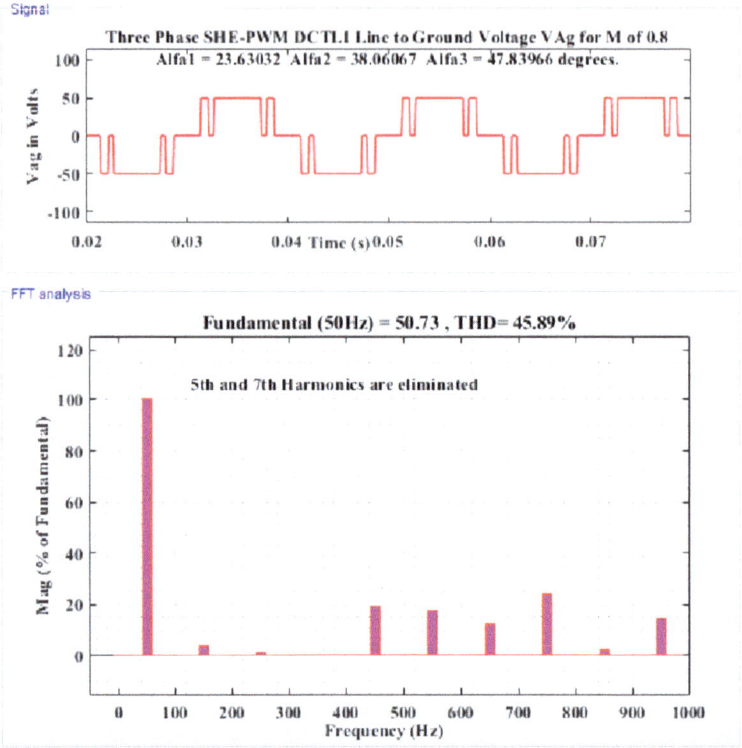

Fig. 3.40 Three-phase SHE-PWM DCTLI Line-to-ground voltage (top) and harmonic spectrum with 5th and 7th harmonics eliminated

Outputs of U9 and U10 are given to OR gate U12. Output of U8, U11, U9 and U12, respectively, forms the gate drive for switches S1a, S2a, S3a and S4a, respectively. Gate drive for switches in phase B and C is derived in the same way by comparison with the reference sine waves for phase B and C, respectively.

Referring to Fig. 3.34, the principle of operation of the gate drive in Fig. 3.44 can be explained as follows:

Case 1: $0 < \omega.t < \alpha 1$ radians. During this time interval of the reference sine wave for phase A, the following operation takes place:

1. Comparators X1, X2 and X3 are LOW, HIGH and LOW and that of X4, X5 and X6 are LOW, HIGH and LOW, respectively.
2. Output of logic gates U1 to U9, respectively, are all LOW. Outputs of logic gates U10, U11 and U12 are all HIGH.
3. Vgs1a, Vgs2a, Vgs3a and vgs4a are, respectively, LOW, HIGH, LOW and HIGH.
4. Switches S2a and S4a are only ON, and the output voltage is 0 volts.

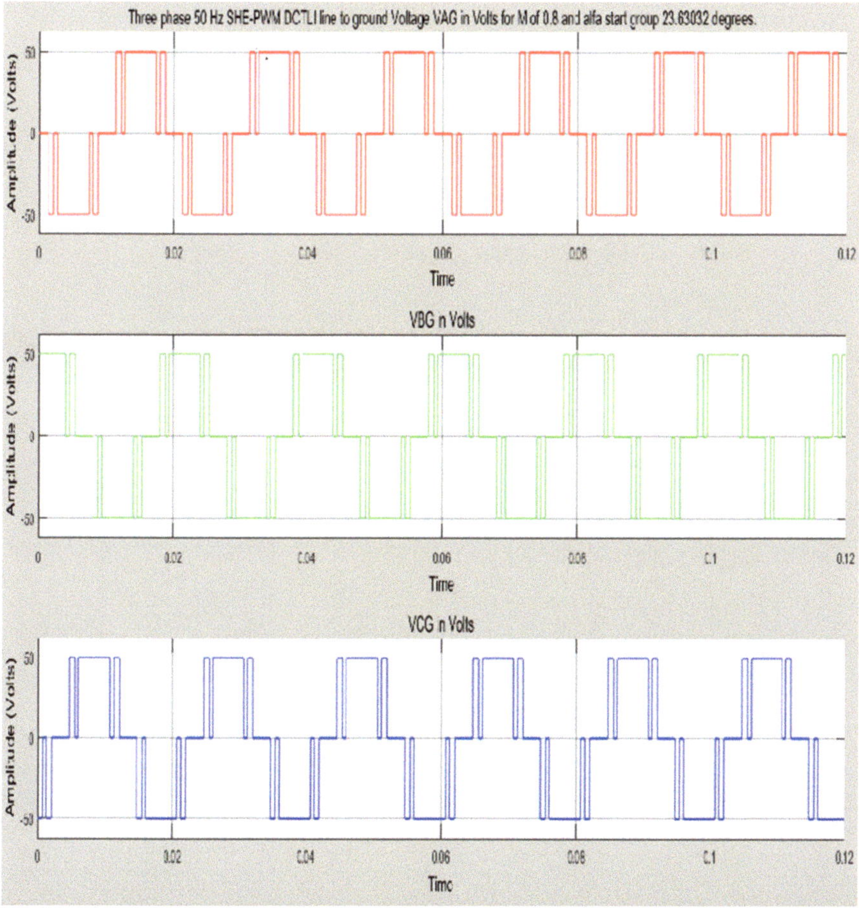

Fig. 3.41 Three-phase SHE-PWM Line-to-ground voltage for M of 0.8 and alfa start group of 23.63032 degrees

Case 2: $\alpha 1 <= \omega.t < \alpha 2$ radians. During this time interval of the reference sine wave for phase A, the following operation takes place:

1. Comparators X1, X2 and X3 are HIGH, HIGH and LOW and that of X4, X5 and X6 are LOW, HIGH and LOW, respectively.
2. Outputs of logic gates U1, U5 and U8 are HIGH and U2, U3, U4, U6, U7 and U9 are LOW.
3. Outputs of logic gates U10, U11 and U12 are LOW, HIGH and LOW, respectively.
4. Vgs1a, Vgs2a, Vgs3a and vgs4a are, respectively, HIGH, HIGH, LOW and LOW.
5. Switches S1a and S2a are only ON, and the output voltage is +Vdc/2 volts.

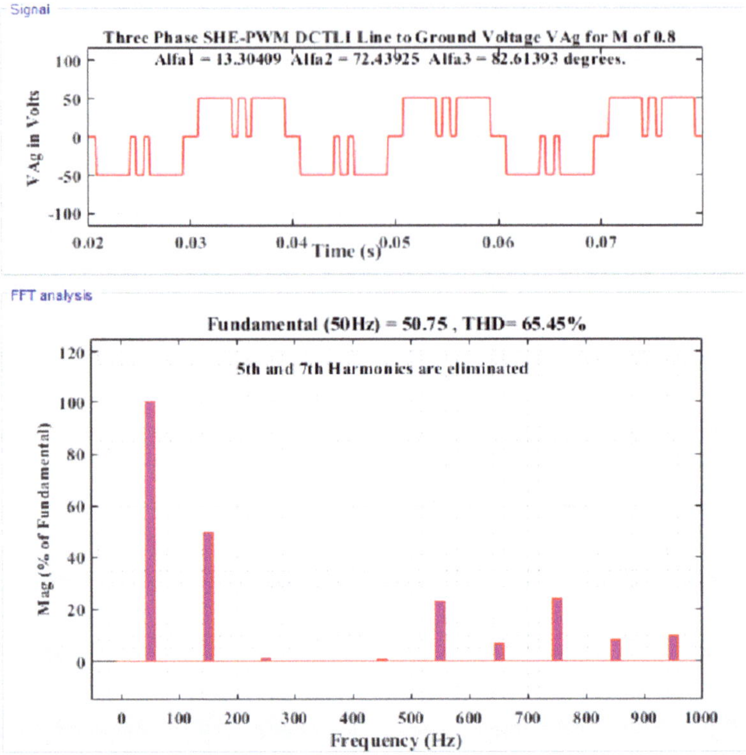

Fig. 3.42 Three-phase SHE-PWM DCTLI line-to-ground voltage (top) and harmonic spectrum with 5th and 7th harmonics eliminated (bottom)

Case 3: $\alpha 2 <= \omega.t < \alpha 3$ radians. During this time interval of the reference sine wave for phase A, the following operation takes place:

1. Comparators X1, X2 and X3 are HIGH, LOW and LOW and that of X4, X5 and X6 are LOW, HIGH and LOW, respectively.
2. Outputs of logic gates U1 to U9 are all LOW.
3. Outputs of logic gates U10, U11 and U12 are all HIGH.
4. Vgs1a, Vgs2a, Vgs3a and vgs4a are, respectively, LOW, HIGH, LOW and HIGH.
5. Switches S2a and S4a are only ON, and the output voltage is 0 volts.

Case 4: $\alpha 3 <= \omega.t < \pi/2$ radians. During this time interval of the reference sine wave for phase A, the following operation takes place:

1. Comparators X1, X2 and X3 are HIGH, LOW and HIGH and that of X4, X5 and X6 are LOW, HIGH and LOW, respectively.
2. Outputs of logic gates U1, U3, U4, U6, U7 and U9 are LOW and that of U2, U5 and U8 are HIGH.

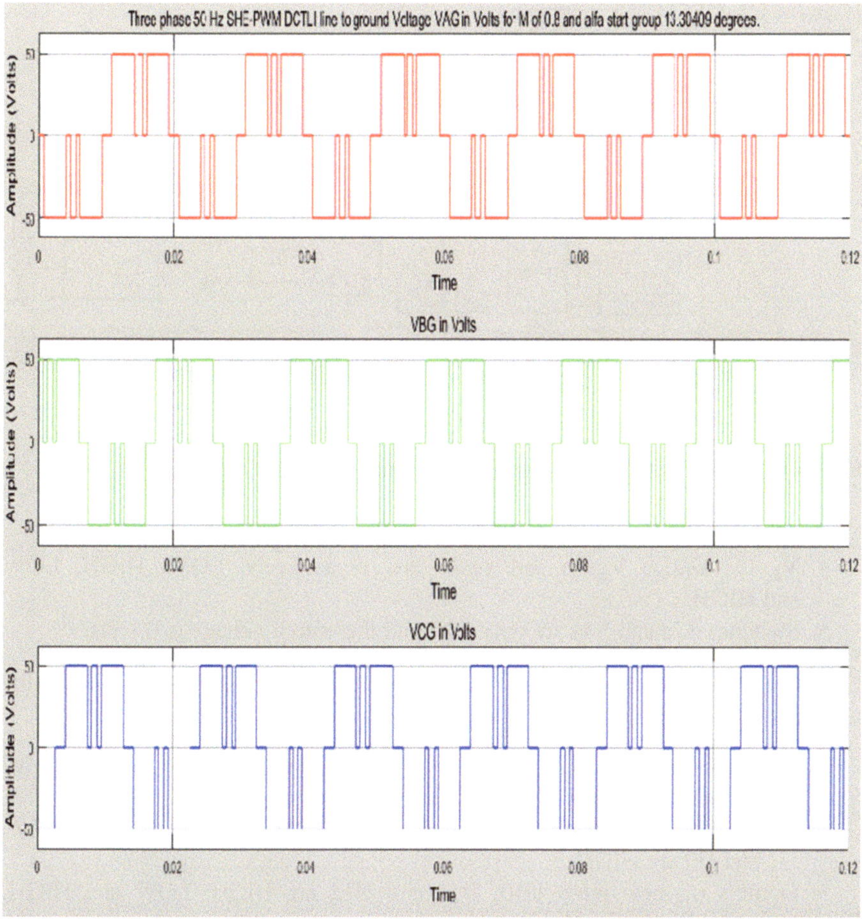

Fig. 3.43 Three-phase SHE-PWM DCTLI line-to-ground voltage for M of 0.8 and alfa start group of 13.30409 degrees

3. Outputs of logic gates U10, U11 and U12 are LOW, HIGH and LOW, respectively.
4. Vgs1a, Vgs2a, Vgs3a and vgs4a are, respectively, HIGH, HIGH, LOW and LOW.
5. Switches S1a and S2a are only ON, and the output voltage is +Vdc/2 volts.

Case 5: $\pi < \omega.t < (\pi + \alpha 1)$ radians. During this time interval of the reference sine wave for phase A, the following operation takes place:

1. Comparators X1, X2 and X3 are LOW, HIGH and LOW and that of X4, X5 and X6 are LOW, HIGH and LOW, respectively.
2. Output of logic gates U1 to U6, U8 and U9 are LOW and that of U7 is HIGH.

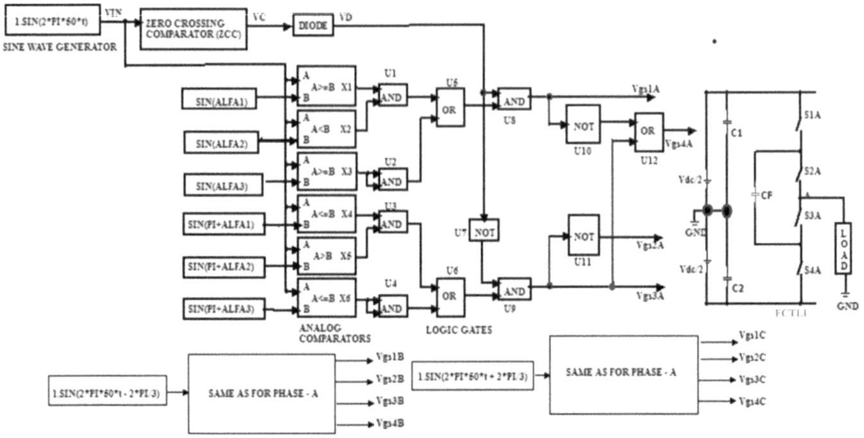

Fig. 3.44 Three-phase FCTL SHE-PWM inverter gate drive model block diagram

3. Output of logic gates U10, U11 and U12 are all HIGH.
4. Vgs1a, Vgs2a, Vgs3a and vgs4a are, respectively, LOW, HIGH, LOW and HIGH.
5. Switches S2a and S4a are only ON, and the output voltage is 0 volts.

Case 6: $(\pi + \alpha 1) <= \omega.t < (\pi + \alpha 2)$ radians. During this time interval of the reference sine wave for phase A, the following operation takes place:

1. Comparators X1, X2 and X3 are LOW, HIGH and LOW and that of X4, X5 and X6 are HIGH, HIGH and LOW, respectively.
2. Outputs of logic gates U1, U2, U4, U5 and U8 are LOW and that of U3, U6, U7 and U9 are HIGH.
3. Outputs of logic gates U10, U11 and U12 are HIGH, LOW and HIGH, respectively.
4. Vgs1a, Vgs2a, Vgs3a and vgs4a are, respectively, LOW, LOW, HIGH and HIGH.
5. Switches S3a and S4a are only ON, and the output voltage is –Vdc/2 volts.

Case 7: $(\pi + \alpha 2) <= \omega.t < (\pi + \alpha 3)$ radians. During this time interval of the reference sine wave for phase A, the following operation takes place:

1. Comparators X1, X2 and X3 are LOW, HIGH and LOW and that of X4, X5 and X6 are HIGH, LOW and LOW, respectively.
2. Outputs of logic gates U1 to U6, U8 and U9 are LOW and that of U7 is HIGH.
3. Outputs of logic gates U10, U11 and U12 are all HIGH.
4. Vgs1a, Vgs2a, Vgs3a and vgs4a are, respectively, LOW, HIGH, LOW and HIGH.
5. Switches S2a and S4a are only ON, and the output voltage is 0 volts.

Case 8: $(\pi + \alpha 3) <= \omega.t < 3\pi/2$ radians. During this time interval of the reference sine wave for phase A, the following operation takes place:

1. Comparators X1, X2 and X3 are LOW, HIGH and LOW and that of X4, X5 and X6 are HIGH, LOW and HIGH, respectively.
2. Outputs of logic gates U1, U2, U3, U5 and U8 are LOW and that of U4, U6, U7 and U9 are HIGH.
3. Outputs of logic gates U10, U11 and U12 are HIGH, LOW and HIGH, respectively.
4. Vgs1a, Vgs2a, Vgs3a and vgs4a are, respectively, LOW, LOW, HIGH and HIGH.
5. Switches S3a and S4a are only ON, and the output voltage is –Vdc/2 volts.

A model of the three-phase SHE-PWM FCTLI is developed using SIMULINK [11]. This model is shown in Fig. 3.45 (Model file: EXAMPLE 3_5), and the model subsystems are shown in Fig. 3.46a, b. Three-phase 50 Hz sine wave AC voltage is used as a reference. The values of $\alpha 1$, $\alpha 2$ and $\alpha 3$ for M value of 0.5 and 0.8 to eliminate 5th and 7th harmonics from the line-to-ground voltages of FCTLI are shown in Table 3.1. The values of switching angles corresponding to M value of 0.5 are used in the model. The inverter switching frequency is 50 Hz. The DC link voltage of FCTLI is 100 volts. The three-phase FCTLI NMOSFET switches are from power electronics library in the Power Systems Specialized Technology block set. The reference sine waves for the three phase are generated using Embedded MATLAB function and the sine of switching angles for $\alpha 1$, $\alpha 2$ and $\alpha 3$ and that for $(\pi + \alpha 1)$, $(\pi + \alpha 2)$ and $(\pi + \alpha 3)$ are generated using another Embedded MATLAB function. The source codes are given in Program segment 3.2 and 3.3 in the model file EXAMPLE 3_5. In Fig. 3.36, the switching angles $\alpha 1$, $\alpha 2$ and $\alpha 3$ in degrees are entered in order in the constant block separated by a space. The selector block selects or directs the first output $\alpha 1$, second output $\alpha 2$ and third output $\alpha 3$. The model subsystem to generate gate pulse for phase A switches of inverter is shown in Fig. 3.37a, b. Here as shown in Fig. 3.37a, reference sine wave for phase A is given to relay block which forms zero-crossing comparator (ZCC) giving square wave output when the input sine wave crosses zero and goes positive and negative, respectively. The ZCC output is compared with zero using a relational operator block which forms the comparator. The ZCC and comparator together operate as a diode. This comparator output is HIGH when the ZCC input is greater than or equal to zero or else its output is LOW. Reference sine wave for phase A $(1.\sin(2*\pi*50*t))$ is compared with sine of $\alpha 1$, $\alpha 2$, $\alpha 3$, $(\pi + \alpha 1)$, $(\pi + \alpha 2)$ and $(\pi + \alpha 3)$ in six relational operator blocks which form the comparators X1 to X6, as shown in Fig. 3.46b. The resulting output of each comparator X1 to X6 and that of ZCC and diode is connected to logic gates U1 to U12 in the same way shown in Fig. 3.44. The output of U8, U11, U9 and U12 forms the gate pulse for switches S1a, S2a, S3a and S4a in phase A.

Similar method is used along with reference sine waves for phase B and C to generate gate pulse for the four switches in the respective phase.

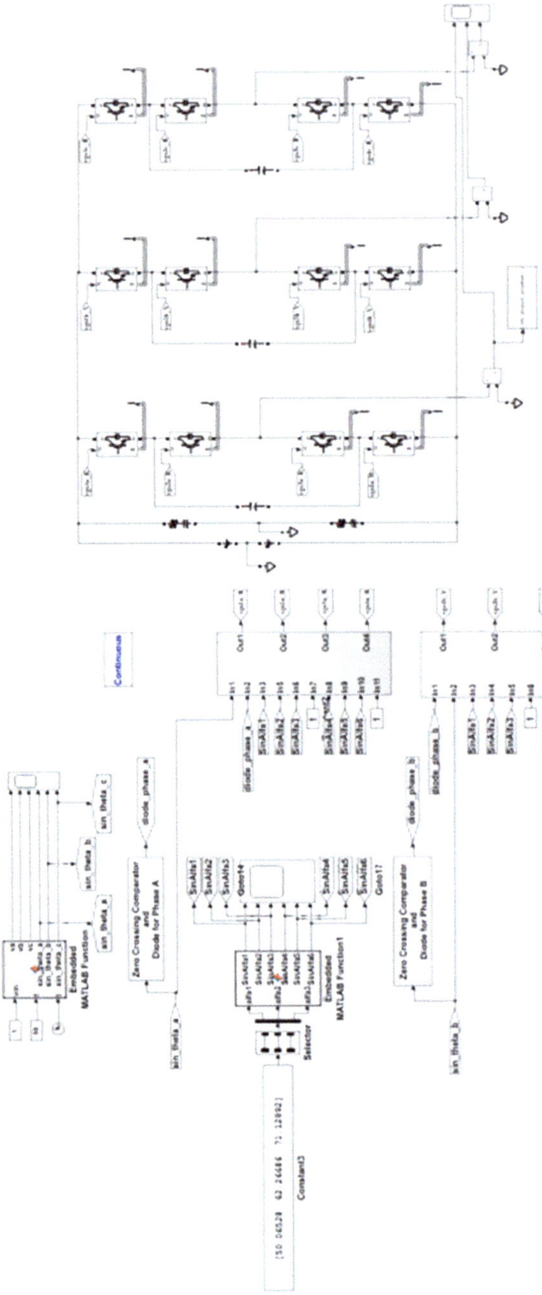

Fig. 3.45 Model of three-phase SHE-PWM FCTLI with three switching angles

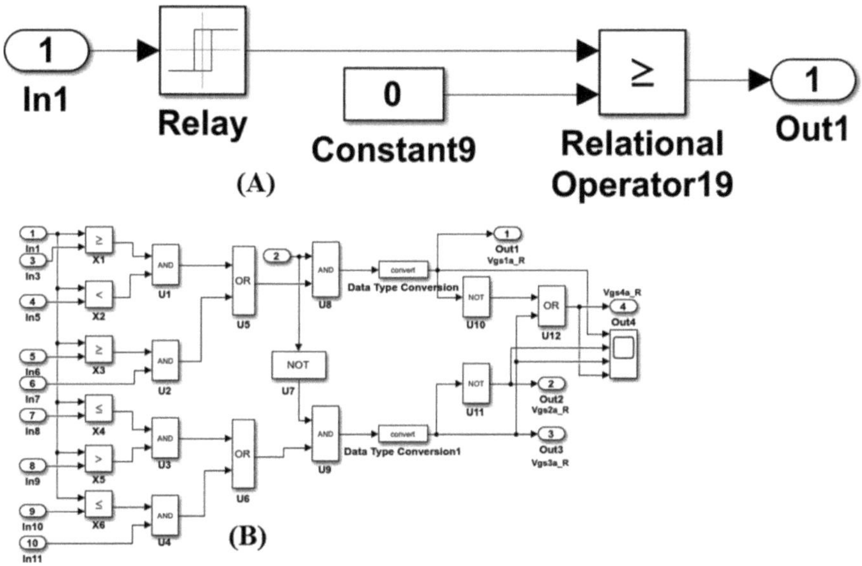

Fig. 3.46 Three-phase SHE-PWM FCTLI model subsystems: (**a**) Zero crossing comparator and diode. (**b**) Gate drive for FCTLI switches in Phase A

3.5.4 Simulation Results

The model simulation of the three-phase SHE-PWM FCTLI was carried out using ode45 (Dormand-Prince) variable step solver. The inverter switching frequency is 50 Hz and DC link voltage is 100 V. The switching angles in Table 3.1 are used to eliminate 5th and 7th harmonics from the line-to-ground voltage. The harmonic spectrum of line-to-ground voltage and the three phase line-to-ground voltage for M values of 0.5 and 0.8 in the respective order are shown in Figs. 3.47 to 3.52, respectively.

3.5.5 Discussion of Results

Referring to the harmonic spectrum for three-phase SHE-PWM DCTLI shown in Figs. 3.38, 3.40 and 3.42, it is seen that the 5th and 7th harmonics are eliminated from the line-to-ground voltage. Also it is seen that the value of the peak fundamental line-to-ground voltage V_1 given by 2*Vdc*M/(π) for the DC link voltage of 100 V and modulation index M value of 0.5 and 0.8 works out to 31.847 V and 50.955 V. The displayed values in Figs. 3.38, 3.40 and 3.42 are 31.99 V, 50.73 V and 50.75 V, respectively. Similarly referring to the harmonic spectrum for three-phase SHE-PWM FCTLI shown in Figs. 3.47, 3.49 and 3.51, it is seen that the 5th and 7th harmonics are eliminated from the line-to-ground voltage. Also the

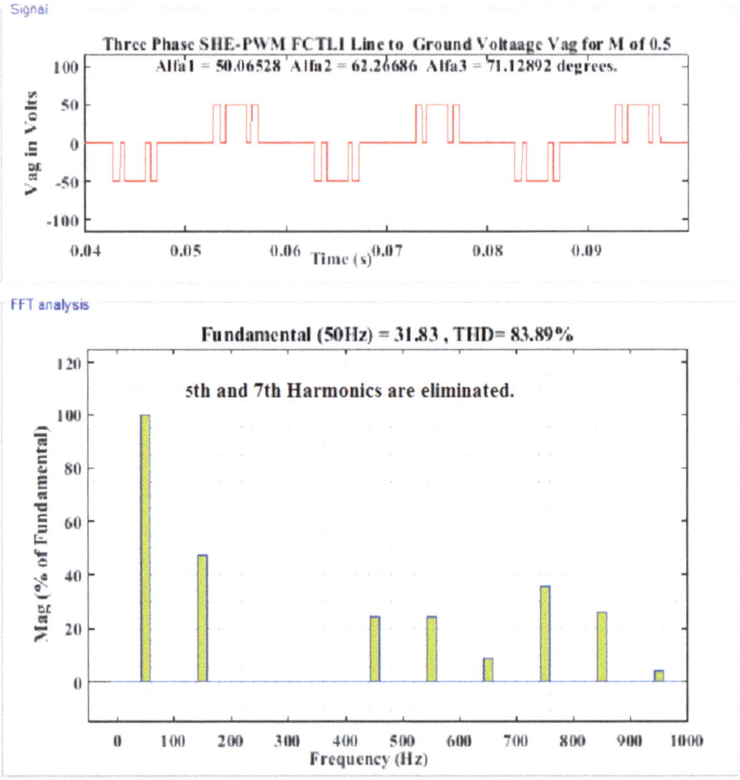

Fig. 3.47 Three-phase SHE-PWM FCTLI line-to-ground voltage (top) and harmonic spectrum with 5th and 7th harmonics eliminated (bottom)

displayed value for peak fundamental line-to-ground voltage V_1 for the given modulation index M value of 0.5 and 0.8 and DC link voltage of 100 V in Figs. 3.47, 3.49 and 3.51 are found to be 31.83 V, 50.93 V and 50.94 V. The discrepancy from the calculated and model values in both cases is very small.

3.6 Case Study: Single-Phase Diode-Clamped and Flying Capacitor Five-Level Inverter

The single-phase diode-clamped five-level (DCFL) and flying capacitor five-level (FCFL) inverter are shown in Fig. 3.53, and their switching tables are shown in Tables 3.6 and 3.7. The The DC link voltage is Vdc volts, and the individual cell voltage is E = Vdc/(m−1) volts where m is the number of line-to-ground output voltage level. Thus for a five-level inverter, E is Vdc/4.

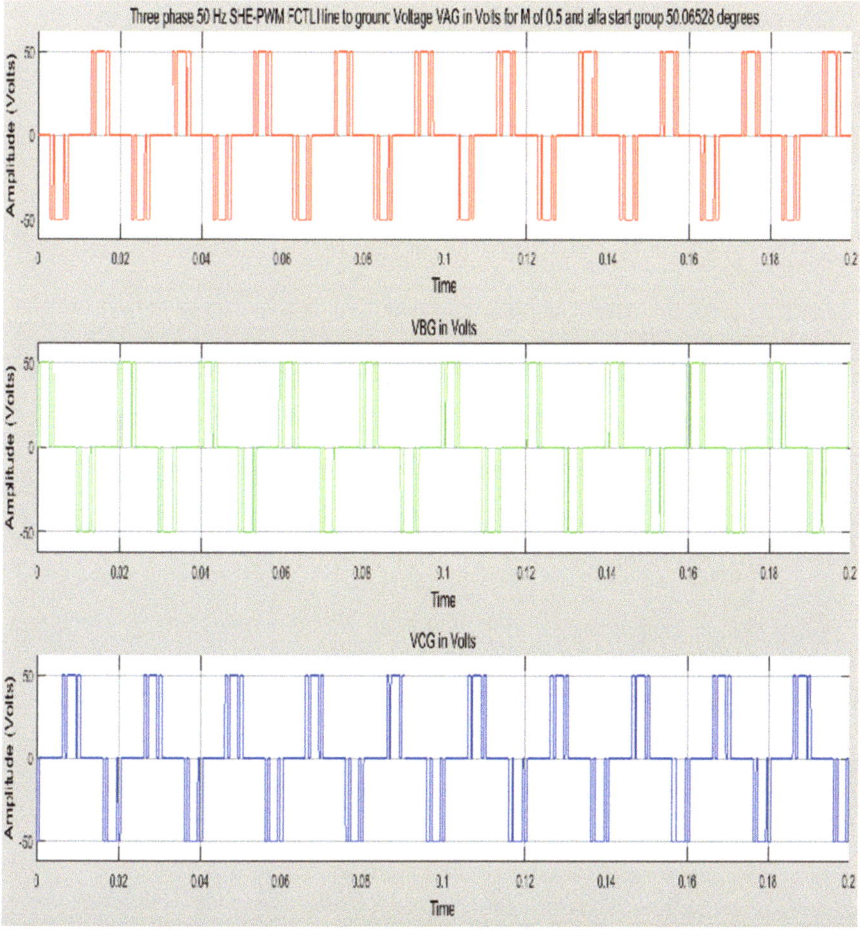

Fig. 3.48 Three-phase SHE-PWM FCTLI line-to-ground voltage for M of 0.5 and alfa start group of 50.06528 degrees

The voltage waveform Vlg of a five-level inverter having two switching angles $\alpha 1$ and $\alpha 2$ is shown in Fig. 3.54. The five voltage levels are +2E, +E, 0, -E and -2E where E is Vdc/4 volts. Based on the derivation for Fourier series expression shown in Sect. 3.4 for Fig. 3.27, assuming that there are N switching angles $\alpha 1$ to αN with ascending values in the range $0 \leq \omega.t < \pi/2$ $\quad 0 < = \omega.t < \pi/2$.

$< \pi/2$, Fourier expression for Vlg can be expressed as follows:

$$V_{lg}(\omega.t) = \sum_{n=1,3,5,7...}^{\infty} \left[\frac{4 * E}{n.\pi} * \left\{ \sum_{k=1}^{N} (-1)^{k+1} * \cos(n * \alpha_k) \right\} * \sin(n.\omega.t) \right] \quad (3.31)$$

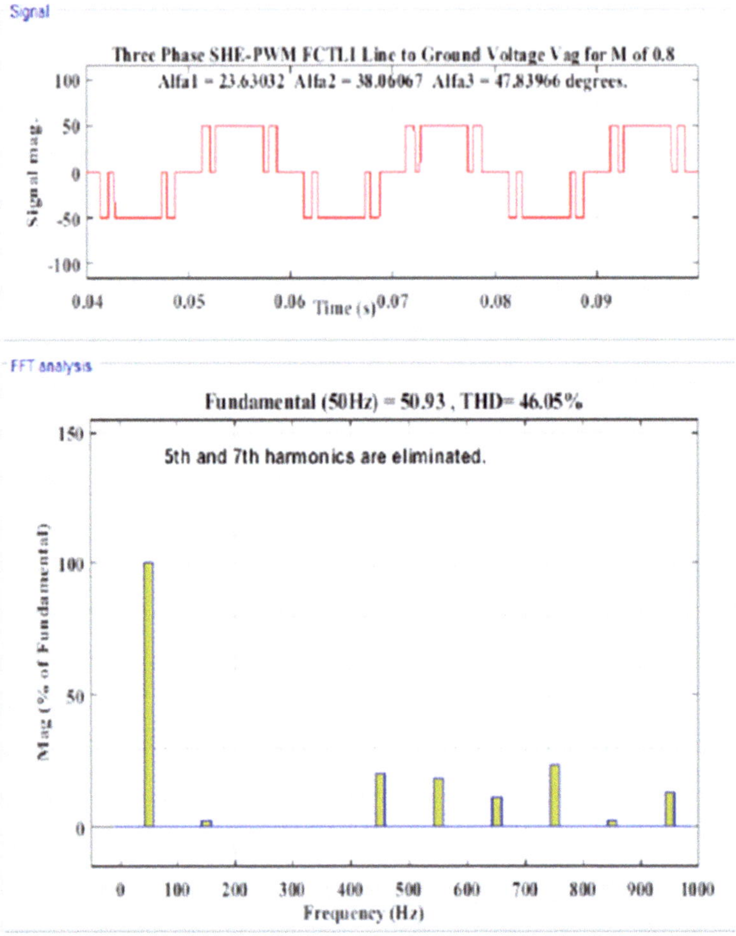

Fig. 3.49 Three-phase SHE-PWM FCTLI line-to-ground voltage (top) and harmonic spectrum with 5th and 7th harmonics eliminated (bottom)

From Eq.3.31, to eliminate the 11th, 13th and 15th harmonics from the line-to-ground voltage of five-level inverter, the following set of equations must be solved:

$$\left.\begin{aligned} \cos(\alpha_1) - \cos(\alpha_2) + \cos(\alpha_3) - \cos(\alpha_4) &= M \\ \cos(11.\alpha_1) - \cos(11.\alpha_2) + \cos(11.\alpha_3) - \cos(11.\alpha_4) &= 0 \\ \cos(13.\alpha_1) - \cos(13.\alpha_2) + \cos(13.\alpha_3) - \cos(13.\alpha_4) &= 0 \\ \cos(15.\alpha_1) - \cos(15.\alpha_2) + \cos(15.\alpha_3) - \cos(15.\alpha_4) &= 0 \end{aligned}\right\} \quad (3.32)$$

In Eq.3.32, $M = (\pi.V_1/4.\mathrm{E})$, where V_1 is the fundamental component of the line-to-ground voltage. Also from Eq.3.31, it is seen that all even harmonics and DC components are zero.

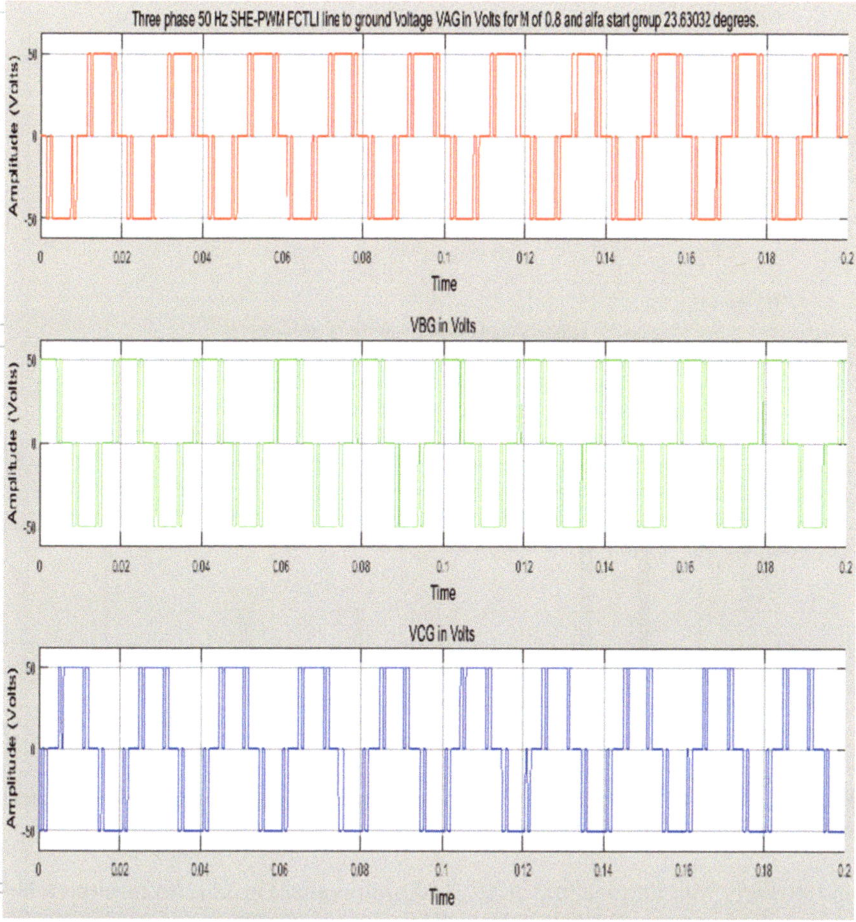

Fig. 3.50 Three-phase SHE-PWM FCTLI line-to-ground voltage for M of 0.8 and alfa start group of 23.63032 degrees

3.6.1 Single-Phase SHE-PWM Diode-Clamped Five-Level Inverter

Single-phase diode-clamped five-level inverter (DCFLI) topology is shown in Fig. 3.53a. To eliminate a set of three odd harmonics from the line-to-ground voltage Vlg, four switching angles $\alpha 1$, $\alpha 2$, $\alpha 3$ and $\alpha 4$ are selected. The desired line-to-ground voltage waveform is shown in Fig. 3.55. The SHE-PWM DCFLI gate drive block diagram is shown in Fig. 3.56. Referring to Fig. 3.56a, the operation of gate drive is explained below:

Comparators are marked from X1 to X16. Here reference sine wave 1*sin (2*pi*f*t) which is the first input A is compared with second input B which are sin(alfa1), in X1, X2, sin(alfa2) in X3, X4, sin(alfa3) in X5, X6, sin(alfa4) in X7, X8

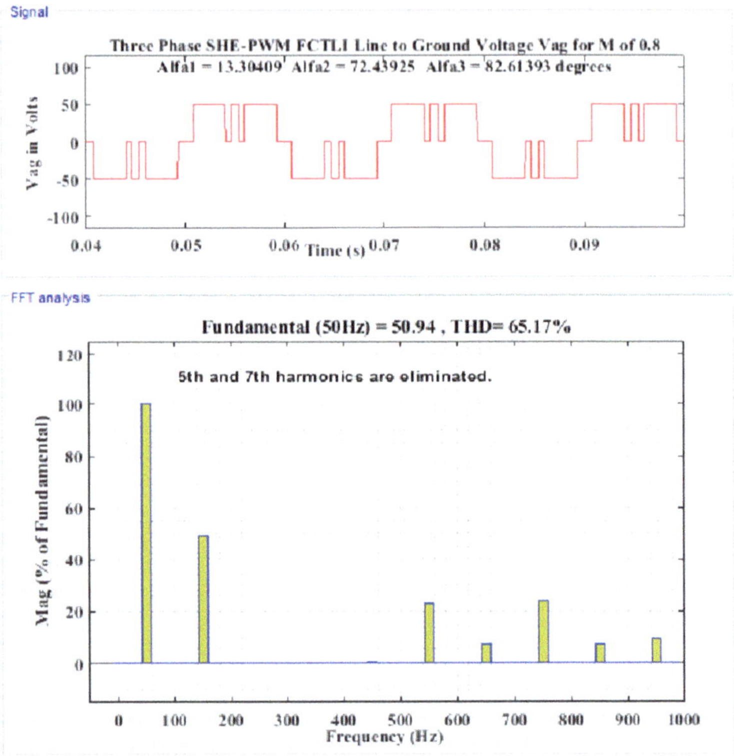

Fig. 3.51 Three-phase SHE-PWM FCTLI line-to-ground voltage (Top) and harmonic spectrum with 5th and 7th harmonics eliminated (bottom)

and similarly with sin(pi+alfa1) in X9, X10, sin(pi+alfa2) in X11, X12, sin(pi+alfa3) in X13, X14 and sin(pi+alfa4) in X15, X16, respectively. The comparators perform the following functions:

Comparators marked A > =B (A > B) gives output HIGH when the first input A is greater than or equal to (greater than) second input B, or else its output is LOW.

Comparators marked A < =B (A < B) gives output HIGH when the first input A is less than or equal to (less than) second input B, or else its output is LOW.

Outputs of comparators X1, X8, X9, X16 are given to single input AND gates U1, U5, U6 and U10. Outputs of comparator pairs X2-X3, X4-X5, X6-X7, X10-X11, X12-X13 and X14-X15 are given to two input AND gates U2, U3, U4, U7, U8 and U9, respectively. The sine wave reference voltage is given to a zero-crossing comparator (ZCC) and diode whose output signal VD is HIGH during the positive half cycle and LOW during the negative half cycle of the reference sine wave input voltage. The output of AND gates U1 to U5 and the ZCC-diode output VD are given to two input AND gates marked U11 to U15 in the respective order. The inverted output of ZCC diode using NOT gate U21 and the output of AND gates U6 to U10 are given to two input AND gates U16 to U20 in the respective order. The outputs

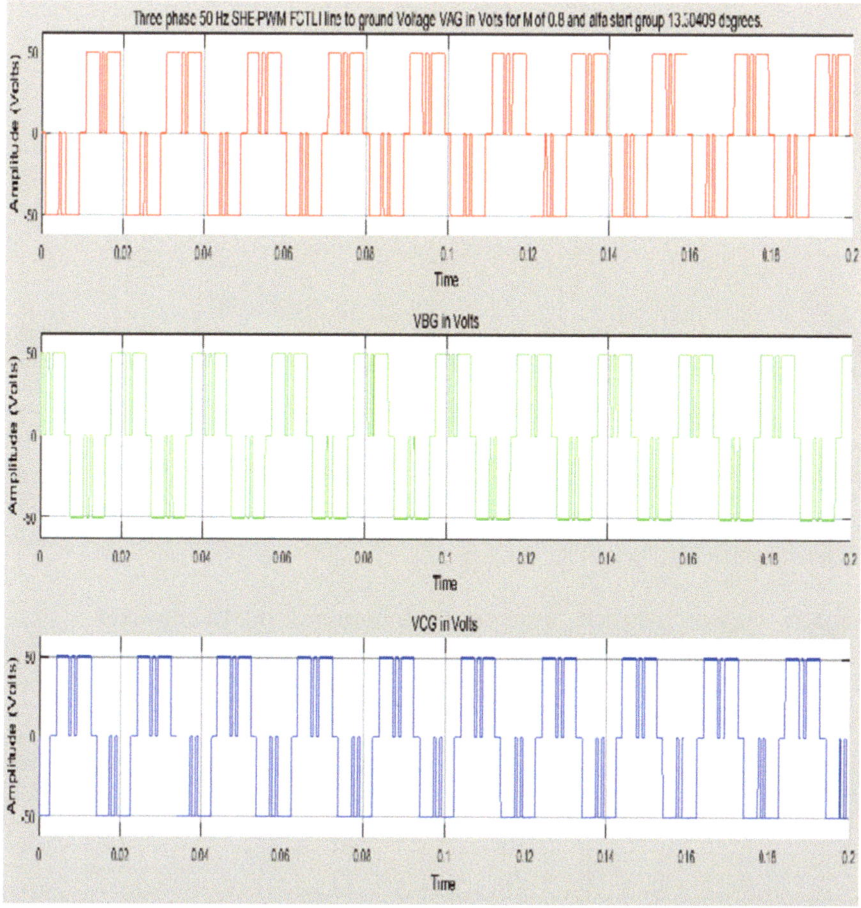

Fig. 3.52 Three-phase SHE-PWM FCTLI line-to-ground voltage for M of 0.8 and alfa start group of 13.30409 degrees

Y0 to Y4 of U11 to U15 take HIGH or LOW during the positive half cycle only, and Y5 to Y9 of U16 to U20 take the above values only during negative half cycle. Thus Y0 is HIGH during the interval $0 <= \omega.t < \alpha 1$, Y1 is HIGH during the interval $\alpha 1 <= \omega.t < \alpha 2$ and so on. During the negative half cycle of the reference sine wave input voltage, Y5 is HIGH during the interval $\pi <= \omega.t < (\pi + \alpha 1)$, Y6 is HIGH during the interval $(\pi + \alpha 1) <= \omega.t < (\pi + \alpha 2)$ and so on. Now to get the shape of the waveform shown in Fig. 3.55 for the line-to-ground voltage, Table 3.8 is prepared using Table 3.6.

In Table 3.8, Vlg column is filled based on Fig. 3.55. For example, during the interval from $\alpha 1$ to $\alpha 2$, gate drive logic output Y1 is HIGH, and the Vlg output is +Vdc/4. Thus S1a, S2a, S3a, S4a are, respectively, marked 0, 1, 1 and 1 from Table 3.6. In this way, all columns are filled referring to Fig. 3.55 and Table 3.6. The logic expression for the gate drive of individual switch is given from Table 3.8.

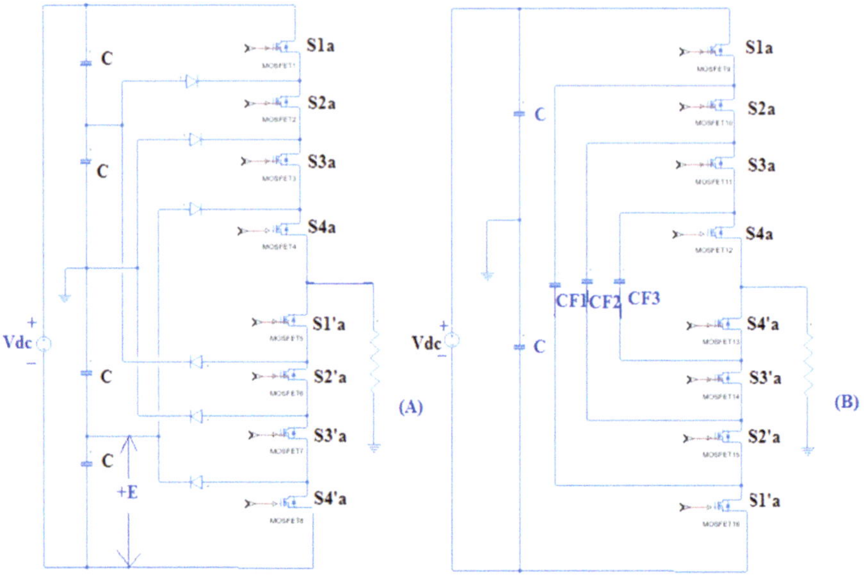

Fig. 3.53 Single-phase five-level inverter: (**a**) Diode clamped and (**b**) Flying capacitor

Table 3.6 Switching truth table for single-phase DCFL inverter

Sl. no.	S1a	S2a	S3a	S4a	S1'a	S2'a	S3'a	S4'a	Vlg (V)
1	1	1	1	1	0	0	0	0	+Vdc/2
2	0	1	1	1	1	0	0	0	+Vdc/4
3	0	0	1	1	1	1	0	0	0
4	0	0	0	1	1	1	1	0	-Vdc/4
5	0	0	0	0	1	1	1	1	-Vdc/2

Table 3.7 Switching truth table for single-phase FCFL inverter

Sl. no.	S1a	S2a	S3a	S4a	S4'a	S3'a	S2'a	S1'a	Vlg (V)
1	1	1	1	1	0	0	0	0	+Vdc/2
2	1	1	1	0	1	0	0	0	+Vdc/4
3	1	1	0	0	1	1	0	0	0
4	1	0	0	0	1	1	1	0	-Vdc/4
5	0	0	0	0	1	1	1	1	-Vdc/2

$$\left.\begin{aligned} S_{1a} &= (Y4 \cup Y2)\\ S_{2a} &= (Y4 \cup Y2 \cup Y1 \cup Y3) = (S_{1a} \cup Y1 \cup Y3)\\ S_{3a} &= (Y0 \cup Y1 \cup Y2 \cup Y3 \cup Y4 \cup Y5) = (S_{2a} \cup Y0 \cup Y5)\\ S_{4a} &= (Y0 \cup Y1 \cup Y2 \cup Y3 \cup Y4 \cup Y5 \cup Y6 \cup Y8) = (S_{3a} \cup Y6 \cup Y8)\\ S'_{1a} &= (\sim S_{1a}); S'_{2a} = (\sim S_{2a}); S'_{3a} = (\sim S_{3a}); S'_{4a} = (\sim S_{4a}) \end{aligned}\right\} \tag{3.33}$$

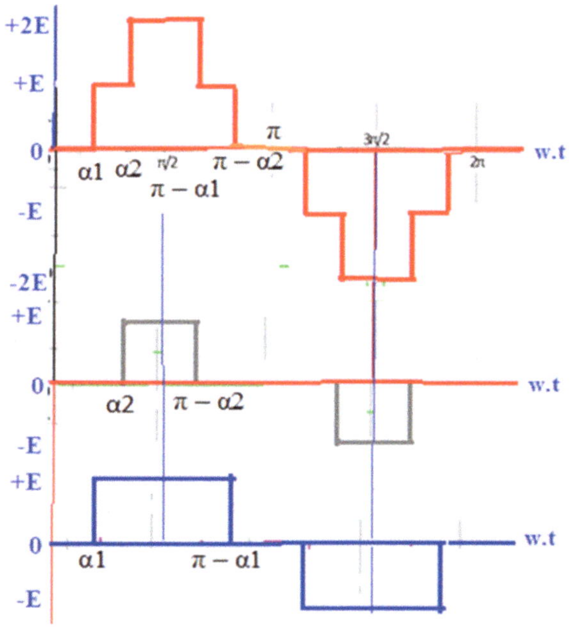

Fig. 3.54 Five-level output voltage wavefrom formed by summation of two three level voltage waveforms

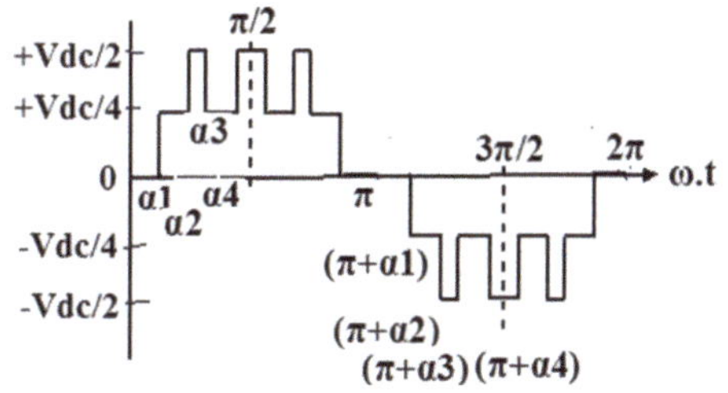

Fig. 3.55 Line-to-ground voltage waveform of a diode clamped five-level inverter

In Eq. 3.33, symbol ∪ stands for logical OR, and symbol ~ stands for logical NOT operation, respectively. Equation 3.33 is shown in Fig. 3.56b.

3.6.2 Model of Single-Phase SHE-PWM Diode-Clamped Five-Level Inverter

A model of the single-phase DCFLI is shown in Fig. 3.57 (Model file: CASE_STUDY_EX3_1). It is found by solving Eq. 3.32 that the four switching angles in degrees to eliminate 11th, 13th and 15th harmonics from the line-to-ground voltage of five-level inverter for an M value of 1.2 are $\alpha 1 = 4.0164$, $\alpha 2 = 30.372$, $\alpha 3 = 31.9023$ and $\alpha 4 = 40.6685$ degrees. To find the 11th, 13th and 15th harmonic and THD contributions, a model of this SHE-PWM DCFLI is developed in Simulink

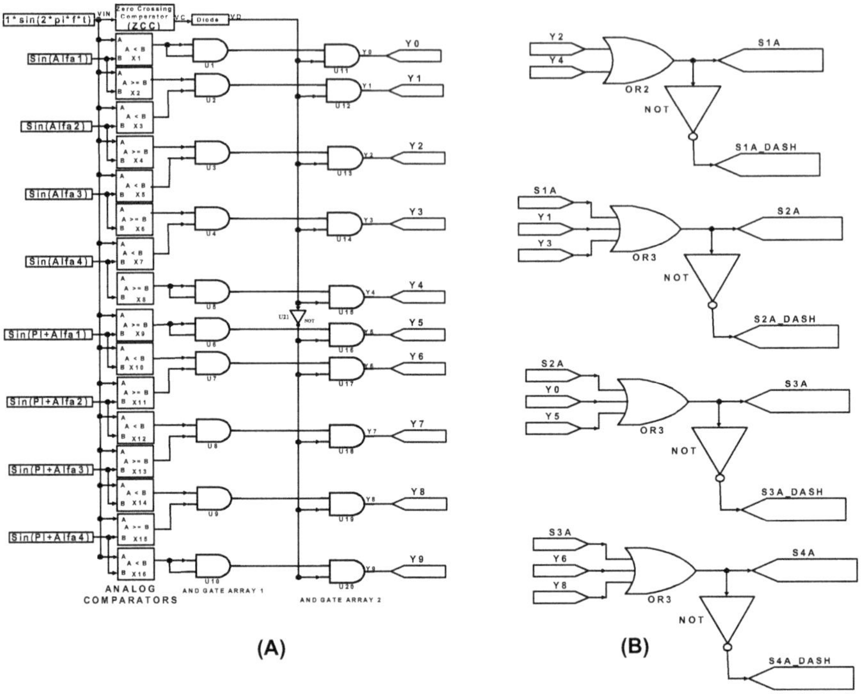

Fig. 3.56 Single-phase SHE-PWM DCFLI and FCFLI Gate Drive: (**a**) Comparators and logic gates, (**b**) Gate drive using OR logic for DCFLI

[11]. The Embedded MATLAB function generates the reference sine wave voltage (1*sin(2*π*50*t) using source code in Program Segment 3.2 in the model file CASE_STUDY_EX3_1. In the constant block, $\alpha 1$ to $\alpha 4$ are entered with space. The selector block along with Demux block directs $\alpha 1$ to $\alpha 4$ in the order. The eight Fcn blocks generate sine of $\alpha 1$ to $\alpha 4$ and that of ($\pi + \alpha 1$) to ($\pi + \alpha 4$). These sine values of $\alpha 1$ to ($\pi + \alpha 4$) are compared with reference sine wave voltage in 16 relational operator blocks which form X1 to X16 in Fig. 3.56. The output of relational operator blocks is given to AND gates as shown in Fig. 3.56. The relay block with switch on and off points zero and output +5 and −5, along with relational operator block which compares relay output with constant zero forms the zero-crossing comparator (ZCC) and diode. The outputs of the above AND gates and that of ZCC are given to a second set of AND gates which give gate drive logic outputs Y0 to Y7 which are connected using OR logic gates as per Eq. 3.33 to generate gate pulses for the semiconductor switches of the five-level inverter. The DC voltage source, input capacitor, diodes and eight NMOSFET switches of DCFLI shown in Fig. 3.57 are from Simscape/electrical and semiconductors and converters library. A gain constant of ten along with gate driver is used for each NMOSFET switch, so that the gate pulse amplitude is greater than the threshold voltage of the switch. The RMS, THD and Fourier Transform blocks are also shown in Fig. 3.57.

Table 3.8 Gate drive switching truth table for single-phase SHE-PWM DCFLI

Gate drive logic outputs												Switch status				
Sl. no.	Y0	Y1	Y2	Y3	Y4	Y5	Y6	Y7	Y8	Y9	HIGH duration	S1a	S2a	S3a	S4a	Vlg (V)
1	1	0	0	0	0	0	0	0	0	0	0 to $\alpha 1$	0	0	1	1	0
2	0	1	0	0	0	0	0	0	0	0	$\alpha 1$ to $\alpha 2$	0	1	1	1	+Vdc/4
3	0	0	1	0	0	0	0	0	0	0	$\alpha 2$ to $\alpha 3$	1	1	1	1	+Vdc/2
4	0	0	0	1	0	0	0	0	0	0	$\alpha 3$ to $\alpha 4$	0	1	1	1	+Vdc/4
5	0	0	0	0	1	0	0	0	0	0	$\alpha 4$ to $\pi/2$	1	1	1	1	+Vdc/2
6	0	0	0	0	0	1	0	0	0	0	π to $\pi + \alpha 1$	0	0	1	1	0
7	0	0	0	0	0	0	1	0	0	0	$\pi + \alpha 1$ to $\pi + \alpha 2$	0	0	0	1	-Vdc/4
8	0	0	0	0	0	0	0	1	0	0	$\pi + \alpha 2$ to $\pi + \alpha 3$	0	0	0	0	-Vdc/2
9	0	0	0	0	0	0	0	0	1	0	$\pi + \alpha 3$ to $\pi + \alpha 4$	0	0	0	1	-Vdc/4
10	0	0	0	0	0	0	0	0	0	1	$\pi + \alpha 4$ to $3.\pi/2$	0	0	0	0	-Vdc/2

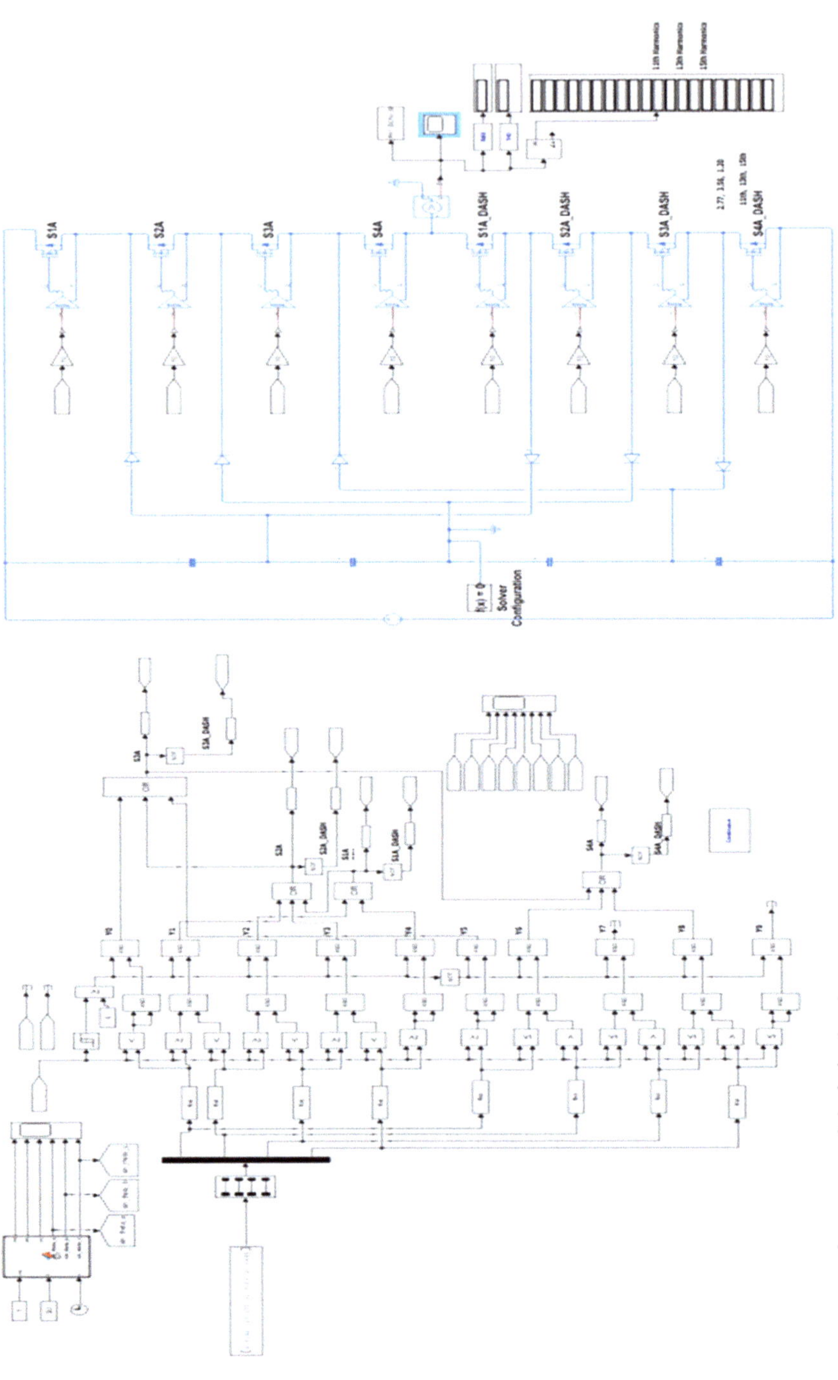

Fig. 3.57 Model of single-phase SHE-PWM DCFL inverter

3.6.3 Simulation Results

The model simulation of single-phase SHE-PWM DCFLI is carried out using ode15s (stiff/NDF) solver in Simulink [11]. The input frequency is 50 Hz, and the DC link voltage is 100 volts. Values of $\alpha1$ to $\alpha4$ given above are used to eliminate 11th, 13th and 15th harmonics from the five-level line-to-ground voltage waveform. The line-to-ground voltage waveform of single-phase SHE-PWM DCFLI is shown in Fig. 3.58. The RMS value, THD and harmonic components from fundamental up to 1000 Hz are shown in Fig. 3.59. Results are presented in Table 3.9.

3.6.4 Single-Phase SHE-PWM Flying Capacitor Five-Level Inverter

Single-phase flying capacitor five-level inverter (FCFLI) topology is shown in Fig. 3.53b. To eliminate a set of three odd harmonics from the line-to-ground voltage Vlg, four switching angles $\alpha1$, $\alpha2$, $\alpha3$ and $\alpha4$ are selected. The desired line-to-ground voltage waveform is shown in Fig. 3.60. The SHE-PWM FCFLI gate drive comparator and logic gate arrangement are shown in Fig. 3.56a. Referring to Fig. 3.56a, the operation of comparator and logic gates is already explained in Sect. 3.6.1.

Now to get the shape of the waveform shown in Fig. 3.60 for the line-to-ground voltage, Table 3.10 is prepared using Table 3.7.

In Table 3.10, Vlg column is filled based on Fig. 3.60. For example, during the interval from $\alpha1$ to $\alpha2$, gate drive logic output Y1 is HIGH and the Vlg output is +Vdc/4. Thus, S1a, S2a, S3a and S4a are, respectively, marked 1, 1, 1 and 0 from Table 3.7. In this way, all columns are filled referring to Fig. 3.60 and Table 3.7. The logic expression for the gate drive of individual switch is given from Table 3.10.

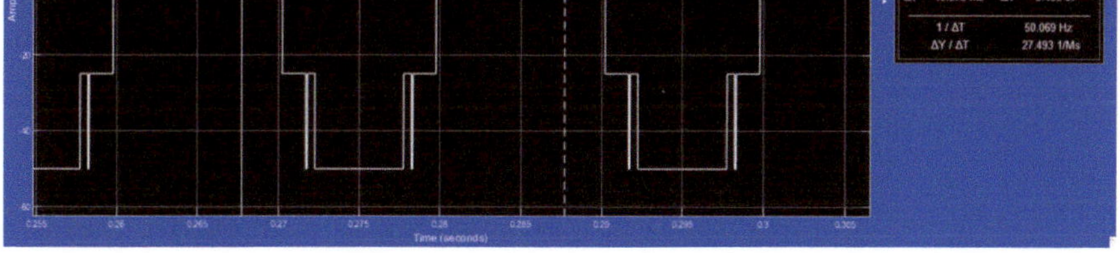

Fig. 3.58 Line-to-Ground Voltage Wavefrom of SHE-PWM FLDCI for alfa start group of 4.0164 Degrees

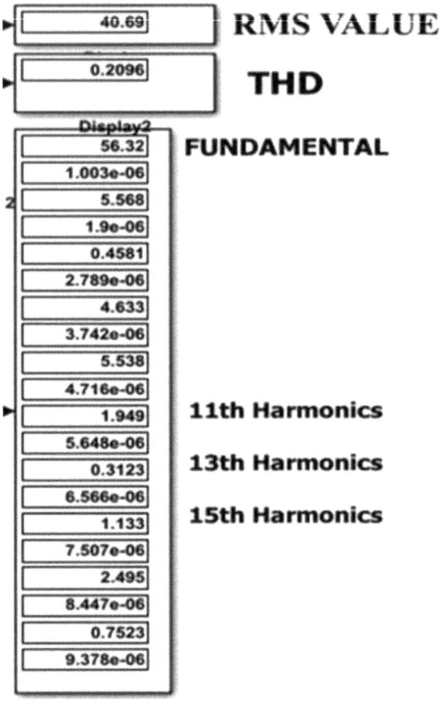

Fig. 3.59 RMS, THD, fundamental 11th, 13th and 15th harmonics of line-to-ground voltage of SHE-PWM DCFLI

Table 3.9 Single-phase SHE-PWM DCFLI simulation results

Sl. no.	RMS value (V)	THD per unit	Fund. V (%)	11th Har. V (%)	13th Har. V (%)	15th Har. V (%)
1)	40.69	0.2096	56.32 (100)	1.949 (3.46)	0.3123 (0.55)	1.133 (2.01)

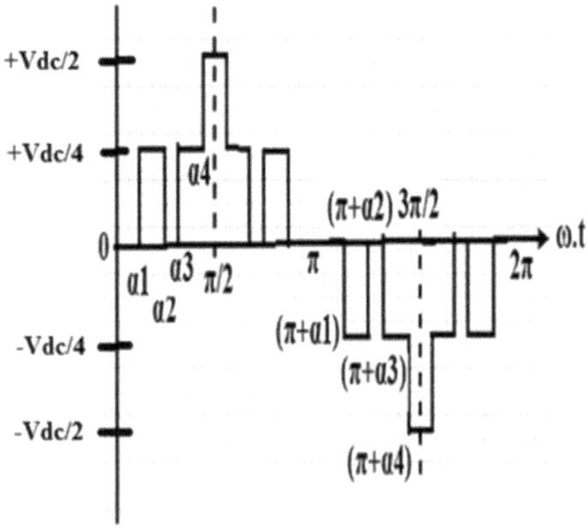

Fig. 3.60 Line-to-ground voltage waveform of a flying capacitor five-level inverter

Table 3.10 Gate drive switching truth table for single-phase SHE-PWM FCFLI

Gate drive logic outputs												Switch status				
Sl. no.	Y0	Y1	Y2	Y3	Y4	Y5	Y6	Y7	Y8	Y9	HIGH duration	S1a	S2a	S3a	S4a	Vlg (V)
1	1	0	0	0	0	0	0	0	0	0	0 to $\alpha 1$	1	1	0	0	0
2	0	1	0	0	0	0	0	0	0	0	$\alpha 1$ to $\alpha 2$	1	1	1	0	+Vdc/4
3	0	0	1	0	0	0	0	0	0	0	$\alpha 2$ to $\alpha 3$	1	1	0	0	0
4	0	0	0	1	0	0	0	0	0	0	$\alpha 3$ to $\alpha 4$	1	1	1	0	+Vdc/4
5	0	0	0	0	1	0	0	0	0	0	$\alpha 4$ to $\pi/2$	1	1	1	1	+Vdc/2
6	0	0	0	0	0	1	0	0	0	0	π to $\pi + \alpha 1$	1	1	0	0	0
7	0	0	0	0	0	0	1	0	0	0	$\pi + \alpha 1$ to $\pi + \alpha 2$	1	0	0	0	−Vdc/4
8	0	0	0	0	0	0	0	1	0	0	$\pi + \alpha 2$ to $\pi + \alpha 3$	1	1	0	0	0
9	0	0	0	0	0	0	0	0	1	0	$\pi + \alpha 3$ to $\pi + \alpha 4$	1	0	0	0	−Vdc/4
10	0	0	0	0	0	0	0	0	0	1	$\pi + \alpha 4$ to $3.\pi/2$	0	0	0	0	−Vdc/2

$$\left.\begin{aligned} S_{1a} &= (Y0 \cup Y1 \cup Y2 \cup Y3 \cup Y4 \cup Y5 \cup Y6 \cup Y7 \cup Y8) = (S_{2a} \cup Y6 \cup Y8) \\ S_{2a} &= (Y0 \cup Y1 \cup Y2 \cup Y3 \cup Y4 \cup Y5 \cup Y7) \\ &= (S_{3a} \cup Y0 \cup Y2 \cup Y5 \cup Y7) \\ S_{3a} &= (Y1 \cup Y3 \cup Y4) = (S_{4a} \cup Y1 \cup Y3) \\ S_{4a} &= (Y4) \\ S'_{1a} &= (\sim S_{1a}); S'_{2a} = (\sim S_{2a}); S'_{3a} = (\sim S_{3a}); S'_{4a} = (\sim S_{4a}) \end{aligned}\right\} \tag{3.34}$$

In Eq. 3.34, symbol ∪ stands for logical OR, and symbol ~ stands for logical NOT operation, respectively. Equation 3.34 is shown in Fig. 3.61.

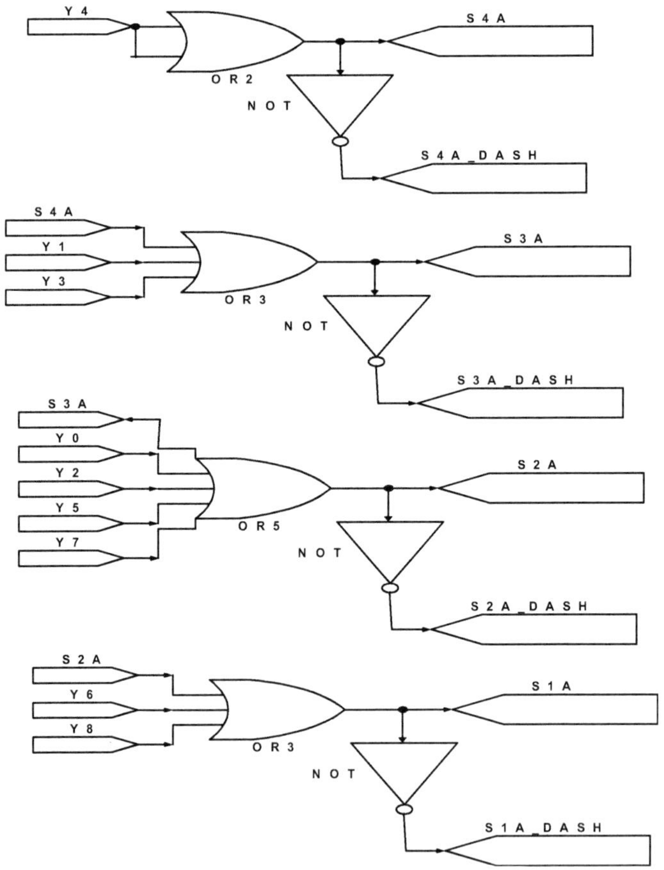

Fig. 3.61 Gate drive using OR logic for FCFLI

3.6.5 Model of Single-Phase SHE-PWM Flying Capacitor Five-Level Inverter

A model of the single-phase FCFLI is shown in Fig. 3.62 (Model file: CASE_STUDY_EX3_2). It is found by solving Eq. 3.32 that the four switching angles in degrees to eliminate 11th, 13th and 15th harmonics from the line-to-ground voltage of five-level inverter for an M value of 1.2 are $\alpha 1 = 4.0164$, $\alpha 2 = 30.372$, $\alpha 3 = 31.9023$ and $\alpha 4 = 40.6685$ degrees. To find the 11th, 13th and 15th harmonic and THD contributions, a model of this SHE-PWM FCFLI is developed in.

Simulink [11]. This model is shown in Fig. 3.62. The Embedded MATLAB function generates the reference sine wave voltage (1*sin(2*π*50*t) using source code in Program Segment 3.2 in the model file CASE_STUDY_EX3_2. The development of this model is already explained in Sect. 3.6.2. Here gate drive logic outputs Y0 to Y7 are developed as presented in Sect. 3.6.2. These gate drive logic Y0 to Y7 are connected as per Eq. 3.34, as shown in Fig. 3.61. The input frequency is 50 Hz, and the DC link voltage is 100 volts. Flying capacitors are from Simscape—Electrical—Specialised Power Systems block set. The RMS, THD and Fourier transform blocks are also shown in Fig. 3.62.

3.6.6 Simulation Results

The model simulation of the single-phase SHE-PWM FCFLI is carried out using ode15s (stiff/NDF) solver in Simulink [11]. The input frequency is 50 Hz, and the DC link voltage is 100 volts. Values of $\alpha 1$ to $\alpha 4$ given above are used to eliminate 11th, 13th and 15th harmonics from the five-level line-to-ground voltage waveform. The line-to-ground voltage waveform of single-phase SHE-PWM FCFLI is shown in Fig. 3.63. The RMS value, THD and harmonic components from fundamental up to 1000 Hz are shown in Fig. 3.64. Results are presented in Table 3.11.

3.6.7 Discussion of Results

Five-level waveform for line-to-ground voltage with different wave shapes is obtained for the SHE-PWM DCFLI and FCFLI. In both cases 11th, 13th and 15th harmonics are eliminated or reduced to a minimum compared to other odd harmonics such as 3rd, 5th, 7th, 9th, etc. In the case of DCFLI topology, the THD of line-to-ground voltage is around 1.3% lower than that for FCFLI topology. Thus, DCFLI line-to-ground voltage waveform is preferable to that of FCFLI waveform from the point of view of THD. The amplitude of fundamental component of line-to-ground voltage is greater for DCFLI compared to that of FCFLI topology. For three-phase DCFLI and FCFLI, comparison of sine of $\alpha 1$ to $\alpha 4$ and $(\pi + \alpha 1)$ to $(\pi + \alpha 4)$ are

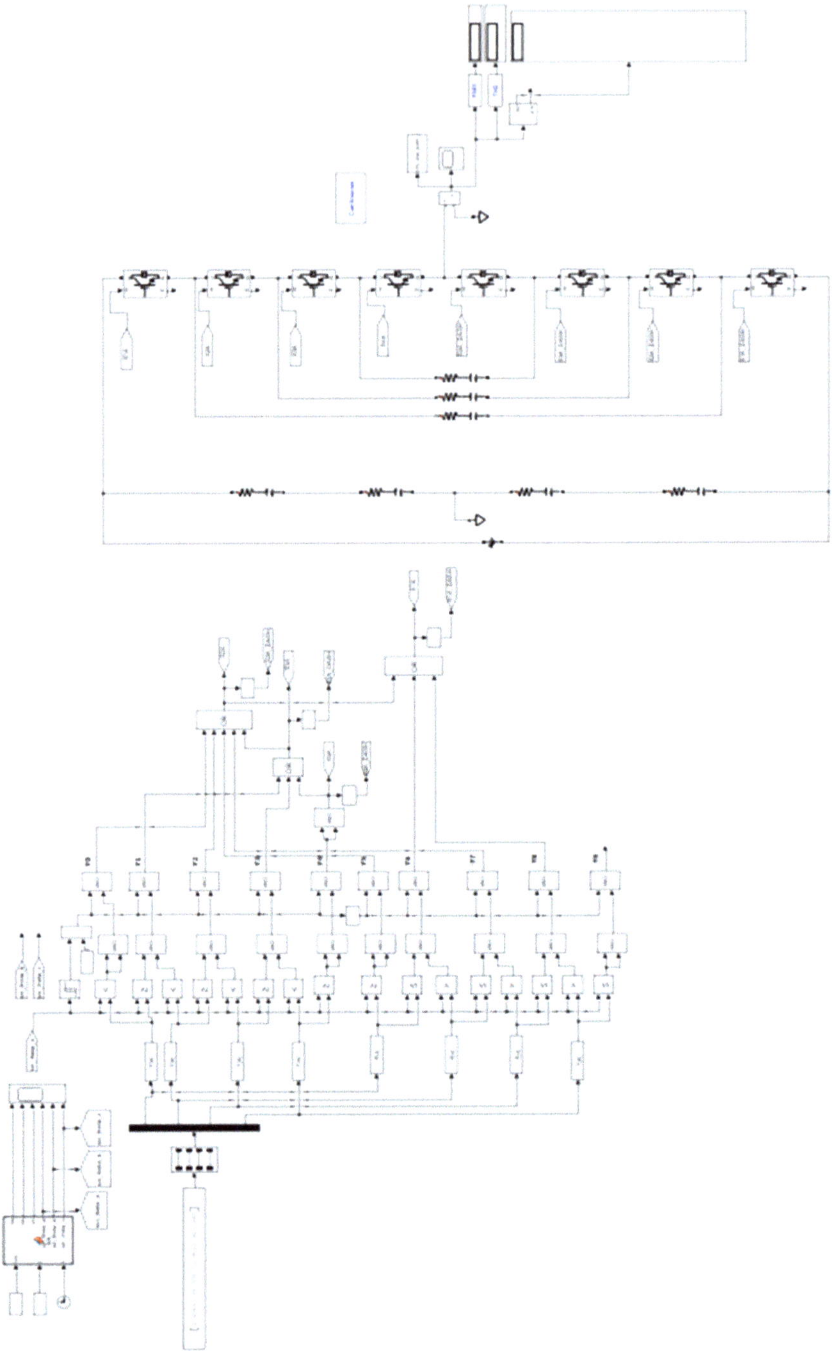

Fig. 3.62 Model of single-phase SHE-PWM FCFL inverter

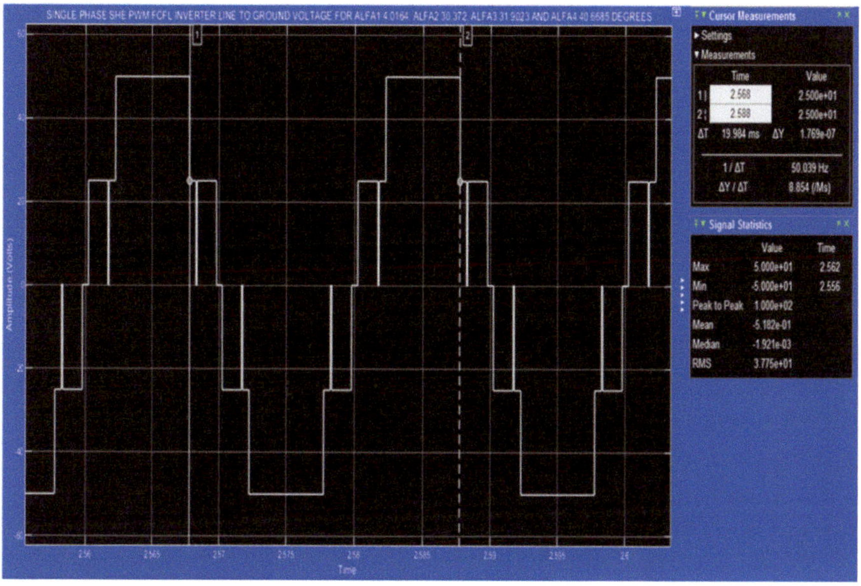

Fig. 3.63 Line-to-ground voltage waveform of a SHE-PWM FCFLI for alfa start group of 4.0164 degrees

Fig. 3.64 RMS, THD, fundamental 11th, 13th and 15th harmonic component of line-to-ground voltage of SHE-PWM FCFLI

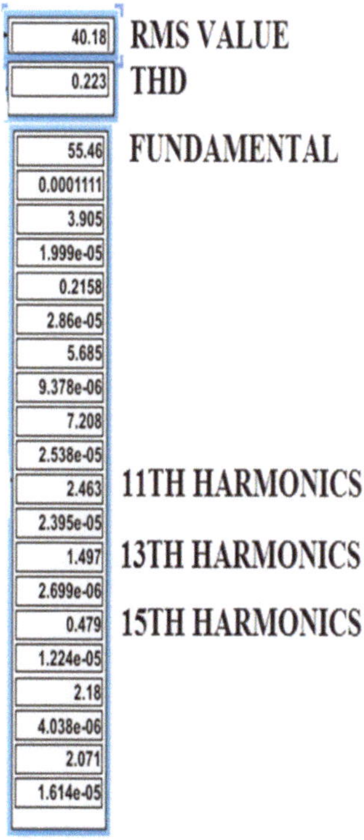

Table 3.11 Single-phase SHE-PWM FCFLI simulation results

Sl. no.	RMS value (V)	THD per unit	Fund. V (%)	11th Har. V (%)	13th Har. V (%)	15th Har. V (%)
1)	40.18	0.223	55.46 (100)	2.463 (4.44)	1.497 (2.7)	0.479 (0.863)

made with 1*sin(2*pi*f*t-2*π/3) for phase B and with 1*sin(2*pi*f*t + 2*π/3) for phase C in separate comparators, and the gate pulse for phase B and C switches is developed as shown in Figs. 3.57 and 3.62.

3.7 Conclusions

In this chapter, models for selective harmonic elimination of three-phase unipolar, bipolar, cascade H-bridge, DCTLI and FCTLI are presented. In all the above cases, 5th and 7th harmonic elimination from the line-to-ground voltage is considered. Simulation results indicate that the 5th and 7th harmonics are eliminated from the line-to-ground voltage successfully and that the peak fundamental line to ground voltage displayed by simulation closely agree with the value calculated using the formula for the given parameters. A real-time HIL simulation of the gate drive for the three-phase unipolar SHE-PWM inverter is also presented. Although there are several methods available for calculating the switching angles to eliminate a given set of harmonics from line-to-ground voltage, the method using "solve" command is presented here. Case studies of single-phase SHE-PWM DCFLI and FCFLI with four switching angles and two different wave shapes for line-to-ground voltage are presented from which it is concluded that the wave shape for DCFLI topology has a lower THD for line-to-ground voltage compared to that for FCFLI topology, with odd harmonics 11th, 13th and 15th eliminated or else minimised in both cases.

References

1. H.S. Patel and R.G. Hoft: "Generalized Techniques of Harmonic Elimination and Voltage Control in Thyristor Inverters: Part I-Harmonic Elimination", IEEE Transactions on Industry Applications, Vol. IA-9, No.3, pp. 310–317. May/June 1973.
2. H.S. Patel and R.G. Hoft: ""Generalized Techniques of Harmonic Elimination and Voltage Control in Thyristor Inverters: Part II—Voltage Control Technique", IEEE Transactions on Industry Applications, Vol.IA-10, No.5, pp. 666–673, September/October 1974.
3. P.N. Enjeti, P.D. Ziogas and J.F. Lindsay: "Programmed PWM Techniques to Eliminate Harmonics: A Critical Evaluation" IEEE Transactions on Industry Applications", vol. 26, No.2, pp. 302–316, March/April 1990.
4. J. Pontt, J. Rodriguez, R. Huerta and J. Pavez: "A Mitigation Method for Non-Eliminated Harmonics of SHEPWM Three-Level Multipulse Three-Phase Active Front End Converter", IEEE ISIE, Vol.1, pp. 258–263, 2003.

5. J. Pontt, J. Rodriguez, R. Huerta and J. Pavez: “Mitigatio-spiepr A3B2 h 0pt-n of Noneliminated Harmonics of SHEPWM Three-Level Multipulse Three-Phase Active Front End Converters with Low Switching Frequency for Meeting Standard IEEE-519–92”, IEEE PESC, pp. 531–536, 2003.
6. J. Pontt, J. Rodriguez and R. Huerta: “Mitigation of Noneliminated Harmonics of SHEPWM Three-Level Multipulse Three-Phase Active Front End Converters with Low Switching Frequency for Meeting Standard IEEE-519-92”, IEEE Transactions on Power Electronics, Vol. 19, No. 6, November 2004, pp. 1594–1600, November 2004.
7. V.G. Agelidis, A.I. Balouktsis and C. Cossar: “On Attaining the Multiple Solutions of Selective Harmonic Elimination PWM Three-Level Waveforms Through Function Minimization”, IEEE Transactions on Industrial Electronics, Vol. 55, No. 3, pp. 996–1004, March 2008.
8. J.N. Chiasson, L.M. Tolbert, K.J. McKenzie and Z. Du: “A Complete Solution to the Harmonic Elimination Problem”, IEEE Transactions on Power Electronics, Vol. 19, No.2, pp. 491–499, March 2004.
9. J.R. Espinoza, G. Joos, J. Guzman, L. Moran and R.P. Burgos: “Selective harmonic Elimination and Current/Voltage Control in Current/Voltage Source Topologies”, IEEE IECON, pp. 318–323, 1999.
10. J.R. Espinoza, G. Joos, J. Guzman, L. Moran and R.P. Burgos: “Selective harmonic Elimination and Current/Voltage Control in Current/Voltage Source Topologies”, IEEE Transactions on Industrial Electronics, Vol.48, No.1, pp. 71–81, February 2001.
11. The Mathworks Inc.: www.mathworks.com: “MATLAB/SIMULINK”, Version R2020a, February 2020.
12. Manitoba HVDC Research Center: “PSCAD V4.2,” 2006.
13. Powersim Inc.: “PSIM Demo V8.0”, 2008.
14. J. Chiasson, L. Tolbert, K. Mckenzie and Z. Du: “Eliminating Harmonics in a Multilevel Converter using Resultant Theory”, IEEE PESC, Cairns, Australia, June 2002, pp. 503–508.
15. Narayanaswamy. P.R.Iyer: “Model for A Three Phase Unipolar Selective Harmonic Elimination PWM Inverter”, Poster, Research Convezatione, School of Electrical and Information Engineering, The University of Sydney, Redfern, NSW, 2008.
16. P. Enjeti and J.F. Lindsay: “Solving Nonlinear Equations of Harmonic Elimination PWM In Power Control”, Electronics Letters, Vol. 23, No.12, June 1987. pp. 656–657.
17. V.G. Agelidis, A. Balouktsis I. Balouktsis and C. Cossar: “Multiple Sets of Solutions for Harmonic Elimination PWM Bipolar Waveforms:Analysis and Experimental Verfication”, IEEE Transactions on Power Electronics, Vol.21, No.2, March 2006, pp. 415–421.
18. Toshiji Kato: “Sequential Homotopy-Based Computation of Multiple Solutions for Selected Harmonic Elimination in PWM Inverters”, IEEE Transactions On Circuits And Systems—I: Fundamental Theory And Applications, Vol. 46, No. 5, May 1999, pp. 586–593.
19. Zhong Du, Leon M. Tolbert, John N. Chiasson, and Burak Özpineci: “A Cascade Multilevel Inverter Using a Single DC Source”, IEEE Conference Record, 2006, pp. 426–429.
20. John N. Chiasson, Leon M. Tolbert, Keith J. McKenzie and Zhong Du: “A Unified Approach to Solving the Harmonic Elimination Equations in Multilevel Converters’, IEEE Transactions on Power Electronics, Vol. 19, No. 2, March 2004, pp. 478–490.
21. H.L. Liu, G.H. Cho and S.S. Park: “Optimal PWM design for High Power Three-level Inverter Through Comparative Studies”, IEEE Transactions on Power Electronics, Vol.10, No.1, January 1995. pp. 38–40.
22. C. Feng, J. Liang and V.G. Agelidis: “Modified Phase Shifted PWM Control for Flying Capacitor Multilevel Converters”, IEEE Transactions on Power Electronics, Vol. 22, No.1, January 2007, pp. 178–185.
23. S. Khomfoi and L.M. Tolbert: “Multilevel Power Converters”, in Power Electronics Handbook, Editor: M.H. Rashid, Elsevier, 2011, Chapter 17, pp. 455–463.
24. Clark Hochgraf, Robert Lasseter, Deepak Divan and T.A. Lipo: “Comparison of Multilevel Inverters for Static VAR Compensation”; IEEE; 1994; pp. 921–928.

25. J.S. Lai and F.Z. Peng: "Multilevel convertors – A New Breed of Power Convertors"; IEEE Transactions on Industry Applications; Vol.32; No.3; May/June 1996; pp. 509–517.
26. F.Z. Peng, J.S. Lai and J.W. Mckeever and J. Van Coevering "A Multilevel Voltage Source Inverter with Separate DC Sources for Static Var Generation"; IEEE Transactions Industry Applications; Vol. 32; No. 5; Sept/Oct. 1996; pp. 1130–1133.
27. Jose Rodriguez, J.S. Lai and F.Z. Peng: "Multilevel Inverters: A Survey of Topologies Control and Applications", IEEE Transactions on Industrial Electronics Vol.49, No.4, August 2002, pp. 724–738.
28. Pradeep M. Bhagwat and V.R. Stefanovic: "Generalized Structure of a Multilevel PWM Inverter"; IEEE Transactions on Industry Applications; Vol.IA-19; No.6; November/December 1983; pp. 1057–1069.
29. Akira Nabae, Isao Takahashi and Hirofumi Akagi: "A Neutral-Point Clamped PWM Inverter", IEEE Transactions on Industry Applications, Vol.IA-17, No.5, September/October 1981, pp. 518–523.
30. Narayanaswamy P R Iyer: "Power Electronic Converters: Interactive Modelling Using Simulink", CRC Press, USA, 2018, Chapter 9, pp. 207–251.
31. Narayanaswamy P R Iyer: "MATLAB/Simulink Modules for Modeling and Simulation of Power Electronic Converters and Electric Drives", M.E. (Research) thesis, University of Technology Sydney, NSW, Australia, Chapter 12, 2006.

Chapter 4
Space Vector Pulse-Width Modulation of Three-Phase Two-Level and Multilevel Inverters

4.1 Introduction

The space vector pulse-width modulation (SVPWM) also known as space vector modulation (SVM) is considered as a better pulse-width modulation (PWM) technique as it has advantages over sine PWM in relation to better utilisation of dc bus voltage, reduced switching frequency for semiconductor switches and minimum torque ripples with induction motor (IM) load. The SVM is a vector technique and is found suitable for vector control method used in IM drives [1–5]. At present, modulation control strategies are implemented by digital technique using digital signal processor (DSP) and SVM is a suitable technique for digital implementation [2]. SVM is based on the polar representation of eight possible output voltages of a three-phase voltage source inverter (VSI) in a two-dimensional x-y plane [1–6]. This chapter presents the modelling of space vector-modulated three-phase two-level and three-phase diode-clamped three-level inverters (DCTLI). Also two case studies are presented one for the SVPWM of three-phase flying capacitor three-level inverter (FCTLI) and another for the SVM of three-phase two-level inverter to eliminate even order harmonics from line-to-line output voltage.

4.2 Space Vector Modulation of Three-Phase Two-Level Inverter

Figure 4.1 shows a three-phase two-level inverter. Such a three-phase inverter has six non-zero active states and two zero states. The six non-zero active state vectors V1 to V6 and the two zero vectors Vo and V7 at the centre are shown in the two axis

Supplementary Information The online version contains supplementary material available at https://doi.org/10.1007/978-3-031-62784-2_4.

N. P. R. Iyer, *Inverters and AC Drives*, Power Systems,
https://doi.org/10.1007/978-3-031-62784-2_4

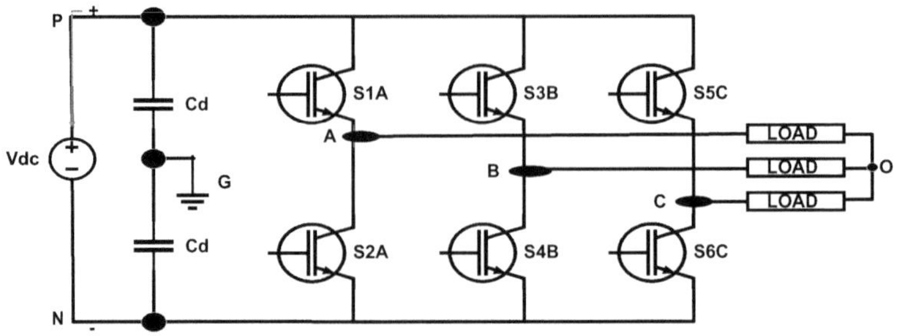

Fig. 4.1 Three phase two level inverter

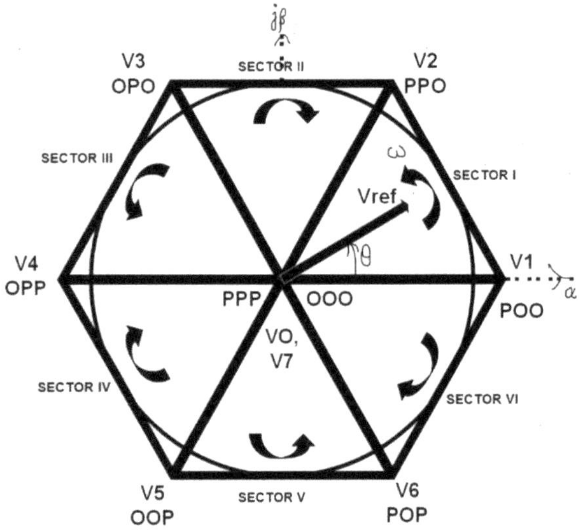

Fig. 4.2 Space vector diagram for three phase two level inverter

Table 4.1 Three-phase two-level inverter switching state

Sl. no.	S1A	S2A	V_{AN} (t)	S3B	S4B	V_{BN} (t)	S5C	S6C	V_{CN} (t)	Switch state	
1	1	0	+Vdc	1	0	+ Vdc	1	0	+ Vdc	p	P
2	0	1	0	0	1	0	0	1	0	O	O

1—Switch ON; 0—Switch OFF

α-β plane in Fig. 4.2. The operation of the two-level inverter in Fig. 4.1 can be explained by switching state defined in Table 4.1 [3].

In Table 4.1, when the upper switch in phase A, B or C is ON, the inverter output voltage V_{AN}, V_{BN} or V_{CN} is +Vdc and the corresponding switch state is denoted by P, and when the lower switch in phase A, B or C is ON, the inverter output voltage V_{AN}, V_{BN} or V_{CN} is zero, and the corresponding switch state is denoted by O. Thus, there are six rotating active space vectors V1 to V6 designated as P O O, P P O, O P O, O

P P, O O P and P O P and two stationary space vectors VO and V7 designated as O O O and P P P. The three instantaneous load phase voltages V_{AO}(t), V_{BO}(t) and V_{CO}(t) of inverter can be represented in the two-dimensional α-β axis by suitable transformation matrix defined in Eq. 4.1 [3].

$$\begin{bmatrix} V_\alpha \\ V_\beta \end{bmatrix} = \frac{2}{3} * \begin{bmatrix} 1 & \frac{-1}{2} & \frac{-1}{2} \\ 0 & \frac{\sqrt{3}}{2} & \frac{-\sqrt{3}}{2} \end{bmatrix} * \begin{bmatrix} V_{AO}(t) \\ V_{BO}(t) \\ V_{CO}(t) \end{bmatrix} \tag{4.1}$$

A voltage space vector *V(t)* in the *α-β* plane can be expressed as follows:

$$\overrightarrow{V(t)} = V_\alpha(t) + jV_\beta(t) \tag{4.2}$$

Using Eqs. 4.1 and 4.2, *V(t)* can be expressed as follows:

$$\overrightarrow{V(t)} = \frac{2}{3} * \left[V_{AO}(t) + \ a.V_{BO}(t) + \ a^2.V_{CO}(t)\right] \tag{4.3}$$

In Eq. 4.3, a is 1/_120 or e^{j120} and a^2 is 1/_240 or e^{j240}.
Also referring to Fig. 4.1, we have the following:

$$\begin{aligned} V_{AO}(t) &= V_{AN}(t) + V_{ON}(t) \\ V_{BO}(t) &= V_{BN}(t) + V_{ON}(t) \\ V_{CO}(t) &= V_{CN}(t) + V_{ON}(t) \end{aligned} \tag{4.4}$$

Using Eq.4.4 in 4.3, noting that $(1 + a + a^2 = 0)$ and simplifying, we have the following:

$$\overrightarrow{V(t)} = \frac{2}{3} * \left[V_{AN}(t) + \ a.V_{BN}(t) + \ a^2.V_{CN}(t)\right] \tag{4.5}$$

Thus Eq.4.1 can also be expressed as follows:

$$\begin{bmatrix} V_\alpha \\ V_\beta \end{bmatrix} = \frac{2}{3} * \begin{bmatrix} 1 & \frac{-1}{2} & \frac{-1}{2} \\ 0 & \frac{\sqrt{3}}{2} & \frac{-\sqrt{3}}{2} \end{bmatrix} * \begin{bmatrix} V_{AN}(t) \\ V_{BN}(t) \\ V_{CN}(t) \end{bmatrix} \tag{4.6}$$

Voltages $V_{AN}(t)$, $V_{BN}(t)$ and $V_{CN}(t)$ are defined in Table 4.1.

Now assume that in Fig. 4.1, switches S1A, S4B and S5C are only ON. This corresponds to the voltage vector V6 in Fig. 4.2 and switch state P O P. Using the voltage values in Table 4.1, Eq. 4.6 can be expressed as follows:

Table 4.2 Space vector table for three-phase two-level inverter

Sl. no.	Space vector	Switch state	ON switches	Vector value
1	*VO, V7*	O O O P P P	S2A, S4B, S6C S1A, S3B, S5C	$VO = V7 = 0$
2	*V1*	P O O	S1A, S4B, S6C	$V1 = \frac{2}{3} * V_{dc}e^{j0}$
3	*V2*	P P O	S1A, S3B, S6C	$V2 = \frac{2}{3} * V_{dc}e^{\frac{j\pi}{3}}$
4	*V3*	O P O	S2A, S3B, S6C	$V3 = \frac{2}{3} * V_{dc}e^{\frac{j2\pi}{3}}$
5	*V4*	O P P	S2A, S3B, S5B	$V4 = \frac{2}{3} * V_{dc}e^{j\pi}$
6	*V5*	O O P	S2A, S4B, S5C	$V5 = \frac{2}{3} * V_{dc}e^{\frac{j4\pi}{3}}$
7	*V6*	P O P	S1A, S4B, S5C	$V6 = \frac{2}{3} * V_{dc}e^{\frac{j5\pi}{3}}$

$$\begin{bmatrix} V_\alpha \\ V_\beta \end{bmatrix} = \frac{2}{3} * \begin{bmatrix} 1 & \frac{-1}{2} & \frac{-1}{2} \\ 0 & \frac{\sqrt{3}}{2} & \frac{-\sqrt{3}}{2} \end{bmatrix} * \begin{bmatrix} V_{dc} \\ 0 \\ V_{dc} \end{bmatrix} \tag{4.7}$$

From Eq.4.7,

$$\vec{V}(t) = \overrightarrow{V_6} = V_\alpha + jV_\beta = \frac{2}{3} * V_{dc} * \left(\frac{1}{2} - \frac{j\sqrt{3}}{2}\right) = \frac{2}{3} * V_{dc}e^{\frac{j5\pi}{3}} \tag{4.8}$$

The voltage space vector *V(t)* for all the remaining switching combination can be similarly calculated. This space vector table is shown in Table 4.2.

The active space vectors *V1* to *V6* and the zero vectors *VO* and *V7* do not rotate in space as their magnitude and position are fixed as shown in Table 4.2. However, V_{ref} shown in Fig. 4.2 rotates at an angular speed ω rad/sec which is 2*π*f where f is the fundamental frequency of the three-phase inverter output voltage. The angle θ made by V_{ref} to the horizontal α-axis can be expressed as follows:

$$\theta(t) = \int_0^t \omega(t).dt + \theta(0) \tag{4.9}$$

There are six sectors from I to VI at 60-degree intervals as shown in Fig. 4.2. Thus, V_{ref} can be synthesised by two nearby vectors based on which switch states can be selected and gate pulse can be generated for active switches. As V_{ref} rotates and passes through different sectors from I to VI, different switching combinations can be selected in such a way that the output voltage level transition is minimum. When V_{ref} completes one revolution, this corresponds to one-time period, (1/f) seconds or one cycle for the inverter output voltage. Magnitude of the inverter output voltage can be controlled by varying the magnitude or length of V_{ref}.

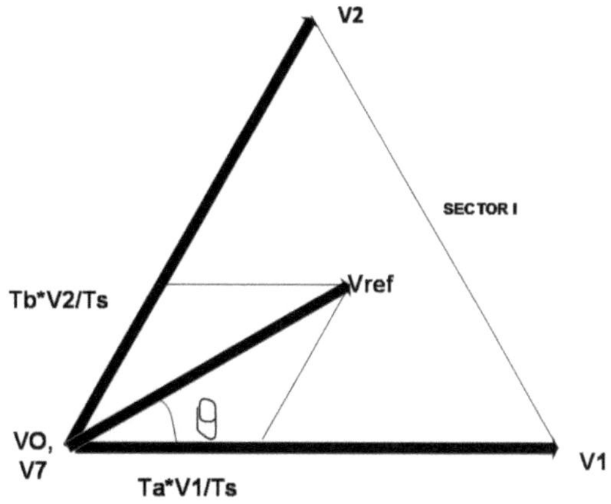

Fig. 4.3 Synthesis of Vref along V1 and V2 vector

The reference voltage V_{ref} can be synthesised by three stationary voltage vectors. The dwell time for stationary vectors represents the ON time or OFF time of the selected switches during the sampling time Ts seconds. The dwell time calculation is based on volt-second principle in which V_{ref} multiplied by Ts is equal to the sum of the individual voltage space vector multiplied by its chosen time interval. Assuming V_{ref} is constant during the small sampling time Ts, it can be synthesised along two active space vectors and one zero vector. Assume V_{ref} lies in Sector I as shown in Fig. 4.3. This can be synthesised along V_1, V_2 and V_O as shown in Fig. 4.3. The volt-second equation can be expressed as follows:

$$\vec{V}_1 * T_a + \vec{V}_2 * T_b + \vec{V}_O * T_O = \vec{V}_{ref} * T_S \tag{4.10}$$

$$T_a + T_b + T_O = T_S \tag{4.11}$$

where T_a, T_b and T_O are the dwell time for space vectors V_1, V_2 and V_O, respectively. Referring to Fig. 4.3, $\vec{V}_{ref}$ can be expressed as $V_{ref}{*}e^{j\theta}$, and from Table 4.2, V_1, V_2 and V_O can be obtained. Substituting these values in Eq. 4.10 and resolving along the real α-axis and imaginary β-axis, we have the following:

$$\frac{2}{3} * V_{dc} * T_a + \frac{2}{3} * V_{dc} * T_b * \cos\left(\frac{\pi}{3}\right) = V_{ref} * \cos(\theta) * T_S \tag{4.12}$$

$$\frac{2}{3} * V_{dc} * T_b * \sin\left(\frac{\pi}{3}\right) = V_{ref} * \sin(\theta) * T_S$$

Solving Eq. 4.12 and using Eq. 4.11, we have the following results:

$$T_a = \frac{\sqrt{3} * V_{ref} * T_S}{V_{dc}} * \sin\left(\frac{\pi}{3} - \theta\right) \tag{4.13}$$

$$T_b = \frac{\sqrt{3} * V_{ref} * T_S}{V_{dc}} * \sin(\theta) \tag{4.14}$$

$$T_O = (T_S - T_a - T_b) \tag{4.15}$$

From Eq. 4.13 and 4.14, it is seen that if θ is π/6 radians, T_a and T_b are equal. If θ is zero radians, T_b is zero, and when θ is π/3 radians, T_a is zero. Thus, T_a is larger than T_b when θ is less than $\pi/6$ radians and vice versa.

For sector II, timing T_a, T_b and T_O are associated with space vectors V2, V3 and VO, respectively, and so on. Also from the absolute value of angle θ, multiple of π/3 radians must be subtracted to get the actual angle θ_dash for the respective sector. This value of θ_dash is defined below:

$$\theta_{dash} = \theta - (k-1) * \frac{\pi}{3} \tag{4.16}$$

where $k = 1,2,3,4,5,6$ for sectors I to VI, respectively, for $0 <= \theta < \pi/3$ radians.

Equation 4.13 and 4.14 can also be expressed in terms of the modulation index m_a as follows:

$$T_a = T_S * m_a * \sin\left(\frac{\pi}{3} - \theta\right) \tag{4.17}$$

$$T_b = T_S * m_a * \sin(\theta) \tag{4.18}$$

$$m_a = \frac{\sqrt{3} * V_{ref}}{V_{dc}} \tag{4.19}$$

where m_a is the modulation index. Also referring to Fig. 4.2 and 4.3, it is clear that the maximum value of V_{ref} can be expressed as follows:

$$V_{ref\_max} = \frac{2 * V_{dc}}{3} * \sin\left(\frac{\pi}{3}\right) = \frac{V_{dc}}{\sqrt{3}} \tag{4.20}$$

Using Eq. 4.20 in 4.19, the modulation index m_a lies in the range 0 to 1. The RMS value of the maximum attainable fundamental component of line-to-line voltage by SVM method can be expressed as follows [3]:

$$V_{LL1\_RMS} = \sqrt{3} * \frac{V_{dc}}{\sqrt{3}} * \frac{1}{\sqrt{2}} = 0.707 * V_{dc} \tag{4.21}$$

4.2.1 Selection of Switching Sequence

For a given V_{ref}, the switching sequence selected must satisfy (a) the transition from one switching state to another state within a given sector should involve only one switch ON and the other OFF in any one inverter leg; and (b) as V_{ref} passes from one sector to another one, there should be no or minimum number of switching. For the

Table 4.3 Seven segment switching scheme for SVM of three-phase two-level inverter

Sl. no.	Sector number	Space vector switch state timing	Space vector switch state timing	Space vector switch state timing	Space vector switch state timing	Space vector switch state timing	Space vector switch state timing	Space vector switch state timing
1	I	*VO* O O O $T_O/4$	*V1* P O O Ta/2	*V2* P P O Tb/2	*V7* P P P $T_O/2$	*V2* P P O Tb/2	*V1* P O O Ta/2	*VO* O O O $T_O/4$
2	II	*VO* O O O $T_O/4$	*V3* O P O Tb/2	*V2* P P O Ta/2	*V7* P P P $T_O/2$	*V2* P P O Ta/2	*V3* O P O Tb/2	*VO* O O O $T_O/4$
3	III	*VO* O O O $T_O/4$	*V3* O P O Ta/2	*V4* O P P Tb/2	*V7* P P P $T_O/2$	*V4* O P P Tb/2	*V3* O P O Ta/2	*VO* O O O $T_O/4$
4	IV	*VO* O O O $T_O/4$	*V5* O O P Tb/2	*V4* O P P Ta/2	*V7* P P P $T_O/2$	*V4* O P P Ta/2	*V5* O O P Tb/2	*VO* O O O $T_O/4$
5	V	*VO* O O O $T_O/4$	*V5* O O P Ta/2	*V6* P O P Tb/2	*V7* P P P $T_O/2$	*V6* P O P Tb/2	*V5* O O P Ta/2	*VO* O O O $T_O/4$
6	VI	*VO* O O O $T_O/4$	*V1* P O O Tb/2	*V6* P O P Ta/2	*V7* P P P $T_O/2$	*V6* P O P Ta/2	*V1* P O O Tb/2	*VO* O O O $T_O/4$

seven-segment switching scheme, sample time T_S is divided into seven segments for the selected voltage vectors [3]. Three, four and five-segment switching schemes are also available [2]. A typical seven-segment switching scheme for SVM of three-phase two-level inverter is tabulated in Table 4.3. From Table 4.3, it is seen that for all odd sectors, the timing sequence is To/4 → Ta/2 → Tb/2 → To/2 → Tb/2 → Ta/2 → To/4, and for all even sectors, it is To/4 → Tb/2 → Ta/2 → To/2 → Ta/2 → Tb/2 → To/4.

To segregate the individual timings Ta, Tb and To from Ts as per pattern shown in Table 4.3, a triangle carrier with period Ts seconds, peak value Ts/2 volts and minimum value zero are used.

This is shown in Figs. 4.4a and 4.4b for sectors I and II. The method shown for sector I is applicable for sectors III and V and that for sector II is applicable for sector IV and VI, where the respective switch state for each sector must be entered from Table 4.3. In Fig. 4.4a, the timing TO/4 is compared with the triangle carrier in a comparator whose output is HIGH when triangle carrier is less than TO/4 or else its output is LOW which is marked A1. Similarly, the cumulative timings (TO/4 + Ta/2), (TO/4 + Ta/2 + Tb/2) and (TO/4 + Ta/2 + Tb/2 + TO/4) are compared with the triangle carrier in separate comparators, and the outputs B1, C1 and D1 are obtained in the same way as for A1. This is applicable to all odd sectors I, III and V. For even sectors II, IV and VI, the order of comparison with triangle carrier changes to TO/4, (TO/4 + Tb/2), (TO/4 + Tb/2 + Ta/2) and (TO/4 + Tb/2 + Ta/2 + TO/4) and the outputs A2, B2, C2 and D2 are obtained in the same way as for A1. The individual switch state dwell time can be obtained as follows:

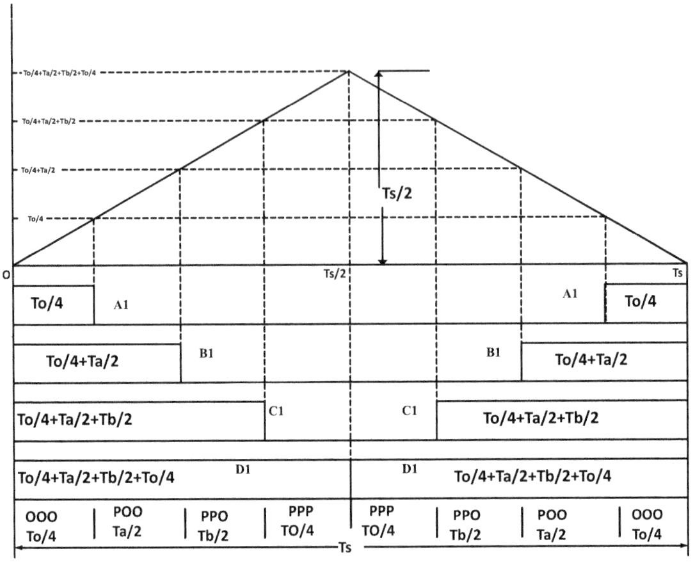

Fig. 4.4a Three phase two level inverter SVM—Seven segment switching scheme for vref in sector I

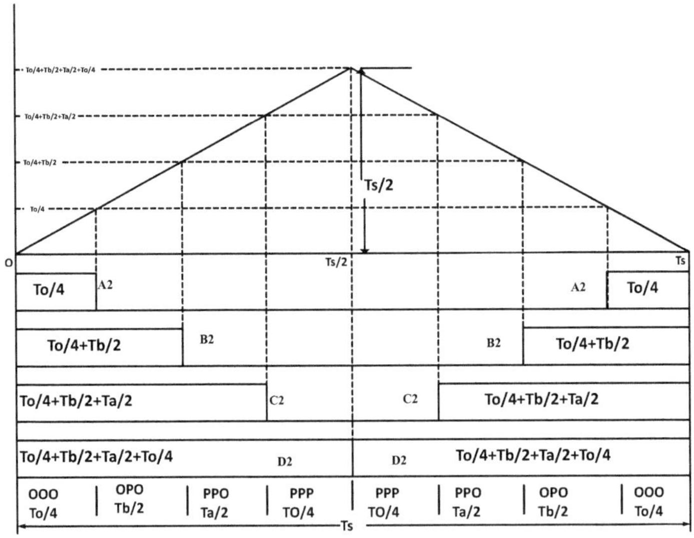

Fig. 4.4b Three phase two level inverter SVM—seven segment switching scheme for Vref in Sector II

$$\left.\begin{aligned} A1 \cap A1 &= \frac{T_O}{4} \\ A1 \varnothing B1 &= \frac{T_a}{2} \\ B1 \varnothing C1 &= \frac{T_b}{2} \\ C1 \varnothing D1 &= \frac{T_O}{4} \end{aligned}\right\} \tag{4.22}$$

$$\left.\begin{aligned} A2 \cap A2 &= \frac{T_O}{4} \\ A2 \varnothing B2 &= \frac{T_b}{2} \\ B2 \varnothing C2 &= \frac{T_a}{2} \\ C2 \varnothing D2 &= \frac{T_O}{4} \end{aligned}\right\} \tag{4.23}$$

In Eqs. 4.22 and 4.23, symbols $\cap$ and $\varnothing$ represent the logical AND and EXCLUSIVE OR operation, respectively. This provides a method to separate the individual dwell time of the inverter switches.

4.2.2 Modelling of Space Vector Modulation of a Three-Phase Two-Level Inverter

The model of SVM of a three-phase two-level inverter developed using SIMULINK [7] is shown in Fig. 4.5 (Model file: EXAMPLE 4_1). The model subsystems are shown in Figs. 4.6a, 4.6b, 4.6c and 4.6d. In this model, seven segment switching sequence shown in Table 4.3 is used. The model parameters are shown in Table 1.1. The space vector and its timing have to be arranged in the order shown in Table 4.3 for odd and even sectors. This is done using the triangle carrier generator whose period is Ts seconds, peak value Ts/2 Volts and minimum value zero volts and using AND and EXCLUSIVE OR logic gates, as explained in Sect. 4.2.1. The model subsystems are explained below:

4.2.2.1 Gate Pulse Timing and Triangle Carrier Generator

This has two Embedded MATLAB function blocks and one triangle carrier generator block as shown in Fig. 4.6a. The first Embedded MATLAB function block has peak input voltage Vim Volts, output frequency of inverter f Hz and time as the input. Although any suitable value of Vim can be used, here V_{ref_max} which is (Vdc/√3) Volts is used. Three-phase output voltages va, vb and vc are generated with peak value Vim and frequency f Hz. This three-phase output voltages are

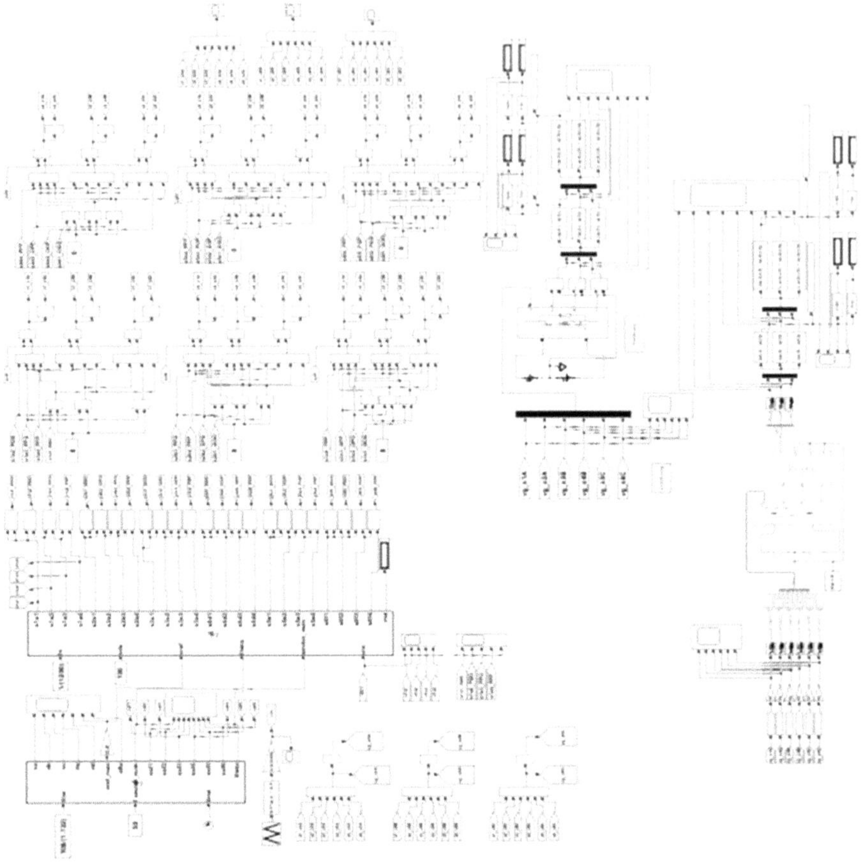

Fig. 4.5 Model of space vector modulation of a three phase two level invertER

resolved into d-q axis components vq and vd using the relation shown in Eq.4.6. Also V_{ref_max} is once again calculated using the relation $V_{ref\_max} = \sqrt{V_q^2 + V_d^2}$. The absolute value of the angle α radians which V_{ref_max} makes with horizontal d-axis is calculated using the relation atan2(vq,vd). The desired V_{ref} value is obtained by multiplying V_{ref_max} with modulation index ma using a gain block shown in Fig. 4.6a. Referring to Fig. 4.2, as V_{ref} pass through sectors I to VI, sector_num 1 to 6 is assigned. When V_{ref} passes through sector I to VI, sector_num goes from 1 to 6, sector switch function ssf1 to ssf6, respectively, goes HIGH and remains LOW otherwise. Also the angle θ which V_{ref} makes within each sector is calculated using Eq.4.16, where θ_dash is replaced by θ and θ is replaced by α. For example, if α lies within the range zero to π/3, sector_num is set to 1, θ is calculated using Eq.4.16, ssf1 is set HIGH (logic 1), or else ssf1 is LOW (logic 0). This is illustrated in Program Segment 4.1 in the model file EXAMPLE 4_1.

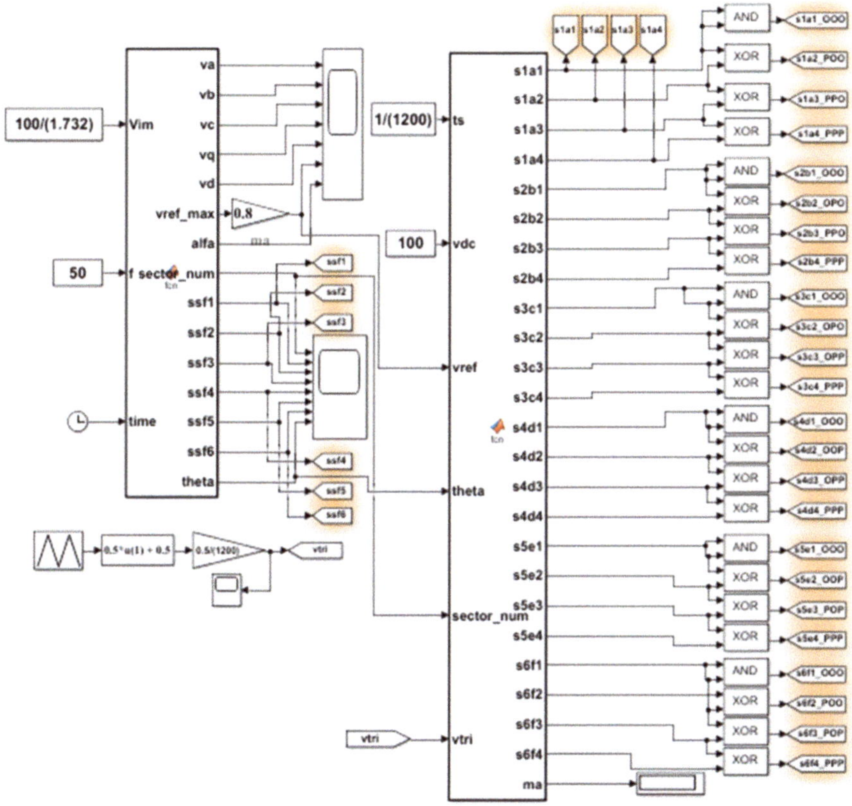

Fig. 4.6a Three phase two level VSI SVM—gale pulse timing arid triangle carrier generator

In the second Embedded MATLAB function, sample time ts, DC link voltage Vdc, reference voltage V_{ref}, sector number sector_num, angular position of V_{ref} within each sector theta and the triangle carrier vtri are given as inputs. Modulation index ma is calculated using Eq. 4.19. For each sector_num from 1 to 6, timing values for ta, tb and t0 are calculated from ts and theta for the respective sector given as inputs, using Eq. 4.17, 4.18 and 4.15. Then the cumulative sector timing magnitudes are calculated for odd and even sectors as in Table 4.3, Figs. 4.4a and 4.4b. These cumulative sector timing magnitudes are then compared with vtri using if-then-else statement and the square pulse output whose pulse width represents the cumulative timings for odd sectors I, III, V and for even sectors II, IV and VI, as in Figs. 4.4a and 4.4b are obtained. For example, in Fig. 4.6a, timing s1a2 for sector I is HIGH if vtri is less than or equal to (0.25*t0 + 0.5*ta), or else s1a2 is LOW. Similarly for sector II, timing s2b2 is HIGH if vtri is less than or equal to (0.25*t0 + 0.5*tb), or else s2b2 is LOW. This is shown in Program Segment 4.2 in the model file EXAMPLE 4_1.

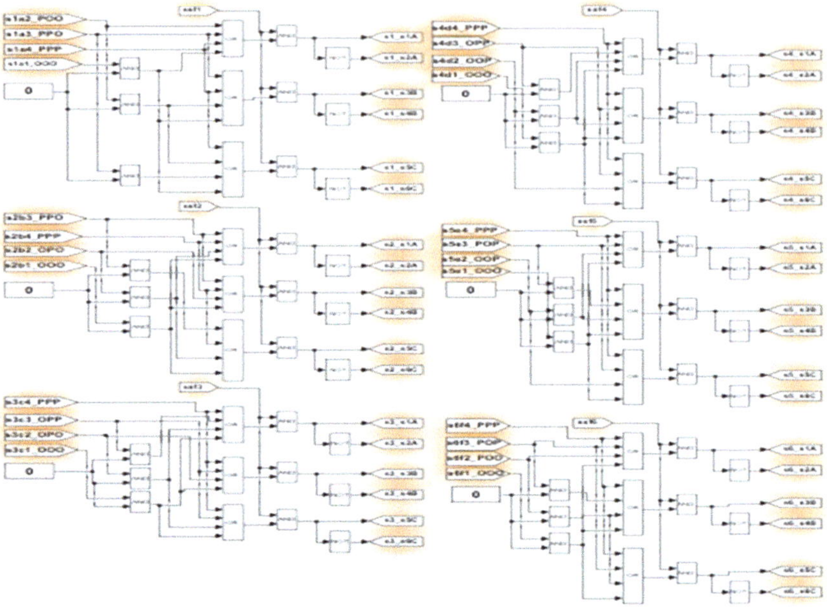

Fig. 4.6b Three phase two level VSI SVM—gate pulse timing sectorwise

Fig. 4.6c Three phase two level VSI SVM—Gate pulse for individual upper and lower switches

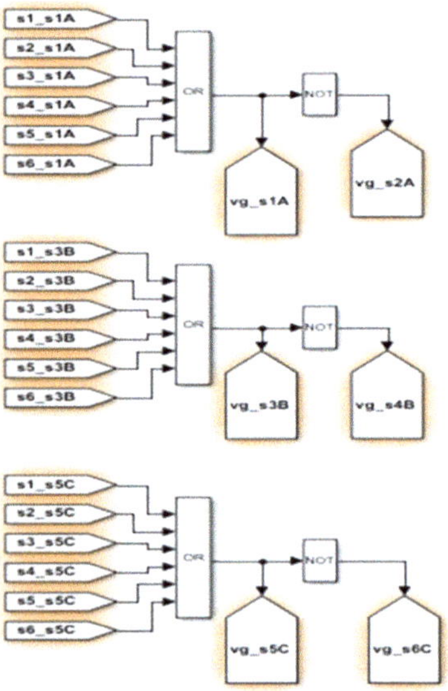

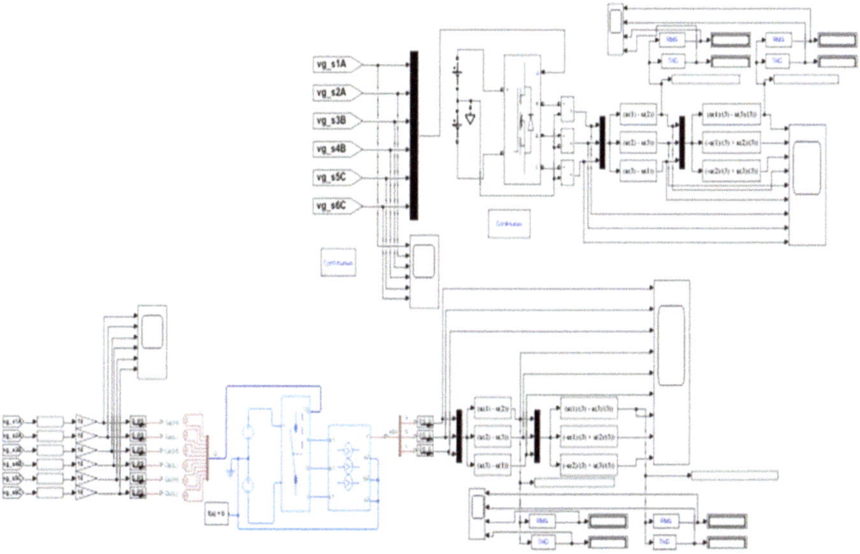

Fig. 4.6d Three phase two level VSI SVM—Modal of VSI using power system specialized technology power electronics components (top) and power system simscape semiconductor components (bottom)

The cumulative sector timings obtained for odd and even sectors are then given to AND and EXCLUSIVE OR gates as in Eqs. $_{4.22}$ and $_{4.23}$, respectively, and the individual timings ta, tb and t0 are obtained for all six sectors. This is shown in Fig. 4.6a. The triangle carrier generator generates the triangle carrier with period Ts (1/1200) seconds and peak value +/−1 Volt. This is then multiplied using a function block with constant (0.5*u(1) + 0.5) and then by another gain block with multiplication constant $0.5*T_S$ (0.5/1200) to obtain triangle carrier as specified in Sect. 4.2.2. This is shown in Fig. 4.6a.

4.2.2.2 Gate Pulse Timing Sectorwise

The individual gate pulse timings from Fig. 4.6a for the switches of the three-phase VSI shown in Fig. 4.1 have to be distributed to the six switches sectorwise according to switch state P or O. This is shown in Fig. 4.6b. Consider sector I when ssf1 is HIGH and ssf2 to ssf6 are all LOW. From Fig. 4.6b, switch timing sequence in order is s1a1_OOO (To/4 Sec), s1a2_POO (Ta/2 Sec), s1a3_PPO (Tb/2 Sec) and s1a4_PPP (To/4 Sec), respectively. For the switch S1A in Fig. 4.1, switch state in order is O (To/4 Sec), P (Ta/2 Sec), P (Tb/2 Sec) and P (To/4 Sec), respectively. To obtain switch state O from the respective switch timing s1a1_OOO, the s1a1_OOO is ANDed with constant block with input value zero. The output state O from this AND gate along with s1a2_POO, s1a3_PPO and s1a4_PPP directly are applied to a four input OR gate. The output of this four input OR gate is ANDed with ssf1 pulse using a two input AND gate and the resulting output of this AND gate is dwell time for switch S1A and its inverted output using NOT gate forms the dwell time for the switch S2A when V_{ref} is in Sector I. These outputs are marked s1_s1A and s1_s2A, respectively. Similarly for the switch S3B and S4B, s1a1_OOO and s1a2_POO are each ANDed with zero using two input AND gate, and these two outputs along with s1a3_PPO and s1a4_PPP directly are applied to the four input OR gate. For the switch S5C and S6C, s1a1_OOO, s1a2_POO and s1a3_PPO are ANDed with zero using three separate two input AND gates, and these three outputs along with s1a4_PPP directly are applied to the four input OR gate. The remaining procedure is the same as for S1A and S2A. The gate pulse for switches in phase B and C are marked s1_s3B, s1_s4B, s1_s5C and s1_s6C, respectively. This procedure for the remaining sectors II to VI is repeated, and the dwell timings for the six switches of the three-phase VSI shown in Fig. 4.1 are obtained sectorwise. This is shown in Fig. 4.6b.

4.2.2.3 Gate Pulse for Individual Upper and Lower Switches of Three-Phase Two-Level Inverter

Here gate pulse timing for the individual upper switches S1A, S3B and S5C in Fig. 4.1 is obtained as explained above. The gate pulse for the lower switches S2A, S4B and S6C is obtained by inverting the corresponding upper switch gate pulse using NOT gate. To obtain the upper switch gate pulse vg_s1A for the switch S1A, s1_s1A, s2_s1A, s3_s1A, s4_s1A, s5_s1A and s6_s1A obtained for all the six sectors are ORed together using six-input OR gate, and this OR gate output gives vg_s1A. This vg_s1A is inverted using a NOT gate, and the resulting output is vg_s2A for the lower switch in phase A shown in Fig. 4.1. Similar procedure is adopted to derive gate pulse for the upper and lower switches in phase B and C shown in Fig. 4.1. This is shown in Fig. 4.6c.

4.2.2.4 Three-Phase Two-Level Voltage Source Inverter

Here models for three-phase VSI are developed (1) using power systems specialised technology power electronics components block set and (2) power systems Simscape semiconductor components block set. For the first one using power electronics components block set, three-phase inverter is from Universal Bridge block which forms the three-phase inverter. The six gate pulse from vg_s1A to vg_s6C is given to a six input MUX, the output of which is given to the gate input of Universal Bridge. The DC link voltage using two battery sources in series form 100 volts, the midpoint of which is grounded. The line-to-line voltage and line-to-neutral voltages are obtained using Fcn blocks, as given in Eqs. 1.3 and 1.4.

For the model using power system Simscape semiconductor components, six-pulse gate multiplexer and converter blocks are used. The converter block forms the three-phase inverter. The six gate pulse vg_s1A to vg_s6C are given to gain blocks with multiplication constant ten so that the gate pulse amplitude will be greater than the semiconductor switch gate threshold voltage. The output of these six

Table 4.4 Three-phase two-level space vector PWM inverter simulation results

Sl. No	Ma	VLL(rms) V	VLL1(rms) V	VLN(rms) V	VLN1(rms) V	THD of VLL	THD of VLN
1	0.2	35.7695	14.2128	20.5962	8.1600	2.3095	2.3175
2	0.4	50.1649	28.0085	29.1630	16.3978	1.4859	1.4707
3	0.6	61.9689	42.6668	35.8234	24.6639	1.0533	1.0534
4	0.8	70.8146	55.9675	40.82	32.3148	0.7752	0.7718
5	1.0	79.8184	70.7814	46.1036	40.8637	0.5212	0.5224

gain blocks forms the gate drive which is applied to the gate terminal of the converter through the six-pulse gate multiplexer. Line-to-line and line-to-neutral voltages are obtained as per Eqs. 1.3 and 1.4. Two DC voltage sources are connected in series to form the DC link voltage of 100 volts. The two VSI models are shown in Fig. 4.6d.

4.3 Simulation Results

The simulation of the SVM of three-phase two-level VSI is carried out using ode23tb (stiff/TR-BDF2) solver [7]. The data shown in Table 1.1 are used. Simulation results are tabulated in Table 4.4. The simulation results of line-to-line and line-to-neutral voltage for Ma of 0.8 and Mf of 24 p.u. are also shown in Fig. 4.7a–m. The harmonic spectrum of line-to-line voltage VAB and line-to-neutral voltage VAO for Ma of 0.8 and Mf of 24 p.u. are shown in Fig. 4.8a–d.

4.4 Discussion of Results

The model of space vector PWM three-phase two-level VSI is presented using seven-segment switching scheme. The simulation results for Ma of 0.8 are shown in Fig. 4.7 and 4.8, and for all other values for Ma are tabulated in Table 4.4. For Ma of 1.0, it is seen from Table 4.4 that the RMS value of fundamental component of line-to-line voltage is 70.7814 volts which well agrees with that obtained using Eq. 4.21.

4.5 Space Vector Modulation of Three-Phase Diode-Clamped Three-Level Inverter

The three phase Diode Clamped Three Level Inverter (DCTLI) schematic also known as Neutral Point Clamped (NPC) converter is shown in Fig. 4.9. The semiconductor switches S1A to S4C can be either NPN BJT, IGBT or NMOSFET. The two diodes connected to the junction of S1A-S2A and S3A-S4A form the

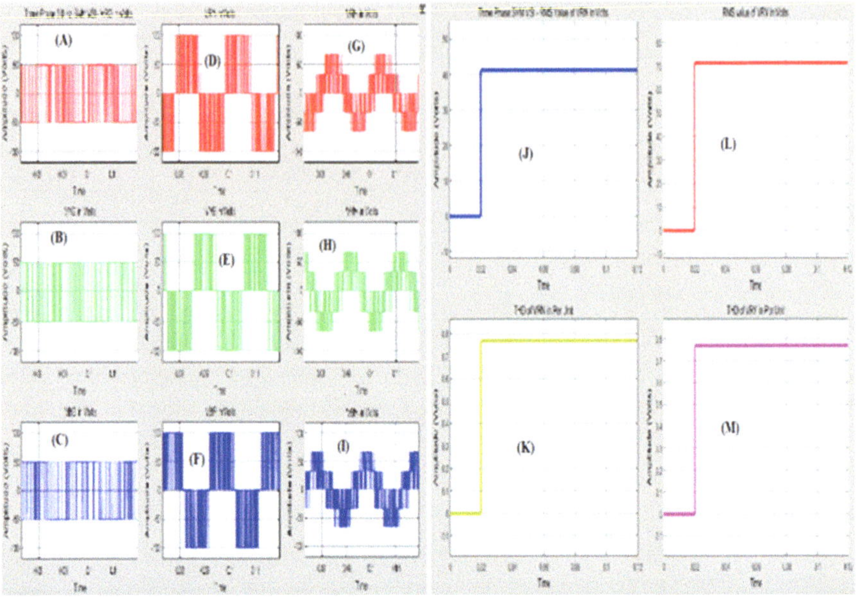

Fig. 4.7 Three phase space vector modulated inverter simulation results: (**a**–**c**) line to ground voltage VRG, VYG and VBG (**d**–**f**) line to line voltage VRY, VYB and VBR (**g**–**i**) line to neutral voltage VRN, VYN and VBN (**j**, **k**) RMS value and THD of VRN. (**l**, **m**) RMS value and THD of YRY

clamping diodes. Similar clamping diodes are connected to phase B and C. The switch state for phase A of Fig. 4.9 is tabulated in Table 4.5. Similar table applies to switches in phase B and C. From Table 4.5, it is seen that the line-to-ground voltage $V_{AG}(t)$ has three levels +Vdc/2, zero and –Vdc/2 volts. The line-to-ground voltage in phase B and C differ by a phase angle of 2π/3 and 4π/3 radians from that of phase A. The line-to-line voltage $V_{AB}(t)$ has five levels +Vdc, +Vdc/2, zero, –Vdc/2 and –Vdc, respectively.

The space vector diagram showing division of sectors and regions for a three-phase NPC inverter is shown in Fig. 4.10. In Fig. 4.10, there are zero vector Vo, small vectors V1 to V6, medium vectors V7 to v12 and large vectors V13 to V18. The respective switch states of zero, small, medium and large vectors are marked in Fig. 4.10. Thus, O O O, P P P, N N N form the zero vector; P O O, O N N, P P O, O O N, O P O, N O N, O P P, N O O, O O P, N N O, P O P, O N O form the small vectors; P O N, O P N, N P O, N O P, O N P, P N O form the medium vectors; and P N N, P P N, N P N, N P P, N N P and P N P form the large vectors.

The three-phase instantaneous load voltage of NPC inverter $V_{AO}(t)$, $V_{BO}(t)$ and $V_{CO}(t)$ can be transformed to α-β axis by suitable transformation matrix. The analysis shown in Eqs. 4.1 to 4.6 applies here as well. Thus selecting small vector O O N for V2, P O N for V7 and N P N for V15, we have the following:

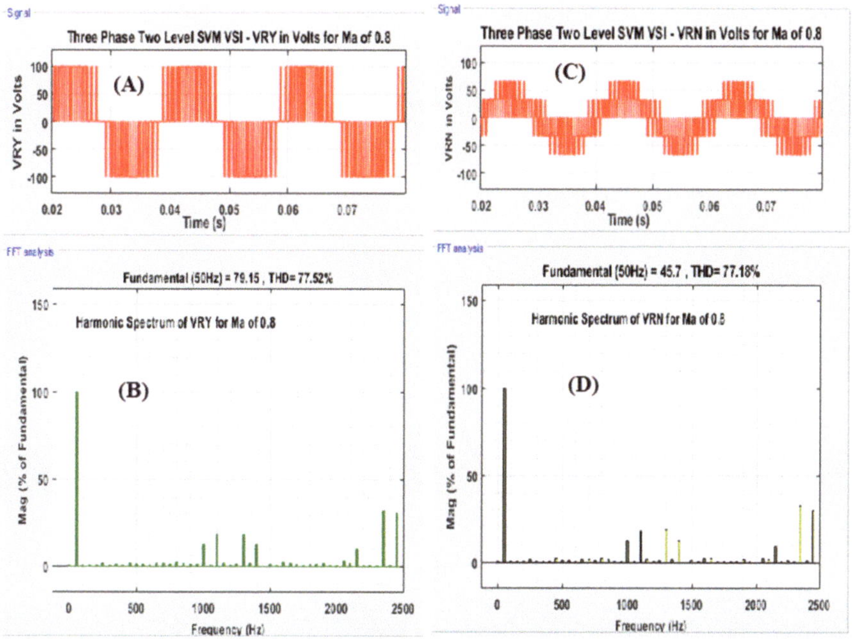

Fig. 4.8 Three phase SVM VSI simulation results: (**a**) line to line voltage VRY. (**b**) harmonic spectrum of VRY. (**c**) line to neutral voltage VRN and (**d**) harmonic spectrum of VRN

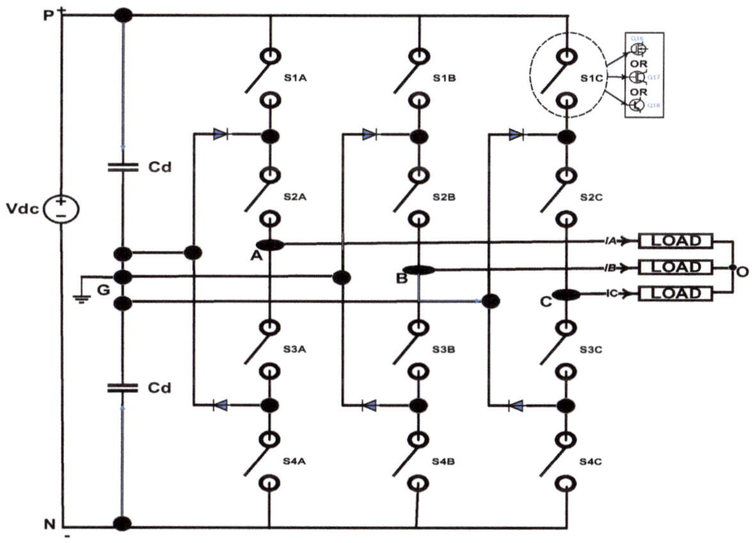

Fig. 4.9 Three phase diode clamped three level inverter

Table 4.5 Three-phase DCTLI switching state for phase A

Sl. no.	S1A S2A S3A S4A	$V_{AG}(t)$ volts	$V_{AN}(t)$ volts	Switch state
1	1 1 0 0	$+V_{dc}/2$	$+V_{dc}$	P
2	0 1 1 0	0	$+V_{dc}/2$	O
3	0 0 1 1	$-V_{dc}/2$	0	N

1: Switch ON and 0: Switch OFF

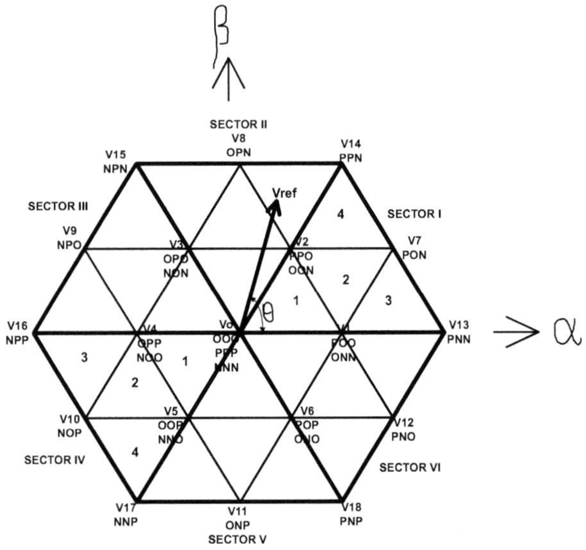

Fig. 4.10 Space vector diagram of three phase NPC inverter

$$\begin{bmatrix} V_\alpha \\ V_\beta \end{bmatrix} = \frac{2}{3} * \begin{bmatrix} 1 & \frac{-1}{2} & \frac{-1}{2} \\ 0 & \frac{\sqrt{3}}{2} & \frac{-\sqrt{3}}{2} \end{bmatrix} * \begin{bmatrix} \frac{V_{dc}}{2} \\ \frac{V_{dc}}{2} \\ 0 \end{bmatrix} \tag{4.24}$$

From Eq. 4.24,

$$\vec{V}(t) = \overrightarrow{V_2} = V_\alpha + jV_\beta = \frac{2}{3} * V_{dc} * \left(\frac{1}{4} + \frac{j\sqrt{3}}{4}\right) = \frac{V_{dc}}{3} * e^{\frac{j\pi}{3}} \tag{4.25}$$

$$\begin{bmatrix} V_\alpha \\ V_\beta \end{bmatrix} = \frac{2}{3} * \begin{bmatrix} 1 & \frac{-1}{2} & \frac{-1}{2} \\ 0 & \frac{\sqrt{3}}{2} & \frac{-\sqrt{3}}{2} \end{bmatrix} * \begin{bmatrix} V_{dc} \\ \frac{V_{dc}}{2} \\ 0 \end{bmatrix} \tag{4.26}$$

From Eq. 4.26,

$$\vec{V}(t)=\overrightarrow{V_7}=V_\alpha+jV_\beta=\frac{1}{\sqrt{3}}*V_{dc}*\left(\frac{\sqrt{3}}{2}+\frac{j}{2}\right)=\frac{V_{dc}}{\sqrt{3}}*e^{\frac{j\pi}{6}} \tag{4.27}$$

$$\begin{bmatrix}V_\alpha\\V_\beta\end{bmatrix}=\frac{2}{3}*\begin{bmatrix}1 & \frac{-1}{2} & \frac{-1}{2}\\ 0 & \frac{\sqrt{3}}{2} & \frac{-\sqrt{3}}{2}\end{bmatrix}*\begin{bmatrix}0\\V_{dc}\\0\end{bmatrix} \tag{4.28}$$

From Eq. 4.28,

$$\vec{V}(t)=\overrightarrow{V_{15}}=V_\alpha+jV_\beta=\frac{2}{3}*V_{dc}*\left(\frac{-1}{2}+\frac{j\sqrt{3}}{2}\right)=\frac{2*V_{dc}}{3}*e^{\frac{j2\pi}{3}} \tag{4.29}$$

All the space vector magnitude and direction can be similarly checked using their switch state. All zero vector (Vo) has magnitude zero, small vectors (V1 to V6) have magnitude Vdc/3, medium vectors (V7 to V12) have magnitude Vdc/√3 and large vectors (V13 to V18) have magnitude 2*Vdc/3 volts, respectively. The reference voltage vector V_{ref} rotates counterclockwise, and the time it takes to complete one revolution is (1/f) seconds where f Hz is the output frequency of the inverter. In Fig. 4.10, there are six triangular sectors I to VI each with four regions 1–4 as marked.

4.5.1 Dwell Time Calculation for Voltage Space Vectors

The dwell time for each voltage vectors in the four different regions can be calculated by volt-second balance principle. Thus in a given triangle region, the sum of the product of each voltage vector and its chosen time interval is equal to the product of Vref and sampling time Ts. Vref can be synthesised along the nearest available voltage vector. The reference voltage vector Vref in the four regions in Sector I are illustrated in Fig. 4.11a–d. The voltage space vectors in the four regions of Fig. 4.11 are defined in Eq. 4.30.

$$V_O=0$$

$$V_1=\frac{V_{dc}.e^{j0}}{3}$$

$$V_2=\frac{V_{dc}.e^{\frac{j\pi}{3}}}{3}$$

$$V_7=\frac{V_{dc}.e^{\frac{j\pi}{6}}}{\sqrt{3}}$$

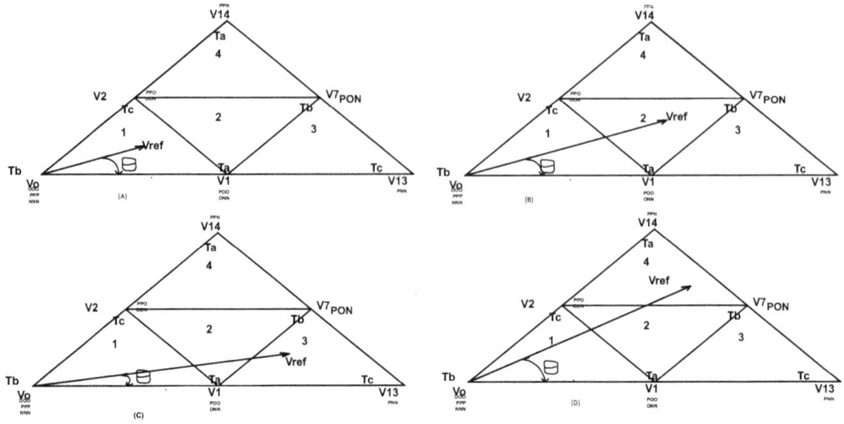

Fig. 4.11 Voltage vector Vref in the four regions in Sector I: (**a**) Region 1 (**b**) Region 2 (**c**) region 3 and (**d**) Region 4

$$V_{13} = \frac{2 * V_{dc}.e^{j0}}{3}$$

$$V_{14} = \frac{2 * V_{dc}.e^{\frac{j\pi}{3}}}{3} \tag{4.30}$$

The voltage space vectors in the four regions of Fig. 4.11 are defined in Eq. 4.30.

Also in any region of a given sector, the following equation connecting dwell time and sample time is valid:

$$T_a + T_b + T_c = T_S \tag{4.31}$$

The dwell time for each voltage vector in the four regions is given in Fig. 4.11. The dwell time for voltage vectors in Sector I in the four regions can be calculated as shown below:

Referring to Fig. 4.11a in the region 1, the volt-second balance equation can be expressed as follows:

$$\vec{V}_O * T_b + \vec{V}_1 * T_a + \vec{V}_2 * T_c = \vec{V}_{ref} * T_S \tag{4.32}$$

Using Eq.4.30 in Eq.4.32 and simplifying,

$$\frac{V_{dc} * T_a}{3} + \frac{V_{dc} * T_c}{3} * \left[cos\,\frac{\pi}{3} + jsin\,\frac{\pi}{3}\right] = V_{ref} * [\cos(\theta) + j\sin(\theta)] * T_S \tag{4.33}$$

Equating real and imaginary part of Eq. 4.33 and simplifying,

$$\frac{V_{dc} * T_a}{3} + \frac{V_{dc} * T_c}{6} = V_{ref} * \cos(\theta) * T_S \tag{4.34}$$

$$\frac{V_{dc} * T_c}{2 * \sqrt{3}} = V_{ref} * sin(\theta) * T_S \tag{4.35}$$

From Eq. 4.35, we have

$$T_c = 2 * m_a * T_S * sin(\theta) \tag{4.36}$$

where

$$m_a = \frac{\sqrt{3} * V_{ref}}{V_{dc}} \tag{4.37}$$

m_a is known as the Modulation Index.

Multiplying Eq. 4.34 by ($\sqrt{3}/V_{dc}$), using Eq.4.36 and 4.37 and simplifying, we have

$$T_a = \sqrt{3} * m_a * T_S * cos(\theta) - m_a * T_S * sin(\theta) \tag{4.38}$$

Equation 4.38 simplifies to the following:

$$T_a = 2 * m_a * T_S * sin\left(\frac{\pi}{3} - \theta\right) \tag{4.39}$$

Using Eq.4.31, 4.37 and 4.39, Tb is calculated below:

$$T_b = T_S - 2 * m_a * T_S * \left[sin\left(\frac{\pi}{3} - \theta\right) - sin(\theta)\right] \tag{4.40}$$

$$i.e. T_b = T_S - 2 * m_a * T_S * sin\left(\frac{\pi}{3} + \theta\right) \ldots .(4.41) \tag{4.41}$$

Referring to Fig. 4.11b in the region 2, the volt-second balance equation can be expressed as follows:

$$\vec{V}_1 * T_a + \vec{V}_7 * T_b + \vec{V}_2 * T_c = \vec{V}_{ref} * T_S \tag{4.42}$$

$$\frac{V_{dc} * T_a}{3} + \frac{V_{dc} * T_b}{\sqrt{3}} * \left[cos\frac{\pi}{6} + jsin\frac{\pi}{6}\right] + \frac{V_{dc} * T_c}{3} * \left[cos\frac{\pi}{3} + jsin\frac{\pi}{3}\right]$$
$$= V_{ref} * [\cos(\theta) + j\sin(\theta)] * T_S \tag{4.43}$$

Multiplying Eq.4.43 by ($\sqrt{3}/V_{dc}$), then equating real and imaginary parts, we have

$$\frac{T_a}{\sqrt{3}} + \frac{\sqrt{3} * T_b}{2} + \frac{T_c}{2 * \sqrt{3}} = m_a * T_S * \cos(\theta) \tag{4.44}$$

$$\frac{T_b}{2} + \frac{T_c}{2} = m_a * T_S * \sin(\theta) \tag{4.45}$$

Using Eq. 4.31 and 4.45, we have

$$T_a = T_S - 2 * m_a * T_S * \sin(\theta) \tag{4.46}$$

Multiplying Eq. 4.44 by √3 and simplifying, we have

$$\frac{3 * T_b}{2} + \frac{T_c}{2} = \sqrt{3} * m_a * T_S * \cos(\theta) + 2 * m_a * T_S * \sin(\theta) - T_S \tag{4.47}$$

Solving Eq. 4.45 and 4.47, we have

$$T_b = 2 * m_a * T_S * \sin\left(\frac{\pi}{3} + \theta\right) - T_S \tag{4.48}$$

$$T_c = T_S - 2 * m_a * T_S * \sin\left(\frac{\pi}{3} - \theta\right) \tag{4.49}$$

Referring to Fig. 4.11c in the region 3, the volt-second balance equation can be expressed as follows:

$$\vec{V}_1 * T_a + \vec{V}_7 * T_b + \vec{V}_{13} * T_c = \vec{V}_{ref} * T_S \tag{4.50}$$

$$\frac{V_{dc} * T_a}{3} + \frac{V_{dc} * T_b}{\sqrt{3}} * \left[\cos\frac{\pi}{6} + j\sin\frac{\pi}{6}\right] + \frac{2 * V_{dc} * T_c}{3} = V_{ref} * [\cos(\theta) + j\sin(\theta)] * T_S \tag{4.51}$$

Multiplying Eq. 4.51 by (√3/V_{dc}), then equating real and imaginary parts, we have

$$\frac{T_a}{\sqrt{3}} + \frac{\sqrt{3} * T_b}{2} + \frac{2 * T_c}{\sqrt{3}} = m_a * T_S * \cos(\theta) \tag{4.52}$$

$$\frac{T_b}{2} = m_a * T_S * \sin(\theta) \tag{4.53}$$

$$T_b = 2 * m_a * T_S * \sin(\theta) \tag{4.54}$$

Multiplying Eq. 4.52 by √3, using Eq.4.53 and rearranging, we have

$$T_a + 2 * T_c = \sqrt{3} * m_a * T_S * \cos(\theta) - 3 * m_a * T_S * \sin(\theta) \tag{4.55}$$

Using Eq.4.31 in Eq.4.55 and rearranging,

$$T_c = \sqrt{3} * m_a * T_S * cos(\theta) - m_a * T_S * sin(\theta) - T_S \tag{4.56}$$

$$T_c = 2 * m_a * T_S * sin\left(\frac{\pi}{3} - \theta\right) - T_S \tag{4.57}$$

Using Eq.4.31, 4.54 and 4.56, T_a can be expressed as follows:

$$T_a = T_S - m_a * T_S * sin(\theta) - \sqrt{3} * m_a * T_S * cos(\theta) + T_S \tag{4.58}$$

$$T_a = 2 * T_S - 2 * m_a * T_S * sin\left(\frac{\pi}{3} + \theta\right) \tag{4.59}$$

Referring to Fig. 4.11d in the region 4, the volt-second balance equation can be expressed as follows:

$$\vec{V}_{14} * T_a + \vec{V}_7 * T_b + \vec{V}_2 * T_c = \vec{V}_{ref} * T_S \tag{4.60}$$

$$\frac{2 * V_{dc} * T_a}{3} * \left[cos\frac{\pi}{3} + jsin\frac{\pi}{3}\right] + \frac{V_{dc} * T_b}{\sqrt{3}} * \left[cos\frac{\pi}{6} + jsin\frac{\pi}{6}\right] + \frac{V_{dc} * T_c}{3} * \left[cos\frac{\pi}{3} + jsin\frac{\pi}{3}\right] = V_{ref} * [\cos(\theta) + j\sin(\theta)] * T_S \tag{4.61}$$

Multiplying Eq.4.61 by ($\sqrt{3}/V_{dc}$), then equating real and imaginary parts, we have

$$\frac{T_a}{\sqrt{3}} + \frac{\sqrt{3} * T_b}{2} + \frac{T_c}{2 * \sqrt{3}} = m_a * T_S * cos(\theta) \tag{4.62}$$

$$T_a + \frac{T_b}{2} + \frac{T_c}{2} = m_a * T_S * sin(\theta) \tag{4.63}$$

Multiplying Eq.4.62 by $\sqrt{3}$, subtracting from Eq.4.63 and simplifying, we have

$$T_b = 2 * ma * T_S * sin\left(\frac{\pi}{3} - \theta\right) \tag{4.64}$$

Substituting for Ta from Eq. 4.31 in Eq. 4.63 and simplifying, we have

$$\frac{T_b}{2} + \frac{T_c}{2} = T_S - m_a * T_S * sin(\theta) \tag{4.65}$$

Subtracting Eq. 4.63 from Eq. 4.65, we have

$$T_a = 2 * m_a * T_S * sin(\theta) - T_S \tag{4.66}$$

Using Eq. 4.64 and 4.65, we have

$$T_c = 2 * T_S - 2 * m_a * T_S * \sin(\theta) - 2 * ma * T_S * \sin\left(\frac{\pi}{3} - \theta\right) \tag{4.67}$$

$$T_c = 2 * T_S - 2 * m_a * T_S * \sin\left(\frac{\pi}{3} + \theta\right) \tag{4.68}$$

The dwell time for the voltage vectors in the four regions in Sector I is tabulated in Table 4.6. For sectors II to VI, the angle θ must be replaced by θ_dash defined in Eq. 4.16.

4.5.2 *Sector and Region Identification for Reference Voltage* $\mathbf{V_{ref}}$

The length of the reference voltage Vref determines the magnitude of the output voltage of the NPC inverter. As the reference voltage Vref rotates counterclockwise at angular frequency ωrad/sec corresponding to the output frequency of this inverter, Vref passes through all the six sectors I to VI. To determine the current sector location for Vref, the three-phase line-to-neutral output voltage of this inverter is transformed to its α-β axis equivalent using Eq. 4.1. Then the absolute value of angle θ is calculated using the relation $\tan^{-1}(V_\beta/V_\alpha)$. Depending on the value of θ in the range from 0 to 2π at intervals of π/3 radians, sector number I to VI applies, as shown in Fig. 4.10.

Depending on the magnitude of reference voltage Vref, the tip of Vref can lie in region 1 alone in all the six sectors or it can lie in region 2, 3 and 4 in all the six sectors at any instant of the sampling time ts. Determining the region location for the tip of Vref at any sampling instant is required to calculate the dwell time and switch state for the switches in that region. The region location for Vref can be determined as follows:

Referring to the four triangles in Sector I shown in Fig. 4.12a–d, it is seen that OF = OA*sin(π/3) = $V_{dc}/(2*\sqrt{3})$, AG = AC*sin(π/3) = $V_{dc}/(2*\sqrt{3})$ and tan (β) = [V_{ref}*sin(θ) / (V_{ref}*cos(θ)—V_{dc}/3)]. Using this data, the region location of V_{ref} can be determined as follows:

- The tip P of V_{ref} is in region 1, if 0 < = V_{ref}*sin(θ) < = [V_{dc}/ (2*√3)] AND β > 2*pi/3 radians.
- The tip P of V_{ref} is in region 2, if 0 < = V_{ref}*sin(θ) < = V_{dc}/ (2*√3) AND π/3 < β < = 2*π/3 radians.
- The tip P of V_{ref} is in region 3, if V_{ref}*cos(θ) > V_{dc}/3 AND β < = π/3 radians.
- The tip P of V_{ref} is in region 4, if V_{ref}*sin(θ) > [V_{dc}/ (2*√3)].

Table 4.6 Three-phase DCTLI: Dwell time for V_{ref} in Sector I

Region number	Voltage vector	Timing Ta	Voltage vector	Timing T_b	Voltage vector	Timing Tc
1	*V1*	$2.m_a.T_S * sin\left(\frac{\pi}{3} - \theta\right)$	*V0*	$T_S * \left[1 - 2.m_a.sin\left(\frac{\pi}{3} + \theta\right)\right]$	*V2*	$2.\ m_a.\ T_S *\ sin\ (\theta)$
2	*V1*	$T_S * [1 - 2.\ m_a.\ sin\ (\theta)]$	*V7*	$T_S * \left[2 * m_a * sin\left(\frac{\pi}{3} + \theta\right) - 1\right]$	*V2*	$T_S * \left[1 - 2.m_a.sin\left(\frac{\pi}{3} - \theta\right)\right]$
3	*V1*	$2.T_S * \left[1 - m_a.sin\left(\frac{\pi}{3} + \theta\right)\right]$	*V7*	$2.\ m_a.\ T_S *\ sin\ (\theta)$	*V13*	$2.m_a.T_S * \left[sin\left(\frac{\pi}{3} - \theta\right) - 1\right]$
4	*V14*	$T_S * [2.\ m_a.\ sin\ (\theta) - 1]$	*V7*	$2.m_a.T_S * sin\left(\frac{\pi}{3} - \theta\right)$	*V2*	$2.T_S * \left[1 - m_a.sin\left(\frac{\pi}{3} + \theta\right)\right]$

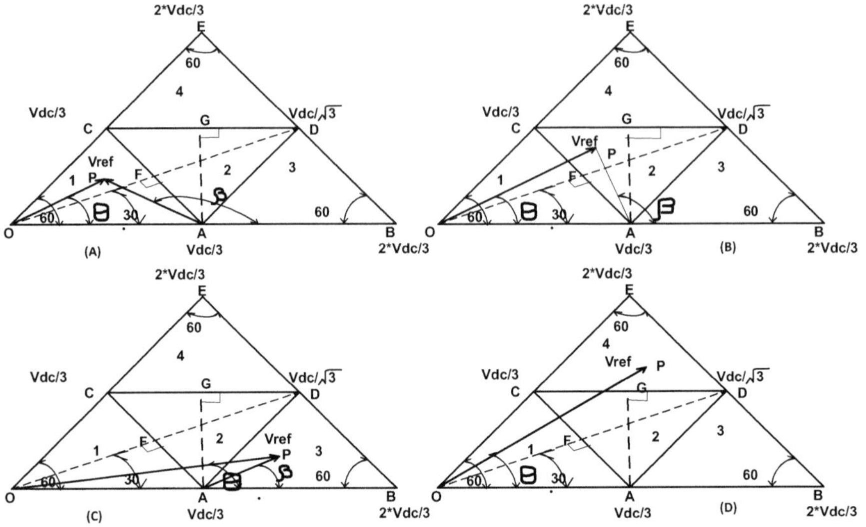

Fig. 4.12 Region identification for Vref: (**a**) Vref in Region 1 (**b**) Vref in Region 2 (**c**) Vref in Region 3 and (**d**) Vref in Region 4

In the above algorithm, θ is the value of Θ_{dash} defined in Eq. 4.16.

4.5.3 Selection of Switching Sequence

Switching sequence is so selected to minimise number of switching operation, transients and losses involved when the switch state changes. For example, changing the switch state from P to O in phase A involves switching off S1A and switch on S3A with switch S2A remaining on. Thus there is only one switch off and one switch on. If switch state changes from P to N in phase A, then switches S1A and S2A should be switched off, and S3A and S4A should be turned on. This involves switching off two switches and switching on two other switches which adds to switching transients and losses. While changing switch state, if two combinations of switching scheme is available, the one with minimum number of switching operation has to be selected [3]. Here seven-segment switching sequence is used [3]. In Fig. 4.10, the following switch sequence is used:

1. For odd sectors I, III and V, following sequence:
   - Ta/4-→Tc/2-→Tb/2-→Ta/2-→Tb/2-→Tc/2-→Ta/4 for regions 1, 2 and 3.
   - Tc/4-→Tb/2-→Ta/2-→Tc/2-→Ta/2-→Tb/2-→Tc/4 for region 4.
2. For even sectors II, IV and VI, the following:
   - Ta/4-→Tb/2-→Tc/2-→Ta/2-→Tc/2-→Tb/2-→Ta/4 for regions 1, 2 and 3.
   - Tc/4-→Ta/2-→Tb/2-→Tc/2-→Tb/2-→Ta/2-→Tc/4 for region 4.

Table 4.7 Switching sequence for Sector I of three-phase DCTLI

Sector number Region number	Space vector Switch state Timing	Space vector Switch state Timing	Space vector Switch state Timing	Space vector Switch state Timing	Space vector Switch state Timing	Space vector Switch state Timing	Space vector Switch state Timing
I 1	*V1* O N N Ta/4	*V2* O O N Tc/2	*V0* O O O Tb/2	*V1* P O O Ta/2	*V0* O O O Tb/2	*V2* O O N Tc/2	*V1* O N N Ta/4
I 2	*V1* O N N Ta/4	*V2* O O N Tc/2	*V7* P O N Tb/2	*V1* P O O Ta/2	*V7* P O N Tb/2	*V2* O O N Tc/2	*V1* O N N Ta/4
I 3	*V1* O N N Ta/4	*V13* P N N Tc/2	*V7* P O N Tb/2	*V1* P O O Ta/2	*V7* P O N Tb/2	*V13* P N N Tc/2	*V1* O N N Ta/4
I 4	*V2* O O N Tc/4	*V7* P O N Tb/2	*V14* P P N Ta/2	*V2* P P O Tc/2	*V14* P P N Ta/2	*V7* P O N Tb/2	*V2* O O N Tc/4

The same method of timing Ta, Tb and Tc shown in Fig. 4.11 for the voltage vectors in the four regions of sector I applies to all the sectors from II to VI. The switching sequence tables for three-phase NPC inverter are shown in Tables 4.7 to 4.12.

To segregate the individual timings Ta, Tb and Tc from Ts as per pattern shown in Tables 4.7 to 4.12, a triangle carrier with period Ts seconds, peak value Ts/2 volts and minimum value zero is used. This is shown in Figs. 4.13 and 4.14 for sector I region 1, 2, 3 and region 4, respectively. The method shown for sector I is applicable for sector III and V and that for sector II is applicable for sector IV and VI, where the respective switch state for each sector must be entered from Tables 4.7 to 4.12. In Fig. 4.13, the timing Ta/4 is compared with the triangle carrier in a comparator whose output is HIGH when triangle carrier is less than Ta/4 or else its output is LOW which is marked A1. Similarly, the cumulative timings (Ta/4 + Tc/2), (Ta/4 + Tc/2 + Tb/2) and (Ta/4 + Tc/2 + Tb/2 + Ta/4) are compared with the triangle carrier in separate comparators, and the outputs B1, C1 and D1 are obtained in the same way as for A1. In the same way in Fig. 4.14, Tc/4, (Tc/4 + Tb/2), (Tc/4 + Tb/2 + Ta/2) and (Tc/4 + Tb/2 + Ta/2 + Tc/4) are compared with triangle carrier in separate comparators, and the outputs A2, B2, C2 and D2 are obtained. This is applicable to all odd sectors, I, III and V. For even sectors II, IV and VI, the order of comparison with triangle carrier changes to Ta/4, (Ta/4 + Tb/2), (Ta/4 + Tb/2 + Tc/2) and (Ta/4 + Tb/2 + Tc/2 + Ta/4), and the outputs A3, B3, C3 and D3 are obtained in the same way as for A1 for regions 1, 2 and 3 of sector I. For region 4 of Sector II, IV and VI, the order of comparison with triangle carrier changes to Tc/4, (Tc/4 + Ta/2), (Tc/4 + Ta/2 + Tb/2) and (Tc/4 + Ta/2 + Tb/2 + Tc/4) and the outputs A4, B4, C4 and D4 are obtained in the same way as for A1 of Sector I. The switch timing for sector I, III and V is given below:

Table 4.8 Switching sequence for Sector II of three-phase DCTLI

Sector number Region number	Space vector Switch state Timing	Space vector Switch state Timing	Space vector Switch state Timing	Space vector Switch state Timing	Space vector Switch state Timing	Space vector Switch state Timing	Space vector Switch state Timing
II 1	*V2* O N N Ta/4	*V0* O O N Tb/2	*V3* O O O Tc/2	*V2* P O O Ta/2	*V3* O O O Tc/2	*V0* O O N Tb/2	*V2* O N N Ta/4
II 2	*V2* O O N Ta/4	*V8* O P N Tb/2	*V3* O P O Tc/2	*V2* P P O Ta/2	*V3* O P O Tc/2	*V8* O P N Tb/2	*V2* O O N Ta/4
II 3	*V2* O O N Ta/4	*V8* O P N Tb/2	*V14* P P N Tc/2	*V2* P P O Ta/2	*V14* P P N Tc/2	*V8* O P N Tb/2	*V2* O O N Ta/4
II 4	*V3* N O N Tc/4	*V15* N P N Ta/2	*V8* O P N Tb/2	*V3* O P O Tc/2	*V8* O P N Tb/2	*V15* N P N Ta/2	*V3* N O N Tc/4

Table 4.9 Switching sequence for Sector III of three-phase DCTLI

Sector number Region number	Space vector Switch state Timing	Space vector Switch state Timing	Space vector Switch state Timing	Space vector Switch state Timing	Space vector Switch state Timing	Space vector Switch state Timing	Space vector Switch state Timing
III 1	*V3* N O N Ta/4	*V4* N O O Tc/2	*V0* O O O Tb/2	*V3* O P O Ta/2	*V0* O O O Tb/2	*V4* N O O Tc/2	*V3* N O N Ta/4
III 2	*V3* O P O Ta/4	*V4* O P P Tc/2	*V9* N P O Tb/2	*V3* N O N Ta/2	*V9* N P O Tb/2	*V4* O P P Tc/2	*V3* O P O Ta/4
III 3	*V3* O P O Ta/4	*V15* N P N Tc/2	*V9* N P O Tb/2	*V3* N O N Ta/2	*V9* N P O Tb/2	*V15* N P N Tc/2	*V3* O P O Ta/4
III 4	*V4* O P P Tc/4	*V9* N P O Tb/2	*V16* N P P Ta/2	*V4* N O O Tc/2	*V16* N P P Ta/2	*V9* N P O Tb/2	*V4* O P P Tc/4

$$\begin{aligned} A1 \cap A1 &= \frac{T_a}{4} \\ A1 \varnothing B1 &= \frac{T_c}{2} \\ B1 \varnothing C1 &= \frac{T_b}{2} \\ C1 \varnothing D1 &= \frac{T_a}{4} \end{aligned} \tag{4.69}$$

Table 4.10 Switching sequence for sector IV of three-phase DCTLI

Sector number Region number	Space vector Switch state Timing	Space vector Switch state Timing	Space vector Switch state Timing	Space vector Switch state Timing	Space vector Switch state Timing	Space vector Switch state Timing	Space vector Switch state Timing
IV 1	*V4* N O O Ta/4	*V0* N N N Tb/2	*V5* N N O Tc/2	*V4* O P P Ta/2	*V5* N N O Tc/2	*V0* N N N Tb/2	*V4* N O O Ta/4
IV 2	*V4* N O O Ta/4	*V10* N O P Tb/2	*V5* N N O Tc/2	*V4* O P P Ta/2	*V5* N N O Tc/2	*V10* N O P Tb/2	*V4* N O O Ta/4
IV 3	*V4* N O O Ta/4	*V10* N O P Tb/2	*V16* N P P Tc/2	*V4* O P P Ta/2	*V16* N P P Tc/2	*V10* N O P Tb/2	*V4* N O O Ta/4
IV 4	*V5* N N O Tc/4	*V17* N N P Ta/2	*V10* N O P Tb/2	*V5* O O P Tc/2	*V10* N O P Tb/2	*V17* N N P Ta/2	*V5* N N O Tc/4

Table 4.11 Switching sequence for sector V of three-phase DCTLI

Sector number Region number	Space vector Switch state Timing	Space vector Switch state Timing	Space vector Switch state Timing	Space vector Switch state Timing	Space vector Switch state Timing	Space vector Switch state Timing	Space vector Switch state Timing
V 1	*V5* N N O Ta/4	*V6* O N O Tc/2	*V0* O O O Tb/2	*V5* O O P Ta/2	*V0* O O O Tb/2	*V6* O N O Tc/2	*V5* N N O Ta/4
V 2	*V5* N N O Ta/4	*V6* O N O Tc/2	*V11* O N P Tb/2	*V5* O O P Ta/2	*V11* O N P Tb/2	*V6* O N O Tc/2	*V5* N N O Ta/4
V 3	*V5* N N O Ta/4	*V17* N N P Tc/2	*V11* O N P Tb/2	*V5* O O P Ta/2	*V11* O N P Tb/2	*V17* N N P Tc/2	*V5* N N O Ta/4
V 4	*V6* O N O Tc/4	*V11* O N P Tb/2	*V18* P N P Ta/2	*V6* P O P Tc/2	*V18* P N P Ta/2	*V11* O N P Tb/2	*V6* O N O Tc/4

$$
\begin{aligned}
A2 \cap A2 &= \frac{T_c}{4} \\
A2 \varnothing B2 &= \frac{T_b}{2} \\
B2 \varnothing C2 &= \frac{T_a}{2} \\
C2 \varnothing D2 &= \frac{T_c}{4}
\end{aligned}
\tag{4.70}
$$

Table 4.12 Switching sequence for Sector VI of three-phase DCTLI

Sector number Region number	Space vector Switch state Timing	Space vector Switch state Timing	Space vector Switch state Timing	Space vector Switch state Timing	Space vector Switch state Timing	Space vector Switch state Timing	Space vector Switch state Timing
VI 1	*V6* O N O Ta/4	*V0* O O O Tb/2	*V1* O N N Tc/2	*V6* P O P Ta/2	*V1* O N N Tc/2	*V0* O O O Tb/2	*V6* O N O Ta/4
VI 2	*V6* P O P Ta/4	*V12* P N O Tb/2	*V1* O N N Tc/2	*V6* O N O Ta/2	*V1* O N N Tc/2	*V12* P N O Tb/2	*V6* P O P Ta/4
VI 3	*V6* P O P Ta/4	*V12* P N O Tb/2	*V18* P N P Tc/2	*V6* O N O Ta/2	*V18* P N P Tc/2	*V12* P N O Tb/2	*V6* P O P Ta/4
VI 4	*V1* P O O Tc/4	*V13* P N N Ta/2	*V12* P N O Tb/2	*V1* O N N Tc/2	*V12* P N O Tb/2	*V13* P N N Ta/2	*V1* P O O Tc/4

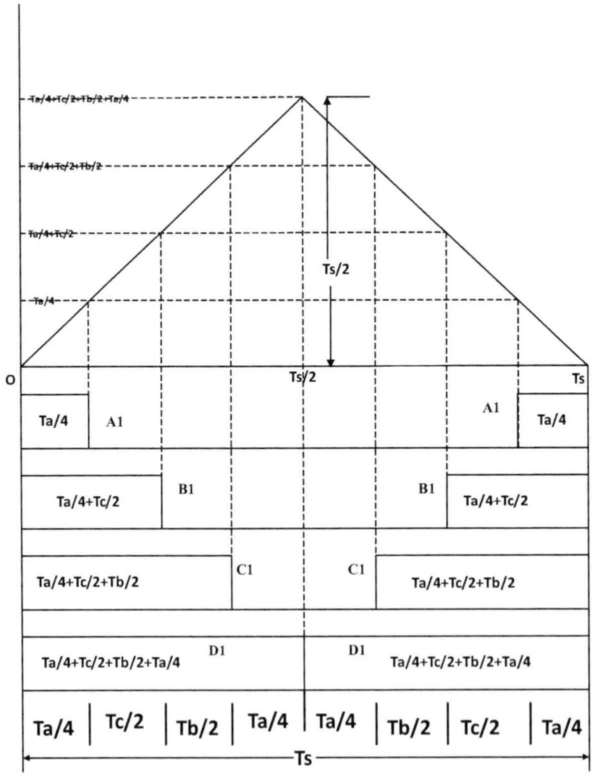

Fig. 4.13 Three phase DCTLI SVM—seven segment switching scheme for Vref in Sector I and Region 1, 2 and 3

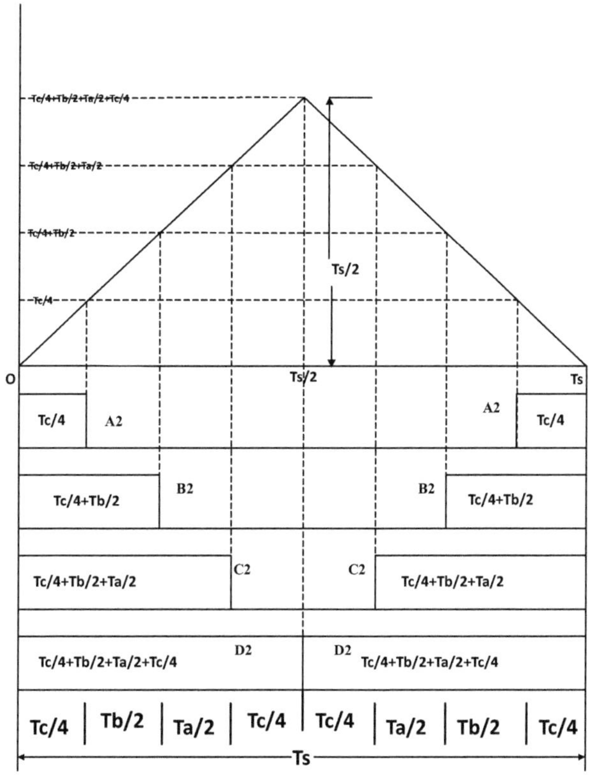

Fig. 4.14 Three phase DCTLI SVM—seven segment switching scheme for Vref in Sector I and Region 4

In Eqs. 4.69 and 4.70, symbols ∩ and ∅ represent the logical AND and EXCLUSIVE OR operation, respectively. The same equation applies to Sectors II, IV and VI region 1, 2 and 3 by replacing A1 to D1 with A3 to D3 and for region 4 by replacing A2 to D2 with A4 to D4. This provides a method to separate the individual dwell time of the DCTLI switches.

4.5.4 Modelling of Space Vector Modulation of a Three-Phase Diode-Clamped Three-Level Inverter

The model of SVM of a three-phase DCTLI developed using SIMULINK [7] is shown in Fig. 4.15 (Model file: EXAMPLE 4_2). The model subsystems are shown in Figs. 4.16a, 4.16b-g, 4.16h, 4.16i. In this model, seven-segment switching sequence shown in Tables 4.7 to 4.12 is used. The model parameters are shown in Table 1.1. The space vector and its timing have to be arranged in the order shown in Tables 4.7 to 4.12 for odd and even sectors. This is done using the triangle carrier

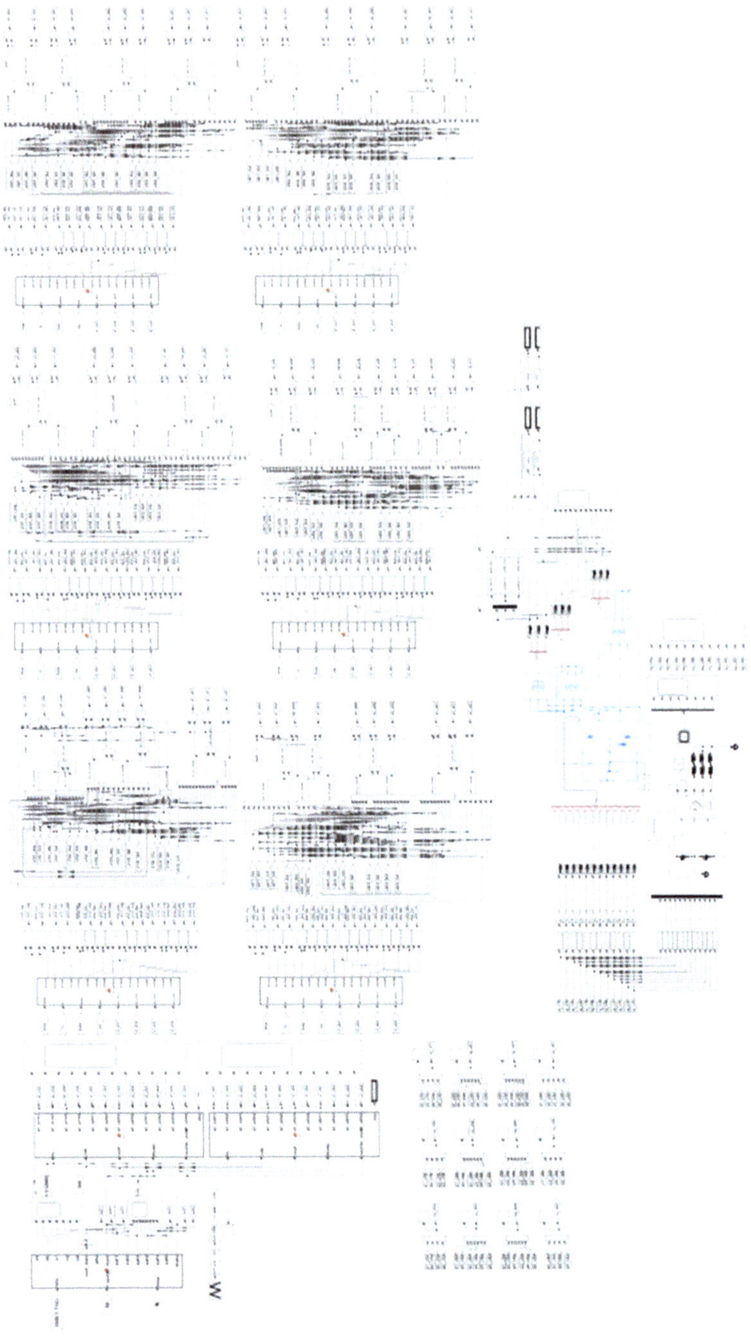

Fig. 4.15 Model of space vector modulation of three phase diode clamped three level inverter

generator whose period is Ts seconds, peak value Ts/2 volts and minimum value zero volts and using AND and EXCLUSIVE OR logic gates, as explained in Sect. 4.5.3. The model subsystems are explained below:

4.5.4.1 Gate Pulse Timing and Triangle Carrier Generator

This has three Embedded MATLAB function blocks and one triangle carrier generator block as shown in Fig. 4.16a. The first Embedded MATLAB function block has peak input voltage Vim Volts, output frequency of inverter f Hz and time as the input. Although any suitable value of Vim can be used, here V_{ref_max} which is (Vdc/√3) volts is used. Three-phase output voltages va, vb and vc are generated with peak value Vim and frequency f Hz. These three-phase output voltages are resolved into d-q axis components vq and vd using the relation shown in Eq. 4.6. Also V_{ref_max} is once again calculated using the relation $V_{ref\_max} = \sqrt{V_q^2 + V_d^2}$. The absolute value of the angle α radians which V_{ref_max} makes with horizontal d-axis is calculated using the relation atan2(vq,vd). The desired V_{ref} value is obtained by multiplying V_{ref_max} with modulation index ma using a gain block shown in

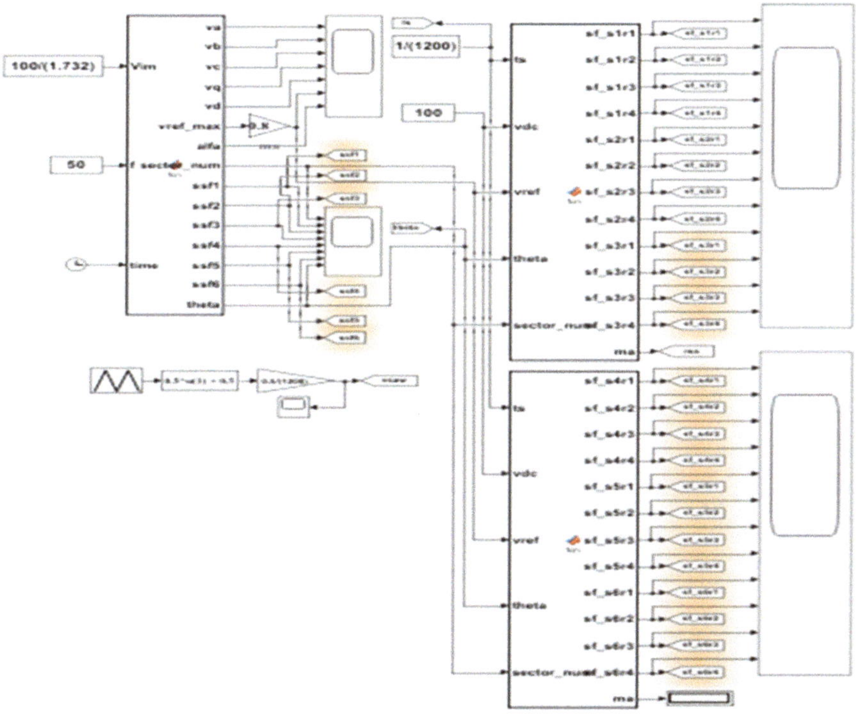

Fig. 4.16a Three phase DCTLI SVM—gate pulse timing and triangle carrier generator

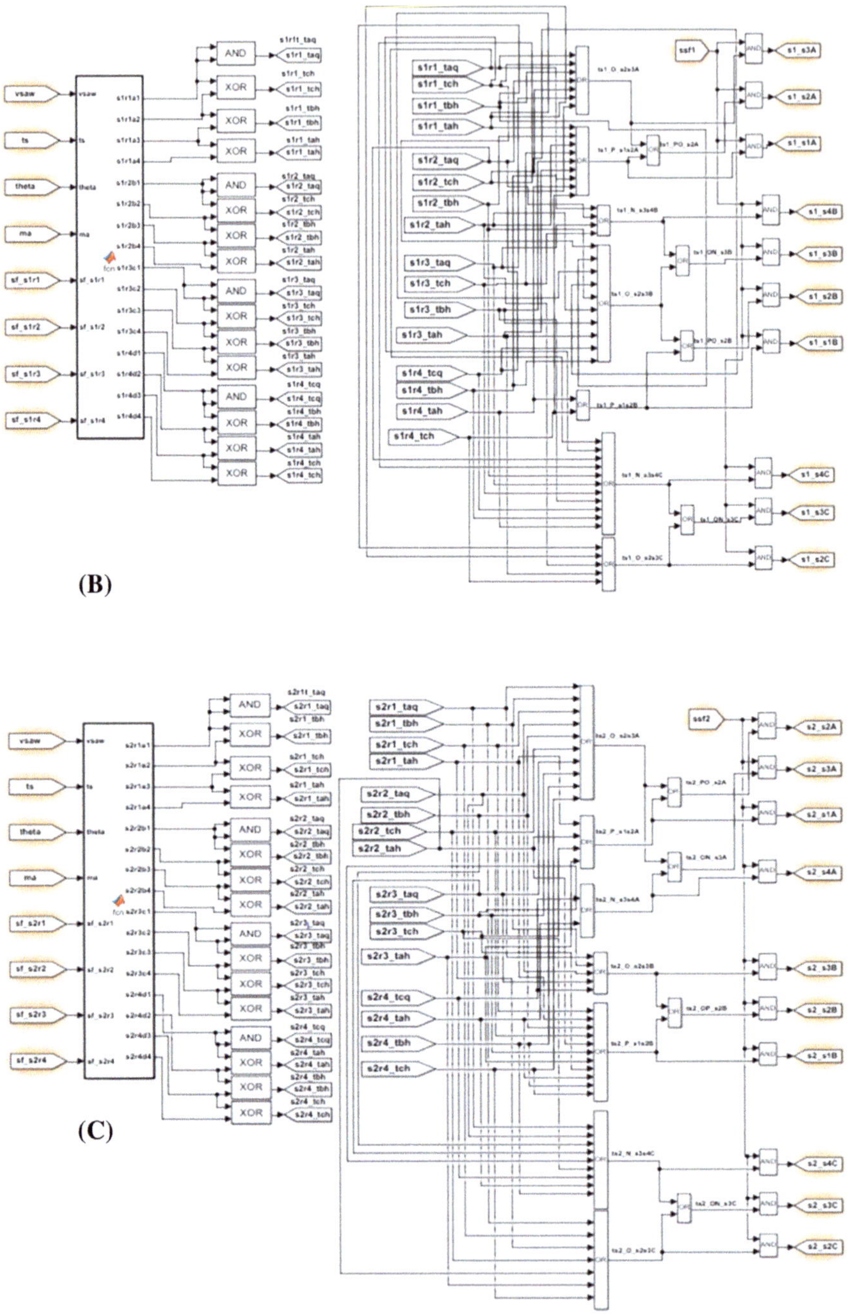

Fig. 4.16b-g Three phase DCTLI SVM—gate pulse timing sectorwise

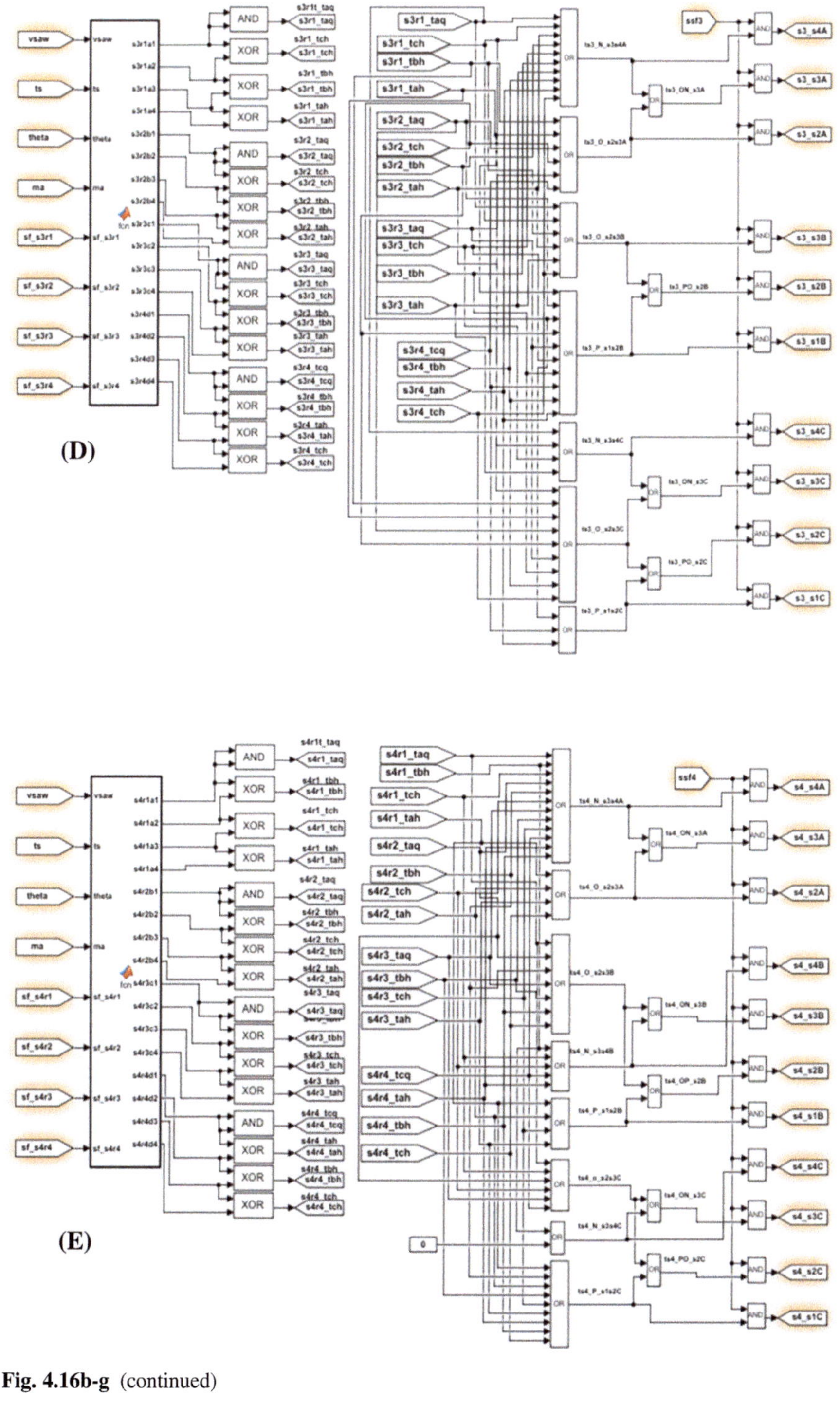

Fig. 4.16b-g (continued)

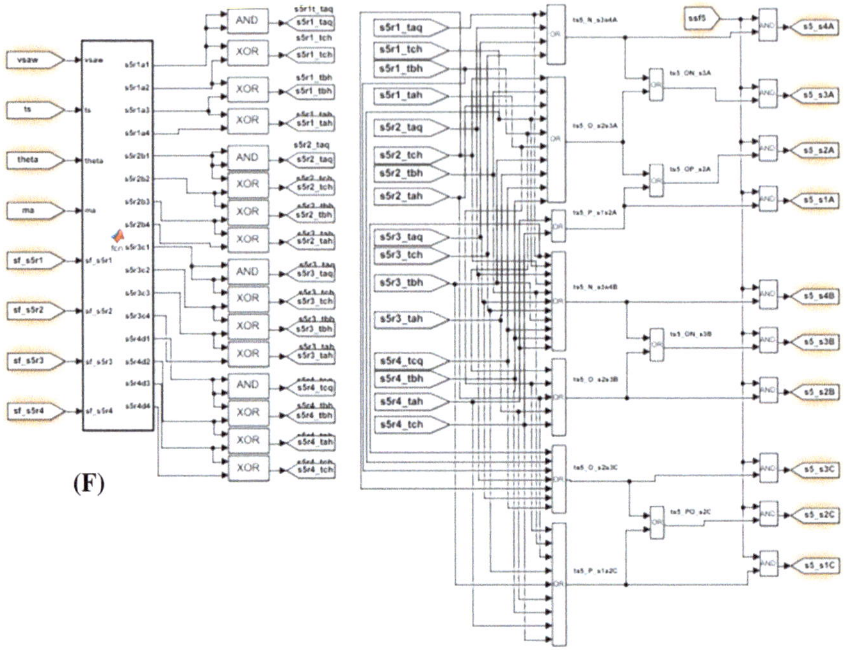

(F)

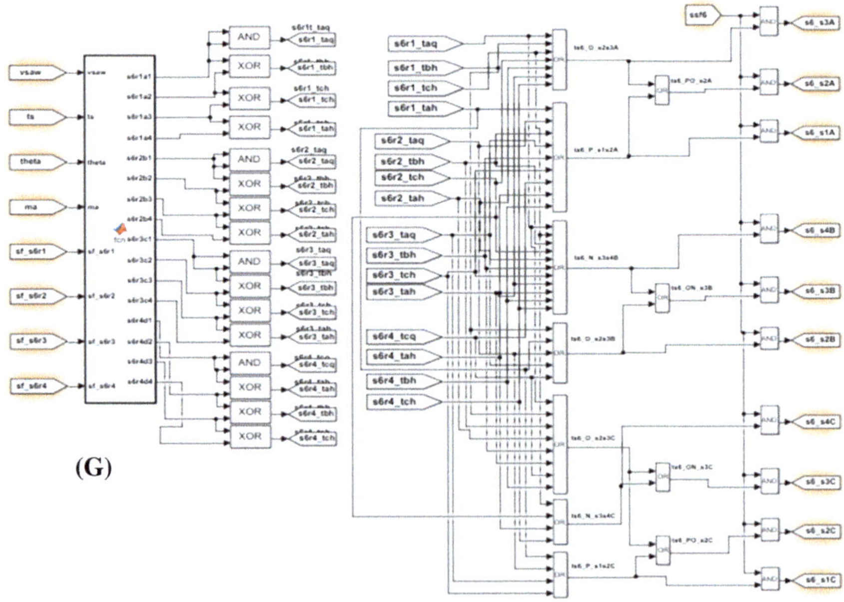

(G)

Fig. 4.16b-g (continued)

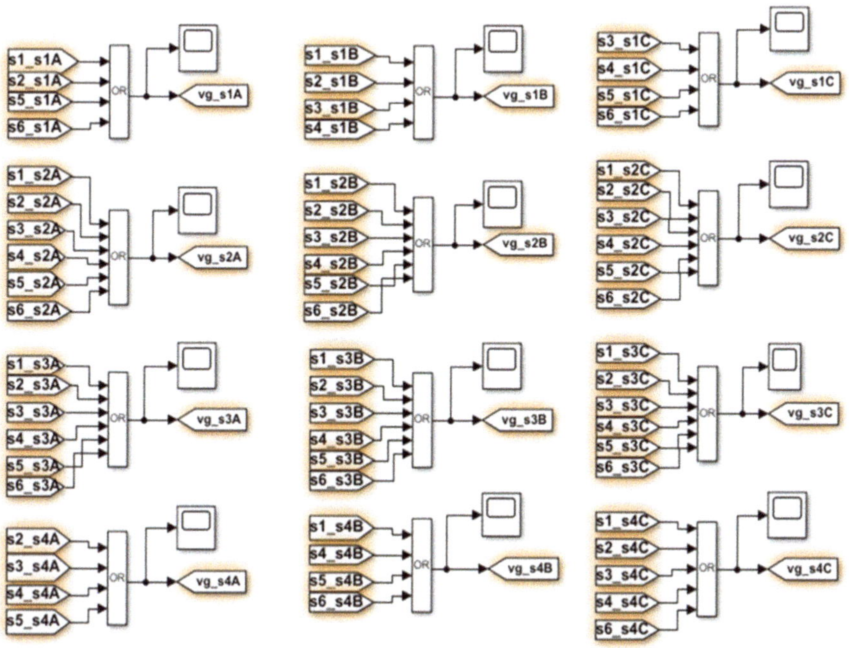

Fig. 4.16h Three phase DCTLI SVM—gate pulse for individual upper and lower switches

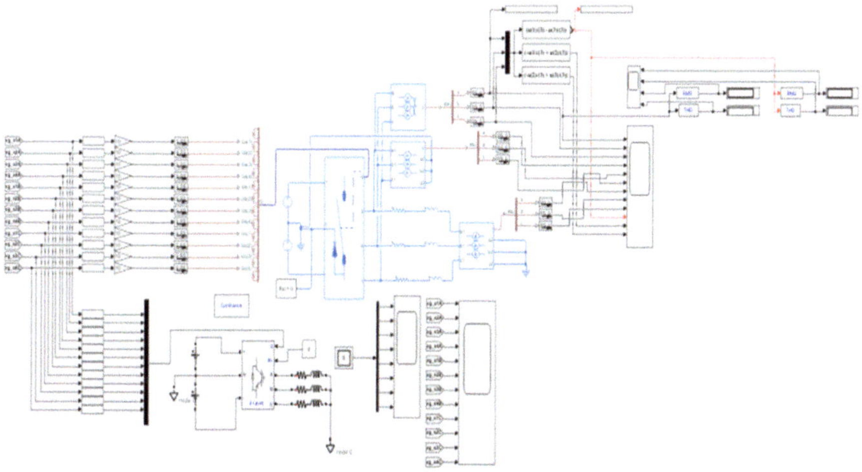

Fig. 4.16i Three phase DCTLI SVM—simscape power systems semiconductor block set (top) and power systems specialized technology power electronics block set (bottom)

Fig. 4.16a. Referring to Fig. 4.10, as V_{ref} pass through sectors I to VI, sector_num 1 to 6 is assigned. When V_{ref} passes through sector I to VI, sector_num goes from 1 to 6; sector switch function ssf1 to ssf6, respectively, goes HIGH and remains LOW otherwise. Also the angle θ which V_{ref} makes within each sector is calculated using Eq. 4.16, where θ_dash is replaced by θ and θ is replaced by α. For example, if α lies within the range zero to π/3, sector_num is set to 1, θ is calculated using Eq. 4.16, ssf1 is set HIGH (logic 1), or else ssf1 is LOW (logic 0). This is illustrated in Program Segment 4.1 in the model file EXAMPLE 4_2.

In the second and third Embedded MATLAB function, sample time T_S, DC link voltage Vdc, reference voltage V_{ref}, angular position of V_{ref} within each sector theta and sector number sector_num are given as inputs. Modulation index ma is calculated using Eq. 4.37. For each sector_num from 1 to 6, the corresponding switch function for sector and region is set to HIGH when Vref is in that sector and region and set to LOW when Vref is NOT in that sector and region. Sector-Region switch function is represented in the form sf_smrn where m takes values from 1 to 6 to represent sector number and n takes values from 1 to 4 to represent region number. Thus when Vref is in sector I and region 1, sf_s1r1 is set to HIGH, and when Vref leaves region 1, sf_s1r1 is set to LOW. Here for Vref, sector identification is done using sector_num, and sector switch functions ssf1 to ssf6, and region identification is done using the algorithm presented in Sect. 4.5.2. This is shown in Program Segment 4.3 and 4.4 in the model file EXAMPLE 4_2 for all the six sectors each having four regions. The triangle carrier generator vsaw generates the triangle carrier with period ts (1/1200) seconds and peak value +/−1 volt. This is then multiplied using a function block with constant (0.5*u(1) + 0.5) and then by another gain block with multiplication constant 0.5*T_S (0.5/1200) to obtain triangle carrier as specified in Sect. 4.2.2. This is shown in Fig. 4.16a.

4.5.4.2 Gate Pulse Timing Sectorwise

Gate pulse timing generators for sector I to VI are shown in Fig. 4.16b-g, respectively. Each timing generator has one Embeded MATLAB Function which calculates timing Ta, Tb and Tc according to the sector and region location of Vref, using equation in Table 4.6. Then cumulative sector timing for each region in a given sector is calculated as explained in Sect. 4.5.3 for Sector I, region 1 to 4. This cumulative sector timing is compared with vsaw for regions 1 to 3 as shown in Fig. 4.13 and for region 4 as shown in Fig. 4.14 and the output A1 to D1 for region 1 to 3 and A2 to D2 for region 4 for odd Sectors I, III, V, A3 to D3 for regions 1 to 3 and A4 to D4 for region 4 for even sectors II, IV and VI are obtained. This cumulative sector timing outputs are given to AND and EXCLUSIVE-OR logic gates as given in Eqs. 4.69 and 4.70 to obtain individual dwell timing for each switch state. This is shown in Program segment 4.5 in the model file EXAMPLE 4_2 for sector I, region 1 to 4. Similar programme applies to sector II to VI which are shown in Program segment 4.6 to 4.10 in the model file EXAMPLE 4_2. In Fig. 4.16b for sector I, the inputs to Embedded MATLAB function are vsaw, Ts, theta, ma, sf_s1r1, sf_s1r2, sf_s1r3 and sf_s1r4, and the outputs are s1r1a1 to s1r1a4, s1r2b1 to s1r2b4, s1r3c1 to s1r3c4 and s1r4d1 to s1r4d4 which correspond to the cumulative time duration in regions 1 to 4 for Sector I. These cumulative timings for each region are given to AND and EXCLUSIVE-OR gates as per Eq. 4.69 and 4.70 to obtain the individual dwell time for each region. These dwell times are as marked in Table 4.7 for each region from left to right. In Fig. 4.16b, these dwell times are represented as s1r1_taq, s1r1_tch, s1r1_tbh, s1r1_tah for region 1, s1r2_taq, s1r2_tch, s1r2_tbh, s1r2_tah for region 2, s1r3_taq, s1r3_tch, s1r3_tbh, s1r3_tah for region 3 and s1r4_tcq, s1r4_tbh, s1r4_tah, s1r4_tch for region 4 of Sector I, where the additional suffix q stands for quarter (1/4th) and h stands for half (½). Similar symbol is used for individual dwell times for the four regions in other sectors from II to VI as shown in Tables 4.8 to 4.12.

In Program segment 4.5 in the model file EXAMPLE 4_2, if sf_s1r1 is HIGH (logic 1), then Vref is in region 1 of Sector I. Timing Ta, Tb and Tc are calculated as per equations in the first row of Table 4.6. Then cumulative timing (0.25*Ta), (0.25*Ta + 0.5*Tc), (0.25*Ta + 0.5*Tc + 0.5*Tb) and (0.25*Ta + 0.5*Tc + 0.5*Tb + 0.5*Ta) are individually compared with vsaw. If vsaw is less than or equal to the cumulative timing, then in the order s1r1a1, s1r1a2, s1r3a3 or s1r4a4, respectively, goes HIGH, or else their value remains LOW. Similar comparison for Vref applies to other three regions of Sector 1 and all the four regions in the other sectors from II to VI. For the order in which cumulative timing has to be carried out for the four regions in each sector, Tables 4.7 to 4.12 must be referred.

The next task is to separate the timing for “P”, “O” and “N” output for each region of a given sector for the three phase A, B and C. For Sector I and region 1 to 4 for phase A, referring to Table 4.7, the following logic for the gate pulse holds good:

$$\begin{aligned} ts1\_O\_s2s3A = {} & s1r1\_taq \cup s1r1\_tch \cup s1r1\_tbh \cup s1\_r2\_taq \cup s1r2\_tch \\ & \cup s1\_r3\_taq \cup s1r4\_tcq \end{aligned} \tag{4.71}$$

$$\begin{aligned} ts1\_P\_s1s2A = {} & s1r1\_tah \cup s1r2\_tbh \cup s1r2\_tah \cup s1r3\_tch \cup s1r3\_tbh \\ & \cup s1r3\_tah \cup s1r4\_tbh \cup s1r4\_tah \cup s1r4\_tch \end{aligned} \tag{4.72}$$

In Eqs. 4.71 and 4.72, the symbol ∪ represents logical OR operation. Equation 4.71 is for “O” and 4.72 is for “P” output of phase A in sector I. Also in sector I, region 1 to 4 there are no “N” output for phase A. In this way, the timing for “P”, “O” and “N” output for the three-phase A, B and C can be obtained for all six sectors by referring to Tables 4.7 to 4.12. These are shown in Fig. 4.16b-g.

Using Table 4.5, Eq. 4.71 and 4.72 for “O” and “P” output, the gate pulse for switch S2A in Fig. 4.9 from region 1 of Sector I can be obtained as follows:

$$ts1\_PO\_s2A = ts1\_O\_s2s3A \cup ts1\_P\_s1s2A \tag{4.73}$$

The gate pulse for S1A and S3A in Fig. 4.9 from region 1 of Sector I is given by Eqs. 4.72 and 4.71, respectively. There is no N switching state in Fig. 4.9 from region 1 of Sector I. These gate pulse for respective switches are ANDed with sector switch function ssf1 for sector I, and the resulting output pulse is marked s1_s1A, s1_s2A and s1_s3A, respectively. In this way, all four regions and six sectors are completed.

4.5.4.3 Gate Pulse for Individual Upper and Lower Switches of Three-Phase DCTLI

The gate pulse for the 12 switches in the three-phase DCTLI of Fig. 4.9, when Vref rotates through all the six sectors, can be determined by collecting the gate pulse for a particular switch in all six sectors and ORing them together using OR gate. This is shown in Fig. 4.16h. To get the gate pulse for switch S1A of Fig. 4.9, s1_s1A, s2_s1A, s5_s1A and S6_S1A are given to an OR gate, and the output of this OR gate is the required gate pulse vg_s1A. Sector III and IV do not contribute gate pulse timing for switch S1A. In this way, the gate pulse timing for all the 12 switches is determined.

4.5.4.4 Three-Phase Three-Level Neutral Point Clamped Voltage Source Inverter

Here models for three-phase DCTLI are developed (1) Simscape power systems semiconductor components block set and (2) using power systems specialised technology power electronics components block set. This is shown in Fig. 4.16i. For the model using Simscape Power system semiconductor components, 12-pulse gate multiplexer and converter blocks are used. The converter block forms the three-phase DCTLI. The 12 gate pulse vg_s1A to vg_s4C are given to gain blocks with multiplication constant ten so that the gate pulse amplitude will be greater than the semiconductor switch gate threshold voltage. The output of these 12 gain blocks form the gate drive which is applied to the gate terminal of the converter through the 12-pulse gate multiplexer. Line-to-line voltage is obtained using line voltage sensor, and line-to-neutral voltage is obtained as per Eq. 1.4, using Fcn blocks. Two DC voltage sources are connected in series to form the DC link voltage of 100 volts.

For the second one using power electronics components block set, three-phase three-level NPC inverter is from three-level inverter block. The 12 gate pulse from vg_s1A to vg_s4C are given to a 12 input MUX, the output of which is given to the gate input of three-level inverter block. The DC link voltage using two battery sources in series form 100 volts, the midpoint of which is grounded. The line-to-ground voltage is obtained using voltage measurement block.

4.6 Simulation Results

The simulation of the SVM of three-phase DCTLI is carried out using ode23tb (stiff/TR-BDF2) solver [7]. The data shown in Table 1.1 are used. In addition, a three-phase balanced R-L load of 16.4 Ω and 14.2 m.H. is used. Simulation results are tabulated in Table 4.13. The simulation results of line-to-line and line-to-neutral voltage for Ma of 0.8 and Mf of 24 p.u. are also shown in Fig. 4.17a–l. The RMS

Table 4.13 Three-phase DCTLI SVPWM: Simulation results

Sl. no	Ma	VLL(rms) V	VLL1(rms) V	VLN(rms) V	VLN1(rms) V	THD of VLL	THD of VLN
1	0.2	24.31	12.36	14.09	7.198	1.69	1.6773
2	0.4	34.27	26.45	19.78	15.29	0.8208	0.8184
3	0.6	44.41	40.19	25.46	22.99	0.4692	0.4740
4	0.8	58.67	54.45	33.83	31.38	0.3965	0.3969
5	1.0	70.83	68.22	40.89	39.44	0.2771	0.2769

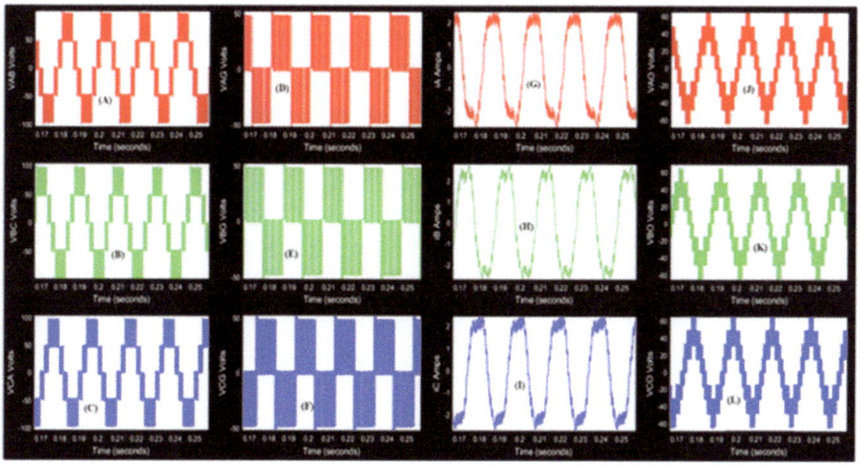

Fig. 4.17 Three phase SVPWM DCTLI simulation results: (**a–c**) line to line voltage (**d–f**) line to ground voltage (**g–i**) load current and (**j–l**) line to neutral voltage

value and THD of line-to-line and line-to-neutral voltages are shown in Fig. 4.18a, b. The harmonic spectrum of line-to-line voltage VAB and line-to-neutral voltage VAO for Ma of 0.8 and Mf of 24 p.u. is shown in Fig. 4.19a, b.

4.7 Discussion of Results

The model of three-phase SVPWM DCTLI using seven-segment switching scheme is presented. The simulation results for Ma of 0.8 are shown in Figs. 4.17, 4.18 and 4.19 and for all other values for Ma are tabulated in Table 4.13. The Ma value of 1 corresponds to voltage vector V7 which is $(Vdc/\sqrt{3})e^{j\pi/6}$. This corresponds to fundamental RMS line-to-line voltage of $Vdc/\sqrt{2}$ which is 70.7 V. Referring to Table 4.13, this value by simulation is 68.22 V which closely agrees with theoretically calculated value. For Ma of 0.5 and below the reference voltage Vref passes through region 1 of sector I to VI and hence as in a conventional three-phase two-

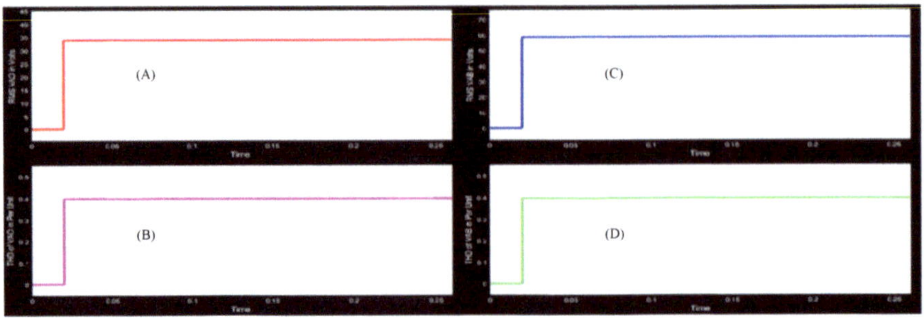

Fig. 4.18 Three phase SVPWM DCTLI simulation results: (**a**, **b**) RMS value and THD of line to neutral voltage (**c**, **d**) RMS value and THD of line to line voltage

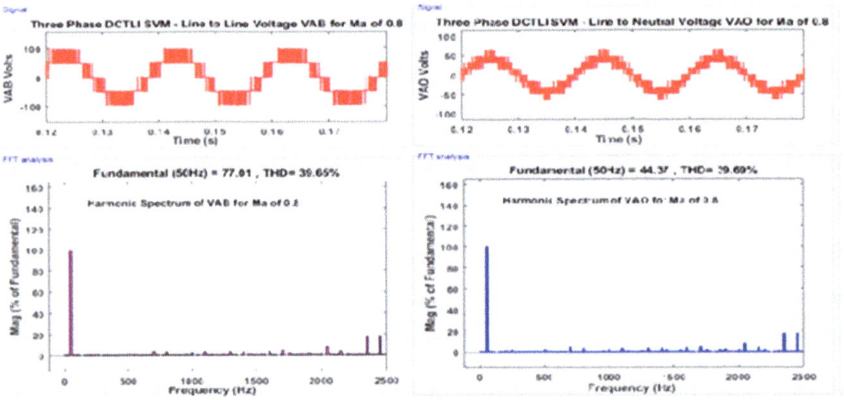

Fig. 4.19 Three phase SVPWM DCTLI simulation results: line to nLine voltage VAB (top left), harmonic spectrum of VAB (bottom left), line to neutral voltage VAO (top right) and harmonic spectrum of VAO (bottom right)

level inverter, the line-to-line voltage has only three levels. The seven-segment switching scheme shown in Tables 4.7 to 4.12 is only one of the many possible switching schemes.

4.8 Case Study: Three-Phase Flying Capacitor Three-Level Inverter

The three-phase flying capacitor three-level inverter (FCTLI) configuration is shown in Fig. 4.20. In this inverter circuit, Cd forms the DC link capacitor, and CF connected to the switches in each phase forms the flying capacitor. The voltage across CF is +Vdc/2 volts where Vdc is the DC link voltage. The switch state for

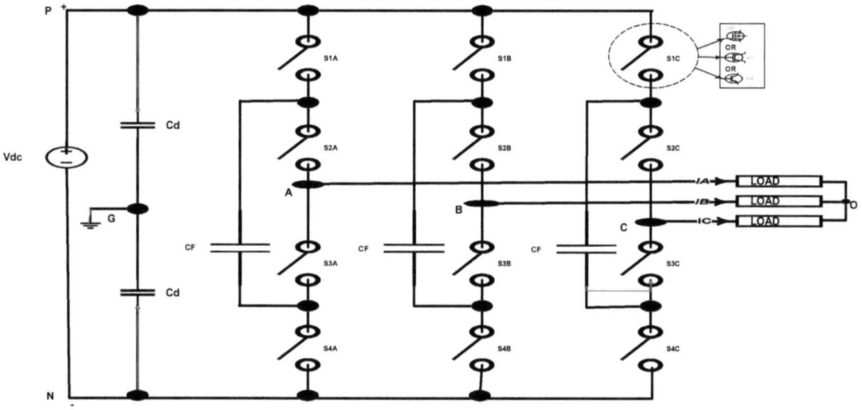

Fig. 4.20 Three phase flying capacitor three level inverter

Table 4.14 Three-phase FCTLI switching state for phase A

Sl. no.	S1A S2A S3A S4A	$V_{AG}(t)$ volts	$V_{AN}(t)$ volts	Switch state
1	1 1 0 0	$+V_{dc}/2$	$+V_{dc}$	P
2	1 0 1 0	0	$+V_{dc}/2$	O
3	0 1 0 1	0	+Vdc/2	O
3	0 0 1 1	$-V_{dc}/2$	0	N

1: Switch ON and 0: Switch OFF

phase A is shown in Table 4.14. It is seen from Table 4.14 that when S1A and S3A are ON or when S2A and S4A are ON, the output voltage VAG(t) is zero. When S1A and S2A are ON, the output voltage VAG(t) is +Vdc/2 and when S3A and S4A are ON, VAG(t) is –Vdc/2 volts [8–10]. The respective switch states "P", "O" and "N" are marked in Table 4.14. Similar table applies to phase B and C. The space vector polygon shown in Fig. 4.10 applies to three-phase FCTLI as well. The only difference here is the "O" state for the switches in the respective phase. Here the "O" state shown in serial number 2 of Table 4.14 is used. The rest of the analysis is the same as for three-phase DCTLI shown in Sect. 4.5.

4.8.1 Modelling of Space Vector Modulation of a Three-Phase Flying Capacitor Three-Level Inverter

The model of SVM of a three-phase FCTLI developed using SIMULINK [7] is shown in Fig. 4.21 (CASE_STUDY_EX4_1). The model subsystem for gate pulse generator for sector I only is shown in Fig. 4.22a to indicate the change caused by "O" state of the switches, and the three-phase FCTLI model subsystem is shown in Fig. 4.22b. The gate pulse generator and triangle carrier subsystem shown in

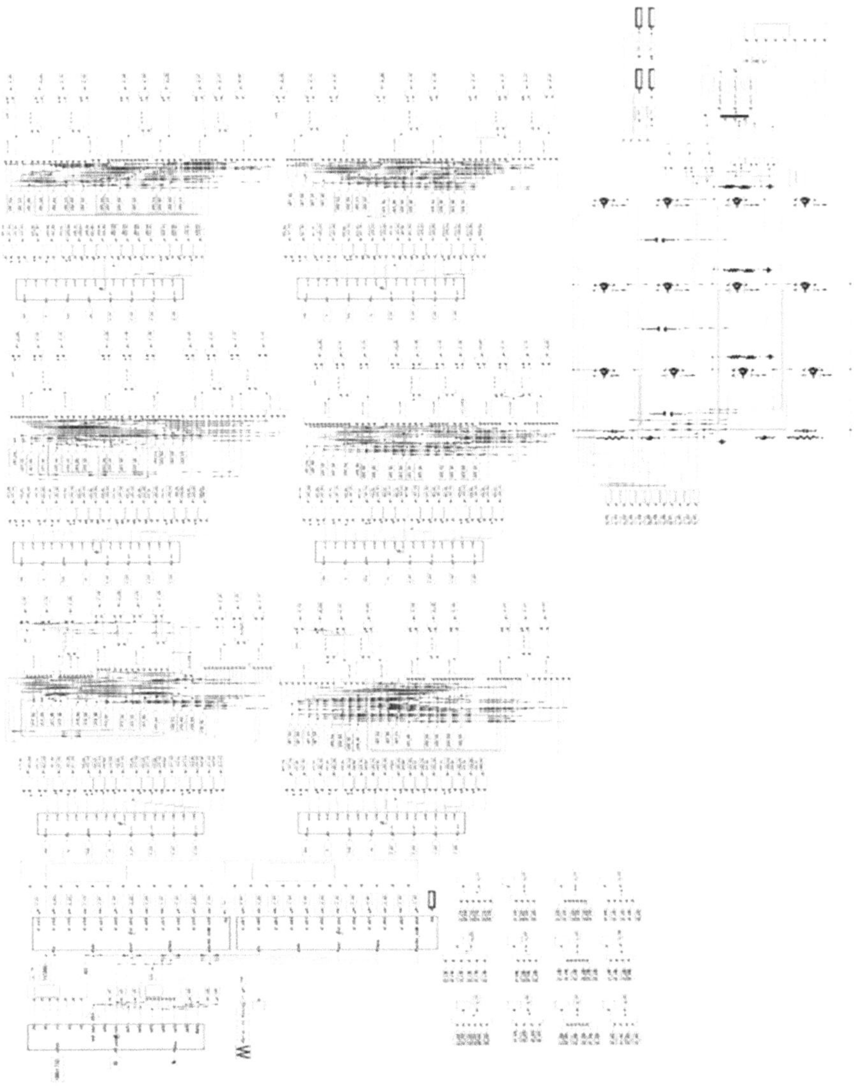

Fig. 4.21 Model of space vector modulation of a three phase flying capacitor three level inverter

Fig. 4.16a are applicable here as well. The gate pulse generator sectorwise subsystem shown in Fig. 4.16b-g and the subsystem named gate pulse for individual upper and lower switches of DCTLI shown in Fig. 4.16h will change here due to the change caused by "O" state of switches in this three-phase FCTLI model. In this model, seven-segment switching sequence shown in Tables 4.7 to 4.12 is used. The model parameters are shown in Table 1.1. Program segments 4.1 to 4.9 are shown in the model file CASE_STUDY_EX4_1. The comparison of cumulative sector timing for each region with vsaw is the same as for DCTLI model explained above.

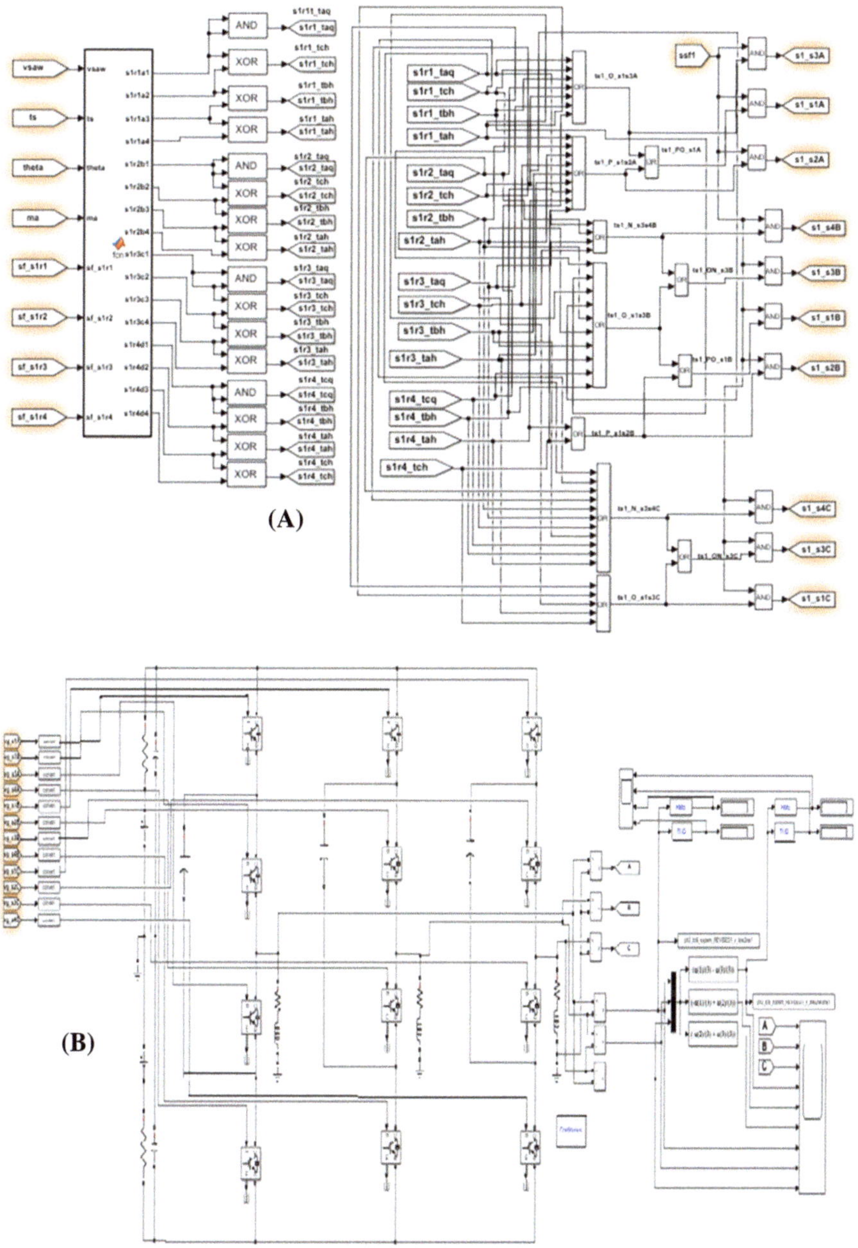

Fig. 4.22 Three phase SVPWM FCTLI—(**a**) gate pulse generator for sector I and (**b**) three phase FCTLI model

Referring to Table 4.7, region 1, the relevant change for the "O" state of switches in phase A of FCTLI is given below:

$$ts1\_O\_s1s3A = s1r1\_taq \cup s1r1\_tch \cup s1r1\_tbh \cup s1\_r2\_taq \cup s1r2\_tch \cup s1\_r3\_taq \cup s1r4\_tcq \tag{4.74}$$

The equation for the "P" state of switches in phase A of FCTLI remains the same as given in Eq. 4.72. The "PO" switch state equation given in 4.73 changes as given below:

$$ts1\_PO\_s1A = ts1\_O\_s1s3A \cup ts1\_P\_s1s2A \tag{4.75}$$

The gate pulse for S2A and S3A in Fig. 4.20 from region 1 of sector I is given by Eq. 4.72 and 4.74 respectively. There is no N state in Fig. 4.20 from region 1 of sector I. These gate pulses for respective switches are ANDed with sector switch function ssf1 for sector I, and the resulting output pulse is marked s1_s1A, s1_s2A and s1_s3A, respectively, as shown in Fig. 4.22a. In this way, all four regions and six sectors are completed. The three-phase FCTLI model shown in Fig. 4.22b is developed using IGBT/diode block, capacitors all from power electronics and power systems specialised technology block set. The individual gate pulse marked vg_s1A to vgs4C is given to the respective gate input of IGBT switches. The DC link voltage is 100 V. The DC link capacitor Cd is 0.1 F, and flying capacitor CF is 1.2e-3 F. The R-L load per phase is 14.2 Ω and 16.5 m.H.

4.8.2 Simulation Results

The simulation of the SVM of three-phase FCTLI is carried out using ode23tb (stiff/TR-BDF2) solver [7]. The data shown in Table 1.1 are used. Simulation results are tabulated in Table 4.15. The simulation results of line-to-line and line-to-neutral voltage for Ma of 0.8 and Mf of 24 p.u. are also shown in Fig. 4.23a–l. The RMS value and THD of line-to-line and line-to-neutral voltages are shown in Fig. 4.24a–d. The harmonic spectrum of line-to-line voltage VAB and line-to-neutral voltage VAO for Ma of 0.8 and Mf of 24 p.u. is shown in Fig. 4.25a–b.

4.8.3 Discussion of Results

The model of three-phase SVPWM FCTLI using seven-segment switching scheme is presented. The simulation results for Ma of 0.8 are shown in Figs. 4.23 to 4.25 and for all other values for Ma are tabulated in Table 4.14. The Ma value of 1 corresponds to voltage vector V7 which is $(Vdc/\sqrt{3})e^{j\pi/6}$. This corresponds to fundamental RMS

Table 4.15 Three-phase FCTLI SVPWM: Simulation results

Sl. no.	Ma	VLL(rms) V	VLL1(rms) V	VLN(rms) V	VLN1(rms) V	THD of VLL	THD of VLN
1	0.2	28.54	15.51	16.81	8.835	1.46	1.51
2	0.4	38.79	30.55	22.62	18.1	0.7567	0.7447
3	0.6	46.51	41.89	26.61	24.3	0.5191	0.4765
4	0.8	61.47	56.9024	35.4	32.7652	0.3876	0.3792
5	1.0	74.95	68.18	41.86	39.49	0.4132	0.3541

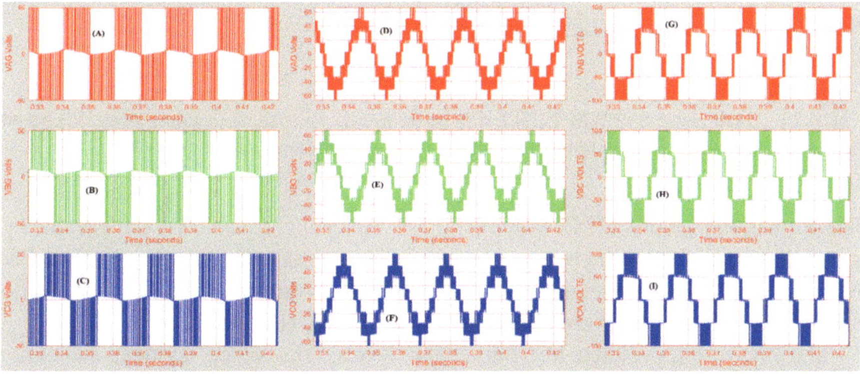

Fig. 4.23 Three phase SVPWM FCTLI simulation results: (**a**–**c**) line to ground voltage VAG, VBG, VCG in volts. (**d**–**f**) line to neutral voltage VAO, VBO, VCO in volts. (**g**–**i**) line to line voltage VAB, VBC, VCA in volts

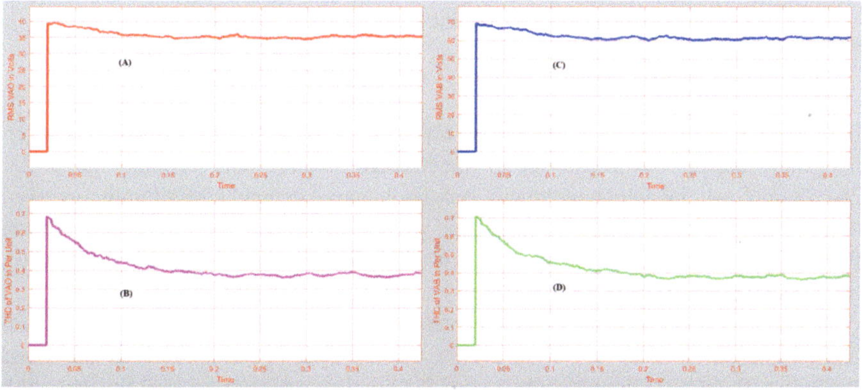

Fig. 4.24 Three phase SVPWM FCTLI simulation results: (**a**, **b**) RMS value and THD of line to neutral voltage VAO, (**c**, **d**) RMS value and THD of line to line voltage VAB

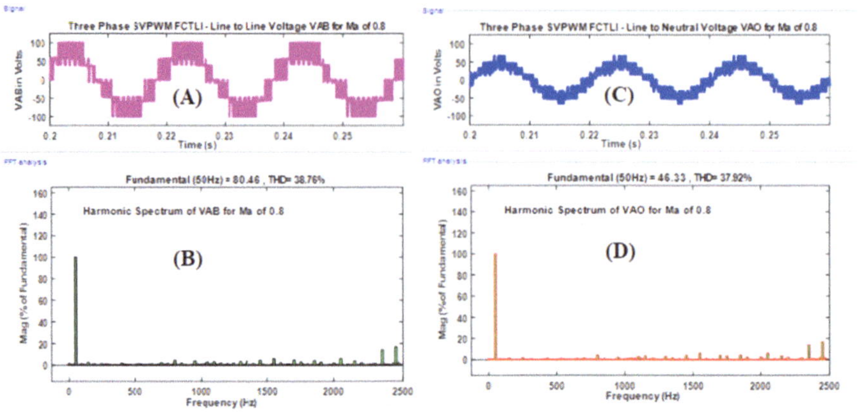

Fig. 4.25 Three phase SVPWM FCTLI simulation results: (**a**, **b**) line to line voltage VAB and harmonic spectrum (**c**, **d**) line to neutral voltage VAO and harmonic spectrum

line-to-line voltage of Vdc/√2 which is 70.7 V. Referring to Table 4.14, this value by simulation is 68.18 V which closely agrees with theoretically calculated value. For Ma of 0.5 and below the reference voltage Vref passes through region 1 of sector I to VI and hence as in a conventional three-phase two-level inverter.

The line-to-line voltage has only three levels. The seven-segment switching scheme shown in Tables 4.7 to 4.12 is only one of the many possible switching schemes.

4.9 Case Study: Space Vector Modulation of a Three-Phase Two-Level Inverter with Eeven Order Harmonics Elimination

In the case of three-phase two-level SVM, it is seen from Fig. 4.8 that even harmonics are present in the line-to-line voltage for harmonic numbers of 20, 22, 26, 28 and so on. Most technical standards have stringent regulation on even harmonics [3]. In this section, a method proposed for even harmonic elimination from line-to-line voltage of three-phase two-level inverter is examined in detail using modeling technique [3, 7].

In Table 4.3, consider Sector II whose switch timing entry can be diagrammatically represented as in Fig. 4.26. Figure 4.26a is type A sequence with starts and finishes with OOO. Figure 4.26b is type B sequence which starts and finishes with PPP [3]. Now consider the type A and type B sequence for Segment V in Table 4.3. This is shown diagrammatically in Fig. 4.27. Comparing Fig. 4.26a with Fig. 4.27b or Fig. 4.26b with Fig. 4.27a, it is seen that VAB(ω.t) = -VAB((ω.t + π), as second and fifth segments are displaced by π radians or Ts/2 s which avoid second

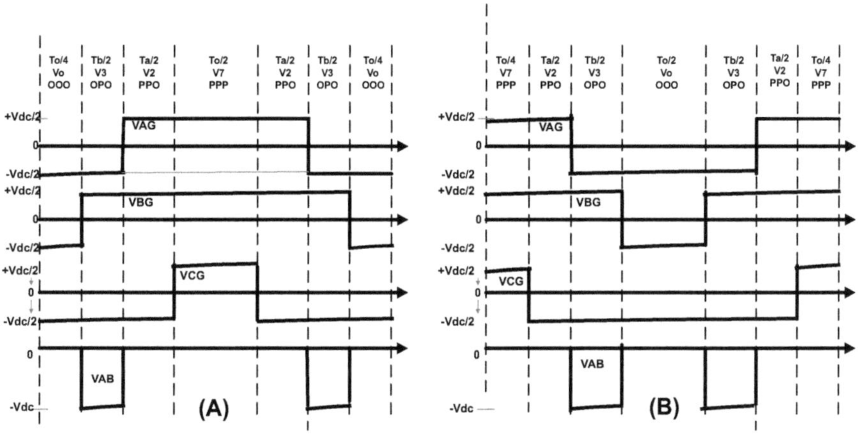

Fig. 4.26 Two valid switching sequence for Vref in Sector II—(**a**) type-a and (**b**) type-b Switching Sequence

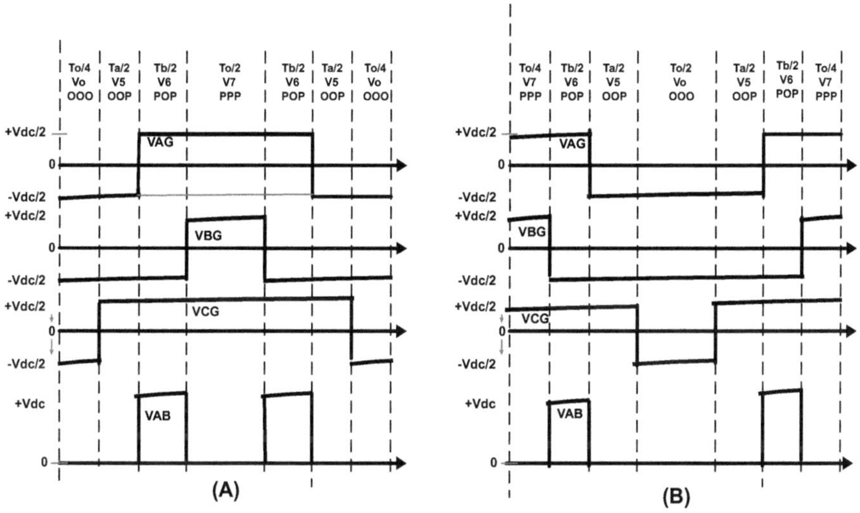

Fig. 4.27 Two valid switching sequence for Vref in Sector V—(**a**) type-a and (**b**) type-b switching sequence

harmonics in VAB [3]. Also changing from Type-A to Type-B switching sequence involves additional switching. It is clear from Fig. 4.26a, b or Fig. 4.27a, b, referring to VAG waveform, switch S1A is switched ON only once during Type-A sequence and twice during Type-B sequence in one sampling period Ts. Similar argument holds good for all other switches in phase B and C. By alternatively using Type-A and Type-B switching, it is possible to make line-to-line voltage of three-phase inverter half wave symmetric [3]. The space vector polygon using both Type-A and

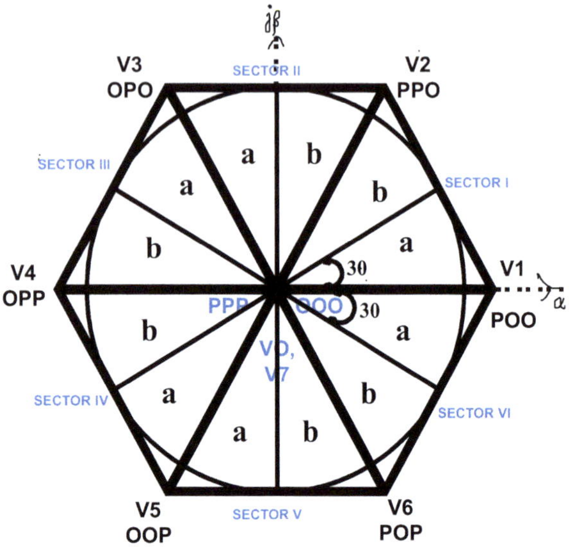

Fig. 4.28 Space vector diagram for three phase two level inverter with type-A and type-B switching sequence

Type-B switching is shown in Fig. 4.28, where region marked "a" uses Type-A switching and region marked "b" uses Type-B switching. The Type-A and Type-B switching sequence to eliminate even order harmonics for SVM of three-phase two-level inverter is summarised in Table 4.16. From Table 4.15, the individual gate pulse timing for the upper switches of three-phase inverter is summarised in Table 4.17. Gate pulse for the lower switches is inverted gate pulse for the respective upper switches. In Table 4.17, Ta1, Tb1 and Tc1 are the cumulative gate pulse timing for upper.

Switches in phase A, B and C calculated for half the sampling period (Ts/2) from Table 4.16. Now referring to conventional SVM switching table for three phase inverter given in Table 4.3, the gate pulse timing for the upper switches are tabulated in Table 4.18 for a sampling interval of Ts/2 s. From Table 4.17, it is clear that the timing for phase A switch corresponding to sectors IIa-IIb and that for Va-Vb are different. Similarly for phase B and C switches, sector timings Ia-Ib, IVa-IVb and IIIa-IIIb, VIa-VIb are different.

4.9.1 Model of Space Vector Modulation of a Three-Phase Two-Level Inverter with Even Order Harmonics Elimination

The model of SVM of a three-phase two-level inverter with even order harmonics elimination is shown in Fig. 4.29 (Model file: CASE_STUDY_EX4_2). Table 4.17 is used to develop the model where Sap, Sbp and Scp are the three gate pulse for

Table 4.16 Three-phase VSI SVM switching table for even order harmonics elimination

Sl. No.	Sector number	Space vector Switch state Timing	Space vector Switch state Timing	Space vector Switch state Timing	Space vector Switch state Timing	Space vector Switch state Timing	Space vector Switch state Timing	Space vector Switch state Timing
1	Ia	*VO* O O O $T_O/4$	*V1* P O O Ta/2	*V2* P P O Tb/2	*V7* P P P $T_O/2$	*V2* P P O Tb/2	*V1* P O O Ta/2	*VO* O O O $T_O/4$
2	Ib	*V7* P P P $T_O/4$	*V2* P P O Ta/2	*V1* P O O Tb/2	*VO* O O O $T_O/2$	*V1* P O O Tb/2	*V2* P P O Ta/2	*V7* P P P $T_O/4$
3	IIa	*V7* P P P $T_O/4$	*V2* P P O Tb/2	*V3* O P O Ta/2	*VO* O O O $T_O/2$	*V3* O P O Ta/2	*V2* P P O Tb/2	*V7* P P P $T_O/4$
4	IIb	*VO* O O O $T_O/4$	*V3* O P O Tb/2	*V2* P P O Ta/2	*V7* P P P $T_O/2$	*V2* P P O Ta/2	*V3* O P O Tb/2	*VO* O O O $T_O/4$
5	IIIa	*VO* O O O $T_O/4$	*V3* O P O Ta/2	*V4* O P P Tb/2	*V7* P P P $T_O/2$	*V4* O P P Tb/2	*V3* O P O Ta/2	*VO* O O O $T_O/4$
6	IIIb	*V7* P P P $T_O/4$	*V4* O P P Ta/2	*V3* O P O Tb/2	*VO* O O O $T_O/2$	*V3* O P O Tb/2	*V4* O P P Ta/2	*V7* P P P $T_O/4$
7	IVa	*V7* P P P $T_O/4$	*V4* O P P Tb/2	*V5* O O P Ta/2	*VO* O O O $T_O/2$	*V5* O O P Ta/2	*V4* O P P Tb/2	*V7* P P P $T_O/4$
8	IVb	*VO* O O O $T_O/4$	*V5* O O P Tb/2	*V4* O P P Ta/2	*V7* P P P $T_O/2$	*V4* O P P Ta/2	*V5* O O P Tb/2	*VO* O O O $T_O/4$
9	Va	*VO* O O O $T_O/4$	*V5* O O P Ta/2	*V6* P O P Tb/2	*V7* P P P $T_O/2$	*V6* P O P Tb/2	*V5* O O P Ta/2	*VO* O O O $T_O/4$
10	Vb	*V7* P P P $T_O/4$	*V6* P O P Ta/2	*V5* O O P Tb/2	*VO* O O O $T_O/2$	*V5* O O P Tb/2	*V6* P O P Ta/2	*V7* P P P $T_O/4$
11	VIa	*V7* P P P $T_O/4$	*V6* P O P Tb/2	*V1* P O O Ta/2	*VO* O O O $T_O/2$	*V1* P O O Ta/2	*V6* P O P Tb/2	*V7* P P P $T_O/4$
12	VIb	*VO* O O O $T_O/4$	*V1* P O O Tb/2	*V6* P O P Ta/2	*V7* P P P $T_O/2$	*V6* P O P Ta/2	*V1* P O O Tb/2	*VO* O O O $T_O/4$

switches in the upper arm and San, Sbn and Scn are the gate pulse for switches in the lower arm of the three-phase inverter. The various subsystems of the model are (1) Sector identifier and Sector Switch Function Generator, and (2) Gate Pulse Timing Generator and (3) Three-Phase Inverter.

Table 4.17 Three-phase SVM two-level VSI: Gate pulse switching time for upper switches for even order harmonics elimination

Sl. no.	Sector number	Sap	Sbp	Scp
1	Ia	Ta1 = 0.5*(ta + Tb + 0.5*T0)	Tb1 = 0.5*(Tb + 0.5*T0)	Tc1 = 0.5*(0.5*T0)
2	IIa	Ta1 = 0.5*(Tb + 0.5*T0)	Tb1 = 0.5*(ta + Tb + 0.5*T0)	Tc1 = 0.5*(0.5*T0)
3	IIIa	Ta1 = 0.5*(0.5*T0)	Tb1 = 0.5*(ta + Tb + 0.5*T0)	Tc1 = 0.5*(Tb + 0.5*T0)
4	IVa	Ta1 = 0.5*(05*T0)	Tb1 = 0.5*(Tb + 0.5*T0)	Tc1 = 0.5*(ta + Tb + 0.5*T0)
5	Va	Ta1 = 0.5*(Tb + 0.5*T0)	Tb1 = 0.5*(0.5*T0)	Tc1 = 0.5*(ta + Tb + 0.5*T0)
6	VIa	Ta1 = 0.5*(ta + Tb + 0.5*T0)	Tb1 = 0.5*(0.5*T0)	Tc1 = 0.5*(Tb + 0.5*T0)
7	Ib	Ta1 = 0.5*(ta + Tb + 0.5*T0)	Tb1 = 0.5*(ta + 0.5*T0)	Tc1 = 0.5*(0.5*T0)
8	IIb	Ta1 = 0.5*(ta + 0.5*T0)	Tb1 = 0.5*(ta + Tb + 0.5*T0)	Tc1 = 0.5*(0.5*T0)
9	IIIb	Ta1 = 0.5*(0.5*T0)	Tb1 = 0.5*(ta + Tb + 0.5*T0)	Tc1 = 0.5*(ta + 0.5*T0)
10	IVb	Ta1 = 0.5*(05*T0)	Tb1 = 0.5*(ta + 0.5*T0)	Tc1 = 0.5*(ta + Tb + 0.5*T0)
11	Vb	Ta1 = 0.5*(ta + 0.5*T0)	Tb1 = 0.5*(0.5*T0)	Tc1 = 0.5*(ta + Tb + 0.5*T0)
12	VIb	Ta1 = 0.5*(ta + Tb + 0.5*T0)	Tb1 = 0.5*(0.5*T0)	Tc1 = 0.5*(ta + 0.5*T0)

Table 4.18 Three-phase two-level SVM inverter: Gate pulse timing for upper switches for conventional switching

Sl. no.	Sector number	Sap	Sbp	Scp
1	I	Ta1 = 0.5*(ta + Tb + 0.5*T0)	Tb1 = 0.5*(Tb + 0.5*T0)	Tc1 = 0.5*(0.5*T0)
2	II	Ta1 = 0.5*(ta + 0.5*T0)	Tb1 = 0.5*(ta + Tb + 0.5*T0)	Tc1 = 0.5*(0.5*T0)
3	III	Ta1 = 0.5*(0.5*T0)	Tb1 = 0.5*(ta + Tb + 0.5*T0)	Tc1 = 0.5*(Tb + 0.5*T0)
4	IV	Ta1 = 0.5*(05*T0)	Tb1 = 0.5*(ta + 0.5*T0)	Tc1 = 0.5*(ta + Tb + 0.5*T0)
5	V	Ta1 = 0.5*(Tb + 0.5*T0)	Tb1 = 0.5*(0.5*T0)	Tc1 = 0.5*(ta + Tb + 0.5*T0)
6	VI	Ta1 = 0.5*(ta + Tb + 0.5*T0)	Tb1 = 0.5*(0.5*T0)	Tc1 = 0.5*(ta + 0.5*T0)

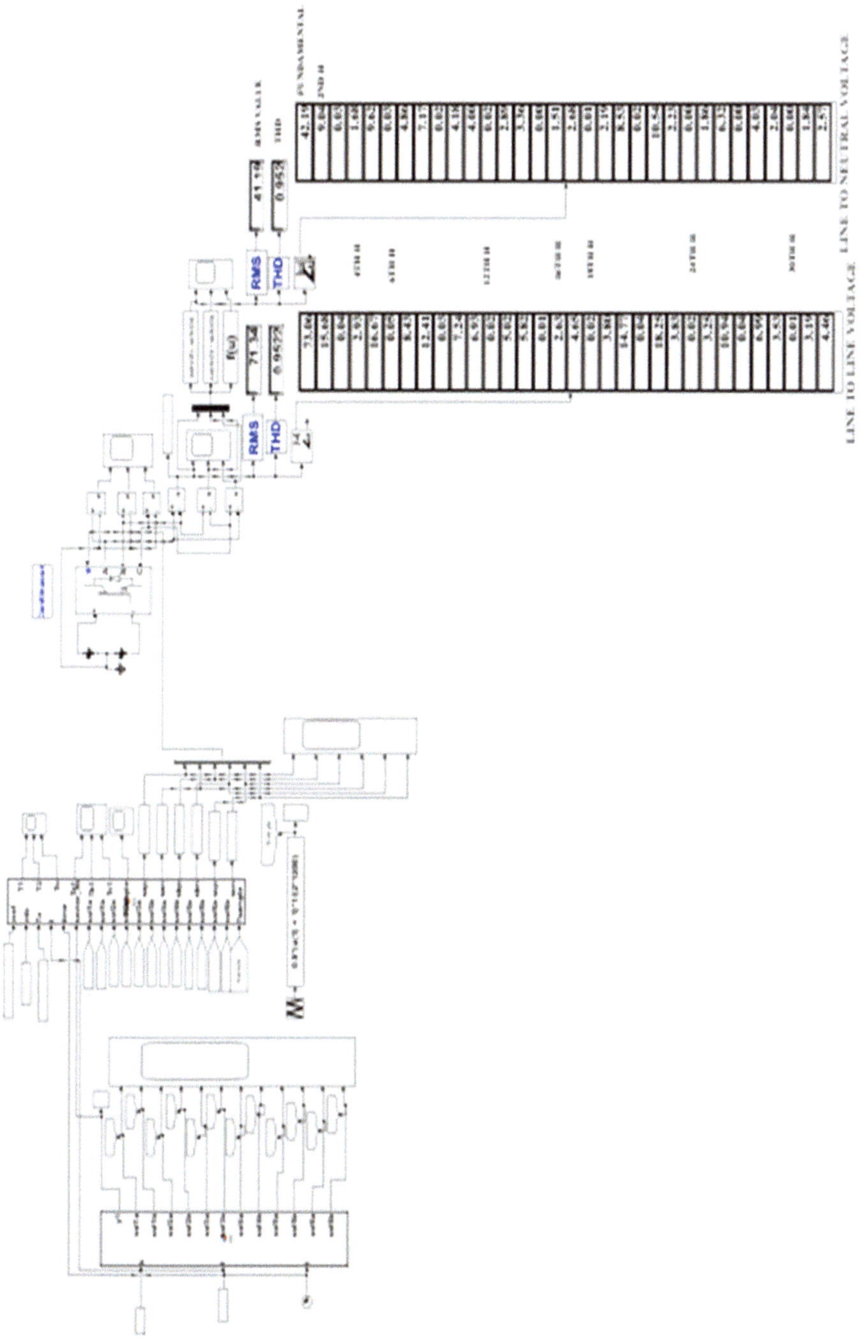

Fig. 4.29 Model of SVM of a three phase two level inverter with even order harmonies elimination

4.9.2 Sector Identifier and Sector Switch Function Generator

Referring to Fig. 4.28, there are six sectors I to VI, and each sector is divided into two subsectors Ia-Ib to VIa-VIb. Vref can lie in any one of the subsectors. These sectors from I to VI are displaced by π/3 radians, and each subsector Ia-Ib to YIa-VIb is displaced by π/6 radians. Embedded MATLAB function with inputs, constant 1 for peakinput voltage, constant 50 for frequency and time module and output sector number y1 and sector switch functions ssf1a-ssf1b to ssf6a-ssf6b as shown in Fig. 4.29. This programme is given under Program Segment 4.11 in the model file CASE_STUDY_EX4_2. Sector output y1 gives sector number 1 to 6 at any instant of time and sector switch functions output ssf1a-ssf1b to ssfVIa-ssfVIb gives location of the tip of reference voltage Vref. For example, when tip of Vref is in Sector Ia, ssf1a is HIGH (logic 1), or else its value is LOW(logic 0).

4.9.3 Gate Pulse Timing Generator

Gate pulse timing for the three-phase inverter switches is generated using another Embedded MATLAB Function. This Embedded MATLAB function has the inputs Vref, Vdc, Tz, f, time t, sector_num, ssf1a, ssf1b to ssf6a, ssf6b, Tsample, and the outputs are T1, T2, T0, Ta1, Tb1, Tc1, Tsample and the six gate pulse sap, san, sbp, sbn, scp and scn. Timings T1, T2 and T0 are defined by Eq. 4.13 to 4.15. Vref is the reference voltage defined in Fig. 4.2, Vdc is the DC link voltage, Tz is the sample time Ts, f is the output frequency of the inverter, sector_num from 1 to 6, sector switch functions from ssf1a, ssf1b to ssf6a, ssf6b and Tsample which is a triangle carrier waveform with period Tz, peak value Tz/2 and minimum value zero. Tsample waveform is shown in Fig. 4.4a, b. The computer programme to generate gate pulse timing is shown under Program Segment 4.12 in the model file CASE_STUDY_EX4_2. In this Program segment 4.12, the symbol (~) represents logical NOT operation. Timing Ta1, Tb1 and Tc1 are calculated as per Table 4.17 along with respective sector switch functions ssf1a-ssf1b to ssf6a-ssf6b. The gate pulse for the upper switches sap, sbp and scp are obtained by comparison of Ta1, Tb1 and Tc1, respectively, with the Tsample waveform. Gate pulse san, sbn and scn for lower switches are obtained by inverting the gate pulse for sap, sbp and scp, respectively. Angle θ_dash is calculated using Eq. 4.16.

The three-phase voltage source inverter model is presented in Sect. 4.2.2.

4.9.4 Simulation Results

The simulation of the three-phase SVM inverter for even harmonics elimination is carried out using ode(23tb stiff/TR-BDF2) solver. The simulation result for the line-

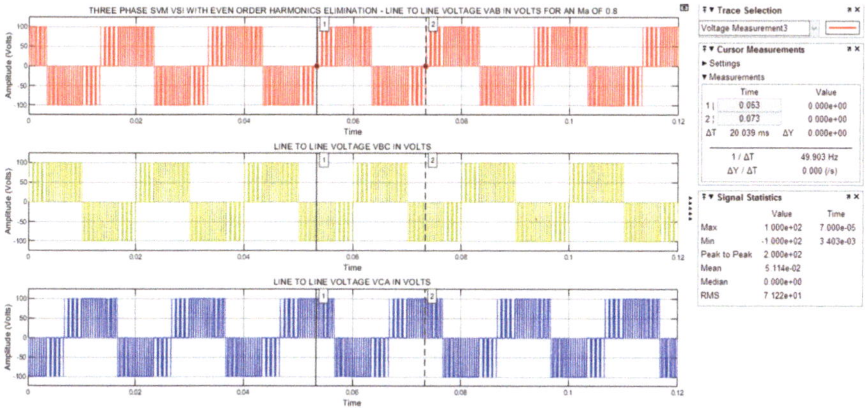

Fig. 4.30 Three phase SVM VSI for even order harmonies elimination—line to line voltage for an A.M. Index of 0.8

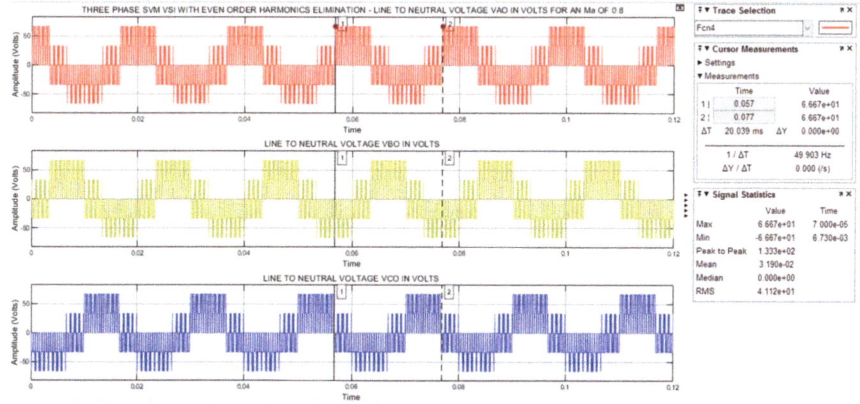

Fig. 4.31 Three phase SVM VSI for even order harmonics elimination—line to neutral voltage for an A.M. Indes of 0.8

to-line voltage VAB for an A.M.index of 0.8 is shown in Fig. 4.30 and that for the line-to-neutral voltage VAO is shown in Fig. 4.31. Simulation results are displayed in Fig. 4.29 and are tabulated in Table 4.19 and in Table 4.20.

4.9.5 Discussion of Results

Referring to Tables 4.17 and 4.18, it is clear that there are differences between three-phase SVM inverter conventional switching and even order harmonics elimination switching. From Tables 4.19 and 4.20, the following points are clear:

Table 4.19 Three-phase SVM VSI with even order harmonics elimination: Harmonic spectrum of line-to-line voltage VAB

Sl. no.	RMS V	THD p.u.	Fund. *V* (%)	4th *V* (%)	6th *V* (%)	12th *V* (%)	16th *V* (%)	18th *V* (%)	24th *V* (%)	30th *V* (%)
(1)	71.34	0.9522	73.06 (100)	2.93 (4.01)	0.05 (0.068)	0.02 (0.027)	2.63 (3.6)	0.02 (0.027)	0.02 (0.027)	0.01 (0.0137)

Table 4.20 Three-phase SVM VSI with even order harmonics elimination: Harmonic spectrum of line-to-neutral voltage VAO

Sl. no.	RMS V	THD p.u.	Fund. *V* (%)	4th *V* (%)	6th *V* (%)	12th *V* (%)	16th *V* (%)	18th *V* (%)	24th *V* (%)	30th *V* (%)
(1)	41.19	0.952	42.19 (100)	1.68 (3.98)	0.03 (0.071)	0.02 (0.047)	1.51 (3.57)	0.01 (0.0237)	0.00 (0)	0.00 (0)

1. The respective RMS and peak fundamental component value for line-to-line voltage VAB and that for line-to-neutral voltage VAO differ only by a small value. The THD values for VAB and VAO are 95.2% .
2. The percentage harmonics values for 6th, 12th, 18th, 24th and 30th are well eliminated for both VAB and VAO, i.e. all even order harmonics in multiples of six are only eliminated from VAB and VAO. In other words, all even order triplen harmonics are well eliminated.
3. However, all other even order harmonic contributions such as 2nd, 4th, 8th, 10th, etc. have a higher value for both VAB and VAO as compared to that for even order harmonics in multiples of six.
4. The THD of VAB and VAO are higher, and their fundamental components are lower for even order harmonics elimination switching as compared to that for conventional switching (Table 4.4).

From Figs. 4.30 and 4.31, the following points are observed:

1. The line-to-line voltage peak values are +100 V and −100 V (+/−Vdc), their balanced RMS values are 71.22 volts for each phase for an A.M. index Ma of 0.8 and frequency 50 Hz.
2. The line-to-neutral voltage peak values are +66.67 V and − 66.67 V (+/−2*Vdc/3); their balanced RMS values are 41.12 volts for each phase for an A.M. index Ma of 0.8 and frequency 50 Hz.

4.10 Conclusions

Models for SVM of three-phase two-level and three-phase DCTLI are presented. The model for SVM of a three-phase two-level inverter presented here is by a rigorous approach. Case studies for model of three-phase FCTLI, that for three-phase two-level SVM inverter with even order harmonics elimination switching, have been successfully presented. Seven-segment switching scheme for SVM is used. The fundamental RMS line-to-line voltage by model simulation for unity modulation index well agree with the theoretically derived results. A new method of determining the region location of reference voltage space vector in the case of three-phase DCTLI and FCTLI is presented. Also by modelling, it is found that even order harmonics elimination switching for a three-phase two-level SVM inverter eliminates even order triplen harmonics which are in multiples of six from the line-to-line and line-to-neutral voltages. In fact, the THD is higher, and the fundamental voltage component is lower for line-to-line and line-to-neutral voltages for even order harmonics elimination switching as compared to that of conventional switching for the three-phase two-level SVM inverter.

References

1. J.F. Silva and S.F. Pinto: "Advanced Control of Switching Power Converters" in "Power Electronics Handbook", Editor: M.H. Rashid, Chapter 36, pp. 1055–1058, Elsevier, 2011.
2. J.R. Espinoza: "Inverters" in "Power Electronics Handbook", Editor: M.H. Rashid, Chapter 15, pp. 372–376, Elsevier, 2011.
3. Bin Wu: "High-Power Converters and AC Drives", IEEE Press, pp. 101–111 and pp. 143–162, 2006.
4. Pradeep M. Bhagwat and V.R. Stefanovic: "Generalized Structure of a Multilevel PWM Inverter"; IEEE Transactions on Industry Applications; Vol. IA-19; No. 6; November/December 1983; p.p. 1057–1069.
5. Akira Nabae, Isao Takahashi and Hirofumi Akagi: "A Neutral-Point Clamped PWM Inverter", IEEE Transactions on Industry Applications, Vol.IA-17, No.5, September/October 1981, pp. 518–523.
6. Amit Kumar Gupta and Ashwin M. Khambadkone; 'A Space Vector PWM Scheme for Multilevel Inverters Based on Two-Level Space Vector PWM", IEEE Transactions On Industrial Electronics, Vol. 53, No. 5, October 2006, pp.1631–1639.
7. The Mathworks Inc., www.mathworks.com: "MATLAB R2020b, 2020.
8. Jose Rodriguez, J.S. Lai and F.Z. Peng: "Multilevel Inverters: A Survey of Topologies Control and Applications", IEEE Transactions on Industrial Electronics Vol.49, No.4, August 2002, pp. 724–738.
9. Clark Hochgraf, Robert Lasseter, Deepak Divan and T.A. Lipo: "Comparison of Multilevel Inverters for Static VAR Compensation"; IEEE; 1994; p.p. 921–928.
10. J.S. Lai and F.Z. Peng: "Multilevel Converters-A New Breed of Power Converters", IEEE Transactions on Industry Applications, vol. 32, No. 3, May/June 1996, pp.509–516.

Chapter 5
Z-Source and Quasi-Z-Source Three-Phase Two-Level and Multilevel Inverters

5.1 Introduction

The impedance source or Z-source converter employs a unique "X-" shaped impedance network consisting of two inductors and two capacitors to couple the power converter with the voltage source, a feature which can't be found in conventional converter [1]. By incorporating such a Z-source network, it is possible to boost or buck the output voltage for a given input voltage. The Z-source inverter finds application in AC drives and in fuel cells to boost the output voltage [1–5]. By adjusting the boost factor B and modulation index M of the Z-source inverter (ZSI), it is possible to boost or buck the output voltage [1]. Unlike conventional inverter, Z-source inverter permits a shoot-through state in which one or more of the upper and lower switches can be simultaneously turned on. The time duration of the shoot-through state determines the boost factor B of the inverter. To overcome drawbacks of ZSI such as discontinuous inductor currents in boost mode and high capacitor voltages, quasi-ZSI (QZSI) was proposed [7]. Various modulation techniques such as simple boost control (SBC), maximum boost control (MBC), third harmonic injection maximum boost control (THIMBC), third harmonic injection maximum constant boost control (THIMCBC) and space vector modulation (SVM) control are used for the Z-source inverter to boost the output voltage [1–5]. In this chapter, models for selected control and modulation techniques for three-phase ZSI and QZSI are presented with simulation results. Two case studies, one for the SVM of three-phase ZSI and another for that of three-phase QZSI are presented. Also models for three-phase Z-source and quasi-Z-source five-level cascade H-bridge inverter (FLCHBI) and neutral point-clamped (NPC) three-level inverter are presented.

Supplementary Information The online version contains supplementary material available at https://doi.org/10.1007/978-3-031-62784-2_5.

N. P. R. Iyer, *Inverters and AC Drives*, Power Systems,
https://doi.org/10.1007/978-3-031-62784-2_5

5.2 Z-Source Inverter Principle of Operation

The three-phase Z-source inverter circuit is shown in Fig. 5.1, and the equivalent circuit for shoot-through (ST) and non-shoot-through (NST) states are shown in Fig. 5.2a, b. The following points relate to Z-source inverter:

This inverter has a Z-source network comprising of series inductors L1 and L2 and cross-connected capacitors C1 and C2 forming an X shape. Diode D is to prevent reverse current flow. Some of the points relating to three-phase conventional inverter and three-phase Z-source inverter are given below [1–5]:

- Three-phase conventional inverter has six non-zero output voltage switching states (100, 110, 010, 011, 001 and 101) and two zero output voltage switching states (000 and 111).

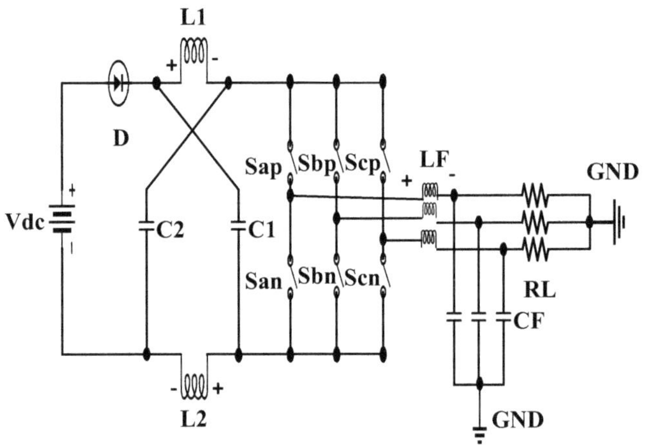

Fig. 5.1 Three phase Z-source inverter

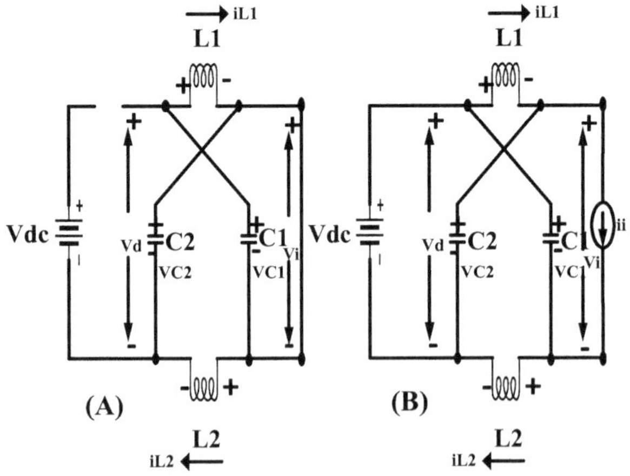

Fig. 5.2 Equivalent circuit for Z-source inverter from DC link side (**a**) shoot through switching state and (**b**) non-shoot through switching state

- Three-phase Z-source inverter output voltage has all the above eight switching states as for a conventional three-phase inverter and a shoot-through zero state, forming a total of nine switching states.
- In a three-phase Z-source inverter, the upper and lower switches in any one arm or a combination of two arms or all the three arms can be simultaneously switched on permitting shoot-through zero state. This state is not permitted in a conventional three-phase inverter as this will cause dangerous short circuit current damaging switching devices.

Z-source inverter analysis is presented below [1]:

In Figs. 5.2a, b, assume that the inductor L1 = L2 = L and capacitor C1 = C2 = C. Then it follows that *VL1* = *VL2* = *VL* and *Vc1* = *Vc2* = *Vc*. Let T be the carrier switching period, T0 be the period of shoot-through (ST) zero switching state and T1 is the period of eight non-shoot-through (NST) switching state. Then it follows that T = T0 + T1.

During the shoot-through interval T0, from Fig. 5.2a, the following equations can be derived:

$$V_L = V_C;\, V_d = 2.V_C = 2.V_L;\, v_i = 0 \tag{5.1}$$

During the non-shoot-through interval T1, from Fig. 5.2b, the following equations can be derived:

$$V_L = (V_{dc} - V_C);\, V_d = V_{dc};\, v_i = (V_C - V_L) = (2.V_C - V_{dc}) \tag{5.2}$$

The average voltage across the inductor over one switching cycle is zero in the steady state which leads to the following equation:

$$\frac{T_O * V_C + T_1 * (V_{dc} - V_C)}{T} = 0$$

$$\frac{V_C}{V_{dc}} = \frac{T_1}{(T_1 - T_O)} \tag{5.3}$$

Using Eqs. 5.1, 5.2 and 5.3, the average dc link voltage across the inverter bridge V_i and peak DC link inverter voltage *vi* can be expressed as follows:

$$V_i = \frac{T_O * 0 + T_1 * (2 * V_C - V_{dc})}{T} = \frac{T_1 * (V_{dc})}{(T_1 - T_O)} = V_C \tag{5.4}$$

$$v_i = (2 * V_C - V_{dc}) = \frac{T * V_{dc}}{(T_1 - T_O)} = B * V_{dc} \tag{5.5}$$

$$B = \frac{T}{((T_1 - T_O))} = \frac{1}{(1 - \frac{2 * T_O}{T})} \geq 1 \tag{5.6}$$

Equation 5.6 is the boost factor B. If v_{ac} is the peak output phase to neutral voltage of the inverter and M is the modulation index, then using Eq. 5.5, v_{ac} can be expressed as follows:

$$v_{ac} = \frac{M * v_i}{2} = \frac{M * B * V_{dc}}{2} \tag{5.7}$$

For a conventional three-phase sine triangle carrier PWM inverter, peak output phase voltage $v_{ac} = (M^* V_{dc}/2)$. B is always greater than 1. The buck-boost factor $B_b = M^*B$ can be adjusted suitably to buck or boost the output voltage of the Z-source inverter [1]. Also using Eqs. 5.1, 5.3 and 5.6, capacitor voltage can be expressed as follows:

$$V_{C1} = V_{C2} = V_C = \left(\frac{1 - \frac{T_O}{T}}{1 - \frac{2.T_O}{T}} \right) * V_{dc} = (1 - D_O) * B * V_{dc} \tag{5.8}$$

where $D_O = (T_O/T)$ is the shoot-through duty cycle.

An approximate method to select the value of inductor and capacitor for the Z-source network is given below:

Referring to Fig. 5.2a, b, the average inductor current over one switching period T is zero. This is expressed as follows:

$$\frac{V_C * D_O * T}{L} + \frac{(V_{dc} - V_i) * (1 - D_O) * T}{2 * L} = 0 \tag{5.9}$$

$$V_C = \frac{(1 - D_O) * (V_i - V_{dc})}{2 * D_O} \tag{5.10}$$

In Eqs. 5.9 and 5.10, $(1\text{-}D_O) = D_1$ is the non-shoot-through duty cycle. The average value of inverter bridge voltage V_i is given by Eq. 5.4.

$$(V_i - V_{dc}) = \frac{D_O * V_{dc}}{(1 - 2 * D_O)} \tag{5.11}$$

Using Eqs. 5.10 and 5.11, the average capacitor voltage V_C can be expressed as follows:

$$V_C = \frac{(1 - D_O) * V_{dc}}{2 * (1 - 2 * D_O)} \tag{5.12}$$

Now $V_C = V_L$ during the shoot-through period T_O. Also $V_L = L^*(di_L/dt) = L^*(\Delta i_L/\Delta t)$ and $\Delta i_L = r_L{}^* i_L$ where r_L is the inductor current ripple factor and Δt is $D_O{}^*T$. Also in the steady state when there is no change in the capacitor voltage V_C, inductor current $i_L = (P/V_{dc})$ where P is the power delivered to the load by the DC source V_{dc}. Using these relations in Eq. 5.12 and simplifying, inductor L can be expressed as follows:

$$L = \frac{V_{dc}^2 * (1 - D_O) * D_O * T}{2 * r_L * P * (1 - 2 * D_O)} \tag{5.13}$$

The capacitor value C can be calculated as follows:

During the shoot-through interval T_O, capacitor current i_C can be expressed as follows:

$$C * \frac{dV_C}{dt} = i_L \tag{5.14}$$

During the non-shoot-through time interval $T_1 = (T\text{-}T_O)$, capacitor current i_C can be expressed as follows:

$$C * \frac{dV_C}{dt} = i_{dc} - i_L \tag{5.15}$$

where i_{dc} is the current supplied by the DC source V_{dc}. The average capacitor voltage over one switching period T is zero. This can be expressed as follows:

$$\frac{i_L * D_O * T}{C} + \frac{(i_{dc} - i_L) * (1 - D_O) * T}{C} = 0 \tag{5.16}$$

$$i_L = C * \frac{dV_C}{dt} = \frac{i_{dc} * (1 - D_O)}{(1 - 2 * D_O)} \tag{5.17}$$

Equation 5.17 also represents the capacitor current ($C*dV_C/dt$) during shoot-through interval D_O*T. This can also be expressed as ($C*\Delta V_C/\Delta t$) = ($C*r_C*V_C/\Delta t$), where r_C is the capacitor voltage ripple factor and Δt is D_O*T. Also in the steady state when the change in inductor current is zero, $V_C = V_{dc}$ and $i_{dc} = (P/V_{dc})$. Using these values in Eq. 5.17 and simplifying,

$$C = \frac{P * D_O * (1 - D_O) * T}{r_C * V_{dc}^2 * (1 - 2 * D_O)} \tag{5.18}$$

Using Eqs. 5.13 and 5.18, Z-source network L and C values for a given V_{dc}, T, D_O and P can be determined.

Similarly output filter inductor Lf and capacitor Cf are determined using the relation $f_O = 1/(2\pi*\text{sqrt}(Lf*Cf))$ where f_O is the cut-off frequency.

5.2.1 Simple Boost Control

Simple boost control is used to control the shoot-through duty-ratio D_O. Figure 5.3 illustrates the simple boost control. In this method, two straight line voltages *Vp* and *Vn* corresponding to the positive and negative peak value of the three-phase reference sine wave modulating signals are used along with traditional sinusoidal PWM to control the shoot-through duty-ratio D_O, simultaneously maintaining the six active switching states. The shoot-through duty ratio for simple boost control is $D_O = (1\text{-M})$, where M is the amplitude modulation index (Vsin/Vtri) as defined for a conventional sinusoidal PWM inverter. Thus no shoot through occurs when M takes the value one and the Z-source inverter behaves like a conventional inverter. Using 5.6 and 5.7, the peak output phase voltage of inverter v_{ac} for simple boost control can be expressed as follows [1, 2]:

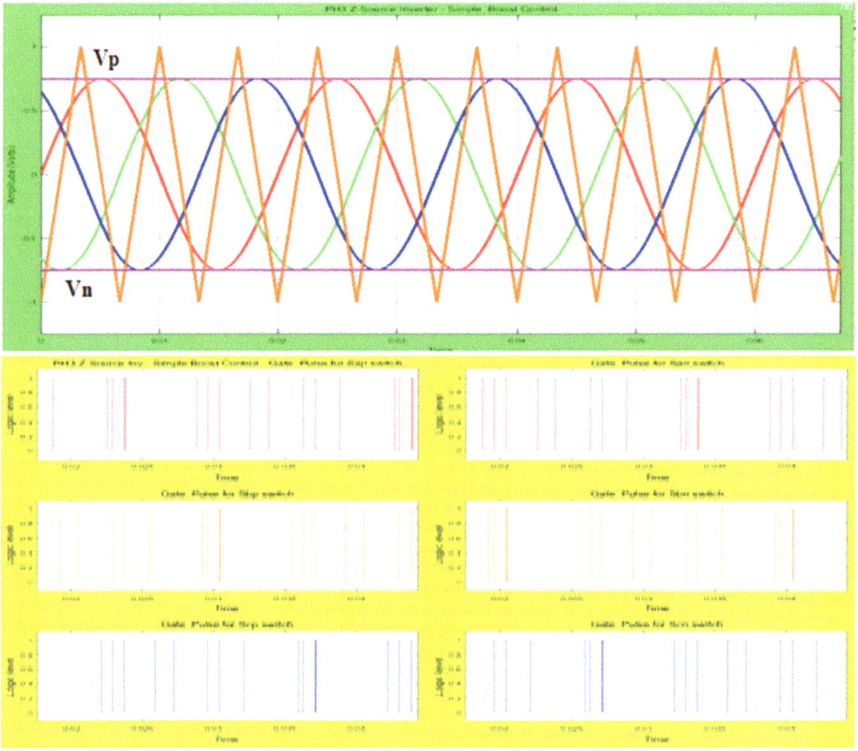

Fig. 5.3 Simple boost control waveforms. Three Phase sine wave modulating signals and triangle carrier (Top) and six gate pulse for the three phase inverter switches (bottom)

Table 5.1 Three-phase Z-source inverter parameters for simple boost control

Name	Value	Unit
Z-source inductor L1 = L2	432	μH
Z-source capacitor C1 = C2	163	μF
Output filter inductor Lf	200	μH
Output filter capacitor Cf	31.67	μF
Shoot-through duty ratio *Do*	0.25	–
Modulation index M (1-D_O)	0.75	–
DC input voltage	48	Volts
Power output P	100–1000	Watts
Inverter output frequency	50	Hz
Maximum boost factor B	2	–
Triangle carrier frequency	20	kHz
Output filter cut-off frequency	2	kHz
Inductor current ripple factor r_L	<10%	–
Capacitor voltage ripple factor r_C	< 10%	–
Load resistor R_L per phase	10	Ω

$$v_{ac} = \frac{M}{(2 * M - 1)} * \frac{V_{dc}}{2} \tag{5.19}$$

Now consider a three-phase Z-source inverter delivering power to a three-phase AC motor load. The Z-source inductance and capacitance are determined using Eqs. 5.13 and 5.18. The parameters are tabulated in Table 5.1. As the boost factor B is 2, this gives a shoot-through duty ratio D_O value of 0.25 using Eq. 5.6. In the following section, models are developed using various boost control techniques for three-phase Z-source inverter, and the simulation results are presented.

5.2.2 *Model of Three-Phase Z-Source Inverter with Simple Boost Control*

A model of the three-phase Z-source inverter with simple boost control is shown in Fig. 5.4 (Model file: EXAMPLE 5_1). The parameters are shown in Table 5.1. As the maximum boost factor B is 2, shoot-through duty ratio D_O is 0.25, and the modulation index M is 0.75. Three sine wave function generators from sources library generating three-phase sine wave AC modulating signals having a frequency of 50 Hz and peak value 0.75 corresponding to that of M are compared with a triangle carrier having a positive and negative peak of +1 V and − 1 V and frequency 20 kHz using three relational operator blocks which form the comparators as in a conventional three-phase sine PWM inverter. When the triangle carrier wave is less than or equal to the respective sine wave modulating signal, the output of that comparator goes HIGH, or else its output is LOW. The output of the three comparators corresponding to each phase is a, b and c, respectively. These outputs a, b and c

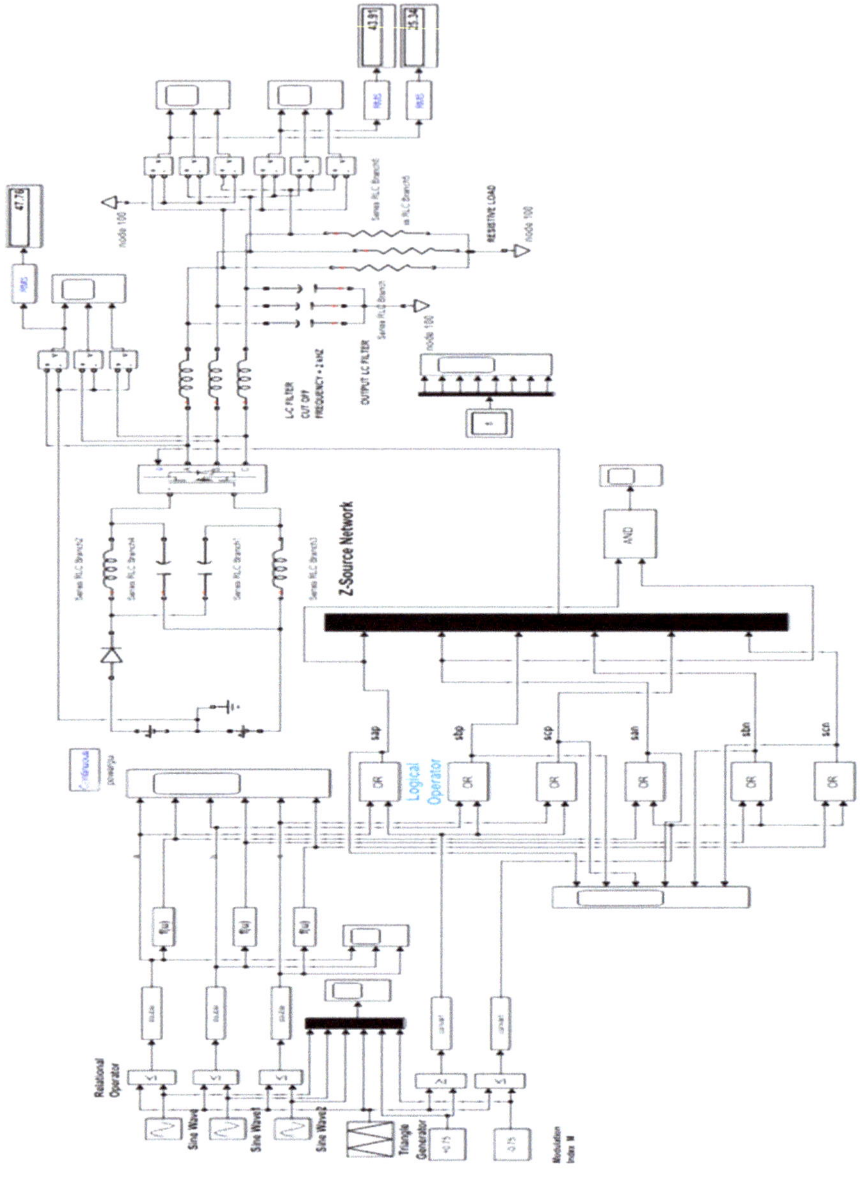

Fig. 5.4 Three phase Z-source inverter with simple boost control

are inverted using Fcn blocks using the relation (1—u(1)). The outputs are a_bar, b_bar and c_bar, respectively. Gate pulse a, b, c, a_bar, b_bar and c_bar forms the PWM gate drive for the upper and lower switches of the three-phase inverter. As these gate pulses correspond to conventional three-phase sine PWM inverter, no shoot-through state can be observed. To introduce shoot-through state, triangle carrier is compared with a constant block +0.75 which corresponds to Vp. The output of relational operator block which forms comparator goes HIGH when triangle carrier exceeds +0.75, or else its output is LOW. This output of this relational operator block is logically ORed with a, b and c signals derived earlier, using three separate OR gates. The output of these three OR gates forms Sap, Sbp and Scp, respectively. Similarly, the triangle carrier is compared with a constant block −0.75 which corresponds to Vn. The output of relational operator block which forms comparator goes HIGH when triangle carrier falls below −0.75, or else its output is LOW. The output of this relational operator block is logically ORed with a_bar, b_bar and c_bar signals derived earlier, using three separate OR gates. The output of these three OR gates forms San, Sbn and Scn, respectively. The gate pulses Sap, Sbp, and Scp and San, Sbn and Scn, respectively, form the gate drive for the upper and lower arm switches of the three-phase inverter. The six gate pulses thus generated provide the necessary shoot-through state. The gate pulses Sap and San corresponding to inverter switches in phase A are given to an AND gate whose output is used to measure the ST period T0.

5.2.3 Simulation Results

The simulation of the three-phase Z-source inverter with simple boost control is carried out using ode23tb (stiff/TR-BDF2) solver [6]. The data shown in Table 5.1 are used. The simulation results for the line-to-neutral output voltage, line-to-line output voltage, line-to-ground output voltage, Z-source inductor currents and capacitor voltages, three-phase load currents and inverter bridge DC voltage Vi are shown in Figs. 5.5 to 5.8, respectively. The shoot-through gate pulse for measurement of ST period T_O is shown in Fig. 5.9. In Fig. 5.9, measurement of shoot-through period T_O is made for one triangle carrier period commencing 0.190 Sec. to 0.19005 Sec. in three different locations. These values are, respectively, 3.165e-6 Sec., 6.276e-6 Sec.

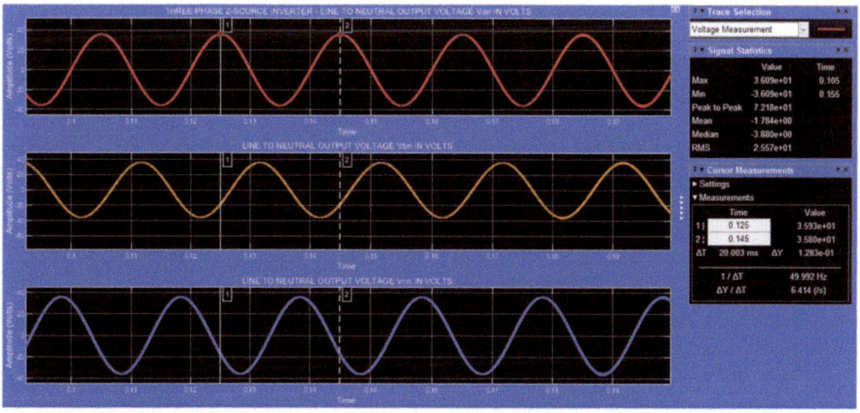

Fig. 5.5 Three phase ZSI with simple boost control—line to neutral output voltages

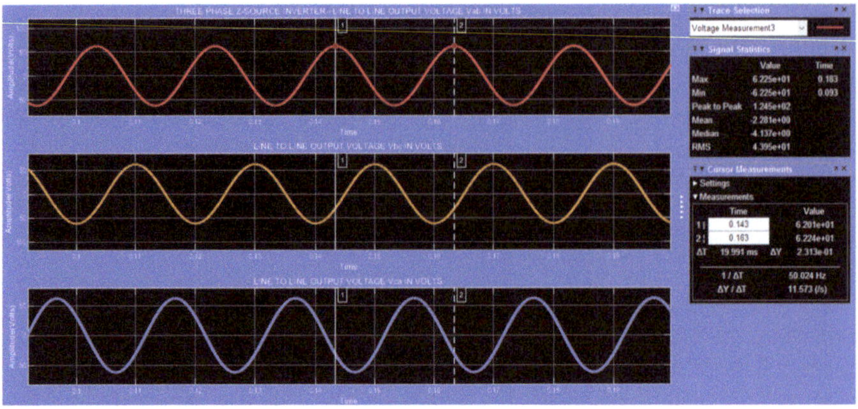

Fig. 5.6 Three phase ZSI with simple boost control—line to line output voltages

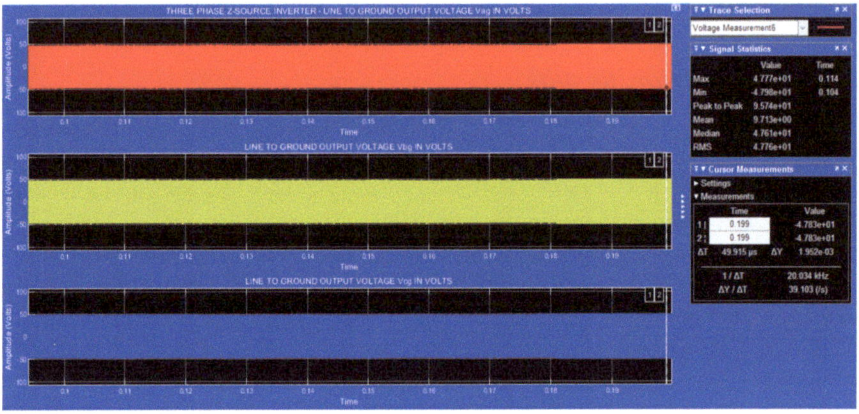

Fig. 5.7 Three phase ZSI with simple boost control—line to ground output voltgaes

and 3.165e-6 Sec. giving a total value of 12.606e-6 Sec. for T_O. The D0 value is 0.252. The simulation results are tabulated in Table 5.2.

5.2.4 Discussion of Results

The model of a three-phase Z-source inverter with simple boost control is developed and simulation results presented. The model values for B and G and model and simulation values for *D*o, $V_{ac(peak)}$, $V_{i(mean)}$, $v_{i(peak)}$ and $V_{C(mean)}$ almost closely well agree with the theoretically calculated values. In Fig. 5.9, cursor location at the end of 0.19005 Sec. is shown. Also from Fig. 5.7, it is seen that the RMS value of line-to-ground voltage is 47.76 V, whereas for a conventional inverter, this value is 24 V which is half the DC source voltage. This gives a boost factor *B* of 1.99.

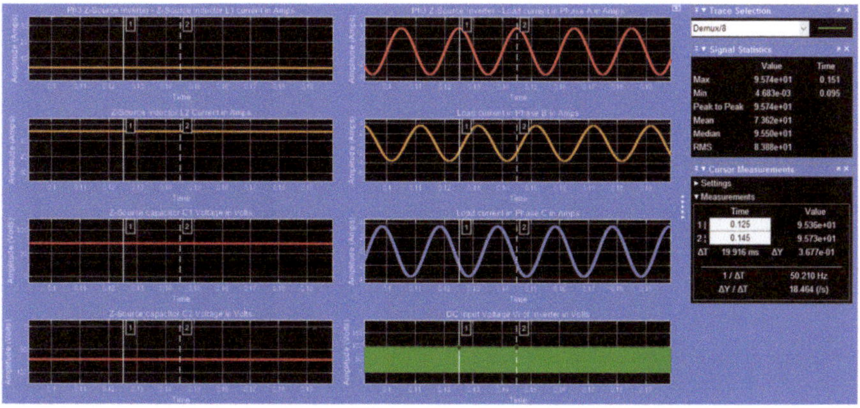

Fig. 5.8 Three phase ZSI with simple boost control—inductor currents iL1, iL2, capacitor voltages vC1, vC2 (Left column top to bottom), three phase load currents, inverter bridge DC link voltage (right column top to bottom)

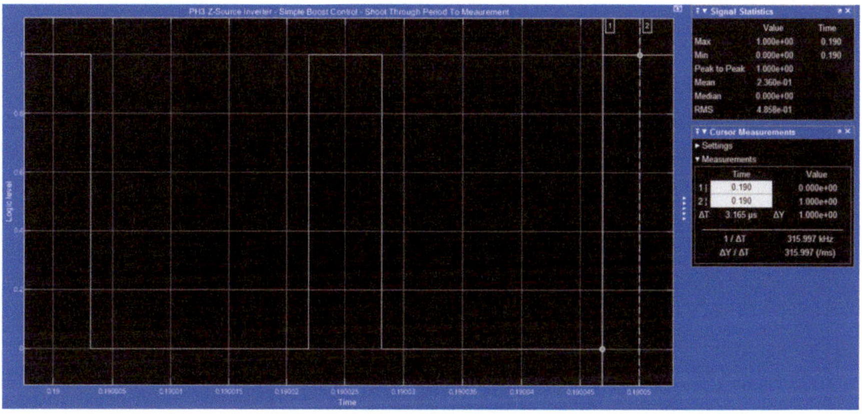

Fig. 5.9 Three phase Z-source inverter with simple boost control—shoot through period T0 measurement

Table 5.2 Three-phase Z-source inverter simple boost control:—Simulation results

S. no.	D_O	B	G B X M	V_{ac} (volts) peak	V_i volts avg./mean	v_i volts peak	V_C volts avg./mean	Remarks
(1)	0.25 [0.252]	2 (2.016)	1.5 (1.512)	36 (36.765) [36.09]	72 (72.38) [71.67]	96 (96.768) [95.74]	72 (72.38) [71.67]	Theoretical (model calculations) [Simulation results]

5.2.5 Maximum Boost Control

In the Z-source inverter, the product M*B represents the voltage gain and B*V_{dc} represents the voltage stress on the switching device. For any given modulation index M, B has to be maximised to increase the voltage gain which requires the shoot-through duty ratio D_O as large as possible. Figure 5.10 shows the maximum boost control (MBC). In this method, conventional sine PWM technique is used to maintain the six active switching states, and in addition, the triangle carrier is compared with the maximum and minimum value of three-phase sine wave modulating signals *va*, *vb* and *vc* to introduce shoot-through switching state. When the triangle carrier goes above the maximum value or goes below the minimum value of three-phase sine wave modulating signals, shoot-through states are introduced whose time period varies with each carrier cycle. Thus average value of the shoot-through time period T_O is calculated for the period from the minimum/maximum value to the maximum/minimum value of any two consecutively available three-phase sine wave modulating signals. This time period is $\theta = \omega.t = \pi/3$ radians where ω is the angular frequency of the sine wave modulating signals [2–4].

Referring to Fig. 5.10, shoot-through duty cycle repeats every π/3 radians, i.e. from the maximum of one modulating sine wave to the next minimum of another modulating sine wave. The total shoot-through time period and duty ratio $T_O(\theta)$ and $D_O(\theta)$ are derived below for the period when V_a goes maximum to the period V_c goes minimum, the duration of which is π/3 radians. The average value of D_O is obtained by integrating the expression for $D_O(\theta)$ and dividing by π/3.

In Fig. 5.10, the lines AB, BC and CD form the sides of the triangle carrier whose period is T Sec. The positive and negative peaks of the triangle carrier are +1 V and -1 V, respectively. The coordinates of the points are A(0, −1), B(T/2, +1), C(T, −1) and D(3 T/2, +1), respectively. Then the equation of the line AB, BC and CD can be expressed as follows:

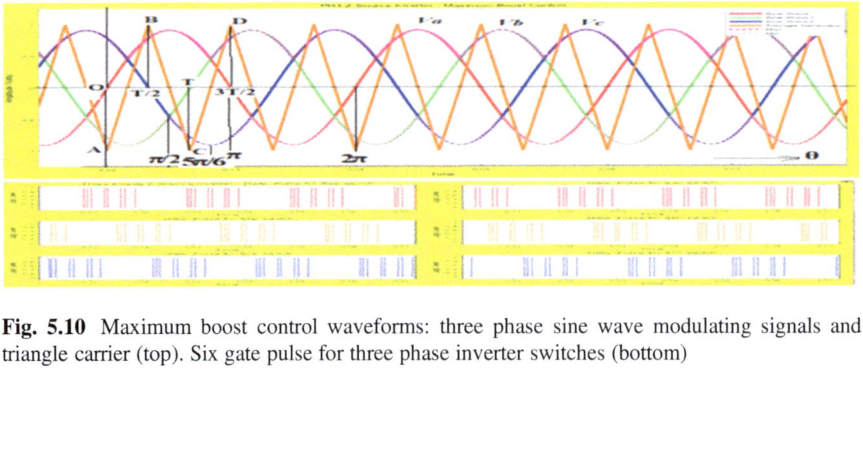

Fig. 5.10 Maximum boost control waveforms: three phase sine wave modulating signals and triangle carrier (top). Six gate pulse for three phase inverter switches (bottom)

$$\text{Equation of line AB} = \frac{(x-0)}{\left(\frac{T}{2}-0\right)} = \frac{(y-(-1))}{(+1-(-1))} \text{ which simplifies to } x$$

$$= \frac{T^{*}y}{4} + \frac{T}{4} \tag{5.20}$$

$$\text{Equation of line BC} = x = \frac{-T^{*}y}{4} + \frac{3^{*}T}{4} \tag{5.21}$$

$$\text{Equation of line CD} = x = \frac{T^{*}y}{4} + \frac{5^{*}T}{4} \tag{5.22}$$

When line AB cuts V_a = M*sin(θ), the point of intersection θ_1 can be expressed as follows:

$$\theta_1 = \frac{T * M * \sin(\theta)}{4} + \frac{T}{4} \tag{5.23}$$

When line BC cuts V_a = M*sin(θ), the point of intersection θ_2 can be expressed as follows:

$$\theta_2 = \frac{-T * M * \sin(\theta)}{4} + \frac{3 * T}{4} \tag{5.24}$$

When line BC cuts V_c = M*sin(θ + 2π/3), the point of intersection θ_3 can be expressed as follows:

$$\theta_3 = \frac{-T * M * \sin\left(\theta + \frac{2\pi}{3}\right)}{4} + \frac{3 * T}{4} \tag{5.25}$$

When line CD cuts V_c = M*sin(θ + 2π/3), the point of intersection θ_4 can be expressed as follows:

$$\theta_4 = \frac{T * M * \sin\left(\theta + \frac{2\pi}{3}\right)}{4} + \frac{5 * T}{4} \tag{5.26}$$

From Eqs. 5.23 and 5.24, the first shoot-through duration is given below:

$$(\theta_2 - \theta_1) = \frac{-T * M * \sin(\theta)}{2} + \frac{T}{2} \tag{5.27}$$

From Eqs. 5.25 and 5.26, the second shoot-through time duration is given below:

$$(\theta_4 - \theta_3) = \frac{T * M * \sin\left(\theta + \frac{2\pi}{3}\right)}{2} + \frac{T}{2} \tag{5.28}$$

The total shoot-through period T_O (θ) can be obtained by adding Eqs. 5.27 and 5.28 which is given below:

$$T_O(\theta) = \frac{T}{2} * M * sin\left(\theta + \frac{2\pi}{3}\right) - \frac{T}{2} * \ M * sin(\theta) + T \tag{5.29}$$

Dividing Eq. 5.29 by T and simplifying, D_O (θ) can be expressed as follows:

$$D_O(\theta) = \frac{2 - \left[M * sin(\theta) - M * sin\left(\theta + \frac{2\pi}{3}\right)\right]}{2} \tag{5.30}$$

Equation 5.30 has to be integrated for the instant commencing when V_a goes maximum (π/2) and the instant ending when V_c goes minimum (5π/6) and divided by the period π/3 which is (5π/6—π/2) to get the average shoot-through duty ratio D_O .

$$D_O = \frac{3}{2\pi} * \left[\int_{\frac{\pi}{2}}^{\frac{5\pi}{6}} \{2 - M * \sin(\theta) + M * \sin(\theta + \frac{2\pi}{3})\} * d\theta\right] \tag{5.31}$$

Integrating Eq. 5.31 and applying limits, D_O can be expressed as follows:

$$D_O = \frac{\left(2 * \pi - 3 * \sqrt{3} * M\right)}{2 * \pi} \tag{5.32}$$

Using Eq. 5.6, equation for B can be expressed as follows:

$$B = \frac{\pi}{\left(3 * \sqrt{3} * M - \pi\ \right)} \tag{5.33}$$

Voltage gain G can be expressed as follows:

$$G = \frac{V_{ac}}{\left(\frac{V_{dc}}{2}\right)} = B * M = \frac{\pi * M}{\left(3 * \sqrt{3} * M - \pi\ \right)} \tag{5.34}$$

5.2.6 *Model of Three-Phase Z-Source Inverter with Maximum Boost Control*

The model of the three-phase ZSI with MBC is shown in Fig. 5.11 (Model file: EXAMPLE 5_2). Now consider the three-phase ZSI whose parameters are given in Table 5.1. The maximum value of B is 2, and this gives a value of 0.907 for M using Eq. 5.33. Three sine wave function generators from sources library generating three-phase sine wave AC modulating signals having a frequency of 50 Hz and peak value

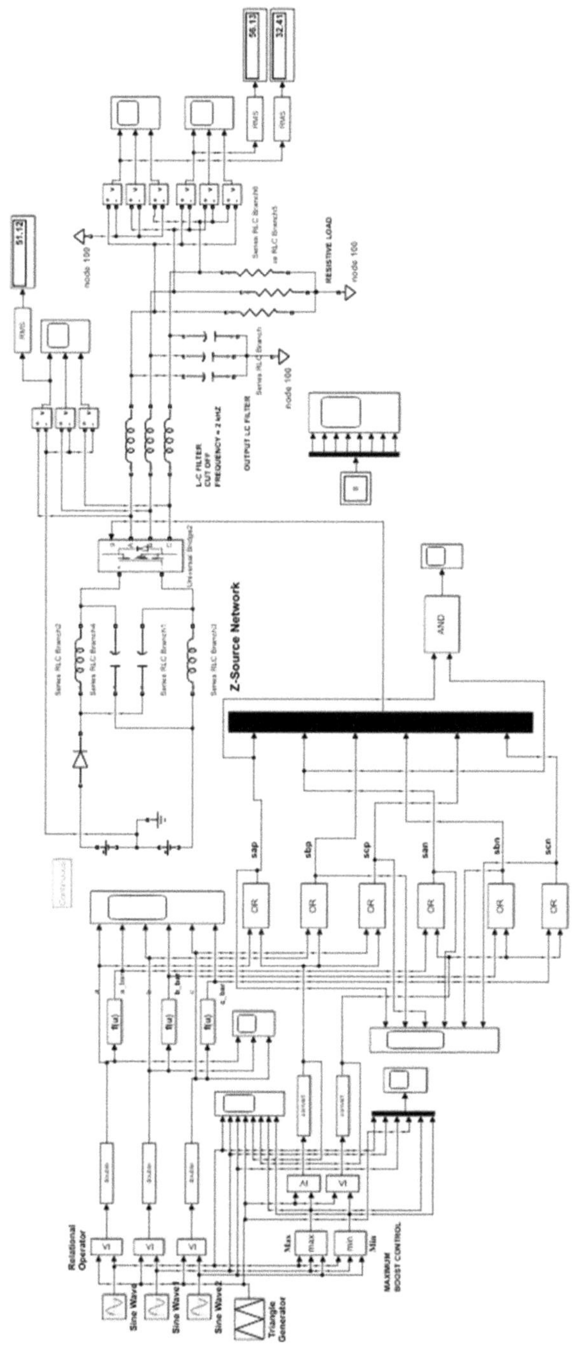

Fig. 5.11 Three phase Z-source inverter—maximum boost control

0.907 corresponding to that of M are compared with a triangle carrier having a positive and negative peak of +1 V and −1 V and frequency 20 kHz using three relational operator blocks which form the comparators as in a conventional three-phase sine PWM inverter. When the triangle carrier wave is less than or equal to the respective sine wave modulating signal, the output of that comparator goes HIGH else its output is LOW. The output of the three comparators corresponding to each phase is a, b and c, respectively. These outputs a, b and c are inverted using Fcn blocks using the relation (1—u(1)). The respective outputs are a_bar, b_bar and c_bar, respectively. Gate pulses a, b, c, a_bar, b_bar and c_bar form the PWM gate drive for the upper and lower switches of the three-phase inverter. As these gate pulses correspond to conventional three-phase sine PWM inverter, no shoot-through state can be observed. To introduce shoot-through state using MBC, triangle carrier is compared with the maximum of the three-phase sine wave modulating signals V_a, V_b and V_c. Maximum of V_a, V_b and V_c is determined using MinMax block in the Math operations library by selecting function Max and number of input ports three for the inputs V_a, V_b and V_c. The output of relational operator block which forms comparator goes HIGH when triangle carrier exceeds the maximum of V_a, V_b and V_c, or else its output is LOW. This output of this relational operator block is logically ORed with a, b and c signals derived earlier, using three separate OR gates. The output of these three OR gates form Sap, Sbp and Scp, respectively. Similarly, the triangle carrier is compared with the minimum of the three-phase sine wave modulating signals V_a, V_b and V_c. Minimum of V_a, V_b and V_c is determined using MinMax block in the Math operations library by selecting function Min and number of input ports three for the inputs V_a, V_b and V_c. The output of relational operator block which forms comparator goes HIGH when triangle carrier falls below the minimum of V_a, V_b and V_c, or else its output is LOW. The output of this relational operator block is logically ORed with a_bar, b_bar and c_bar signals derived earlier, using three separate OR gates. The output of these three OR gates forms San, Sbn and Scn, respectively. The gate pulses Sap, Sbp, Scp and San, Sbn and Scn, respectively, form the gate drive for the upper and lower arms of the three-phase inverter. The six gate pulses thus generated provide the necessary shoot-through state. The gate pulses Sap and San corresponding to inverter switches in phase A are given to an AND gate whose output is used to measure the ST period T0.

5.2.7 *Simulation Results*

The simulation of the three-phase Z-source inverter with maximum boost control is carried out using ode23tb (stiff/TR-BDF2) solver [6]. The data shown in Table 5.1 are used, with the M value replaced with 0.907. The simulation results for the line-to-neutral output voltage, line-to-line output voltage, line-to-ground output voltage, Z-source inductor currents and capacitor voltages, three-phase load currents and inverter bridge DC link voltage Vi are shown in Figs. 5.12 to 5.15, respectively. The shoot-through gate pulse for measurement of ST period T_O is shown in Fig. 5.16. In

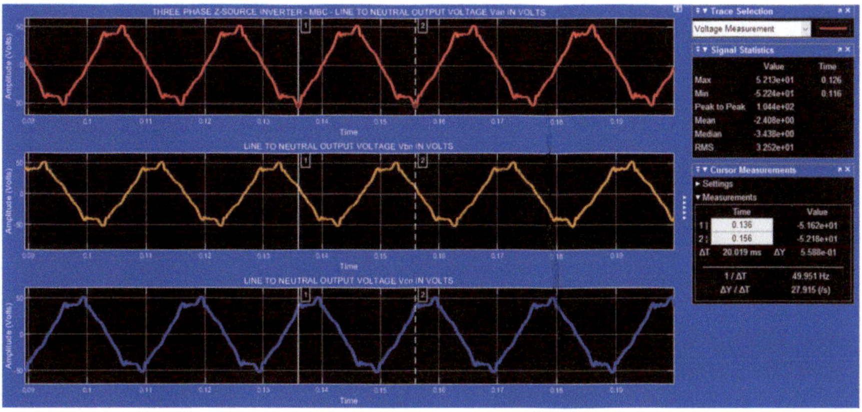

Fig. 5.12 Three phase ZSI with maximum boost control—line to neutral output voltage

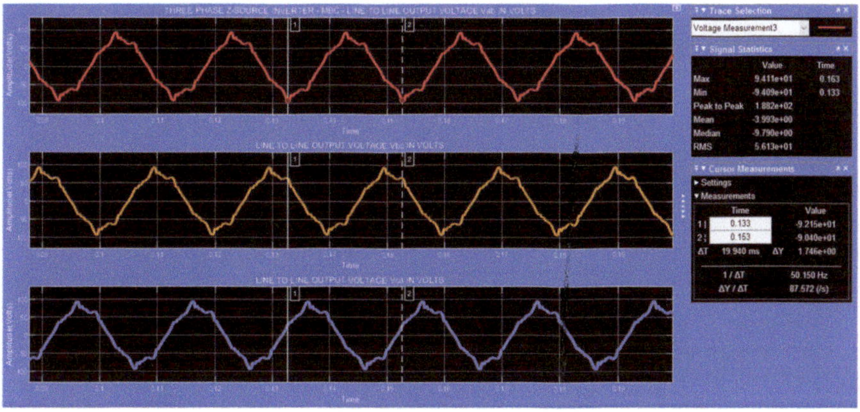

Fig. 5.13 Three phase ZSI with maximum boost control—line to line output voltage

Fig. 5.16, measurement of shoot-through period T_O is made for one triangle carrier period commencing 0.170 Sec. to 0.17005 Sec in three different locations. These values are 2.652e-6 Sec., 5.252e-6 Sec. and 2.756e-6 Sec., respectively, giving a total value of 10.66e-6 Sec. for T_O. The D0 value is 0.2132. The simulation results are tabulated in Table 5.3.

5.2.8 Discussion of Results

The model of a three-phase Z-Source inverter with maximum boost control is developed and simulation results presented. The model/simulation values for D_O, B and G differ from theoretical values by 14.72%, 12.8% and 8.56%, respectively.

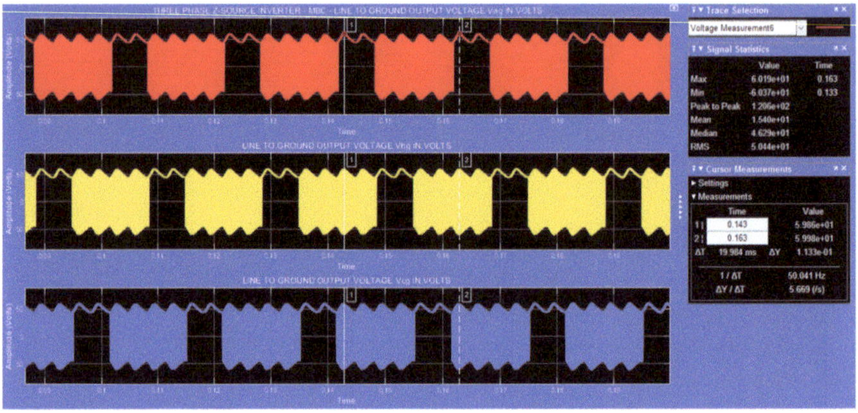

Fig. 5.14 Three phase ZSI with maximum boost control—line to ground output voltage

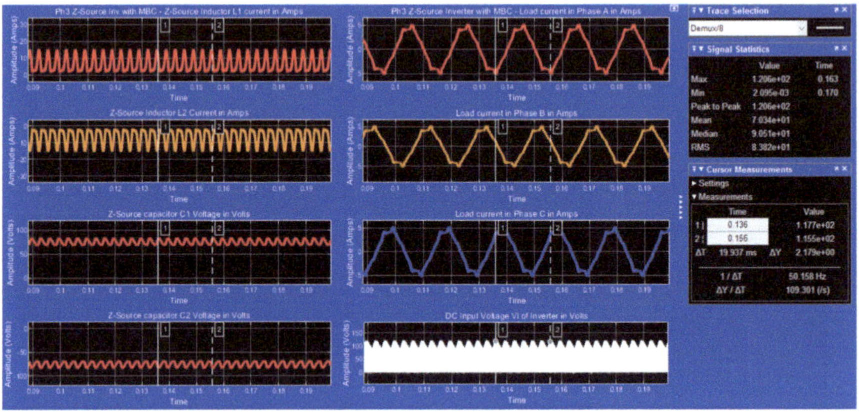

Fig. 5.15 Three phase ZSI with maximum boost control—inductor currents iL1, iL2, capacitor voltages vC1, vC2 (left column top to bottom), three phase load currents, inverter bridge DC link voltage (right Column top to bottom)

Model and simulation value for $V_{ac(peak)}$ differ from theoretical value by 8.56% and -19.7%, respectively. Model values for $V_{i(mean)}$, $v_{i(peak)}$ and $V_{C(mean)}$ differ from their theoretical values by 8.55%, 14.4% and 8.55%, respectively. Simulation results for $V_{i(mean)}$ and $V_{C(mean)}$ closely agree with theoretical values. Simulation results for $V_{ac(peak)}$ and $v_{i(peak)}$ differ by -19.74% and -25.6%, respectively. In Fig. 5.16, cursor location is shown in one location at the start of 0.17 s. Also from Fig. 5.14, it is seen that the RMS value of line-to-ground voltage is 50.44 V, whereas for a conventional inverter, this value is 24 V which is half the DC source voltage. This gives a boost factor B of 2.1016.

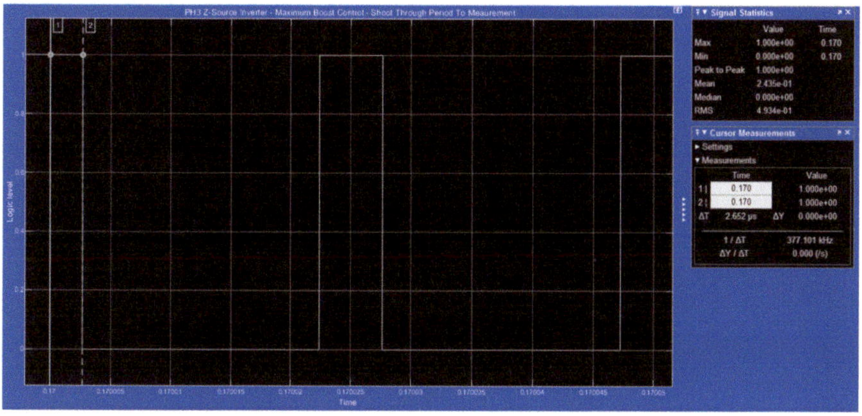

Fig. 5.16 Three phase ZSI with maximum boost control—shoot through period to measurement

Table 5.3 Three-phase Z-source inverter maximum boost control—Simulation results

S. No.	D_O	B	G B X M	V_{ac} (volts) peak	V_i volts avg./mean	v_i volts peak	V_C volts avg./mean	Remarks
(1)	0.25 [0.2132]	2 (1.7435)	1.814 (1.6587)	43.536 (39.81) [52.13]	72 (65.84) [70.34]	96 (82.14) [120.6]	72 (65.84) [76.17]	Theoretical (model calculations) [Simulation results]

5.2.9 Third Harmonic Injection Maximum Boost Control

Third harmonic injection maximum boost control (THIMBC) is shown in Fig. 5.17. In this Method, conventional sine PWM technique is used to maintain the six active switching states, and in addition, the triangle carrier is compared with the maximum and minimum value of three-phase third harmonic injection sine wave modulating signals vtha, vthb and vthc to introduce shoot-through switching state. When the triangle carrier goes above the maximum value or goes below the minimum value of three-phase third harmonic injection (THI) sine wave modulating signals, shoot-through states are introduced whose time period varies with each carrier cycle. Thus, average value of the shoot-through time period T_O is calculated for the period from the minimum/maximum value to the maximum/minimum value of any two consecutively available three-phase THI sine wave modulating signals. This time period is $\theta = \omega.t = \pi/3$ radians where ω is the angular frequency of the THI sine wave modulating signals. If M is the amplitude of the fundamental frequency sine wave, then a harmonic sine wave whose frequency is three times the frequency of the fundamental sine wave and amplitude (M/6) is added or injected to the fundamental

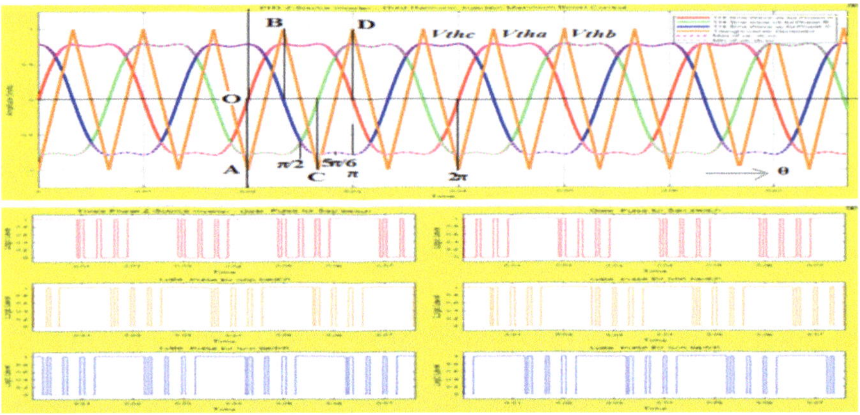

Fig. 5.17 Third harmonic injection maximum boost control: three phase third harmonic sine wave with triangle carrier (top). Six gate pulse for three phase inverter switches (bottom)

frequency sine wave. The general expression for the THI sine wave is given below [2–4].

$$Vth_K = M * \sin(\theta + \beta_K) + \frac{M}{6} * \sin(3\theta + \gamma_K) \tag{5.35}$$

where $\beta_K = 0, \frac{-2\pi}{3}, \frac{+2\pi}{3}$ and $\gamma_K = 0, -2\pi, +2\pi$ for $K = a, b, c$, respectively.

Following the derivation shown in Sect. 5.2.5 and noting the expression for the THI sine wave, D_O can be expressed as follows:

$$D_O = \frac{3}{2\pi} * \left[\int_{\frac{\pi}{2}}^{\frac{5\pi}{6}} \left\{ 2 - M * sin(\theta) + M * sin\left(\theta + \frac{2\pi}{3}\right) - \frac{M}{6} * sin(3\theta) + \frac{M}{6} * \sin(3\theta + 2\pi) \right\} * d\theta \right] \tag{5.36}$$

Integrating Eq. 5.36 and applying limits, D_O can be expressed as follows:

$$D_O = \frac{(2 * \pi - 3 * \sqrt{3} * M)}{2 * \pi} \tag{5.37}$$

Equation for B can be expressed as follows:

$$B = \frac{\pi}{(3 * \sqrt{3} * M - \pi)} \tag{5.38}$$

Voltage gain G can be expressed as follows:

$$G = \frac{V_{ac}}{\left(\frac{V_{dc}}{2}\right)} = B * M = \frac{\pi * M}{\left(3 * \sqrt{3} * M - \pi\right)} \tag{5.39}$$

5.2.10 Model of Three-Phase Z-Source Inverter with Third Harmonic Injection Maximum Boost Control

The model of the three-phase ZSI with THIMBC is shown in Fig. 5.18 (Model file: EXAMPLE 5_3). Now consider the three-phase ZSI whose parameters are given in Table 5.1. The maximum value of B is 2, and this gives a value of 0.907 for M using Eq. 5.38. The Embedded MATLAB function generates three-phase THI sine wave defined in Eq. 5.35. The source code for generating this three-phase THI sine wave is shown in Program segment 5.1 in the model file EXAMPLE 5_3 which are then compared with a triangle carrier having a positive and negative peak of +1 V and −1 V and frequency 20 kHz using three relational operator blocks which form the comparators as in a conventional three-phase sine PWM inverter. When the triangle carrier wave is less than or equal to the respective THI sine wave modulating signal, the output of that comparator goes HIGH else its output is LOW. The output of the three comparators corresponding to each phase is a, b and c, respectively. These outputs a, b and c are inverted using Fcn blocks using the relation (1—u(1)). The respective outputs are a_bar, b_bar and c_bar, respectively. Gate pulses a, b, c, a_bar, b_bar and c_bar form the PWM gate drive for the upper and lower switches of the three-phase inverter. As these gate pulses correspond to conventional three-phase sine PWM inverter, no shoot-through state can be observed. To introduce shoot-through state using THIMBC, triangle carrier is compared with the maximum of the three-phase THI sine wave modulating signals V_{tha}, V_{thb} and V_{thc}. Maximum of V_{tha}, V_{thb} and V_{thc} is determined using MinMax block in the Math operations library by selecting function Max and number of input ports three for the inputs V_{tha}, V_{thb} and V_{thc}. The output of relational operator block which forms comparator goes HIGH when triangle carrier exceeds the maximum of V_{tha}, V_{thb} and V_{thc}, or else its output is LOW. This output of this relational operator block is logically ORed with a, b and c signals derived earlier, using three separate OR gates. The output of these three OR gates form Sap, Sbp and Scp, respectively. Similarly the triangle carrier is compared with the minimum of the three-phase THI sine wave modulating signals V_{tha}, V_{thb} and V_{thc}. Minimum of V_{tha}, V_{thb} and V_{thc} is determined using MinMax block in the Math operations library by selecting function Min and number of input ports three for the inputs V_{tha}, V_{thb} and V_{thc}. The output of relational operator block which forms comparator goes HIGH when triangle carrier falls below the minimum of V_{tha}, V_{thb} and V_{thc}, or else its output is LOW. The output of this relational operator block is logically Ored with a_bar, b_bar and c_bar signals derived earlier using three

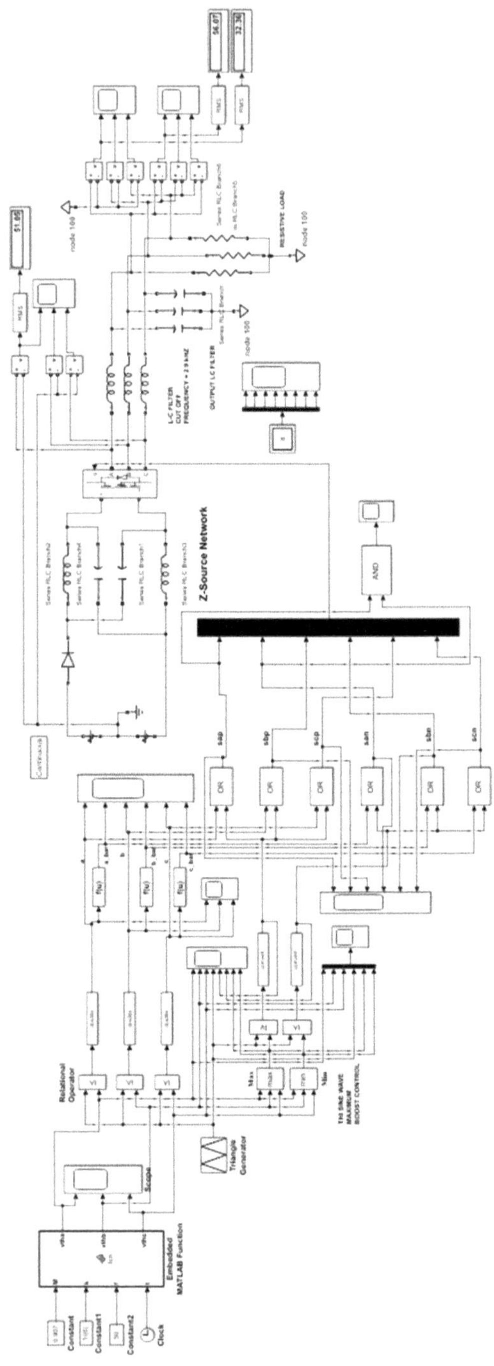

Fig. 5.18 Three phase Z-source inverter—third harmonic injection maximum boost control

separate OR gates. The output of these three OR gates forms San, Sbn and Scn, respectively. The gate pulses Sap, Sbp, Scp and San and Sbn and Scn, respectively, form the gate drive for the upper and lower arms of the three-phase inverter. The six gate pulses thus generated provide the necessary shoot-through state. The six gate pulses thus generated provide the necessary shoot-through state.

5.2.11 Simulation Results

The simulation of the three-phase Z-source inverter with THIMBC is carried out using ode23tb (stiff/TR-BDF2) solver [6]. The data shown in Table 5.1 are used, with the M value replaced with 0.907. The simulation results for the line-to-neutral output voltage, line-to-line output voltage, line-to-ground output voltage, Z-source inductor currents and capacitor voltages, three-phase load currents and inverter bridge DC link voltage Vi are shown in Fig. 5.19 to 5.22, respectively. The shoot-through gate pulse for To measurement is shown in Fig. 5.23. In Fig. 5.23, measurement of shoot-through period T_O is made for one triangle carrier period commencing 0.170 Sec. to 0.17005 Sec in three different locations. These values are, respectively, 2.698e-6 Sec., 5.396e-6 Sec. and 2.698e-6 Sec., respectively, giving a total value of 10.792e-6 Sec. for T_O. The D0 value is 0.214. The simulation results are tabulated in Table 5.4.

5.2.12 Discussion of Results

The model of a three-phase Z-source inverter with THIMBC is developed and simulation results presented. The model/simulation values for D_O, B and G differ

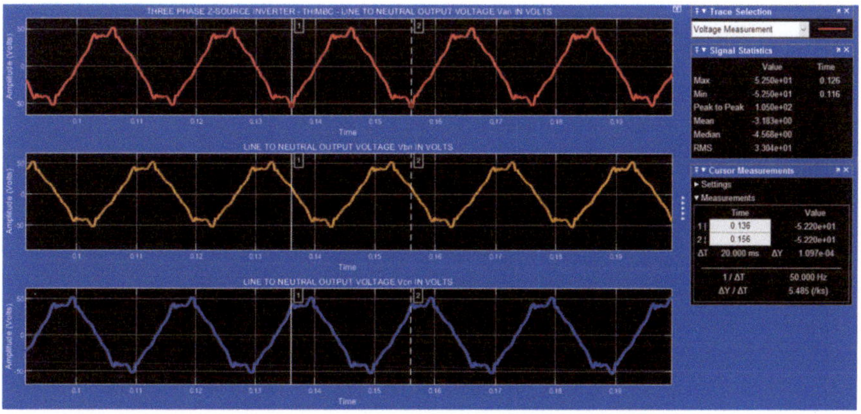

Fig. 5.19 Three phase ZSI with THIMBC—line to neutral output voltage

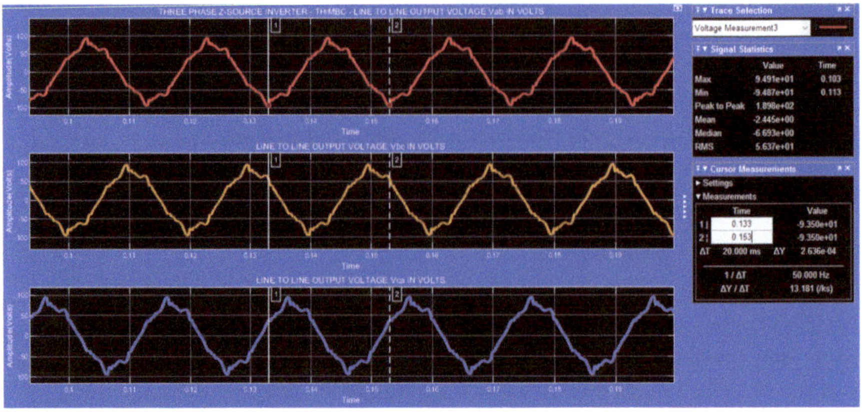

Fig. 5.20 Three phase ZSI with THIMBC—line to line output voltage

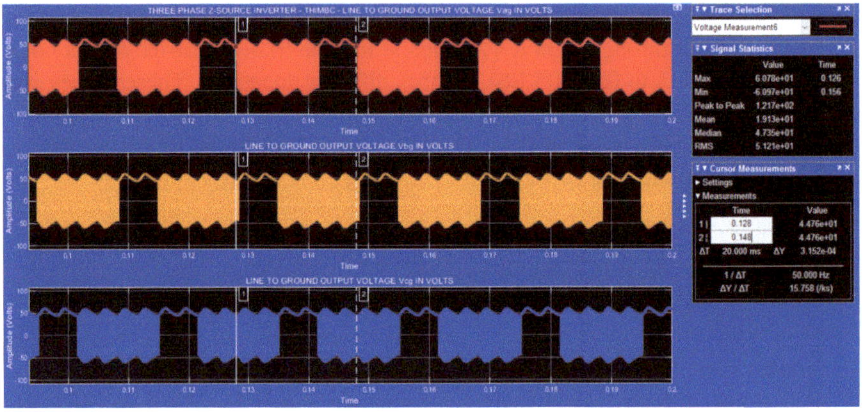

Fig. 5.21 Three phase ZSI with THIMBC—line to ground output voltage

from theoretical values by 13.6%, 11.95% and 7.9%, respectively. Model values for $V_{ac(peak)}$, $V_{i(mean)}$, $v_{i(peak)}$ and $V_{C(mean)}$ differ from their theoretical values by 7.94%, 7.96%, 11.9% and 7.96%, respectively. Simulation results for $V_{i(mean)}$ and $V_{C(mean)}$ closely agree with theoretical values. Simulation results for $V_{ac(peak)}$, $V_{i(mean)}$, $v_{i(peak)}$ and $V_{C(mean)}$ differ from the theoretical values by −20.6%, 13%, −26.8%, and −6.25%, respectively. In Fig. 5.23, cursor is shown in one location at the end of 0.17005 s. Also from Fig. 5.21, it is seen that the RMS value of line-to-ground voltage is 51.21 V where as for a conventional inverter, this value is 24 V which is half the DC source voltage. This gives a boost factor B of 2.134.

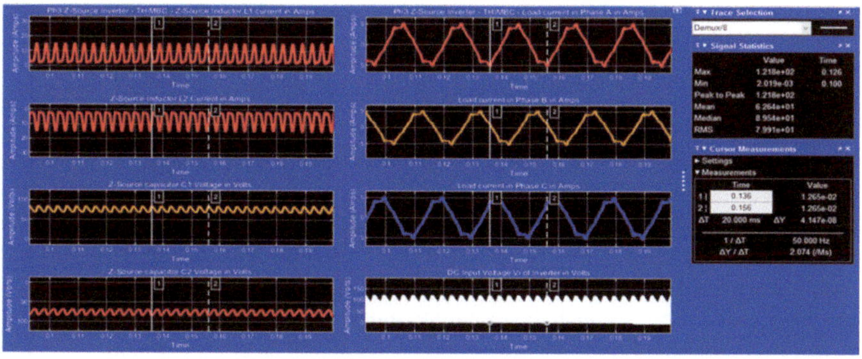

Fig. 5.22 Three phase ZSI with THIMBC—inductor currents iL1, iL2, capacitor voltages vC1, vC2 (left column top to bottom), three phase load currents, inverter bridge DC link voltage (right column top to bottom)

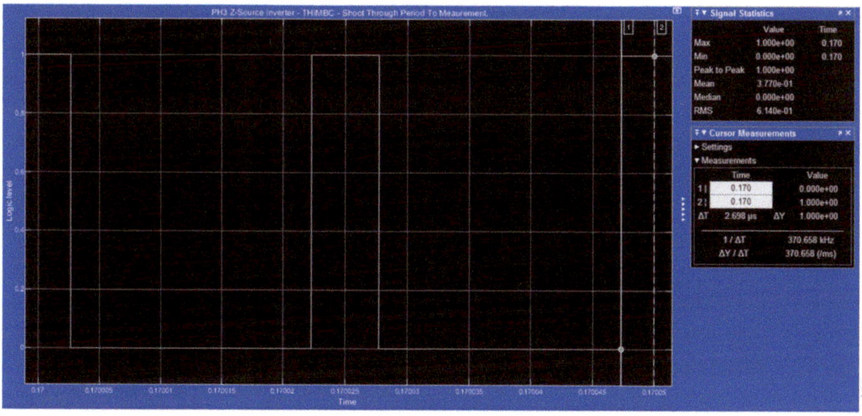

Fig. 5.23 Three phase ZSI with THIMBC—shoot through period to measurement

Table 5.4 Three-phase Z-source inverter THIMBC—Simulation results

S. No.	D_O	B	G B X M	V_{ac} (volts) peak	V_i volts avg./ Mean	v_i volts peak	V_C volts avg./ mean	Remarks
(1)	0.25 [0.216]	2 (1.761)	1.814 (1.67)	43.536 (40.08) [52.5]	72 (66.27) [62.64]	96 (84.53) [121.8]	72 (66.27)) [76.5]	Theoretical (model calculations) [Simulation results]

5.2.13 Third Harmonic Injection Maximum Constant Boost Control

To reduce the cost and size of Z-source source network, it is necessary to maintain the shoot—through duty ratio constant with each carrier switching cycle. This results in a constant boost factor B and a constant voltage gain for a given modulation index. The aim is to maintain the boost factor B as large as possible and constant throughout each carrier switching cycle. Referring to Eq. 5.30, the part of the time varying shoot-through duty ratio which is a function of θ takes the maximum value (√3*M), and the average shoot-through duty ratio D_O thus becomes [1 – (√3*M/2)]. One way to achieve this shoot-through duty ratio is by using a third harmonic injected modulating sine wave as defined in Eq. 5.35. Here, a constant positive and negative reference voltage +Vp and -Vn with magnitude (+√3*M/2) and (−√3*M/2), respectively, are used. In this method, conventional THI sine PWM technique is used to maintain the six active switching states, and in addition, the triangle carrier is compared with +Vp and –Vn reference values. When the triangle carrier goes above the reference value +Vp or goes below the reference value -Vn, shoot-through states are introduced. Thus, average value of the shoot-through time period T_O is calculated for the period from the minimum/maximum value to the maximum/minimum value of any two consecutively available three-phase THI sine wave modulating signals. This time period is $\theta = \omega.t = \pi/3$ radians where ω is the angular frequency of the THI sine wave modulating signals. Maximum constant boost control (MCBC) is shown in Fig. 5.24. Referring to Fig. 5.24, shoot-through duty cycle repeats every π/3 radians, i.e. from the maximum of one modulating THI sine wave to the next minimum another modulating THI sine wave. The total shoot-through time period and duty ratio $T_O(\theta)$ and $D_O(\theta)$ are derived below for the period when V_a goes maximum to the period V_c goes minimum, the duration of which is π/3 radians. The average value of D_O is obtained by integrating the expression for $D_O(\theta)$ and dividing by π/3.

In Fig. 5.24, the lines AB, BC and CD form the sides of the triangle carrier whose period is *T* Sec. The positive and negative peaks of the triangle carrier are +1 V and

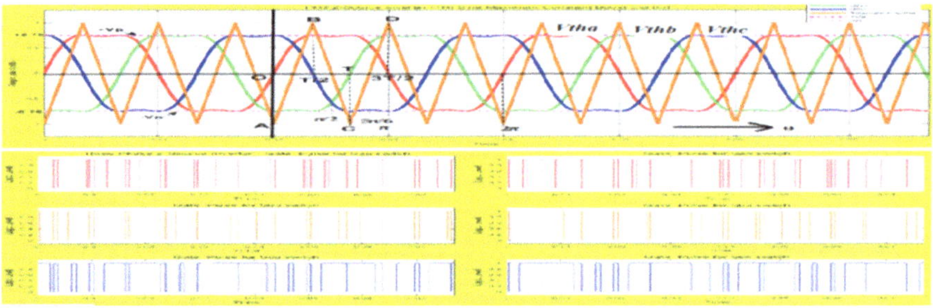

Fig. 5.24 Three phase Z-source inverter with THI sine maximum constant boost control (MCBC)—three phase THI sine modulating signals with triangle carrier (top) and gate pulse for six inverter switches (bottom)

-1 V, respectively. The coordinates of the points are A(0, -1), B(T/2, +1), C(T, -1) and D(3 T/2, +1), respectively. Then the equation of the line AB, BC and CD can be expressed as in Eqs. 5.20 to 5.22. Reference voltages *Vp* and *Vn* are expressed as in Eqs. 5.40 and 5.41:

$$\mathrm{y} = \frac{\sqrt{3} * M}{2} \tag{5.40}$$

$$\mathrm{y} = \frac{-\sqrt{3} * M}{2} \tag{5.41}$$

Let line AB cuts *Vp* at θ1 which can be expressed as follows:

$$\theta_1 = \frac{T}{4} * \left(\frac{\sqrt{3} * M}{2}\right) + \frac{T}{4} \tag{5.42}$$

Let line BC cuts *Vp* at θ2 which can be expressed as follows:

$$\theta_2 = \frac{-T}{4} * \left(\frac{\sqrt{3} * M}{2}\right) + \frac{3 * T}{4} \tag{5.43}$$

Thus from Eqs. 5.42 and 5.43, first shoot-through pulse width be expressed as follows:

$$(\theta_2 - \theta_1) = \frac{-T * \sqrt{3} * M}{4} + \frac{T}{2} \tag{5.44}$$

Let BC cuts *Vn* at θ3 which can be expressed as follows:

$$\theta_3 = \frac{T}{4} * \left(\frac{\sqrt{3} * M}{2}\right) + \frac{3 * T}{4} \tag{5.45}$$

Let CD cuts *Vn* at θ4 which can be expressed as follows:

$$\theta_4 = \frac{-T}{4} * \left(\frac{\sqrt{3} * M}{2}\right) + \frac{5 * T}{4} \tag{5.46}$$

Thus from Eqs. 5.45 and 5.46, second shoot-through pulse width be expressed as follows:

$$(\theta_4 - \theta_3) = \frac{-T * \sqrt{3} * M}{4} + \frac{T}{2} \tag{5.47}$$

The shoot-through time interval $T_O(\theta)$ can be obtained by adding Eqs. 5.44 and 5.47. This is given below:

$$T_O(\theta) = \frac{-T * \sqrt{3} * M}{2} + T \tag{5.48}$$

The shoot-through duty ratio $D_O(\theta)$ can be expressed as follows:

$$D_O(\theta) = \left[1 - \left(\frac{\sqrt{3} * M}{2}\right)\right] \tag{5.49}$$

Equation 5.49 has to be integrated for the instant commencing when V_a goes maximum (π/2) and the instant ending when V_c goes minimum (5π/6) and divided by the period π/3 which is (5π/6 – π/2) to get the average shoot-through duty ratio D_O. This is given below:

$$D_O(\theta) = \frac{3}{\pi} * \int_{\frac{\pi}{2}}^{\frac{5\pi}{6}} \left[1 - \left(\frac{\sqrt{3} * M}{2}\right)\right] * d\theta \tag{5.50}$$

$$D_O = \left[1 - \left(\frac{\sqrt{3} * M}{2}\right)\right] \tag{5.51}$$

In this case as the triangle carrier cuts reference voltage lines Vp and –Vn parallel to the x-axis, the shoot-through period will be uniform for each carrier cycle. This is clear from Eqs. 5.49 and 5.51. Equation for B can be expressed as follows:

$$B = \frac{1}{(\sqrt{3} * M - 1)} \tag{5.52}$$

Voltage gain G can be expressed as follows:

$$G = \frac{V_{ac}}{\left(\frac{V_{dc}}{2}\right)} = B * M = \frac{M}{(\sqrt{3} * M - 1)} \tag{5.53}$$

5.2.14 Model of Three-Phase Z-Source Inverter with Third Harmonic Injection Maximum Constant Boost Control

The model of the three-phase ZSI with third harmonic injection maximum constant boost control (THIMCBC) is shown in Fig. 5.25 (Model file: EXAMPLE 5_4). Now consider the three-phase ZSI whose parameters are given in Table 5.1. The

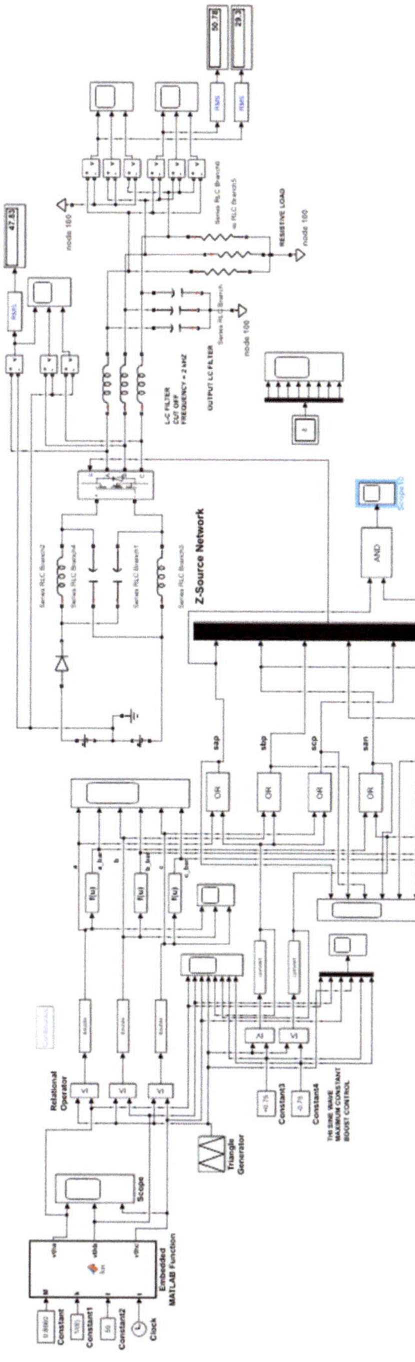

Fig. 5.25 Three phase Z-source inverter—third harmonic injection maximum constant boost control

maximum value of B is 2, and this gives a value of 0.866 for M using Eq. 5.52. The Embedded MATLAB function generates three-phase THI sine wave defined in Eq. 5.35. The source code for generating this three-phase THI sine wave is shown in Program segment 5.1 in the model file EXAMPLE 5_4, which are then compared with a triangle carrier having a positive and negative peak of +1 V and −1 V, 20 kHz frequency using three relational operator blocks which form the comparators as in a conventional three-phase sine PWM inverter. When the triangle carrier wave is less than or equal to the respective THI sine wave modulating signal, the output of that comparator goes HIGH, or else its output is LOW. The outputs of the three comparators corresponding to each phase is a, b and c, respectively. These outputs a, b and c are inverted using Fcn blocks using the relation (1 − u(1)). The respective outputs are a_bar, b_bar and c_bar, respectively. Gate pulses a, b, c, a_bar, b_bar and c_bar form the PWM gate drive for the upper and lower switches of the three-phase inverter. As these gate pulses correspond to conventional three-phase sine PWM inverter, no shoot-through state can be observed. To introduce shoot-through state, triangle carrier is compared with a constant block +0.75 (+√3*M/2) which corresponds to Vp. The output of relational operator block which forms comparator goes HIGH when triangle carrier exceeds +0.75 (+√3*M/2), or else its output is LOW. This output of this relational operator block is logically ORed with a, b and c signals derived earlier using three separate OR gates. The output of these three OR gates forms Sap, Sbp and Scp, respectively. Similarly, the triangle carrier is compared with a constant block −0.75 (−√3*M/2) which corresponds to Vn. The output of relational operator block which forms comparator goes HIGH when triangle carrier falls below −0.75 (−√3*M/2), or else its output is LOW. The output of this relational operator block is logically ORed with a_bar, b_bar and c_bar signals derived earlier, using three separate OR gates. The output of these three OR gates forms San, Sbn and Scn, respectively. The gate pulses Sap, Sbp, Scp and San and Sbn and Scn, respectively, form the gate drive for the upper and lower arms of the three-phase inverter. The six gate pulses thus generated provide the necessary shoot-through state.

5.2.15 Simulation Results

The simulation of the three-phase Z-source inverter with THIMCBC is carried out using ode23tb (stiff/TR-BDF2) solver [6]. The data shown in Table 5.1 are used, with the M value replaced with 0.866. The simulation results for the line-to-neutral output voltage, line-to-line output voltage, line-to-ground output voltage, Z-source inductor currents and capacitor voltages, three-phase load currents and inverter bridge DC link voltage Vi are shown in Figs. 5.26 to 5.29, respectively. The gate pulse for shoot-through period T_O measurement is shown in Fig. 5.30. In Fig. 5.30, measurement of shoot-through period T_O is made for one triangle carrier period commencing 0.190 Sec. to 0.19005 Sec in three different locations. These values are,

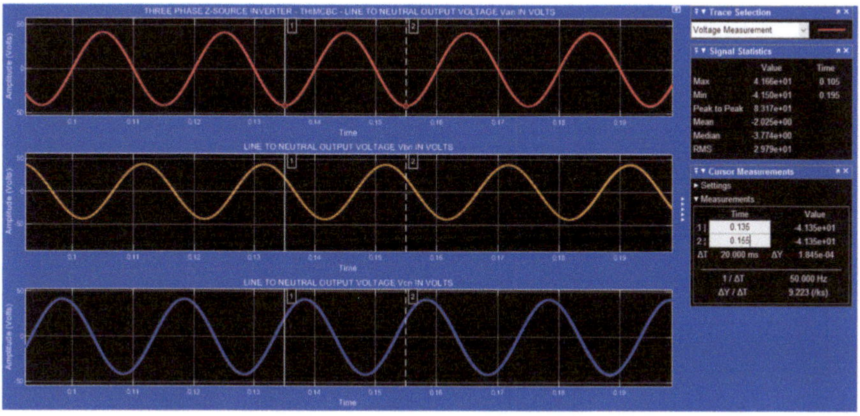

Fig. 5.26 Three phase ZSI with THIMCBC—line to neutral output voltage

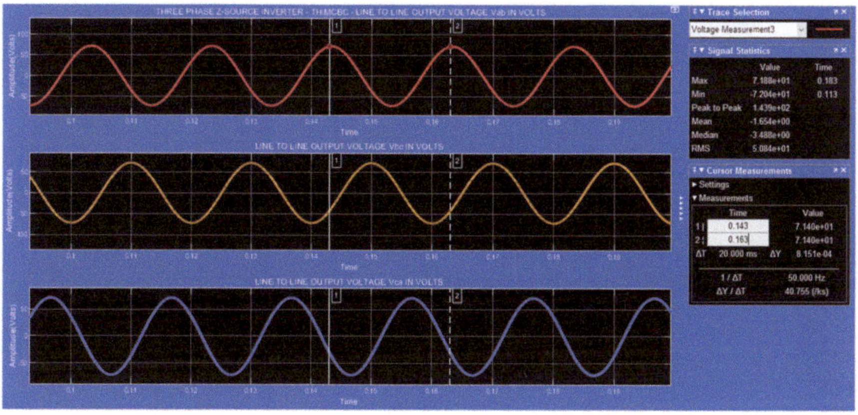

Fig. 5.27 Three phase ZSI with THIMCBC—line to line voltage

respectively, 3.146e-6 Sec., 6.240e-6 Sec. and 3.146e-6 Sec., respectively, giving a total value of 12.532e-6 Sec. for T_O. This gives D0 value of 0.2506. The simulation results are tabulated in Table 5.5.

5.2.16 Discussion of Results

The model of a three-phase Z-source inverter with THIMCBC is developed, and simulation results are presented. The model/simulation values for D_O, B and G differ from theoretical values by a narrow margin. Both model and simulation values for $V_{ac(peak)}$, $V_{i(mean)}$, $v_{i(peak)}$ and $V_{C(mean)}$ closely agree with their respective theoretical values. In Fig. 5.30, cursor is shown in one location at the end of 0.19005 Seconds.

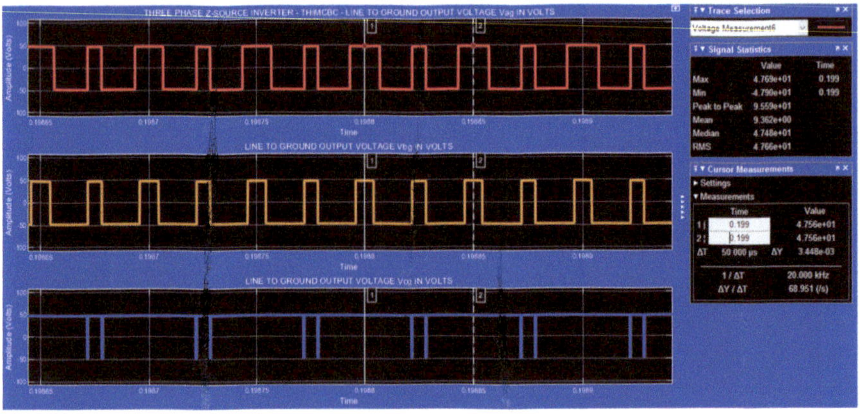

Fig. 5.28 Three phase ZSI with THIMCBC—line to ground voltage

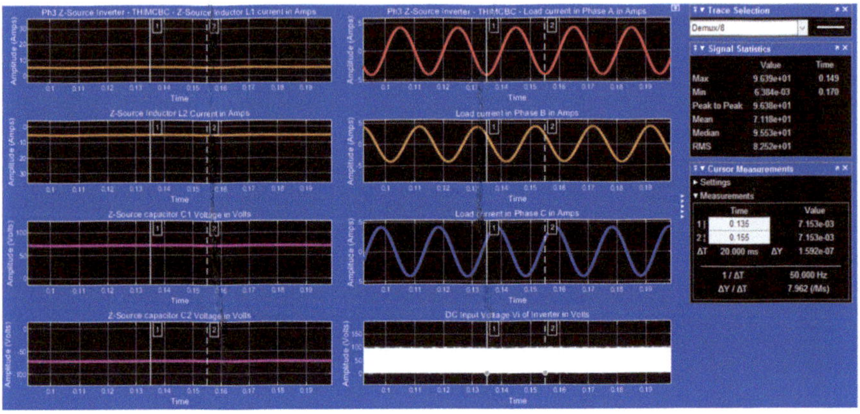

Fig. 5.29 Three phase ZSI with THIMCBC—inductor currents iL1, iL2, capacitor voltages vC1, vC2 (left column top to bottom), three phase load currents, inverter bridge DC link voltage (right column top to bottom)

Also from Fig. 5.28, it is seen that the RMS value of line-to-ground voltage is 47.66 V, whereas for a conventional inverter, this value is 24 V which is half the DC source voltage. This gives a boost factor B of 1.986.

5.3 Three-Phase Quasi-Z-Source Inverter

The quasi-Z-source inverter (QZSI) configuration is shown in Fig. 5.31. The QZSI can be operated in the non-shoot-through (NST) mode (six active state vectors and two zero state vectors) and in the shoot-through (ST) mode where the switches in the

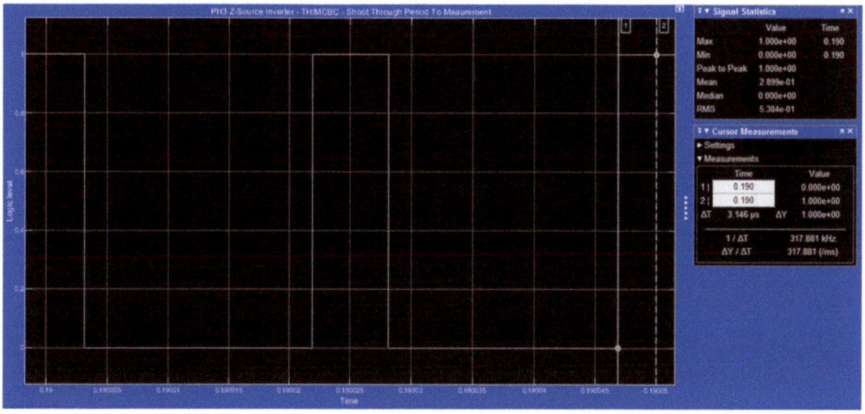

Fig. 5.30 Three phase ZSI with THIMCBC—shoot through period to measurement

Table 5.5 Three-phase Z-source inverter THIMCBC—Simulation results

S. No.	D_O	B	G B X M	V_{ac} (volts) peak	V_i volts avg./mean	v_i volts peak	V_C volts avg./mean	Remarks
(1)	0.25 [0.2506]	2 (2.005)	1.732 (1.735)	41.57 (41.64) [41.66]	72 (72.12) [71.18]	96 (98.52) [96.4]	72 (72.12) [71.7]	Theoretical (model calculations) [Simulation results]

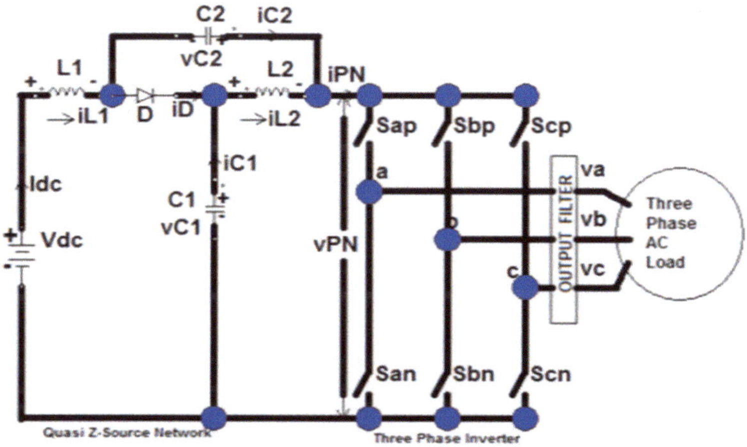

Fig. 5.31 Three phase quasi Z-source inverter

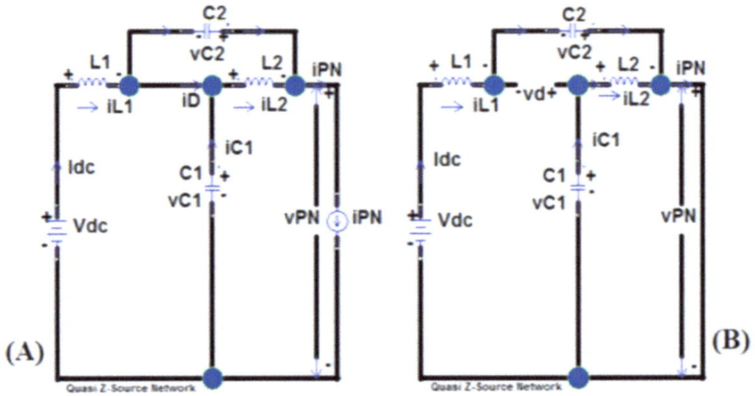

Fig. 5.32 Three phase quasi z-source inverter: (**a**) non-shoot through state and (**b**) shoot through state

upper and lower legs of one or all the three phase are switched simultaneously providing a direct current shoot-through path to the ground thus boosting the output voltage of inverter compared to the DC input voltage. This QZSI can also be operated in the buck mode when the output voltage will be less than the input DC voltage. The equivalent circuit of the QZSI in the NST and ST modes are shown in Fig. 5.32a, b, respectively. During NST state, the inverter acts as a current source delivering power to the load, and during ST state, the upper and lower switches of one or more legs of inverter are short circuited and thereby opening the diode switch. The circuit analysis during NST and ST state is given below [7].

Referring to Fig. 5.32a, during the NST state, the following equations apply [7]:

$$v_{L1} = (V_{dc} - V_{C1}) \tag{5.54}$$

$$v_{L2} = (V_{C1} - v_{PN}) = -V_{c2} \tag{5.55}$$

$$v_{PN} = (V_{C1} + V_{C2}) \tag{5.56}$$

$$v_D = 0 \tag{5.57}$$

$$i_{C1} = (i_{PN} - I_{L1}) \tag{5.58}$$

$$i_{C2} = (i_{PN} - I_{L2}) \tag{5.59}$$

Referring to Fig. 5.32b, during the ST state, the following equations apply [7]:

$$v_{L1} = (V_{dc} + V_{C2}) \tag{5.60}$$

$$v_{L2} = V_{C1} \tag{5.61}$$

$$v_{PN} = 0 \tag{5.62}$$

$$v_D = (V_{C1} + V_{C2}) \tag{5.63}$$

$$i_{C1} = (I_{L2}) \tag{5.64}$$

$$i_{C2} = (I_{L1}) \tag{5.65}$$

The carrier switching period T, NST period T_1, ST period T_O and shoot-through duty-ratio D_O are as defined in Sect. 5.2. The average capacitor voltage across C1 and C2 over one carrier switching period T is zero. This results in the following equations:

$$(1 - D_O) * T^*(i_{PN} - I_{L1}) * \frac{1}{C_1} + D_O{}^* T^* I_{L2}{}^* \frac{1}{C_1} = 0 \tag{5.66}$$

$$(1 - D_O) * T^*(i_{PN} - I_{L2}) * \frac{1}{C_2} + D_O{}^* T^* I_{L1}{}^* \frac{1}{C_2} = 0 \tag{5.67}$$

Subtracting Eq.5.67 from 5.66 results in the following:

$$I_{L1} = I_{L2} \tag{5.68}$$

If P is the input power, then

$$\mathrm{I_{L1} = I_{L2} = (P/V_{dc})} \tag{5.69}$$

The average inductor current flowing through L1 and L2 over one carrier switching period is zero which results in the following set of equations:

$$(V_{dc} - V_{C1}) * (1 - D_O) * T * \frac{1}{L_1} + (V_{dc} + V_{C2}) * D_O * T * \frac{1}{L_1} = 0 \tag{5.70}$$

$$- V_{C2} * (1 - D_O) * T * \frac{1}{L_2} + V_{C1} * D_O * T * \frac{1}{L_2} = 0 \tag{5.71}$$

Simplifying Eqs. 5.70 and 5.71, the following equations result:

$$V_{C1} = \frac{(V_{dc} + D_O * V_{C2})}{(1 - D_O)} \tag{5.72}$$

$$V_{C2} = \frac{D_O * V_{C1}}{(1 - D_O)} \tag{5.73}$$

Using Eq. 5.73 in Eq. 5.72 and simplifying,

$$V_{C1} = \frac{V_{dc} * (1 - D_O)}{(1 - 2 * D_O)} \tag{5.74}$$

Using Eq. 5.74 in Eq. 5.73 and simplifying,

$$V_{C2} = \frac{V_{dc} * D_O}{(1 - 2 * D_O)} \tag{5.75}$$

The peak inverter bridge DC link voltage is given below:

$$v_{PN(peak)} = (V_{C1} + V_{C2}) = \frac{V_{dc}}{(1 - 2 * D_O)} = B * V_{dc} \tag{5.76}$$

The average inverter bridge DC link voltage is given below:

$$v_{PN(avg)} = \frac{V_{dc} * (1 - D_O)}{(1 - 2 * D_O)} = B * (1 - D_O) * V_{dc} \tag{5.77}$$

where $B = [1/(1\text{–}2^*D_O)]$ is the boost factor. In the buck mode, the QZSI is operated entirely in the NST mode. The ST duty ratio D_O is zero, and from Eqs. 5.74 and 5.75, $V_{C1} = V_{dc}$ and $V_{C2} = 0$, respectively. The peak line-to-neutral output voltage is defined below:

$$v_{ln(peak)} = \frac{M * v_{PN(peak)}}{2} = \frac{M * V_{dc}}{2} \tag{5.78}$$

where M is the Amplitude Modulation Index.

If the QZSI is operated in the ST mode, the peak line-to-neutral output voltage can be expressed as follows:

$$v_{ln(peak)} = \frac{M * v_{PN(peak)}}{2} = \frac{M * B * V_{dc}}{2} \tag{5.79}$$

The boost factor B is always greater than unity, and the line-to-neutral output voltage is boosted compared to that of a conventional inverter.

A method to select the values of inductors and capacitors for the quasi-Z-source network is given below [8]:

During the ST period, referring to Fig. 5.32b, the following state space Eq. 5.80 applies [8].

The inductance and capacitance values of quasi-Z-source network are selected according to current ripple and voltage ripple, respectively. Thus, the following general expressions can be written as in Eq. 5.81.

$$\begin{bmatrix} \frac{dv_{C1}}{dt} \\ \frac{dv_{C2}}{dt} \\ \frac{di_{L1}}{dt} \\ \frac{di_{L2}}{dt} \end{bmatrix} = \begin{bmatrix} 0 & 0 & 0 & \frac{-1}{C_1} \\ 0 & 0 & \frac{-1}{C_2} & 0 \\ 0 & \frac{1}{L_1} & 0 & 0 \\ \frac{1}{L_2} & 0 & 0 & 0 \end{bmatrix} * \begin{bmatrix} v_{C1} \\ v_{C2} \\ i_{L1} \\ i_{L2} \end{bmatrix} + V_{dc} * \begin{bmatrix} 0 \\ 0 \\ \frac{1}{L_1} \\ 0 \end{bmatrix} \tag{5.80}$$

$$L_x * \frac{di_{Lx}}{dt} = L_x * \frac{\Delta i_{Lx}}{\Delta t} = v_{Lx}, C_x * \frac{dv_{Cx}}{dt} = C_x * \frac{\Delta v_{Cx}}{\Delta t} = i_{Cx}\ and\ x \in 1,2. \quad (5.81)$$

During the shoot-through period, referring to Eqs. 5.80 and 5.81, the following equations can be written:

$$\begin{aligned} \Delta v_{C1} &= \frac{i_{L2} * D_0 * T}{C_1}, \Delta v_{C2} = \frac{i_{L1} * D_0 * T}{C_2}, \\ \Delta i_{L1} &= \frac{(v_{C2} + V_{dc}) * D_0 * T}{L_1}, \Delta i_{L2} = \frac{v_{C1} * D_0 * T}{L_2} \end{aligned} \quad (5.82)$$

In Eq. 5.81, $\Delta v_{Cx} = r_C * v_{Cx}$ and $\Delta i_{Lx} = r_L * i_{Lx}$ where r_C and r_L are the capacitor and inductor voltage and current ripple ratio, respectively. Using this relation along with Eqs. 5.69, 5.74, 5.75 and 5.82, the following equations can be derived [8]:

$$\begin{aligned} C_1 &\geq \frac{P * D_0 * (1 - 2 * D_0) * T}{r_{C1} * (1 - D_0) * V_{dc}^2}, C_2 \geq \frac{P * (1 - 2 * D_0) * T}{r_{C2} * V_{dc}^2}, \\ L_1 = L_2 &\geq \frac{V_{dc}^2 * (1 - D_0) * D_0 * T}{P * r_L * (1 - 2 * D_0)}. \end{aligned} \quad (5.83)$$

The various boost control methods are (1) simple boost control (SBC), (2) maximum boost control (MBC), (3) third harmonic injection maximum boost control (THIMBC) and (4) third harmonic injection maximum constant boost control (THIMCBC). These are already explained in Sects. 5.2.1, 5.2.5, 5.2.9 and 5.2.13.

5.3.1 Model of Three-Phase Quasi-Z-Source Inverter Using Simple Boost Control

A model of the three-phase quasi-Z-source inverter (QZSI) with simple boost control is shown in Fig. 5.33 (Model file: EXAMPLE 5_5). The parameters are shown in Table 5.6. As the maximum boost factor *B* is 2, shoot-through duty ratio D_O is 0.25, and the modulation index *M* is 0.75. The model parameters for quasi-Z-source network are selected as per Eq. 5.83. The model operation is the same as explained in Sect. 5.2.2.

5.3.2 Simulation Results

The simulation of the three-phase QZSI with simple boost control is carried out using ode15s (stiff/NDF) solver [6]. The data shown in Table 5.6 are used. The simulation results for the line-to-neutral output voltage, line-to-line output voltage,

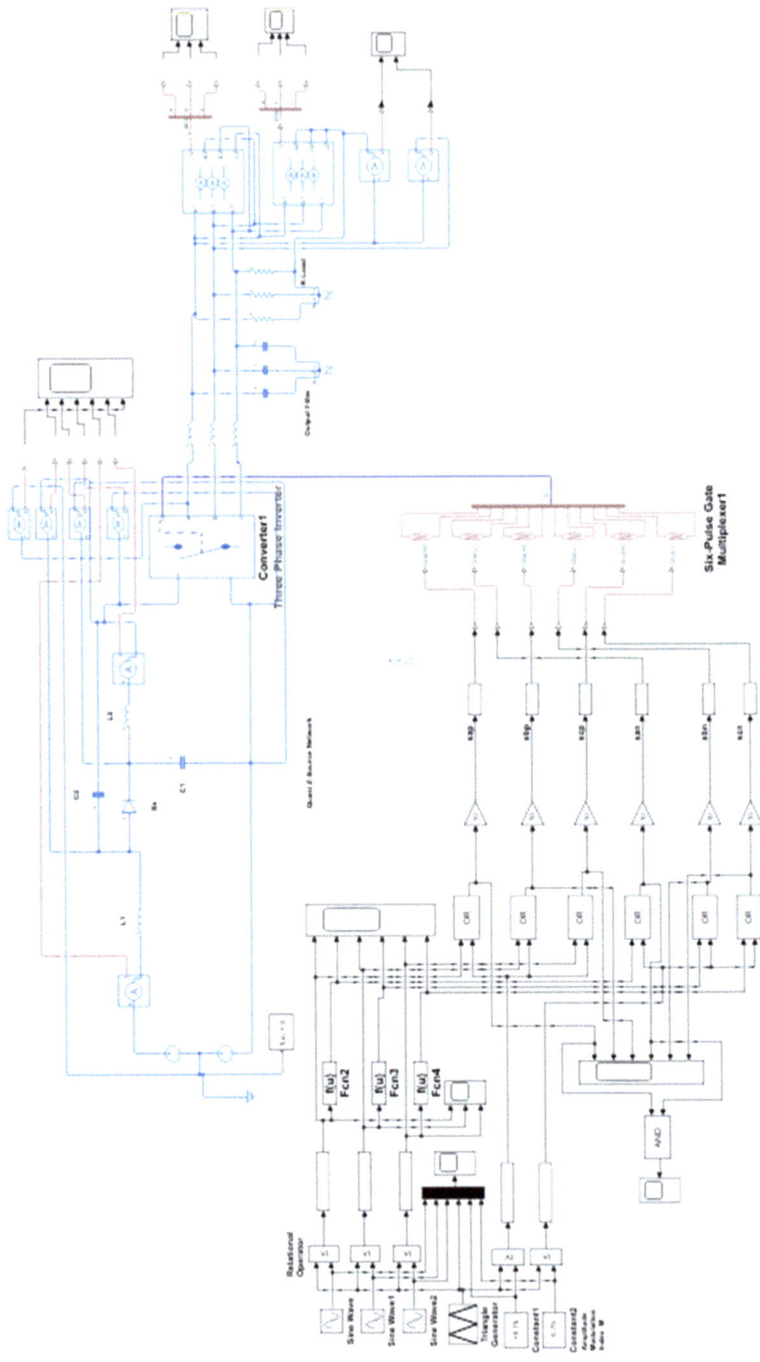

Fig. 5.33 Model of three phase quasi Z-source inverter—simple boost control

Table 5.6 Three-phase QZSI with simple boost control—Model parameters

Name	Value	Unit
Z-source inductor L1, L2	200E-6, 200E-6	H
Z-source capacitor C1, C2	20E-6, 50E-6	F
Output filter inductor Lf	200E-6	H
Output filter capacitor Cf	50.661E-9	F
Shoot-through duty ratio Do	0.25	–
Modulation index M (1-D_O)	0.75	–
DC input voltage	48	Volts
Power output P	400–1000	Watts
Inverter output frequency	50	Hz
Maximum boost factor B	2	–
Triangle carrier frequency	100E3	Hz
Output filter cut-off frequency	10E3	Hz
Inductor current ripple factor r_L	< 10%	–
Capacitor voltage ripple factor r_C	< 10%	–
Load resistor R_L per phase	10	Ω

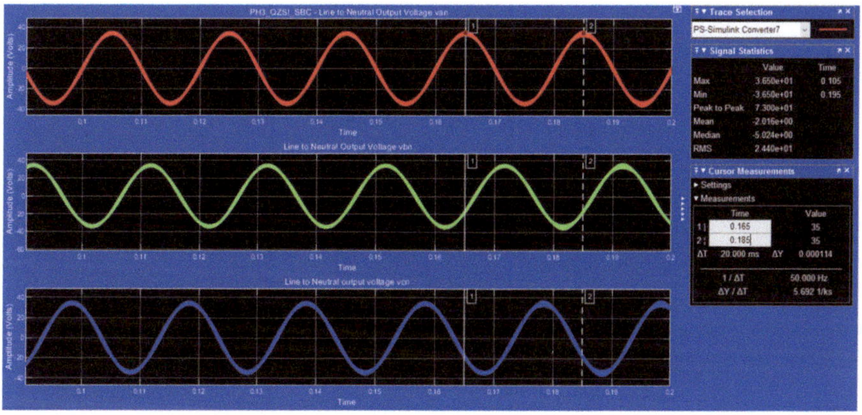

Fig. 5.34 Three phase QZSI with simple boost control—line to neutral output voltage

quasi-Z-source inductor currents, capacitor voltages, inverter bridge DC link voltage v_{PN} and line-to-ground output voltage for phase A are shown in Figs. 5.34 to 5.36, respectively. The shoot-through gate pulse for To measurement is shown in Fig. 5.37. In Fig. 5.37, measurement of shoot-through period T_O is made for one triangle carrier period commencing 0.190 Sec. to 0.19001 Sec in three different locations. These values are 624.109e-9 Sec., 1.258e-6 Sec. and 624.109e-9 Sec., respectively, giving a total value of 2.506e-6 Sec. for T_O. The simulation results are tabulated in Table 5.7.

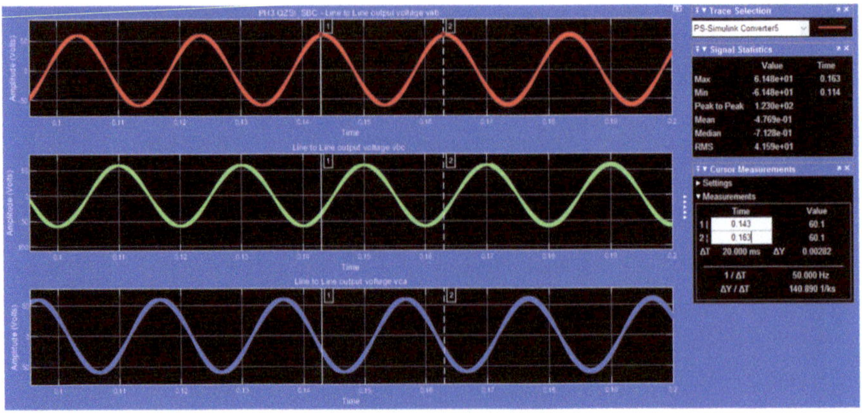

Fig. 5.35 Three phase QZSI with simple boost control—line to line output voltage

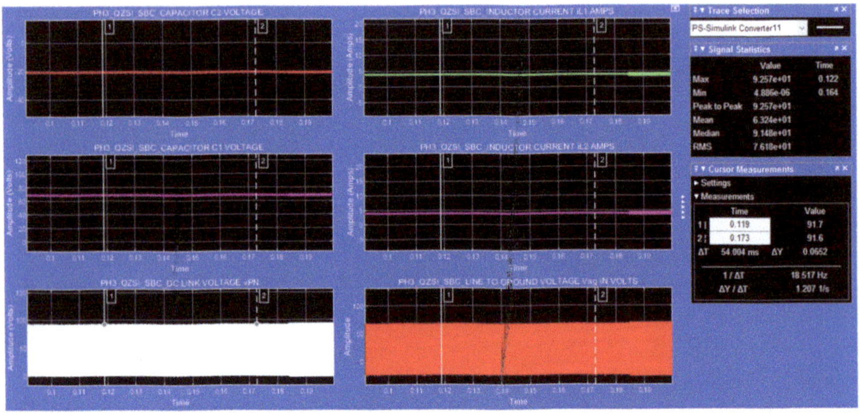

Fig. 5.36 Three phase QZSI with simple boost control—capacitor voltages vC2, vC1 and inverter DC link voltage vPN (left column top to bottom), inductor currents iL1, iL2 and line to ground output voltage vag (right column top to bottom)

5.3.3 Discussion of Results

The model values calculated, and simulation results closely agree with the respective theoretical values for D_O, B, G, V_{C1}, V_{C2}, v_{ln}, v_{PN}, I_{L1} and I_{L2}. The line-to-ground RMS output voltage is found to be 43.51 V from Fig. 5.36, whereas this value is 24 V which is half the DC source voltage for a conventional inverter, giving a boost factor B of 1.813.

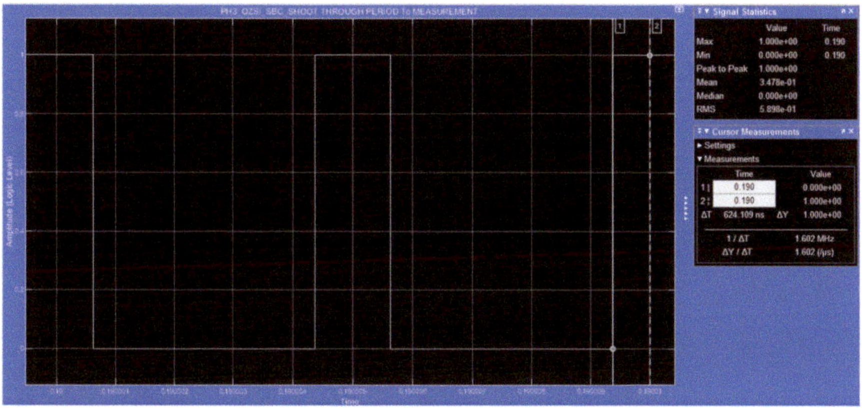

Fig. 5.37 Three phase QZSI with SBC—shoot through period To measurement

5.3.4 Model of Three-Phase Quasi-Z-Source Inverter Using Maximum Boost Control

The model of the three-phase QZSI with MBC is shown in Fig. 5.38 (Model file: EXAMPLE 5_6). The three-phase QZSI parameters are given in Table 5.6. The maximum value of B is 2, and this gives a value of 0.907 for M using Eq. 5.33 derived in Sect. 5.2.5. The model operation is the same as explained in Sect. 5.2.6.

5.3.5 Simulation Results

The simulation of the three-phase QZSI with MBC is carried out using ode23tb (stiff/TR-BDF2) solver [6]. The data shown in Table 5.6 are used with M value 0.907. The simulation results for the line-to-neutral output voltage, line-to-line output voltage, quasi-Z-source inductor currents and capacitor voltages, inverter bridge DC link voltage vPN and line-to-ground voltage are shown in Figs. 5.39 to 5.41, respectively. The shoot-through gate pulse for To measurement is shown in Fig. 5.42. In Fig. 5.42, measurement of shoot-through period T_O is made for one triangle carrier period commencing 0.196 Sec. to 0.19601 Sec in three different locations. These values are, respectively, 337.562e-9 Sec., 1.626e-6 Sec. and 347.792e-9 Sec., respectively, giving a total value of 2.31e-6 Sec. for T_O. The simulation results are tabulated in Table 5.8.

Table 5.7 Three-phase quasi-Z-source inverter simple boost control—simulation results

S. no.	D_O	B	G B X M	V_{C1} volts mean	V_{C2} volts mean	$I_{L1,}$, I_{L2} amps (mean)	V_{In} volts peak	v_{PN} volts mean	v_{PN} volts peak	Remarks
(1)	0.25 [0.2506]	2 (2.005)	1.5 (1.5025)	72 (70.03) [68.90]	24 (22.03) [20.82]	[3.88] [3.88]	36 (34.5) [36.5]	72 72.12 [63.24]	96 (92.06) [92.57]	Theoretical (model calculations) [Simulation results]

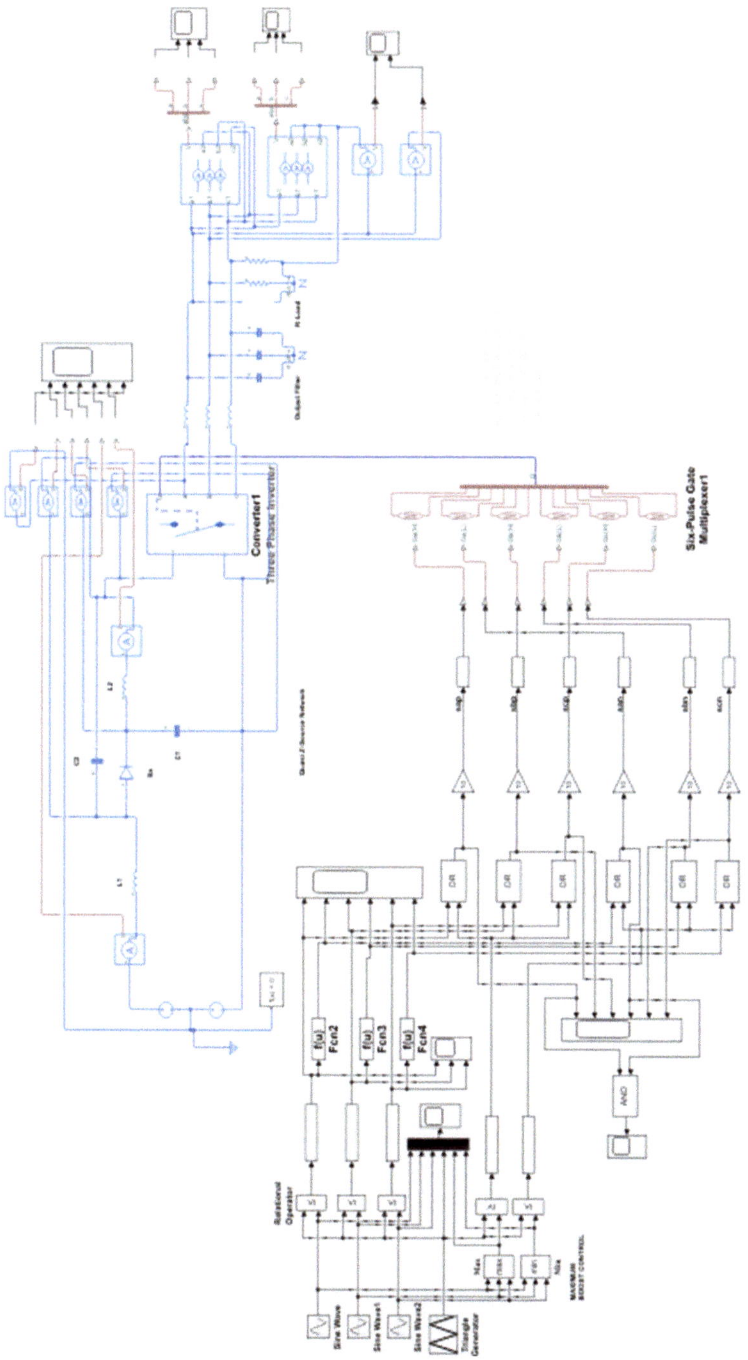

Fig. 5.38 Model of three phase quasi z-source inverter—maximum boost control

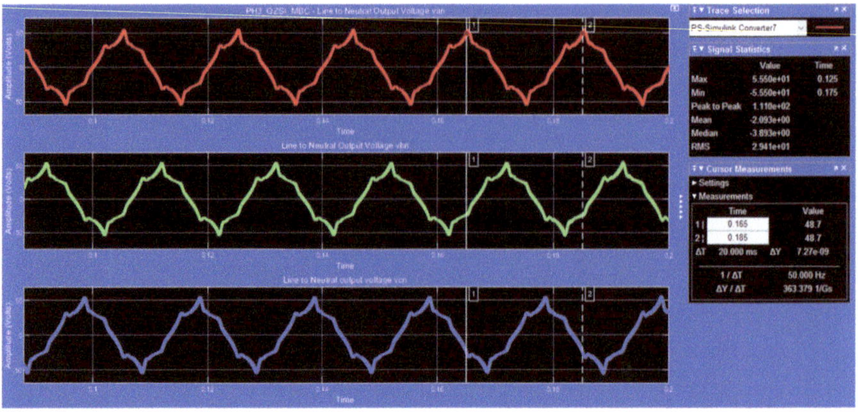

Fig. 5.39 Three phase QZSI with MBC—line to neutral output voltage

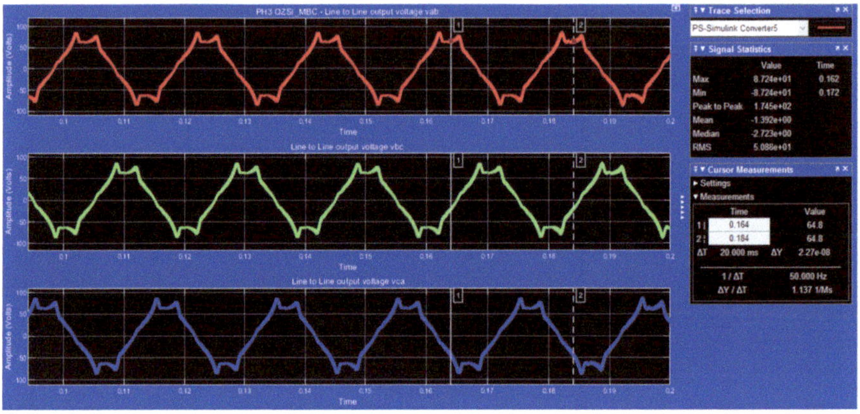

Fig. 5.40 Three phase QZSI with MBC—line to line output voltage

5.3.6 Discussion of Results

The model values calculated closely agree with respective theoretical values for *B*, *G*, V_{C1}, V_{C2}, v_{ln} and v_{PN}. Simulation results for I_{L1} and I_{L2} agree. The simulation results for Do closely agree with theoretical value. Simulation results for V_{C1}, V_{C2}, $v_{ln(peak)}$, *vPN(mean)* and *vPN(peak)* differ from respective theoretical value by −5.2%, −15.6%, 27.5%, −13.8% and 22.3% . The line-to-ground RMS output voltage is found to be 43.2 V from Fig. 5.41, whereas this value is 24 V which is half the DC source voltage for a conventional inverter, giving a boost factor B of 1.8.

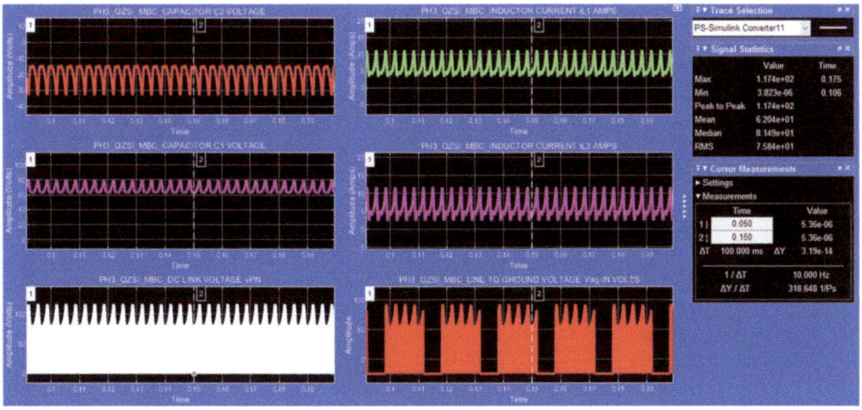

Fig. 5.41 Three phase QZSI with maximum boost control—capacitor voltages vC2, vC1 and inverter DC link voltage vPN (left column top to bottom), inductor currents iL1, iL2 and line to ground output voltage vag (right column top to bottom)

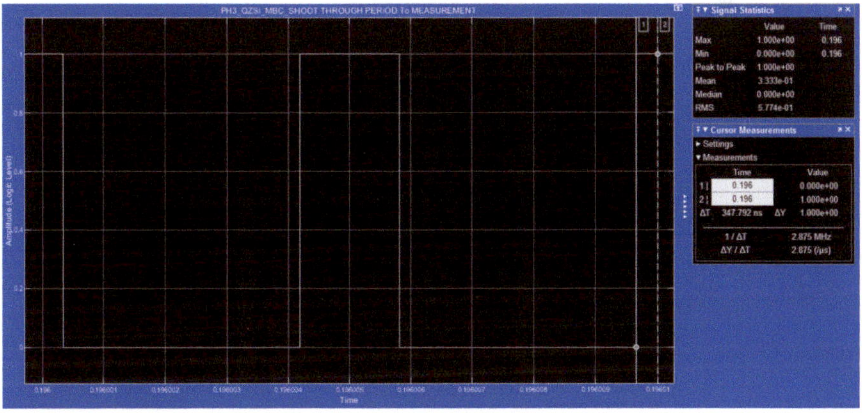

Fig. 5.42 Three phase QZSI with MBC—shoot through period to measurement

5.3.7 *Model of Three-Phase Quasi-Z-Source Inverter with Third Harmonic Injection Maximum Boost Control*

The model of the three-phase QZSI with THIMBC is shown in Fig. 5.43 (Model file: EXAMPLE 5_7). The three-phase QZSI parameters are given in Table 5.6. The maximum value of B is 2, and this gives a value of 0.907 for M using Eq. 5.38 derived in Sect. 5.2.9. The model operation is the same as explained in Sect. 5.2.10.

Table 5.8 Three-phase quasi-Z-source inverter maximum boost control—simulation results

S. no.	D_O	B	G B X M	V_{C1} volts mean	V_{C2} volts mean	$I_{L1,}$, I_{L2} amps (mean)	V_{In} volts peak	v_{PN} volts mean	v_{PN} volts peak	Remarks
(1)	0.2499 [0.231]	1.9995 (1.86)	1.8135 (1.73)	72 (68.66) [68.25]	24 (20.62) [20.25]	[5.936] [5.936]	43.52 (41.51) [55.5]	72 (68.66) [62.04]	95.98 (89.3) [117.4]	Theoretical (model calculations) [Simulation results]

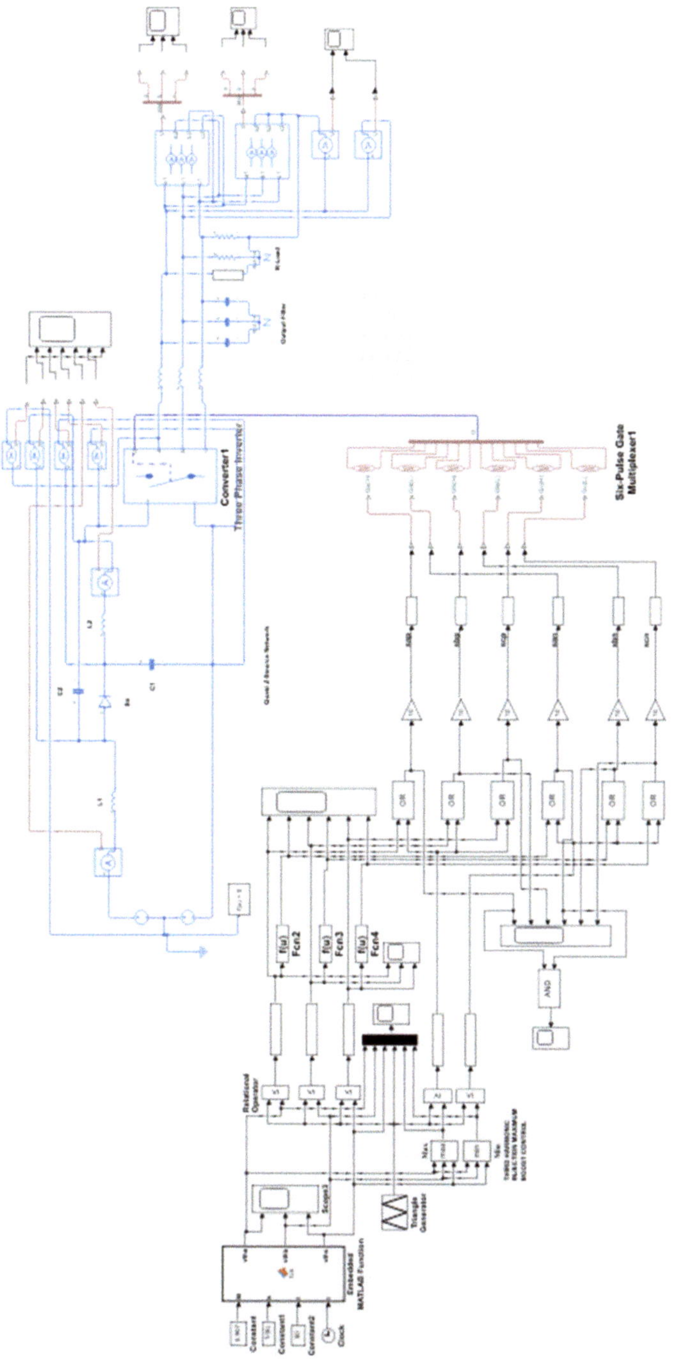

Fig. 5.43 Model of three phase quasi Z-source inverter—third harmonic injection maximum boost control

5.3.8 Simulation Results

The simulation of the three-phase QZSI with THIMBC is carried out using ode23tb (stiff/TR-BDF2) solver [6]. The data shown in Table 5.6 are used with M value 0.907. The simulation results for the line-to-neutral output voltage, line-to-line output voltage, quasi-Z-source inductor currents, capacitor voltages, inverter bridge DC link voltage v_{PN} and line-to-ground output voltage Vag for phase A are shown in Figs. 5.44 to 5.46, respectively. The shoot-through gate pulse for T_O measurement is shown in Fig. 5.47. In Fig. 5.47, measurement of shoot-through period T_O is made for one triangle carrier period commencing 0.196 Sec. to 0.19601 Sec in three different locations. These values are, respectively, 561.418e-9 Sec., 1.183e-6 Sec. and 561.253e-9 Sec., respectively, giving a total value of 2.305e-6 Sec. for T_O. The simulation results are tabulated in Table 5.9.

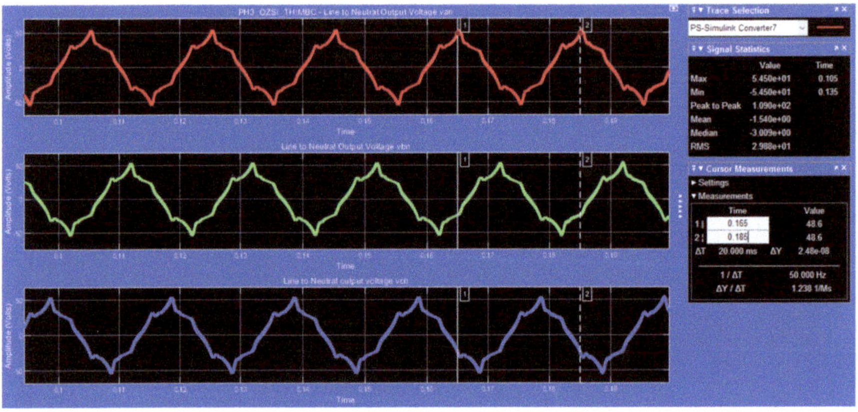

Fig. 5.44 Three phase QZSI with THIMBC—line to neutral output voltage

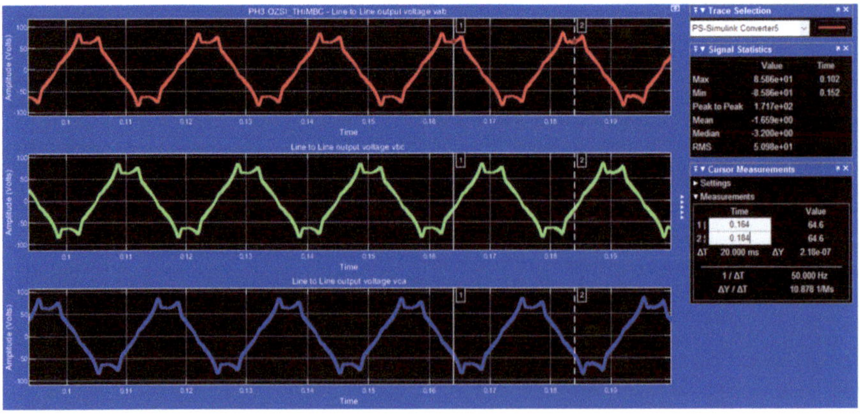

Fig. 5.45 Three phase QZSI with THIMBC—line to line output voltage

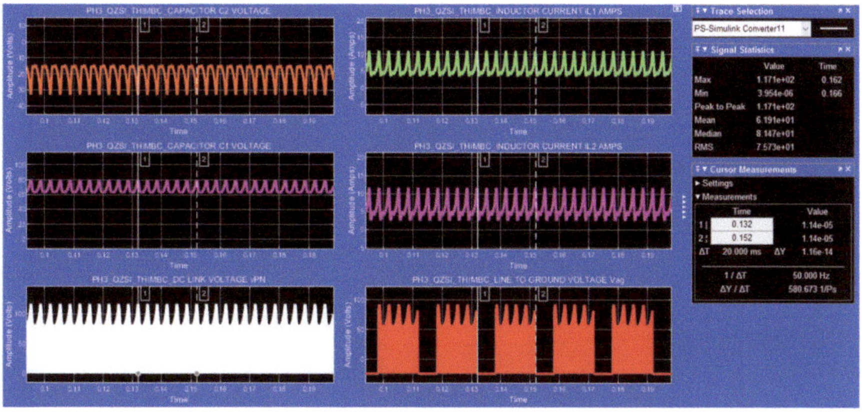

Fig. 5.46 Three phase QZSI with THIMBC—capacitor voltage vC2, vC1 and inverter bridge DC link voltage vPN (left column top to bottom), inductor currents iL1, iL2 and line to ground voltage *Vag* for phase a (right column top to bottom)

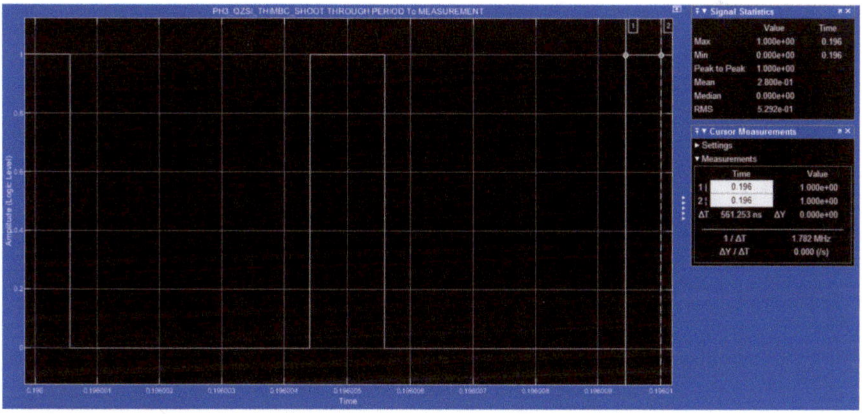

Fig. 5.47 Three phase QZSI with THIMBC—shoot through period xo measurement

5.3.9 Discussion of Results

The model values calculated closely agree with respective theoretical values for B, G, V_{C1}, V_{C2}, v_{ln} and v_{PN}. Simulation results for I_{L1} and I_{L2} agree. The simulation results for Do closely agree with theoretical value. Simulation results for *Vc1*, *Vc2*, *Vln(Peak)*, *vPN(Mean)* and *vPN(Peak)* differ from the respective theoretical values by −5.16%, −15.45%, 25.23%, −14.01% and 22%. The RMS value of line-to-ground voltage Vag for phase A is 43.61 V, whereas for a conventional inverter, this value is 24 V which is half the DC source voltage. This gives a boost factor B of 1.817.

Table 5.9 Three-phase quasi-Z-source inverter THIMBC: Simulation results

S. no.	D_O	B	G B X M	V_{C1} volts mean	V_{C2} volts mean	$I_{L1,}$, I_{L2} amps (mean)	V_{ln} volts peak	v_{PN} volts mean	v_{PN} volts peak	Remarks
(1)	0.2499 [0.2305]	1.9995 (1.855)	1.8135 (1.726)	72 (68.52) [68.28]	24 (20.52) [20.29]	[5.960] [5.967]	43.52 (41.424) [54.50]	72 (68.52) [61.91]	95.98 (89.04) [117.1]	Theoretical (model calculations) [Simulation results]

5.3.10 Model OF Three-Phase Quasi-Z-Source Inverter with Third Harmonic Injection Maximum Constant Boost Control

The model of the three-phase QZSI with third harmonic injection maximum constant boost control (THIMCBC) is shown in Fig. 5.48 (Model file: EXAMPLE 5_8). Now consider the three-phase QZSI whose parameters are given in Table 5.6. The maximum value of B is 2, and this gives a value of 0.866 for M using Eq. 5.52 derived in Sect. 5.2.13. The model operation is the same as explained in Sect. 5.2.14.

5.3.11 Simulation Results

The simulation of the three-phase QZSI with THIMCBC is carried out using ode23tb (stiff/TR-BDF2) solver [6]. The data shown in Table 5.6 are used with M value 0.866. The simulation results for the line-to-neutral output voltage, line-to-line output voltage, quasi-Z-source inductor currents, capacitor voltages, inverter bridge DC link voltage v_{PN} and line-to-ground voltage for phase A are shown in Figs. 5.49 to 5.51, respectively. The shoot-through gate pulse for T_O measurement is shown in Fig. 5.52. In Fig. 5.52, measurement of shoot-through period T_O is made for one triangle carrier period commencing 0.190 Sec. to 0.19001 Sec in three different locations. These values are, respectively, 624.603e-9 Sec., 1.259e-6 Sec. and 624.603e-9 Sec., respectively, giving a total value of 2.507e-6 Sec. for T_O. The simulation results are tabulated in Table 5.10.

5.3.12 Discussion of Results

The model values calculated closely agree with respective theoretical values for B, G, V_{C1}, V_{C2}, v_{ln} and v_{PN}. The simulation results for I_{L1} and I_{L2} well agree. The simulation results for Do, V_{C1}, *vln(Peak)* and *vPN(Peak)* closely agree with theoretical values. The simulation results for V_{C2} and vPN(Mean) differ from the theoretical values by −16.5% and −11.5%, respectively. The RMS value of the line-to-ground voltage is 43.44 V from Fig. 5.51, whereas this value is 24 V which is half the DC source voltage for a conventional inverter. This gives a boost factor B of 1.81.

5.4 Case Study: Space Vector Modulation of a Three-Phase Z-Source Inverter

The space vector modulation (SVM) of three-phase two-level inverter presented in Sects. 4.2 and 4.9 of Chap. 4 is used here. Only six segments I to VI at π/3 radians interval shown in Fig. 4.2 are considered. Table 4.3 shows the switching sequence

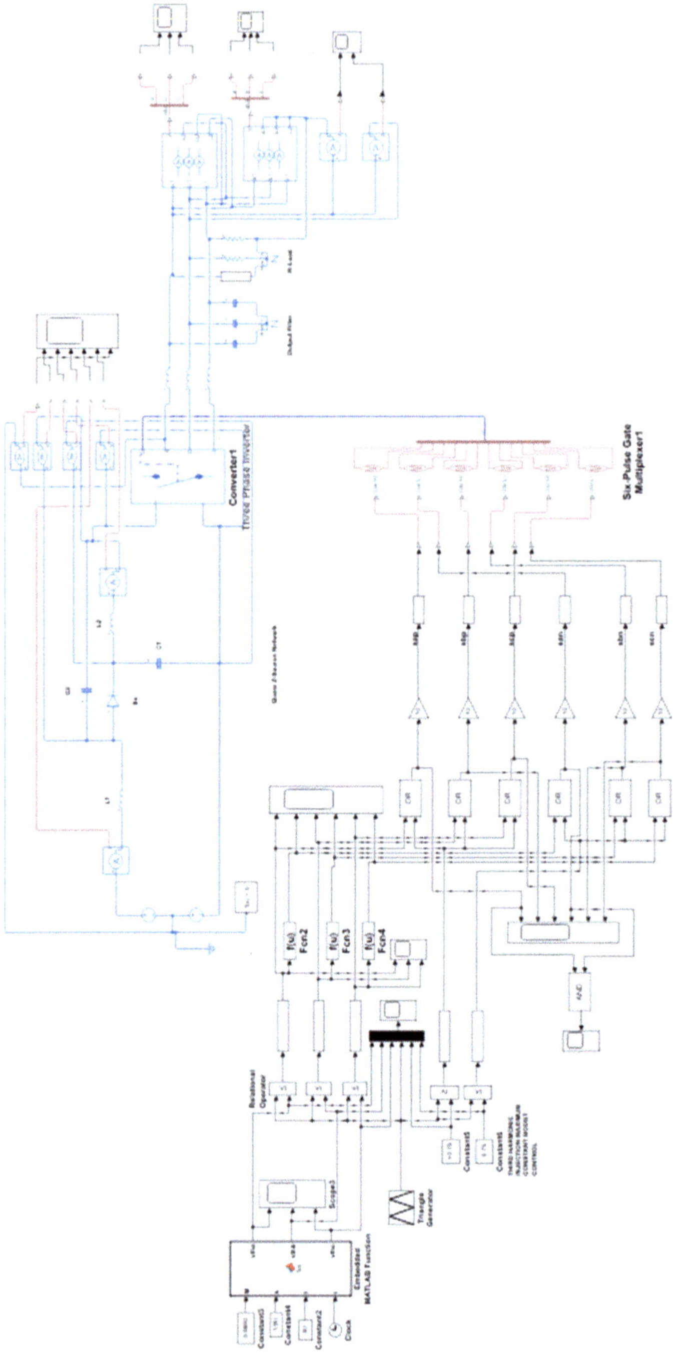

Fig. 5.48 Model of three phase quasi Z-source inverter—third harmonic injection maximum constant boost control

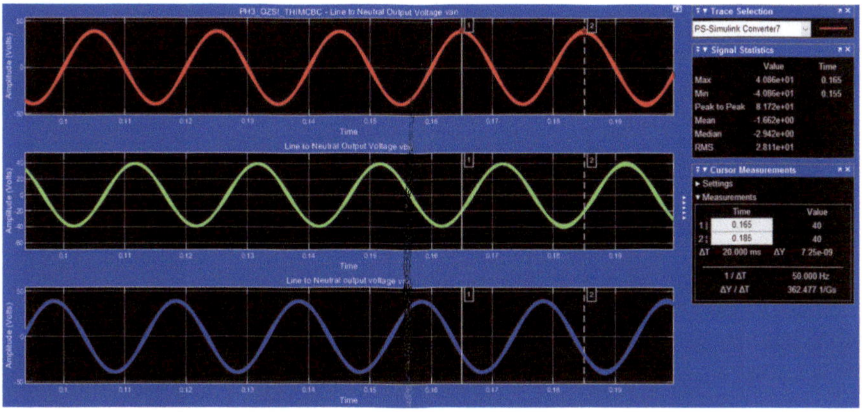

Fig. 5.49 Three phase QZSI with THIMCBC—line to neutral output voltage

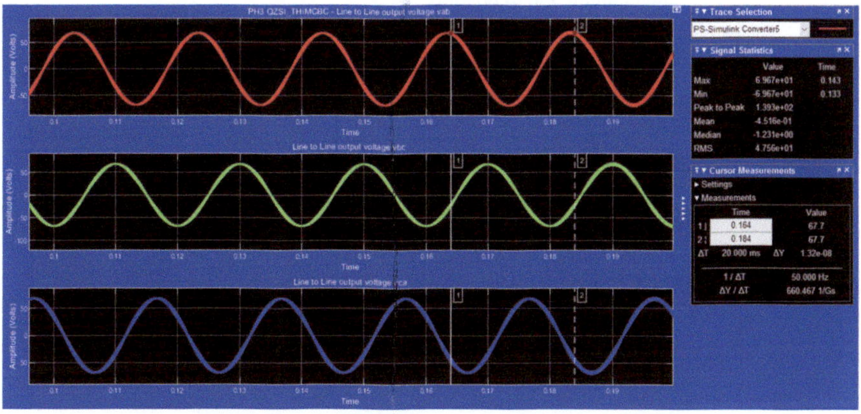

Fig. 5.50 Three phase QZSI with THIMCBC—line to line output voltage

table, and Table 4.17 gives the gate pulse timing or dwell time for the upper switches of the inverter. The method of generating the shoot-through state for SVM of three-phase Z-source inverter is presented below:

Initially consider that the cumulative switching time for all six sectors of phase A is calculated as Ta1 Seconds using Ta, Tb and To values as per Table 4.17. Now assume a shoot-through constant Ksh between 0 and 1. Now reduce the timing for switch states VO and V7 to a value To1 = (To—Ksh*To). Now calculate the new cumulative switching time Ta2 using Ta, Tb and new value of To1 as per Table 4.17. Ta2 is less than Ta1. Now compare Ta1 and Ta2 with Tsample in two comparators. The comparator comparing Ta1 and Tsample gives HIGH output when Ta1 is greater than or equal to Tsample, or else its output is LOW. Similarly, the comparator comparing Ta2 and Tsample gives HIGH output when Ta2 is less than or equal to Tsample, or else its output is LOW. This is shown in Fig. 5.53. The shaded regions Tsh1 and Tsh2 in Fig. 5.53 give the shoot-through interval for one sampling period.

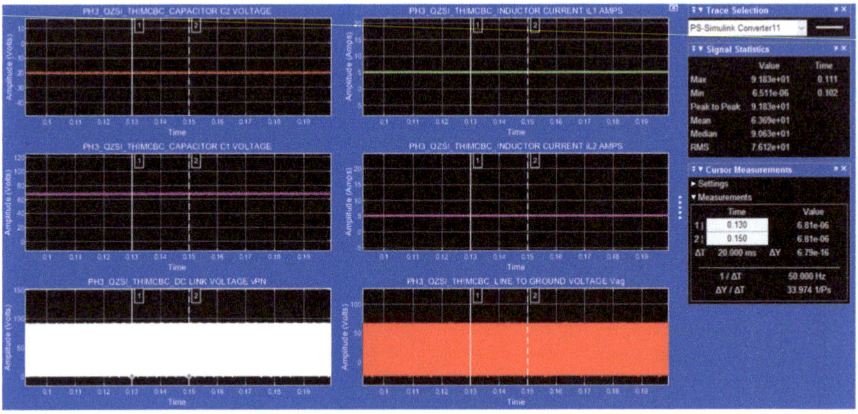

Fig. 5.51 Three phase QZSI with THIMCBC—capacitor voltage vC1, vC2 and inverter DC link bridge voltage vPN (letft column top to bottom), inductor currents iL1, iL2 and inverter line to ground voltage *Vag* (right column top to bottom)

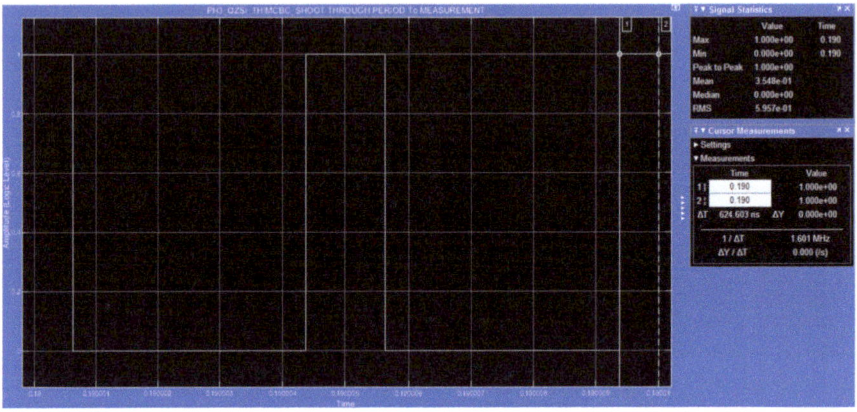

Fig. 5.52 Three phase QZSI with THIMCBC—shoot through period To measurement

5.4.1 *Model of Space Vector Modulation of a Three-Phase Z-Source Inverter*

The model of SVM of a three-phase Z-source inverter is shown in Fig. 5.54 (Model file: CASE_STUDY_EX5_1). The various subsystems are (1) sector identifier, (2) sector switch function generator, (3) triangle carrier generator, (4) gate pulse timing generator, (5) Z-source network, (6) three-phase inverter and (7) L-C filter and load.

Table 5.10 Three-phase quasi-Z-source inverter THIMCBC:—Simulation results

S. no.	D_O	B	G B X M	V_{C1} volts mean	V_{C2} volts mean	$I_{L1,}$, I_{L2} amps (mean)	V_{In} volts peak	v_{PN} volts mean	v_{PN} volts peak	Remarks
(1)	0.25 [0.2507]	2 (2.007)	1.732 (1.736)	72 (72.18) [68.03]	24 (24.15) [20.03]	[5.112] [5.112]	41.568 (41.664) [40.86]	72 (72.18) [63.69]	96.00 (96.336) [91.83]	Theoretical (model calculations) [Simulation results]

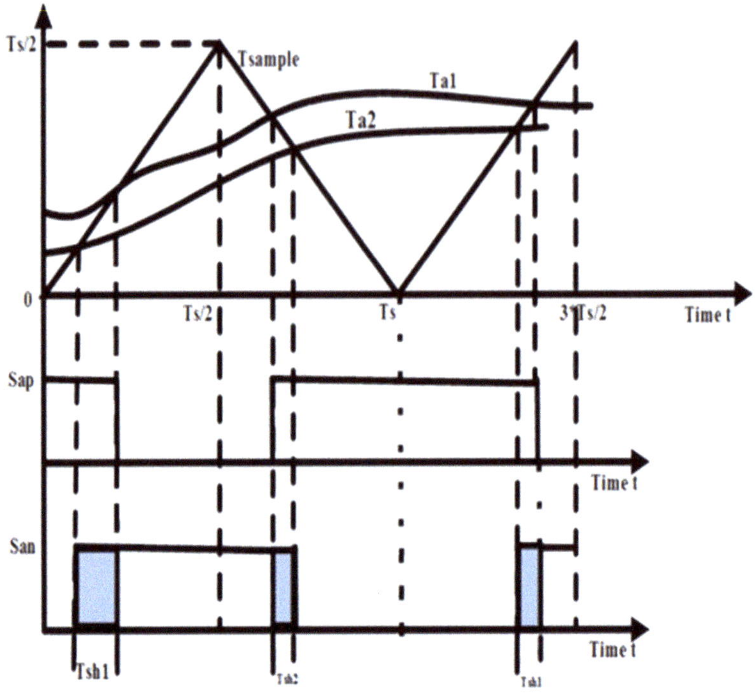

Fig. 5.53 Diagram illustrating shoot through state in a three phase SVPWM Z-source inverter

5.4.2 Sector Identifier

Referring to Fig. 4.2 of Chap. 4, there are six sectors, and Vref can lie in any one of the six sectors. These sectors from I to VI are displaced by π/3 radians. Embedded MATLAB function with inputs constant 1 for peak input voltage, constant 50 for frequency and time module and output sector y1 in Fig. 5.54 is used for identifying the sectors. This programme is given under Program Segment 5.2 in the model file CASE_STUDY_EX5_1. Sector output y1 gives the sector number at any instant of time.

5.4.3 Sector Switch Function Generator

The sector output giving sector number is given as input to another Embedded MATLAB function which gives sector switch functions ssf1 to ssf6 as outputs. For example, when Vref is in Sector I, ssf1 output is HIGH (logic 1), or else its output is LOW (logic 0). The computer programme to generate this sector switch functions is given under Program Segment 5.3 in the model file CASE_STUDY_EX5_1. In Program Segment 5.3, output y1 to y6 correspond to ssf1 to ssf6.

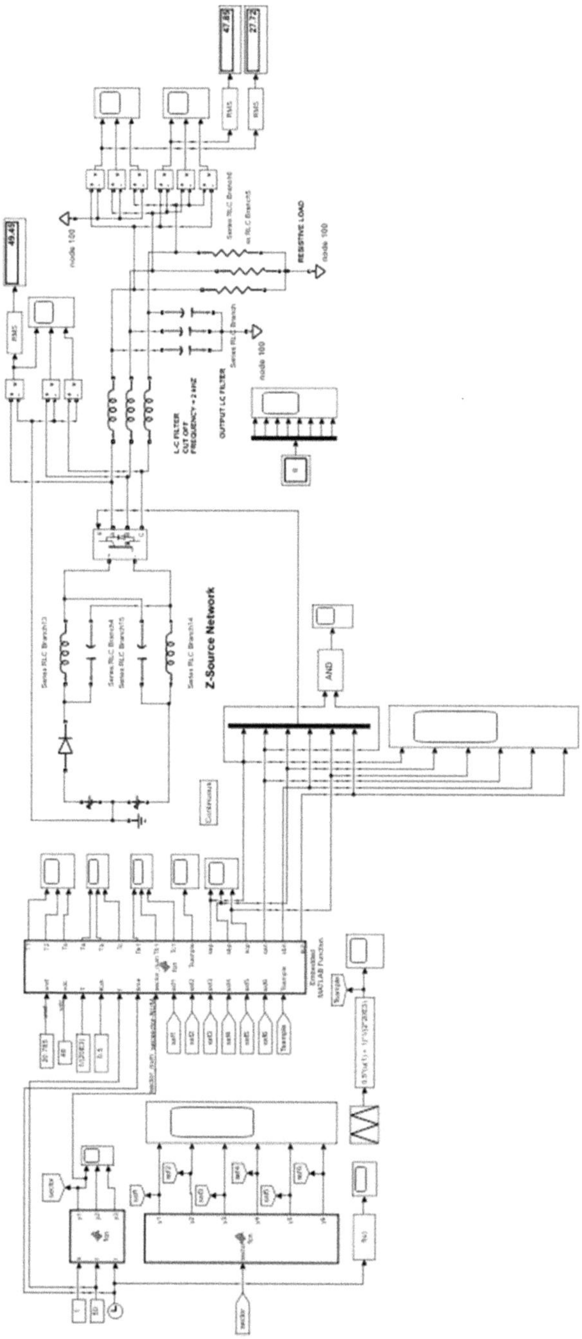

Fig. 5.54 Model of space vector modulation of a three phase Z-source inverter

5.4.4 Gate Pulse Timing Generator

Gate pulse timing for the three-phase inverter switches is generated using another Embedded MATLAB Function. This Embedded MATLAB function has the inputs Vref, Vdc, T, Ksh, f, time t, sector_num, ssf1 to ssf6 and Tsample, and the outputs are T1, T2, To, Ta, Tb, Tc, Ta1, Tb1, Tc1, Tsample, sap, sbp, scp, san, sbn and scn. Vref is the reference voltage defined in Fig. 4.2, Vdc is the DC link voltage, T is the sample time Ts, Ksh is the shoot-through factor, f is the output frequency of the inverter, sector_num is from 1 to 6, sector switch functions from ssf1 to ssf6 and Tsample which is a triangle carrier waveform with period Ts, and peak value Ts/2 and minimum value zero. Tsample waveform is shown in Fig. 4.4a, b of Chap. 4. Similarly outputs T1, T2 and To correspond timings as defined in Eqs. 4.13, 4.14 and 4.15. Ta, Tb and Tc correspond to Ta1, Tb1 and Tc1 defined in Table 4.17 for conventional switching. Ta1, Tb1 and Tc1 are the cumulative sector timing for the three phase using T1, T2 and To1 in the place of To where To1 is (To—Ksh*To) and sap, sbp, scp, san, sbn and scn are the six gate pulse for the upper and lower switches of inverter shown in Fig. 4.1. The computer programme to generate gate pulse timing is shown under Program Segment 5.4 in the model file CASE_STUDY_EX5_1. In Program segment 5.4, the symbol (~) represents logical NOT operation. Timing Ta, Tb and Tc calculated as per Table 4.17 using T1, T2 and To which are then compared with Tsample to generate gate pulse sap, sbp and scp for the upper switches and timing Ta1, Tb1 and Tc1 calculated as per Table 4.17 using T1, T2 and To1 which are then compared with Tsample to generate gate pulse san, sbn and scn for the lower switches of inverter shown in Fig. 4.1. The method of generating the six gate pulse with shoot-through state is explained in Sect. 5.4. Triangle carrier generator to generate Tsample is explained in Sect. 4.2.1 and in Sect. 4.2.2.1. Z-source network is explained in Sect. 5.2. The three-phase inverter model is already presented in Sect. 4.2.2.4 of Chap. 4. The L-C filter is designed for a cut-off frequency of 2 kHz using the relation fc = 1/(2*π*sqrt(L*C)) where fc is the cut-off frequency and L and C are the inductor and capacitor. Three-phase star-connected resistive load of 10 Ω is used.

5.4.5 Simulation Results

The simulation of the three-phase SVM Z-source inverter is carried out using ode23tb (stiff/TR-BDF2) solver in Simulink [6]. The DC source voltage used for the Z-source network is 48 V. The sampling period Ts is 50 microseconds. The reference voltage Vref is 20.785 V. Inverter output frequency f is 50 Hz. Shoot-through factor Ksh used is 0.5. Modulation index using Eq. 4.19 is 0.75. The simulation results for the three-phase line-to-neutral voltage, line-to-line voltage and line-to-ground voltage are shown in Figs. 5.55 to 5.57. The Z-source network

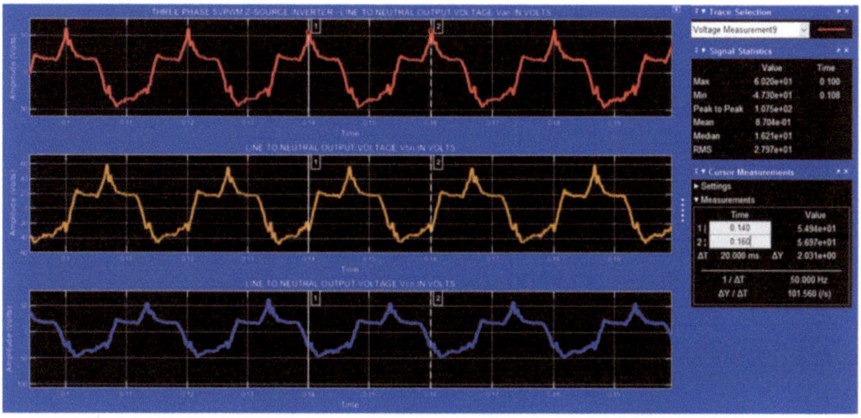

Fig. 5.55 Three phase SVPWM Z-source inverter—line to neutral output voltage

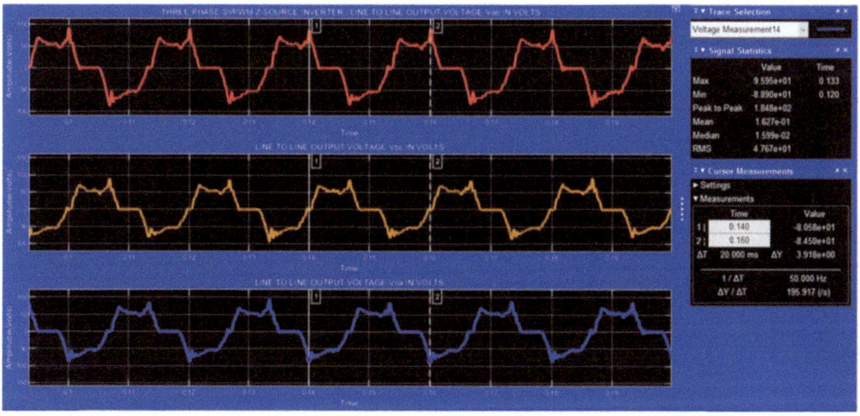

Fig. 5.56 Three phase SVPWM inverter—line to line output voltage

inductor currents, capacitor voltages, load currents and the DC link bridge voltage Vi across inverter legs are shown in Fig. 5.58. The shoot-through gate pulse for Tsh measurement is shown in Fig. 5.59. From Fig. 5.59, measurements are made using cursor to calculate shoot-through period Tsh during the arbitrary interval from 0.07 Sec. to 0.07005 Sec. The value of Tsh1 and Tsh2 are found to be 5.125e-6Sec. and 5.373e-6 Sec. giving a total Tsh value of 10.5e-6 Second. The shoot-through duty ratio Do is found to be 0.21. Simulation results are shown in Table 5.11. The line-to-ground RMS voltage is found to be 51.09 V. With a conventional inverter, this value is 24 V which is half the DC source voltage. This gives a boost factor B of 2.13. This value is shown in Table 5.11 as model calculated value.

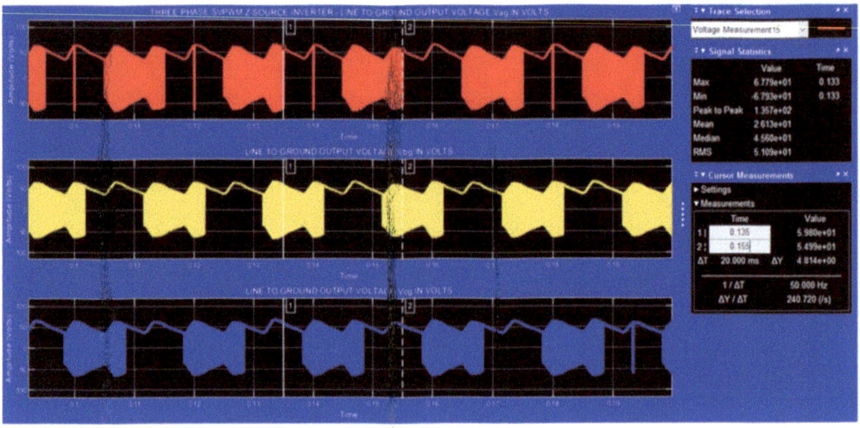

Fig. 5.57 Three phase SVPWM Z-source inverter—line to ground output voltage

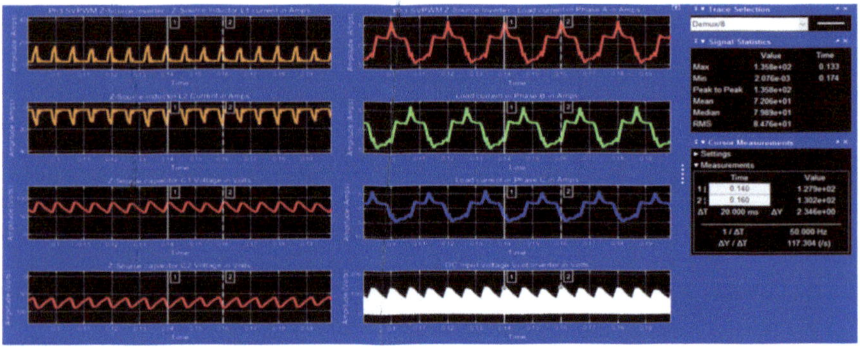

Fig. 5.58 Three phase SVPWM Z-source inverter—inductor currents iL1, iL2, capacitor voltages vC1, vC2 (left column top to bottom), three phase load currents, inveter bridge DC link voltage vi (right column top to bottom)

5.4.6 Discussion of Results

Referring to Fig. 5.59, two shoot-through intervals Tsh1 and Tsh2 are observed for one sampling period Ts. The RMS value of line-to-ground voltage display in Fig. 5.57 is found to be 51.09 volts, whereas for a normal inverter with conventional switching, this value is 24 V which is half the DC link voltage. Thus, the boost factor B is found to be 51.09/24 which is 2.13. Using Do value of 0.21 calculated using measurements from Fig. 5.59, the value of B is found to be 1.724. In Table 5.11, the measured value of peak line-to-neutral output voltage Vac(Peak) and inverter bridge DC link peak voltage $V_{i(Peak)}$ differs from model values by 28.7% and 24.7%, respectively. The measured value of *V*i(Mean), *VC1(Mean)* and *VC2(Mean)* closely agrees with their respective model calculated value. The measured values of IL1 (Mean) and IL2(Mean) are found to be equal.

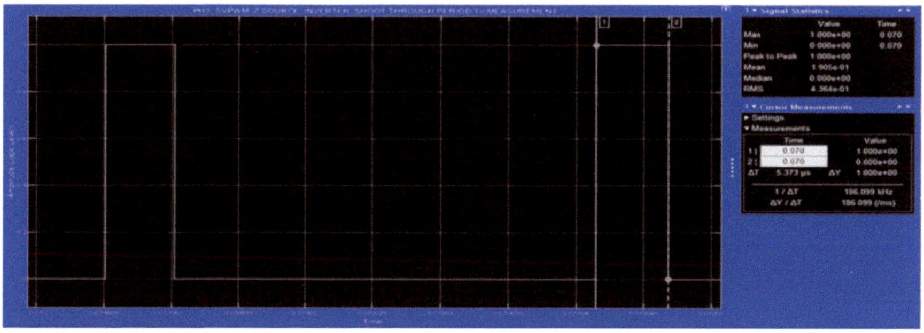

Fig. 5.59 Three phase SVPWM Z-source inverter—shoot through period Tsh measurement

5.5 Case Study: Space Vector Modulation of a Three Phase Quasi-Z-Source Inverter

The space vector modulation (SVM) of three-phase two-level inverter presented in Sect. 4.2 and 4.9 of Chap. 4 is used here. Only six segments I to VI at $\pi/3$ radians interval shown in Fig. 4.2 are considered. Table 4.3 shows the switching sequence table, and Table 4.17 gives the gate pulse timing or dwell time for the upper switches of the inverter. The method of generating the shoot-through state for SVM of three-phase quasi-Z-source inverter is the same as for the SVM of three-phase Z-source inverter presented in Sect. 5.4.

5.5.1 Model of Space Vector Modulation of a Three-Phase Quasi-Z-Source Inverter

The model of SVM of a three-phase quasi-Z-source inverter is shown in Fig. 5.60 (Model file: CASE_STUDY_EX5_2). The various subsystems are (1) sector identifier, (2) sector switch function generator, (3) triangle carrier generator, (4) gate pulse timing generator, (5) quasi-Z-source network, (6) three-phase inverter and (7) three-phase load. The first four and sixth subsystems are already explained in Sect. 5.4.2. Generation of shoot-through period Tsh is presented in Sect. 5.4. These are shown in the model file CASE_STUDY_EX5_2.

5.5.2 Simulation Results

The simulation of the three-phase SVPWM QZSI is carried out using ode15s (stiff/NDF) solver in Simulink [6]. The data for the QZS network shown in Table 5.6 are used. The simulation results for the three-phase line-to-neutral voltage, line-to-line

Table 5.11 Three-phase SVPWM Z-source inverter simulation results

S. no.	D_O	B	G B X M	V_{ac} (volts) peak	V_i volts avg./mean	v_i volts peak	V_C volts avg./mean	Remarks
(1)	(0.2357) [0.2158]	(1.8958) [1.759]	(1.4188) [1.3192]	(45) [103.8 / 2]	(66.2238) [67.34]	(84.432) [72]	(69.3997) [70.45]	(Model calculations) [Simulation results]

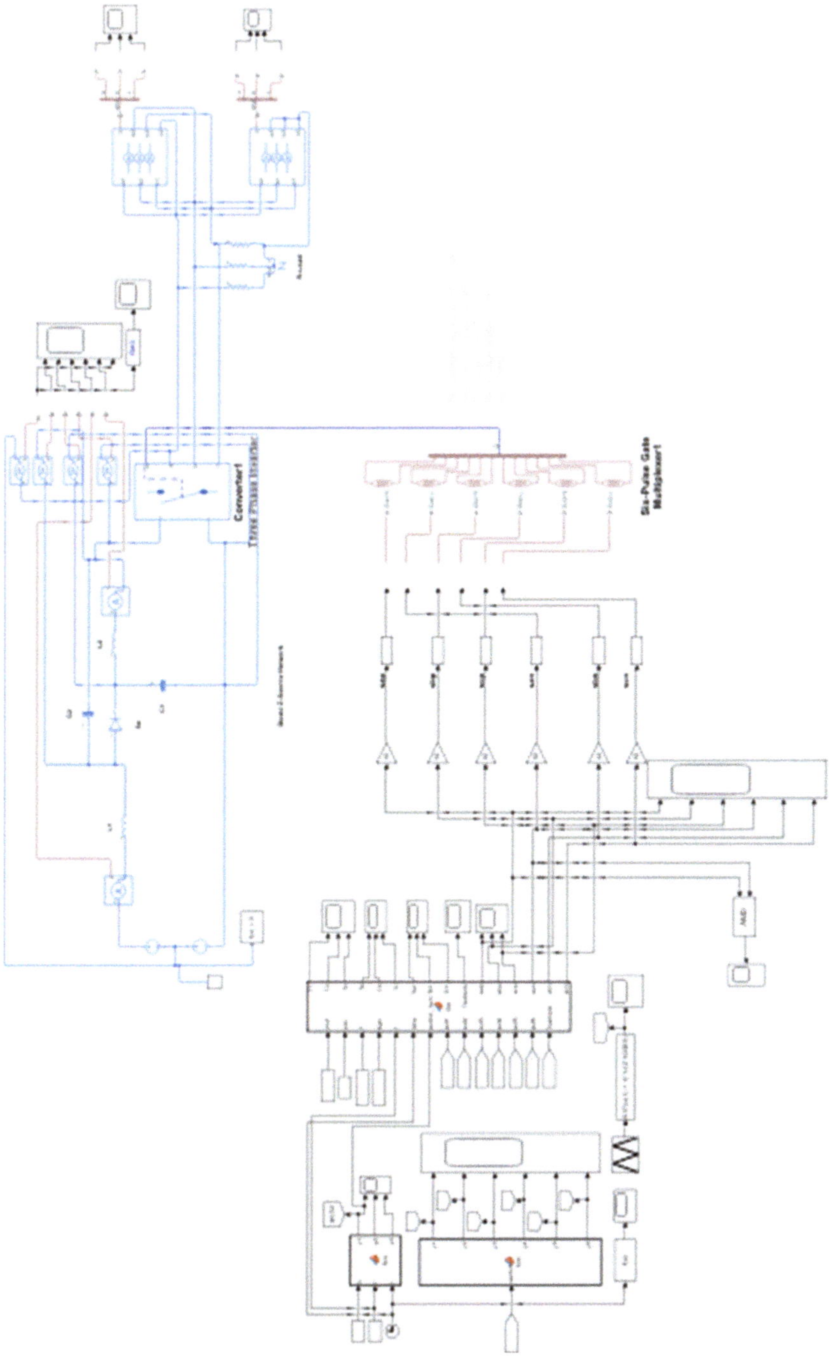

Fig. 5.60 Model of three phase SVPWM quasi Z-source inverter

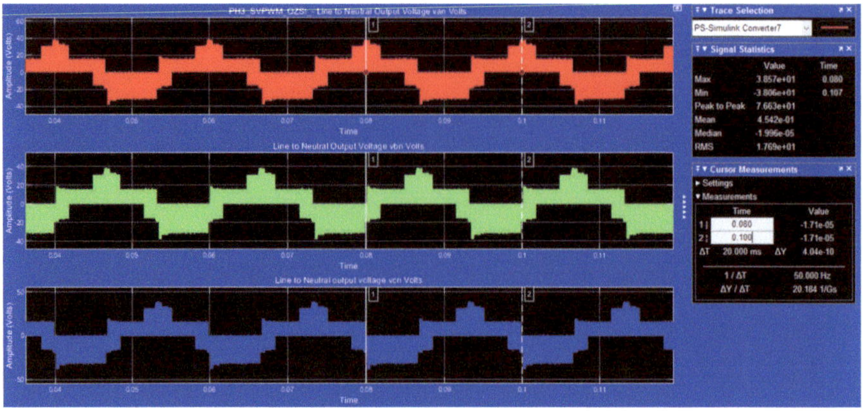

Fig. 5.61 Three phase SVPWM QZSI—line to neutral output voltage

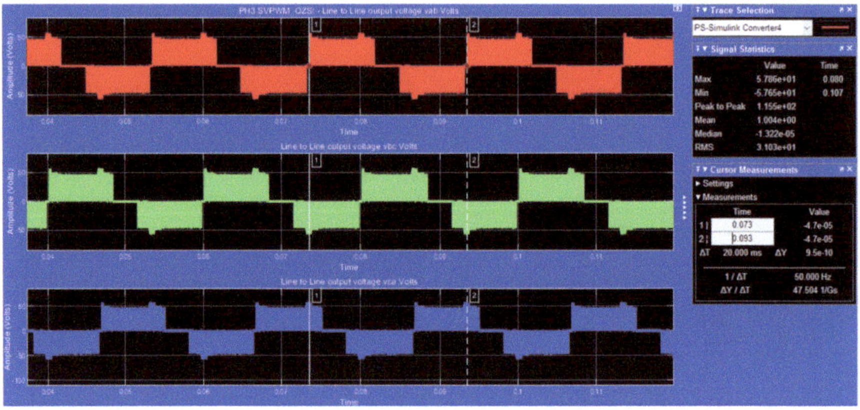

Fig. 5.62 Three phase SVPWM QZSI—line to line output voltage

voltage, two capacitor voltages, DC link voltage *Vi* across inverter legs, two inductor currents and line-to-ground voltage VAG for phase A are shown in Figs. 5.61 to 5.63, respectively. Shoot-through gate pulse waveform for Tsh measurement is shown in Fig. 5.64. For the SVM, a modulation index of 0.8 is assumed which gives a Vref value of 22.1709 V. A shoot-through factor Ksh of 0.5 is assumed. Simulation results are tabulated in Table 5.12. Shoot-through periods are measured for one sampling period from 114.82E-3 Sec. to 114.83E-3 Sec. The values of Tsh1 and Tsh2 are found to be 508.057E-9 Sec. and 508.495E-9 Sec. giving a total shoot-through period Tsh of 1.016E-6 Sec. The shoot-through duty-ratio Do is 0.1016.

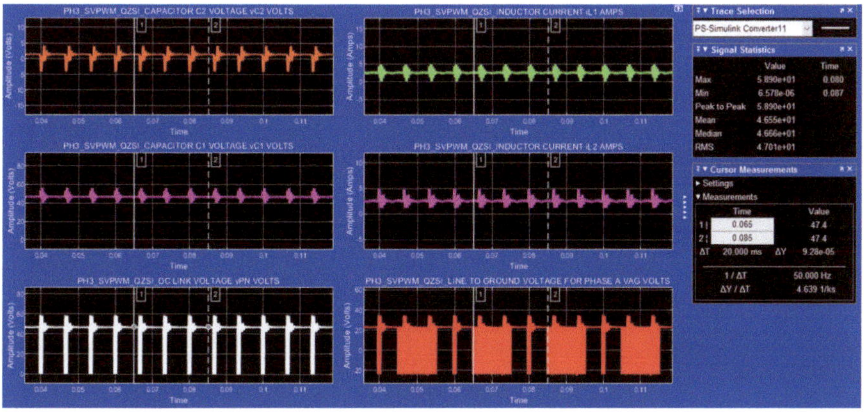

Fig. 5.63 Three phase SVPWM QZSI—capacitor voltages vC1, vC2 and inverter bridge DC link voltage vPN (Top to Bottom), inductor currents iL1, iL2 and line to ground voltage *Vag* for phase A

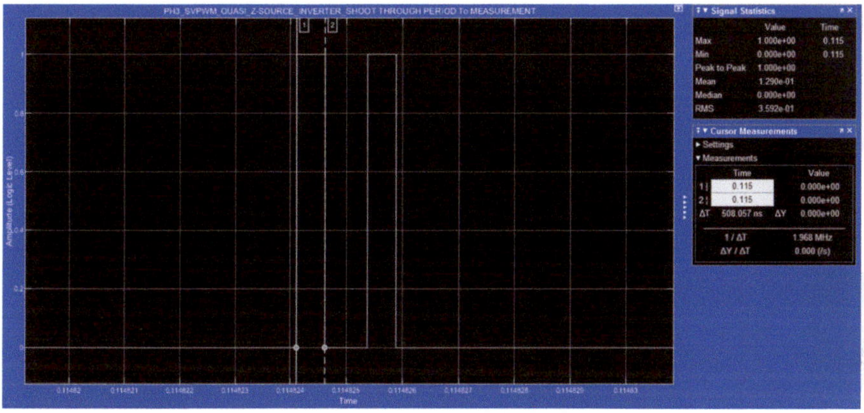

Fig. 5.64 Three phase SVPWM QZSI—shoot through period Tsh measurement

5.5.3 Discussion of Results

Two shoot-through regions Tsh1 and Tsh2 are found as per derivation in Sect. 5.4. The mean values of two inductor currents iL1 and il2 by simulation are found to be the same as per derivation. The model calculations are made as per derivation in Sect. 5.3. The model calculations and simulation results for Vc1(mean), vPN(Peak) and vPN(Mean) are found to agree each other with a small percentage error. The percentage error for the model values and simulation results for vln(Peak) and Vc2 (Mean) are a little higher than that for other measured parameters. The RMS value of line-to-ground voltage VAG is found to be 23.78 V from Fig. 5.63, whereas for a conventional inverter, this value is 24 V which is half the DC source voltage. This

Table 5.12 Three-phase SVPWM quasi-Z-source inverter:—Simulation results

S. no.	D_O	B	G B X M	V_{C1} volts mean	V_{C2} volts mean	$I_{L1,}$, I_{L2} amps (mean)	V_{In} volts peak	v_{PN} volts peak	v_{PN} volts mean	Remarks
(1)	[0.1016]	[1.255]	[1.004]	(54.12) [46.97]	(6.12) [1.073]	[2.6] [2.609]	(24.1) [38.57]	(60.24) [58.9]	(54.12) [46.55]	(Model Calculations) [simulation Results]

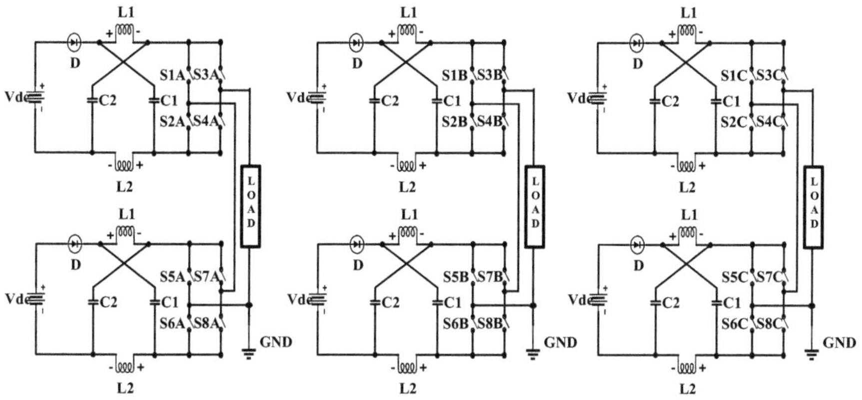

Fig. 5.65 Three phase Z-source five level cascade H-bridge inverter

gives a boost factor B of 0.99 which is very close to unity. This boost factor 0.99 implies that the three-phase boost SVPWM QZSI inverter is working as a conventional SVPWM QZSI in the normal mode.

5.6 Three-Phase Z-Source Five-Level Cascade H-Bridge Inverter

The topology of the three-phase five-level cascade H-bridge Z-source inverter is shown in Fig. 5.65. The analysis of the Z-source inverter is already presented in Sect. 5.2. The same analysis for the Z-source network holds good here for each cell. The bottom and top cell in each phase is connected in series so as to get a five-level voltage output across the load. The DC source voltage for both cells is each Vdc volts. The operation of the five-level cascade H- bridge inverter is already presented in Sect. 2.13 of Chap. 2 [9, 10].

5.6.1 Model of Three-Phase Z-Source Cascade H-Bridge Inverter

The model of the three-phase Z-source cascade H-bridge inverter is shown in Fig. 5.66 (Model file: EXAMPLE 5_9). The model subsystem for phase A is shown in Fig. 5.67. The Z-source network parameters are the same as given in Table 5.1. Here, SBC technique is used to generate gate pulse which is already explained in Sect. 5.2.1. Here ST duty-ratio Do is 0.25, and the modulation index M is 0.75. Triangle carrier has a peak of +/−1 volt and frequency 20 kHz. Two triangle

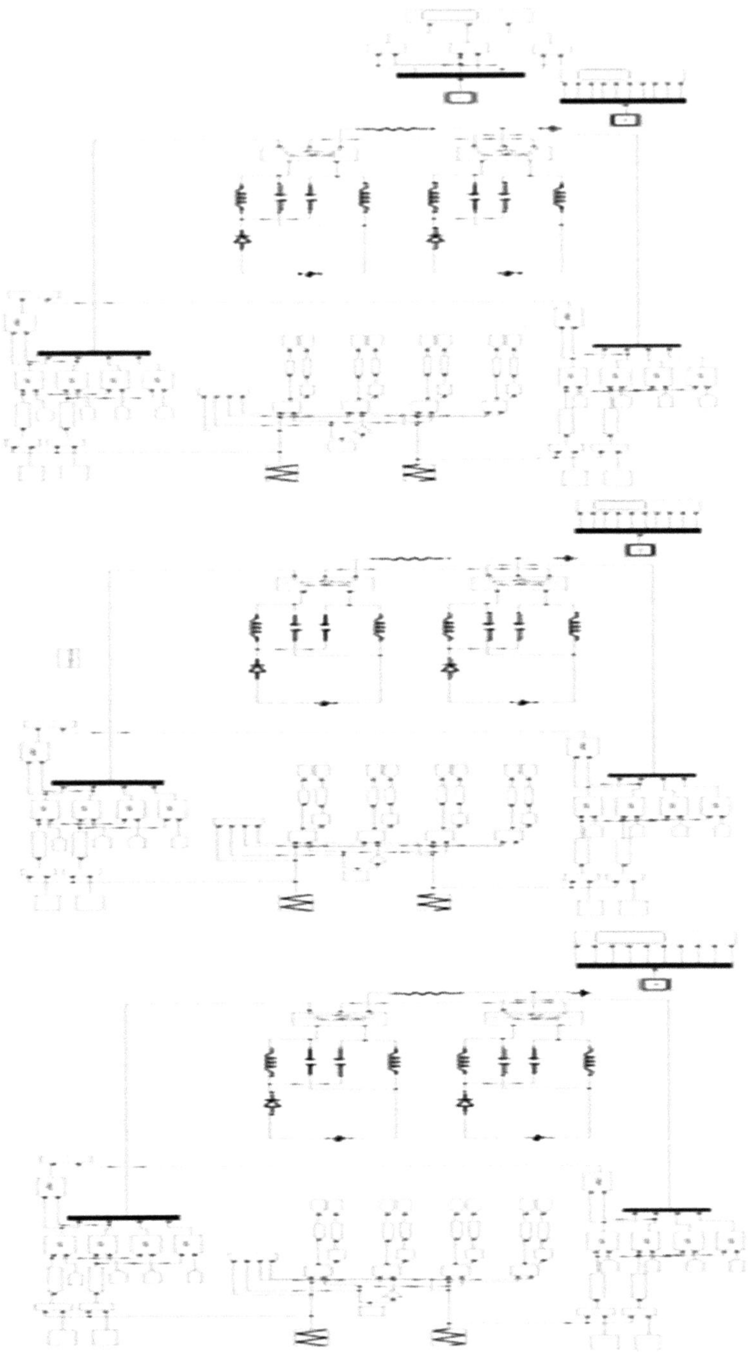

Fig. 5.66 Model of three phase Z-source cascade H-bridge inverter

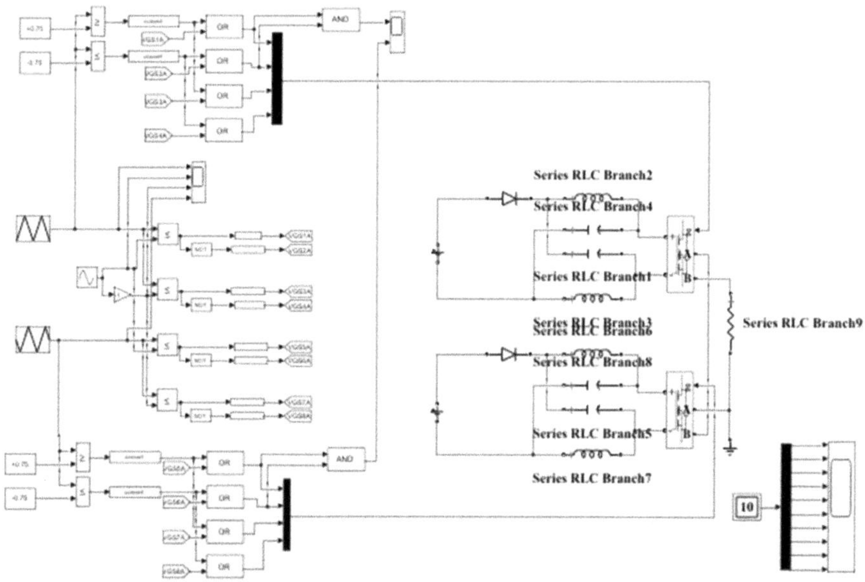

Fig. 5.67 Three phase Z-source cascade H-bridge inverter—model subsystem for phase A

carriers having the same peak value and frequency with a phase shift of zero and $\pi/2$ radians are used, the former for the upper and the later for the lower H-bridge switches. The gate pulse for the switches S1A, S2A, S3A and S4A in the upper H-bridge and S5A, S6A, S7A and S8A for the lower H-bridge is first generated as for the conventional cascade H-bridge inverter presented in Sect. 2.13.1.

To introduce shoot-through state, triangle carrier with zero-phase shift is compared with a constant block +0.75 which corresponds to Vp. The output of relational operator block which forms comparator goes HIGH when triangle carrier exceeds +0.75, or else its output is LOW. This output of this relational operator block is logically ORed with VGS1A and VGS3A using two separate OR gates. The output of these two OR gates forms VGS1AU and VGS3AU for the upper H-bridge switches S1A and S3A, respectively. Similarly, the triangle carrier with zero phase shift is compared with a constant block −0.75 which corresponds to Vn. The output of relational operator block which forms comparator goes HIGH when triangle carrier falls below −0.75, or else its output is LOW. The output of this relational operator block is logically ORed with VGS2A and VGS4A using two separate OR gates. The output of these two OR gates forms VGS2AU and VGS4AU, respectively, for the switches S2A and S4A in the upper H-bridge. The four gate pulses VGS1AU, VGS2AU, VGS3AU and VGS4AU through the four input Mux form the gate drive and provide necessary shoot-through state for the upper H-bridge. The gate pulse VGS1AU and VGS2AU corresponding to switches in phase A are given to an AND gate whose output is used to measure the ST period T0 for the upper H-bridge.

For the gate drive of lower H-bridge switches, the procedure mentioned above for the upper H-bridge switches is repeated with the triangle carrier replaced with the one having π/2 radians phase shift. The four gate pulse VGS5AL, VGS6AL, VGS7AL and VGS8AL thus generated form the gate drive for the lower H-bridge switches. These are taken through a four input Mux and given to the gate input of lower H-bridge and provide necessary shoot-through state for the lower H-bridge. The gate pulse VGS5AL and VGS6AL are given to an AND gate whose out- put gives shoot-through period To for the lower H-Bridge. For the switches in phase B and C, the two triangle carriers having zero and π/2 phase shift remain the same, and comparison is made with sine wave modulating signals with the same peak value and frequency having a phase of -2π/3 and + 2π/3 radians, respectively.

5.6.2 Simulation Results

The simulation of the three-phase Z-source FLCHBI is carried out using ode23tb (stiff/TR-BDF2) solver. Simulation results for the two capacitor voltages of Z-source network for the upper and lower H-bridge, DC link bridge voltage for the upper and lower H-bridge, line voltage across each individual H-bridge, load voltage and load current of cascade H-bridge relating to phase A are shown in Fig. 5.68. The line-to-line output voltage of cascaded H-bridge inverter is shown in Fig. 5.69. The shoot-through period To measurement for the upper and lower H-bridge is shown in Fig. 5.70. The total shoot-through period To for the upper and lower H-bridges

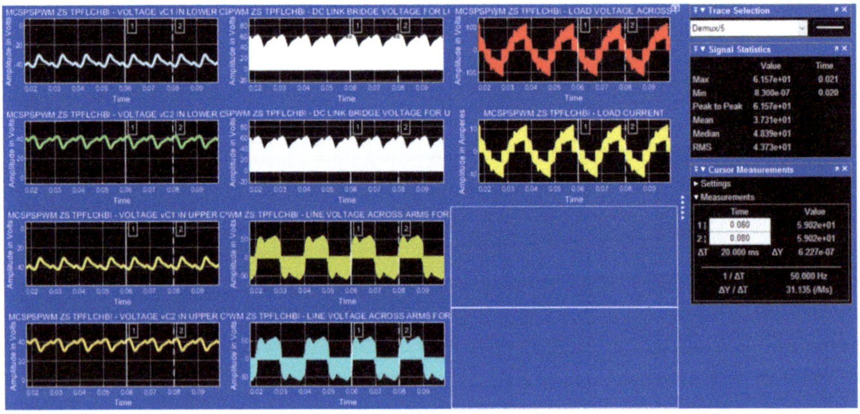

Fig. 5.68 Three phase Z-source FLCHBI simulation results—capcitor voltages vC1, vC2 for the lower and upper H-bridges (letft column top to bottom), DC link bridge voltage for lower and upper H-bridges, line voltage across lower and upper H-bridges (middle column top to bottom), load voltage and load current of FLCHBI (right column top to bottom)

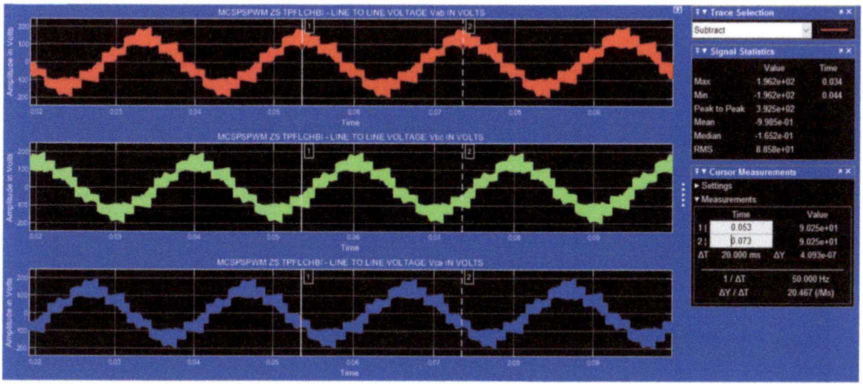

Fig. 5.69 Three phase Z-source FLCHBI—line to line output voltage

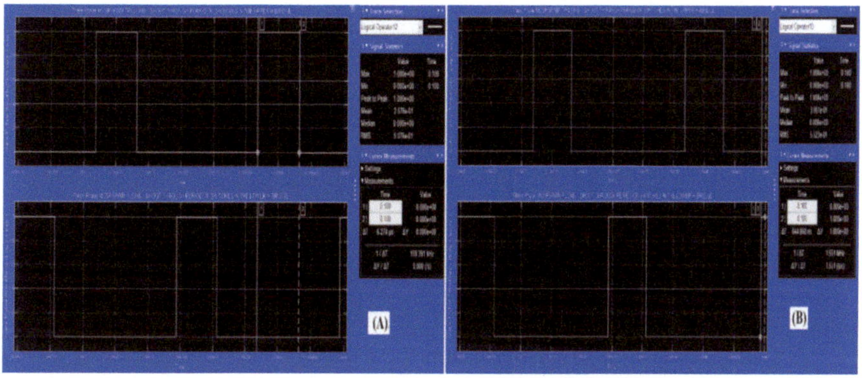

Fig. 5.70 Three phase Z-source FLCHBI shoot through period to measurement: (**a**) To measurement for upper H-bridge, (**b**) To measurement for lower H bridge

from Fig. 5.70 is found to be 12.548e-6 Sec. and 12.5e-6 Sec., respectively. This gives a Do of 0.2509 for the upper H-bridge and 0.25 for the lower H-bridge. The simulation results for the upper and lower H-bridge are tabulated in Table 5.13.

5.6.3 Discussion of Results

The theoretical and model values for Do, B, G, *Vi(Mean)*, *Vc1(Mean)* and *Vc2 (Mean)* closely agree for both the upper and lower H-bridge. The simulation results DC link bridge voltage for upper and lower H-bridge *Vi(Peak)* differ from theoretically calculated value by a large percentage. The line-to-line peak voltage of lower and upper H-bridge when added together closely agrees with the peak value of load voltage *Van(Peak)*.

Table 5.13 Three-phase Z-source FLCHBI: Simulation results

S. no.	D_O	B	G B X M	V_i volts avg./mean	v_i volts peak	V_{C1}, V_{C2} volts avg/mean	*VLL* volts peak	Load voltage van(peak) *V*	Remarks
(1)	(0.25) [0.2509]	(2.00) [2.007]	(1.5) [1.505]	(36) [37.17]	(48) [61.57]	(36, 36) [−37.90, 37.90]	[57.25]	[113.1]	(Model calculations) [Simulation results] Upper H-bridge
(2)	(0.25) [0.25]	(2.00) [2.00]	(1.5) [1.5]	(36) [37.31]	(48) [61.57]	(36, 36) [−37.90, 37.90]	[57.25]	[113.1]	(Model calculations) [Simulation results] Lower H-bridge

5.7 Three-Phase Quasi-Z-Source Five-Level Cascade H-Bridge Inverter

The topology of the three-phase quasi-Z-source (QZS) five-level cascade H-bridge inverter (FLCHBI) is shown in Fig. 5.71. The analysis of the quasi-Z-source inverter is presented in Sect. 5.3. The same analysis for the quasi-Z-source network holds good here for each cell. The bottom and top cells in each phase are connected in series so as to get a five-level voltage output across the load. The DC source voltage for both cells is each Vdc volts. The operation of the five-level cascade H-bridge inverter is already presented in Sect. 2.13 of Chap. 2 [9, 10].

5.7.1 Model of Three-Phase Quasi-Z-Source Five-Level Cascade H-Bridge Inverter

The model of the three-phase quasi-Z-source cascade H-bridge inverter is shown in Fig. 5.72 (Model file: EXAMPLE 5_10). The model subsystem for phase A is shown in Fig. 5.73. The quasi-Z-source network parameters are the same as given in Table 5.6. Here, SBC technique is used to generate gate pulse which is already explained in Sect. 5.2.1. Here ST duty-ratio Do is 0.25, and the modulation index M is 0.75. Triangle carrier has a peak of +/−1 volt and frequency 100 kHz. Two triangle carriers having the same peak value and frequency with a phase shift of zero and π/2 radians are used, the former for the upper and the later for the lower H-bridge switches. The gate pulse for the switches S1A, S2A, S3A and S4A in the upper H-bridge and S5A, S6A, S7A and S8A in the lower H-bridge is first generated as for

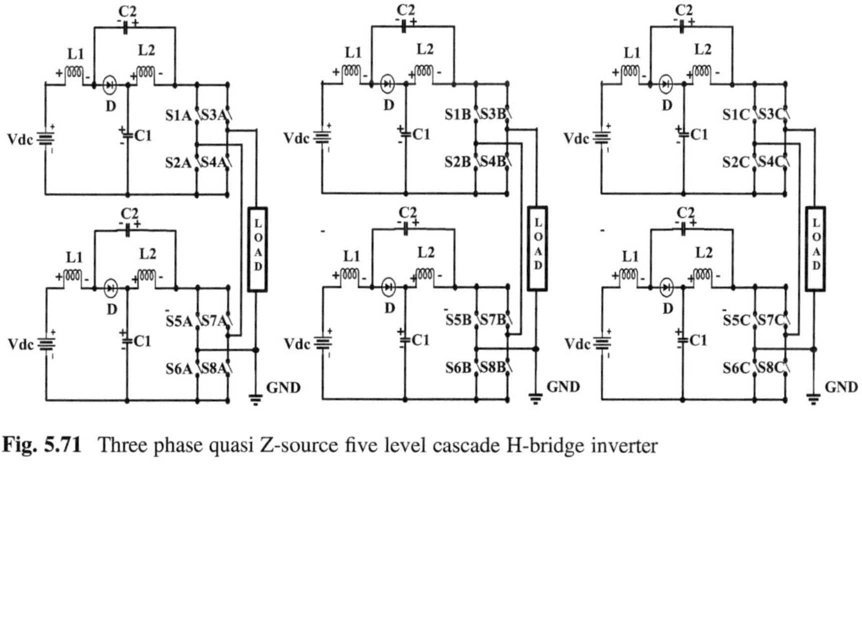

Fig. 5.71 Three phase quasi Z-source five level cascade H-bridge inverter

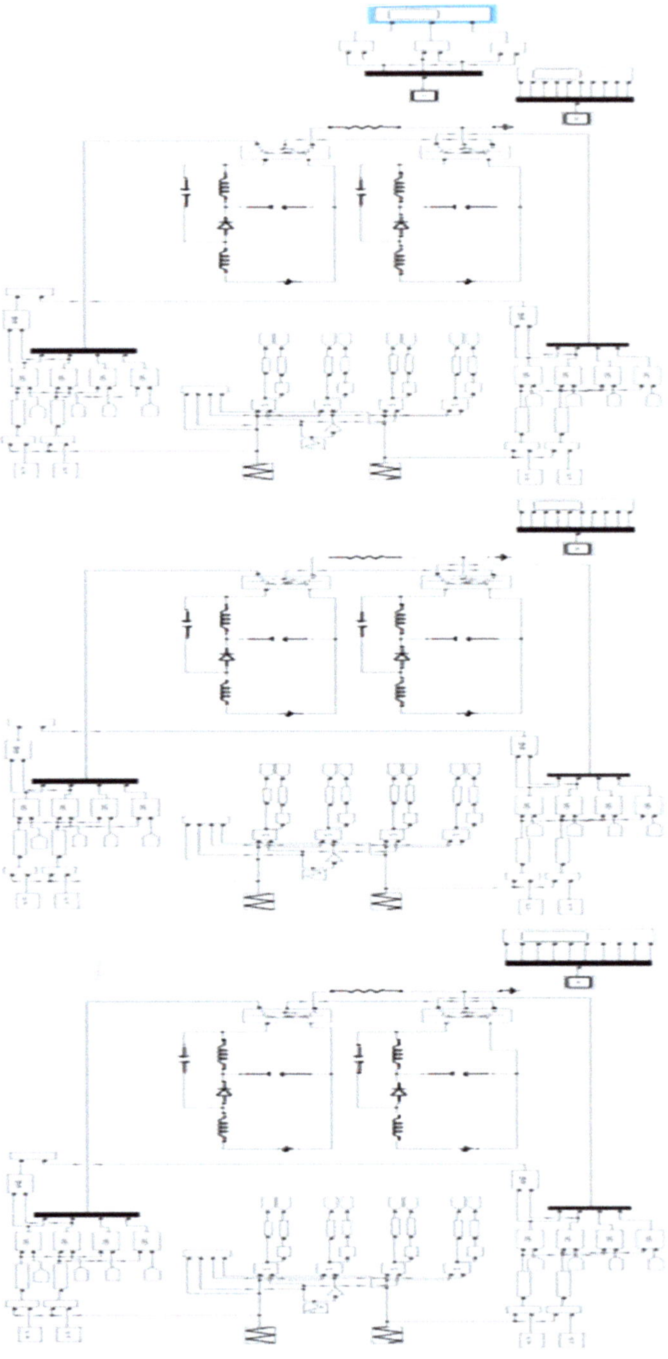

Fig. 5.72 Model of three phase quasi Z-source five level cascade H-bridge inverter

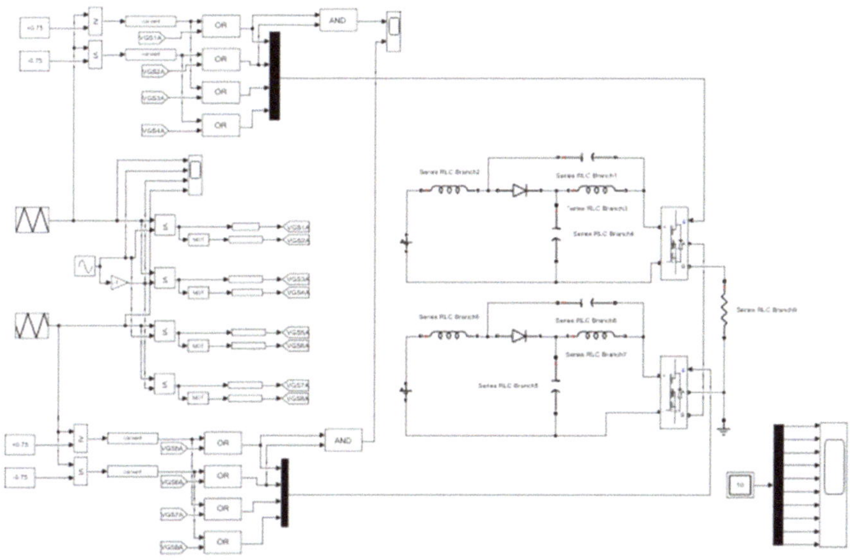

Fig. 5.73 Three phase QZS FLCHBI—model subsystem for phase A

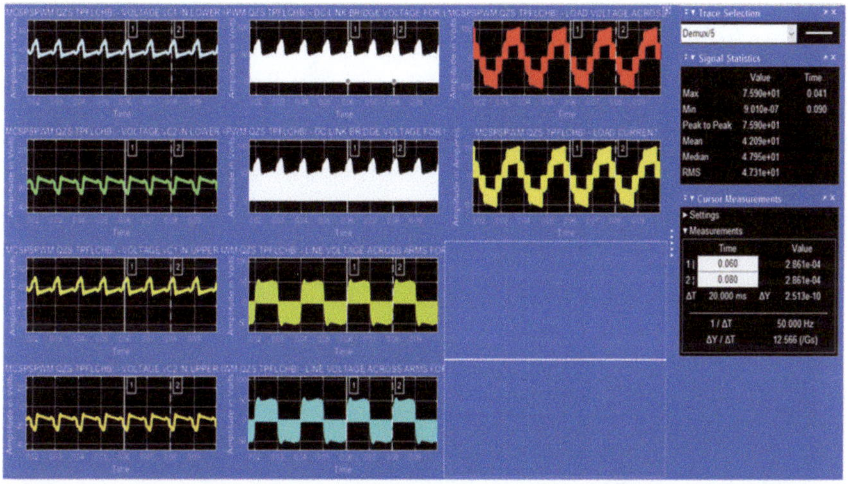

Fig. 5.74 Three phase quasi Z-source FL CHBI simulation results: capacitor voltages vC1, vC2 for the upper and lower H-bridges (Left Column top to bottom), DC link bridge voltage of upper and lower H-bridge, line voltages of upper and lower H-bridge (middle column top to bottom), load voltage and load current of cascade H-bridge inverter (right column top to bottom)

the conventional cascade H-bridge inverter presented in Sect. 2.13.1. The method of generating the gate pulse VGS1AU to VGS4AU for the upper H-bridge and VGS5AL to VGS8AL for the lower H-bridge and the respective shoot-through period To measurement are the same as presented in Sect. 5.6.1.

5.7.2 Simulation Results

The simulation of the three-phase quasi-Z-source FLCHBI is carried out using ode23tb (stiff/TR-BDF2) solver. Simulation results for the two capacitor voltages of quasi-Z-source network for the upper and lower H-bridge, DC link bridge voltage for the upper and lower H-bridge, line voltage across each individual H-bridge, load voltage and load current of cascade H-bridge relating to phase A are shown in Fig. 5.74. The line-to-line output voltage of cascaded H-bridge inverter is shown in Fig. 5.75. The shoot-through period To measurement for the upper and lower H-bridge are shown in Fig. 5.76. The total shoot-through period To for the upper and lower H-bridges from Fig. 5.76 is found to be 2.494e-6 Sec. and 2.493e-6 Sec., respectively. This gives a Do of 0.2494 for the upper H-bridge and 0.2493 for the lower H-bridge. The simulation results for the upper and lower H-bridge are tabulated in Table 5.14.

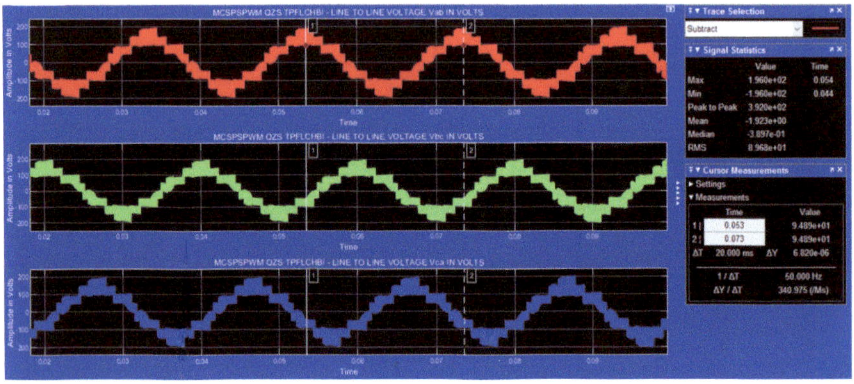

Fig. 5.75 Three phase quasi Z-source FLCHBI—line to line output voltage

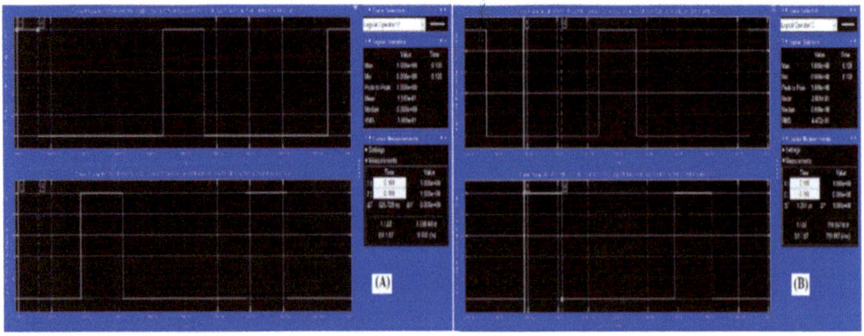

Fig. 5.76 Three phase quasi Z-source FLCHBI—(**a**) shoot through period To measurement for upper H-bridge and (**b**) shoot through period To measurement for lower H-bridge

Table 5.14 Three-phase quasi-Z-source FLCHBI: Simulation results

S. no.	D_O	B	G B X M	V_i volts avg./mean	v_i volts peak	V_{C1}, V_{C2} volts avg./mean	*VLL* volts peak	Load voltage van (peak) *V*	Remarks
(1)	(0.25) [0.2494]	(2.00) [1.995]	(1.5) [1.496]	(36) [42.09]	(48) [75.90]	(36, 12) [38.46, −14.46]	[54.32]	[103.6]	(Model calculations) [Simulation results] Upper H-bridge
(2)	(0.25) [0.2493]	(2.00) [1.994]	(1.5) [1.497]	(36) [42.03]	(48) [75.90]	(36, 12) [38.46, −14.46]	[54.32]	[103.6]	(Model calculations) [Simulation results] Lower H-bridge

5.7.3 Discussion of Results

The theoretical and model values for Do, B, G, *Vi(Mean)*, *Vc1(Mean)* and *Vc2 (Mean)* closely agree for both the upper and lower H-bridge. The simulation results for DC link bridge voltage for upper and lower H-bridge *Vi(Peak)* differ from theoretically calculated value by a large percentage. The line-to-line peak voltage for upper and lower H-bridge when added together closely agrees with peak value of load voltage *Van(Peak)*.

5.8 Three-Phase Z-Source Neutral Point-Clamped Three-Level Inverter

The topology of the three-phase Z-source neutral point-clamped (NPC) three-level inverter is shown in Fig. 5.77. The equivalent circuits for the NST and full shoot-through (FST) state of the NPC inverter are shown in Figs. 5.78a, b. During NST period, the upper and lower switches of each phase act as current source with midpoint grounded. During FST, diodes D1 and D2 are open, and all the switches from S1x to S4x ($x = A, B, C$) in any one or more phase are short circuited. The relevant equations referring to Fig. 5.78a, b using KVL are given below [11, 12]:

From Fig. 5.78a, during NST, the following equations for vL1 and vL2 apply:

$$v_{L1} = v_{C1} - v_i \tag{5.84}$$

$$v_{L1} = 2 * V_{dc} - v_{C2} \tag{5.85}$$

$$v_{L2} = v_{C2} - v_i \tag{5.86}$$

From Fig. 5.78b, during FST, the following equations for v_{L1} and v_{L2} apply:

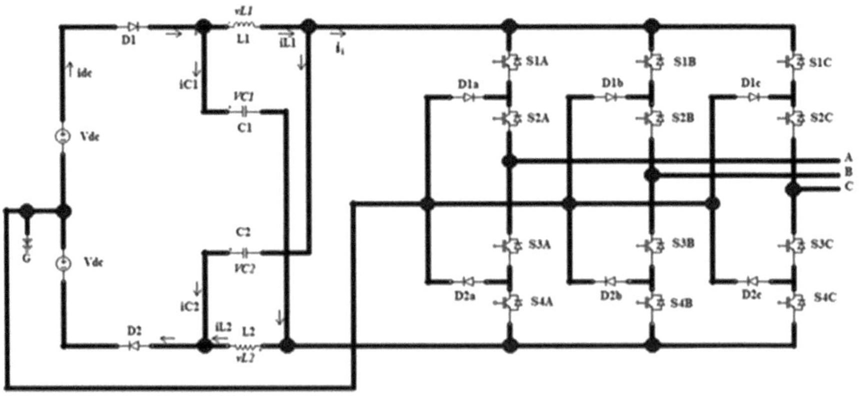

Fig. 5.77 Three phase neutral point clamped three level inverter

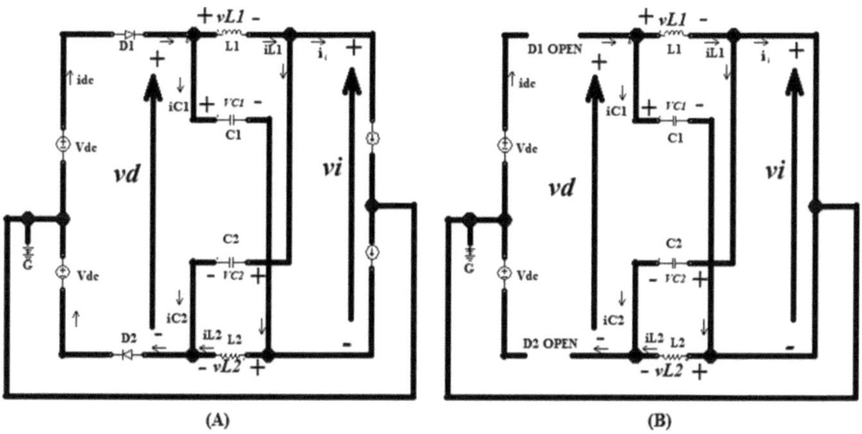

Fig. 5.78 Three phase NPC inverter equivalent circuit: (**a**) non-shoot-through state and (**b**) full shoot-through state

$$v_{L1} = v_{C1} \tag{5.87}$$

$$v_{L2} = v_{C2} \tag{5.88}$$

The average current through L1 and L2 for one carrier switching period T is zero. This leads to the following equations:

$$(v_{C1} - v_i) * (1 - D) * (T/L_1) + v_{C1} * D * (T/L_1) = 0 \tag{5.89}$$

$$(v_{C2} - v_i) * (1 - D) * (T/L_2) + v_{C2} * D * (T/L_1) = 0 \tag{5.90}$$

From Eqs. 5.89 to 5.90, the following:

$$v_{C1} = v_{C2} = v_i * (1 - D) \tag{5.91}$$

In Eqs. 5.89–5.91, D is the full shoot-through duty ratio (T0/T) where T0 is the full shoot-through period.

Also using Eqs. 5.85 and 5.87, the average inductor current through L1 for one carrier switching period T can also be expressed as follows:

$$(2 * V_{dc} - v_{C2}) * (1 - D) * (T/L_1) + v_{C1} * D * (T/L_1) = 0 \tag{5.92}$$

Using Eq. 5.91 in 5.92 and simplifying,

$$v_{C1} = v_{C2} = \frac{2 * V_{dc} * (1 - D)}{(1 - 2 * D)} \tag{5.93}$$

$$v_i = \frac{2 * V_{dc}}{(1 - 2 * D)} \tag{5.94}$$

The AC output phase voltage v_x ($x = A, B, C$) and the boost factor B can be expressed as follows:

$$v_x = \frac{M * v_i}{2} = \frac{M * V_{dc}}{(1 - 2 * D)} \tag{5.95}$$

$$B = \frac{1}{(1 - 2 * D)} \tag{5.96}$$

5.8.1 *Model of a Three-Phase Z-Source Neutral Point-Clamped Three-Level Inverter*

The model of the three-phase Z-source NPC inverter is shown in Fig. 5.79 (Model file: EXAMPLE 5_11). The gate drives for phase A switches are shown in Fig. 5.80. The data in Table 5.1 are used to develop the model. Here, the modulation index *M* is taken as 0.8. The diodes and the three-phase three-level NPC inverter are from specialised power systems and power electronics blockset. The L-C components, DC voltage source and ground connection are from specialised power systems blockset. The gate drive for phase A is shown in Fig. 5.80. In Fig. 5.80, the triangle carrier generates triangle carrier signal, with peak value +/− 1 volt and frequency 20 kHz. This is given as input to an interpreted MATLAB Fcn block with multiplication constant [(1 + u(1)*0.5] whose output is a triangle carrier wave with peak value 1 volt, minimum value zero and frequency 20 kHz corresponding to the chosen carrier frequency. This interpreted MATLAB Fcn output is added with −1 using Sum12 block. The output of Sum12 block is a triangle carrier with peak value zero, minimum value −1 and frequency 20 kHz. The modulating sine wave signal with amplitude 0.8 V corresponding to modulation index value M and frequency 50 Hz corresponding to that of inverter output frequency is generated using sine wave1 function generator. These waveforms are shown in Fig. 5.81. In Fig. 5.81, the upper triangle carrier corresponds to Embedded MATLAB Fcn output, lower triangle carrier corresponds to Sum12 block output, and the modulating sine wave corresponds to sine wave1 function generator output. Referring to Fig. 5.80, the upper triangle carrier is compared with sine wave modulating function in the relational operator 1 comparator block, the output of which is HIGH when the former value is less than or equal to the later value, or else its output is LOW. The relational operator 1 block output corresponds to the gate pulse for switch S1A, and its inverted output using NOT gate corresponds to that of switch S3A. Similarly, the lower triangle

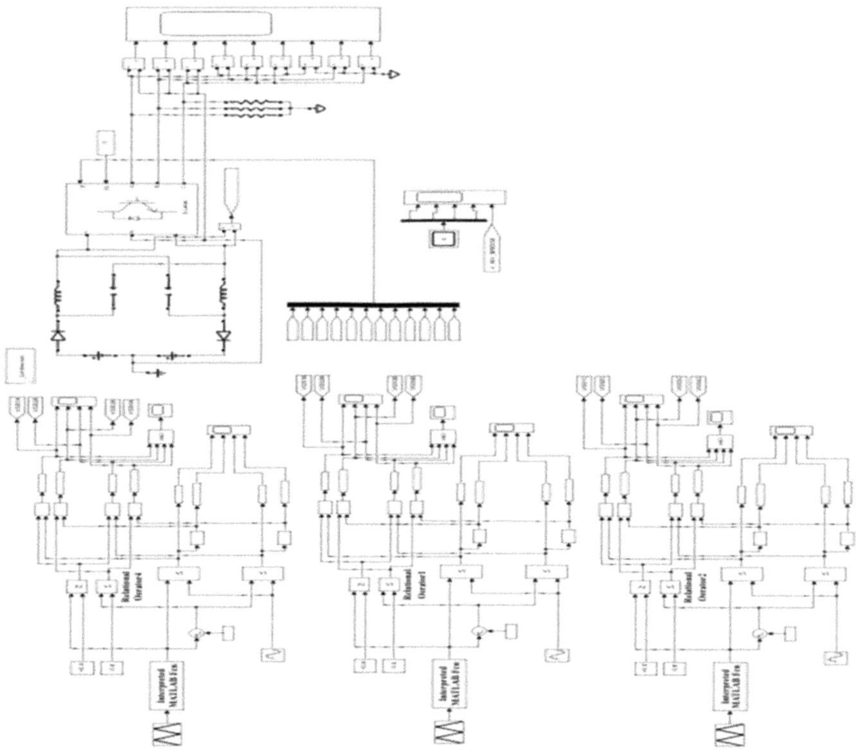

Fig. 5.79 Model of a three phase Z-source NPC inverter

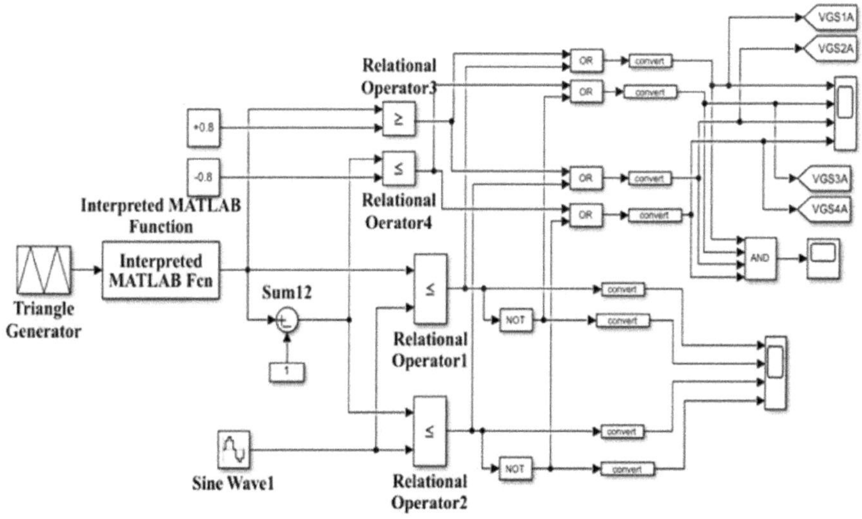

Fig. 5.80 Phase A gate drive model of a three phase Z-source NPC three level inverter

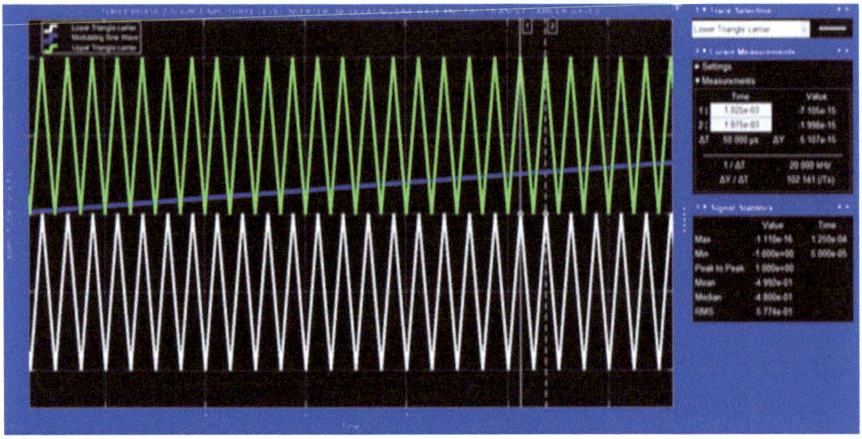

Fig. 5.81 Three phase Z-source NPC inverter—modulating Sine ware signal and two triangle carrier signals

carrier is compared with sine wave modulating function in the relational operator2 comparator block, the output of which is HIGH when the former value is less than or equal to the later value, or else its output is LOW. The relational operator2 block output corresponds to the gate pulse for switch S2A, and its inverted output using NOT gate corresponds to that of switch S4A. The gate pulse generated in this way never produces shoot-through state.

To add shoot-through state, two constant values corresponding to positive and negative modulation index M, +0.8 and −0.8, are compared with upper triangle and lower triangle carriers using relational operator3 and relational operator4 comparator blocks, respectively. When the upper triangle carrier goes above +0.8, relational operator3 block goes HIGH, or else its output is LOW. Similarly, when the lower triangle carrier goes below −0.8, relational operator4 comparator block goes HIGH, or else its output is low. The gate pulse for switch S1A and S2A generated as mentioned above is each logically ORed with relational operator3 comparator output using two separate OR gates and their respective outputs form the new gate drive for switch S1A and S2A. Similarly, the NOT gate output corresponding to gate pulse of switch S3A and S4A mentioned above are each logically ORed with relational operator4 comparator output using two separate OR gates and their respective output form the new gate drive for switches S3A and S4A, respectively. The gate pulse generated in this way introduces shoot-through state.

5.8.2 *Simulation Results*

The simulation of the three-phase Z-source NPC three-level inverter is carried out using ode 15 s (Stiff/NDF) solver in Simulink [6]. The modulation index M is 0.8. The data shown in Table 5.1 are used. The simulation results for the line-to-ground,

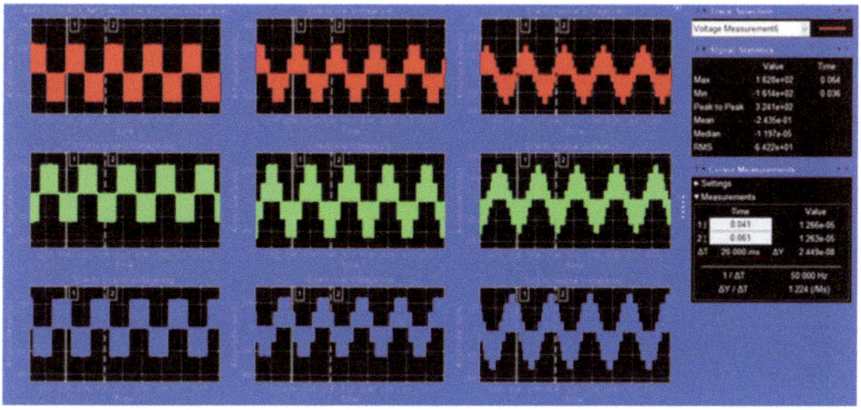

Fig. 5.82 Three phase Z-source NPC inverter—line to ground voltage (left column top to bottom), line to line voltage (middle column top to bottom) and line to neutral voltage (right column top to bottom)

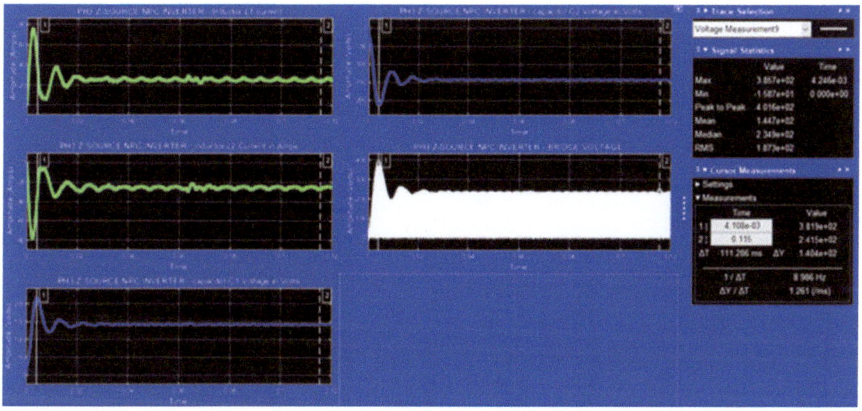

Fig. 5.83 Three phase Z-source NPC inverter—current through Inductor L1, current through inductor L2, Voltage across capacitor C1, Voltage across capacitor C2. inverter bridge voltage vi

line-to-line and line-to-neutral voltages are shown in Fig. 5.82. The simulation results for the two capacitor voltages, two inductor currents and inverter bridge voltage are shown in Fig. 5.83. The full shoot-through period measurement is shown in Fig. 5.84. From Fig. 5.84, the T0 measured for the one carrier period duration from 119e-3 Sec. to 119.05e-3 Sec. is found to be 9.991e-6 s. This gives a shoot-through duty-ratio D value of 0.1998. The simulation results are tabulated in Table 5.15.

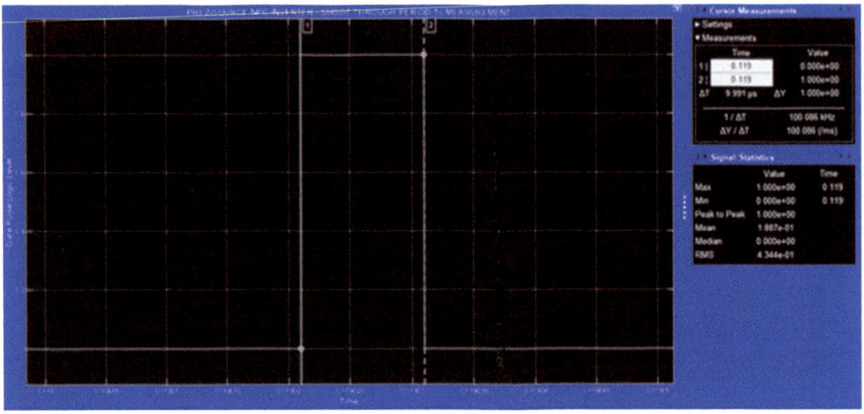

Fig. 5.84 Three phase Z-source NPC inverter—full shoot through period T0 measurement

Table 5.15 Three-phase Z-source NPC inverter—Model simulation results

S. no.	D	B	G B X M	V_i volts avg./mean	V_{C1}, V_{C2} volts	V_{AN} volts RMS	Remarks
(1)	(0.2) [0.1998]	(1.67) [1.665]	(1.336) [1.332]	(160) [144.7]	(128, 128) [148, – 148]	(64.13) [64.22]	(Model calculations) [Simulation results]

5.8.3 *Discussion of Results*

The model values for shoot-through duty-ratio D, boost factor B and gain G very closely agree with simulation results. The two capacitor voltages vc1 and vc2 by simulation are found to be equal which agree with theoretical derivation. The inverter bridge voltage v_i and AC output phase voltage v_{an} by simulation agree with theoretically derived values, the former having a percentage error of around 9.37%.

5.9 Three-Phase Quasi-Z-Source Neutral Point-Clamped Three-Level Inverter

The topology of the three-phase quasi-Z-source (QZS) NPC three-level inverter is shown in Fig. 5.85 (Model file: EXAMPLE 5_12). There are three equivalent circuits for this QZS network configuration. The NST active state when either of the switch pair S1x–S2x or S3x–S4x (x = A, B, C) is only ON delivering power to the load, NST zero state when switch pair S2x–S3x are only ON applying zero

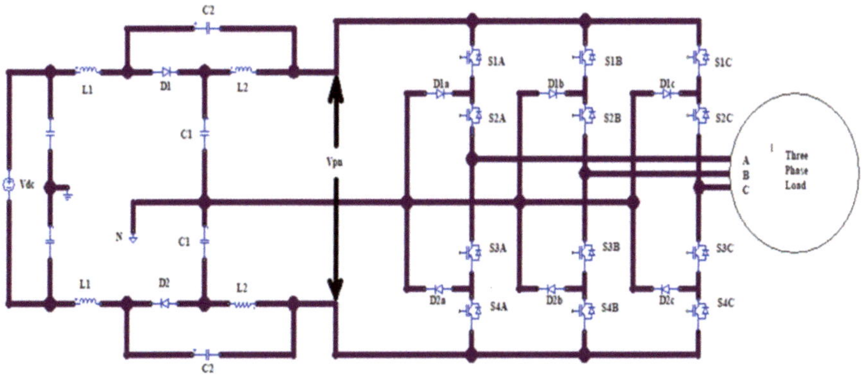

Fig. 5.85 Three phase quasi Z-source NPC three level inverter

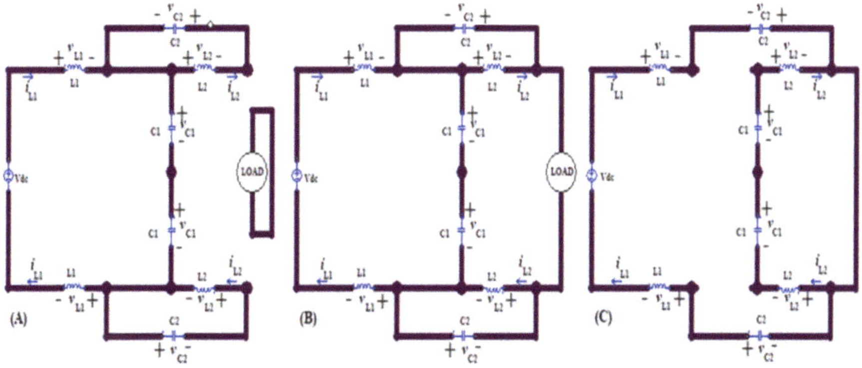

Fig. 5.86 Three phase quasi Z-source NPC three level inverter (**a**) NST zero state, (**b**) NST active state and (**c**) FST state

voltage to the load, FST when all switches S1x to S4x in any one or more phase are ON forming a full short circuit state [13]. These equivalent circuits are shown in Fig. 5.86.

Assume that NST zero state duty ratio is Dz, NST active state duty ratio is Da and FST duty ratio is D0. Then (Dz + Da + D0) = 1.

Applying KVL to Fig. 5.86a, the following equations:

$$2 * v_{L1} + 2 * v_{C1} = V_{dc} \tag{5.97}$$

$$v_{L2} + v_{C2} = 0 \tag{5.98}$$

From Fig. 5.86b, the following equations:

$$2 * v_{L1} + 2 * v_{C1} = V_{dc} \tag{5.99}$$

$$v_{L2} + v_{C2} = 0 \tag{5.100}$$

From Fig. 5.86c, the following equations:

$$2 * v_{L1} - 2 * v_{C2} = V_{dc} \tag{5.101}$$

$$2 * v_{L2} - 2 * v_{C1} = 0 \tag{5.102}$$

Equations for NST active state and NST zero state are one and the same, and they can be combined having a time duration (Da*T + Dz*T) = (1-D0)*T, where T is the carrier switching period.

During the NST period (1-D0)*T, from Eqs. 5.97 to 5.100, the following:

$$v_{L1} = \frac{V_{dc}}{2} - v_{C1} \tag{5.103}$$

$$v_{L2} = -v_{C2} \tag{5.104}$$

During the FST period D0*T, from Eqs. 5.101 and 5.102, the following:

$$v_{L1} = \frac{V_{dc}}{2} + v_{C2} \tag{5.105}$$

$$v_{L2} = v_{C1} \tag{5.106}$$

The average current through inductors L1 and L2 for one switching period T is zero. This can be expressed as follows:

$$\left(\frac{V_{dc}}{2} - v_{C1}\right) * (1 - D_0) * (T/L_1) + \left(\frac{V_{dc}}{2} + v_{C2}\right) * D_0 * (T/L_1) = 0 \tag{5.107}$$

$$(- v_{C2}) * (1 - D_0) * (T/L_2) + (v_{C1}) * D_0 * (T/L_2) = 0 \tag{5.108}$$

Simplifying Eqs. 5.107 and 5.108, the following:

$$v_{C1} * (D_0 - 1) + v_{C2} * D_0 = \frac{-V_{dc}}{2} \tag{5.109}$$

$$v_{C1} * D_0 + v_{C2} * (D_0 - 1) = 0 \tag{5.110}$$

Solving Eqs. 5.109 and 5.110, the following:

$$v_{C1} = \frac{\begin{vmatrix} \frac{-V_{dc}}{\genfrac{}{}{0pt}{}{2}{0}} & \frac{D_0}{(D_0-1)} \end{vmatrix}}{\begin{vmatrix} \frac{(D_0-1)}{D_0} & \frac{D_0}{(D_0-1)} \end{vmatrix}} = \frac{V_{dc} * (1 - D_0)}{(2 - 4 * D_0)} \tag{5.111}$$

$$v_{C2} = \frac{\begin{vmatrix} \frac{(D_0-1)}{D_0} & \frac{-V_{dc}}{\genfrac{}{}{0pt}{}{2}{0}} \end{vmatrix}}{\begin{vmatrix} \frac{(D_0-1)}{D_0} & \frac{D_0}{(D_0-1)} \end{vmatrix}} = \frac{V_{dc} * D_0}{(2 - 4 * D_0)} \tag{5.112}$$

Now in the steady state when inductor currents are constants, referring to Fig. 5.86a, b, the peak DC link voltage V_{pn} can be expressed as follows:

$$V_{pn} = 2 * (v_{C1} + v_{C2}) \tag{5.113}$$

Using Eqs. 5.111 and 5.112, Eq. 5.113 can be expressed as follows:

$$V_{pn} = \frac{V_{dc}}{(1 - 2 * D_0)} \tag{5.114}$$

$$B = \frac{1}{(1 - 2 * D_0)} \tag{5.115}$$

The average value of the output phase voltage v_{an} is given below:

$$v_{an} = \frac{M * V_{pn}}{2} = \frac{M * V_{dc}}{2 * (1 - 2 * D_0)} \tag{5.116}$$

5.9.1 Model of a Three-Phase Quasi-Z-Source Neutral Point-Clamped Three-Level Inverter

The model of a three-phase QZS NPC three-level inverter is shown in Fig. 5.87 (Model file: EXAMPLE 5_12). The data are shown in Table 5.6. Here the three-phase NPC three-level inverter is from Simscape—electrical—semiconductors and converters blockset. The DC voltage source and all other passive components are from Simscape—electrical blockset. The gate drive model is the same as shown in Fig. 5.80. The model development for implementing shoot-through state is the same as explained in Sect. 5.8.1. The two triangle carriers and the sine wave modulating singals are the same as shown in Fig. 5.81 except that here the two triangle carriers have a frequency of 100 kHz. The modulating sine wave frequency is 50 Hz with an amplitude of 0.8 volts.

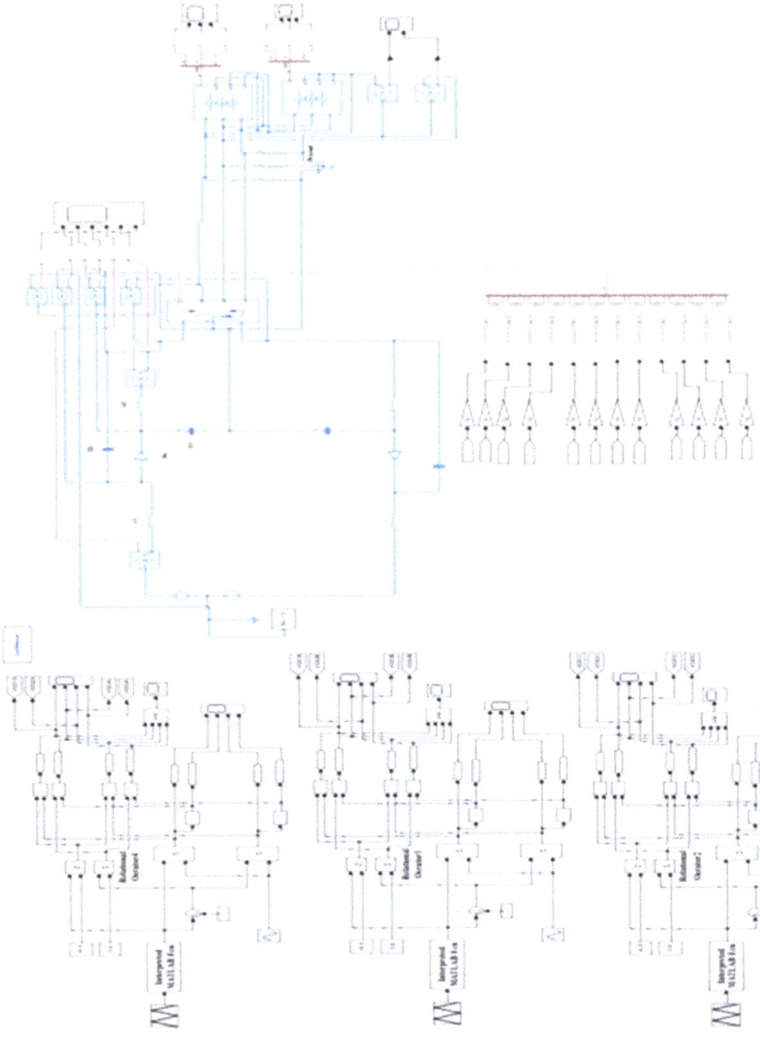

Fig. 5.87 Three phase quasi Z-source NPC three level inverter

5.9.2 Simulation Results

The simulation of the three-phase quasi-Z-source NPC three-level inverter is carried out using fixed-step ode1be (Backward Euler) solver in Simulink [6]. The modulation index M is 0.8. The data shown in Table 5.6 are used. The output filter is eliminated. The simulation results for the line-to-line and line-to-neutral voltages are shown in Figs. 5.88 and 5.89. The simulation results for the two capacitor voltages, inverter bridge voltage, two inductor currents and the line-to-ground voltage are shown in Fig. 5.90. The full shoot-through period measurement is shown in

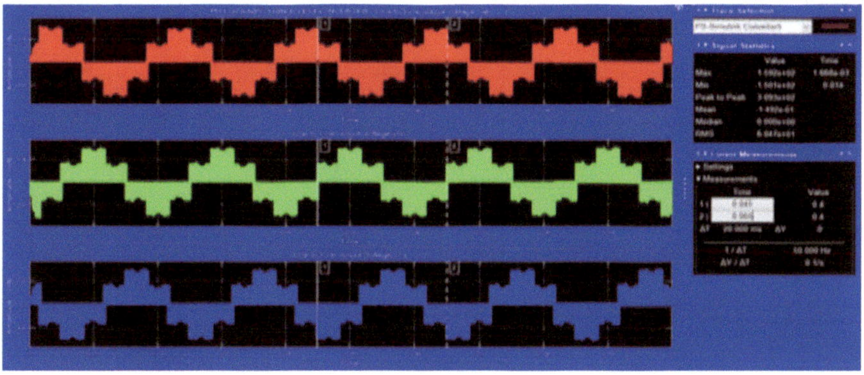

Fig. 5.88 Three phase QZS NPC three level inverter—line to line voltage vab, Vbc Vea

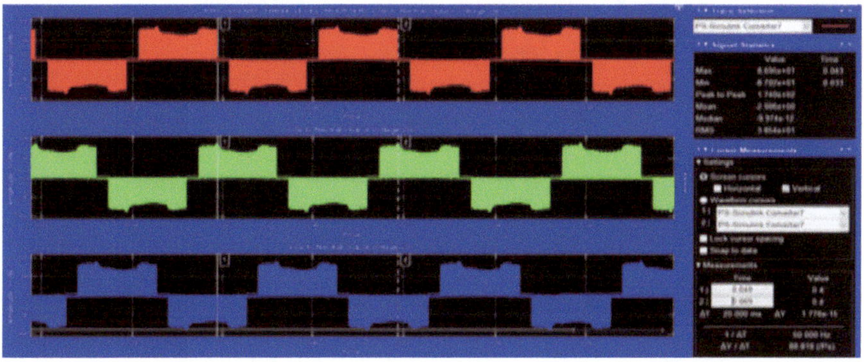

Fig. 5.89 Three phase QZS NPC three level inverter—line to neutral voltage van, vbn and ven

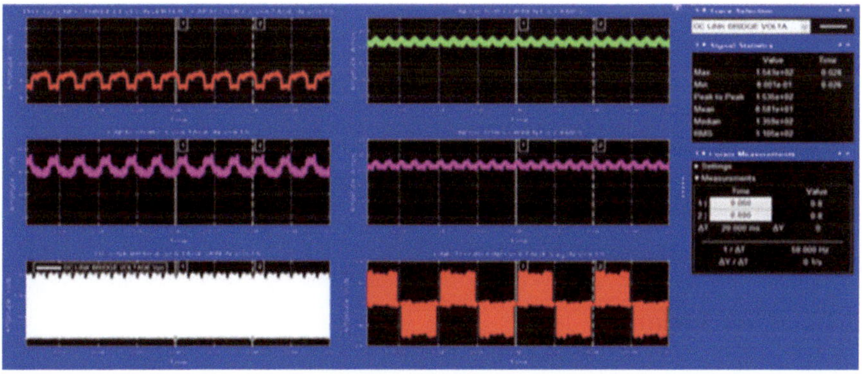

Fig. 5.90 Three phase QZS NPC three level inverter—capacitor voltage vC2, capacitor voltage vC1, inverter DC link bridge voltage Vpn (left column tap to bottom), inductor currents iL1, iL2 and Line to ground voltage vag (right column top to bottom)

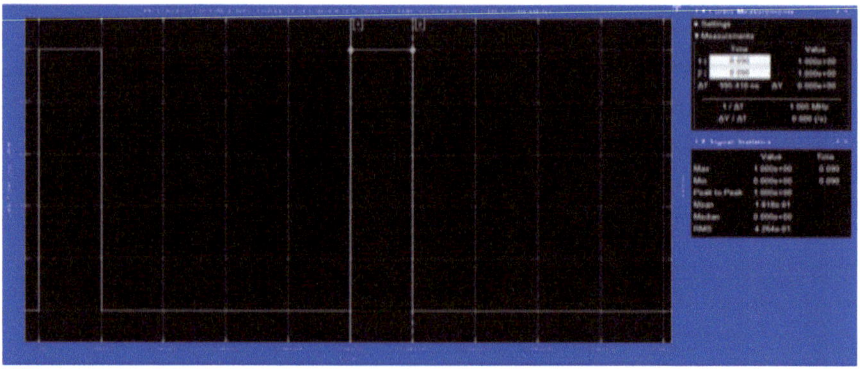

Fig. 5.91 Three phase QZS NPC three level inverter—shoot through period T0 measurement

Table 5.16 Three-phase QZS NPC three-level inverter—Simulation results

S. no.	*Do*	*B*	*G* B X M	V_{pn} volts avg./ mean	V_{C1}, V_{C2} volts	V_{AN} volts *RMS*	Remarks
(1)	(0.2) [0.2]	(1.67) [1.67]	(1.336) [1.336]	(80) [85.81]	(32, 8) [43.26, −19.28]	(32) [38.54]	(Model calculations) [Simulation results]

Fig. 5.91. From Fig. 5.91, the T0 measured for one carrier period duration from 90e-3 Sec. to 90.01e-3 Sec. is found to be 2e-6 Sec. This gives a shoot-through duty-ratio D_O value of 0.2. The simulation results are tabulated in Table 5.16.

5.9.3 *Discussion of Results*

The calculated values for *D*o, *B* and *G* well agree with that from model simulation. The calculated values for V_{pn} and V_{an} differ from model values by 7.25% and 20.3%, respectively. The calculated values for v_{c1} and v_{c2} differ from the model values by a large percentage.

5.10 Conclusions

Models for three-phase Z-source and quasi-Z-source inverter using SBC, MBC, THIMBC and THIMCBC techniques are presented. For Z-source inverter using SBC and THIMCBC, inductor currents are found to be uniform, and all the simulation results closely agree with respective theoretical values. For the ZSI using MBC and THIMBC, inductor currents are discontinuous or non-uniform. The equations for DO, B and G for MBC, THIMBC and THIMCBC are derived from fundamentals. For the three-phase QZSI, it is seen that the voltage stress on capacitor C1 is almost three times higher than that on C2. Inductor currents iL1 and iL2 are uniform and equal.

Designed for the same power output P, DC input voltage, inductor currents ripple ratios and capacitor voltages ripple ratios, the size of the two inductors and two capacitors for QZS network is much smaller compared to that for ZS network. This makes QZSI design economical compared to ZSI. Two case studies one for the SVPWM of three-phase ZSI and another for that of QZSI are presented. A new method of using shoot-through factor Ksh is developed to generate shoot-through state in ZSI and QZSI with space vector modulation. Also three-phase Z-source and quasi-Z-source inverter models with (a) five-level cascade H-bridge topology and (b) three-level NPC topology both with SBC technique are presented with simulation results.

References

1. F.Z. Peng: "Z-Source Inverter", IEEE Transactions on Industry Applications, Vol. 39, No. 2, March/April 2003, pp. 504—510.
2. F.Z. Peng, M. Shen and Z. Qian: "Maximum Boost Control of the Z-Source Inverter", IEEE Transactions on Power Electronics, Vol. 20, No. 4, July 2005, pp. 833—838.
3. M. Shen, J. Wang, Alan Joseph, F.Z. Peng, L.M. Tolbert and D.J. Adams: "Constant Boost Control of the Z-Source Inverter to minimize current ripple and voltage stress," IEEE Transactions on Industry Applications, Vol.42, No.3, May/June 2006, pp. 770–778.
4. M. Shen, J. Wang, Alan Joseph, F.Z. Peng, L.M. Tolbert and D.J. Adams: "Maximum Constant Boost Control of the Z-Source Inverter", IEEE Industry Applications Society Conference, 2004, pp. 142–147.
5. F.Z. Peng, Alan Joseph, Jin Wang, M. Shen, Lithua Chen, Zhiguo Pan, E.O. Riviera and Yi Huang: "Z-Source Inverter for Motor Drives", IEEE Transactions on Power Electronics, Vol. 20, NO. 4, JULY 2005, pp. 857—863.
6. The Mathworks Inc.: www.mathworks.com: "MATLAB/SIMULINK", R2021a, 2021.
7. Y. Li, J. Anderson, F.Z. Peng and D. Lu: "Quasi-Z-Source Inverter for Photovoltaic Power Generation Systems', IEEE APEC, February 2009, Washington DC, USA, pp. 918—924.
8. Shuo Liu, Baoming Gay, X. Jiang, Haitham Abu Rub, F.Z. Peng: "A Novel Indirect Quasi Z-Source Matix Converter Applied to Induction Motor Drives"; IEEE Energy Conversion Congress and Exposition; Denver, CO, USA, September 15–19, 2013; pp. 2440–2444.
9. Surin Khomfoi and Leon M. Tolbert. 2011. "Multilevel Power Converters" in Power Electronics Handbook, Ed. M.H. Rashid, 455–484, Elsevier, Butterworth-Heinemann.
10. N P R Iyer: "Power Electronic Converters: Interactive Modeling using Simulink", CRC Press, USA, Chapter 9 and 10, pp. 243–249 and pp. 315–319., 2018.
11. Liu Yushan etal.: "Impedance Source Power Electronic Converters", Chapter 12, John Wiley and Sons, 2016.
12. Martin Stempfle, Steffen Bintz, Julian Wölfle, Jörg Roth-Stielow: "Adaptive closed-loop state control system for a three-level neutral-point-clamped Z-source inverter", IET Electrical Systems in Transportation, 2016, Vol. 6, Iss. 1, pp. 12–19.
13. Oleksandr Husev, Carlos Roncero-Clemente, Enrique Romero-Cadaval, Dmitri Vinnikov, Tanel Jalakas: "Three-level three-phase quasi-Z-source neutral-point-clamped inverter with novel modulation technique for photovoltaic application", Electric Power Systems Research, Elsevier, 2016, pp. 10-21.

Chapter 6
Switched Inductor, Switched Capacitor and Diode-Assisted Z-Source and Quasi Z-Source Three-Phase Inverter

6.1 Introduction

The Z-source inverter (ZSI) has advantage over the conventional inverter in the sense that the former topology can boost the DC link voltage of inverter or boost or buck the inverter output phase voltage whereas for the latter topology, the output phase voltage of inverter is always less than the input DC source voltage [1]. The ZSI has discontinuous input current, high voltage appearing on both the Z-source network capacitors. The quasi Z-source inverter (QZSI) has advantages over the classical ZSI. The QZSI carries a continuous input current and has a common earthing for DC voltage source and DC link bus, low voltage across one of the capacitors of the quasi Z-source network, smaller size of passive components, reduced component cost, simple design and enhanced efficiency [2–4]. These features were examined by model simulation in Chap. 5.

The boost factor and the ability to boost the input DC voltage are low for ZSI and QZSI. To further increase the boost factor and enhance the boost ability, switched inductor ZSI (SL-ZSI), switched inductor (SL) and switched capacitor (SC) QZSI (SL and SC-QZSI) were proposed [5–7]. It is seen that the two inductor currents and the two capacitor voltages are continuous with minimum ripple and the voltage stress on the capacitors is low. In addition active switched QZSI (AS-QZSI), extended active switched QZSI (EAS-QZSI) and diode-assisted QZSI (DA-QZSI) are proposed [8, 11]. This active switched topology increases the boost factor, minimises input current ripple as well as voltage ripple across quasi Z-source network capacitors and reduces voltage stress on the inverter switches [8, 11].

Supplementary Information The online version contains supplementary material available at https://doi.org/10.1007/978-3-031-62784-2_6.

N. P. R. Iyer, *Inverters and AC Drives*, Power Systems,
https://doi.org/10.1007/978-3-031-62784-2_6

In this chapter, models for SL-ZSI, SL-QZSI, ASC-QZSI and DAEB-QZSI are presented with simulation results for selected boost control techniques, and a comparison of the performance is made with that of classical ZSI and QZSI. In addition four case studies are presented each for the space vector modulation (SVM) of three-phase SL-QZSI, ASC-QZSI, DAEB-QZSI and SL-ZSI.

6.2 Switched Inductor Z-Source Three-Phase Inverter

The classical ZSI has a low boost factor. In order to increase the boost factor further and to obtain higher DC link bridge voltage and output phase voltage for the inverter, switched inductor ZSI (SL-ZSI) is used [5, 6]. This has higher boost factor and voltage gain compared to the classical ZSI. This is achieved by proper choice of the shoot through duty ratio D_O which is T_O/T where T_O is the shoot-through and T is the carrier switching period. A circuit schematic of the SL-ZSI is shown in Fig. 6.1. The equivalent circuits of the SL-ZSI for the shoot-through (ST) and non-shoot-through (NST) period are shown in Fig. 6.2a, b, respectively. During the ST period, diodes D1 and D4 are open, D2 and D3 are closed, and the inductors L1 and L2 are connected in parallel. During the NST period, diodes D1 and D4 are closed, D2 and D3 are open, and the inductors L1 and L2 are connected in series. The analysis is given below:

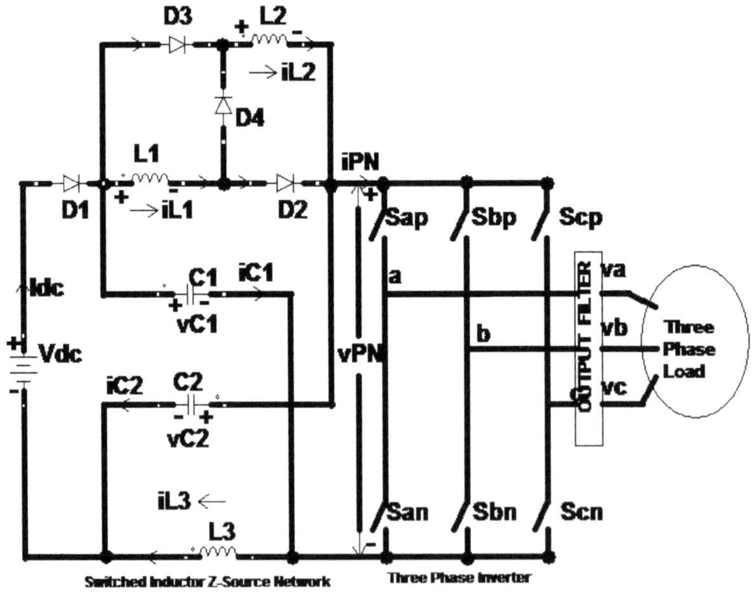

Fig. 6.1 Switched inductor Z-source three phase inverter

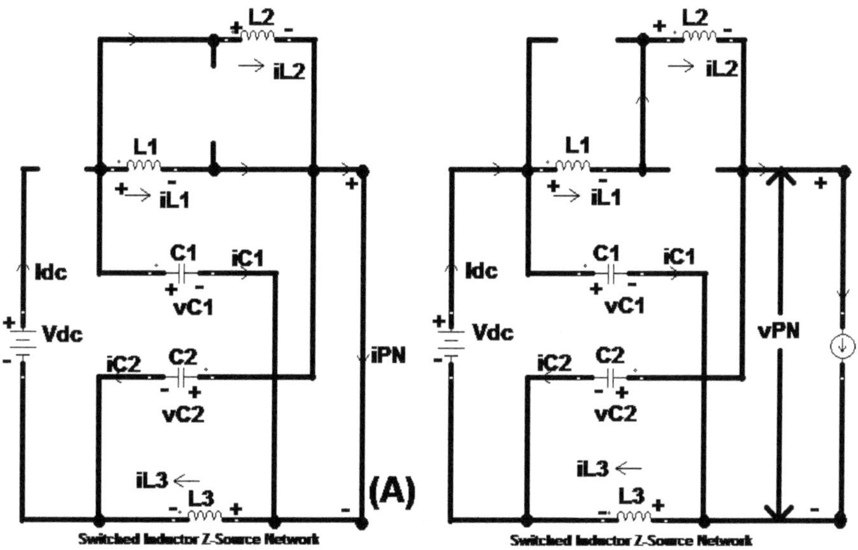

Fig. 6.2 Switched inductor three-phase ZSI: (**a**) shoot-through state and (**b**) non-shoot-through state

During the ST period, the following equations are valid:

$$v_{L1} = V_{C1} \tag{6.1}$$

$$v_{L2} = V_{C1} \tag{6.2}$$

$$v_{L1} = v_{L2} \tag{6.3}$$

$$v_{L3} = V_{C2} \tag{6.4}$$

During the NST period, the following equations are valid assuming $L1 = L2 = L$:

$$v_{L1} = v_{L2} = v_L = \left(\frac{V_{\text{dc}}}{2} - \frac{V_{C2}}{2}\right) \tag{6.5}$$

$$v_{L3} = (V_{C2} - v_{\text{PN}}) = (V_{\text{dc}} - V_{C1}) \tag{6.6}$$

From Eq. 6.6, v_{PN} can be expressed as follows:

$$v_{\text{PN}} = (V_{C1} + V_{C2} - V_{\text{dc}}) \tag{6.7}$$

The average current through inductor L1 over one switching period is zero which leads to the following equation:

$$\left(\frac{V_{C1}{}^*D_O{}^*T}{L_1}\right) + \frac{\left(\frac{V_{dc}}{2} - \frac{V_{C2}}{2}\right) * (1 - D_O) * T}{L_1} = 0 \tag{6.8}$$

$$i.e.\ V_{C1} = \frac{(1 - D_O)^*}{2^* D_O}(V_{C2} - V_{dc}) \tag{6.9}$$

Similarly the average current through inductor L3 over one switching period is zero which leads to the following equation:

$$\left(\frac{V_{C2}{}^* D_O{}^* T}{L_3}\right) + \frac{(V_{dc} - V_{C1}) * (1 - D_O) * T}{L_3} = 0 \tag{6.10}$$

$$i.e.\ V_{C2} = \frac{(1 - D_O)^*}{D_O}(V_{C1} - V_{dc}) \tag{6.11}$$

Using Eq. 6.11 in Eq. 6.9 and simplifying, V_{C1} can be expressed as follows:

$$V_{C1} = \frac{V_{dc}{}^*(1 - D_O)}{(1 - 2^* D_O - D_O{}^2)} \tag{6.12}$$

Using Eq. 6.12 in Eq. 6.11 and simplifying, V_{C2} can be expressed as follows:

$$V_{C2} = \frac{V_{dc}{}^*(1 - D_O) * (1 + D_O)}{(1 - 2^* D_O - D_O{}^2)} = \frac{V_{dc}{}^*(1 - D_O{}^2)}{(1 - 2^* D_O - D_O{}^2)} \tag{6.13}$$

Using Eqs. 6.12 and 6.13 in Eq. 6.7, the mean inverter bridge DC link voltage v_{PN} can be expressed as follows:

$$v_{PN} = \frac{V_{dc}{}^*(1 + D_O)}{(1 - 2^* D_O - D_O{}^2)} \tag{6.14}$$

$$B = \frac{v_{PN}}{V_{dc}} = \frac{(1 + D_O)}{(1 - 2^* D_O - D_O{}^2)} \tag{6.15}$$

In Eq. 6.15, B is the boost factor.

6.2.1 Model of Switched Inductor Z-Source Three-Phase Inverter with Simple Boost Control

The simple boost control (SBC) technique is explained in Sect. 5.2.1. The model of the SL-ZSI is shown in Fig. 6.3 (model file: EXAMPLE6_1). The parameters shown in Table 5.1 are used except for the values of Z-source network inductor L3 which is made half the value of L1. Modulation index M is 0.75. Boost factor B is calculated by using Eq. 6.15. The method of generating the gate pulse for SBC technique is already explained in Sect. 5.2.2. Here the shoot through period T_O is

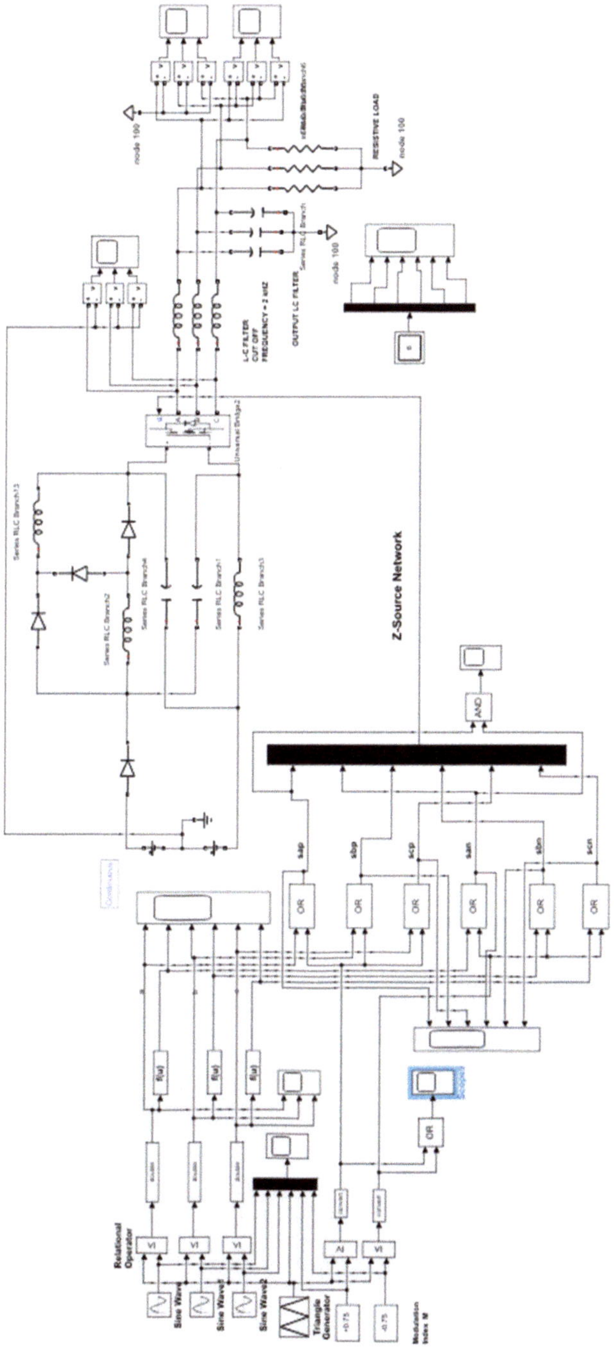

Fig. 6.3 Mode of three-phase switched inductor Z-source inverter with simple boost control

directly measured from the gate drive model without going to the gate pulse waveforms. Here the constant blocks relating to +0.75 and −0.75 are compared with triangle carrier using two relational operator blocks which form the comparators, and the output of these two comparators is given to OR gate, the output of which gives the shoot-through period T_O. Also the gate pulse Sap and San corresponding to the upper and lower switches in Phase A of the inverter are given to a two-input AND gate, the output of which gives the shoot- through period T_O.

6.2.2 Simulation Results

The simulation of the three-phase SL-ZSI with SBC is carried out using ode23tb (stiff/TR-BDF2) solver [9]. The data shown in Table 5.1 are used with the Z-source network inductor value L3 changed to 216e-6 H. The simulation results for the line to neutral output voltage, line to line output voltage, line to ground output voltage and Z-source inductor currents and capacitor voltages, three-phase load currents and inverter bridge DC link voltage V_{pn} are shown in Figs. 6.4 to 6.7 respectively. The shoot-through period T_O measurement is shown in Fig. 6.8. In Fig. 6.8, measurement of shoot-through period T_O is made for one triangle carrier period commencing 0.190 sec. to 0.19005 sec. in three different locations. These values are, respectively, 3.123e-6 sec., 6.297e-6 sec. and 3.123e-6 sec. respectively giving a total value of 12.543e-6 sec. for T_O using AND gate and 3.119e-6 sec., 6.289e-6 sec. and 3.119e-6 sec. giving a total value of 12.527e-6 sec. for To using OR gate. The value of To using AND gate is considered here. The D0 value is 0.2508. The simulation results are tabulated in Table 6.1.

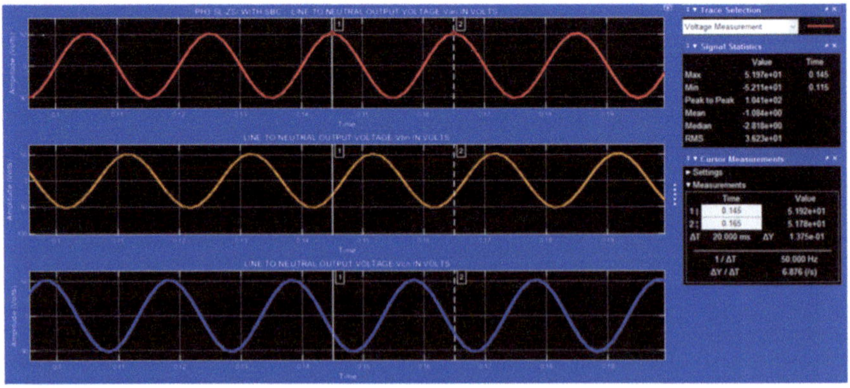

Fig. 6.4 Three-phase SL-ZSI—line to neutral output voltage

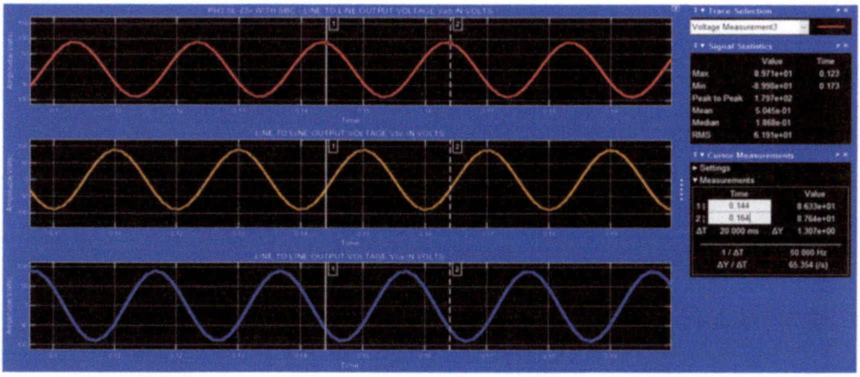

Fig. 6.5 Three-phase SL-ZSI—line to line output voltage

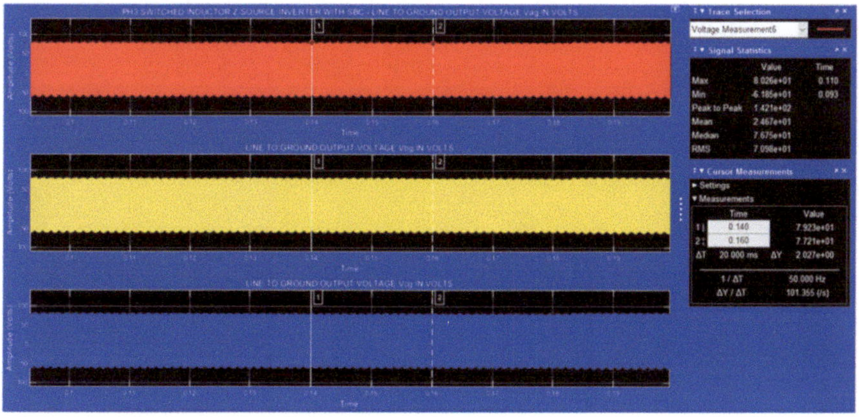

Fig. 6.6 Three-phase SL-ZSI—line to ground output voltage

6.2.3 Discussion of Results

The model values and simulation results in Table 6.1 for D_O, B, G, *Vac(peak)*, *Vpn (mean)*, *vpn(peak)*, *VC1(mean)* and *VC2(mean)* agree with the theoretical values. Comparing the result with that shown in Table 5.2 in Chap. 5, it is seen that for the same D_O and SBC technique, the values of B and G are increased and this has boosted the value of *Vac(peak)*, *Vpn(mean)*, *vpn(peak)*, *VC1(mean)* and *VC2(mean)*. Inductor currents are continuous. Also the Root Mean Square (RMS) value of line to ground voltage from Fig. 6.6 is found to be 70.98 V, whereas for a conventional inverter, this value is 24 V which is half the DC source voltage. This gives a boost factor B of 2.9575.

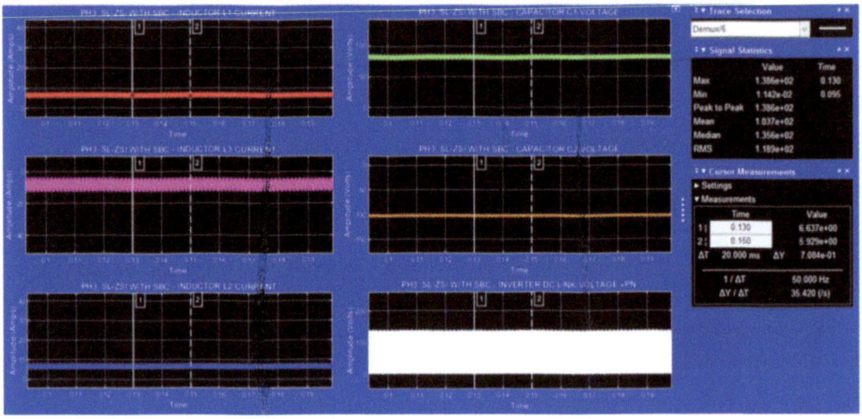

Fig. 6.7 Three-phase SL-ZSI—inductor currents iL1, iL3 and iL2 (left column, top to bottom), capacitor voltages vC1 and vC2 and inverter bridge DC link voltage V_{PN} (right column, top to bottom)

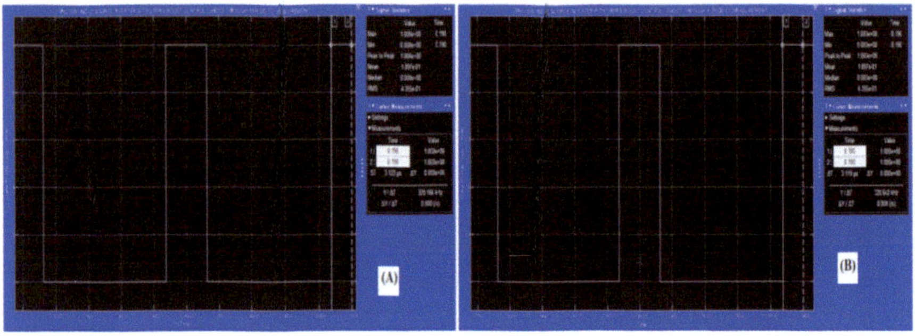

Fig. 6.8 Three-phase SL-ZSI using simple boost control—shoot through period To measurement (**a**) using AND gate and (**b**) using OR gate

6.2.4 Model of Switched Inductor Z-Source Three-Phase Inverter with Third Harmonic Injection Maximum Constant Boost Control

The maximum constant boost control technique is presented in Sect. 5.2.13. The model of the SL-ZSI with third harmonic injection maximum constant boost control (THIMCBC) is shown in Fig. 6.9 (model file: EXAMPLE6_2). Now consider the three-phase SL-ZSI whose parameters are given in Table 5.1. Inductor value L3 is made half the value of inductor L1. The value of M is taken as 0.866 and using Eq. 5.51, D_O is 0.25. The Embedded MATLAB function generates three-phase THI sine wave defined in Eq. 5.35. The source code for generating this three-phase THI sine wave is shown in Program segment 5.1. Boost factor B is calculated by using Eq. 6.15. The method of generating the gate pulse for THIMCBC technique is

Table 6.1 Three-phase SL-ZSI with simple boost control—simulation results

S. no.	D_O	B	G B X M	V_{ac} (Volts) Peak	V_{PN} Volts Avg/mean	v_{PN} Volts Peak	V_{C1} Volts Avg/mean	V_{C2} Volts Avg/mean	Remarks
1)	0.25 [0.2508]	2.8571 (2.8721)	2.143 (2.152)	51.4286 (51.648) [51.97]	102.86 (103.285) [103.7]	137.1429 (137.8611) [138.6]	82.2857 (82.5756) [81.93]	102.8571 (103.2855) [102.1]	Theoretical (Model Calculations) [Simulation Results]

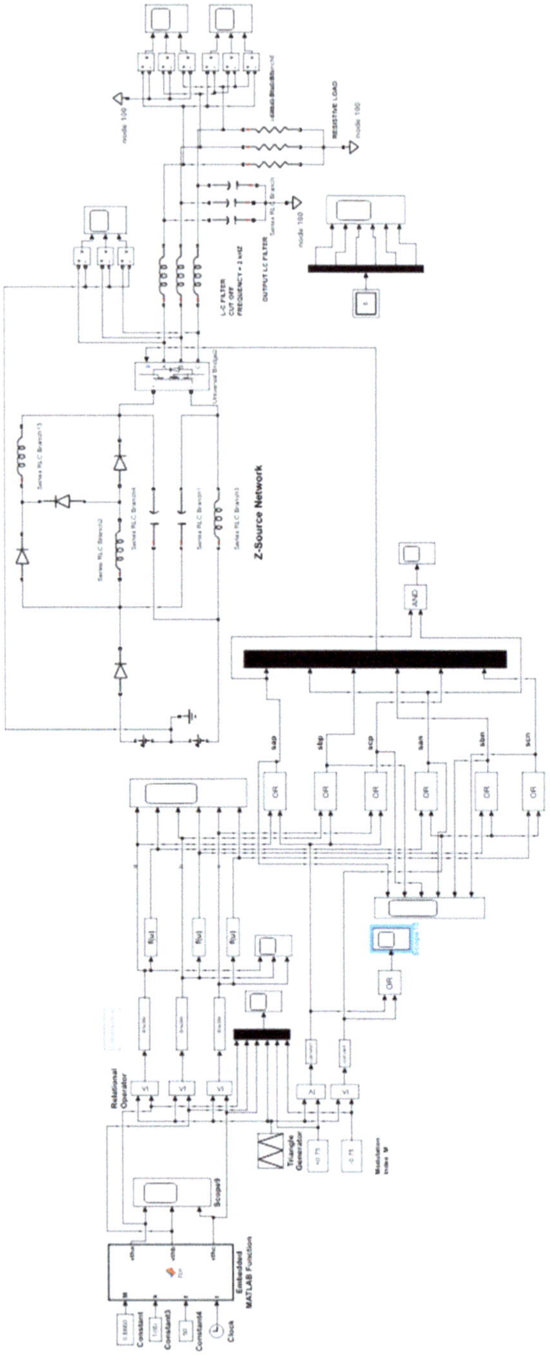

Fig. 6.9 Three-phase switched inductor Z-source inverter with third harmonic injection maximum constant boost control

already explained in Sect. 5.2.13. Here the shoot-through period T_O is directly measured from the gate drive model without going to the gate pulse waveforms. Here the constant blocks relating to $+\sqrt{3}*M/2$ (+0.75) and $-\sqrt{3}*M/2$ (−0.75) are compared with triangle carrier using two relational operator blocks which form the comparators, and the output of these two comparators is given to OR gate, the output of which gives the shoot through period T_O. In addition the gate pulse corresponding to switches Sap and San in Phase A of the inverter is given to a two-input AND gate, the output of which gives the shoot-through period T_O.

6.2.5 Simulation Results

The simulation of the three-phase SL-ZSI with THIMCBC is carried out using ode23tb (stiff/TR-BDF2) solver [10]. The data shown in Table 5.1 are used with the Z-source network inductor value L3 changed to 216e-6 H. The simulation results for the line to neutral output voltage, line to line output voltage, line to ground output voltage and Z-source inductor currents, capacitor voltages and inverter bridge DC link voltage *Vpn* are shown in Figs. 6.10 to 6.13 respectively. The shoot-through period T_O measurement is shown in Fig. 6.14. In Fig. 6.14, measurement of shoot-through period To is made for one triangle carrier period commencing 0.190 sec. to 0.19005 sec. in three different locations. These values are, respectively, 3.167e-6 sec., 6.284e-6 sec. and 3.167e-6 sec. giving a total value of 12.618e-6 sec. for T_O using AND gate and 3.197e-6 sec., 6.237e-6 sec. and 3.145e-6 sec. giving a total value of 12.579e-6 sec. for To using OR gate. The value of To using AND gate is considered here. The D0 value is 0.252. The simulation results are tabulated in Table 6.2.

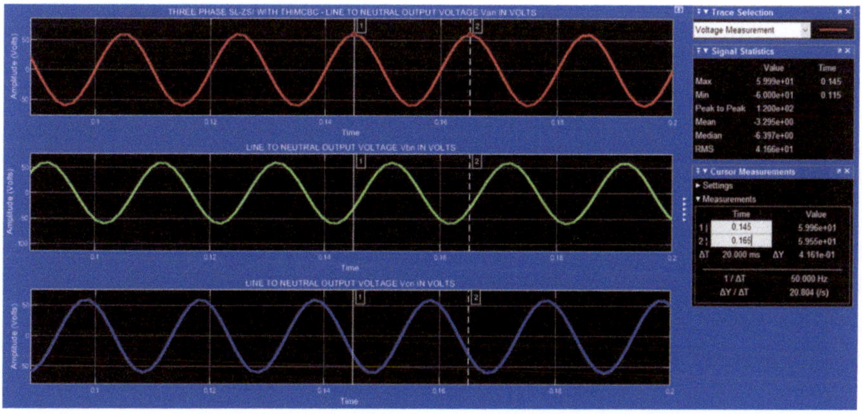

Fig. 6.10 Three-phase SL-ZSI with THIMCBC—line to neutral output voltage

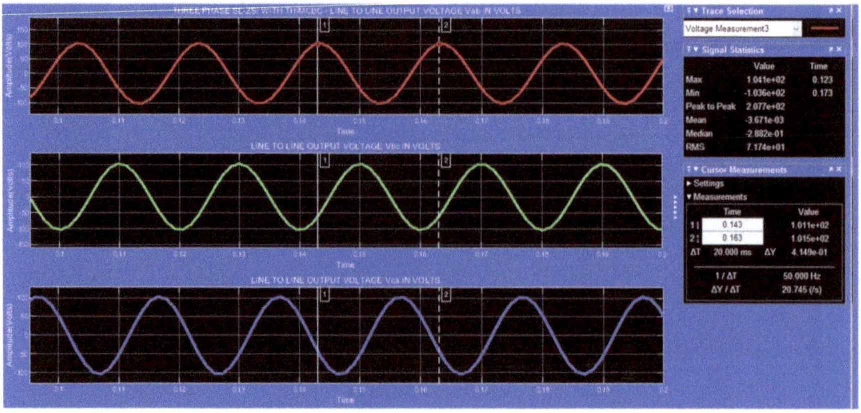

Fig. 6.11 Three-phase SL-ZSI with THIMCBC—line to line output voltage

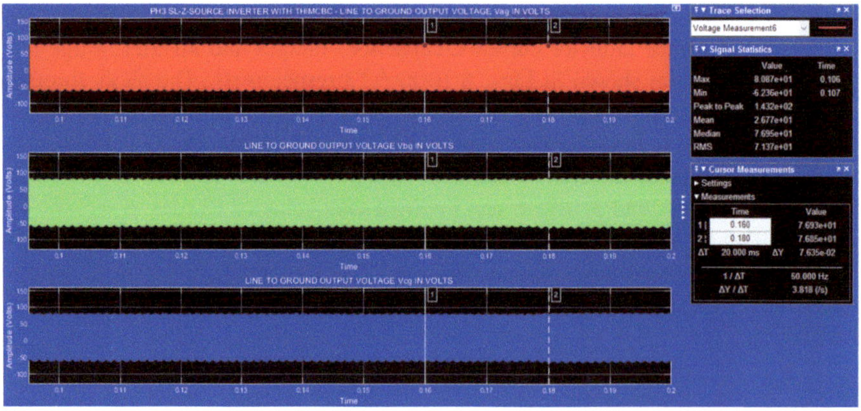

Fig. 6.12 Three-phase SL-ZSI with THIMCBC—line to ground output voltage

6.2.6 Discussion of Results

The model values and simulation results in Table 6.2 for D_O, B, G, *Vac(peak)*, *Vpn (mean)*, *vpn(peak)*, *VC1(mean)* and *VC2(mean)* agree with the theoretical values. Comparing the result with that shown in Table 5.5 in Chap. 5, it is seen that for the same D_O and THIMCBC technique, the values of B and G are increased and this has boosted the value of *Vac(peak)*, *Vpn(mean)*, *vpn(peak)*, *VC1(mean)* and *VC2(mean)*. Inductor currents are continuous. Also from Fig. 6.12, the RMS value of line to ground output voltage is 71.37 V, whereas for a conventional inverter, this value is 24 V which is half the DC source voltage. This gives a boost factor B of 2.97.

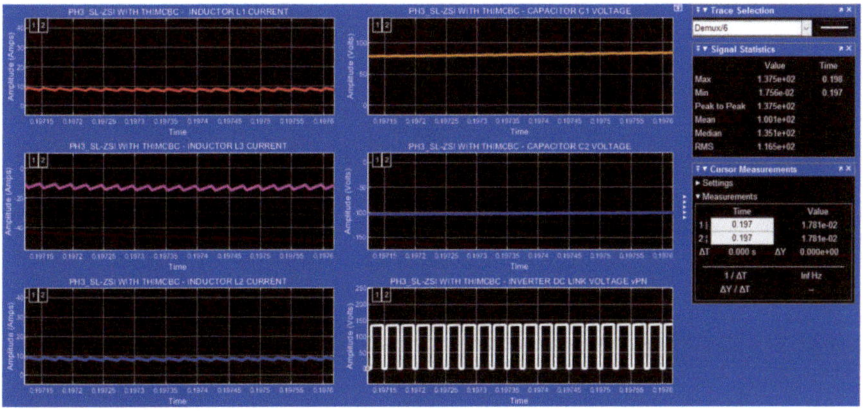

Fig. 6.13 Three-phase SL-ZSI with THIMCBC—inductor currents iL1, iL3 and iL2 (left column, top to bottom), capacitor voltages vC1 and vC2 and inverter bridge DC link voltage V_{PN} (right column, top to bottom)

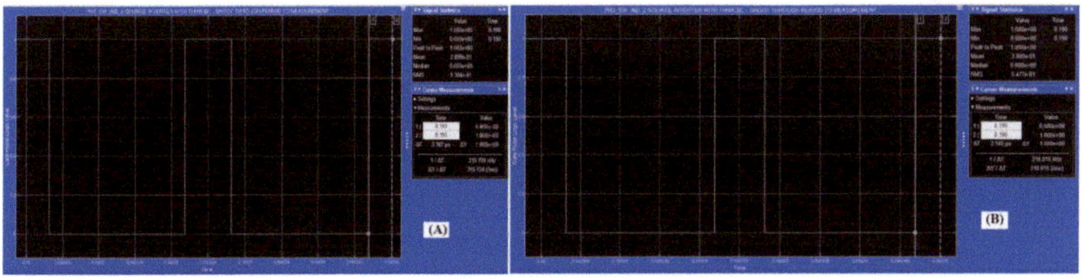

Fig. 6.14 Three-phase SL-ZSI with THIMCBC—shoot- through period To measurement (**a**) using AND gate and (**b**) using OR gate View mode NOT enabled.

6.3 Switched Inductor Quasi Z-Source Three-Phase Inverter

The classical quasi Z-source inverter (QZSI) was proposed to overcome the drawbacks of the classical Z-source inverter (ZSI) such as huge inrush current, high voltage appearing on Z-source network capacitors, discontinuous input current, high voltage stress on inverter switches and capacitors and no common ground between inverter and the DC voltage source [2–4]. QZSI draws continuous input current from the DC voltage source and inverter switches, and one of the Z-source capacitors has a lower voltage stress and has a common ground point for inverter and DC voltage source. This continuous input current makes QZSI suitable for photovoltaic power generation [3, 4]. The boost factor for the ZSI and QZSI is still low due to narrow shoot-through shoot throughrange, and enhancing the boost factor even further enhances the output voltage and reduces the voltage stress [5–9].

To enhance the boost factor even further than that possible by classical ZSI and QZSI, switched inductor (SL) and switched capacitor (SC) QZSI (SL/SC-QZSI)

Table 6.2 Three-phase SL-ZSI with THIMCBC—simulation results

S. no.	D_O	B	G B X M	V_{ac} (Volts) Peak	V_{PN} Volts Avg/mean	v_{PN} Volts Peak	V_{C1} Volts Avg/mean	V_{C2} Volts Avg/mean	Remarks
1)	0.25 [0.252]	2.8571 (2.8948)	2.4742 (2.507)	59.383 (60.1660) [60.00]	102.86 (103.93) [101.7]	137.143 (138.9516) [139.3]	82.286 (83.0158) [82.0]	102.86 (103.9358) [102.2]	Theoretical (Model Calculations) [Simulation Results]

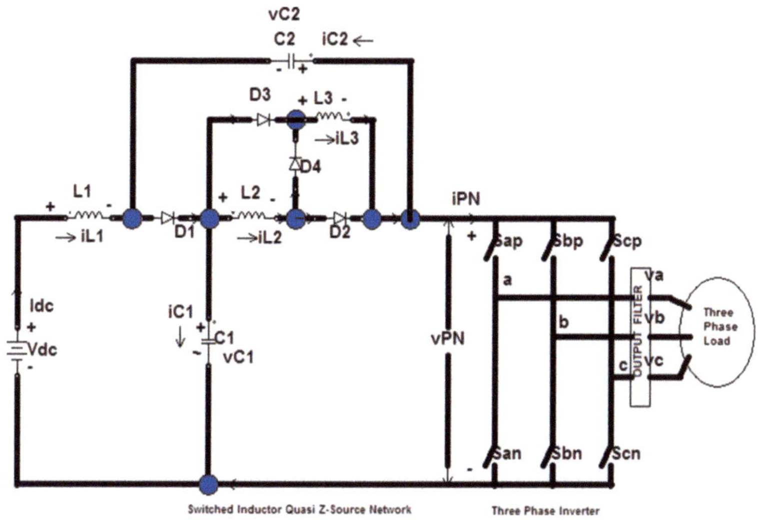

Fig. 6.15 Switched inductor quasi Z-source three-phase inverter

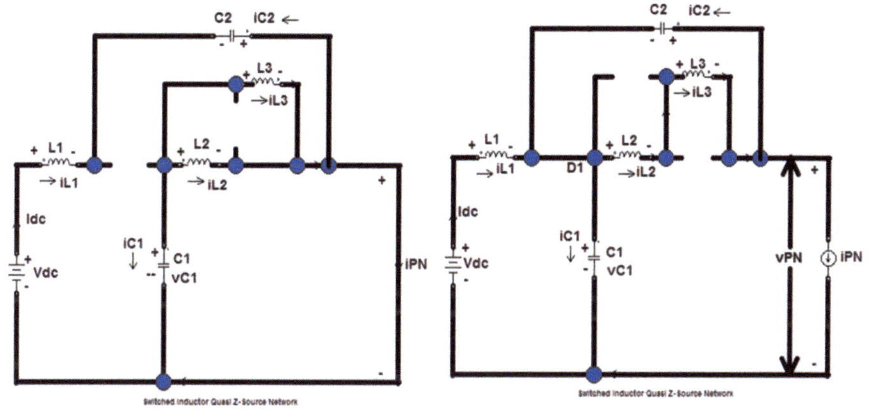

Fig. 6.16 Switched inductor quasi Z-source three-phase inverter—(a) shoot-through state and (b) non-shoot-through state

topologies were proposed [6–8]. This is achieved by proper choice of the shoot-through duty ratio D_O which is T_O/T where T_O is the shoot-through period and T is the carrier switching period. The analysis of a SL-QZSI topology is presented below [7].

The switched inductor quasi Z-source inverter (SL-QZSI) topology is shown in Fig. 6.15. Here in addition to the quasi Z-source network inductors L1 and L2 and diode D1, additional inductor L3 and additional diodes D2, D3 and D4 are provided. The inductors L2 and L3 are equal. The equivalent circuit for the shoot-through period and the non-shoot-through period is shown in Fig. 6.16a, b, respectively.

During the ST period, the following equations are valid:

$$v_{L2} = v_{L3} = V_{C1} \tag{6.16}$$

$$v_{L1} = (V_{dc} + V_{C2}) \tag{6.17}$$

$$i_{L2} = i_{L3} \tag{6.18}$$

$$v_{PN} = 0 \tag{6.19}$$

$$v_{L1} = (V_{dc} - V_{C1}) \tag{6.20}$$

$$v_{L2} = v_{L3} = \left(\frac{V_{C1}}{2} - \frac{v_{PN}}{2}\right) \tag{6.21}$$

$$v_{PN} = (V_{C1} + V_{C2}) \tag{6.22}$$

Equations are valid assuming L2 = L3 = L :
During the NST period, the following

$$v_{L2} = v_{L3} = \frac{-V_{C2}}{2} \tag{6.23}$$

During NST period Eqs. 6.20 to 6.23 are valid. The average current through inductors L1 and L2 over one switching cycle is zero which results in the following equations:

$$\frac{(V_{dc} + V_{C2}) * D_O{}^*T}{L_1} + \frac{(V_{dc} - V_{C1}) * (1 - D_O) * T}{L_1} = 0 \tag{6.24}$$

$$i.e.\ \ V_{C1} = \frac{V_{dc}}{(1 - D_O)} + \frac{D_O{}^* V_{C2}}{(1 - D_O)} \tag{6.25}$$

$$\frac{(V_{C1}) * D_O{}^*T}{L_2} + \left(\frac{-V_{C2}}{2}\right) * \frac{(1 - D_O) * T}{L_2} = 0 \tag{6.26}$$

$$i.e.\ \ V_{C2} = \frac{2^* V_{C1}{}^* D_O}{(1 - D_O)} \tag{6.27}$$

Using Eq. 6.27 in Eq. 6.25 and simplifying,

$$V_{C1} = \frac{V_{dc}{}^*(1 - D_O)}{(1 - 2^* D_O - D_O{}^2)} \tag{6.28}$$

Using Eq. 6.28 in Eq. 6.27 and simplifying,

$$V_{C2} = \frac{2^* V_{dc}{}^* D_O}{(1 - 2^* D_O - D_O{}^2)} \tag{6.29}$$

Using Eqs. 6.22, peak inverter bridge DC link voltage v_{PN} can be expressed as follows:

$$v_{\mathrm{PN}} = \frac{V_{\mathrm{dc}}{}^{*}(1 + D_O)}{\left(1 - 2^{*}D_O - D_O{}^2\right)} \tag{6.30}$$

$$B = \frac{v_{\mathrm{PN}}}{V_{\mathrm{dc}}} = \frac{(1 + D_O)}{\left(1 - 2^{*}D_O - D_O{}^2\right)} \tag{6.31}$$

In Eq. 6.31, B is the boost factor.

6.3.1 *Model of Switched Inductor Quasi Z-Source Three-Phase Inverter with Simple Boost Control*

The SBC technique is explained in Sect. 5.2.1. The model of the SL-QZSI using SBC is shown in Fig. 6.17 (model file: EXAMPLE6_3). The value of L2 and L3 is equal and is each 200e-6 H. The value of inductor L1 is 100e-6 H. All other parameters are as shown in Table 5.6. Modulation index M is 0.75 and this gives a value of 0.25 for $\mathrm{D_O}$. Boost factor B is calculated using Eq. 6.31. The method of generating the gate pulse for SBC technique is already explained in Sect. 5.2.2. Here the constant blocks relating to +0.75 and −0.75 are compared with triangle carrier using two relational operator blocks which form the comparators, and the output of these two comparators is given to OR gate, the output of which gives the shoot through period $\mathrm{T_O}$. Also the gate pulse for the upper switch Sap and lower switch San of Phase A is given to an AND gate, the output of which gives the shoot through period $\mathrm{T_O}$.

6.3.2 *Simulation Results*

The simulation of the three-phase SL-QZSI with SBC is carried out using ode23tb (stiff/TR-BDF2) solver [9]. The data shown in Table 5.6 are used. The simulation results for the line to neutral output voltage, line to line output voltage and Z-source inductor currents, capacitor voltages, inverter bridge DC link voltage V_{pn} and line to ground output voltage V_{ag} for Phase A are shown in Figs. 6.18 to 6.20, respectively. The shoot-through shoot through period $\mathrm{T_O}$ measurement is shown in Fig. 6.21. In Fig. 6.21 measurement of shoot-through period $\mathrm{T_O}$ is made for one triangle carrier period commencing 0.190 sec. to 0.19001 sec. in three different locations. These values are, respectively, 628.304e-9 sec., 1.246e-6 sec. and 618.170e-9 sec. respectively giving a total value of 2.492e-6 sec. for $\mathrm{T_O}$ using AND gate and 621.340e-9 sec., 1.253e-6 sec. and 621.340e-9 sec. giving a total value of 2.495e-6 sec. for To using OR gate. The To value using AND gate is considered here. The $\mathrm{D_O}$ value is 0.2492. The simulation results are tabulated in Table 6.3.

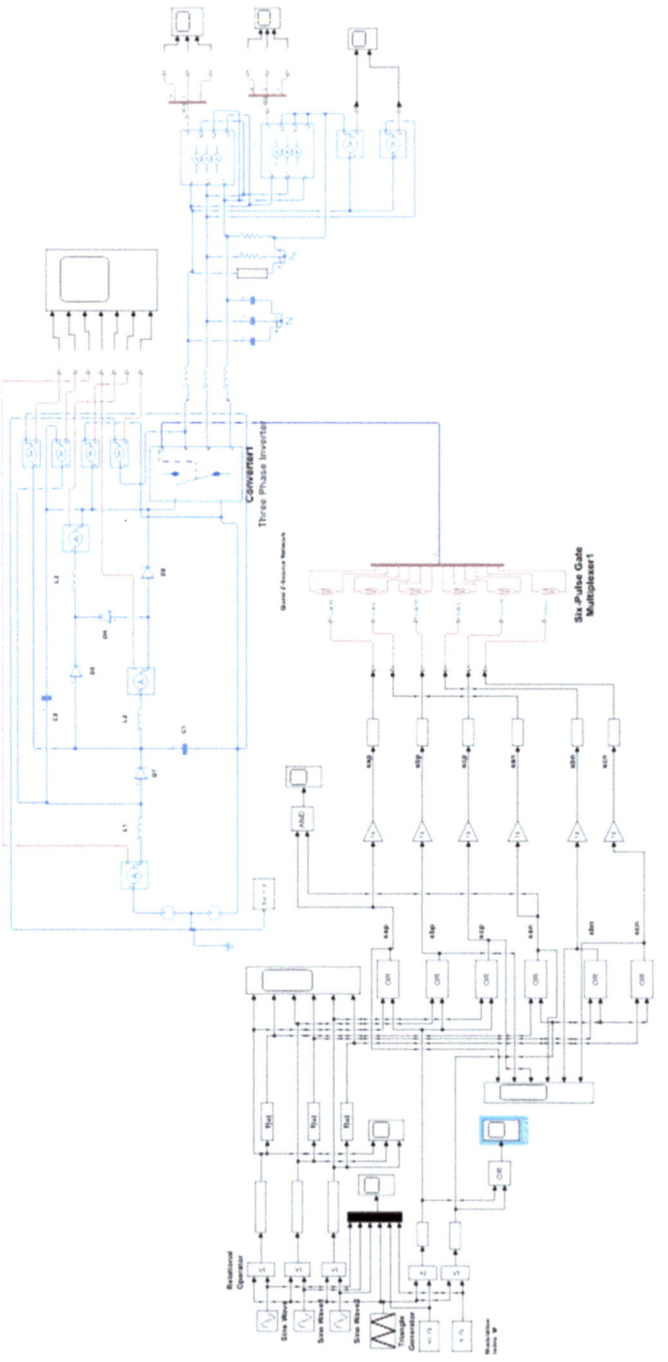

Fig. 6.17 Switched inductor quasi Z-source three-phase inverter with simple boost control

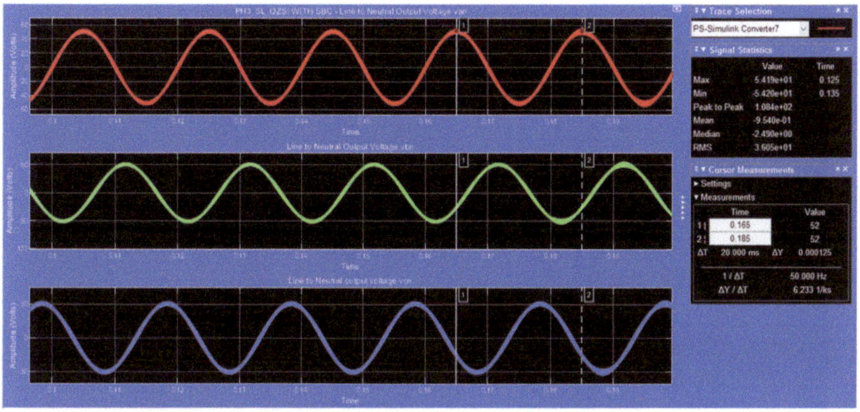

Fig. 6.18 SL-QZSI with SBC—three-phase line to neutral output voltage

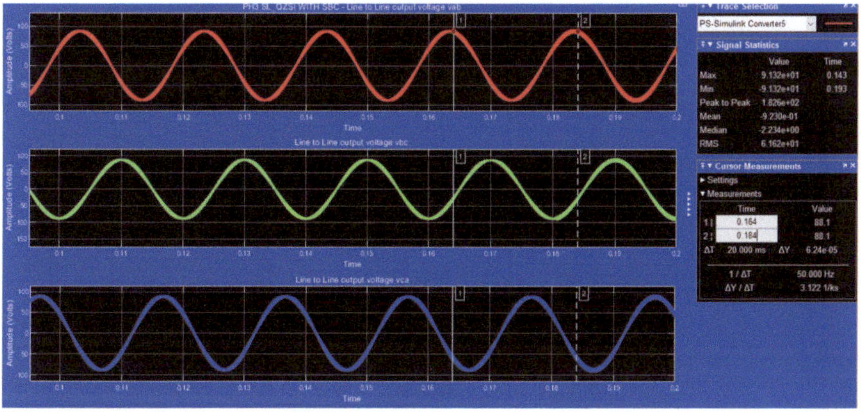

Fig. 6.19 SL-QZSI with SBC—three-phase line to line output voltage

6.3.3 Discussion of Results

The model values and simulation results in Table 6.3 for D_O, B, G, *Vac(peak), Vpn (mean), vpn(peak), VC1(mean)* and *VC2(mean)* closely agree with the theoretical values. Comparing the result with that shown in Table 5.7 in Chap. 5, it is seen that for the same D_O and SBC technique, the values of B and G are increased and this has boosted the value of *Vac(peak)*, *Vpn(mean)*, *vpn(peak)*, *VC1(mean)* and *VC2(mean)*. Inductor currents are continuous. Also from Fig. 6.20, the RMS value of line to ground output voltage *Vag* is 70.80 V, whereas for a conventional inverter, this value is 24 V which is half the DC source voltage. This gives a boost factor B of 2.95.

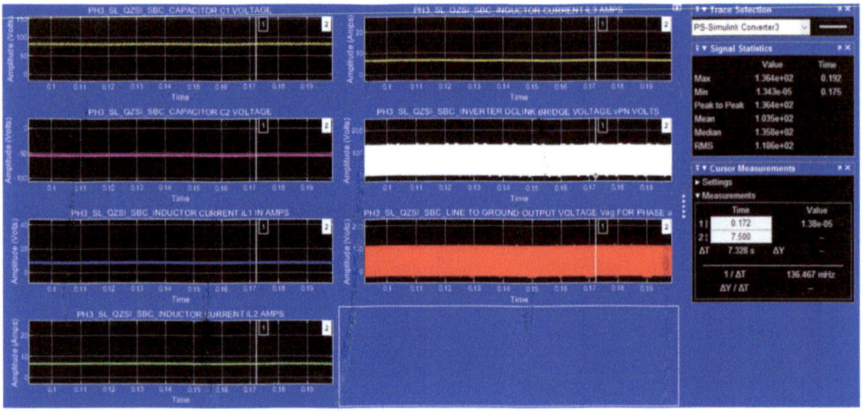

Fig. 6.20 SL-QZSI with SBC—capacitor voltages vC1 and vC2 and inductor currents iL1 and iL2 (left column, top to bottom), inductor current iL3, inverter bridge DC link voltage V_{PN} and line to ground output voltage V_{ag} (right column, top to bottom)

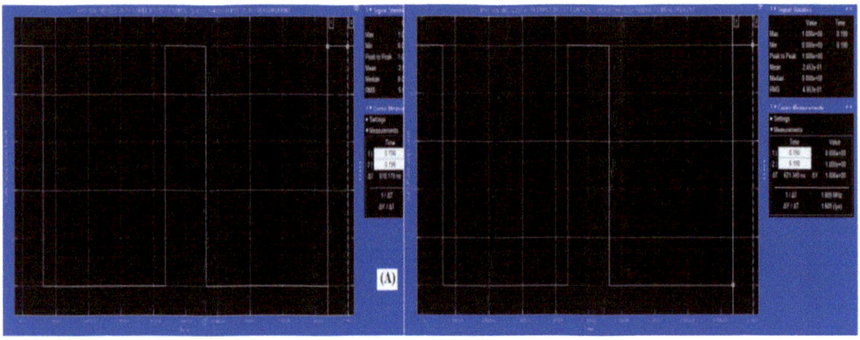

Fig. 6.21 SL-QZSI with SBC—shoot-through period to measurement (**a**) using AND gate and (**b**) using OR gate

6.3.4 *Model of Switched Inductor Quasi Z-Source Three-Phase Inverter with Third Harmonic Injection Maximum Constant Boost Control*

The third harmonic injection maximum constant boost control technique is presented in Sect. 5.2.13. The model of the SL-QZSI with THIMCBC is shown in Fig. 6.22 (model file: EXAMPLE6_4). Now consider the three-phase SL-QZSI whose data are given in Sect. 6.3.1. The value of M is taken as 0.866, using Eq. 5.51 and D_O 0.25. The Embedded MATLAB function generates three-phase THI sine wave defined in Eq. 5.35. The source code for generating this three-phase THI sine wave is shown in Program segment 5.1. Boost factor B is calculated by using Eq. 6.31. The method of generating the gate pulse for THIMCBC technique is already

Table 6.3 Three-phase SL-QZSI with simple boost control—simulation results

S. no.	D_O	B	G B X M	V_{ac} (Volts) Peak	V_{PN} Volts Avg/mean	v_{PN} Volts Peak	V_{C1} Volts Avg/mean	V_{C2} Volts Avg/mean	Remarks
1)	0.25 [0.2492]	2.8571 (2.8423)	2.143 (2.1340)	51.43 (51.2164) [54.19]	102.86 (102.433) [103.5]	137.143 (136.4316) [136.4]	82.2857 (82.00) [81.77]	54.86 (54.4328) [53.95]	Theoretical (Model Calculations) [Simulation Results]

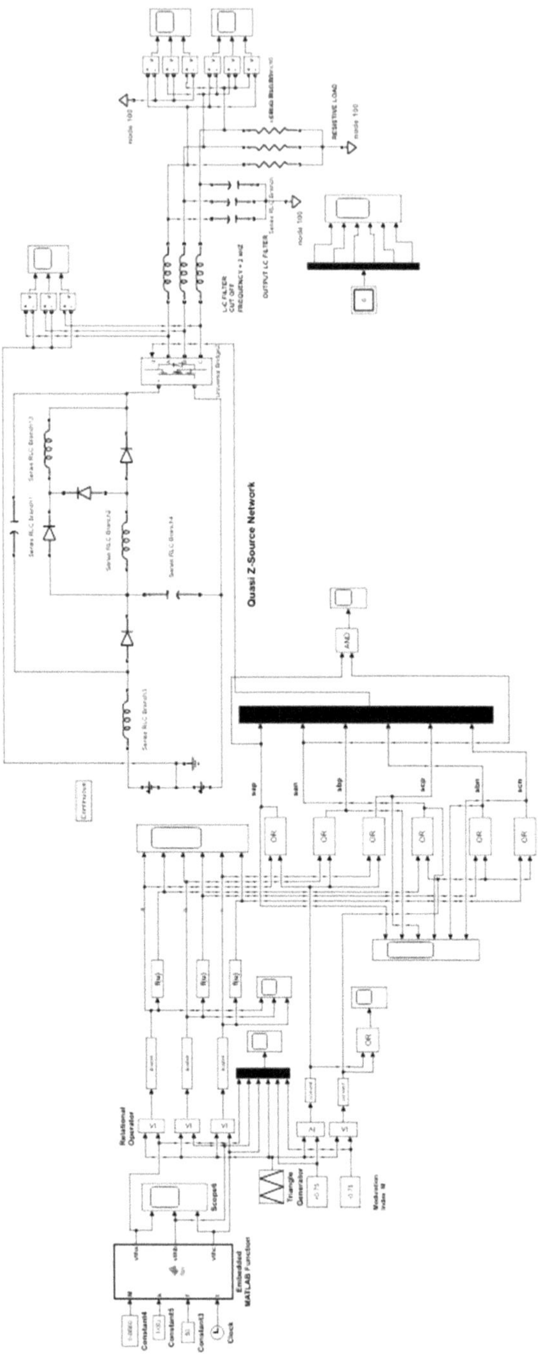

FIG. 6.22 Switched inductor quasi Z-source three-phase inverter with maximum constant boost control

explained in Sect. 5.2.14. Here the shoot-through period T_O is directly measured from the gate drive model without going to the gate pulse waveform. Here the constant blocks relating to $+\sqrt{3}*M/2$ (+0.75) and $-\sqrt{3}*M/2$ (−0.75) are compared with triangle carrier using two relational operator blocks which form the comparators, and the output of these two comparators is given to OR gate, the output of which gives the shoot-through period T_O. In addition the gate pulse corresponding to switches Sap and San in Phase A of inverter is given to AND gate whose output gives shoot-through period To.

6.3.5 Simulation Results

The simulation of the three-phase SL-QZSI with THIMCBC is carried out using ode23tb (stiff/TR-BDF2) solver [9]. The data shown in Sect. 6.3.1 are used. The simulation results for the line to neutral output voltage, line to line output voltage, line to ground output voltage and quasi Z-source inductor currents, capacitor voltages and inverter bridge DC link voltage V_{pn} are shown in Figs. 6.23 to 6.26, respectively. The shoot-through period T_O measurement is shown in Fig. 6.27a, b, the former using AND gate and the latter using OR gate. In Fig. 6.27a, b, measurement of shoot through period T_O is made for one triangle carrier period commencing 0.190 sec. to 0.19001 sec in three different locations. These values are, respectively, 632.655e-9 sec., 1.255e-6 sec. and 622.284e-9 sec. respectively giving a total value of 2.509e-6 sec. for T_O using AND gate and 630.174e-6 sec., 1.26e-6 sec. and 630.174e-6 sec. giving a total value of 2.52e-6 sec. for To using OR gate. The To value using AND gate is considered here. The simulation results are tabulated in Table 6.4.

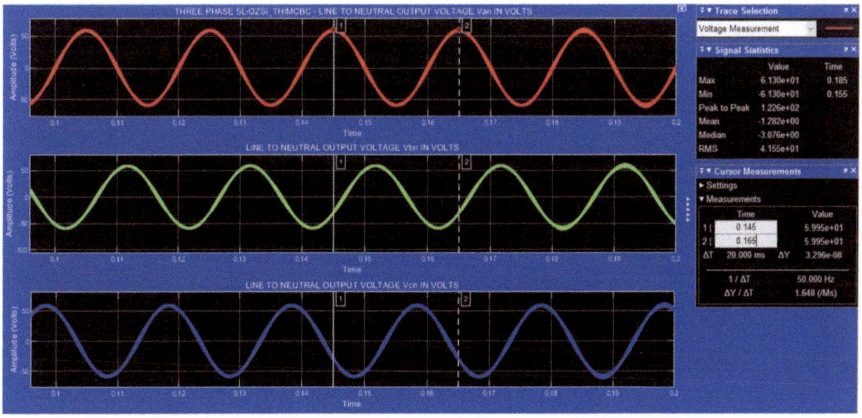

Fig. 6.23 Three-phase SL-QZSI with THIMCBC—line to neutral output voltage

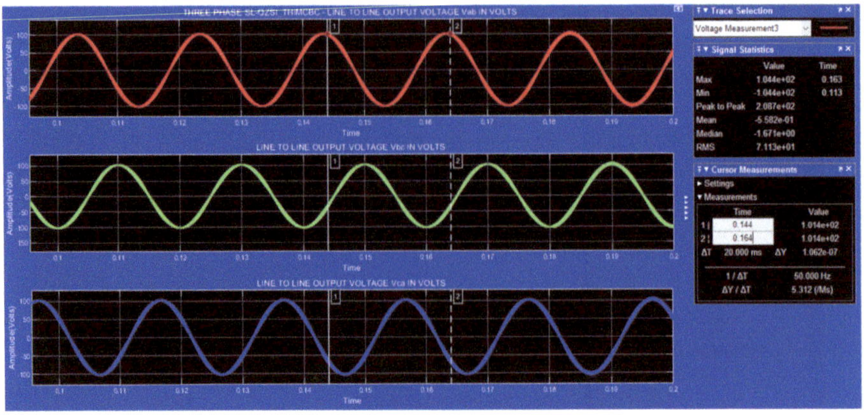

Fig. 6.24 Three-phase SL-QZSI with THIMCBC—line to line output voltage

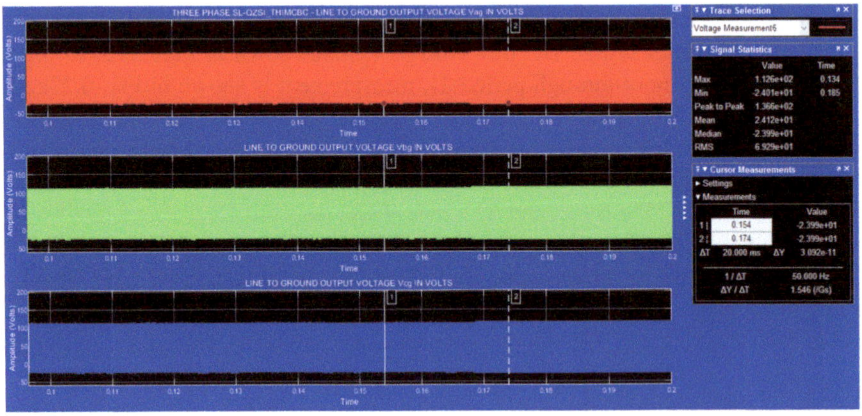

Fig. 6.25 Three-phase SL-QZSI with THIMCBC—line to ground output voltage

6.3.6 Discussion of Results

The model values and simulation results for D_O, B, G, *Vac(peak)*, *Vpn(mean)*, *vpn (peak)*, *VC1(mean)* and *VC2(mean)* agree with the theoretical values. Comparing the result with that shown in Table 5.10 in Chap. 5, it is seen that for the same D_O and THIMCBC technique, the values of B and G are increased and this has boosted the value of *Vac(peak)*, *Vpn(mean)*, *vpn(peak)*, *VC1(mean)* and *VC2(mean)*. Current through inductors L1, L2 and L3 is continuous. Also from Fig. 6.25, the RMS value of the line to ground output voltage is 69.29 volts, whereas for a conventional inverter, this value is 24 V which is half the DC source voltage. This gives a boost factor B of 2.887.

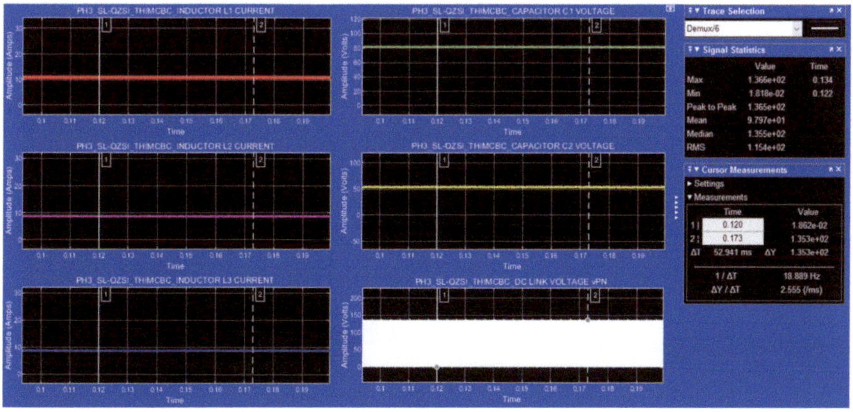

Fig. 6.26 Three-phase SL-QZSI with THIMCBC—inductor currents *iL1, iL2* and *iL3* (left column, top to bottom), capacitor voltages *vC1* and *vC2* and inverter bridge DC link voltage V_{PN} (right column, top to bottom)

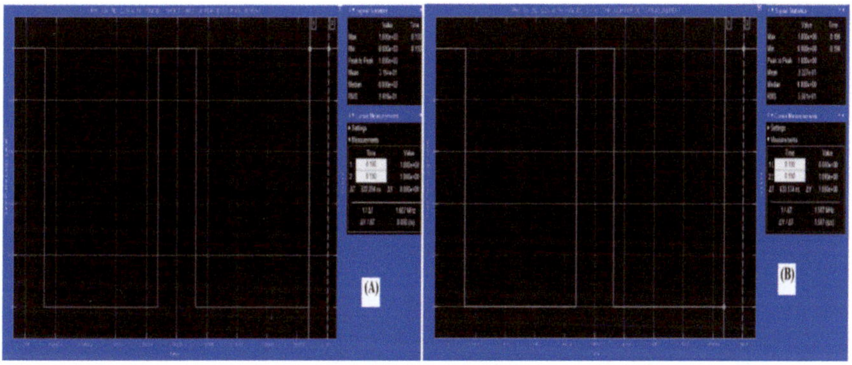

Fig. 6.27 Three-phase switched inductor QZSI with THIMCBC—shoot-through period To measurement using (**a**) AND gate and (**b**) OR gate

6.4 Active Switched Capacitor Quasi Z-Source Three-Phase Inverter

Defining an active electronic on-off switch as a transistor/MOSFET switch driven by an external gate pulse, active switched capacitor (ASC) quasi Z-source inverter (ASC-QZSI) is reported [8, 12]. This ASC-QZSI topology offers higher boost factor, greater inversion ability and lower voltage stress across inverter switches compared to that of QZSI and SL-QZSI [8]. In this section, ASC-QZSI topology is analysed in detail, model is developed, and simulation results are presented. Figure 6.28 shows the ASC-QZSI topology [8]. By controlling the shoot-through (ST) duty ratio D_O, the desired boost factor B can be achieved. The equivalent circuit of the ASC-QZSI

Table 6.4 Three-phase SL-QZSI with THIMCBC—simulation results

S. no.	D_O	B	G B X M	V_{ac} (Volts) Peak	V_{PN} Volts Avg/mean	v_{PN} Volts Peak	V_{C1} Volts Avg/mean	$V_{C2\ 55.3393}$ Volts Avg/mean	Remarks
1)	0.25 [0.2509]	2.8571 (2.8740)	2.4743 (2.4889)	59.383 (59.7329) [61.30]	102.8571 (103.3393) [97.97]	137.1429 (137.9513) [136.6]	82.2857 (82.6120) [81.72]	54.8571 (55.3393) [53.91]	Theoretical (Model Calculations) [Simulation Results]

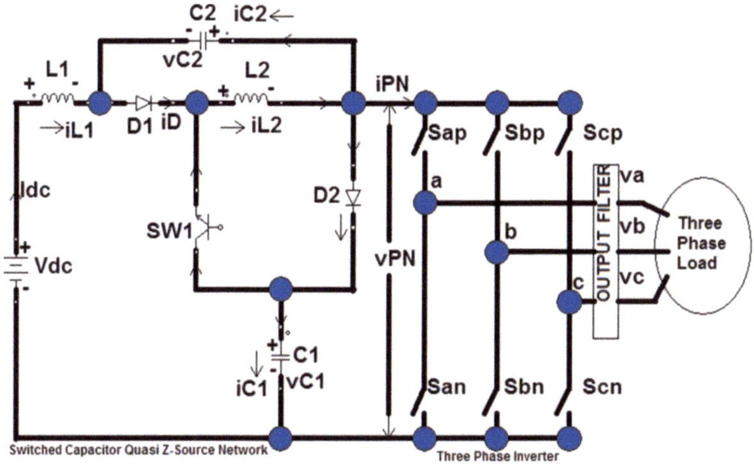

Fig. 6.28 Active switched capacitor quasi Z-source three-phase inverter

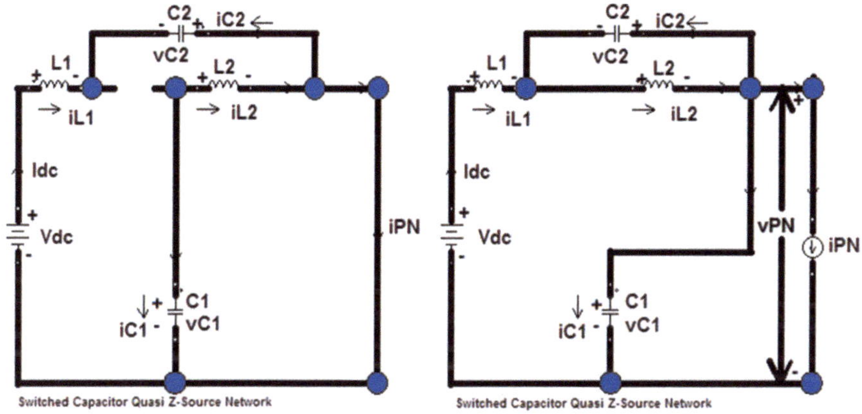

Fig. 6.29 Active switched capacitor quasi Z-source three-phase inverter—(**a**) shoot-through state and (**b**) non-shoot-through state

for the shoot-through (ST) period and for the non-shoot-through (NST) period is shown in Fig. 6.29a, b, respectively [8]. The analysis is given below:

During the ST period, the following equations are valid:

$$v_{L1} = (V_{dc} + V_{C2}) \tag{6.32}$$

$$v_{L2} = V_{C1} \tag{6.33}$$

$$i_{L1} = -i_{C2} \tag{6.34}$$

$$i_{L2} = -i_{C1} \tag{6.35}$$

During the NST period, the following equations are valid:

$$V_{C1} = v_{PN} \tag{6.36}$$

$$V_{C2} = -v_{L2} \tag{6.37}$$

$$v_{L1} = (V_{dc} - V_{L2} - V_{C1}) = (V_{dc} - V_{C1} + V_{C2}) \tag{6.38}$$

$$i_{L2} = (i_{L1} + i_{C2}) = (i_{C1} + i_{C2} + i_{PN}) \tag{6.39}$$

The average current through inductor L1 over one carrier switching period is zero which can be expressed as follows:

$$\frac{(V_{dc} + V_{C2}) * D_O{}^* T}{L_1} + \frac{(V_{dc} - V_{C1} + V_{C2}) * (1 - D_O) * T}{L_1} = 0 \tag{6.40}$$

$$i.e.\ \ V_{C1} = \frac{V_{dc}}{(1 - D_O)} + \frac{V_{C2}}{(1 - D_O)} \tag{6.41}$$

Similarly the average current through inductor L2 over one carrier switching period is zero which can be expressed as follows:

$$\frac{(V_{C1}) * D_O{}^* T}{L_2} + \frac{(-V_{C2}) * (1 - D_O) * T}{L_2} = 0 \tag{6.42}$$

$$i.e.\ \ V_{C2} = \frac{V_{C1}{}^* D_O}{(1 - D_O)} \tag{6.43}$$

Using Eq. 6.43 in Eq. 6.41 and simplifying,

$$V_{C1} = \frac{V_{dc}{}^*(1 - D_O)}{(D_O{}^2 - 3^* D_O + 1)} \tag{6.44}$$

Using Eq. 6.44 in Eq. 6.43 and simplifying,

$$V_{C2} = \frac{V_{dc}{}^* D_O}{(D_O{}^2 - 3^* D_O + 1)} \tag{6.45}$$

The peak DC link bridge voltage of the inverter v_{PN} is equal to V_{C1} which can be expressed as follows:

$$v_{PN} = \frac{V_{dc}{}^*(1 - D_O)}{(D_O{}^2 - 3^* D_O + 1)} \tag{6.46}$$

From Eq. 6.46, boost factor B can be expressed as follows:

$$B = \frac{(1 - D_O)}{(D_O{}^2 - 3^* D_O + 1)} \tag{6.47}$$

If M is the modulation index, the peak line to neutral phase voltage v_{ac} can be expressed as follows:

$$v_{\mathrm{ac}} = \frac{(M^* B^* V_{\mathrm{dc}})}{2} \tag{6.48}$$

6.4.1 *Modelling of Active Switched Capacitor Quasi Z-Source Three-Phase Inverter Using Simple Boost Control*

The SBC technique is explained in Sect. 5.2.1. The model of the ASC-QZSI is shown in Fig. 6.30 (model file: EXAMPLE6_5). The data shown in Table 5.6 are used. Modulation index M is 0.75 and this gives a value of 0.25 for D_O. Boost factor B is calculated by using Eq. 6.47. The method of generating the gate pulse for SBC technique is already explained in Sect. 5.2.2. Here the shoot-through period T_O is directly measured from the gate drive model without going to the gate pulse waveform. Here the constant blocks relating to +M(+0.75) and –M(−0.75) are compared with triangle carrier using two relational operator blocks which form the comparators, and the output of these two comparators is given to OR gate, the output of which gives the shoot through period T_O. Also the gate pulse corresponding to switches Sap and San in Phase A of the inverter is given to an AND gate, the output of which also gives the shoot through period To. Any one output either from OR gate or AND gate can be used. The output of this AND gate multiplied by gain block with gain multiplier value of ten forms the gate pulse for the NPN transistor switch SW1 in Fig. 6.28. The gate pulse amplitude must be greater than the gate threshold voltage for switch SW1 used in the model.

6.4.2 *Simulation Results*

The simulation of the three-phase ASC-QZSI with SBC is carried out using ode23tb (stiff/TR-BDF2) solver [9]. The data shown in Table 5.6 are used. The simulation results for the line to neutral output voltage, line to line output voltage and quasi Z-source inductor currents, capacitor voltages, inverter bridge peak DC link voltage V_{pn} and line to ground output voltage *Vag* are shown in Figs. 6.31 to 6.33 respectively. The shoot-through period T_O measurement is shown in Fig. 6.34a, b, the former using AND gate and the latter using OR gate. In Fig. 6.34, measurement of shoot through period T_O is made for one triangle carrier period commencing

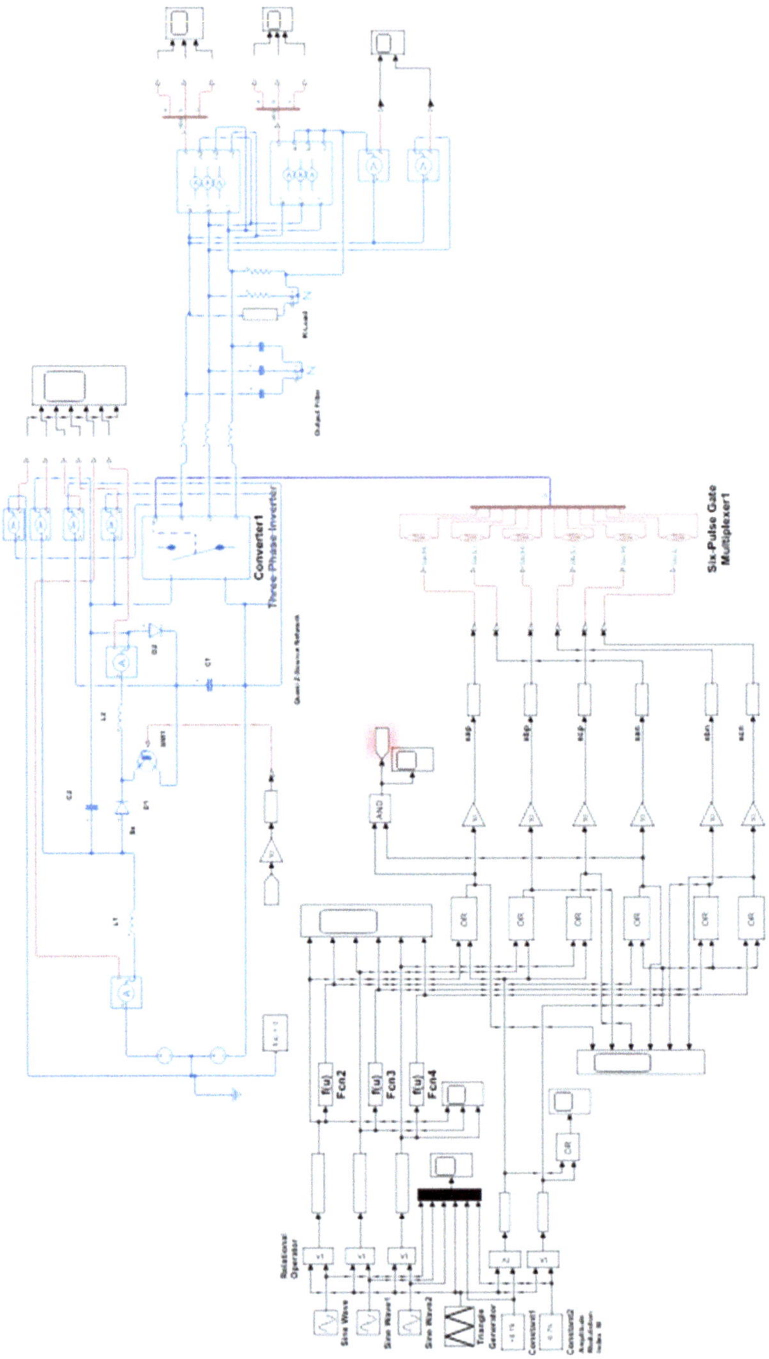

Fig. 6.30 Model of active switched capacitor three-phase quasi Z-source inverter—simple boost control

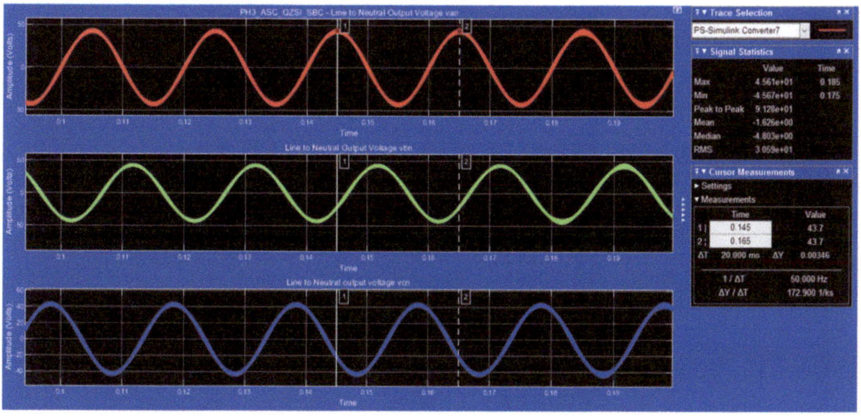

Fig. 6.31 Three-phase ASC-QZSI with SBC—line to neutral output voltage

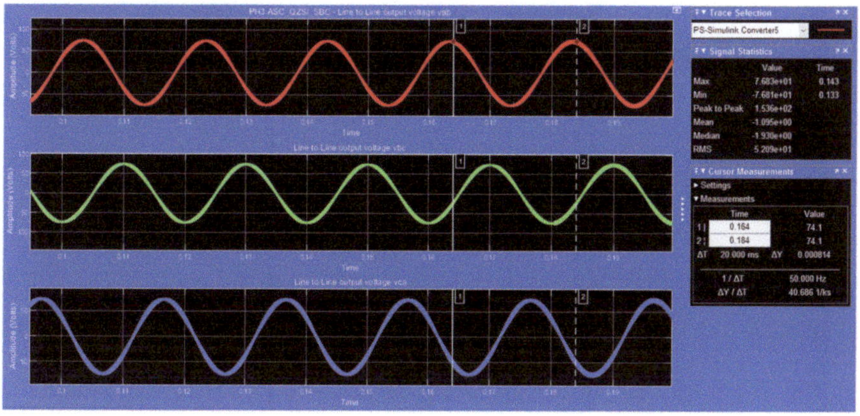

Fig. 6.32 Three-phase ASC-QZSI with SBC—line to line output voltage

0.190 sec. to 0.19001 sec. in three different locations. These values are, respectively, 621.031e-9 sec., 1.262e-6 sec. and 621.031e-9 sec. giving a total value of 2.504e-6 sec. for T_O using AND gate and 629.147e-9 sec., 1.258e-6 sec. and 626.294e-9 sec. giving a total value of 2.513e-6 sec. for To using OR gate. The value using AND gate is used here. The simulation results are tabulated in Table 6.5.

6.4.3 Discussion of Results

The model values and simulation results in Table 6.5 for D_O, B, G, *Vac(peak)*, *Vpn (mean)*, *vpn(peak)*, *VC1(mean)* and *VC2(mean)* agree with the theoretical values. Comparing the result with that shown in Table 5.7 in Chap. 5, it is seen that for the

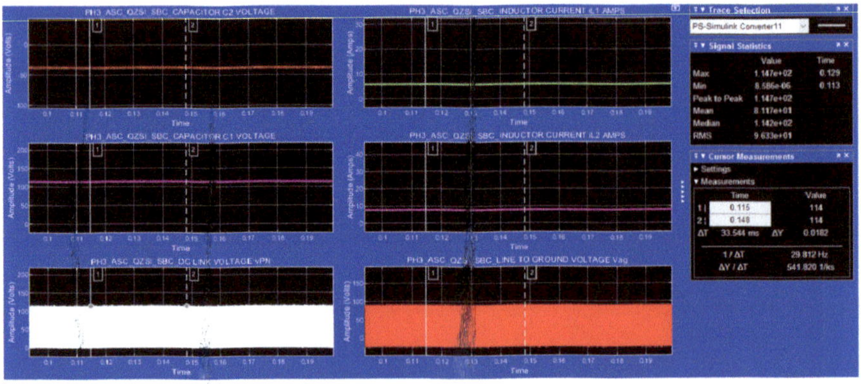

Fig. 6.33 Three-phase ASC-QZSI with SBC—capacitor voltages vC2 and vC1 and inverter bridge DC link voltage V_{PN} (left column, top to bottom), inductor currents *iL1* and *iL2* and line to ground voltage *Vag* (right column, top to bottom)

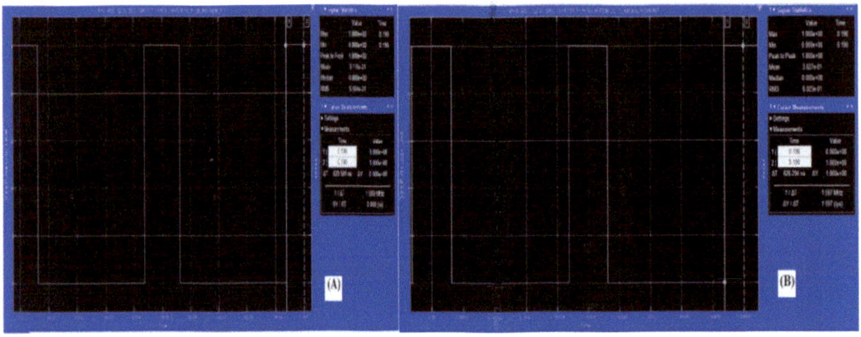

Fig. 6.34 Three-phase ASC-QZSI with SBC—shoot through period To measurement (**a**) using AND gate and (**b**) using OR gate

same D_O and SBC technique, the values of B and G are increased and this has boosted the value of *Vac(peak)*, *Vpn(mean)*, *vpn(peak)*, *VC1(mean)* and *VC2(mean)*. Current through inductors L1 and L2 is continuous. Also from Fig. 6.33, the line to ground RMS output voltage is found to be 58.24 V, whereas for a conventional inverter, this value is 24 V which is half the DC source voltage. This gives a boost factor B of 2.427. The V_{C2} value in Table 6.5 is shown with a minus sign which means that the polarity of V_{C2} is reversed from the one shown in Fig. 6.28.

Table 6.5 Three-phase ASC-QZSI with SBC—simulation results

S. no.	D_O	B	G B X M	V_{ac} (Volts) Peak	V_{PN} Volts Avg/mean	v_{PN} Volts Peak	V_{C1} Volts Avg/mean	V_{C2} Volts Avg/mean	Remarks
1)	0.25 [0.2504]	2.40 (2.4064)	1.8 (1.8039)	43.2 (43.2924) [45.61]	86.40 (86.5849) [81.17]	115.20 (115.5081) [114.7]	115.20 (115.5081) [114.1]	38.40 (38.5849) [−37.76]	Theoretical (Model Calculations) [Simulation Results]

6.4.4 Modelling of Active Switched Capacitor Quasi Z-Source Three-Phase Inverter Using Third Harmonic Injection Maximum Constant Boost Control

The maximum constant boost control technique is presented in Sect. 5.2.13. The model of the ASC-QZSI with third harmonic injection maximum constant boost control (THIMCBC) is shown in Fig. 6.35 (model file: EXAMPLE6_6). Now consider the three-phase ASC-QZSI whose data are given in Table 5.6. The value of M is taken as 0.866 and using Eq. 5.51, D_O is 0.25. The Embedded MATLAB function generates three-phase THI sine wave defined in Eq. 5.35. The source code for generating this three-phase THI sine wave is shown in Program segment 5.1. Boost factor B is calculated by using Eq. 6.47. The method of generating the gate pulse for THIMCBC technique is already explained in Sect. 5.2.14. Here the shoot-through period T_O is directly measured from the gate drive model without going to the gate pulse waveform. Here the constant blocks relating to +√3*M/2 (+0.75) and −√3*M/2 (−0.75) are compared with triangle carrier using two relational operator blocks which form the comparators, and the output of these two comparators is given to OR gate, the output of which gives the shoot-through period T_O. Also the gate pulse corresponding to upper and lower switches Sap and San of Phase A is given to an AND gate, the output of which gives the shoot through period To. Both methods are shown in Fig. 6.35. The output of this AND gate multiplied by gain block with gain multiplier value of ten forms the gate pulse for the NPN transistor switch SW1 in Fig. 6.28. The gate pulse amplitude must be greater than the gate threshold voltage of switch SW1 used in the model.

6.4.5 Simulation Results

The simulation of the three-phase ASC-QZSI with THIMCBC is carried out using ode23tb (stiff/TR-BDF2) solver [9]. The data shown in Table 5.6 are used. The simulation results for the line to neutral output voltage, line to line output voltage and quasi Z-source inductor currents, capacitor voltages, inverter bridge DC link voltage *Vpn* and line to ground output voltage *Vag* for Phase A are shown in Figs. 6.36 to 6.38 respectively. The shoot through period T_O measurement is shown in Fig. 6.39. In Fig. 6.39, measurement of shoot through period T_O is made for one triangle carrier period commencing 0.190 sec. to 0.19001 sec. in three different locations. These values are 633.054e-9 sec., 1.246e-6 sec. and 623.006e-9 sec., respectively, giving a total value of 2.502e-6 sec. for T_O using AND gate and 635.154e-9 sec., 1.26e-6 sec. 614.991e-9 sec., respectively, giving a total value of 2.51e-6 sec. for To using OR gate. The values using AND gate are considered here. The D_O value is 0.2502. The simulation results are tabulated in Table 6.6.

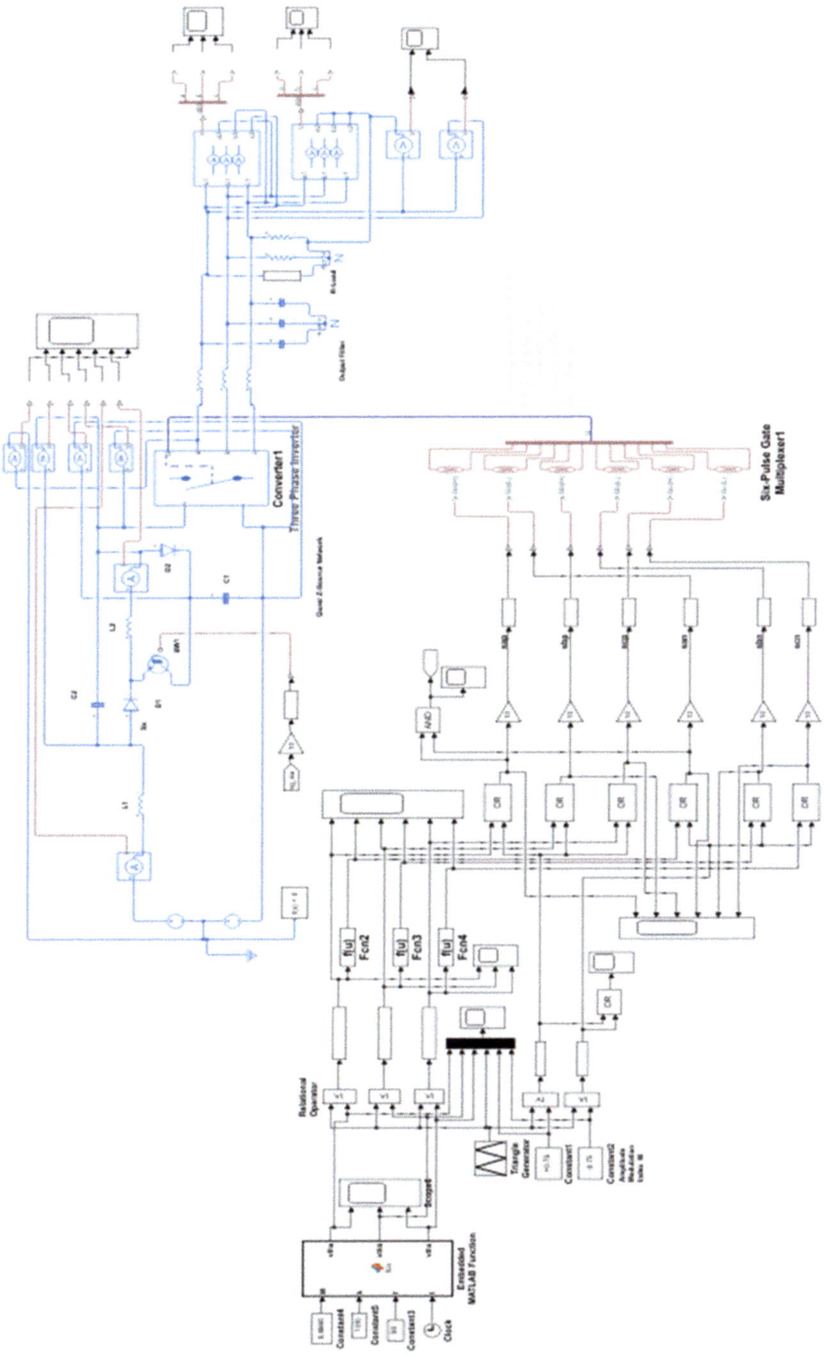

Fig. 6.35 Model of active switched capacitor quasi Z-source three-phase inverter—THI maximum constant boost control

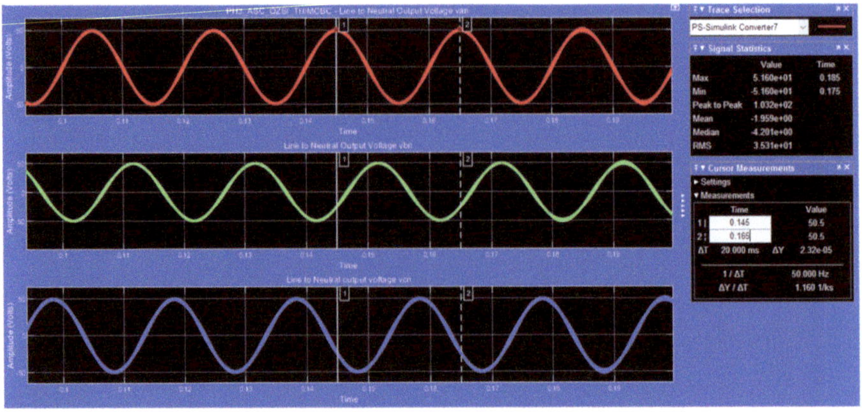

Fig. 6.36 Three-phase ASC-QZSI with THIMCBC—line to neutral output voltage

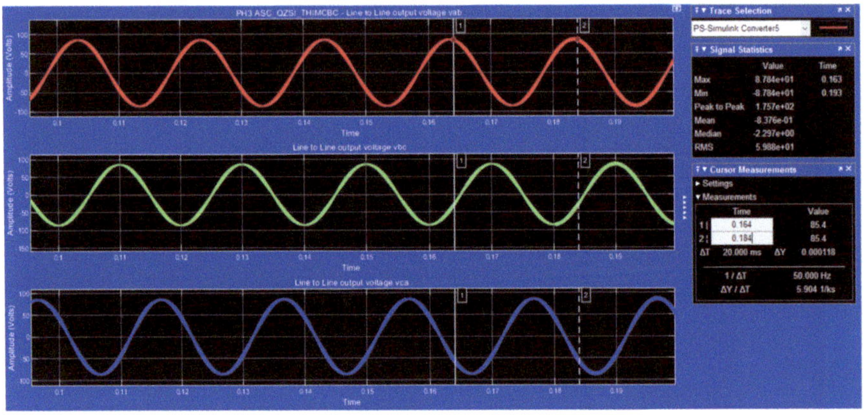

Fig. 6.37 Three-phase ASC-QZSI with THIMCBC—line to line output voltage

6.4.6 Discussion of Results

The model values and simulation results in Table 6.6 for D_O, B, G, *Vac(peak)*, *Vpn (mean)*, *vpn(peak)*, *VC1(mean)* and *VC2(mean)* agree with the theoretical values. Comparing the result with that shown in Table 5.10 in Chap. 5, it is seen that for the same D_O and THIMCBC technique, the values of B and G are increased and this has boosted the value of *Vac(peak)*, *Vpn(mean)*, *vpn(peak)*, *VC1(mean)* and *VC2(mean)*. Current through inductors L1 and L2 is continuous. Also from Fig. 6.38, it is seen that the RMS value of line to ground output voltage *Vag* for Phase A is 57.18 V, whereas for a conventional inverter, this value is 24 V which is half the DC source voltage. This gives a boost factor B of 2.38. The value of V_{C2} in Table 6.6 is entered with a minus sign which means that the polarity of *vC2* is reversed from the one shown in Fig. 6.28.

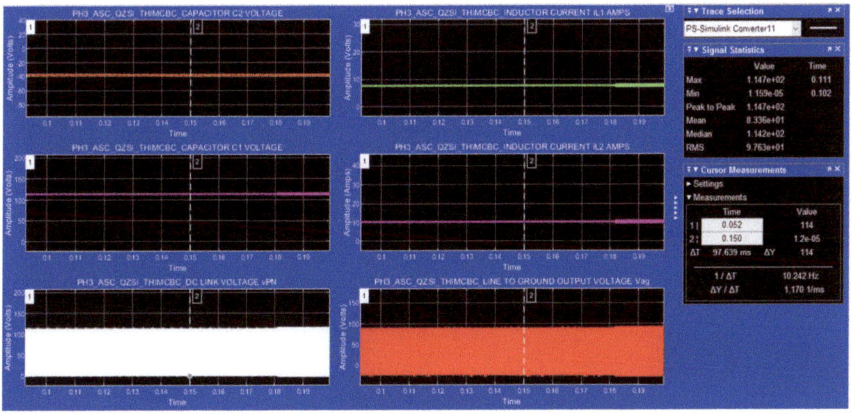

Fig. 6.38 Three-phase ASC-QZSI with THIMCBC—capacitor voltages vC2 and vC1 and inverter bridge DC link voltage V_{PN} (left column, top to bottom), inductor currents *iL1* and *iL2* and line to ground voltage for Phase A *Vag* (right column, top to bottom)

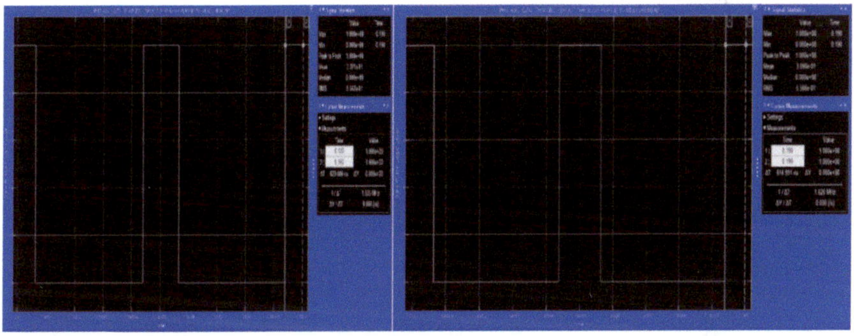

Fig. 6.39 Three-phase ASC-QZSI with THIMCBC—shoot through period To measurement (**a**) using AND gate and (**b**) using OR gate

6.5 Diode-Assisted Extended Boost Quasi Z-Source Three-Phase Inverter

The topology of the diode-assisted extended boost QZSI (DA-QZSI) is shown in Fig. 6.40, and the equivalent circuits for ST and NST state are shown in Fig. 6.41a, b, respectively [8, 11]. This has one additional inductor and capacitor compared to classical QZSI topology.

A higher boost factor is achieved by properly controlling the shoot-through duty ratio D_O. All the conventional methods of boost control technique used for classical QZSI are applicable to this DAEB-QZSI topology. The analysis is given below:

During the ST period, the following equations are valid:

Table 6.6 Three-phase ASC-QZSI with THIMCBC—simulation results

S. no.	D_O	B	G B X M	V_{ac} (Volts) Peak	V_{PN} Volts Avg/mean	v_{PN} 6Volts Peak	V_{C1} Volts Avg/mean	V_{C2} Volts Avg/mean	Remarks
1)	0.25 [0.2502]	2.4 (2.4032)	2.0784 (2.0812)	49.8816 (49.9482) [51.6]	86.4 (86.4923) [83.36]	115.2 (115.3538) [114.7]	115.2 (115.3538) [114.2]	38.4 (38.4923) [−37.77]	Theoretical (Model Calculations) [Simulation Results]

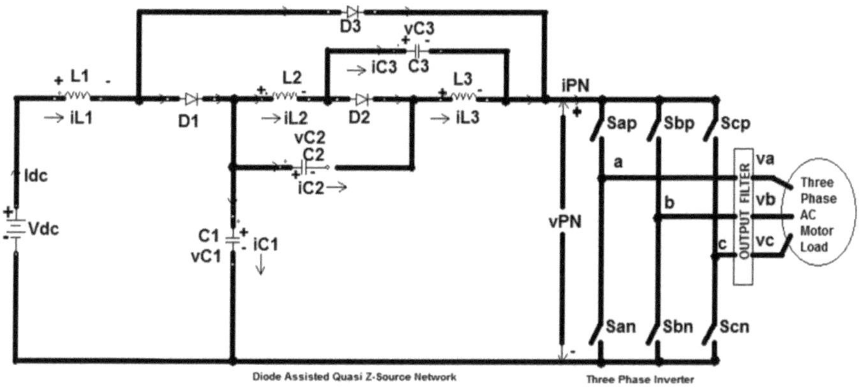

Fig. 6.40 Diode assisted extended boost quasi Z-source three phase inverter

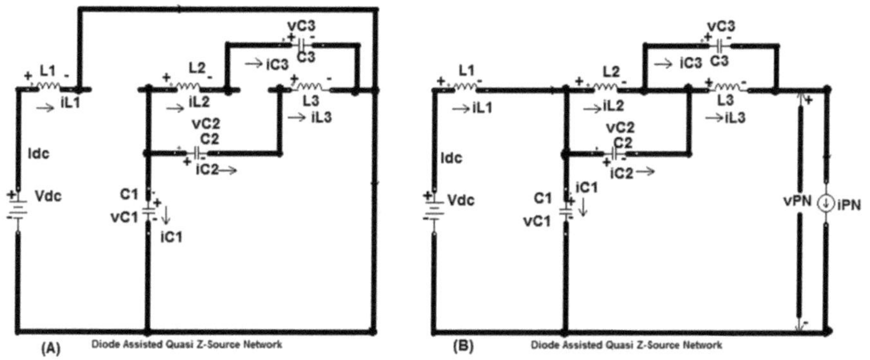

Fig. 6.41 Diode assisted extended boost quasi Z-source three phase inverter (**a**) shoot- through state and (**b**) non-shoot- through state

$$v_{L1} = V_{\text{dc}} \tag{6.49}$$

$$v_{L2} = (V_{C1} - V_{C3}) \tag{6.50}$$

$$v_{L3} = (V_{C1} - V_{C2}); \tag{6.51}$$

During the NST period, the following equations are valid:

$$v_{L1} = (V_{\text{dc}} - V_{C1}) \tag{6.52}$$

$$v_{L2} = V_{C2} \tag{6.53}$$

$$v_{L3} = V_{C3} \tag{6.54}$$

$$v_{\text{PN}} = (V_{\text{dc}} - v_{L1} - v_{L2} - v_{L3}) = (V_{C1} - V_{C2} - V_{C3}) \tag{6.55}$$

The average current through inductor L1 over one carrier switching period is zero which results in the following equations:

$$\frac{(V_{dc}{}^*D_O{}^*T)}{L_1} + \frac{(V_{dc} - V_{C1}) * (1 - D_O) * T}{L_1} = 0 \tag{6.56}$$

$$i.e. \quad V_{C1} = \frac{V_{dc}}{(1 - D_O)} \tag{6.57}$$

Similarly average current through inductor L2 over one carrier switching period is zero which results in the following equations:

$$\frac{(V_{C1} - V_{C3}) * D_O{}^*T}{L_2} + \frac{V_{C2}{}^*(1 - D_O) * T}{L_2} = 0 \tag{6.58}$$

$$(V_{C1} - V_{C3}) * D_O + V_{C2}{}^*(1 - D_O) = 0 \tag{6.59}$$

The average current through inductor L3 over one carrier switching period is zero which results in the following equations:

$$\frac{(V_{C1} - V_{C2}) * D_O{}^*T}{L_3} + \frac{V_{C3}{}^*(1 - D_O) * T}{L_3} = 0 \tag{6.60}$$

$$(V_{C1} - V_{C2}) * D_O + V_{C3}{}^*(1 - D_O) = 0 \tag{6.61}$$

Using Eq. 6.57 in Eq. 6.61 and simplifying,

$$V_{C2}{}^*D_O - V_{C3}{}^*(1 - D_O) = \frac{V_{dc}{}^*D_O}{(1 - D_O)} \tag{6.62}$$

Similarly using Eq. 6.57 in Eq. 6.59 and simplifying,

$$V_{C3}{}^*D_O - V_{C2}{}^*(1 - D_O) = \frac{V_{dc}{}^*D_O}{(1 - D_O)} \tag{6.63}$$

Subtracting Eq. 6.62 from 6.63 and simplifying,

$$V_{C2} = V_{C3} \tag{6.64}$$

Using Eq. 6.64 in Eq. 6.59, then using Eq. 6.57 and simplifying,

$$V_{C2} = V_{C3} = \frac{V_{C1}{}^*D_O}{(2^*D_O - 1)} = \frac{-V_{dc}{}^*D_O}{(1 - 3^*D_O + 2^*D_O^2)} \tag{6.65}$$

The minus sign in Eq. 6.65 indicate that the polarity of the voltage across capacitors C2 and C3 is reversed from the one shown in Fig. 6.40.

Using Eqs. 6.55, 6.57 and 6.65, the peak value of DC link bridge voltage v_{PN} across the inverter can be expressed as follows:

$$v_{PN} = \frac{V_{C1}}{(1 - 2^*D_O)} = \frac{V_{dc}}{(1 - 3^*D_O + 2^*D_O^2)} \tag{6.66}$$

The boost factor B for DA-QZSI can be expressed as follows:

$$B = \frac{1}{(1 - 3^*D_O + 2^*D_O^2)} \tag{6.67}$$

The voltage gain G is given below:

$$G = B^*M = \frac{M}{(1 - 3^*D_O + 2^*D_O^2)} \tag{6.68}$$

The peak value of line to neutral phase voltage can be expressed as follows:

$$v_{ac} = G^* \frac{V_{dc}}{2} = B^*M^* \frac{V_{dc}}{2} \tag{6.69}$$

6.5.1 Modelling of Diode-Assisted Extended Boost Quasi Z-Source Three-Phase Inverter Using Simple Boost Control

The SBC technique is explained in Sect. 5.2.1. The model of the DAEB-QZSI with SBC is shown in Fig. 6.42 (model file: EXAMPLE6_7). The data shown in Table 5.6 are used, with inductor L1 value changed to 100e-6 H. Modulation index M is 0.75 and this gives a value of 0.25 for D_O. Boost factor B is calculated by using Eq. 6.67. The method of generating the gate pulse for SBC technique is already explained in Sect. 5.2.2. Here the shoot through period T_O is directly measured from the gate drive model without going to the gate pulse waveform. Here the constant blocks relating to +M(+0.75) and –M(−0.75) are compared with triangle carrier using two relational operator blocks which form the comparators, and the output of these two comparators is given to OR gate, the output of which gives the shoot through period T_O. Shoot through period is also measured by connecting the gate pulse corresponding upper and lower switches Sap and San of inverter in Phase A to an AND gate.

Fig. 6.42 Diode-assisted extended boost quasi Z-source three-phase inverter with simple boost control

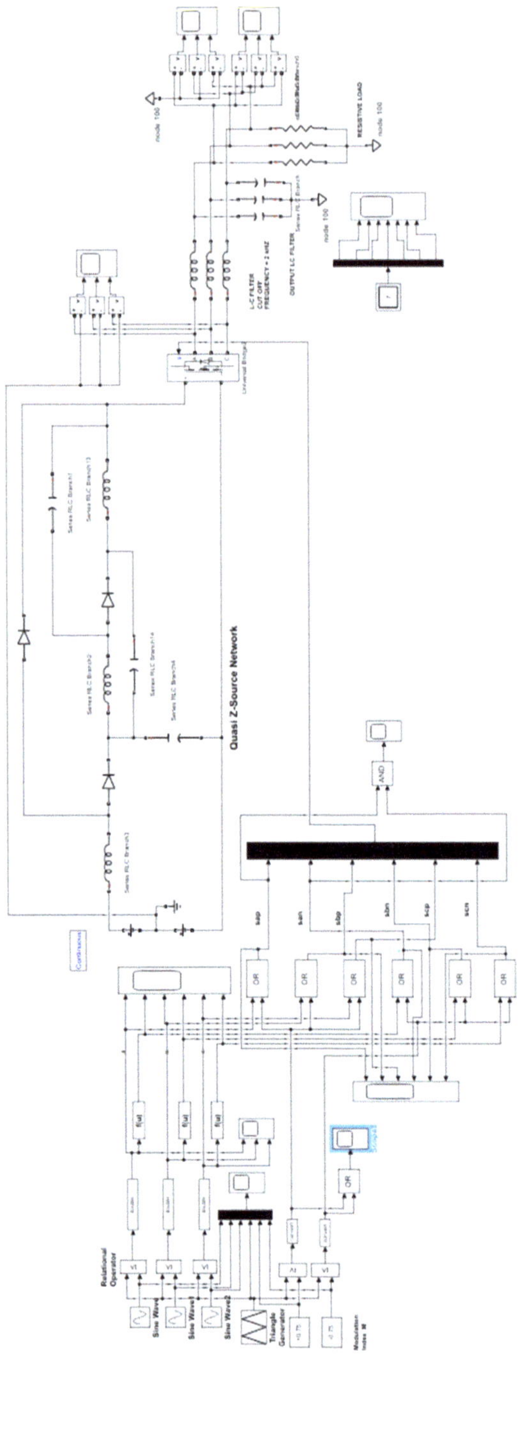

6.5.2 Simulation Results

The simulation of the three-phase DAEB-QZSI with SBC is carried out using ode23tb (stiff/TR-BDF2) solver [9]. The data shown in Table 5.6 are used. The simulation results for the line to neutral output voltage, line to line output voltage, line to ground output voltage and quasi Z-source inductor currents, capacitor voltages and inverter bridge peak DC link voltage V_{pn} are shown in Figs. 6.43 to 6.46, respectively. The shoot-through period T_O measurement is shown in Fig. 6.47. In Fig. 6.47, measurement of shoot-through period T_O is made for one triangle carrier period commencing 0.190 sec. to 0.19001 sec. in three different locations. These

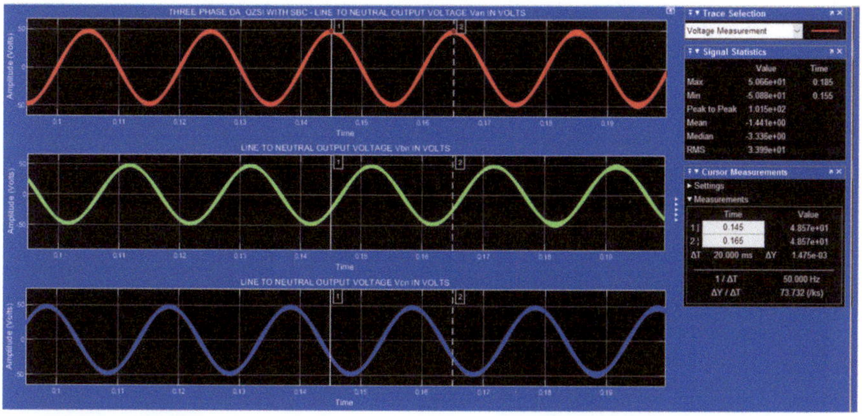

Fig. 6.43 Three-phase diode-assisted extended boost QZSI with SBC—line to neutral output voltage

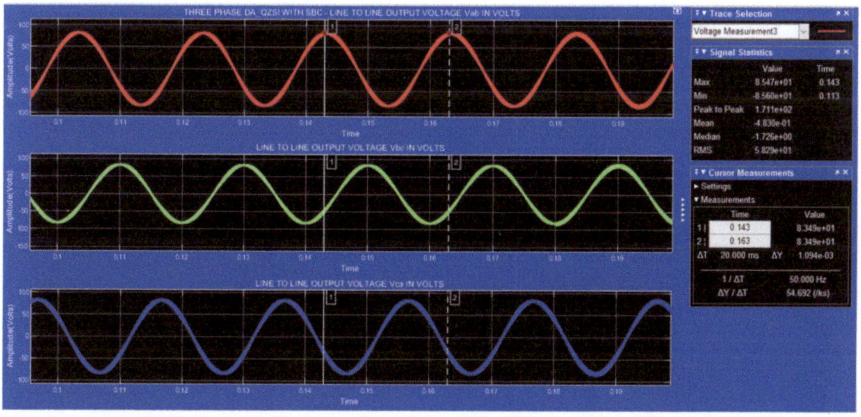

Fig. 6.44 Three-phase diode-assisted extended boost QZSI with SBC—line to line output voltage

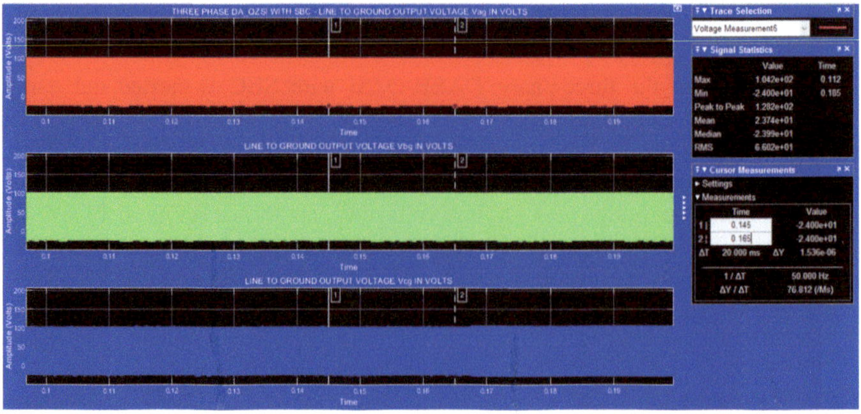

Fig. 6.45 Three-phase diode-assisted extended boost QZSI with SBC—line to ground output voltage

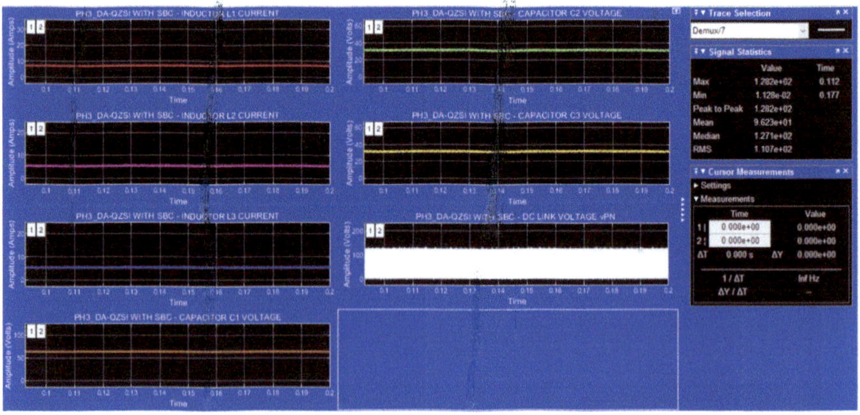

Fig. 6.46 Three-phase diode-assisted extended boost QZSI with SBC—inductor currents *iL1, iL2* and *iL3* and capacitor voltage *vC1* (left column, top to bottom), capacitor voltages *vC2* and *vC3* and inverter bridge DC link voltage V_{PN} (right column, top to bottom)

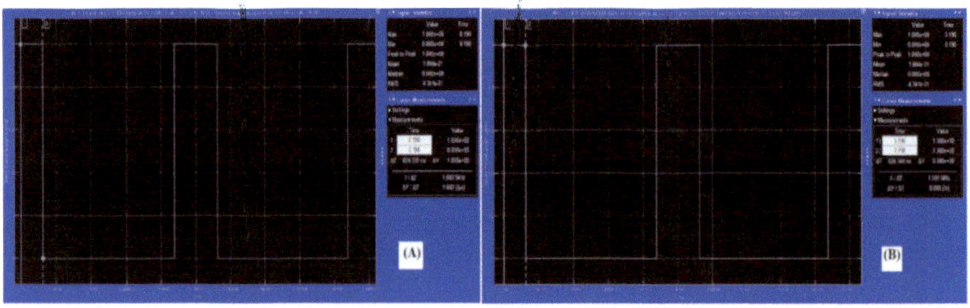

Fig. 6.47 Three-phase diode-assisted QZSI with SBC—shoot through period To measurement (**a**) using AND gate and (**b**) using OR gate

values are, respectively, 624.339e-9 sec., 1.249e-6 sec. and 624.339e-9 sec. giving a total value of 2.497e-6 sec. for T_O using AND gate and 628.340e-9 sec., 1.257e-6 sec. and 618.040e-9 sec. giving a total value of 2.503e-6 sec. for To using OR gate. The values using AND gate is considered here. The simulation results are tabulated in Table 6.7.

6.5.3 Discussion of Results

The model values and simulation results in Table 6.7 for D_O, B, G, *Vac(peak)*, *Vpn (mean)*, *vpn(peak)*, *VC1(mean)*, *VC2(mean)* and *VC3(mean)* agree with the respective theoretical values. Comparing the result with that shown in Table 5.7 in Chap. 5, it is seen that for the same D_O and SBC technique, the values of B and G are increased and this has boosted the value of *Vac(peak)*, *Vpn(mean)*, *vpn(peak)*, *VC2 (mean)* and *VC3(mean)*. Current through inductors L1, L2 and L3 is continuous. Also from Fig. 6.45, the RMS value of line to ground voltage for Phase A is 66.02 V, whereas for a conventional inverter, this is 24 V which is half the DC source voltage. This gives a boost factor B of 2.75.

6.5.4 Modelling of Diode-Assisted Extended Boost Quasi Z-Source Three-Phase Inverter Using Third Harmonic Injection Maximum Constant Boost Control

The maximum constant boost control technique is presented in Sect. 5.2.13. The model of the DAEB-QZSI with THIMCBC is shown in Fig. 6.48 (model file: EXAMPLE6_8). Now consider the three-phase DAEB-QZSI whose data are given in Table 5.6. The value of M is taken as 0.866 using Eq. 5.51, with D_O 0.25. The Embedded MATLAB function generates three-phase THI sine wave defined in Eq. 5.35. The source code for generating this three-phase THI sine wave is shown in Program segment 5.1. Boost factor B is calculated by using Eq. 6.67. The method of generating the gate pulse for THIMCBC technique is already explained in Sect. 5. 2.14. Here the shoot through period T_O is directly measured from the gate drive model without going to the gate pulse waveform. Here the constant blocks relating to +√3*M/2 (+0.75) and −√3*M/2 (−0.75) are compared with triangle carrier using two relational operator blocks which form the comparators, and the output of these two comparators is given to OR gate, the output of which gives the shoot through period T_O. Also shoot through period is measured by connecting the upper and lower switches Sap and San of inverter to an AND gate. The output of AND gate gives the shoot through period To.

Table 6.7 Three-phase DA-QZSI with SBC—simulation results

S. no.	D_O	B	G B X M	V_{ac} (Volts) Peak	V_{PN} Volts Avg/mean	v_{PN} Volts Peak	V_{C1}, V_{C2}, V_{C3} Volts Avg/mean	Remarks
1)	0.25 [0.2497]	2.6667 (2.66)	2.0 (1.9976)	48.0 (47.644) [50.66]	96.0 (95.88) [96.2]	128.0 (127.79) [128.2]	64.0, 32.0, 32.0 (63.97, 31.91, 31.91) [63.71, 31.68, 31.68]	Theoretical (Model Calculations) [Simulation Results]

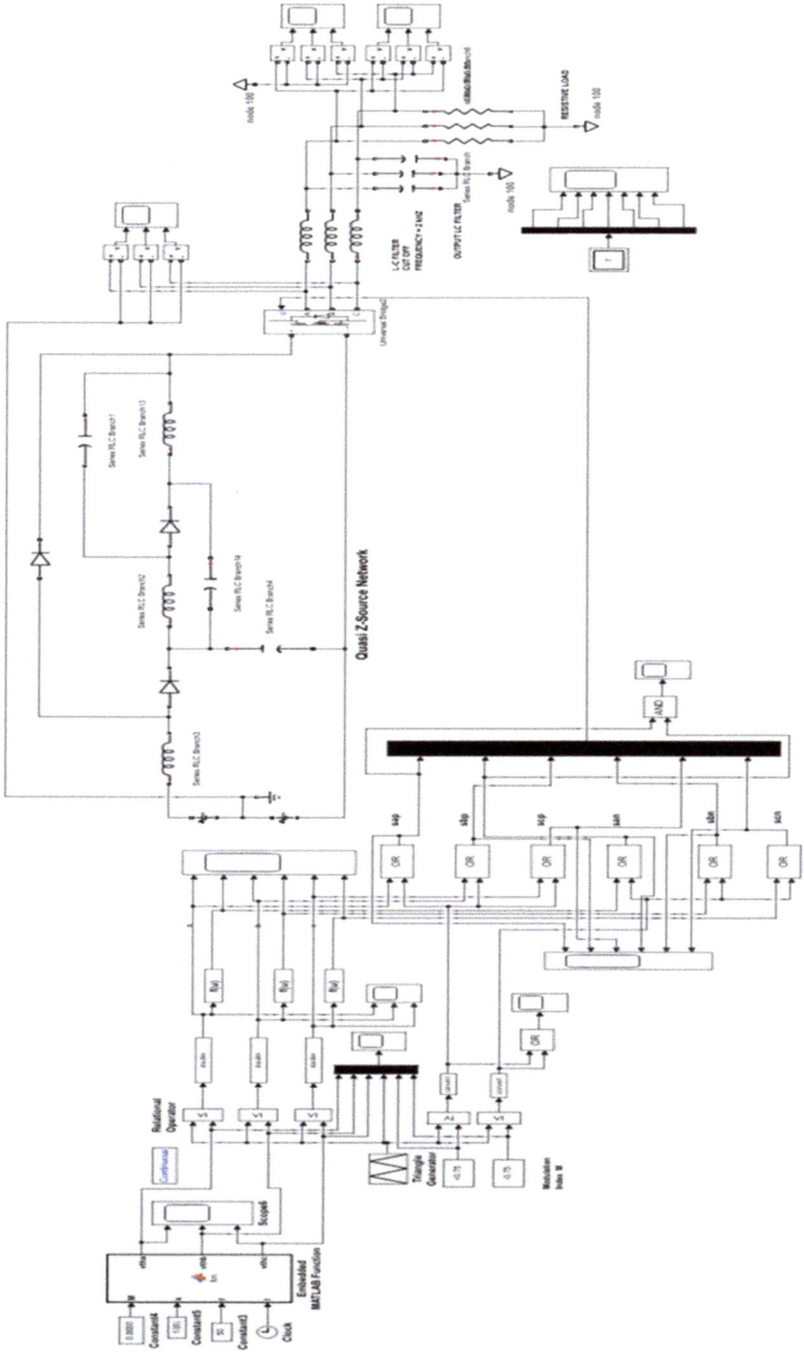

Fig. 6.48 Diode-assisted extended boost quasi Z-source three-phase inverter with third harmonic injection maximum constant boost control

6.5.5 Simulation Results

The simulation of the three-phase DA-QZSI with THIMCBC is carried out using ode15s (stiff/NDF) solver [9]. The data shown in Table 5.6 are used with inductor L1 changed to 100e-6 H. The simulation results for the line to neutral output voltage, line to line output voltage, line to ground output voltage and quasi Z-source inductor currents, capacitor voltages and inverter bridge peak DC link voltage V_{pn} are shown in Figs. 6.49 to 6.52, respectively. The shoot-through period T_O measurement is shown in Fig. 6.53. In Fig. 6.53, measurement of shoot through period T_O is made for one triangle carrier period commencing 0.190 sec. to 0.19001 sec. in three different locations. These values are, respectively, 631.448e-9 sec., 1.252e-6 sec. and 620.924e-9 sec. giving a total value of 2.503e-6 sec. for T_O using AND gate and

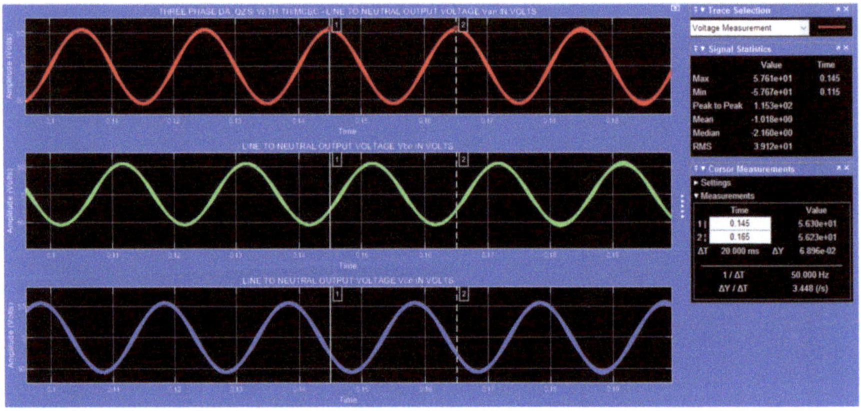

Fig. 6.49 Three-phase DA extended boost QZSI with THIMCBC—line to neutral output voltage

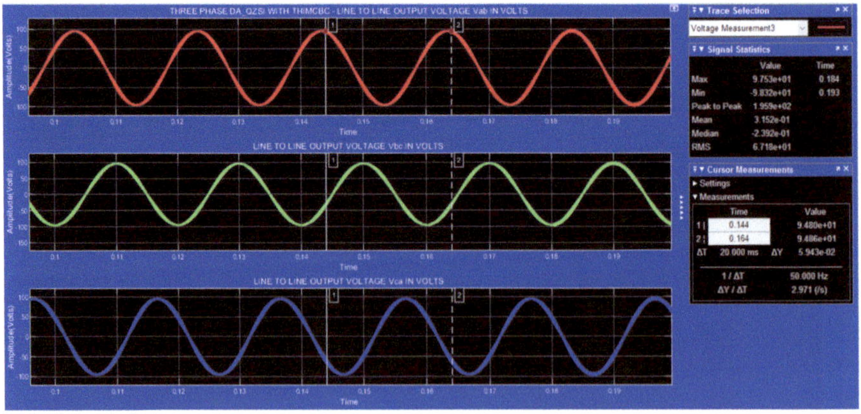

Fig. 6.50 Three-phase DA extended boost QZSI with THIMCBC—line to line output voltage

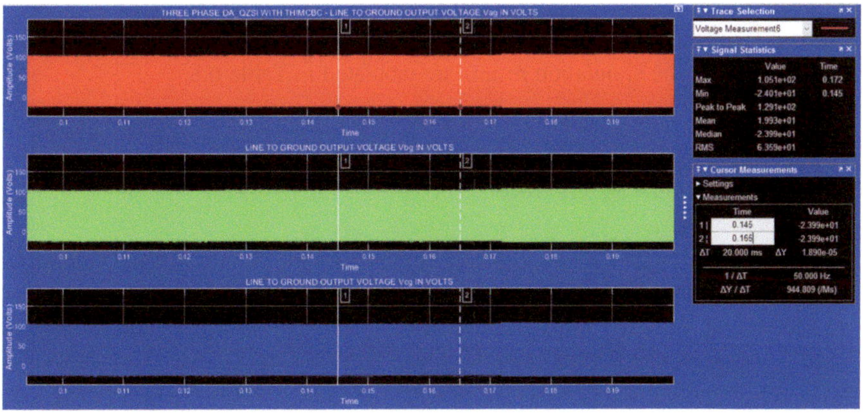

Fig. 6.51 Three-phase DA extended boost QZSI with THIMCBC—line to ground output voltage

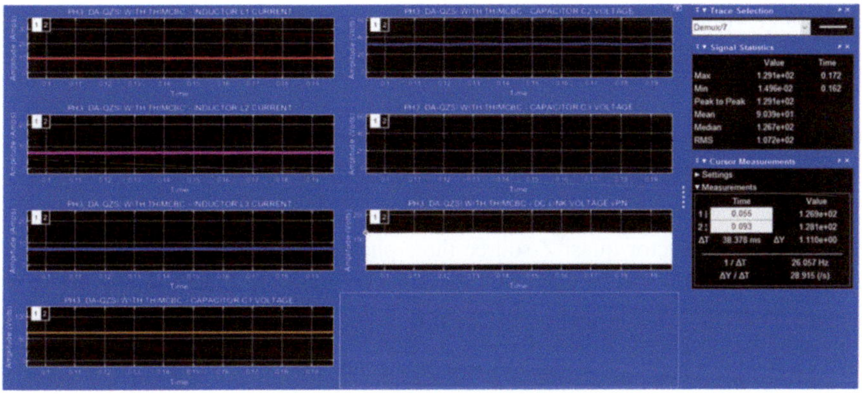

Fig. 6.52 Three-phase DA extended boost QZSI with THIMCBC—inductor currents *iL1, iL2* and *iL3* and capacitor voltage *vC1* (left column, top to bottom), capacitor voltages *vC2* and *vC3* and inverter bridge DC link voltage V_{PN} (right column, top to bottom)

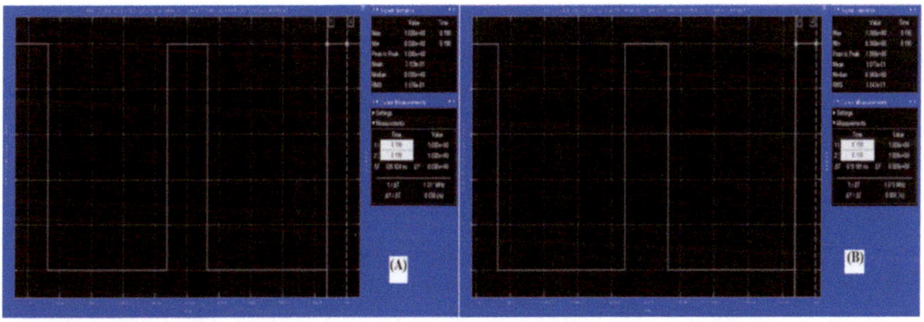

Fig. 6.53 Three-phase DA extended boost QZSI with THIMCBC—shoot-through period to measurement (**a**) using AND gate and (**b**) using OR gate

629.178e-9 sec., 1.258e-6 sec. and 619.191e-9 sec. giving a total value of 2.506e-6 sec. for To using OR gate. The values using AND gate are considered here. The D_O value is 0.2503. The simulation results are tabulated in Table 6.8.

6.5.6 *Discussion of Results*

The model values and simulation results in Table 6.8 for D_O, B, G, *Vac(peak)*, *Vpn (mean)*, *vpn(peak)*, *VC1(mean)*, *VC2(mean)* and *VC3(mean)* agree with the respective theoretical values. Comparing the result with that shown in Table 5.10 in Chap. 5, it is seen that for the same D_O and THIMCBC technique, the values of B and G are increased and this has boosted the value of *Vac(peak)*, *Vpn(mean)*, *vpn (peak)*, *VC2(mean)* and *VC3(mean)*. Current through inductors L1, L2 and L3 is continuous. Also from Fig. 6.51, the RMS value of line to ground voltage for Phase A is 63.59 V, whereas for a conventional inverter, this value is 24 V which is half the DC source voltage. This gives a boost factor B of 2.649.

6.6 Case Study: Space Vector Modulation of a Switched Inductor Quasi Z-Source Three-Phase Inverter

The switched inductor quasi Z-source three-phase inverter (SL-QZSI) topology is shown in Fig. 6.15. The ST and NST equivalent circuits are shown in Fig. 6.16a, b. The derivations for the capacitor voltages *Vc1* and *Vc2*, inverter bridge voltage V_{PN} and boost factor B are derived in Eqs. 6.28 to 6.31. The advantages of SL-QZSI over classical ZSI and QZSI are presented in Sect. 6.3. In this section model for space vector modulation (SVM) of a three-phase SL-QZSI is presented with simulation results. The results are compared with that obtained using SBC and THIMCBC techniques.

6.6.1 *Model of Space Vector Modulation of a Switched Inductor Quasi Z-Source Three-Phase Inverter*

The model of space vector modulation (SVM) of a SL-QZSI is shown in Fig. 6.54 (model file: CASE_STUDY_EX6_1). The SL-QZSI model is already presented in Sect. 6.3. The SVM subsystem consists of (1) sector identifier, (2) sector switch function generator and (3) gate pulse timing generator. These are presented in Program Segments 5.1, 5.2 and 5.3 in Sect. 5.4 of Chap. 5. The quasi Z-source parameters are the same as presented in Sect. 6.3.1. The inverter output frequency is

Table 6.8 Three-phase DA-QZSI with THIMCBC—simulation results

S. no.	D_O	B	G B X M	V_{ac} (Volts) Peak	V_{PN} Volts Avg/mean	v_{PN} Volts Peak	V_{C1}, V_{C2}, V_{C3} Volts Avg/mean	Remarks
1)	0.25 [0.2503]	2.6667 (2.6709)	2.309 (2.313)	55.42 (55.5128) [57.61]	96.0 (96.115) [90.31]	128.0 (128.205) [129.1]	64.0, 32.0, 32.0 (64.0256, 32.089, 32.089) [63.69, 31.63, 31.63]	Theoretical (Model Calculations) [Simulation Results]

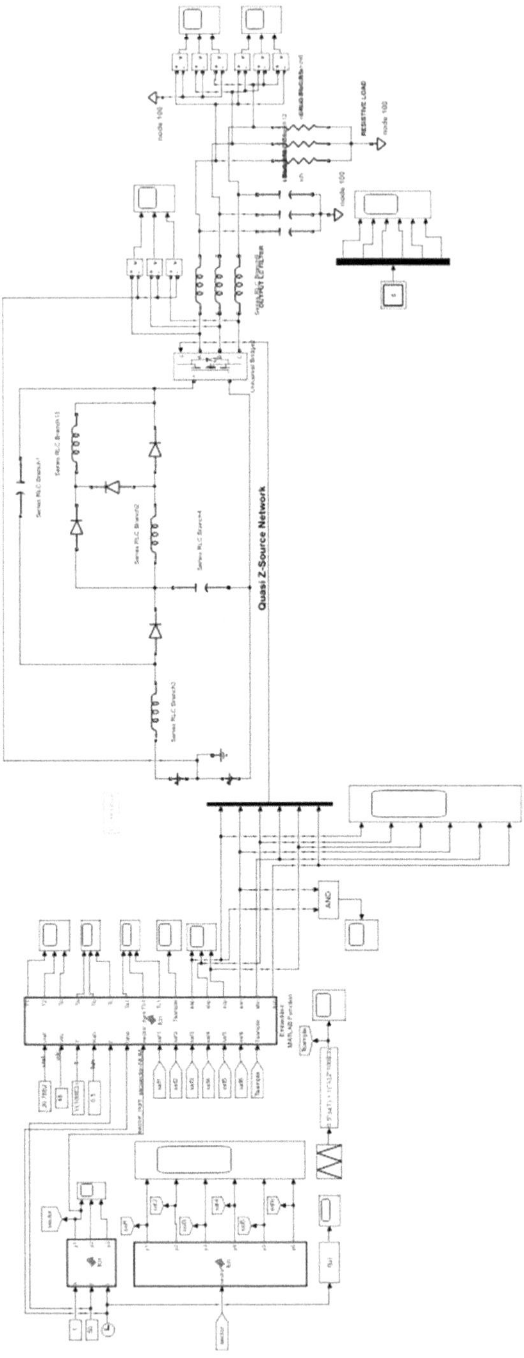

Fig. 6.54 Model of space vector modulation of a switched inductor quasi Z-source three-phase inverter

50 Hz and the sampling frequency is 100 kHz. The modulation index Ma is 0.75. This gives a Vref value of 20.7852 volts. The DC link input voltage is 48 volts. A shoot-through factor Ksh of 0.5 is used.

6.6.2 Simulation Results

The simulation of the Space Vector Pulse Width Modulation (SVPWM) SL-QZSI is carried out using ode23tb (stiff/TR-BDF2) solver in Simulink [9]. The simulation results for the three-phase line to neutral output voltage, line to line output voltage, line to ground output voltage and inductor currents, capacitor voltages and DC link bridge voltage are shown in Figs. 6.55 to 6.58, respectively. The gate pulse for the

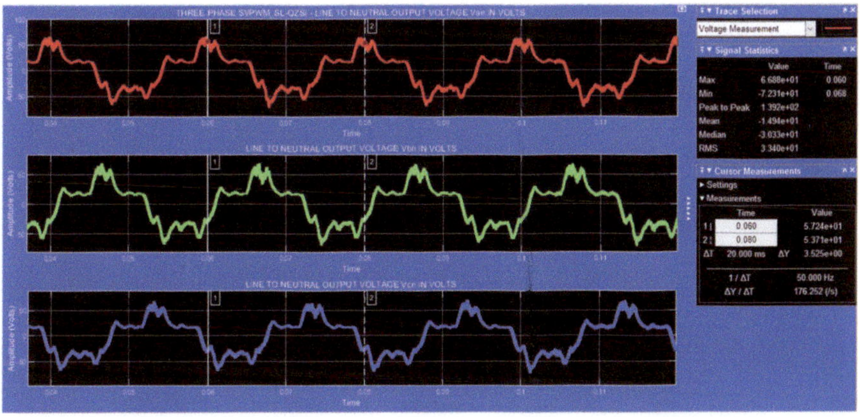

Fig. 6.55 Three-phase SVPWM SL-QZSI—line to neutral output voltage

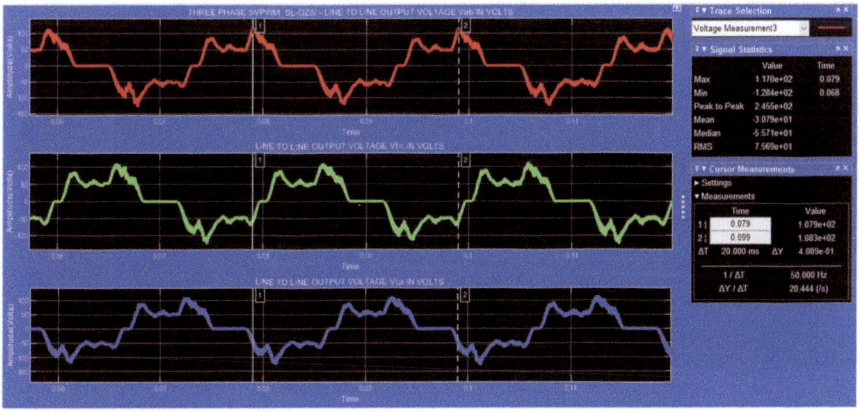

Fig. 6.56 Three-phase SVPWM SL-QZSI—line to line output voltage

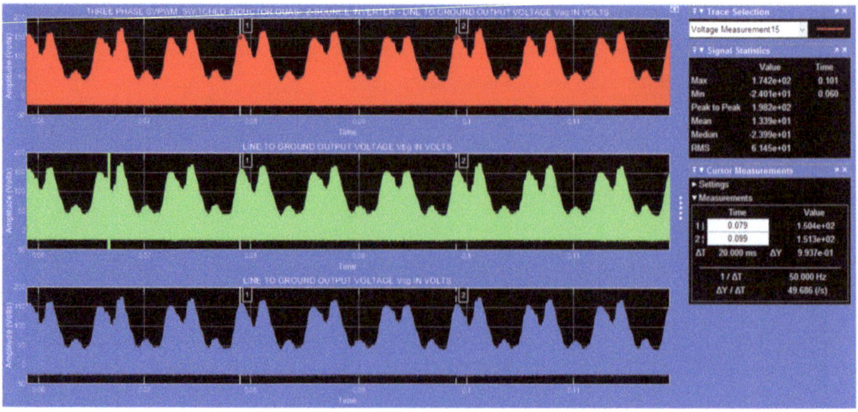

Fig. 6.57 Three-phase SVPWM SL-QZSI—line to ground output voltage

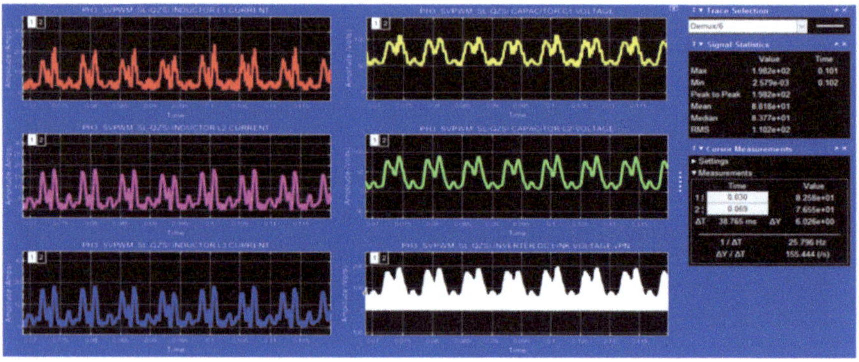

Fig. 6.58 Three-phase SVPWM SL-QZSI—inductor currents *iL1*, *iL2* and *iL3* (left column, top to bottom), capacitor voltages vC1 and vC2 and inverter bridge DC link voltage V_{PN} (right column, top to bottom)

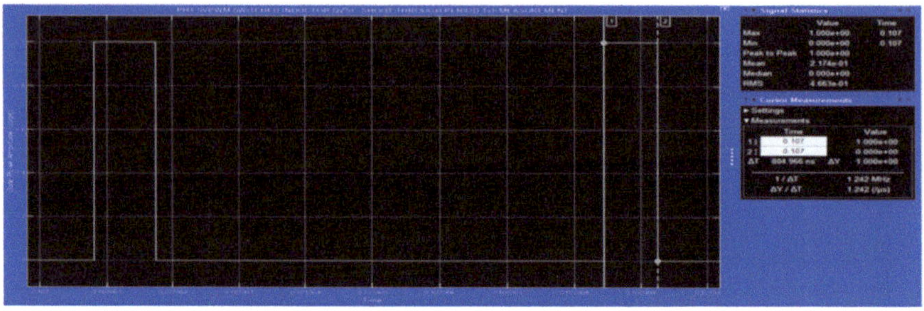

Fig. 6.59 Three-phase SVPWM SL-QZSI—shoot-through period Tsh measurement

Table 6.9 Three-phase SVPWM SL-QZSI—simulation results

S. no.	D_O	B	V_{PN} Volts Peak	V_{PN} Volts Mean	V_{C1} Volts Avg/ mean	V_{C2} Volts Avg/mean	Remarks
1)	(0.2323) [0.174]	(2.56) [1.8883]	(122.86) [198.2]	(94.3214) [88.18]	(76.54) [78.5]	(46.32) [48.6]	(model calculations) [Simulation results]

measurement of shoot-through period is shown in Fig. 6.59. Simulation results are tabulated in Table 6.9. From the simulation results in Fig. 6.57, it is seen that the RMS value of the line to ground output voltage of inverter is 61.45 volts whereas for a conventional inverter, this value is 24 V which is half the DC source voltage. Thus the boost factor B is 2.56. Shoot through period measurement is made using Fig. 6.59 during one sampling period from 0.107 sec. to 0.10701 sec. The two shoot through values Tsh1 and Tsh2 are found to be 935.774e-9 sec. and 804.966e-9 sec., respectively, giving a total shoot-through period Tsh of 1.74e-6 sec. giving a Do value of 0.174. Using B value of 2.56, the calculated value of Do is 0.2323 using Eq. 6.31.

6.6.3 Discussion of Results

The simulation results for Do and B show that their values are lower than that shown in Tables 6.3 and 6.4 for simple and maximum constant boost control. One factor is the choice of shoot through factor Ksh and the other is the modulation index M. By varying Ksh it is possible to vary shoot through period Tsh as is clear from Fig. 5.49 in Sect. 5.4. The V_{PN}(peak) value by model calculation and simulation result differs by a large value, whereas *vPN(mean)*, *vC1(mean)* and *vC2(mean)* closely well agree.

6.7 Case Study: Space Vector Modulation of an Active Switched Capacitor Quasi Z-Source Three-Phase Inverter

The active switched capacitor (ASC) quasi Z-source three-phase inverter (ASC-QZSI) topology is shown in Fig. 6.25. The ST and NST equivalent circuits are shown in Fig. 6.26a, b. The derivations for the capacitor voltages *Vc1* and *Vc2,* inverter bridge voltage V_{PN} and boost factor B are derived in Eqs. 6.44 to 6.47. The advantages of ASC-QZSI over classical ZSI and QZSI are presented in Sect. 6.4. In this section model for space vector modulation (SVM) of a three-phase ASC-QZSI is presented with simulation results. The results are compared with that obtained using SBC and THIMCBC techniques.

6.7.1 Model of Space Vector Modulation of an Active Switched Capacitor Quasi Z-Source Three-Phase Inverter

The model of space vector modulation (SVM) of a three-phase ASC-QZSI is shown in Fig. 6.60 (model file: CASE_STUDY_EX6_2). The AS-QZSI model is already presented in Sect. 6.4. The SVM subsystem consists of (1) sector identifier, (2) sector switch function generator and (3) gate pulse timing generator. These are presented in Program Segments 5.1, 5.2 and 5.3 in Sect. 5.4 of Chap. 5. The quasi Z-source parameters are the same as presented in Sect. 6.4.1. The inverter output frequency is 50 Hz and the sampling frequency is 100 kHz. The modulation index Ma is 0.75. This gives a Vref value of 20.7852 volts. The DC input voltage is 48 volts. A shoot-through factor Ksh of 0.5 is used. The one difference here is the gate pulse generation for the NPN transistor switch SW1 in Fig. 6.25. This SW1 is closed during ST and open during NST period. This shoot-through gate pulse is obtained by connecting the gate pulse for reference Phase A of inverter upper switch Sap and lower switch San to an AND gate. This AND gate output pulse duration for one sampling period gives the shoot through period Tsh. This gate pulse amplitude and that for the three-phase inverter are multiplied by a gain of ten to account for the gate threshold voltage of the semiconductor switches used in the model.

6.7.2 Simulation Results

The simulation of the SVPWM ASC-QZSI is carried out using ode15s (stiff/NDF) solver in Simulink [9]. The simulation results for the three-phase line to neutral output voltage, line to line output voltage and inductor currents, capacitor voltages, inverter bridge DC link voltage and line to ground output voltage for Phase A are shown in Figs. 6.61 to 6.63, respectively. The gate pulse during the shoot through period is shown in Fig. 6.64. Simulation results are tabulated in Table 6.10. From Fig. 6.63, it is seen that the RMS value of line to ground voltage is displayed as 95.62 V whereas for a conventional three-phase inverter, this value is 24 V which is half the DC source voltage. This gives a boost factor B value of 3.98. Shoot through period measurement is made using Fig. 6.64 during one sampling period from 0.106 sec. to 0.10601 sec. The two shoot through values Tsh1 and Tsh2 are found to be 0.712e-6 sec. and 0.712e-6 sec., respectively, giving a total shoot through

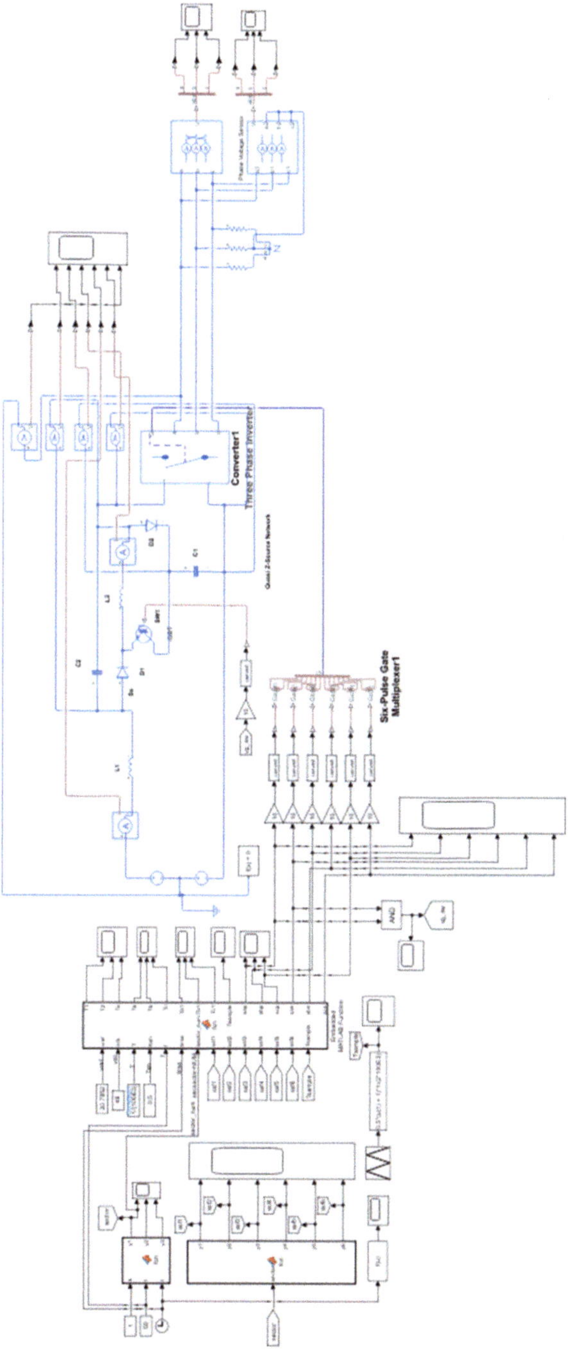

Fig. 6.60 Model of space vector pulse width modulation of an active switched capacitor quasi Z-source three-phase inverter

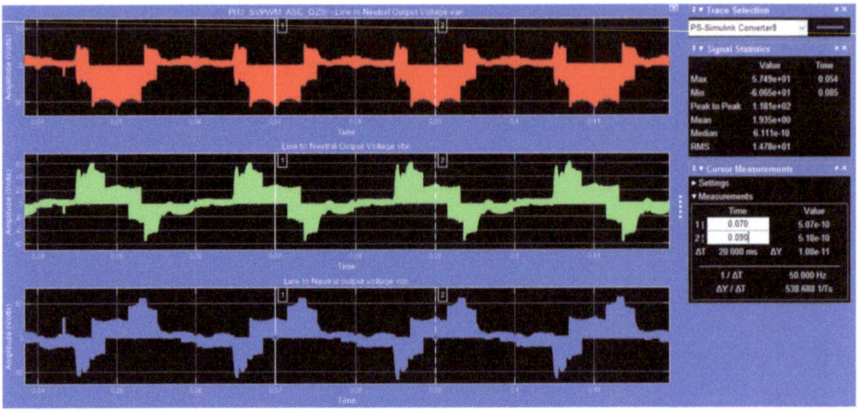

Fig. 6.61 Three-phase SVPWM ASC-QZSI—line to neutral output voltage

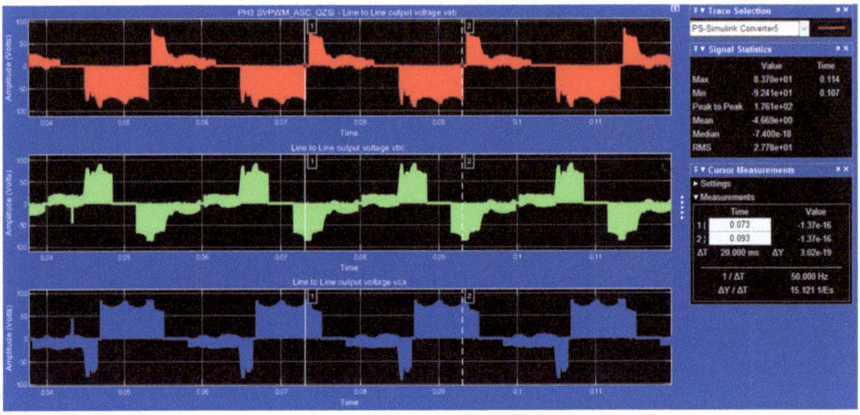

Fig. 6.62 Three-phase SVPWM ASC-QZSI—line to line output voltage

period Tsh of 1.424e-6 sec. This gives a Do value of 0.1424. The two capacitor voltages *Vc1* and *Vc2*, inverter bridge DC link voltage V_{PN} and boost factor B are calculated using Eqs. 6.44 to 6.47, respectively.

6.7.3 Discussion of Results

The simulation results for Do and B show that the value of Do and B is lower than that shown in Tables 6.5 and 6.6 for simple and maximum constant boost control. One factor is the choice of shoot through factor Ksh. By varying Ksh it is possible to

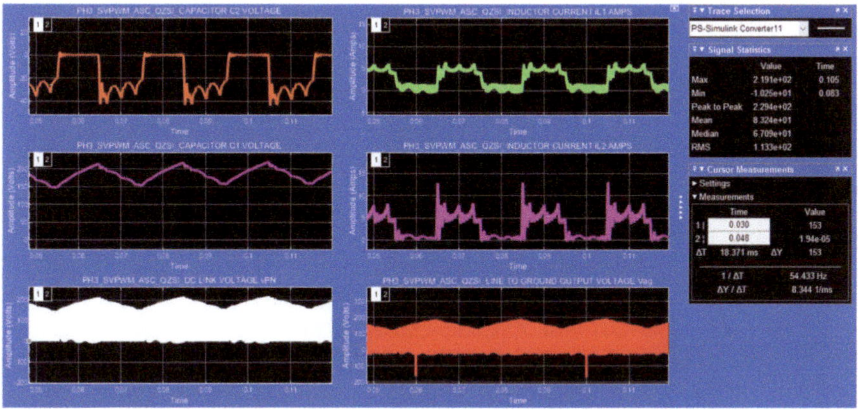

Fig. 6.63 Three-phase SVPWM ASC-QZSI—capacitor voltages vC2 and vC1 and inverter bridge DC link voltage V_{PN} (left column, top to bottom), inductor currents *iL1* and *iL2* and line to ground output voltage for Phase A *Vag* (right column, top to bottom)

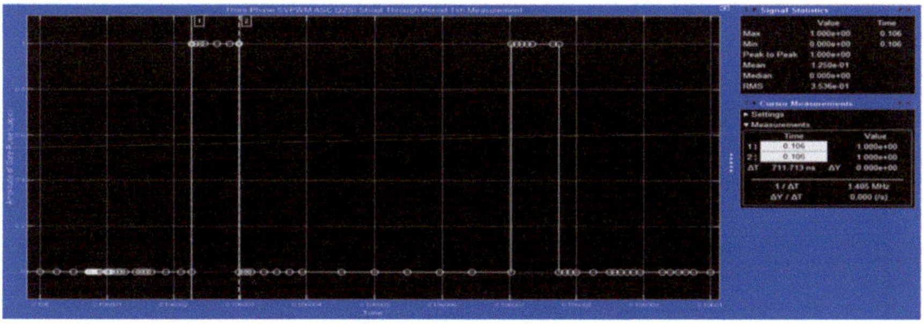

Fig. 6.64 Three-phase SVPWM ASC-QZSI shoot through period Tsh measurement

Table 6.10 Three phase SVPWM ASC-QZSI—simulation results

S. No.	D_O	B	V_{PN} Volts Peak	V_{PN} Volts Mean	V_{C1} Volts Mean	V_{C2} Volts Mean	Remarks
1)	[0.1424] (0.3066)	[1.446] (3.98)	(191.059) [219.1]	(132.48) [83.24]	(191.059) [183.0]	(16.6313) [-15.62]	(Model Calculations) [Simulation Results]

vary shoot through period Tsh as is clear from Fig. 5.49 in Sect. 5.4. The V_{PN} (peak), *Vc1* and *Vc2* values by model calculation and simulation results differ by a small percentage. The value of V_{PN} (Mean) by model calculation and simulation result differs by a large percentage.

6.8 Case Study: Space Vector Modulation of a Diode-Assisted Extended Boost Quasi Z-Source Three-Phase Inverter

The diode-assisted extended boost quasi Z-source three-phase inverter (DAEB-QZSI) topology is shown in Fig. 6.37. The ST and NST equivalent circuits are shown in Fig. 6.38a, b.

The derivations for the capacitor voltages *Vc1, Vc2* and *Vc3*, inverter bridge DC link voltage V_{PN} and boost factor B are derived in Eqs. 6.57 to 6.67. The advantages of DAEB-QZSI over classical ZSI and QZSI are presented in Sect. 6.5. In this section model for space vector modulation (SVM) of a three-phase DAEB-QZSI is presented with simulation results. The results are compared with that obtained using SBC and THIMCBC techniques.

6.8.1 Model of Space Vector Modulation of a Diode-Assisted Extended Boost Quasi Z-Source Three-Phase Inverter

The model of space vector modulation (SVM) of a three-phase DAEB-QZSI is shown in Fig. 6.65 (model file: CASE_STUDY_EX6_3). The DAEB-QZSI model is already presented in Sect. 6.5.

The SVM subsystem consists of (1) sector identifier, (2) sector switch function generator and (3) gate pulse timing generator. These are presented in Program Segments 5.1, 5.2 and 5.3 in Sect. 5.4 of Chap. 5. The data are presented in Sect. 6.5.1. The inverter output frequency is 50 Hz and the sampling frequency is 100 kHz. The modulation index Ma is 0.75. This gives a Vref value of 20.7852 volts. The DC input voltage is 48 volts. A shoot through factor Ksh of 0.5 is used.

6.8.2 Simulation Results

The simulation of the SVPWM DAEB-QZSI is carried out using ode23tb (stiff/TR-BDF2) solver in Simulink [9]. The simulation results for the three-phase line to neutral output voltage, line to line output voltage, line to ground output voltage and inductor currents, capacitor voltages and inverter bridge DC link voltage are shown in Figs. 6.66 to 6.69, respectively. The gate pulse during the shoot-through period is shown in Fig. 6.70. Simulation results are tabulated in Table 6.11. Shoot-through period measurements are made using Fig. 6.70 during one sampling period from 0.0945 sec. to 0.09451 sec. The two shoot-through values Tsh1 and Tsh2 are found to be 1.25e-6 sec. and 1.29e-6 sec., respectively, giving a total shoot-through period Tsh of 2.54e-6 sec. This gives a Do value of 0.254. The three capacitor voltages *Vc1, Vc2* and *Vc3,* inverter bridge DC link voltage V_{PN} and boost factor B are calculated

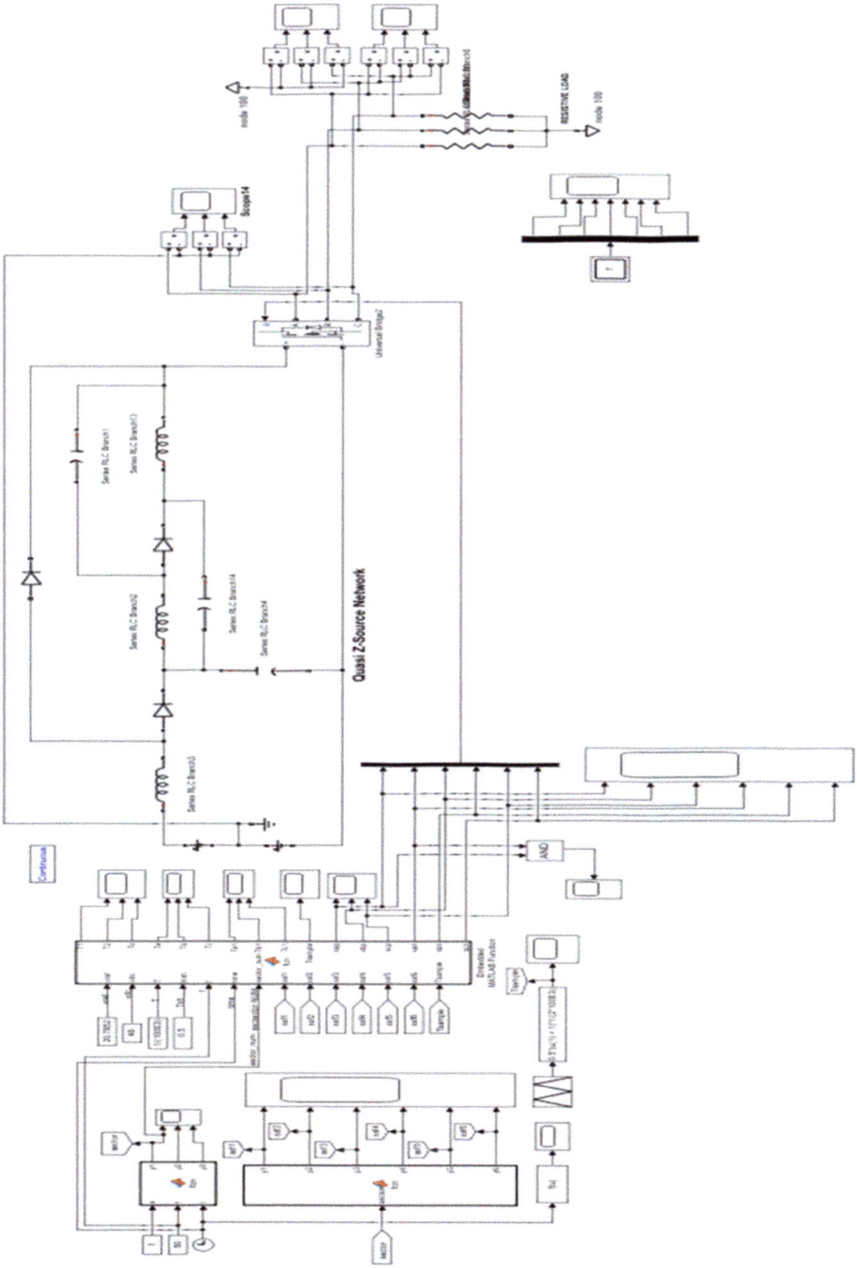

Fig. 6.65 Model of space vector modulation of a diode-assisted extended boost quasi Z-source three-phase inverter

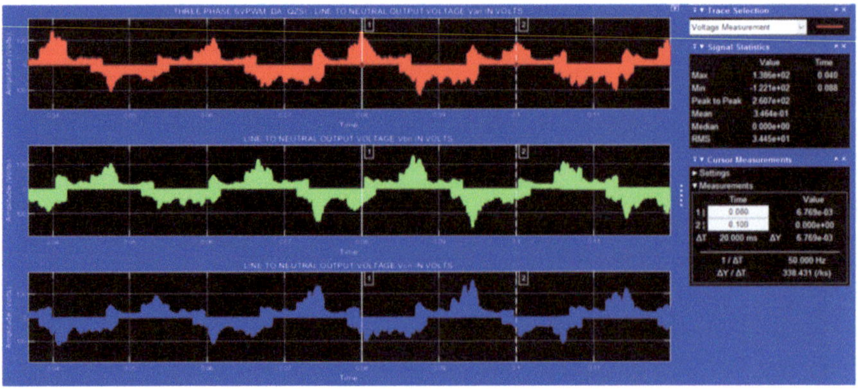

Fig. 6.66 Three-phase DA-extended boost QZSI—line to neutral output voltage

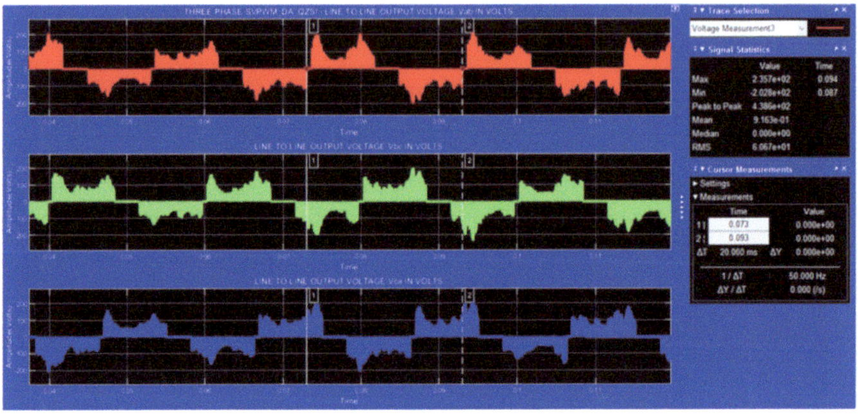

Fig. 6.67 Three-phase DA-extended boost QZSI—line to line output voltage

using Eqs. 6.57 to 6.67, respectively. Also from the line to ground voltage shown in Fig. 6.68, the RMS value is found to be 55.98 V, whereas for a conventional three-phase inverter, this value is half the DC source voltage which is 24 V. This gives a boost factor B of 55.98/24 which is 2.33 and a Do value of 0.2168.

6.8.3 *Discussion of Results*

The simulation results for Do and B show that the value of Do and B is lower than that shown in Tables 6.7 and 6.8 for simple and maximum constant boost control. One factor is the choice of shoot-through factor Ksh. By varying Ksh it is possible to vary shoot-through period Tsh as is clear from Fig. 5.49 in Sect. 5.4. The values of

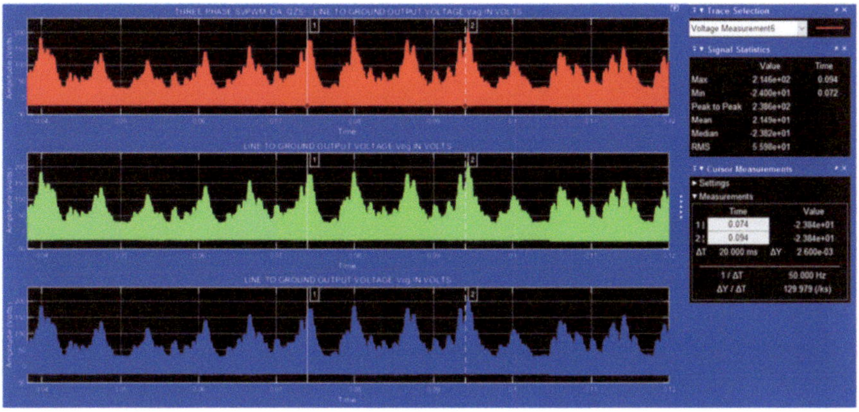

Fig. 6.68 Three-phase DA-extended boost QZSI—line to ground output voltage

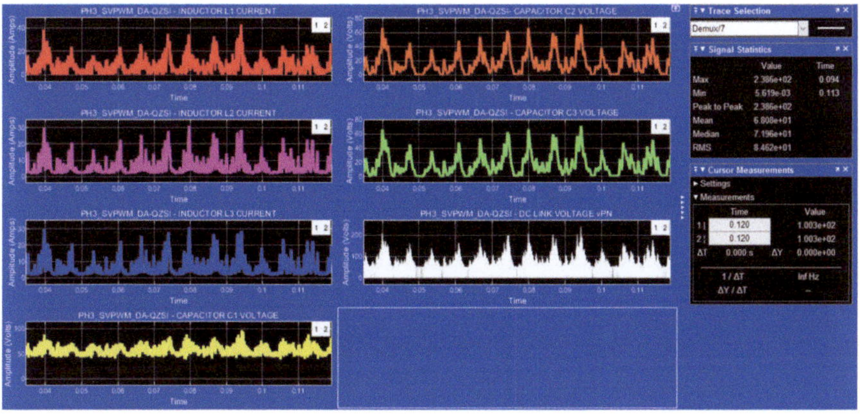

Fig. 6.69 Three-phase DA-extended boost QZSI—inductor currents *iL1, iL2* and *iL3* and capacitor voltage *vC1* (left column, top to bottom), capacitor voltages *vC2* and *vC3* and inverter bridge DC link voltage V_{PN} (right column, top to bottom)

Vc1, Vc2 and *Vc3* by model calculation and simulation results differ by a small percentage. The values of *Vc2* and *Vc3* are found to be equal by simulation which agrees with derivation. For the Do and B values calculated using RMS value of line to ground voltage, the calculated values for *Vc1, Vc2* and *Vc3* closely agree with simulation results. The value of V_{PN} by model calculation and simulation differs by a large percentage.

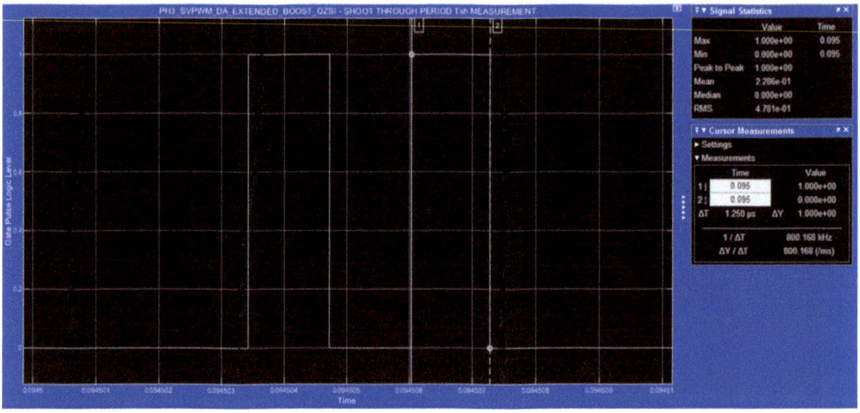

Fig. 6.70 Three-phase SVPWM DA-extended boost QZSI—shoot through period Tsh measurement

Table 6.11 Three-phase SVPWM DA-extended boost QZSI—simulation results

S. no.	D_O	B	V_{PN} Volts Peak	V_{C1} Volts Mean	V_{C2} Volts Mean	V_{c3} Volts Mean	Remarks
1)	[0.2540] (0.2236)	[2.72] (2.33)	(130.7788) (111.837) [238.6]	(64.34) (61.82) [58.09]	(33.2178) (25.00) [19.45]	(33.2178) (25.00) [19.45]	(Model Calculations) [Simulation Results]

6.9 Case Study: Space Vector Modulation of a Switched Inductor Z-Source Three-Phase Inverter

The switched inductor Z-source three-phase inverter (SL-ZSI) topology is shown in Fig. 6.1. The ST and NST equivalent circuits are shown in Fig. 6.2a, b. The derivations for the capacitor voltages Vc1 and Vc2, inverter bridge voltage vPN and boost factor B are derived in Eqs. 6.12 to 6.15. The advantages of SL-ZSI over classical ZSI are presented in Sect. 6.2. In this section model for space vector modulation (SVM) of a three-phase SL-ZSI is presented with simulation results. The results are compared with that obtained with SBC and THIMCBC techniques.

6.9.1 Model of Space Vector Modulation of a Switched Inductor Z-Source Three-Phase Inverter

The model of SVM of a three-phase switched inductor Z-source inverter (SL-ZSI) is shown in Fig. 6.71 (model file: CASE_STUDY_EX6_4). The SL-ZSI model is already presented in Sect. 6.3. The SVM subsystem consists of (1) sector identifier,

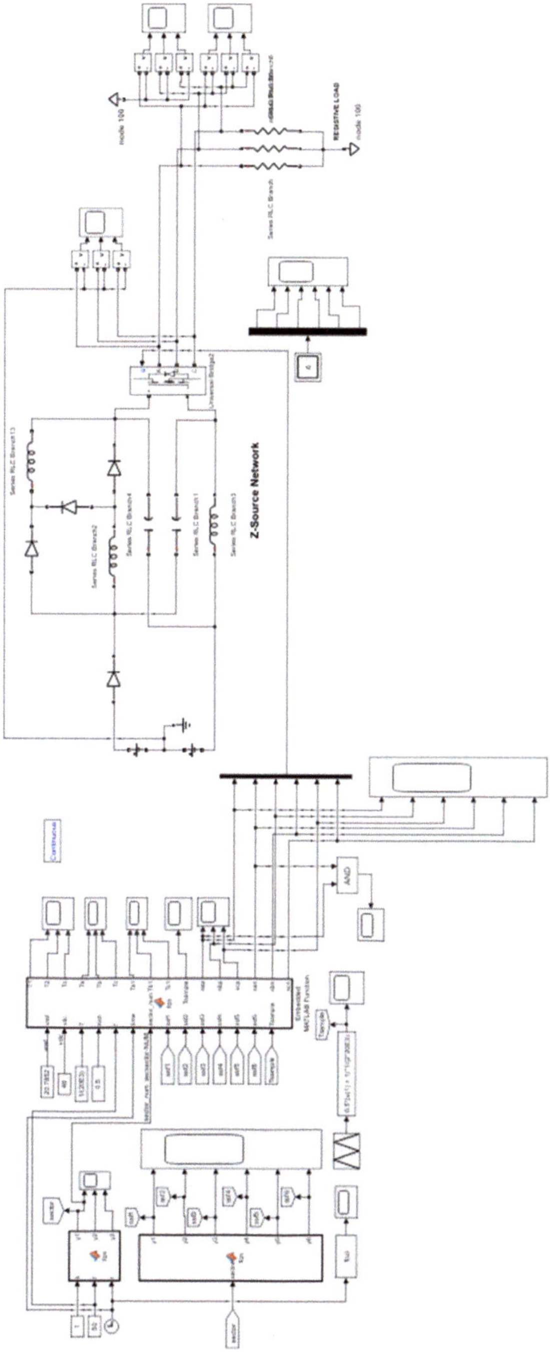

Fig. 6.71 Model of space vector modulation of a three-phase switched inductor Z-source inverter

(2) sector switch function generator and (3) gate pulse timing generator. These are presented in Program segments 5.1, 5.2 and 5.3 in Sect. 5.4 of Chap. 5. The SL-ZSI parameters are the same as presented in Sect. 6.2.1. The inverter output frequency is 50 Hz and the sampling frequency is 20 kHz. The modulation index Ma is 0.75. The DC source voltage is 48 V. This gives a Vref value of 20.7852 volts. A shoot through factor Ksh of 0.5 is used.

6.9.2 Simulation Results

The simulation of the SVPWM SL-ZSI is carried out using ode23tb (stiff/TR-BDF2) solver in Simulink [9]. The simulation results for the three-phase line to neutral output voltage, line to line output voltage, line to ground output voltage and inductor currents, capacitor voltages and inverter bridge DC link voltage are shown in Figs. 6.72 to 6.75, respectively. The gate pulse during the shoot-through period is shown in Fig. 6.76. Simulation results are tabulated in Table 6.12. Shoot-through period measurement is made using Fig. 6.76 during one sampling period from 0.073 sec. to 0.07305 sec. The two shoot-through values Tsh1 and Tsh2 are found to be 3.747e-6 sec. and 3.798e-6 sec., respectively, giving a total shoot- through period Tsh of 7.545e-6 sec. This gives a Do value of 0.1509. The two capacitor voltages $Vc1$ and $Vc2$, inverter bridge DC link voltage V_{PN} and boost factor B are calculated using Eqs. 6.12 to 6.15, respectively. Also from the line to ground voltage in Fig. 6.74, the RMS value is found to be 58.61 V, whereas for a conventional inverter, this value is half the DC source voltage which is 24 V. This gives a boost factor B of 58.61/24 which is 2.442 and a Do value of 0.2242.

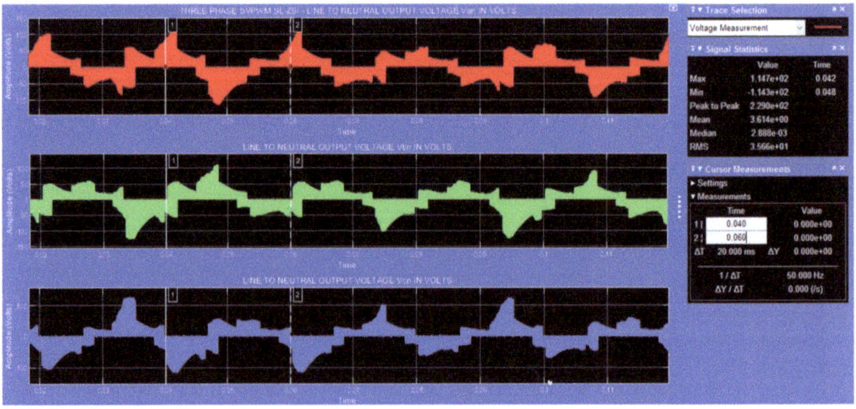

Fig. 6.72 Three-phase SVPWM SL-ZSI—line to neutral output voltage

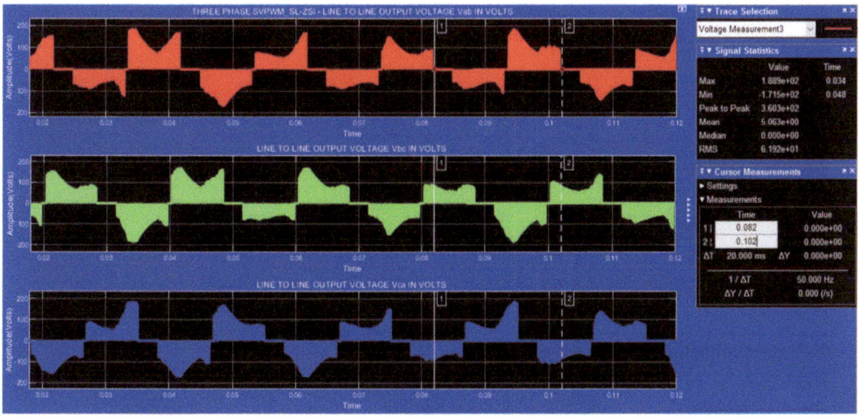

Fig. 6.73 Three-phase SVPWM SL-ZSI—line to line output voltage

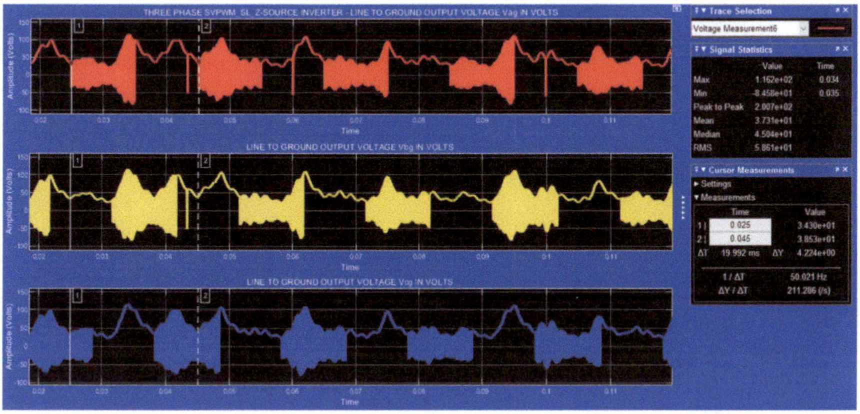

Fig. 6.74 Three-phase SVPWM SL-ZSI—line to ground output voltage

6.9.3 Discussion of Results

The simulation results for Do and B show that the value of Do and B is lower than that shown in Tables 6.1 and 6.2 for simple and maximum constant boost control. One factor is the choice of shoot through factor Ksh. By varying Ksh it is possible to vary shoot through period Tsh as is clear from Fig. 5.49 in Sect. 5.4. The V_{PN}(peak) value by model calculation and by simulation differs by a large percentage. The mean values of *vc1*, *vc2* and V_{PN} by calculation and by model simulation differ by a small value. The mean and peak voltage of *Vc2* by simulation have a minus sign which indicates that the polarity of *Vc2* is reversed from the one shown in Fig. 6.1.

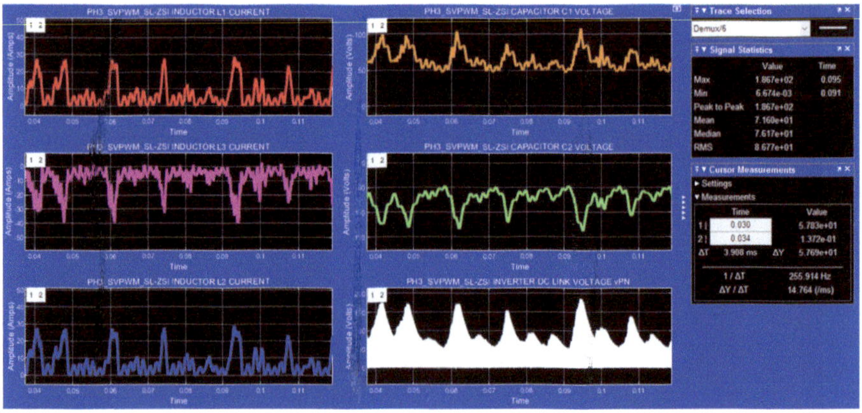

Fig. 6.75 Three-phase SVPWM SL-ZSI—inductor currents *iL1, iL3* and *iL2* (left column, top to bottom), capacitor voltages *vC1* and *vC2* and inverter bridge DC link voltage V_{PN} (right column, top to bottom)

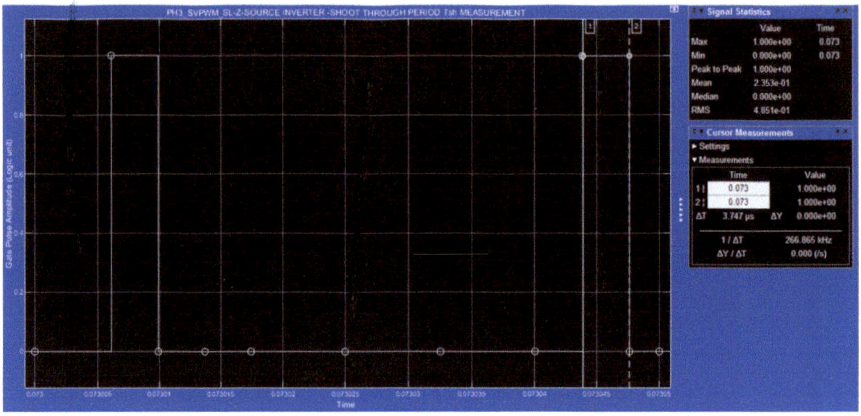

Fig. 6.76 Three-phase SVPWM SL-ZSI—shoot through period Tsh measurement

6.10 Conclusions

Models for SL-ZSI, SL-QZSI, ASC-QZSI and DA-QZSI are presented with simulation results. The analysis and derivation of switched inductor Z-source inverter in Sect. 6.2 are new.

Comparing the performance of the SL-ZSI, SL-QZSI, ASC-QZSI and DAEB-QZSI with that of the respective topology presented in Chap. 5, it is seen that the value of boost factor B, voltage gain G, Z-source network capacitor voltages *VC1* and *VC2* and inverter DC link bridge voltage V_{PN} has all been increased for the same shoot through duty ratio and boost control method. Also with SL, ASC and DAEB topology, it is seen that the inductor currents are continuous. Four case studies are

Table 6.12 Three phase SVPWM SL-ZSI—simulation results

S. No.	D_O	B	V_{PN} Volts Peak	V_{PN} Volts Mean	V_{C1} Volts Peak	V_{C2} Volts Peak	V_{C1} Volts Mean	V_{C2} Volts Mean	Remarks
1)	[0.1509] (0.2242)	[1.704] (2.442)	(81.7898) (117.21) [186.7]	(69.447) (90.93) [71.6]	[108.1]	[-137.3]	(60.34) (74.278) [67.3]	(69.4477) (90.93) [-80.01]	(Model Calculations) [Simulation Results]

presented using space vector modulation for SL, ASC, DAEB-QZSI and SL-ZSI topologies. The results are compared with their respective values for simple boost and maximum constant boost control techniques. In the case of space vector modulation, the shoot through duty ratio Do and boost factor B can be increased by using different shoot through factors for Ksh and also by varying the reference voltage Vref. It is also seen that in the case of all SVPWM SL, ASC, DA-extended boost QZSI and SL-ZSI, shoot through period Tsh for one sampling period varies with time, giving different values for different instants of time. Hence the values shown in the relevant tables are only approximate.

References

1. F.Z. Peng: "Z-Source Inverter", IEEE Transactions on Industry Applications, Vol. 39, No.2, March/April 2003, pp. 504–510.
2. J. Anderson and F. Peng, "Four quasi-Z-source inverters," in Proc. IEEE Power Electronics Specialist Conference., Rhodes, Greece, Jun. 2008, pp. 2743–2749.
3. Y. Li, J. Anderson, F.Z. Peng and D. Liu: "Quasi-Z-Source Inverter for Photovoltaic Power Generation Systems", IEEE APEC, February 2009, Washington DC, USA, pp. 918–924.
4. Ayman Ayad, Stefan Hanafiah, Ralph Kennel: "A Comparison of Quasi-Z-Source Inverter and Traditional Two-Stage Inverter for Photovoltaic Application", PCIM Europe, 19–21 May 2015, Nuremberg, Germany, pp.1580–1587.
5. M. Zhu, K. Yu, and F. L. Luo, "Switched inductor Z-source inverter," IEEE Trans. Power Electronics., Vol. 25, No. 8, pp. 2150–2158, August 2010.
6. Hailong Liu, Yuyao He, Mingyang Zhang Liu: "Quasi-Switched-Inductor Z-Source Inverter", IEEE 8th International Power Electronics and Motion Control Conference (IPEMC-ECCE Asia), 2016.
7. M. K. Nguyen, Y. C. Lim, and G. B. Cho, "Switched-inductor quasi-Z-source inverter," IEEE Trans. Power Electronics., vol. 26, no. 11, pp. 3183–3191, November 2011.
8. Anh-Vu Ho and Tae-Won Chun: "Topologies of Active-Switched Quasi-Z-source Inverters with High-Boost Capability"; Journal of Power Electronics, Vol. 16, No. 5, pp. 1716–1724, September 2016.
9. Mathworks Inc., www.mathworks.com: "MATLAB/SIMULINK", R2020b, September 30, 2020.
10. C. J. Gajanayake, F. L. Luo, H. B. Gooi, P. L. So, and L. K. Siow, "Extended boost Z-source inverters," IEEE Trans. Power Electronics., Vol. 25, No. 10, pp. 2642–2652, Oct. 2010.
11. F.L. Luo and H. Ye: "Power Electronics: Advanced Conversion Technologies", CRC Press, 2010, pp. 460–477.
12. J. Yuan, Y. Yang and Frede Blaabjerg: "A Switched Quasi Z-Source Inverter with Continuous Input Currents", MDPI, Energies, 13(6), March 2020 pp. 1–12.

Chapter 7
Six-Step Inverter-Fed Permanent Magnet Synchronous Motor and Brushless DC Motor Drives

7.1 Introduction

Permanent magnet synchronous motors (PMSM) and brushless DC motors (BLDCM) find wide applications in the industry. In PMSM and BLDCM, the rotor is made of permanent magnets such as neodymium-iron-boron alloy (NdFeB) and thus avoids rotor field winding. The interior permanent magnet synchronous motor supplied by a six-step inverter with electronic commutation is used as variable speed drive in pumps and fans [1]. Permanent magnet brushless DC motors are used in laser printers, hard disc drives and electric vehicles [2, 27]. Electronic switching of the six-step inverter is controlled by the rotor position which is sensed by using either the optical or the Hall effect sensors [2–5]. Many sensorless techniques for inverter switching such as the back e.m.f. detection, indirect flux detection by online reactance measurement (INFORM), Kalman filtering, phase current measurement and conduction state of freewheeling diodes have been reported in the literature. The PMSM has a sinusoidal back e.m.f. and BLDCM has a trapezoidal back e.m.f. The six-step inverter switching used for PMSM drives can be either 180-degree mode or the 120-degree mode, whereas for BLDCM it is 120-degree mode. Modelling of PMSM drives fed by six-step continuous and discontinuous current mode inverter using various computer programming methods and also using six-step inverters with fixed and variable switching angles is reported in the literature [4–15]. Modelling of three-phase inverter-fed BLDCM drive fed by hysteresis and Pulse Width Modulated (PWM) current controllers and speed control above rated speed by advancing the phase angle of stator phase current ahead of back e.m.f. are also reported [20–22, 25, 26].

Supplementary Information The online version contains supplementary material available at https://doi.org/10.1007/978-3-031-62784-2_7.

N. P. R. Iyer, *Inverters and AC Drives*, Power Systems,
https://doi.org/10.1007/978-3-031-62784-2_7

In this chapter models with simulation and experimental results for PMSM drives supplied by six-step continuous and discontinuous current mode inverter, model of a three-phase vector controlled PMSM, case study for a three-phase space vector pulse width modulated (SVPWM) inverter-fed vector controlled PMSM drive and models for three-phase inverter-fed BLDCM drive in open loop and with hysteresis current controller are presented. This is followed by a case study for a three-phase inverter-fed BLDCM drive with hysteresis current control and switching angle advancing facility.

7.2 Six-Step Inverter-Fed Permanent Magnet Synchronous Motor Drive

Three-phase inverter-fed PMSM drive method also applicable to permanent magnet brushless DC motor (PMBLDCM) drive is shown in Fig. 7.1 [1–5]. A BLDCM inverter is driven by generating gate pulse coupled with rotor position. The voltage must be properly applied to the three-phase windings of the BLDCM stator such that the magnetic flux produced by stator coils and that produced by rotor permanent

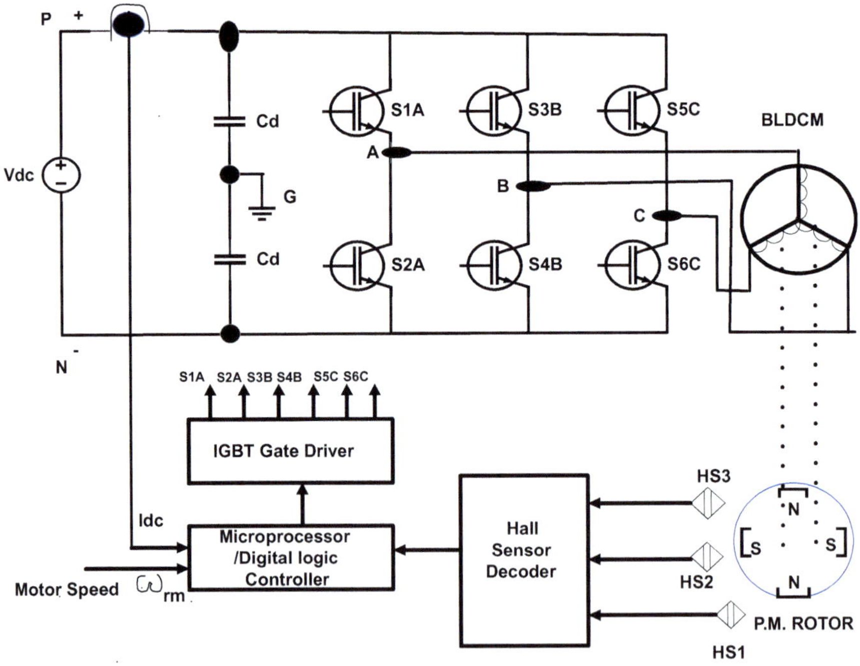

Fig. 7.1 Three-phase inverter-fed permanent magnet brushless DC motor drive

magnets make an angle of $\pi/2$ radians to develop maximum electromagnetic torque. This requires that the BLDCM controller must be able to determine the position of the rotor shaft relative to the stator coils. This rotor position information is detected by (1) using mechanical encoder with photosensors and (2) by Hall effect sensors. Sensorless rotor position detection is also possible by sensing the back e.m.f. of the BLDC motor and the DC input current using voltage and current sensors. Hall effect sensors are mounted on the stator near the air gap of the BLDCM,and when the rotor magnet rotates, the rotor position is detected by the Hall sensor which produces a Hall voltage perpendicular to the direction of the current flow through the Hall element and the rotor magnetic field. This Hall voltage is amplified and digitised by using operational amplifiers (op.amp) comparator and Schmitt trigger. The voltage produced by the Hall sensor is on for 180 electrical degree of rotation and off for the next 180-degree rotation of the rotor. In Fig. 7.1, three Hall sensors are mounted near the stator air gap at 60-degree interval which divides the 360-degree rotation of the rotor into six intervals each of 60 degrees. The process of switching the current to flow through any two phases of the stator winding for every 60-degree rotation of the rotor is called electronic commutation. The switching action of the three-phase inverter is triggered by the signals from the three Hall position sensors mounted at 60-degree intervals on the stator. When properly aligned with stator three-phase windings, these Hall sensors produce digital pulses which can be decoded, and these decoded pulses along with the output of rotor speed sensor and DC link current information from current sensor can be given as input to a microprocessor or else to a digital logic circuit to generate necessary switching pulses for the three-phase inverter.

7.2.1 Analysis of Six-Step Continuous Current Mode Inverter-Fed PMSM Drive

In deriving the dq-axis model of PMSM drives, the following assumptions are made:

(1) Back e.m.f. is sinusoidal. (2) There are equal turns per phase. (3) Rotor flux is concentrated along d-axis. (4) There is NO flux along the q-axis. (5) Core loss is negligible. (6) There is constant rotor flux. (7) Rotor reference frame is used to express stator voltage eqs. (8) The d-q axis stator windings have fixed phase relationship with rotor magnet axis or d-axis. The modelling equations of PMSM all in rotor frame are given below [4–12]:

$$v_{qs} = R_S * i_{qs} + S.\lambda_{qs} + \omega_{re} * \lambda_{ds} \tag{7.1}$$

$$v_{ds} = R_S * i_{ds} + S.\lambda_{ds} - \omega_{re} * \lambda_{qs} \tag{7.2}$$

where v_{ds} and v_{qs} are the dq-axis stator voltages and ω_{re} is the angular speed of the rotor in electrical radians per second. The dq-axis flux linkages are expressed as follows:

$$\lambda_{qs} = L_q * i_{qs} \tag{7.3}$$

$$\lambda_{ds} = L_d * i_{ds} + \lambda_m \tag{7.4}$$

where λ_m is the amplitude of the stator flux linkages established by the permanent magnet, which is a constant. Using Eqs. 7.3 and 7.4 in Eqs. 7.1 and 7.2, Eq. 7.5 follows:

$$\begin{bmatrix} v_{qs} \\ v_{ds} \end{bmatrix} = \begin{bmatrix} R_s + s.L_q & \omega_{re} * L_d \\ -\omega_{re} * L_q & R_s + s.L_d \end{bmatrix} * \begin{bmatrix} i_{qs} \\ i_{ds} \end{bmatrix} + \begin{bmatrix} \omega_{re} * \lambda_m \\ 0 \end{bmatrix} \tag{7.5}$$

In Eq. 7.5, R_s, L_d and L_q are the resistance and inductance of the stator in dq-axis, respectively.

The electromagnetic torque is given by Eq. 7.6:

$$T_{em} = \left[\frac{3}{2}\right] * \left[\frac{P}{2}\right] * \left(\lambda_{ds} * i_{qs} - \lambda_{qs} * i_{ds}\right) \tag{7.6}$$

where P is the number of poles. Using Eqs. 7.3 and 7.4 in Eq. 7.6, Eq. 7.7 follows:

$$T_{em} = \left[\frac{3}{2}\right] * \left[\frac{P}{2}\right] * \left(\lambda_m * i_{qs} + \left(L_d - L_q\right) * i_{qs} * i_{ds}\right) \tag{7.7}$$

The rotor e.m. torque T_{em} and speed are related as follows:

$$T_{em} = J * \frac{d\omega_{rm}}{dt} + D * \omega_{rm} + T_{mech} \tag{7.8}$$

where ω_{rm} is the rotor speed in mechanical radians per second, J is rotor inertia, D is damping, and T_{mech} is the mechanical load torque. The essential modelling equations of the PMSM with sinusoidal back e.m.f. are given in Eqs. 7.1 to 7.8. The phasor diagram of the PMSM is shown in Fig. 7.2 [7]. The vector control of PMSM is derived from its dynamic model. Considering the currents as inputs, the three-phase currents are given below in Eq. 7.9.

$$\left.\begin{aligned} i_{as} &= i_S * \sin(\omega_{re} * t + \delta)) \\ i_{bs} &= i_S * \sin\left(\omega_{re} * t + \delta - \frac{2\pi}{3}\right)) \\ i_{cs} &= i_S * \sin\left(\omega_{re} * t + \delta + \frac{2\pi}{3}\right)) \end{aligned}\right\rangle \tag{7.9}$$

Transforming Eq. 7.9 to its equivalent dq-axis component, Eq. 7.10 follows:

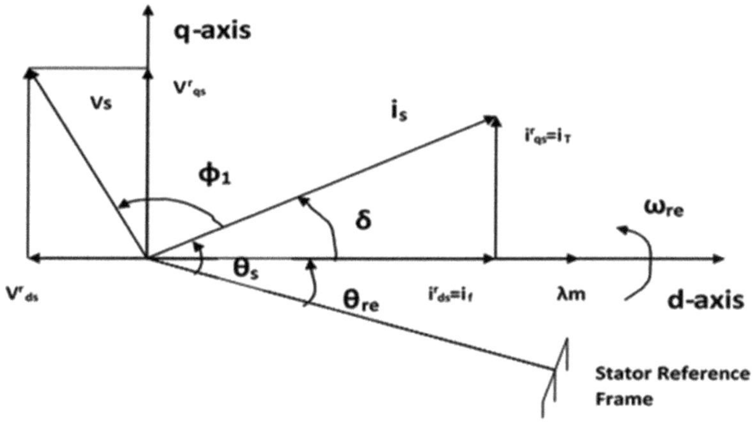

Fig. 7.2 Phasor diagram of PMSM

$$\begin{bmatrix} i_{qs} \\ i_{ds} \end{bmatrix} = \frac{2}{3} * \begin{bmatrix} \cos(\omega_{re} * t) & \cos\left(\omega_{re} * t - \frac{2\pi}{3}\right) & \cos\left(\omega_{re} * t + \frac{2\pi}{3}\right) \\ \sin(\omega_{re} * t) & \sin\left(\omega_{re} * t - \frac{2\pi}{3}\right) & \sin\left(\omega_{re} * t + \frac{2\pi}{3}\right) \end{bmatrix} * \begin{bmatrix} i_S * \sin(\omega_{re} * t + \delta) \\ i_S * \sin\left(\omega_{re} * t + \delta - \frac{2\pi}{3}\right) \\ i_S * \sin\left(\omega_{re} * t + \delta + \frac{2\pi}{3}\right) \end{bmatrix} \tag{7.10}$$

Simplifying Eq. 7.10, the dq-axis stator currents in **rotor reference frame** are obtained as shown in Eq. 7.11.

$$\begin{bmatrix} i_{qs} \\ i_{ds} \end{bmatrix} = i_S * \left[\sin(\delta) \cos(\delta) \right] \tag{7.11}$$

Using Eq. 7.11 in Eq. 7.7, the electromagnetic torque equation is obtained as shown in Eq. 7.12.

$$T_{em} = \frac{3}{2} * \frac{P}{2} * \left[\frac{1}{2} * \left(L_d - L_q\right) * i_S^2 * \sin 2\delta + \lambda_m * i_S * \sin\delta\right] \tag{7.12}$$

For $\delta = \pi/2$, the equation for T_{em} is modified as in Eq. 7.13.

$$T_{em} = \left[\frac{3}{2}\right] * \left[\frac{P}{2}\right] * \lambda_m * i_s \text{ Nw} - \text{meter} \tag{7.13}$$

Thus by maintaining δ, a value of $\pi/2$, the e.m. torque T_{em} can be controlled by varying the stator current, as λ_m is a constant.

The PMSM drive is inverter driven and the sensors provide information on the position of the rotor poles and the dq-axis. By appropriately switching the six-step inverter, it is possible to change v_{qs} and v_{ds} and the rotor position. The line to neutral voltage of a PMSM driven by six-step 180-degree mode inverter can be expressed as in Eq. 7.14 [4–16].

$$V_{as} = \frac{2 * V_{dc}}{\pi} * \left[\sin(\theta_{ev}) + \frac{1}{5} * \sin(5\theta_{ev}) + \frac{1}{7} * \sin(7\theta_{ev}) + \ldots\right] \tag{7.14}$$

In Eq. 7.14, $\boldsymbol{\omega_e.t} = \boldsymbol{\theta_{ev}}$ is the switching angle or the phase angle of the applied voltage. If switching angle is advanced by $\pi/2$ radians by appropriately switching the inverter switches, then θ_{ev} in Eq. 7.14 can be replaced by $(\boldsymbol{\theta_{ev}} + \boldsymbol{\pi/2})$. Equation 7.14 can be written as in Eq. 7.15 [6].

$$V_{as} = \frac{2 * V_{dc}}{\pi} * \left[\cos(\theta_{ev}) + \frac{1}{5} * \cos(5\theta_{ev}) + \frac{1}{7} * \cos(7\theta_{ev}) + \ldots\right] \tag{7.15}$$

In Eqs. 7.14 and 7.15, ω_e is the angular frequency of switching of the inverter in electrical radians per second and is equal to ω_{re}, the angular frequency of the rotor in electrical radians per second. The dq-axis stator voltages in rotor frame can be expressed as in Eq. 7.16.

$$\left.\begin{aligned} V_{qs} &= \sqrt{2} * V_s * \cos(\varphi) \\ V_{ds} &= -\sqrt{2} * V_s * \sin(\varphi) \\ \sqrt{2} * V_s &= \frac{2 * V_{dc}}{\pi} \\ \varphi &= \theta_{ev}(0) - \theta_{re}(0) \end{aligned}\right\} \tag{7.16}$$

The steady state voltage equations neglecting harmonics can be expressed as in Eq. 7.17, by letting $s = d/dt = 0$ in Eq. 7.5.

$$\begin{bmatrix} V_{qs} \\ V_{ds} \end{bmatrix} = \begin{bmatrix} R_s & \omega_{re} * L_d \\ -\omega_{re} * L_q & R_s \end{bmatrix} * \begin{bmatrix} I_{qs} \\ I_{ds} \end{bmatrix} + \begin{bmatrix} \omega_{re} * \lambda_m \\ 0 \end{bmatrix} \tag{7.17}$$

Now letting $\mathbf{L_d} = \mathbf{L_q} = \mathbf{L_s}$ and solving Eq. 7.17, the equation for I_{qs} can be written as in Eq. 7.18.

$$I_{qs} = \frac{R_S}{\left(R_S^2 + \omega_{re}^2 * L_S^2\right)} * \left[V_{qs} - \omega_{re} * \lambda_m - \frac{\omega_{re} * L_S * V_{ds}}{R_S}\right] \tag{7.18}$$

The e.m. torque T_{em} in Eq. 7.7 can be written as in Eq. 7.19.

$$T_{em} = \left(\frac{3 * P}{4}\right) * \frac{R_S * \lambda_m}{\left(R_S^2 + \omega_{re}^2 * L_S^2\right)} * \left[V_{qs} - \omega_{re} * \lambda_m - \frac{\omega_{re} * L_S * V_{ds}}{R_S}\right] \tag{7.19}$$

Using Eq. 7.16 in Eq. 7.19, the torque expression simplifies as in Eq. 7.20.

$$T_{\mathrm{em}} = \left(\frac{3 * P}{4}\right) * \frac{R_{\mathrm{S}} * \lambda_{\mathrm{m}}}{\left(R_{\mathrm{S}}^2 + \omega_{\mathrm{re}}^2 * L_{\mathrm{S}}^2\right)} * \left[\sqrt{2} * V_{\mathrm{S}} * \cos(\varphi) - \omega_{\mathrm{re}} * \lambda_{\mathrm{m}} + \frac{\omega_{\mathrm{re}} * L_{\mathrm{S}} * \sqrt{2} * V_{\mathrm{S}} * \sin(\varphi)}{R_{\mathrm{S}}}\right] \tag{7.20}$$

For maximum e.m. torque T_{em} differentiating Eq. 7.20 with respect to φ and equating to zero, Eq. 7.21 is obtained [4, 6]:

$$\varphi = \varphi_{\mathrm{MT}} = \tan^{-1}\left[\frac{\omega_{\mathrm{re}} * L_{\mathrm{S}}}{R_{\mathrm{S}}}\right] \tag{7.21}$$

where φ_{MT} is the value of φ for maximum e.m. torque.

7.2.2 Modelling of Six-Step Continuous Current Mode Inverter-Fed PMSM Drive

The model of the six-step continuous current (180-degree) mode [16] inverter fed PMSM drive is shown in Fig. 7.3 (model file: EXAMPLE 7_1). The various dialog boxes are shown in Fig. 7.4. The various subsystems are shown in Fig. 7.5a–j. The various subsystems are explained below [11, 12, 17]:

Figure 7.5a, b gives phase advancer and gate drive block. The dialog boxes Three Phase 180 Degree Mode Inverter Gate Drive Data and Three Phase 180 Degree Mode Inverter Data in Fig. 7.4 correspond to Fig. 7.5a and b, respectively. In Fig. 7.5a, an arbitrary constant *K*, rotor speed ω_{re}, time and phase advancer are given as input to the mux. This mux output is given in Fig. 7.5b. In Fig. 7.5b, the outputs R, G and B of Fcn blocks are three-phase sine wave with peak value of 10 and angular frequency of ω_{re} electrical radians per second, with phase advance entered in the phase shifter block of Fig. 7.5a added. The outputs a, b and c of Fcn blocks are logic 1, when each of the three-phase inputs just crosses zero and goes positive and zero when the respective input crosses zero and goes negative. The gate drives a, b and c are given to three-phase continuous current mode inverter block shown in Fig. 7.5c [15]. Gate drives a, b and c are given to the second input u(2) of respective threshold switch blocks, while the first u(1) and third u(3) inputs are at $+V_{\mathrm{dc}}/2$ and $-V_{\mathrm{dc}}/2$ volts, respectively, where V_{dc} is the DC link voltage. The outputs of the switches are $+V_{\mathrm{dc}}/2$ when the respective u(2) input is greater than or equal to the threshold value of 0.5; else the outputs are $-V_{\mathrm{dc}}/2$. The output of the three switches is three-phase line to ground voltages, where ground is the midpoint of the two series-connected sources $+V_{\mathrm{dc}}/2$ and $-V_{\mathrm{dc}}/2$, respectively. Using the SUBTRACT blocks, the respective three-phase line to line voltages are derived. The three line to line voltages are given to VLL to VLN block shown in Fig. 7.5d, and the three line to ground voltages are given to ABC to DQ block shown in Fig. 7.5e. The VLL

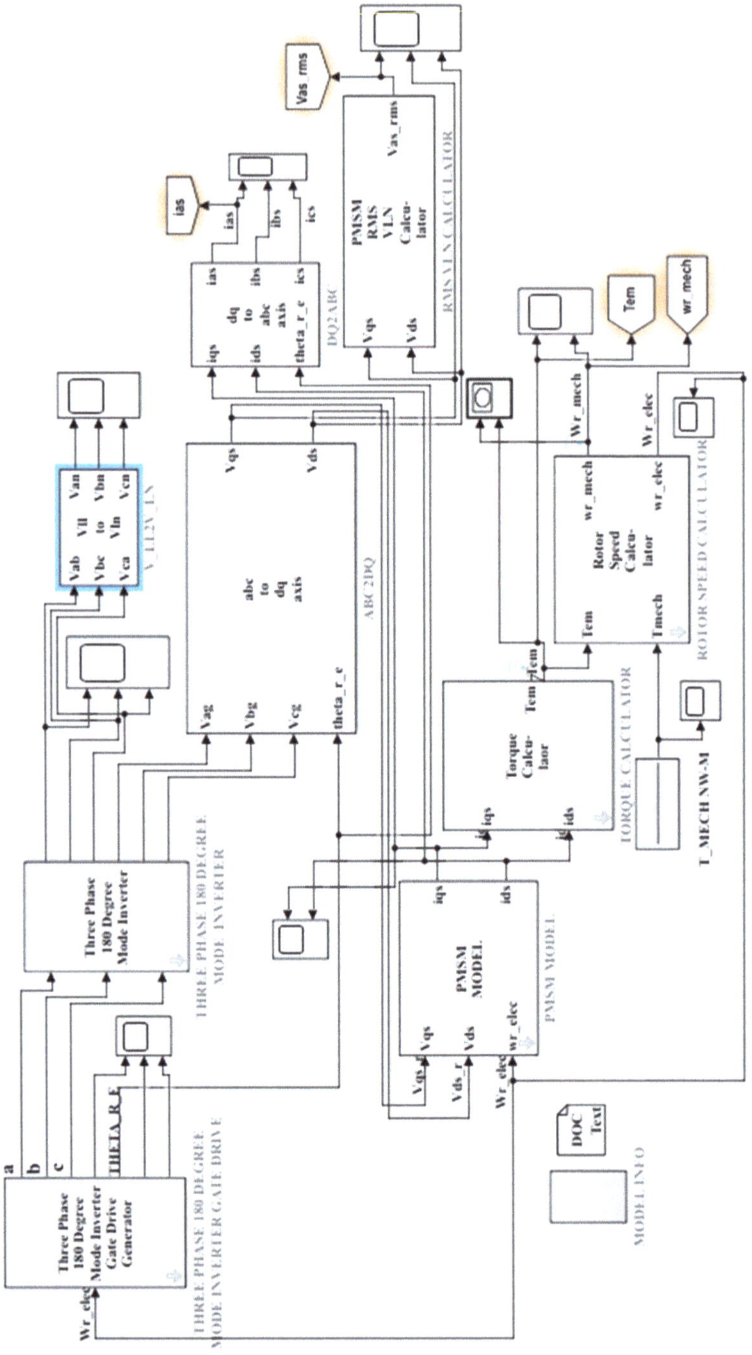

Fig. 7.3 Three-phase six-step continuous current mode inverter-fed PMSM drive model

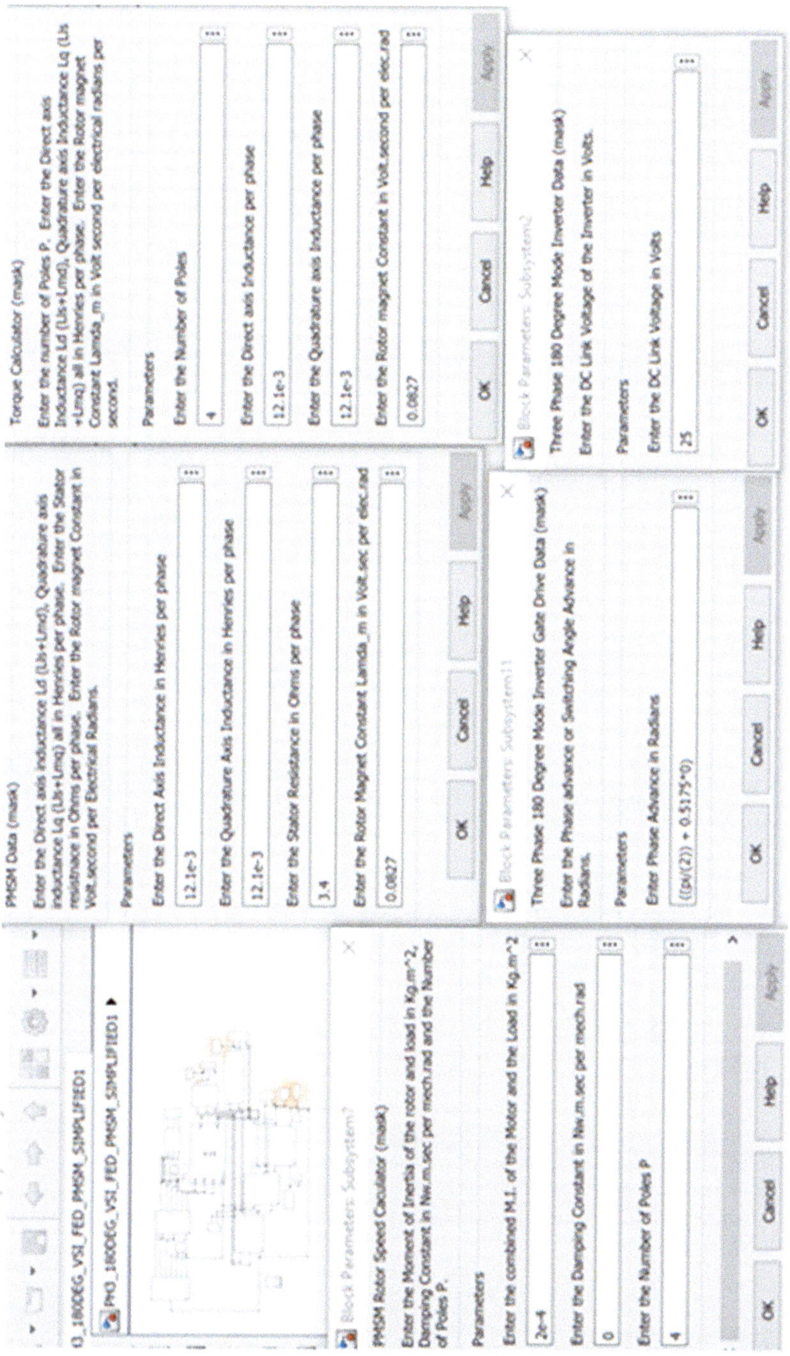

Fig. 7.4 Model of six-step continuous current mode inverter-fed PMSM drive with interactive dialog box

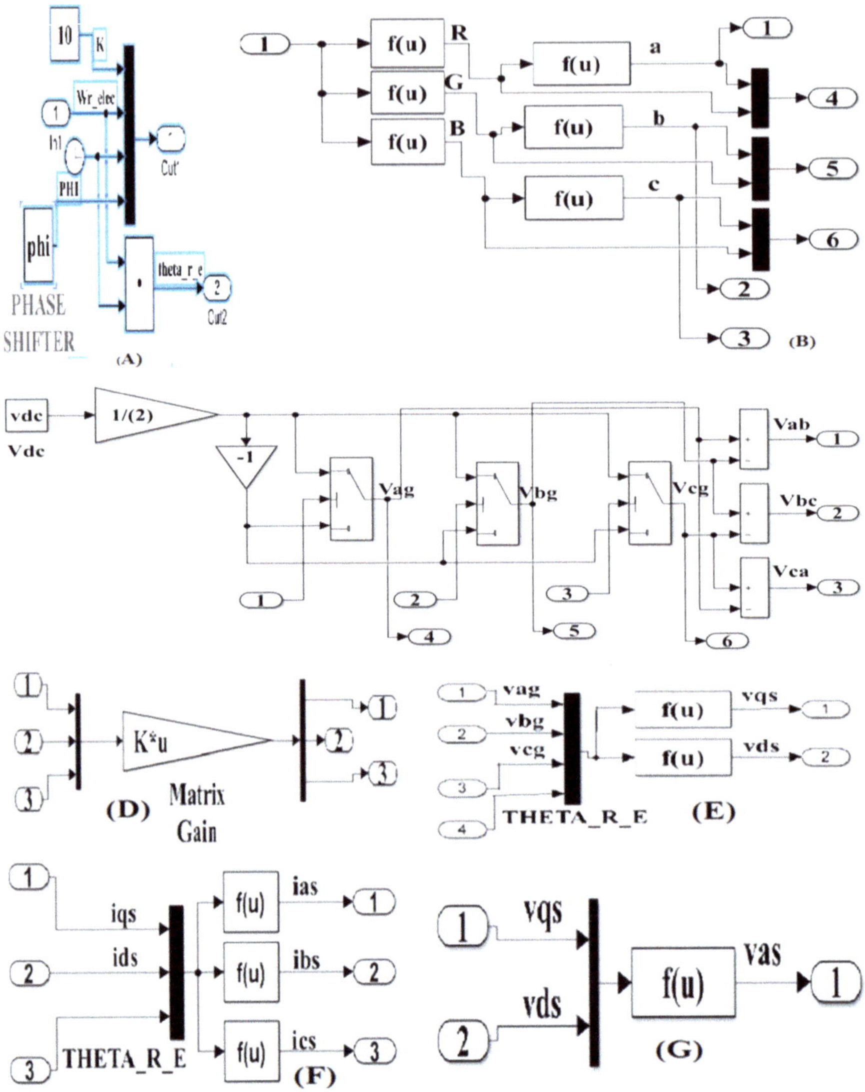

Fig. 7.5 (a, b) Six-step continuous current mode inverter gate drive. (c) Six-step 180-degree mode inverter. (d) Line-to-line to line-to-neutral voltage transformation. (e) abc axis to dq axis voltage transformation. (f) dq axis to abc axis current transformation. (g) dq axis voltage to line to neutral voltage transformation. (h) Three-phase PMSM model. (i) Rotor electromagnetic torque calculator. (j) Rotor speed calculator

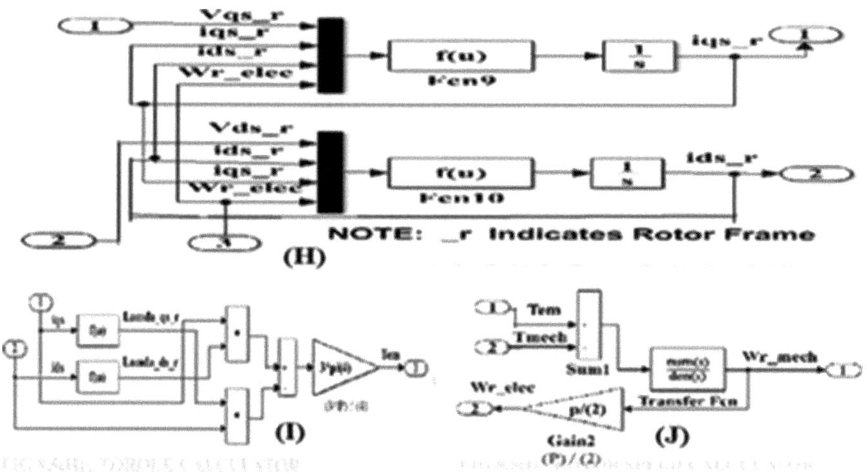

Fig. 7.5 (continued)

to VLN block shown in Fig. 7.5d transforms line to line voltage to line to neutral voltage. The matrix gain block essentially transforms the three-phase line to line voltage input to the mux block to corresponding line to neutral voltage output at the Demux block. The transformation matrix K in Fig. 7.5d is given by Eq. 1.4 in Chap. 1. The ABC to DQ transform block is shown in Fig. 7.5e. The four inputs to the mux are the three-phase line to ground voltages from inverter block and the rotor position θ_{re} in electrical radians. The two Fcn blocks generate V_{qs} and V_{ds} as per the transformation matrix given in Eq. 7.22

$$\begin{bmatrix} V_{\text{qs}} \\ V_{\text{ds}} \end{bmatrix} = \frac{2}{3} * \begin{bmatrix} \cos(\theta_{\text{re}}) & \cos\left(\theta_{\text{re}} - \frac{2\pi}{3}\right) & \cos\left(\theta_{\text{re}} + \frac{2\pi}{3}\right) \\ \sin(\theta_{\text{re}}) & \sin\left(\theta_{\text{re}} - \frac{2\pi}{3}\right) & \sin\left(\theta_{\text{re}} + \frac{2\pi}{3}\right) \end{bmatrix} * \begin{bmatrix} V_{\text{ag}} \\ V_{\text{bg}} \\ V_{\text{cg}} \end{bmatrix} \tag{7.22}$$

Figure 7.5f performs the dq-axis to abc-axis transformation of the stator currents. The three inputs to the mux are i_{qs}, i_{ds} and θ_{re}, and the three Fcn blocks perform dq to abc axis transformation of the stator currents as per the transformation matrix defined in Eq. 7.23.

$$\begin{bmatrix} I_{\text{as}} \\ I_{\text{bs}} \\ I_{\text{cs}} \end{bmatrix} = \begin{bmatrix} \cos(\theta_{\text{re}}) & \sin(\theta_{\text{re}}) \\ \cos\left(\theta_{\text{re}} - \frac{2\pi}{3}\right) & \sin\left(\theta_{\text{re}} - \frac{2\pi}{3}\right) \\ \cos\left(\theta_{\text{re}} + \frac{2\pi}{3}\right) & \sin\left(\theta_{\text{re}} + \frac{2\pi}{3}\right) \end{bmatrix} * \begin{bmatrix} I_{\text{qs}} \\ I_{\text{ds}} \end{bmatrix} \tag{7.23}$$

The phase to neutral voltage calculator block is shown in Fig. 7.5g. The qd-axis stator voltages V_{qs} and V_{ds} are the two inputs to the mux, and the Fcn block calculates the RMS value of the phase to neutral voltage across the stator of PMSM as given in Eq. 7.24:

$$v_{as} = \sqrt{\frac{v_{qs}^2 + v_{ds}^2}{2}} \tag{7.24}$$

The PMSM block is shown in Fig. 7.5h. The top mux along with Fcn 9 block and integrator solves for i_{qs}, while the bottom part of mux, Fcn 10 block and integrator solve for i_{ds}, both defined by Eq. 7.5. The torque calculator block is shown in Fig. 7.5i. With i_{qs} and i_{ds} as inputs to two Fcn blocks, Eqs. 7.3 and 7.4 are solved to obtain qd axis flux linkages. Equation 7.6 is solved using two multipliers, summer and a gain block to obtain electromagnetic torque T_{em}. The rotor speed calculator block is shown in Fig. 7.5j. The load torque T_{mech} is applied externally using the repeating sequence block shown in Fig. 7.3. Equation 7.8 is solved using the Subtract block Sum1 which subtracts T_{mech} from T_{em}, and this output is given to the transfer function block with transfer function $[1/(J_s + D)]$ to obtain rotor speed in mechanical radians per second. This rotor speed is then multiplied by P/2 using gain block to obtain rotor speed ω_{re} in electrical radians per second, where P is the number of poles. Rotor position θ_{re} is obtained by integrating speed ω_{re} using integrator block 1/s or by multiplying ω_{re} with time block.

7.2.3 *Simulation Results*

The simulation of PMSM drive fed by six-step continuous current mode inverter was carried out using ode23tb (stiff/TR-BDF2) solver in Simulink [18, 17]. The data relating to the six-step inverter-fed PMSM is shown in Table 7.1 [4, 6]. To verify Eq. 7.21, a rotor speed of 80 mech rad/sec was selected. The simulation of the six-step continuous current mode inverter-fed PMSM drive was carried out at no load with an inverter phase advance of $\pi/2$ electrical radians, and the rotor electromagnetic (EM) torque T_{em} corresponding to 80 mech rad/sec was found to be 0.19

Table 7.1 Six-step 180-degree mode inverter-fed PMSM drive—model parameters

Sl. no.	Parameter	Value	Unit
1	**Number of phases**	**3**	
2	**Rated voltage vs per phase**	**11.25**	Volts
3	**DC link voltage**	**25**	Volts
4	**Stator resistance per phase R_s**	**3.4**	Ohms
5	**Stator inductance per phase L_s ($L_d = L_q$)**	**12.1**	Millihenries
6	**Number of poles P**	**4**	–
7	**Moment of inertia J**	**$2 \times 10e^{-4}$**	$Kg.m^2$
8	Rotor magnet constant λ_m	0.0827	volt sec/elec rad

Nw-metres. The value of φ_{MT} for maximum torque for a rotor speed of 80 mech rad/sec was found to be 0.5175 electrical radians, using Eq. 7.21. This phase advance of 0.5175 was added to $\pi/2$ in the phase shifter block in Fig. 7.5a. From the plot of torque-time and speed-time curves at no load, the rotor EM torque T_{em} corresponding to speed of 80 mech rad/sec was found to be 0.29 Nw-metres. The mechanical load was set to 0.19 Nw-M with a phase advance of $\pi/2$ electrical radians, and the plot of PMSM stator line to neutral voltage, stator current, e.m. torque, rotor speed, torque-speed curve and load torque T_{mech} are shown in Fig. 7.6a–f, respectively. The mechanical load was then set to 0.29 Nw-M with an inverter phase advance of [$\pi/2$ + 0.5175] electrical radians, and the plot of PMSM stator line to neutral voltage, stator current, e.m. torque, rotor speed, torque-speed curve and load torque T_{mech} are shown in Fig. 7.7a–f, respectively. It is seen that by advancing the phase angle of inverter switching by an additional 0.5175 electrical radians corresponding to φ_{MT}, the e.m. torque of motor has increased by 53% as compared to NO additional phase advance, i.e. by maintaining the phase advance of inverter switching by $\pi/2$ electrical radians, with the rotor speed maintained at 80 mech rad/sec in both cases [6].

7.2.4 Discussion of Results

Simulation results shown in Fig. 7.6c–e and that shown in Fig. 7.7c–e indicate that the rotor electromagnetic torque has increased by around 53% by shifting the phase angle of switching of the inverter from $\pi/2$ to ($\pi/2$ + 0.5175) electrical radians while maintaining the rotor speed at 80 mech rad/sec in both cases. This is in confirmation with Eq. 7.21. Also observation of the stator line to neutral voltage waveforms and stator currents in Figs. 7.6a and b and 7.7a and b indicates that their respective frequencies range from 25.93 Hz to 26.08 Hz which gives a range of 162.84 to 163.78 elec rad/sec. These values closely agree with the rotor speed of 160 elec rad/sec. The error in the measurement can be due to discrete pixel movement of the cursor used for measurement.

7.3 Modelling of Six-Step Discontinuous Current Mode Inverter-Fed PMSM Drive

The model of the discontinuous current mode [16] inverter-fed PMSM drive is shown in Fig. 7.8 (model file: EXAMPLE 7_2). The various dialog boxes are shown in Fig. 7.9. The subsystems relating to six-step 120-degree mode inverter and gate drive are shown in Fig. 7.10a, b. The dialog box Three Phase 120 Degree Mode Inverter and Gate Drive in Fig. 7.9 corresponds to this subsystem. All other subsystems are the same as for six-step 180-degree mode inverter shown in Fig. 7.5d–j. The subsystems shown in Fig. 7.10a, b are explained below [13–17].

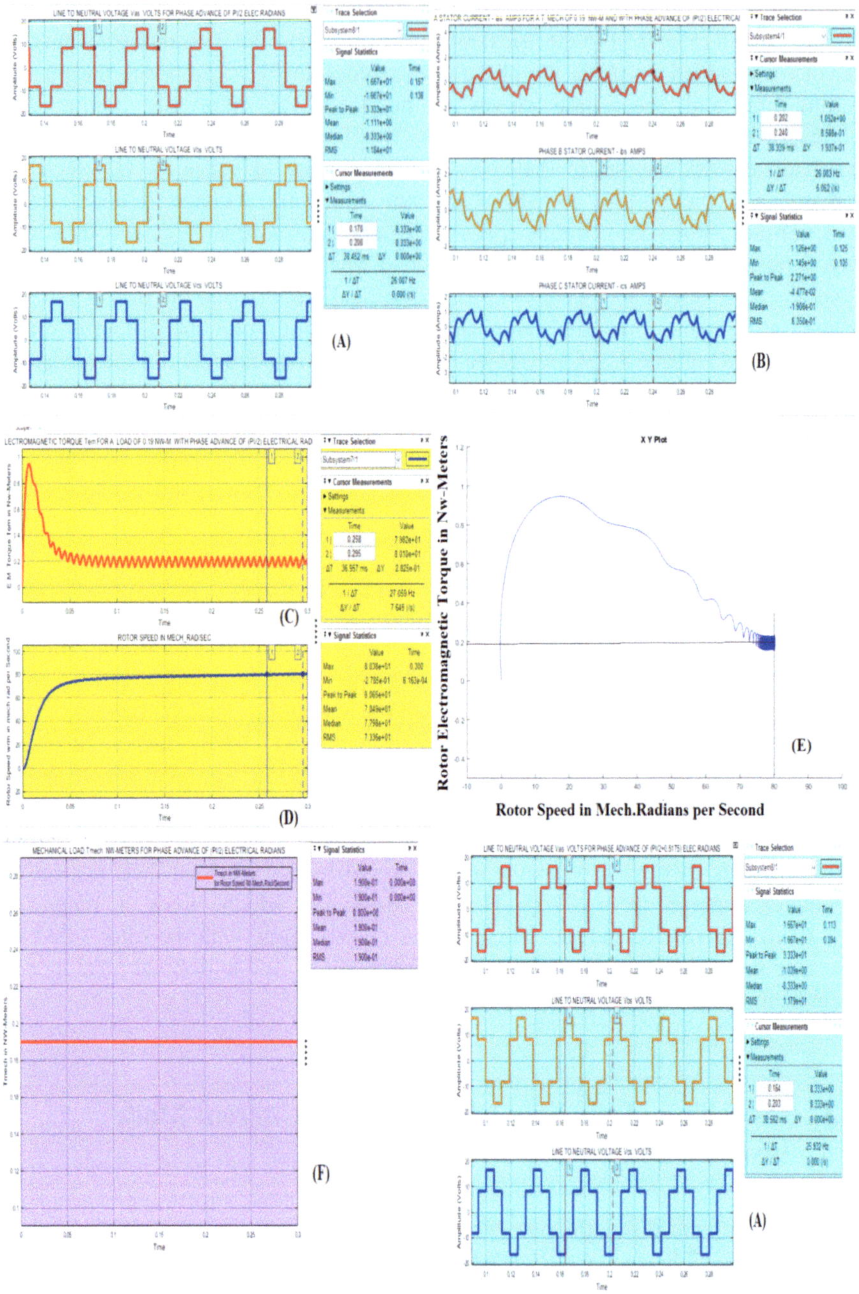

Fig. 7.6 Six-step continuous current mode inverter-fed PMSM drive simulation results: (**a**) Three-phase stator voltages. (**b**) Three-phase stator currents. (**c**) Rotor e.m. torque. (**d**) Rotor speed. (**e**) Rotor torque-speed curve for phase advance pi/2 elec radians, T_{meth} 0.19 Nw-meters and rotor speed of 80 mech.rad / sec.. (**f**) Mechanical load torque for rotor speed of 80 mech rad/sec

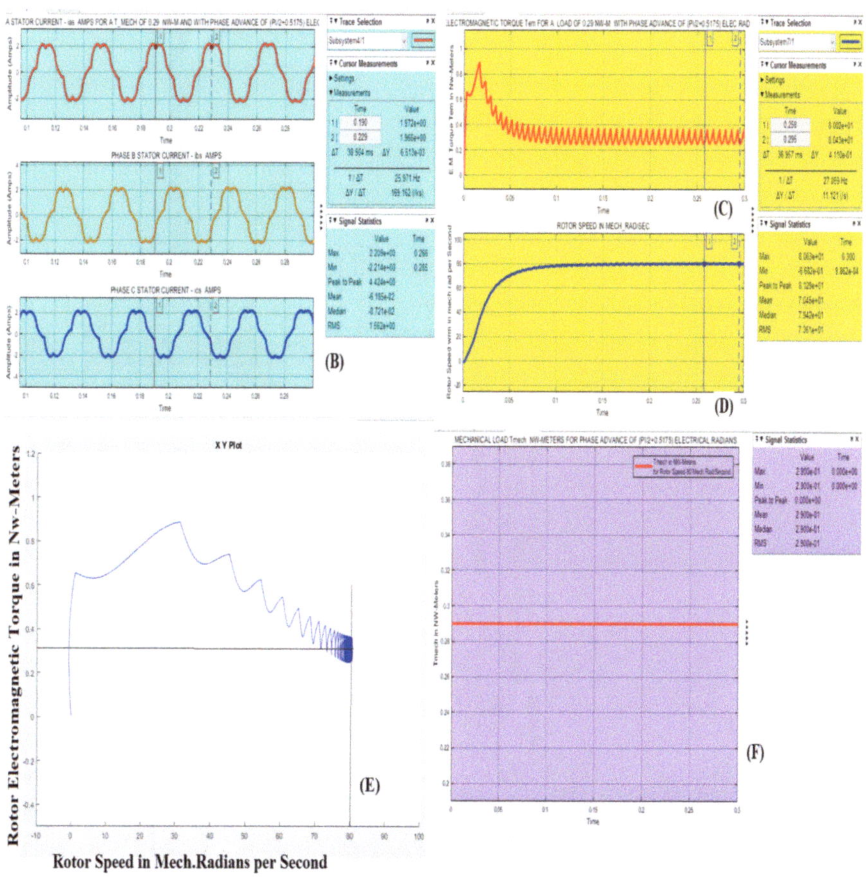

Fig. 7.7 Six-step continuous current-mode inverter-fed PMSM drive simulation results: (**a**) three-phase stator line-to-neutral voltage three-phase stator currents, (**c**) rotor e.m. torque (**d**), rotor speed, (**e**) rotor torque-speed curve for a phase advance of (pi/2 + 0.5175) elec.radans, T_{mech} 0.29 Nw-Meters and rotor speed of 80 mech rad/sec, (**f**) mechanical load torque Tmech for a rotor speed of 80 mech rad/sec

The four inputs to the mux are the arbitrary constant, rotor speed ω_{re}, time and inverter switching angle advance. The three-phase sine wave AC with angular frequency ω_{re} with phase advance phi added is generated using three Fcn blocks, Fcn, Fcn1 and Fcn2. Each of the three-phase AC outputs is then compared in a Schmitt trigger relay comparator used as zero crossing comparators, with output logic 1 and 0, respectively, during the positive and negative half cycle of the input sine wave. The a, b and c outputs of the three relays marked Relay, Relay1 and Relay2 are given to three Fcn blocks, Fcn6, Fcn7 and Fcn8. The three Fcn blocks Fcn6, Fcn7 and Fcn8 subtract the two inputs to their respective mux. The resulting output of these three Fcn blocks is then multiplied by $V_{dc}/2$ using multiplier blocks to

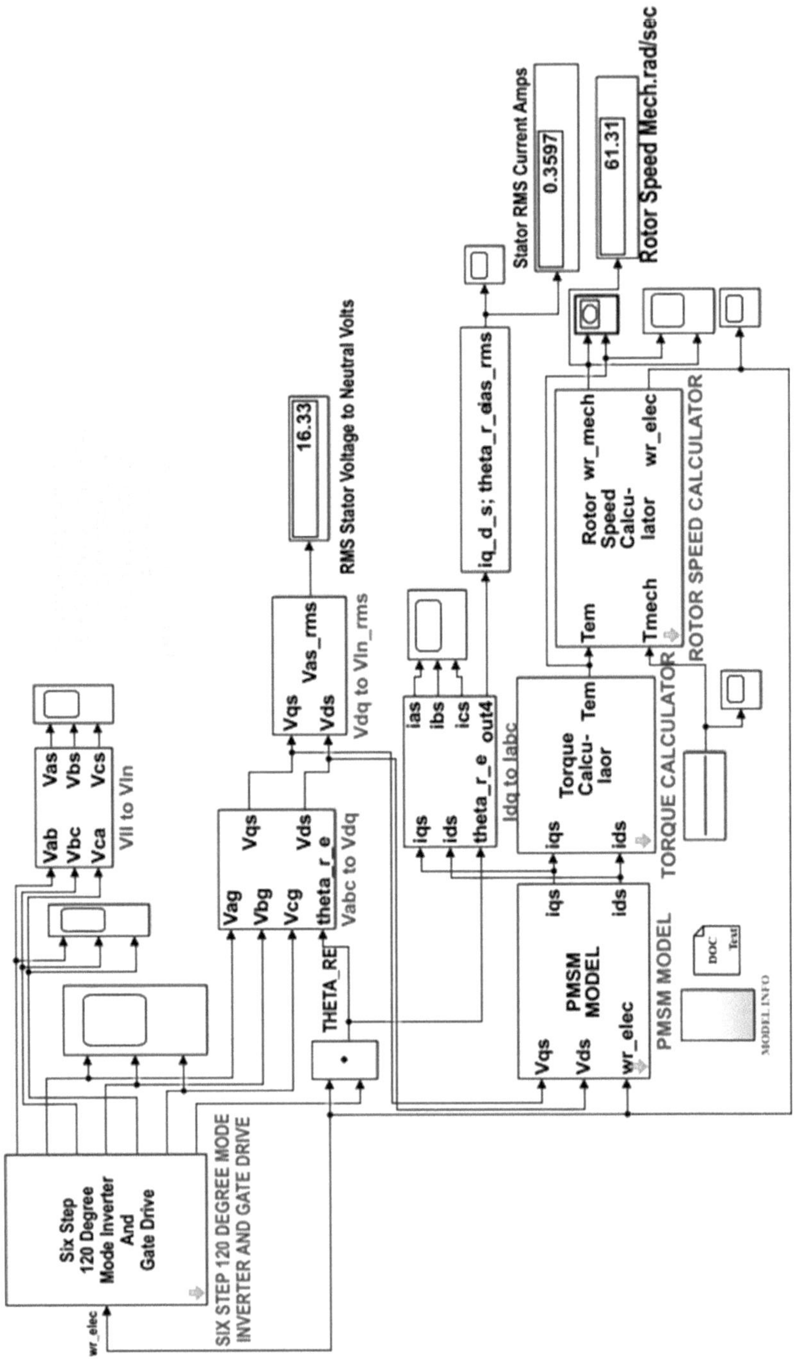

Fig. 7.8 Six-step 120-degree mode inverter-fed PMSM drive

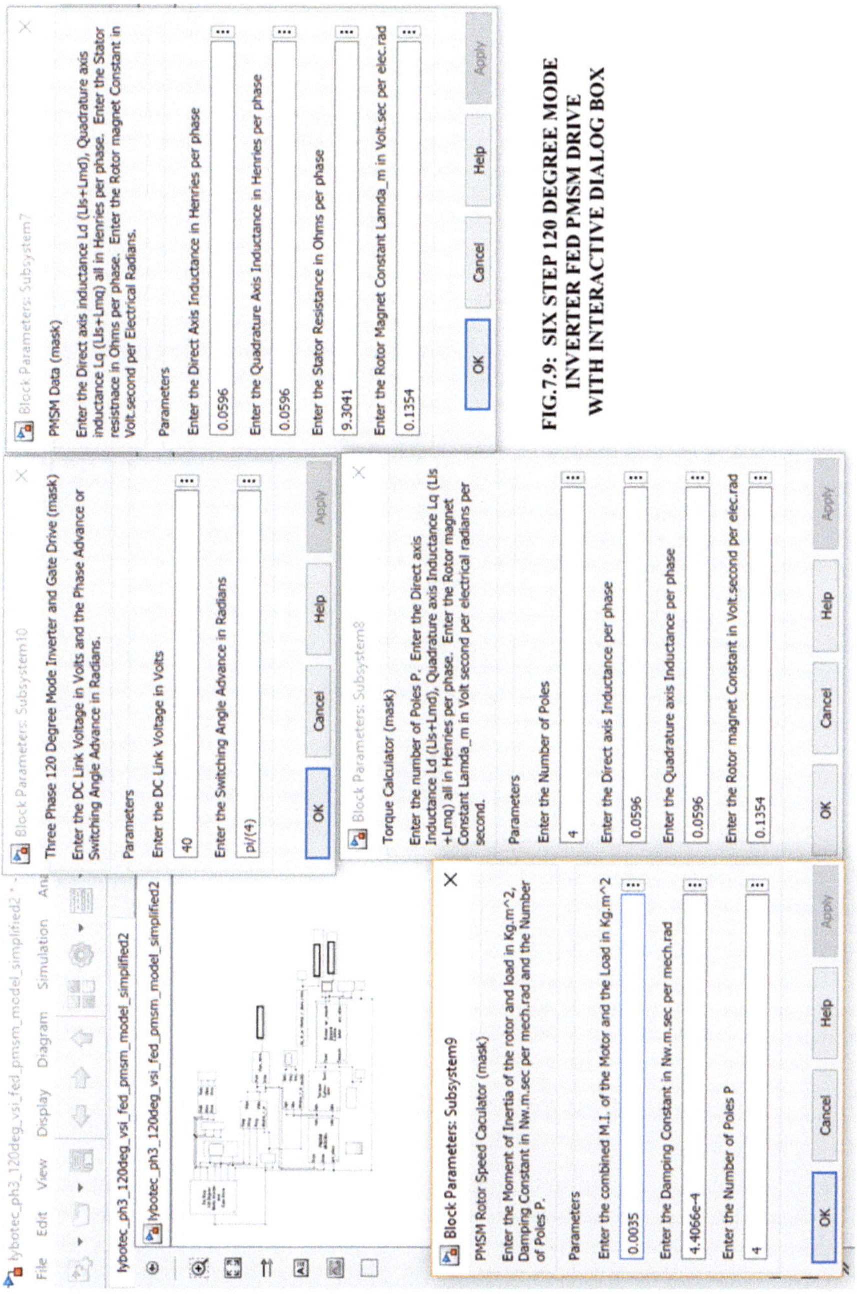

Fig. 7.9 Six-step 120-degree mode inverter-fed PMSM drive with interactive dialog box

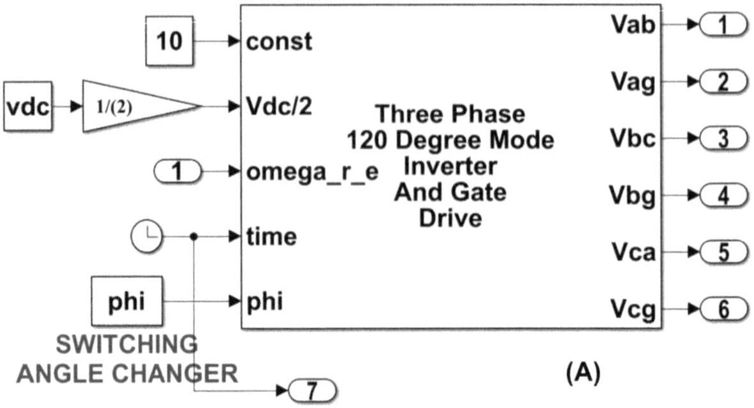

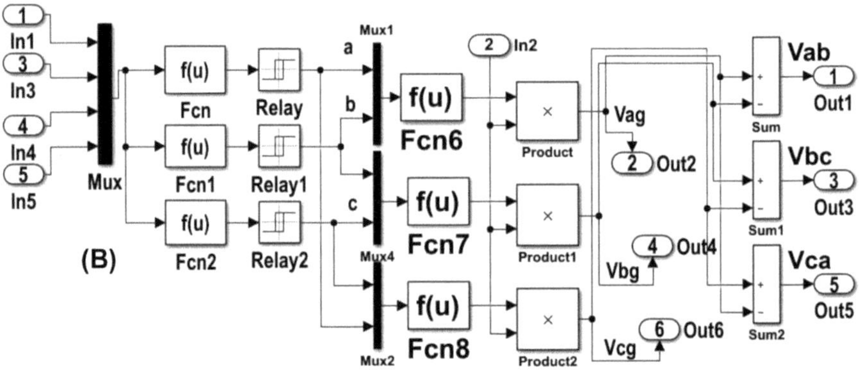

Fig. 7.10 (**a**, **b**) Six-step 120-degree mode gate drive and inverter

generate line to ground voltages V_{ag}, V_{bg} and V_{cg} of the three-phase 120-degree mode inverter, where V_{dc} is the DC link voltage. The three Subtract blocks marked Sum, Sum1 and Sum2 are used to generate the three-phase line to line voltages V_{ab}, V_{bc} and V_{ca}.

7.3.1 Simulation Results

The simulation was carried out using ode15s (stiff/NDF) solver in Simulink [18]. The data relating to the six-step discontinuous current mode Lybotec inverter-fed PMSM drive is shown in Table 7.2. The simulation of this Lybotec inverter-fed PMSM was carried out for phase advance of $\pi/4$ electrical radians. These simulation results are compared with that obtained experimentally in the laboratory. The simulation of line to neutral voltage of PMSM stator, stator current, rotor electromagnetic torque, rotor speed, rotor torque-speed curve for zero mechanical load and inverter switching angle advance of $\pi/4$ electrical radians is shown in Fig. 7.11a–e, as well as the recorded value of the RMS. Line to neutral voltage of the PMSM was found to be 16.33 volts which corresponds to $V_{dc}/\sqrt{6}$ volts [15, 16].

Table 7.2 Six-step 120-degree mode inverter-fed PMSM drive—model parameters

Sl. no.	Parameters	Value	Unit
1	**Number of phases**	**3**	–
2	**No. of poles P**	**4**	–
3	**Stator resistance R_s**	**9.3041**	Ohms/Ph
4	**Stator inductance L_s**	**0.0596**	H/Ph
5	**Rotor magnet constant λ_m**	**0.1354**	V.sec/elec.rad
6	**Damping constant D**	**0.00044066**	Nw.M.sec/mech.rad
7	**Rotor inertia J**	**0.0035**	$Kg.m^2$
8	DC link voltage V_{dc}	40	Volts

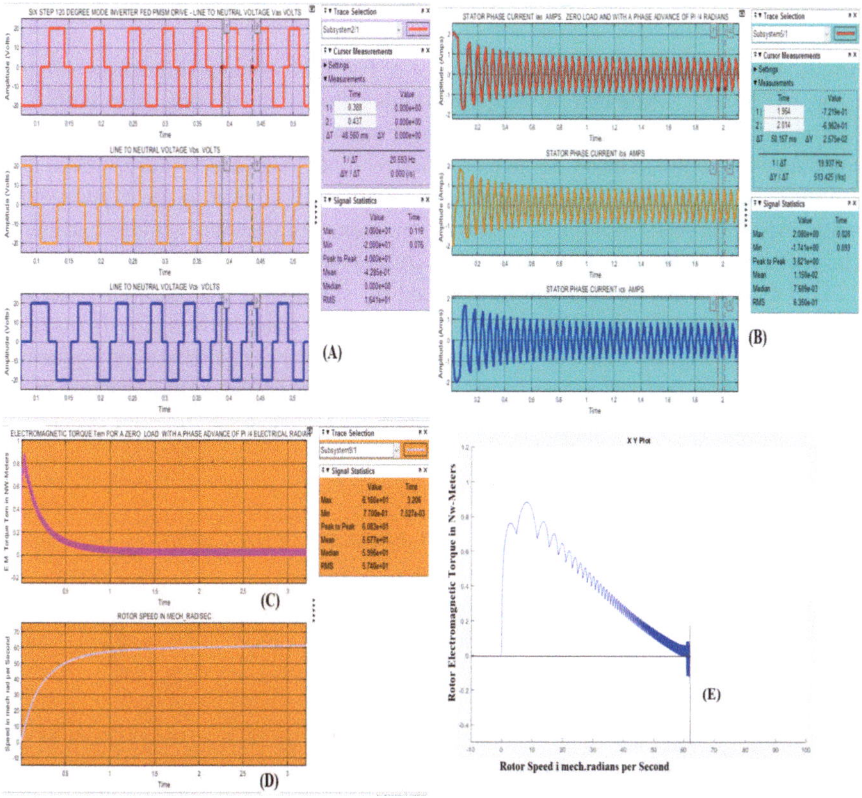

Fig. 7.11 Six-step 120-degree mode inverter-fed PMSM Drive simulation results: (**a**) Three-phase stator line to neutral voltage, (**b**) three-phase stator current, (**c**) rotor e.m. torque, (**d**) rotor speed, and (**e**) rotor torque-speed curve for zero mechanical load and inverter switching phase advance of pi/4 electric radians

7.3.2 Experimental Verification

The performance of the Lybotec six-step inverter-fed PMSM drive was experimentally verified in the laboratories [13–15, 17]. The experimental connection of the six-step Lybotec inverter-fed PMSM is shown in Fig. 7.12. The DC link voltage was maintained at 40 volts. The mechanical load was zero. The experimental readings observed are tabulated as shown in Table 7.3. The experimental set-up is shown in Fig. 7.13a and the oscilloscope waveform of the stator Phase A current is shown in Fig. 7.13b.

7.3.3 Discussion of Results

The RMS value of the line to neutral voltage of PMSM by model simulation is found to be 16.33 volts, while that by experiment this value is 17.09 volts. Also with zero mechanical load and inverter switching angle advance of $\pi/4$ electrical radians, the rotor speed is found to be 61.31 mech rad/sec by model simulation and 54 mech rad/sec by experiment which gives an error of 13.5% with respect to the experimental value. Also it is seen from Fig. 7.11a, b the frequency of the stator voltage and stator currents is 20.53 Hz and 19.93 Hz, respectively, which give a value of 128.93 and 125.16 elec rad/sec whereas the rotor speed by simulation is 122.62 elec rad/sec. Thus the frequency of stator line to neutral voltage and stator current closely matches with the rotor angular frequency. The discrepancy in the measurement of frequency of stator current and stator line to neutral voltage can be due to the discrete pixel movement of the cursor used for measurement. Lybotec six-step inverter provides a sensorless control of PMSM drive.

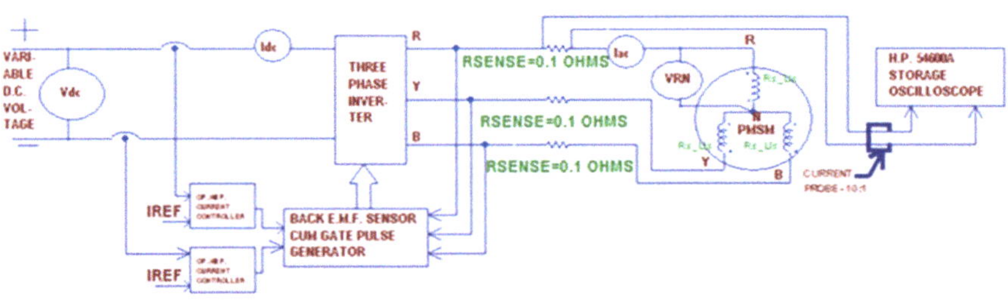

Fig. 7.12 Six-step Lybotec inverter-fed PMSM drive—stator current waveform measurement

Table 7.3 Six-step Lybotec inverter-fed PMSM drive—experimental results

Sl. no.	V_{dc} Volts	I_{dc} Amps	V_{rn} or V_s Volts	I_{ac} Amps	Speed rpm	Speed mech rad/sec	Tmech Nw.m
1	40	0.044	17.09	0.042	517	54	0

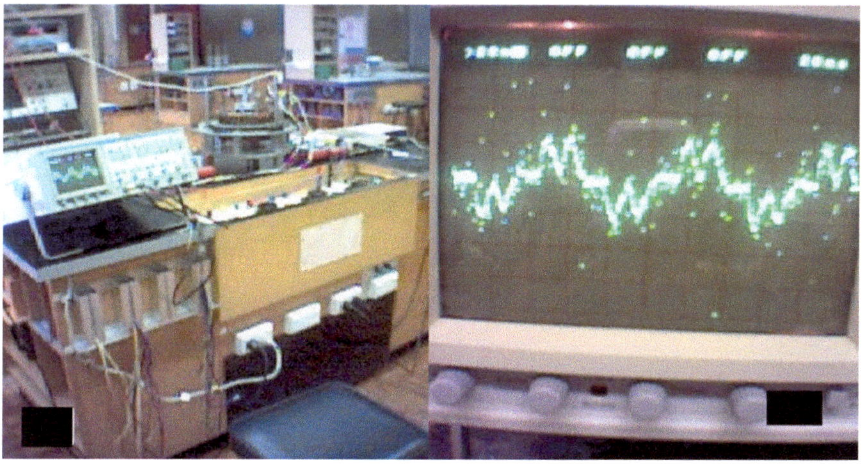

Fig. 7.13 (**a**) Lybotec six-step inverter-fed PMSM drive experimental set-up and (**b**) PMSM stator current waveform in phase A

7.4 Vector Control of Three-Phase Permanent Magnet Synchronous Motor Drive

Conventional vector control of permanent magnet synchronous motor (PMSM) drive requires a rotor speed and position hardware sensors such as Hall effect device or encoder to orient the stator current vector at right angle to the rotor magnet flux. As the rotor speed and position hardware sensors form a significant cost of the whole PMSM drive unit, several sensorless control schemes have been proposed without speed and position hardware sensors, with suitable information on the actual rotor position and speed obtained by computing stator dq-axis currents and voltages. The lack of any hardware sensors mounted on the rotor shaft reduces the system complexity, space and cost and allows the use of such drive for industrial applications. Such schemes have been developed mostly for surface PMSM drives [18–21]. Thus sensorless vector control also known as Field-Oriented Control (FOC) consists of transforming the three-phase stator voltages and stator currents into its equivalent dq-axis voltages and currents to compute the rotor position and speed. The dq-axis stator currents i_{ds} and i_{qs} are responsible for the production of magnetic flux and electromagnetic torque, respectively. Both currents i_{ds} and i_{qs} are separately controlled using PI controller. The reference current i_{ds}* for flux component is taken as zero and that for the torque component i_{qs}* is derived from speed controller. The outer speed loop controls the speed of the motor and generates reference current I_{qs}* for torque controller or q-axis current controller, and the inner current loop controls the magnetic flux and torque of motor. The block diagram of the sensorless vector control scheme of three-phase surface magnet PMSM drive is shown in Fig. 7.14.

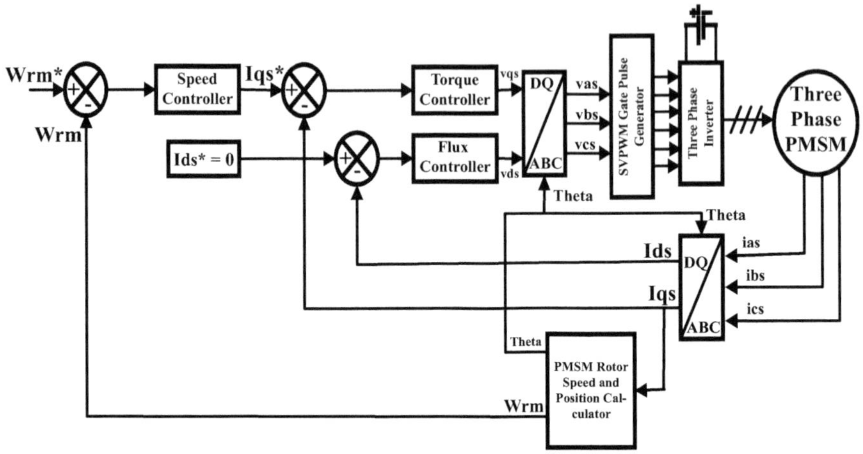

Fig. 7.14 Sensorless vector control scheme of a three-phase permanent magnet synchronous motor drive

Here back e.m.f. sensing method is used. The three-phase currents of the PMSM are transformed to dq-axis currents i_{ds} and i_{qs} using abc to dq transformation block also known as Park transformation. The q-axis current i_{qs} is used to calculate the rotor speed and position. Here three-phase surface magnet PMSM is used where $L_d = L_q = L_s$. The q-axis current i_{qs} is multiplied by the gain block with gain constant [3*P*λ_m/4] to get the electromagnetic torque T_{em}. This T_{em} value, assuming zero mechanical load, is given to a Transfer Fcn block with transfer function [1/(J.s + B)] to get the rotor speed in ω_{rm} in mechanical radians per second which is then multiplied by pole pair using the gain block [P/2] to get the rotor speed ω_{re} in electrical radians per second. This value of ω_{re} is then integrated using integrator block [1/s] to get the rotor position θ_{re} in electrical radians. This value of θ_{re} is given as input to dq- to abc-axis and abc- to dq-axis transformation block. The rotor speed ω_{rm} is then compared with a reference value of rotor speed ω^*_{rm} in a speed comparator whose error output is given to a speed controller to get the reference value I^*_{qs} for the q-axis current. The d-axis reference current I^*_{ds} is zero for vector control. The reference currents I^*_{qs} and I^*_{ds} are then compared with their respective values i_{qs} and i_{ds} obtained previously using abc-axis to dq-axis transformation, in two separate comparators, and the error outputs are respectively given to the torque controller (q-axis current controller) and flux controller (d-axis current controller). The output of these torque and flux controllers gives v_{qs} and v_{ds}, respectively. This vqs and v_{ds} are transformed to abc-axis quantities vas, v_{bs} and v_{cs} using dq-axis to abc-axis transformation or inverse Park transformation block. These three-phase AC voltages are used to generate Space Vector PWM (SVPWM) gate signals which drive the three-phase inverter switches. This three-phase inverter output voltage is fed to the PMSM.

7.4.1 Design of Speed Controller

The speed controller is shown in Fig. 7.15, where PI controller constants K_p and K_i are calculated on the basis of assumption of a suitable crossover frequency and phase margin for the speed controller [19]. In Fig. 7.15, ω^*_{rm} is the reference rotor speed and ω_{rm} is the actual rotor speed. The crossover frequency or bandwidth selected for the speed loop must lie within the rotor speed range in mechanical radians per second.

Referring to speed control loop in Fig. 7.15, constant k is defined below:

$$k = \frac{3 * P * \lambda_m}{4} \tag{7.25}$$

For PI controller design, the loop gain $G(s)*H(s) = GH1(S)$ is equated to unity which is given below:

$$\left(\frac{K_p S + K_i}{S}\right) * \frac{k}{(JS + B)} = 1 \tag{7.26}$$

Let $S = j\omega_{cr}$ be the given gain crossover frequency and PM be the desired phase margin. Then the loop gain $GH1(j\omega_{cr})$ can be expressed as follows:

$$GH1(j\omega_{cr}) = \frac{(j\omega_{cr} * K_p + K_i) * k}{j\omega_{cr} * (J * j\omega_{cr} + B)} \tag{7.27}$$

Using the given phase margin PM and Eq. 7.27, the expression for phase contribution of PI controller follows:

$$\frac{K_p * \omega_{cr}}{K_i} = \tan\left[\text{PM} - \frac{\pi}{2} - \tan^{-1}\left(\frac{J * \omega_{cr}}{B}\right)\right] \tag{7.28}$$

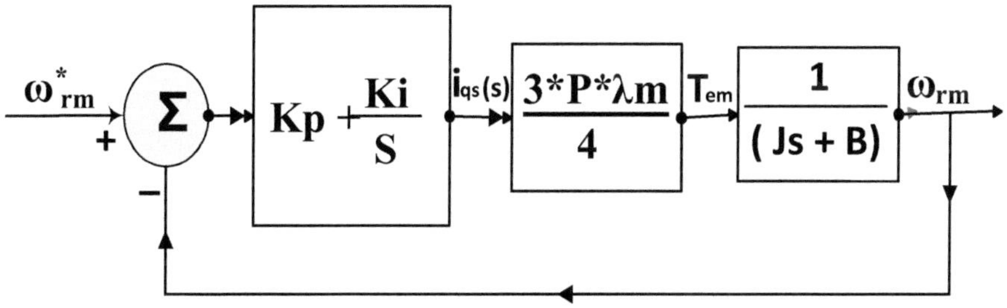

Fig. 7.15 Speed controller for PMSM drive

$$K_{\mathrm{i}} = \frac{K_p * \omega_{\mathrm{cr}}}{\tan\left[\mathrm{PM} - \frac{\pi}{2} - \tan^{-1}\left(\frac{J*\omega_{\mathrm{cr}}}{B}\right)\right]} \tag{7.29}$$

In Eq. 7.29, ω_{k1} is defined below:

$$\omega_{\mathrm{k1}} = \frac{\omega_{\mathrm{cr}}}{\tan\left[\mathrm{PM} - \frac{\pi}{2} - \tan^{-1}\left(\frac{J*\omega_{\mathrm{cr}}}{B}\right)\right]} \tag{7.30}$$

$$K_{\mathrm{i}} = K_{\mathrm{p}} * \omega_{\mathrm{k1}} \tag{7.31}$$

Equating the gain of Eq. 7.27 to unity, we have the following:

$$\frac{k * \left(\sqrt{K_{\mathrm{p}}^2 * \omega_{\mathrm{cr}}^2 + K_{\mathrm{i}}^2}\right)}{\omega_{\mathrm{cr}} * \left(\sqrt{J^2 * \omega_{\mathrm{cr}}^2 + B^2}\right)} = 1 \tag{7.32}$$

Using Eq. 7.31 in Eq. 7.32 and simplifying,

$$K_{\mathrm{p}} = \frac{\omega_{\mathrm{cr}} * \left(\sqrt{J^2 * \omega_{\mathrm{cr}}^2 + B^2}\right)}{k * \left(\sqrt{\omega_{\mathrm{cr}}^2 + \omega_{\mathrm{k1}}^2}\right)} \tag{7.33}$$

Using Eqs. 7.31 and 7.33, the PI controller constants K_{p} and K_{i} for the speed control loop can be evaluated.

7.4.2 *Design of Current Controller*

The PI controller for current loop is designed assuming a suitable crossover frequency and phase margin for the current loop controller [19]. The crossover frequency or bandwidth must be within the range used for the PWM carrier frequency which is much higher than the rotor speed range. The dq-axis current controllers are shown in Fig. 7.16a, b. Assume that the PI controller transfer function for the current control loop is [K_{pi} + (K_{ii}/S)]. Further assume that the given gain crossover frequency and phase margin for the current control loops are ω_{ci} rad/sec and PMi radians, respectively. Thus the loop gain $G(s)*H(s) = GH2(s) = 1$ results in the following equation:

$$\left(\frac{K_{\mathrm{pi}}.S + K_{\mathrm{ii}}}{S}\right) * \frac{1}{(S.L_S + R_S)} = 1 \tag{7.34}$$

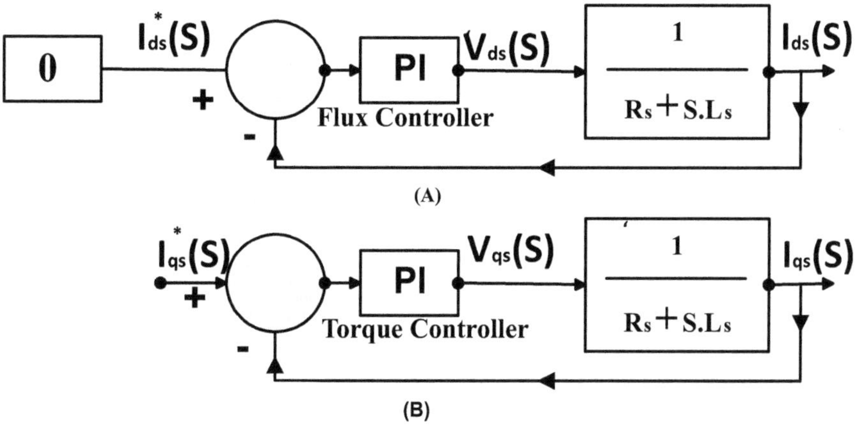

Fig. 7.16 (**a**, **b**) Three-phase PMSM vector control—current loop controllers

$$\mathrm{GH2}(j\omega_{\mathrm{ci}}) = \frac{(j\omega_{\mathrm{ci}} * K_{\mathrm{pi}} + K_{\mathrm{ii}})}{j\omega_{\mathrm{ci}} * (j\omega_{\mathrm{ci}} * L_S + R_S)} \tag{7.35}$$

Using the given phase margin PMi and Eq. 7.35, the expression for phase contribution of PI controller follows:

$$\frac{K_{\mathrm{pi}} * \omega_{\mathrm{ci}}}{K_{\mathrm{ii}}} = \tan\left[\mathrm{PMi} - \frac{\pi}{2} - \tan^{-1}\left(\frac{L_S * \omega_{\mathrm{ci}}}{R_S}\right)\right] \tag{7.36}$$

$$K_{\mathrm{ii}} = \frac{K_{\mathrm{pi}} * \omega_{\mathrm{ci}}}{\tan\left[\mathrm{PMi} - \frac{\pi}{2} - \tan^{-1}\left(\frac{L_S * \omega_{\mathrm{ci}}}{R_S}\right)\right]} \tag{7.37}$$

In Eq. 7.37, ω_{k2} is defined below:

$$\omega_{\mathrm{k2}} = \frac{\omega_{\mathrm{ci}}}{\tan\left[\mathrm{PMi} - \frac{\pi}{2} - \tan^{-1}\left(\frac{L_S * \omega_{\mathrm{ci}}}{R_S}\right)\right]} \tag{7.38}$$

$$K_{\mathrm{ii}} = K_{\mathrm{pi}} * \omega_{\mathrm{k2}} \tag{7.39}$$

Following the same procedure used for deriving Eqs. 7.32 and 7.33, K_{pi} can be expressed as follows:

$$K_{\mathrm{pi}} = \frac{\omega_{\mathrm{ci}} * \left(\sqrt{L_S^2 * \omega_{\mathrm{ci}}^2 + R_S^2}\right)}{\left(\sqrt{\omega_{\mathrm{ci}}^2 + \omega_{\mathrm{k2}}^2}\right)} \tag{7.40}$$

Using Eqs. 7.39 and 7.40, the PI controller constants K_{pi} and K_{ii} for current controller loop can be evaluated.

7.4.3 Model of Three-Phase Vector Controlled Permanent Magnet Synchronous Motor

The model of vector controlled three-phase PMSM is developed using Simulink [18]. The data shown in Table 7.2 are used. The model of vector controlled three-phase PMSM driven by three-phase sine wave AC voltage source obtained by dq-axis to abc-axis transformation is shown in Fig. 7.17 (model file: EXAMPLE 7_3). The PI controller parameters for the speed and current control loops are calculated using equations derived in Sects. 7.4.1 and 7.4.2. This is shown in Program segment 7.1. From the simulation results, the values are tabulated in Table 7.4.

Program segment 7.1

```
%% Vector control of Three Phase PMSM Drive
%% Calculation of Initial Conditions
%% Dr. Narayanaswamy P R Iyer
%% Reference 19
%% PMSM Data
Rs=9.3041; Ls=0.0596; J = 0.0035; p=4;
Lamda_m = 0.1354; B = 4.4066e-4;
%% Steady State Operating Condition
%% Phasor Calculations
Wrm_0=517*2*pi/(60);
Wre=Wrm_0*(p/2);
Tem_0=0.4;
TL_0=Tem_0;
Iqs_0=4*Tem_0/(3*p*Lamda_m); %% eq. 7.7.
Ids_0=0; %%Requirement for vector control of PMSM drive.
Lamda_ds_0=Ls*Ids_0+Lamda_m; %%eq. 7.3 and 7.4.
Lamda_qs_0=Ls*Iqs_0;
E = Lamda_m*Wre;
Is=(-j)*(Ids_0+j*Iqs_0); %%d-axis is 90 degrees behind the a axis.
Eq.8.11.
Ia=Is/1.5;
Va=E+Ia*(Rs+j*Wre*Ls);
theta_0=angle(Va);
VLLrms=abs(Va)*sqrt(3)/(sqrt(2));
Vs=1.5*Va;
Vds_0=(2/3)*real(j*Vs);
Vqs_0=(2/3)*imag(j*Vs);
%% PI controller controller calculations for speed loop.
Wcr=200; %% crossover frequency in rad/s
k=(3*p/4)*Lamda_m; % Iqs to torque gain. Eq. 7.7.
PM=75*pi/180; %% Assumed phase margin in rad/s
```

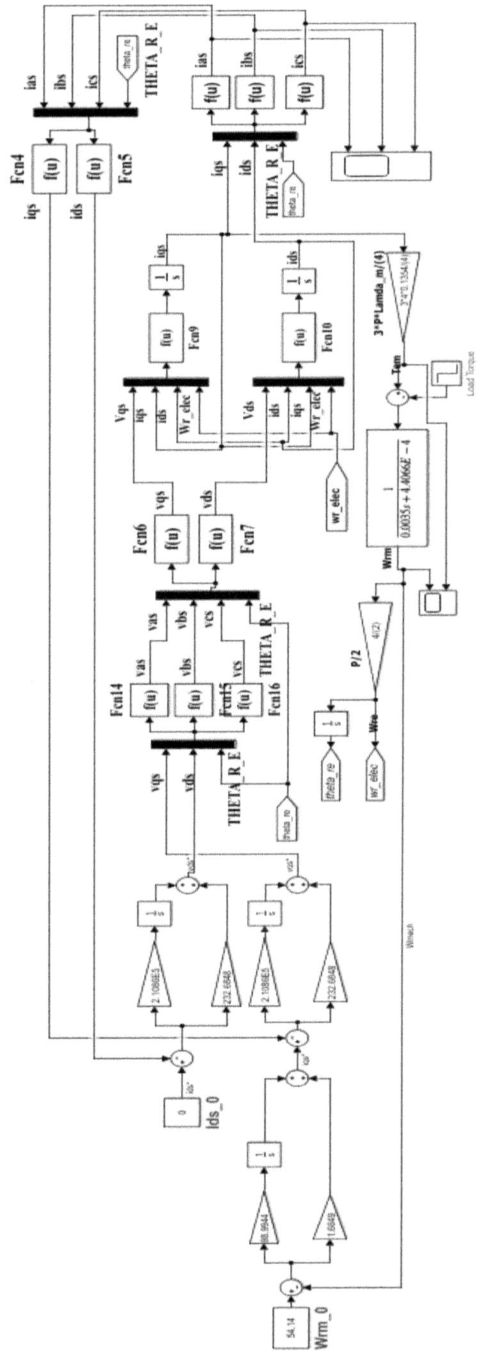

Fig. 7.17 Model of vector-controlled three-phase permanent magnet synchronous motor

Table 7.4 PI speed and current controller parameters for vector controlled three-phase PMSM

	Speed control loop		Current control loop		Rotor speed	Initial stator current
Sl. no.	K_p	K_i	K_{pi}	K_{ii}	mech rad/sec	Amps (RMS)
1	1.6649	88.9944	232.6848	2.1086e + 05	54.140	0.9847

```
Wcr_kp_by_ki = tan(PM - pi/(2) - atan(J*Wcr/(B)));
Wk1 = Wcr/(Wcr_kp_by_ki); %%Eq. 7.30
kp = Wcr*(sqrt(J*J*Wcr*Wcr+B*B))/(k*(sqrt(Wcr*Wcr+Wk1*Wk1)));%%
Eq. 7.33.
ki = kp*Wk1; %% Eq. 7.31.
%% PI in current loop
Wci=20*Wcr;
PMi=PM; %% Phase margin as for speed loop.
Wci_kpi_by_kii = tan(PMi-pi/(2)-atan(Wci*Ls/(Rs)));
Wk2 = Wci/(Wci_kpi_by_kii); %%Eq. 7.38.
kpi = Wci*(sqrt(Wci*Wci*Ls*Ls+Rs*Rs))/(sqrt(Wci*Wci+Wk2*Wk2));%%
Eq. 7.40.
kii = kpi*Wk2; %%Eq. 7.39.
```

Here rotor reference speed ω^*_{rm} which is 54.140 mech rad/sec from Table 7.4 is compared with the actual rotor speed ω_{rm} from rotor speed and position calculator in a comparator, and the rotor output is given to the speed PI controller with K_p and K_i value from Table 7.4. The speed PI controller output gives I^*_{qs}, the q-axis stator reference current. The value of I^*_{ds} is zero. The reference currents I^*_{ds} and I^*_{qs} are given to two separate current comparators. The estimated dq-axis currents i_{ds} and i_{qs} from the PMSM model form the second input to the two current comparators. The PI flux controller and torque controller have the K_{pi} and K_{ii} values given in Table 7.4. The torque and flux controller give outputs v_{qs} and v_{ds}, respectively. Here v_{qs}, v_{ds} and rotor position θ_{re} from rotor speed and position calculator are given as input to a three-input mux, and the abc-axis voltages v_{as}, v_{bs} and v_{cs} are generated using three Fcn blocks as per Eq. 7.23. These three-phase AC voltages v_{as}, v_{bs} and v_{cs} along with θ_{re} are given as input to a four-input mux, and the dq-axis voltages v_{ds} and v_{qs} are calculated using two Fcn blocks as per Eq. 7.22. This value of v_{qs}, v_{ds} and rotor speed ω_{re} in electrical radians per second from rotor speed and position calculator along with output currents i_{ds} and i_{qs} is fed back as input to the two mux to find dq-axis stator currents i_{ds} and i_{qs} using the PMSM model shown in Fig. 7.5h and explained in Sect. 7.2.2. This dq-axis stator current from PMSM model along with θ_{re} is given as input to a three-input mux to generate three-phase stator current of PMSM using two Fcn blocks as per Eq. 7.23. These three-phase stator currents i_{as}, i_{bs} and i_{cs} along with θ_{re} are given as input to a four-input mux to find the estimated value of i_{ds} and i_{qs} from PMSM model using two Fcn blocks as per Eq. 7.22.

For the rotor speed and position calculator, the q-axis current i_{qs} from PMSM model is given as input to a gain block with gain constant [3*P*λ_m/4]. This gain block output is given to a Transfer Fcn block with transfer function [1/(J.s + B)] to get the rotor speed ω_{rm} in mechanical radians per second. The output of Transfer Fcn block is multiplied by a gain block with gain constant [P/2] to get the rotor speed ω_{re} in electrical radians per second. This value of ω_{re} is then integrated using integrator block [1/s] to get the rotor position θ_{re} in electrical radians.

7.4.4 Simulation Results

The model of the three-phase vector controlled PMSM is carried out using ode45 (Dormand-Prince) variable step solver in Simulink [18]. The data shown in Tables 7.2 and 7.4 are used for simulation. Here an external step load of 0.5 Nw-metres for the first 2 seconds and 0.25 Nw-metres for the remaining time is used for simulation. The rotor speed, e.m. torque and three-phase stator currents of PMSM by simulation are shown in Fig. 7.18.

7.4.5 Discussion of Results

From the simulation results shown in Fig. 7.18, it is seen that as the load torque is varied from an initial value of 0.5 Nw-metres to 0.25 Nw-metres in 2 seconds, the rotor speed is initially 54.14 mech rad/sec, rises to 54.4 mech rad/sec at 2 seconds and then falls to 54.14 mech rad/sec. The rotor e.m. torque is initially close to 0.5 newton metres and at 2 seconds, this torque falls close to 0.25 Nw-metres. The PMSM stator current is initially 0.906 A(RMS) and this falls to 0.477 A(RMS) in 2 seconds. The rotor speed is maintained constant irrespective of load variation which is the advantage of vector control.

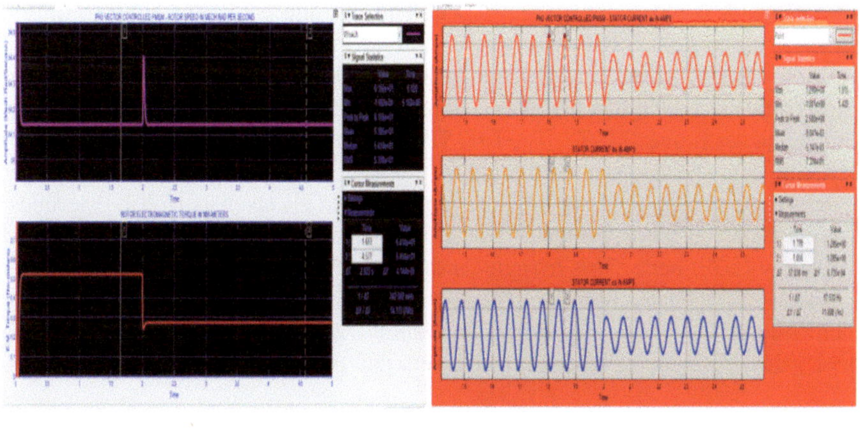

Fig. 7.18 Three-phase vector-controlled pmsm simulaton results: rotor speed in mechanical radians per second (top left), Rotor electromagnetic torque in Nw-Meters (bottom left), three-phase PMSM stator current (right top to bottom)

7.5 Case Study: Model of Three-Phase Space Vector PWM Inverter-Fed Vector Controlled PMSM Drive

The model of the three-phase vector controlled PMSM drive fed by a SVPWM inverter is shown in Fig. 7.19 (model file: CASE_STUDY_EX7_1). Here a vector controlled PMSM model shown in Fig. 7.17 is used as a reference model to generate the gate pulse for the three-phase SVPWM inverter. Here two Embedded MATLAB functions are used to generate the gate pulse and the universal bridge model is used as the three-phase SVPWM inverter. The three-phase output voltage of this inverter is fed to the Permanent Magnet Synchronous Machine model from specialised technology block set. To this machine model, an external step load of 0.5 Nw-M for the first 2 seconds and 0.25 Nw-M for the remaining time is applied. Machine parameters shown in Table 7.2 are used.

Here the dq-axis voltages v_{ds} and v_{qs} from the reference model of three-phase vector controlled PMSM are given to two MinMax blocks to determine their respective maximum values which are given to a two-input mux, and using function block Fcn connected to this mux, the maximum voltage V_{im} is calculated using the relation $V_{im} =$ sqrt(v_{ds}_max*v_{ds}_max + v_{qs}_max*v_{qs}_max). This output voltage V_{im} and rotor speed ω_{re} from rotor speed and position calculator along with the time module t are given as input to the first Embedded MATLAB function. This Embedded MATLAB function generates three-phase voltages v_a, v_b and v_c where $v_a = V_{im}$*sin(ω_{re}*time). The program to calculate maximum reference voltage V_{ref_max}, dq-axis voltage v_d, v_q, sector numbers, angular position θ for the reference voltage V_{ref_max} and sector switch functions from ssf1 to ssf6 is given in the model file CASE_STUDY_EX7_1. The V_{im} calculated above is multiplied by another Fcn block with constant (1.732*V_{im}/V_{dc}) to get modulation index ma. The V_{ref_max} output is multiplied by Fcn block with output ma using Multiplier 1 block to get reference voltage V_{ref}. The reference voltage V_{ref}, DC link voltage 40 V from Table 7.2, sample time T_z, rotor speed ω_{re} from rotor speed and position calculator, time module t, sector number, sector switch functions ssf1 to ssf6, triangle carrier V_{tri} and the position θ for V_{ref} from first Embedded MATLAB function are given as input to the second Embedded MATLAB function. The second Embedded MATLAB function calculates timing T1, T2, T0, Ta1, Tb1 and Tc1 and generates gate pulses sap, sbp, scp, san, sbn and scn for the upper and lower switches of the inverter as per Program segment 4.7 of Chap. 4. Here sample time T_z is taken as 1/(20E3) seconds. The triangle generator generates triangle wave with peak value +/−1 volt and frequency ($1/T_z$) hertz. This triangle carrier output is multiplied by function block Fcn with multiplication constant [0.5*(u(1) + 0.5], and the output of Fcn block is multiplied by a gain block with gain constant [T_z/2] to generate the carrier V_{tri} with peak value (T_z/2) volts, minimum value zero and period T_z seconds.

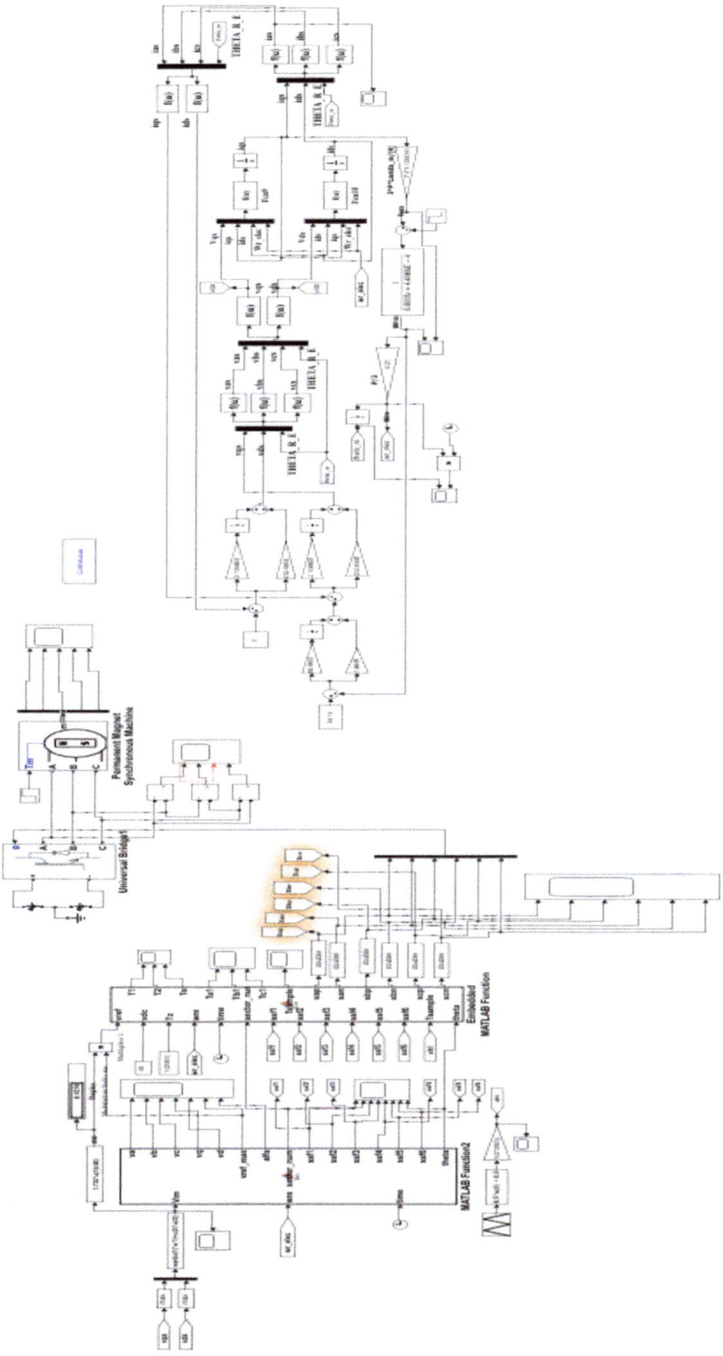

Fig. 7.19 Three-phase SVPWM inverter-fed vector-controlled PMSM drive

7.5.1 Simulation Results

The simulation of the vector controlled three-phase PMSM drive fed by SVPWM inverter is carried out using ode23tb (stiff/TR-BDF2) solver in Simulink [18]. The inverter and PMSM data are from Table 7.2. Carrier frequency is 20 kHz. Simulation results are shown in Figs. 7.20 and 7.21.

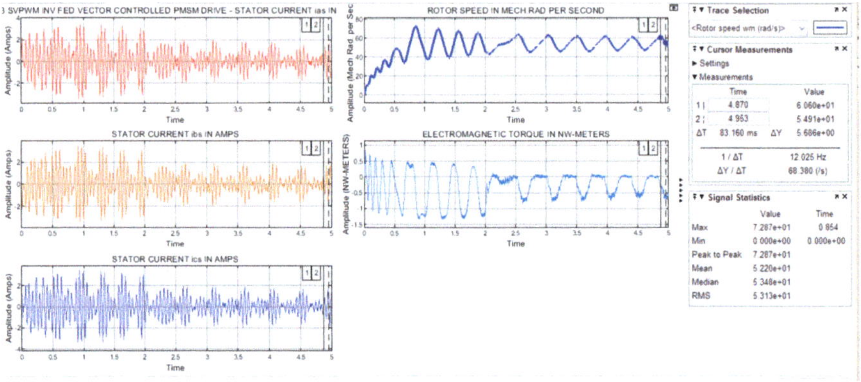

Fig. 7.20 PH3 SVPWM inverter-fed vector-controlled PMSM drive simulation results. Three-phase stator currents (left column top to bottom), rotor speed and e.m. torque (right column top to bottom)

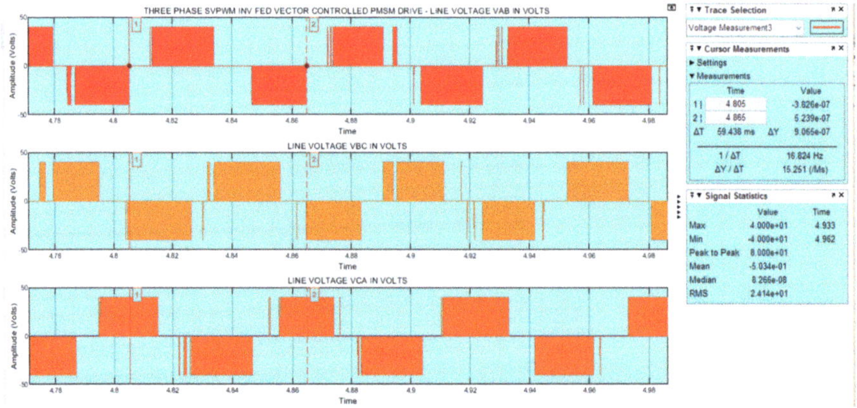

Fig. 7.21 Three-phase SVPWM inverter-fed vector controlled PMSM drive simulation results: line-to-line output voltage of inverter

7.5.2 Discussion of Results

Simulation results shown in Fig. 7.20 indicate a balanced three-phase stator current. The mean rotor speed is 52.2 mech rad/sec which is very close to the set point speed of 54.140 mech rad/sec. The steady state stator current in Fig. 7.20 is close to 0.99 amps as recorded in Table 7.4. Also from Fig. 7.21, the line to line output voltage of three-phase SVPWM inverter is 24.14 volts (RMS) and the frequency is 16.824 Hz. Converting the rotor set point speed of 54.140 mech rad/sec to electrical frequency, this value is 17.233 Hz which closely agrees with the frequency of the line to line output voltage of inverter. The modulation index ma is found to be 0.9258 by simulation. This model demonstrates the accuracy of vector controlled PMSM drive.

7.6 Brushless DC Motor Drives

The permanent magnet (PM) brushless DC motor (PMBLDCM) drive configuration is shown in Fig. 7.1. The brushless DC motor (BLDCM) is basically a non-salient pole permanent magnet machine with concentrated stator winding which produces trapezoidal back electromotive force (EMF). The BLDCM requires rectangular-shaped stator phase currents to produce constant torque. Due to trapezoidal back EMF, the stator inductance variation is non-sinusoidal with respect to the rotor position, and the conventional technique for modelling as for PMSM drive by transforming the three-phase voltages and currents in the abc-axis to the dq-axis is unsuitable for PMBLDCM drive. Therefore modelling of three-phase inverter-fed PMBLDCM drive is done in the three-phase abc-axis using the relevant equations connecting applied phase voltage, trapezoidal back EMFs, phase currents, stator winding resistance and self and mutual inductance parameters. To maintain the stator currents close to the reference current value, either hysteresis current control or pulse width modulation (PWM) technique is used [20–22].

As presented in Sect. 7.1, when the stator coils are energised by the inverter three-phase AC voltage, electromagnetic field is established in the air gap which interacts with the PM rotor field. As a result of this force of interaction, rotor rotates. This requires rotor position information to energise the appropriate stator phase windings to maintain continuous rotation. For this reason Hall effect sensors are used which are embedded in the stator. Based on these three Hall signals, the exact sequence of commutation can be determined. The induced back EMF and stator current waveform of a three-phase inverter-fed BLDCM drive are shown in Fig. 7.22. As seen from Fig. 7.22, at any given interval of rotor angle θ_e, only any two stator phase windings carry current, and there is no current flow in the third phase winding. This requires that the three-phase inverter connected to the PMBLDCM drive has to be switched in the 120-degree mode. This mode of switching is shown in Table 7.5.

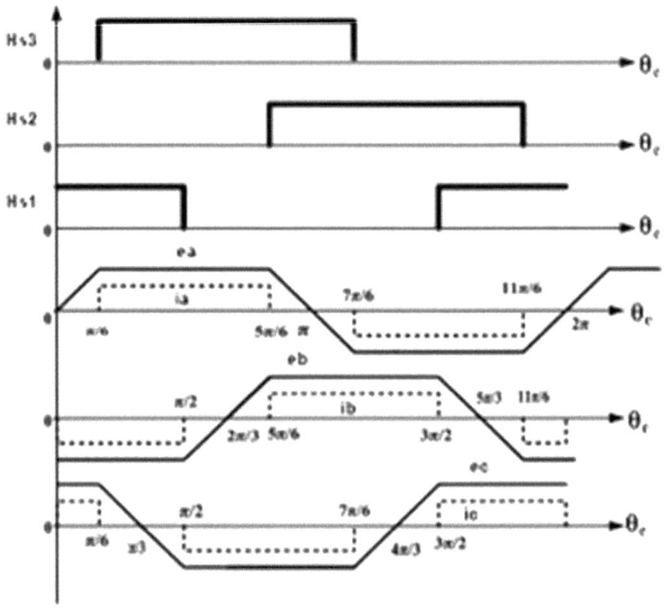

Fig. 7.22 Waveforms of three-phase back EMFs and Stator currents along with Hall sensor signals

Table 7.5 Switching sequence table

Sl. no.	Rotor angle θ_e interval (degrees)	Switches turned ON	Conducting phase a	b	c	Hall sensor output Hs3	Hs2	Hs1
1	30–90	S1A, S4B	+1	–1	0	1	0	1
2	90–150	S1A, S6C	+1	0	–1	1	0	0
3	150–210	S3B, S6C	0	+1	–1	1	1	0
4	210–270	S3B, S2A	–1	+1	0	0	1	0
5	270–330	S5C, S2A	–1	0	+1	0	1	1
6	330–390	S5C, S4B	0	–1	+1	0	0	1

In the case of BLDCM drive, the phase current waveform is a square wave and hence the RMS value is the same as the peak value. For PMSM, RMS current is reduced by a factor of $1/\sqrt{2}$ of the peak current. Thus for a given back e.m.f., the power output of BLDCM is high compared to PMSM drive. Due to commutation at 60-degree interval, torque ripple in BLDCM is high compared to PMSM. High-speed operation of BLDCM drive is possible by advancing the phase current so that it leads the phase back e.m.f. This is achieved by advancing the switching angle of the inverter.

7.6.1 Dynamic Model of Brushless DC Motor

The equivalent circuit a three-phase BLDCM is shown in Fig. 7.23. The coil of each phase has a resistance R_x and self-inductance L_x connected in series with a back e.m.f. source e_x where $x \in$ a, b, c. Each phase coil has also a mutual inductance M_{yz} with the neighbouring phase coils where $y \in$ a, b, a and $z \in$ b, c, c, respectively. The applied voltages across each phase to neutral n are v_x and the current through each phase is i_x where $x \in$ a, b, c. From Fig. 7.23, the voltage equation for each phase can be expressed in the following form [20–22]:

$$\begin{bmatrix} v_a \\ v_b \\ v_c \end{bmatrix} = \begin{bmatrix} R_a & 0 & 0 \\ 0 & R_b & 0 \\ 0 & 0 & R_c \end{bmatrix} * \begin{bmatrix} i_a \\ i_b \\ i_c \end{bmatrix} + \begin{bmatrix} L_a & M_{ab} & M_{ac} \\ M_{ab} & L_b & M_{bc} \\ M_{ac} & M_{bc} & L_C \end{bmatrix} * \begin{bmatrix} i_{a_{dot}} \\ i_{b_{dot}} \\ i_{c_{dot}} \end{bmatrix} + \begin{bmatrix} e_a \\ e_b \\ e_c \end{bmatrix} \tag{7.41}$$

For identical three-phase stator windings, $R_a = R_b = R_c = R_s$, $L_a = L_b = L_c = L_s$, $M_{ab} = M_{bc} = M_{ac} = M$. Also when the stator currents are balanced, $(i_a + i_b + i_c = 0)$. Using these relations, Eq. 7.41 can be expressed as follows:

$$\begin{bmatrix} v_a \\ v_b \\ v_c \end{bmatrix} = \begin{bmatrix} R_s & 0 & 0 \\ 0 & R_s & 0 \\ 0 & 0 & R_s \end{bmatrix} * \begin{bmatrix} i_a \\ i_b \\ i_c \end{bmatrix} + \begin{bmatrix} (L_s - M) & 0 & 0 \\ 0 & (L_s - M) & 0 \\ 0 & 0 & (L_s - M) \end{bmatrix} * \begin{bmatrix} i_{a_{dot}} \\ i_{b_{dot}} \\ i_{c_{dot}} \end{bmatrix} + \begin{bmatrix} e_a \\ e_b \\ e_c \end{bmatrix} \tag{7.42}$$

Letting $(L_s - M) = L$, Eq. 7.42 can be expressed as follows:

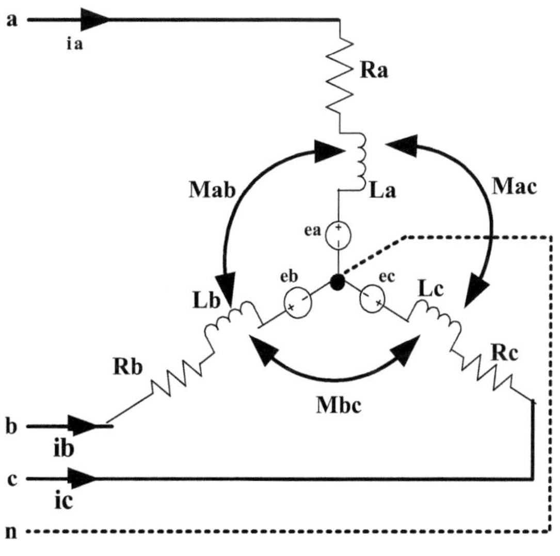

Fig. 7.23 Equivalent circuit of three-phase BLDCM

$$\begin{bmatrix} i_{a_{dot}} \\ i_{b_{dot}} \\ i_{c_{dot}} \end{bmatrix} = \begin{bmatrix} \frac{1}{L} & 0 & 0 \\ 0 & \frac{1}{L} & 0 \\ 0 & 0 & \frac{1}{L} \end{bmatrix} * \left[\begin{bmatrix} v_a \\ v_b \\ v_c \end{bmatrix} - \begin{bmatrix} e_a \\ e_b \\ e_c \end{bmatrix} - \begin{bmatrix} R_s & 0 & 0 \\ 0 & R_s & 0 \\ 0 & 0 & R_s \end{bmatrix} * \begin{bmatrix} i_a \\ i_b \\ i_c \end{bmatrix} \right] \tag{7.43}$$

Assigning diagonal matrix with (L_s-M) term as the L matrix and stator resistance matrix as R_s, Eq. 7.42 can also be expressed in the following form:

$$\overrightarrow{\left[\boldsymbol{i}_{\mathbf{x}_{\mathbf{dot}}} \right]} = \overrightarrow{\left[\boldsymbol{L}^{-1} \right]} * \left\{ - \left[\overrightarrow{\boldsymbol{R}_{\mathbf{S}}} \right] * \left[\overrightarrow{\boldsymbol{i}_{\mathbf{x}}} \right] + \left[\overrightarrow{\boldsymbol{v}_{\mathbf{x}}} - \overrightarrow{\boldsymbol{e}_{\mathbf{x}}} \right] \right\} \tag{7.44}$$

In Eq. 7.44, $i_{x\_dot} = di_x/dt$ where x € a, b, c. Also if P_e is the electric power output, T_{em} is the electromagnetic torque developed, and ω_m is the rotor mechanical speed in radians per second, then P_e can be expressed as follows:

$$P_e = (e_a * i_a + e_b * i_b + e_c * i_c) \tag{7.45}$$

$$T_{em} = \frac{(e_a * i_a + e_b * i_b + e_c * i_c)}{\omega_m} \tag{7.46}$$

If ω_e is the rotor speed in electrical radians per second, then ω_e and T_{em} can also be expressed as follows:

$$\omega_e = \frac{P * \omega_m}{2} \tag{7.47}$$

$$T_{em} = \frac{P * (e_a * i_a + e_b * i_b + e_c * i_c)}{2 * \omega_e} \tag{7.48}$$

In Eq. 7.47 and 7.48, P is the number of poles. If BLDCM is driven by a three-phase 120-degree mode inverter, then from Fig. 7.22, it is seen that for every 60-degree interval, only any two stator phase windings carry current and the current in the third phase winding is zero. If E_P is the peak value of the trapezoidal back EMF and I_P is the peak value of the square wave phase current, then power output and electromagnetic torque can also be expressed as follows:

$$P_e = (E_P) * (I_P) + (-E_P) * (-I_P) = 2 * E_P * I_P \tag{7.49}$$

$$T_{em} = \frac{2 * E_P * I_P}{\omega_m} = 2 * k_e * I_P \tag{7.50}$$

Also if k_T is the torque constant in Nw-metres per ampere, then $T_{em} = k_T * i_a$. In Eq. 7.50, k_e is the back EMF constant expressed as volt sec/rad. The rotor dynamic equation is given below:

$$J * \frac{d\omega_{\mathrm{m}}}{dt} + B * \omega_{\mathrm{m}} = (T_{\mathrm{em}} - T_{\mathrm{L}}) \tag{7.51}$$

In Eq. 7.51, J is the rotor inertia, B is the damping constant, and T_{L} is the load torque.

Also referring to Fig. 7.22, the back e.m.f. equation for reference Phase A can be defined as follows:

$$\begin{aligned} e_{\mathrm{a}} &= \frac{6 * E_{\mathrm{p}} * \theta_{\mathrm{e}}}{\pi} \text{ for } 0 \leq \theta_{\mathrm{e}} < \frac{\pi}{6} \\ &= +E_{\mathrm{p}} \text{ for } \frac{\pi}{6} \leq \theta_{\mathrm{e}} < \frac{5 * \pi}{6} \\ &= \frac{-6 * E_{\mathrm{p}} * \theta_{\mathrm{e}}}{\pi} + 6 * E_{\mathrm{p}} \text{ for } \frac{5 * \pi}{6} \leq \theta_{\mathrm{e}} < \frac{7 * \pi}{6} \\ &= -E_{\mathrm{p}} \text{ for } \frac{7 * \pi}{6} \leq \theta_{\mathrm{e}} < \frac{11 * \pi}{6} \\ &= \frac{6 * E_{\mathrm{p}} * \theta_{\mathrm{e}}}{\pi} - 12 * E_{\mathrm{p}} \text{ for } \frac{11 * \pi}{6} \leq \theta_{\mathrm{e}} < 2 * \pi \end{aligned} \tag{7.52}$$

In Fig. 7.22, for example, corresponding to θ_{e} value of $11*\pi/6$ and $2*\pi$, the coordinates are $(11*\pi/6, -E_{\mathrm{p}})$ and $(2*\pi, 0)$, respectively. The equation of the line joining these two points takes the form shown in Eq. 7.52.

7.6.2 Model of Six-Step 120-Degree Mode Inverter-Fed BLDCM Drive

Based on the mathematical model presented in Sect. 7.6.1, an open loop model of the three-phase inverter-fed BLDCM drive is presented in this section. This model is shown in Fig. 7.24 (model file: EXAMPLE 7_4). In this model no feedback control is used. The various subsystems are (1) six-step 120-degree mode inverter gate drive, (2) three-phase inverter and (3) PMBLDCM model. These are briefly presented below:

7.6.2.1 Six-Step 120-Degree Mode Inverter Gate Drive

The three-phase sine wave AC generator block generates three-phase AC voltage with amplitude 1 volt and frequency corresponding to the angular frequency of rotation of BLDCM rotor in electrical radians per second. These are given to three op.amp zero crossing comparators (ZCC) whose output is +5 V and 0 volts when the relevant input crosses zero and goes positive and negative, respectively. The output

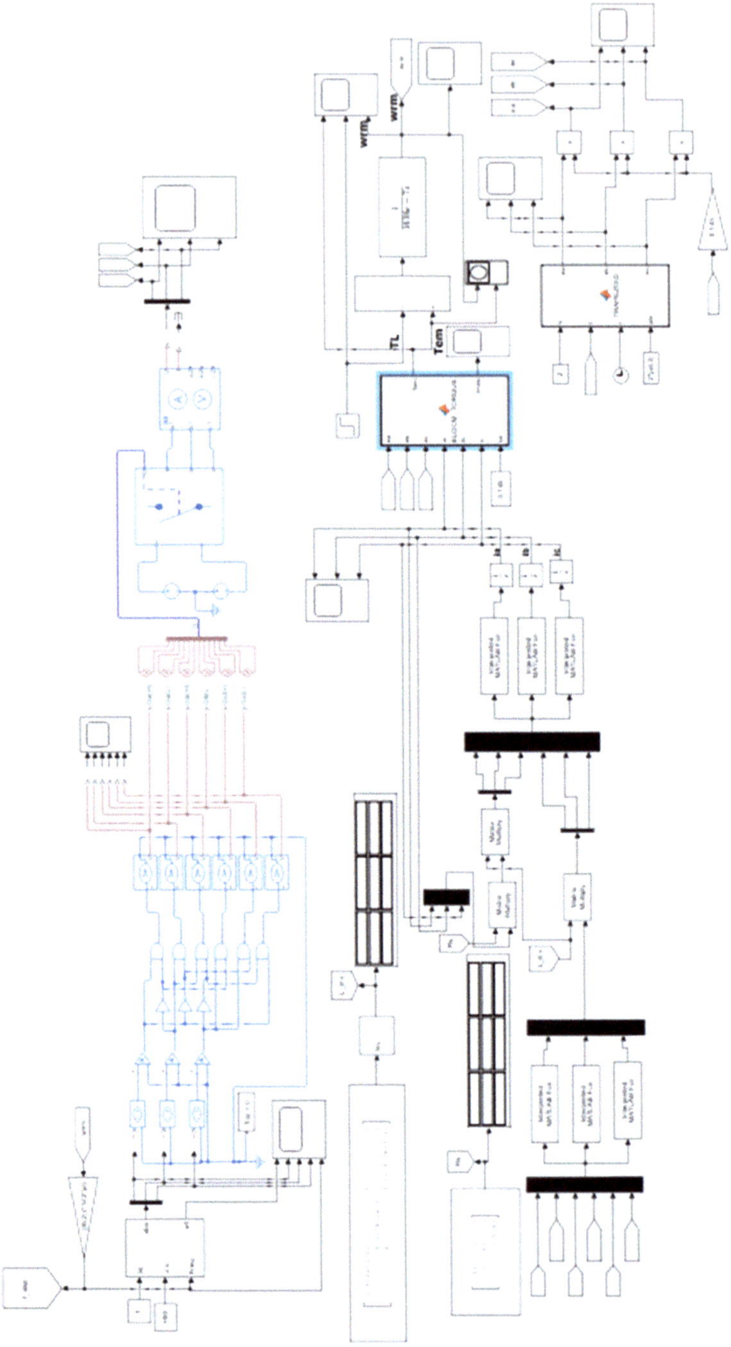

Fig. 7.24 Model of three-phase inverter-fed PMBLDCM drive

of the ZCC for each phase is V_{ga}, V_{gb} and V_{gc}, respectively. These are given to logic gates to generate the required gate pulse for the inverter switches. Referring to Fig. 7.1, the gate pulse logic for the upper switch S1A and lower switch S2A in Phase A are given below:

$$V_{gs1a} = V_{ga} \cap (\sim V_{gb}) \tag{7.53}$$

$$V_{gs2a} = V_{gb} \cap (\sim V_{ga}) \tag{7.54}$$

In Eqs. 7.53 and 7.54, the symbols $\cap$ and $\sim$ stand for logic AND and NOT operation, respectively. Similar gate drive logic applies to the switches in the other two phases. The gate drive for the upper and lower switches in Phase A, B and C in the respective order is given to a six-pulse gate multiplexer.

7.6.2.2 Three-Phase Inverter

The three-phase IGBT inverter [16, 17] is from Simscape Electrical-Semiconductors-converters block set. Two DC voltage sources in series with midpoint grounded from electrical sources block set are used as the input DC link voltage source. Here for the IGBT inverter, gate threshold voltage is set to 3 V. The gate pulse amplitude must be greater than this threshold value for proper operation of the inverter.

7.6.2.3 PMBLDCM Model

The BLDCM model is developed from Eqs. 7.43 to 7.51 [20–22]. The model subsystems are (1) stator phase current generator, (2) electromagnetic torque generator and (3) rotor speed generator. Stator phase current generator is developed using Eq. 7.43 where va, vb and vc are the inverter line to ground output voltages and ea, eb and ec are back e.m.f. in each phase. The stator resistance R_s and stator inductance L matrix are known. The trapezoidal back EMF waveform has to be generated at the electrical rotational angular frequency of the rotor with amplitude E_p which is k_e*ω_m. Assuming an E_p of 1 p.u., the slope of the waveform during the interval $0 <= \theta_e < \pi/6$ is $6/\pi$, and the equation of the line in this range of θ_e is $(6*\theta_e/\pi)$. This equation of the line is approximately $2*\sin(\theta_e)$, taking $\sin(\theta_e) = \theta_e$ in the interval $0 <= \theta_e <= \pi/6$. The error caused by this assumption is very small. Program segment 7.2 in the model file EXAMPLE 7_4 illustrates this method of trapezoidal three-phase back e.m.f. generation. Here three-phase trapezoid back e.m.f. with 1 volt amplitude is generated using Embedded MATLAB function TRAPEZOID given in the model file EXAMPLE 7_4. The inputs to TRAPEZOID function are amplitude constant k with value 2 V, electrical frequency of rotation of rotor felec, time module t and an optional phase shift module phi. Three-phase AC voltages f1, f2 and f3 are generated with amplitude, frequency and phase shift added. For θ_e range in the order defined in Eq. 7.52, back e.m.f. ea. takes values f1, +1, f1, −1 and f1. The same principle holds good for generating eb and ec using f2 and f3. These per unit values of ea, eb and ec are then multiplied by $k_e*\omega_m$ using gain and multiplier blocks to get the actual values of ea, eb and ec.

The diagonal stator resistance matrix R_s and the inductance matrix L ($L_s - M$) elements are entered in two constant blocks row-wise with a space between each element and a semicolon before entering the next row of elements. The L matrix constant block is then inverted using Matrix Divide or Matrix Inverse block whose output is L^{-1}. The vx output of three-phase inverter and ex output from trapezoidal waveform generator are given as input to a six-input Mux. Using three Interpreted MATLAB Function blocks, (vx–ex) for each phase are computed and given as input to a three-input Mux to output a 3 × 1 column vector which is then pre-multiplied with L^{-1} matrix using a Matrix Multiply block to realise a 3 × 1 column vector output [L^{-1}*(vx–ex)]. This output is given to a three-output Demux to realise individual elements of the three rows corresponding to Phase A, B and C. The stator phase current ix (x = A, B, C) assuming to be available is given as input to a three-input Mux to realise ix as a 3 × 1 column vector which is then post-multiplied with stator resistance R_s matrix using Matrix Multiply block to realise a 3 × 1 (R_s*ix) column vector. This R_s*ix output is pre-multiplied with L^{-1} matrix to realise a 3 × 1 (L^{-1}*R_s*ix) column vector. This output is given to a three-output Demux to realise individual elements of the three rows corresponding to Phase A, B and C. The individual elements of [L^{-1}*(vx–ex)] matrix and that of (L^{-1}*R_s*ix) matrix are then given as input to a six-input Mux, and using three Interpreted MATLAB Function blocks, [L^{-1}*(vx–ex–R_s*ix)] is computed for each phase which is then integrated using three integrator blocks 1/s to realise phase currents ia, ib, and ic. ia, ib, and ic outputs are then fed back to the three-input Mux where it was initially assumed to be available. This way Eq. 7.44 is implemented. The electromagnetic torque generator is realised using Embedded MATLAB function BLDCM_TORQUE shown in the model file EXAMPLE 7_4. Here ea, eb, ec, ia, ib, ic, and ke are given as input and the outputs are electromagnetic torque T_{em} and peak value of stator phase current imax. The Program segment 7.3 in the model file EXAMPLE 7_4 shows the method for calculating imax. Here imax is calculated by comparison of each phase current with the other two phase currents. Thus if ia > = ib AND ia > = ic is TRUE, then ia takes the value of imax. Similar comparison is used for ib and ic. The value of T_{em} is calculated using Eq. 7.50.

Rotor speed generator is realised using Sum and Transfer Fcn blocks. The load torque TL is realised using step source. The T_{em} output of electromagnetic torque generator and that of TL are given to a Sum block with a minus sign for TL. The output (T_{em}—TL) is given to a Transfer Fcn block with transfer function 1/(J.s + B) to realise the rotor speed ω_m as per Eq. 7.51.

7.6.3 Simulation Results

The simulation of the model shown in Fig. 7.24 was carried out using fixed step ode1be (Backward Euler) solver in Simulink [18]. The data used for simulation are shown in Table 7.6 [23]. Simulation results for the three-phase line to neutral output voltage of inverter is shown in Fig. 7.25. The three-phase trapezoidal back EMF waveforms are shown in Fig. 7.26. Three-phase stator current waveforms are shown in Fig. 7.27. The electromagnetic torque, load torque and rotor speed waveforms are shown Fig. 7.28. Here the BLDCM drive is started at no load and at 50 milliseconds a full load torque TL of 0.83 Nw-metres is applied.

Table 7.6 PMBLDCM data

Sl. no.	Parameter	Value	Unit
1	Terminal voltage	48	Volts DC
2	Rated speed	2319	RPM
3	Rated torque	0.83	Nw-M
4	Rated current	5.53	Amps
5	Rated power	201	Watts
6	Torque constant k_T	0.15	Nw-metre per ampere
7	Back EMF constant k_e	15.2	Volts per KRPM
8	Stator resistance R_s	0.93	Ohms per phase
9	Stator inductance L (L_S–M)	1.872	Millihenries per phase
10	Rotor inertia J	1638	gm-cm^2
11	Number of poles P	8	–
12	No load speed	3200	RPM

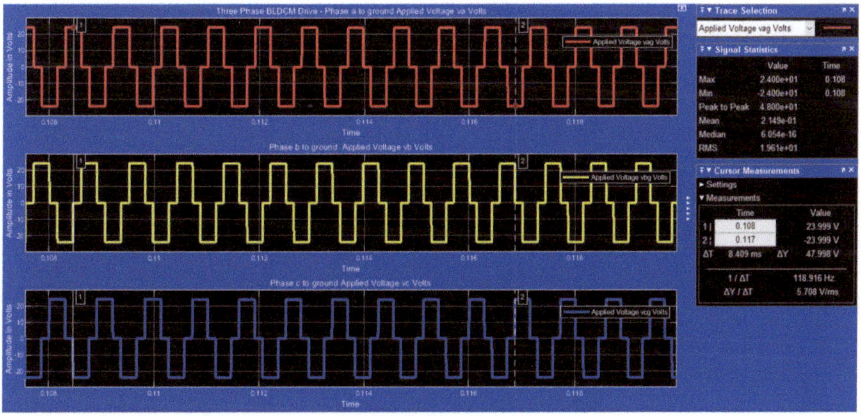

Fig. 7.25 Three-phase inverter-fed PMBLDCM drive—line-to-ground applied voltage va, vb and vc

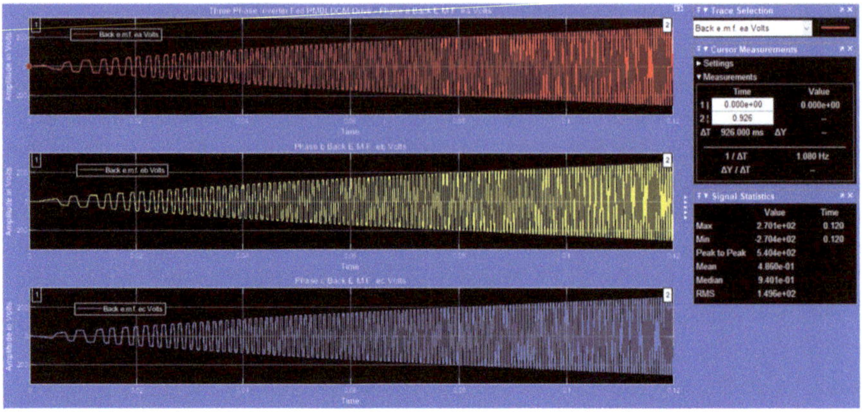

Fig. 7.26 Three-phase inverter-fed BLDCM drive—trapezoidal back e.m.f. ea, eb and ec

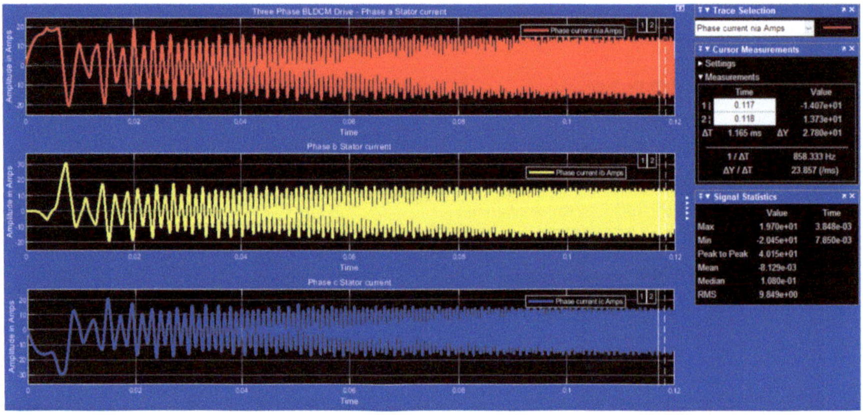

Fig. 7.27 Three-phase inverter-fed BLDCM drive—stator phase current waveforms ia, ib and ic

7.6.4 Discussion of Results

From Fig. 7.25 it is seen that the line to ground voltages va, vb and vc confirm to six-step 120-degree mode inverter switching. The RMS value of line to ground voltage is displayed as 19.63 volts which closely agree with ($V_{dc}/\surd 6$) where V_{dc} is the DC link voltage of the inverter. The trapezoidal back EMF amplitude increases with rotor speed. The phase current amplitude is initially small and increases from 50 milliseconds onwards when the load torque is applied. The load current has a trapezoidal shape with a narrow peak flat region. The electromagnetic torque is high at the time of starting and gradually reduces to a low value, and from $50e^{-3}$ sec onwards when load is applied, the peak electromagnetic torque decreases. The load torque waveform shoots up from 0 to 0.83 Nw-M at $50e^{-3}$ sec. The rotor speed

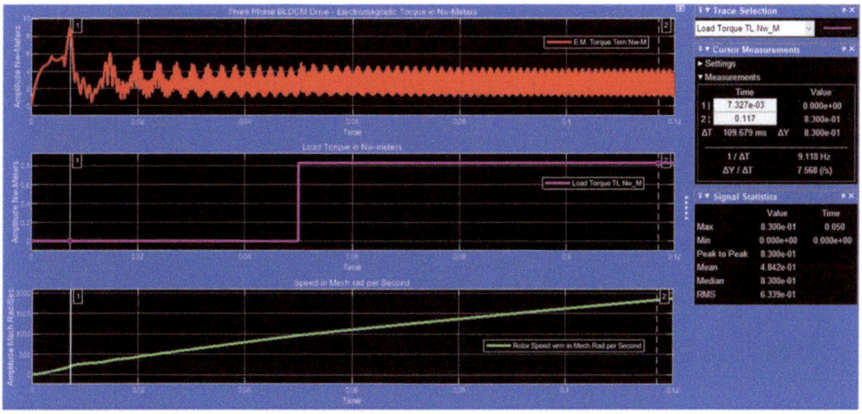

Fig. 7.28 Three-phase inverter-fed BLDCM drive—electromagnetic torque, T_{em} Nw-M; load torque, TL Nw-M; rotor speed in mechanical radians per second

linearly increases from 0 and at $50e^{-3}$ sec takes a slight bend towards the time axis and continues to increase linearly. Here simulation is done open loop. No feedback speed control is used. The rotor speed waveform is theoretical. The back e.m.f. is proportional to rotor speed and adds up for every 2π radians rotation of rotor and hence the stator current and rotor speed increase. Feedback control mechanism is required to limit this rotor speed.

7.6.5 Model of Hysteresis Current Controlled Three-Phase Inverter-Fed PMBLDCM Drive

The hysteresis current control of three-phase PMBLDCM is used to maintain the stator phase current close to the reference current value [25–26]. The hysteresis current controller block diagram is shown in Fig. 7.29. Here the reference speed ω_m* is compared with actual rotor speed ω_m and the error speed is given to a PI controller. The PI controller output is the reference torque T_{em}* which is divided by torque constant K_T to get the reference stator phase current I_{ref}. This rotor speed ω_m is multiplied by pole pair number and then integrated to get the rotor electrical angle θ_e in radians. The I_{ref} and θ_e are given as input to a Current Transformation block whose output is the reference stator phase current i_{abc}* (i_a*, i_b* and i_c*) for each phase. The reference stator phase current values are shown in Table 7.7. The rotor position θ_e is initially set to 0 degrees although any convenient starting value can be assumed. The reference phase current I_{abc}* is compared with the respective actual stator phase current of motor i_{abc} using separate comparators, and the error current outputs are given to a hysteresis current comparator and gate pulse generator along with that of Hall sensor output to generate the gate pulse for the inverter switches. Here a small hysteresis band current δ_i is assumed. Considering Phase A, when reference current

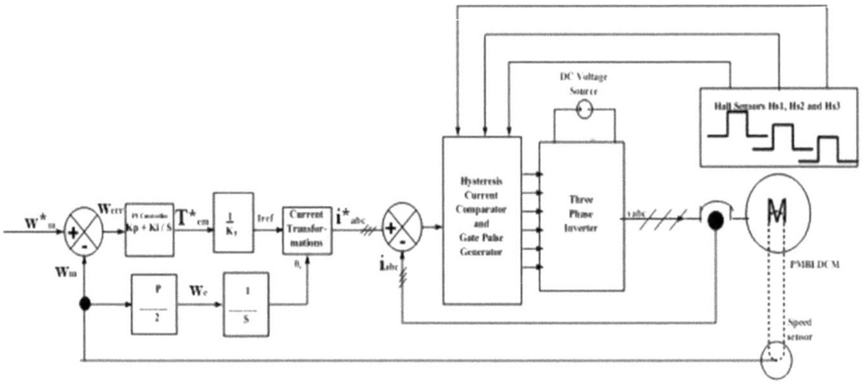

Fig. 7.29 Block diagram of a hysteresis current-controlled three-phase inverter-fed PMBLDCM drive

Table 7.7 Current transformation

Sl. no.	Rotor angle θ_e in electrical degrees	Reference stator phase current in amps I_a*	I_b*	I_C*	Switches ON (Fig. 7.1)
1	$0 \leq \theta_e < 60$	$+I_{ref}$	$-I_{ref}$	0	S1A, S4B
2	$60° \leq \theta_e < 120$	$+I_{ref}$	0	$-I_{ref}$	S1A, S6C
3	$120° \leq \theta_e < 180$	0	$+I_{ref}$	$-I_{ref}$	S3B, S6C
4	$180° \leq \theta_e < 240$	$-I_{ref}$	$+I_{ref}$	0	S2A, S3B
5	$240° \leq \theta_e < 300$	$-I_{ref}$	0	$+I_{ref}$	S2A, S5C
6	$300° \leq \theta_e < 360$	0	$-I_{ref}$	$+I_{ref}$	S4B, S5C

I_a* is positive and increasing, that is when $I_a* >= 0$, if the actual motor current is ia > ($I_a* + \delta_i$), then switch S1A is OFF, and S2A is ON, and when ia < = ($I_a* - \delta_i$), S1A is ON, and S2A is OFF. Similarly when I_a* is negative and decreasing, that is when $I_a* < 0$, if the actual motor current is ia > ($I_a* + \delta_i$), then switch S1A is OFF, and S2A is ON, and when ia < = ($I_a* - \delta_i$), S1A is ON, and S2A is OFF. Similar principle holds good for switches in Phase B and C.

The model of the hysteresis current controlled three-phase BLDCM drive is shown in Fig. 7.30 (model file: EXAMPLE 7_5). The model subsystems are (1) three-phase inverter, (2) BLDC motor, (3) PI controller and (4) hysteresis current controller and (5) gate pulse generator.

Three-phase inverter with DC voltage source is the same as explained in Sect. 7.6.2. [16]. The BLDCM is a built-in model from Simscape Electrical—Electromechanical—Permanent Magnet library. The relevant data for rotor and stator are entered from Table 7.6. Wye wound stator winding connection is selected. Back e.m.f. profile is selected as perfect trapezoid with provision to specify maximum rotor-induced back e.m.f. Rotor angle definition is selected as angle between a-phase magnetic axis and the q-axis. For stator parameterisation option, specify L_s, L_m and M_s. is selected. Zero sequence option is excluded. Rotor inertia and damping are entered zero. However rotor inertia is selected using an external inertia source

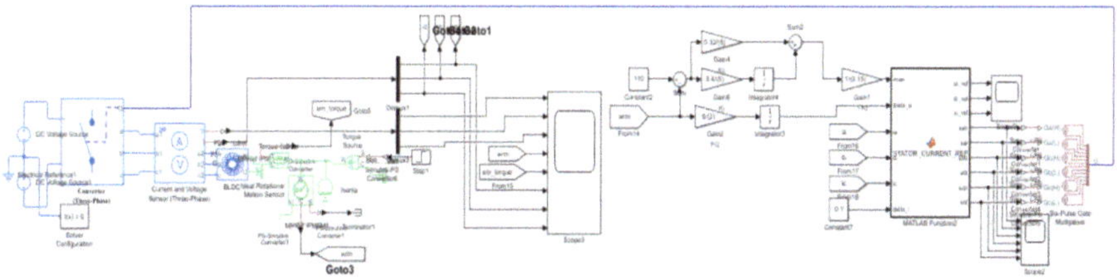

Fig. 7.30 Model of hysteresis current-controlled three-phase inverter-fed PMBLDCM drive

from Simscape Foundation library—Mechanical—Rotational Elements block set where the data for moment of inertia J is entered from Table 7.6. The ideal Torque Source is from Simscape Foundation library—Mechanical—Mechanical sources block set. To this Torque Source, load torque TL which is a step input is connected to terminal S which is the physical signal port. Ideal Torque Sensor and Ideal Rotational Motion Sensor are from Simscape Foundation library—Mechanical—Mechanical Sensors block set, and Mechanical Rotational Reference is from Mechanical Rotational Elements block set. The Torque Source output terminal R is connected to torque sensor terminal R whose output terminal C is given to the rotor terminal R of BLDCM. The inertia block and Ideal Rotational Motion Sensor R terminal are connected in parallel to the output terminal R of Torque Source as shown in Fig. 7.30. The C terminal of BLDCM, Torque Source and that of Ideal Rotational Motion Sensor are connected to rotational mechanical reference. The terminal marked T of Ideal Torque Sensor gives the electromagnetic torque output of motor. Terminals W and A of Ideal Rotational Motion Sensor give the angular velocity ω_m and rotor angle θ_m of motor.

The PI controller parameters are initially derived from the open loop plot of rotor speed ω_m shown in Fig. 7.28 using Ziegler-Nichols algorithm for PI controller tuning [24]. Using this value of K_p and K_i, closed loop response of rotor speed ω_m is obtained, and from this plot, values of K_p and K_i are further tuned so as to get stator reference phase currents i_{abc}* within reasonable limits. These values of K_p and K_i are 0.064 and 0.68, respectively. PI controller is derived using gain, integrator and summer blocks as shown in Fig. 7.30. A constant block is used to enter the reference speed ω_m* which is compared with actual rotor speed ω_m using comparator, and the error output is given to the PI controller.

The hysteresis current controller and gate pulse generator are developed using Embedded MATLAB function STATOR_CURRENT_REF shown in the model file EXAMPLE 7_5. The actual rotor speed ω_m is multiplied by pole pair number and integrated using integrator block to get the rotor electrical angle θ_e. The PI controller output is divided by torque constant K_T to get the maximum value of reference phase current imax. This imax, θ_e, stator phase currents i_{abc} from BLDCM model and assumed value of hysteresis current band δ_i are given as input to the STATOR_CURRENT_REF function block. The outputs of this function block are the stator reference phase currents i_{abc_ref} and gate pulse for the upper and lower switches in each phase of inverter marked sah, sal, sbh, sbl, sch and scl, respectively.

The program to generate the reference phase currents and six-gate pulse are shown in Program Segment 7.4 in the model file EXAMPLE 7_5. Here for every 60° interval for θ_e commencing from zero, stator reference phase current values i_{abc}* are allocated the value as per Table 7.7, and the relevant inverter switches are either set HIGH (logic 1) or LOW (logic 0). The values of switches set HIGH are then multiplied by a gain of ten so that the gate pulse amplitude is greater than the gate threshold voltage of the switch. These gate pulses are given to the gate input of three-phase inverter through a six-pulse gate multiplexer.

7.6.6 Simulation Results

The simulation of the model shown in Fig. 7.30 is carried out using ode15s (stiff/NDF) solver in Simulink [18]. The data shown in Table 7.6 are used. The load torque TL is initially set to zero and steps up to 0.83 Nw-metres at 50 milliseconds. The hysteresis current band is 0.1 amps. The rotor reference speed is set to 110 mech rad/sec. The simulation results are shown in Figs. 7.31 and 7.32, respectively.

7.6.7 Discussion of Results

The rotor mean speed is found to be 118.6 mech rad/sec from Fig. 7.31 which is close to the set point value. The stator reference phase currents follow a quasi square wave as shown in Fig. 7.32. The stator reference current frequency in hertz closely corresponds to the rotor speed in electrical radians per second.

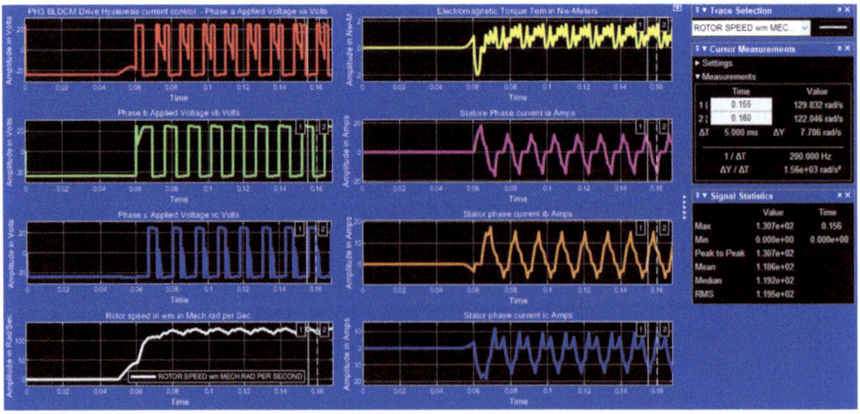

Fig. 7.31 Three-phase hysteresis current-controlled inverter-fed BLDCM drive: three-phase inverter output voltage V_{abe} and rotor speed w_m (left column top to bottom), Electromagnetic torque T_{em} and three-phase stator phase currents i_{abc} (right column top to bottom)

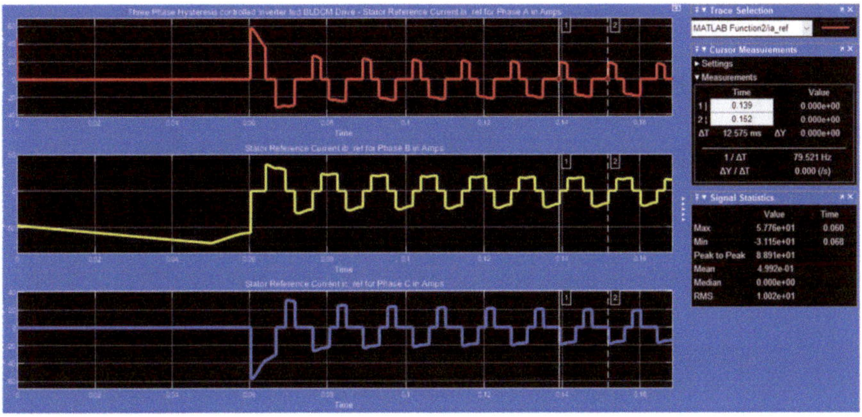

Fig. 7.32 Three-phase hysteresis current-controlled inverter-fed BLDCM drive—stator reference phase currents for phases A, B and C (top to bottom)

7.7 Case Study: Phase Angle Advance Control of Three-Phase PMBLDCM Drive for Extended Speed Range Operation

For many applications such as in electric vehicles, machine tools, high-speed operation, above the rated speed is required [25, 26]. In three-phase PMSM and IM drives, this high-speed range is achieved by field weakening. Here the direct axis component of current i_d responsible for flux production is set to zero or negative. Such a dq-axis transformation of three-phase stator phase currents is difficult with BLDCM due to its square wave nature. In the case of BLDCM, field weakening is achieved by advancing the phase angle of the stator phase currents i_x to a value θ_0 ahead with respect to the back e.m.f. e_x of the relevant phase x [25, 26]. In a BLDCM, two types of induced EMF are generated. One is the transformer induced EMF (TEMF) L^*di_x/dt due to the rate of change of current in the inductive stator phase windings, and the other is the back EMF (BEMF) e_x due to the magnetic field of PM rotor. For speed control below the rated speed, phase currents i_x are in phase with the relevant phase BEMF. For operation above rated speed when the phase currents i_x lead the relevant phase BEMF by an angle θ_0, the stator phase current for a given applied phase voltage v_x increases, and this increases TEMF which counteract or act against the BEMF. This is equivalent to field weakening operation [25, 26]. As a result the rotor speed increases above the rated speed. Advancing the phase angle of phase currents requires advancing the switching angle of the inverter. This in turn advances the phase of relevant applied phase voltage. The BEMF amplitude is proportional to rotor angular speed, and its frequency is that of inverter switching frequency which is also the rotor angular frequency of rotation in electrical radians per second and is independent of the switching angle of the inverter. Mathematically this can be expressed as in Eq. 7.55:

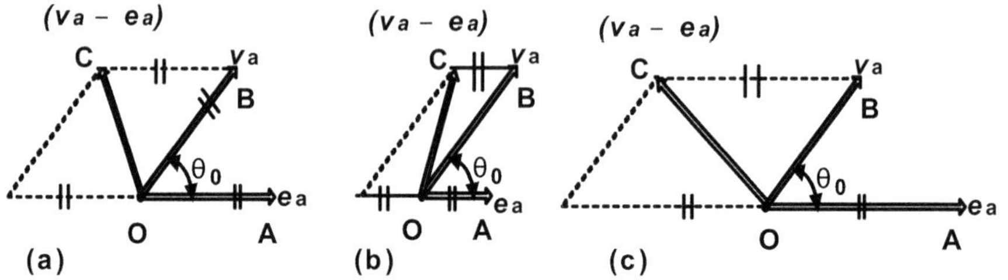

Fig. 7.33 Phasor diagram of va and ea with phase angle advance: (**a**) ea = va, (**b**) ea < va and (**c**) ea > va

$$v_{\mathrm{x}}\left(\theta_{\mathrm{e}}+\theta_{O}-\frac{n*2\pi}{3}\right)=R_{\mathrm{S}}*i_{\mathrm{x}}\left(\theta_{\mathrm{e}}+\theta_{\mathrm{O}}-\frac{n*2\pi}{3}\right)+L$$
$$*\,i_{\mathrm{x_{dot}}}\left(\theta_{\mathrm{e}}+\theta_{O}-\frac{n*2\pi}{3}\right)+e_{\mathrm{x}}\left(\theta_{\mathrm{e}}-\frac{n*2\pi}{3}\right) \tag{7.55}$$

In Eq. 7.55, $\theta_{\mathrm{e}}=\omega_{\mathrm{e}}{*}t$, $i_{\mathrm{x\_dot}}=di_{\mathrm{x}}/dt$ for $x \in$ a, b, c and $n = 0, 1, 2$ for $x =$ a, b, c, respectively.

This is illustrated in Fig. 7.33 for Phase A values. From Fig. 7.33 with phase advance angle θ_0, it is clear that (v_{a}—e_{a}) value is always nonzero and hence stator phase current is either positive increasing or negative decreasing. With no phase angle advance, when va = ea., there is no phase current flow. The BEMF ea. is zero while starting and increases with increasing rotor speed. The detailed analysis of phase angle advance approach and the derivation for calculation of phase advance angle for various rotor speeds are given in the literature reference [25]. The extension of speed range above the reference speed and the nonzero stator phase current are illustrated by modelling in the next section.

7.7.1 Modelling of a Hysteresis Current Controlled Three-Phase Inverter-Fed PMBLDCM Drive with Phase Angle Advance Control

The block diagram of the hysteresis current controlled three-phase inverter-fed PMBLDCM drive is shown in Fig. 7.34. Comparing with the model block diagram shown in Fig. 7.29, the only difference here is that the rotor electrical angle θ_{e} is added with inverter switching angle or phase angle advance θ_0 in a separate adder and the resulting output is given to the Current Transformation block. All other parts of the block diagram are the same as in Fig. 7.29. The operation of the various units of Fig. 7.34 is the same as explained in Sect. 7.6.5.

The model of the hysteresis current controlled three-phase inverter-fed PMBLDCM drive with phase angle advance is shown in Fig. 7.35 (model file: CASE_STUDY_EX7_2). The model subsystems are the same as presented in Sect.

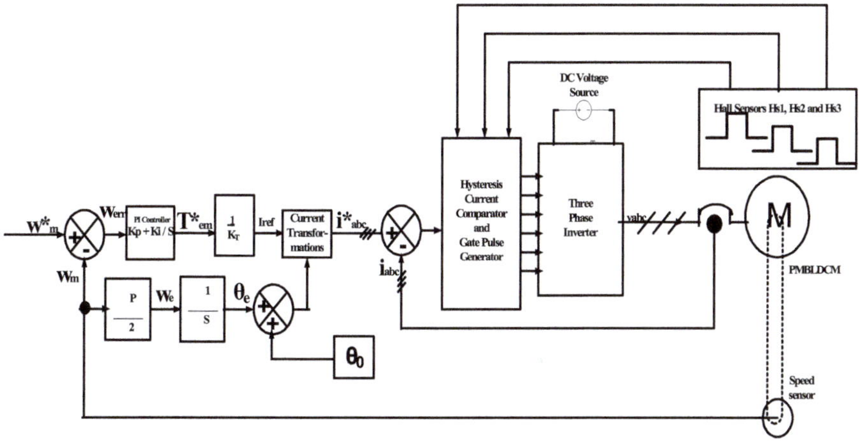

Fig. 7.34 Block diagram of a hysteresis current-controlled three-phase inverter-fed PMBLDCM drive with phase-angle advance control

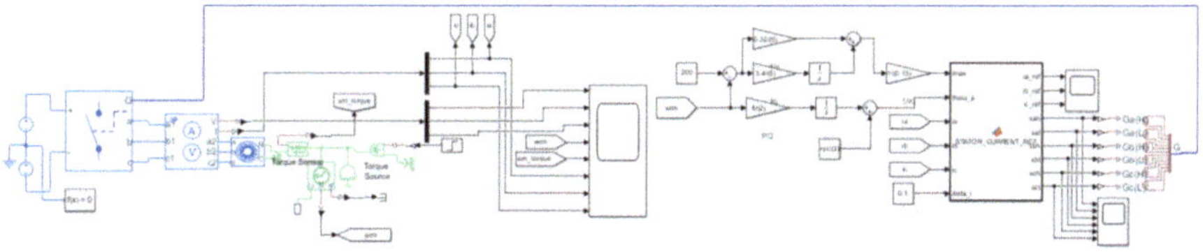

Fig. 7.35 Model of hysteresis current-controlled three-phase inverter-fed PMBLDCM drive with phase-angle advance control

7.6.5. Here the rotor electrical angle θ_e is added with phase angle advance θ_0 in a separate summing unit and is given as theta_e input to the STATOR_CURRENT_REF function block. The same Program Segment 7.4 in the model file CASE_STUDY_EX7_2 is used to allocate the current transformation and to generate the gate pulse for the three-phase inverter.

7.7.2 *Simulation Results*

The simulation of the model shown in Fig. 7.35 is carried out using ode15s (stiff/NDF) solver in Simulink [18]. The data shown in Table 7.6 are used. The load torque TL is initially set to zero and steps up to 0.83 Nw-metres at 50 milliseconds. The hysteresis current band is 0.1 amps. The phase angle advance θ_0 is set to $+\pi/2$ electrical radians. The rotor reference speed is set to 200 mech rad/sec. The simulation results are shown in Figs. 7.36 and 7.37, respectively.

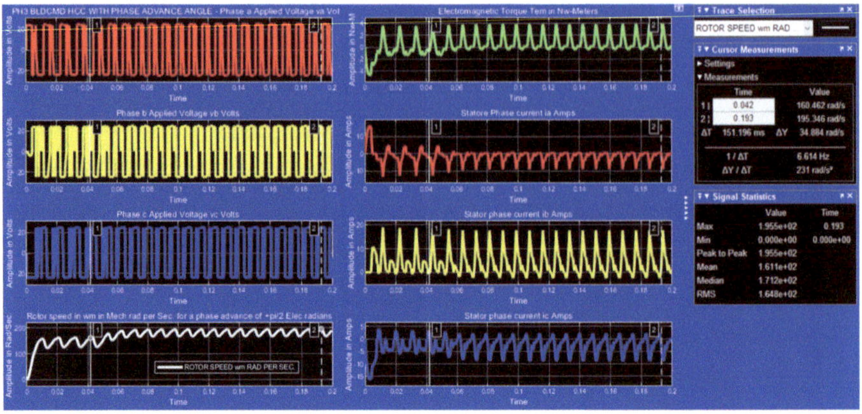

Fig. 7.36 Three-phase hysteresis current-controlled BLDCM drive with phase-angle advance control—three-phase applied voltage vabc and rotor speed (left column top to bottom). Electromagnetic torque T_{em} and three-phase stator currents iabc (right column top to bottom)

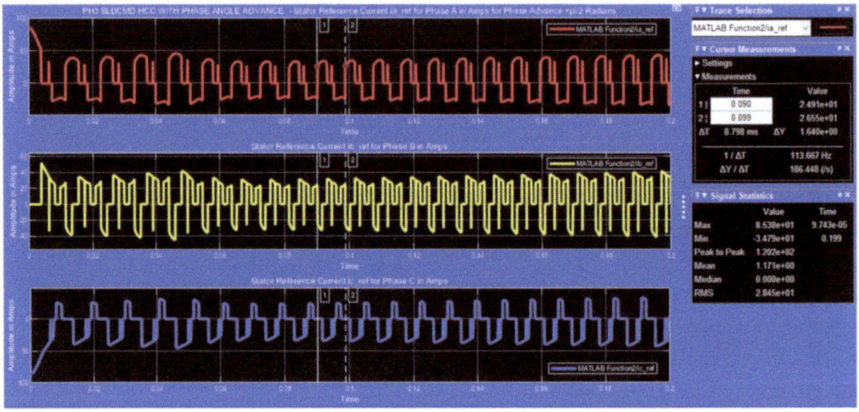

Fig. 7.37 Three-phase hysteresis current-controlled BLDCM drive with phase-angle advance control—stator reference phase current for phases A, B and C (top to bottom)

7.7.3 Discussion of Results

From Fig. 7.36 it is clear that with a phase angle advance of $+\pi/2$ electrical radians, the rotor speed reaches a peak value of 195.5 and a mean value of 161.1 mech rad/sec, whereas from Fig. 7.31 with no phase angle advance, the rotor speed reaches a peak value of 130.7 and a mean speed of 118.6 mech rad/sec. Also with no phase angle advance, it is seen from Fig. 7.31 that the rotor speed, e.m. torque and three stator phase currents are all zero from the start till 50 milliseconds and, when load torque is applied at this instant, start increasing in value. In Fig. 7.36 with phase angle advance, it is seen that the rotor speed, e.m. torque and three-phase stator currents all start increasing from the start and at 50 milliseconds when the load torque is applied, rotor speed and stator phase currents further increase, whereas the e.m. torque decreases in value. Thus speed increase is observed with phase angle advance.

7.8 Conclusions

Models for the six-step continuous and discontinuous conduction mode inverter have been successfully developed and presented with simulation results. The criteria for inverter switching angle advance φ_{MT} for maximum electromagnetic torque of six-step continuous current mode inverter-fed PMSM drive are verified by modelling technique. For the six-step Lybotec inverter-fed PMSM drive which is a six-step discontinuous conduction mode inverter-fed PMSM drive forming a sensorless control technique, the experimental results for stator line to neutral voltage and rotor speed closely agree with that of simulation results, and the error for the latter case is 13.5%. The PMSM stator current comparison by experiment and that by model simulation are not possible as the Lybotec six-step inverter has op.amp. current comparators and current limit switches which are not incorporated in the model as the current controller transfer function is unknown. Vector control of three-phase PMSM when driven by inverter with sinusoidal three-phase sine wave AC power supply as well as when fed by three-phase SVPWM inverter is presented with model simulation results which show the accuracy of vector control. Open loop model for three-phase inverter-fed PMBLDCM drive is presented. This is followed by models for three-phase hysteresis current controlled inverter-fed BLDCM drive with zero and $+\pi/2$ phase angle advance, and the rotor speed increases, and continuous stator current flow by phase angle advance is established by model simulation.

References

1. J.F. Gieras and M. Wing: "Permanent Magnet Motor Technology"; Marcel Dekker Inc.; New York; 2002; pp. 211–212.
2. T. Kenjo and S. Nagamori: "Permanent-Magnet and Brushless DC Motors"; Oxford Press; 1985; pp. 58–64.
3. T.J.E. Miller: "Brushless Permanent and Reluctance Motor Drives"; 1993; pp. 80–83.
4. Paul C. Krause: "Analysis of Electric Machinery"; McGraw Hill Inc.;1986; pp. 178–197.
5. R. Krishnan: "Electric Motor Drives Modeling Analysis and Control"; Prentice Hall Inc.; 2001.
6. P.C. Krause, R.R. Nucera R.J. Krefta and O. Wasynczuk: "Analysis of a Permanent magnet Synchronous Machine Supplied From A 180° Inverter With Phase Control"; IEEE Transactions on Energy Conversion; Vol.EC-2; No.3; September 1987; pp. 423–431.
7. P. Pillay and R. Krishnan: "Modeling, Simulation and Analysis of Permanent– Magnet Motor Drives, Part I: The Permanent magnet Synchronous Motor Drive"; IEEE Transactions on Industry Applications, Vol.25, No.2, March/April 1989; pp. 265–273.

8. Cui Bowen, Zhou Jihua and Ren Zhang: "Modeling and Simulation of Permanent Magnet Synchronous Motor Drives"; Proceedings of the Fifth International Conference on Electrical Machines and Systems; 2001; pp. 905–908.
9. R.R. Nucera, S.D. Sudhoff and P.C. Krause: "Computation of Steady-State Performance of an Electronically Commutated Motor", IEEE Transactions on Industry Applications, Vol.25, No.6, Nov/Dec 1989, pp. 1110—1117.
10. P. Pillay and R. Krishnan: "Modeling of Permanent magnet Motor Drives"; IEEE Transactions on Industrial Electronics; Vol.35; No.4; November 1988; pp. 537–541.
11. Chee-Mun-Ong: "Dynamic Simulation of Electric machinery Using MATLAB/SIMULINK"; Prentice Hall Inc.; 1998.
12. N.P.R. Iyer and J.G. Zhu: Interactive Modeling and Simulation of A Six Step Continuous Current Mode Inverter fed PMSM Drive Using SIMULINK", 17th International Conference on Electrical Machines, Greece, OTM2-4, September 2–5, 2006.
13. N.P.R. Iyer and J.G. Zhu: "Modeling and Simulation of A Six Step Discontinuous Current Mode Inverter Fed Permanent Magnet Synchronous Motor Drive Using SIMULINK", Sixth IEEE International Conference on Power Electronics and Drive Systems (PEDS05), Kuala Lumpur, Malaysia, Nov.-Dec. 2005, pp. 1056–1061.
14. N.P.R. Iyer and J.G. Zhu: "Interactive Modelling of a Six Step Discontinuous Current Mode Inverter fed Permanent Magnet Synchronous Motor Drives", 37th IEEE Power Electronics Specialists Conference, June 18–22, 2020, pp. 261–315.06, Korea, PS 1–94, ISBN 1-4244-9717-7.
15. N.P.R. Iyer and J.G. Zhu: "Modelling and Simulation of a Six Step Discontinuous Conduction Mode MOSFET Inverter Fed Permanent Magnet Synchronous Motor Drive"; Australasian Universities Power Engineering Conference; Melbourne, December 10–13, 2006.
16. N.P.R. Iyer: "Power Electronic Converters: Interactive Modelling Using SIMULINK", CRC Press, FL, USA, 2018, pp. 60–68.
17. N P R Iyer: "MATLAB/SIMULINK Modules for Modeling and Simulation of Power Electronic Converters and Electric Drives", M.E.(Research) thesis, The University of Technology Sydney, NSW, Australia, 2006, Chapter 8, pp. 261–315.
18. The MathWorks Inc., www.mathworks.com: "MATLAB/Simulink, R2017b, 2017.
19. Ned Mohan: "Advanced Electric Drives: Analysis, Control and Modeling Using MATLAB/ Simulink", Ch.10, John Wiley, 2014.
20. A Consoli, A. Raciti and A. Testa: "Sensorless Vector and Speed Control of Brushless Motor Drives", IEEE Transactions on Industrial Electronics, Vol. 41, No. 1, February 1994, pp. 91–96.
21. J. Mat Lazi, Z. Ibrahim, S.N. Mat Isa, A. Anugerah, F.A. Patakor and R. Mustafa: "Sensorless Speed Control of PMSM Drives using dSPACE DS1103 Board", IEEE International Conference on Power and Energy (PECon), 2-5, December 2012, Malaysia, pp. 922–927.
22. P. Pillay and R. Krishnan: "Modeling, Simulation, and Analysis of Permanent-Magnet Motor Drives, Part II: The Brushless DC Motor Drive", IEEE Transactions On Industry Applications, Vol. 25, No. 2, March/April 1989, pp. 274–279.
23. Moog Inc.: "Silencer series Brushless DC Motor", BSG23 Specifications, January 2019.
24. K. Ogata: "Modern Control Engineering", Prentice Hall Inc., NJ, USA, 2010.
25. B. Nguyen and C. Ta, "Phase advance approach to expand the speed range of brushless DC motor," International Conference on Power Electronics and Drive Systems, 2007, pp. 1255–1262.
26. Chau, K. T.: "Electric Vehicle Machines and Drives: Design, Analysis and Application", John Wiley & Sons, Inc., 2015. pp. 69–94.

Chapter 8
Three-Phase Inverter-Fed Induction Motor Drives

8.1 Introduction

To study the transients due to the effect of load and supply frequency variations, dynamic models of three-phase induction motor (IM) are developed using dq0-axis voltage-current and flux linkage equations. The abc-axis to dq0-axis voltage and current transformations are performed using the well-known Park transformation, and the results obtained in dq0-axis can be brought back to abc-axis by inverse Park transformation [1, 2, 5]. The reference frames used for this abc-axis to dq0-axis transformation can be any of arbitrary, synchronous, stator or stationary and rotor. Also to control the output voltage in the case of three-phase inverter-fed IM drive, pulse width modulation (PWM) technique is used.

In this chapter interactive model of the three-phase IM is developed in synchronous reference frame using the well-known flux linkage equations in state space [3, 4]. Then this model is used to study the performance of the IM when fed by three-phase two-level and three-level inverters with (a) sine-triangle carrier PWM (SPWM), (b) third harmonic injection PWM (THIPWM), (c) clipped sine PWM (CSPWM) and (d) multicarrier sine level shift PWM (MCSLSPWM). Models for vector control and direct torque control (DTC) of three-phase IM drive using classical switching table are presented. Four case studies are presented: one for the three-phase space vector pulse width modulated (SVPWM) inverter-fed IM drive, the second for that of vector controlled IM drive fed by three-phase sine wave AC voltage and third and fourth for DTC of three-phase IM drive fed by SVPWM two-level inverter using modified switching table. Finally steady state analysis using equivalent circuit, scalar control and models for closed loop constant stator voltage to frequency ratio control for six-step 180-degree mode inverter-fed and three-phase sine PWM inverter-fed IM drive are presented.

Supplementary Information The online version contains supplementary material available at https://doi.org/10.1007/978-3-031-62784-2_8.

N. P. R. Iyer, *Inverters and AC Drives*, Power Systems,
https://doi.org/10.1007/978-3-031-62784-2_8

8.2 Model of Three-Phase Induction Motor

The model of the three-phase IM represented using dq-axis is shown in Fig. 8.1. The d-axis and q-axis stator and rotor windings are displaced in space by 90 degrees. The angular speed of the rotor is θ_{r_dot} which is $(d\theta_r/dt)$ radians per second. As rotor winding is aligned along the d-axis and q-axis, the mutual inductance between stator and rotor winding is a constant. The model shown in Fig. 8.1 is in the stator or stationary frame of reference. Reference frames are like observer platforms which simplify the system equations. It is advantageous to derive the system model in any arbitrary reference frame rotating at any arbitrary angular speed ω_c radians per second and then transfer this model to any desired reference frame. The relationship between the stationary reference frame marked by dq-axis and arbitrary reference frame marked by d^cq^c-axis is shown in Fig. 8.2 [5]. The relationship between the electrical quantities in the stationary dq reference frame and the arbitrary d^cq^c reference frame is given by Eq. 8.1. In Eq. 8.1, f can be voltage, current or flux linkage.

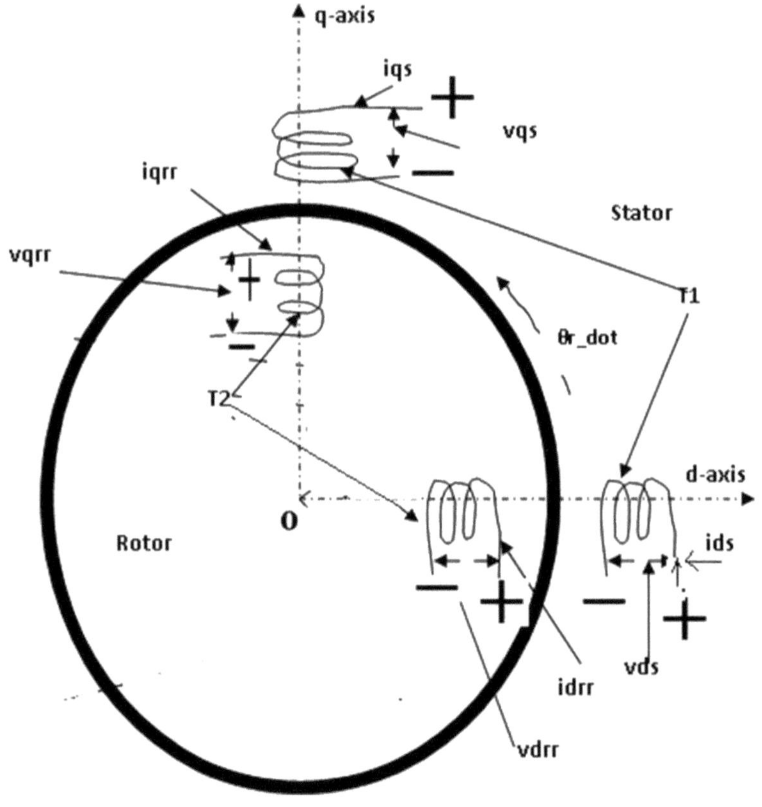

Fig. 8.1 Two-axis representation of three-phase induction motor

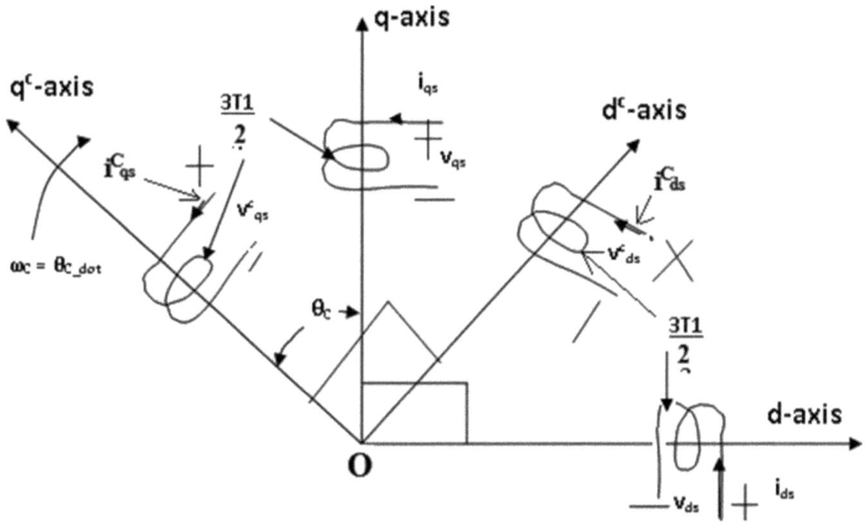

Fig. 8.2 Stationary and arbitrary reference frames

$$\begin{bmatrix} f_{qs} \\ f_{ds} \end{bmatrix} = \begin{bmatrix} \cos\theta_C & \sin\theta_C \\ -\sin\theta_C & \cos\theta_C \end{bmatrix} * \begin{bmatrix} f^C_{qs} \\ f^C_{ds} \end{bmatrix} \tag{8.1}$$

The model of the three-phase IM in the arbitrary reference frame ω_C can be expressed as in Eq. 8.2. In Eq. 8.2, the superfix C above dq-axis stator and rotor voltages and currents corresponds to arbitrary reference frame.

The electromagnetic torque Tem of the induction machine is given in Eq. 8.3.

$$\begin{bmatrix} v^C_{qs} \\ v^C_{ds} \\ v^C_{qr} \\ v^C_{dr} \end{bmatrix} = \begin{bmatrix} (R_S + sL_S) & (\omega_C * L_S) & sL_m & (\omega_C * L_m) \\ (-\omega_C * L_S) & (R_S + sL_S) & (-\omega_C * L_m) & sL_m \\ sL_m & (\omega_C - \omega_{re}) * L_m & (R_r + sL_r) & (\omega_C - \omega_{re}) * L_r \\ -(\omega_C - \omega_{re}) * L_m & sL_m & -(\omega_C - \omega_{re}) * L_r & (R_r + sL_r) \end{bmatrix} * \begin{bmatrix} i^C_{qs} \\ i^C_{ds} \\ i^C_{qr} \\ i^C_{dr} \end{bmatrix} \tag{8.2}$$

In Eq. 8.2, arbitrary reference frame ω_C is replaced by ω_e or ω_S for synchronous or excitation reference frame, ω_r for rotor reference frame and zero for stationary reference frame.

$$T_{\text{em}} = \left[\frac{3}{2}\right] * \left[\frac{P}{2}\right] * L_m * \left[i^C_{qs} * i^C_{dr} - i^C_{ds} * i^C_{qr}\right] \text{ Nw} - \text{Meter} \tag{8.3}$$

8.2.1 Model of Three-Phase Induction Motor in Synchronous Reference Frame Using dq0-Axis Flux Linkage Equations in State Space

The three-phase induction motor model in arbitrary reference frame ω_C, using dq0-axis flux linkage equations in state space is developed as shown in Eqs. 8.4, 8.5, 8.6, and 8.7 [2–5]. By letting ω_C the values of ω_e or ω_S, in Eqs. 8.4, 8.5, 8.6, and 8.7, the flux linkage equations in synchronous (excitation) reference frame can be obtained.

The stator and rotor referred to stator side qd0-axis voltage equations in arbitrary reference frame ω_C are given in Eqs. 8.4 and 8.5.

$$\begin{bmatrix} v_{qs} \\ v_{ds} \\ v_{0s} \end{bmatrix} = \begin{bmatrix} S & \omega_C & 0 \\ -\omega_C & S & 0 \\ 0 & 0 & S \end{bmatrix} * \begin{bmatrix} \lambda_{qs} \\ \lambda_{ds} \\ \lambda_{0s} \end{bmatrix} + r_S * \begin{bmatrix} i_{qs} \\ i_{ds} \\ i_{0s} \end{bmatrix} \tag{8.4}$$

$$\begin{bmatrix} v'_{qr} \\ v'_{dr} \\ v'_{0r} \end{bmatrix} = \begin{bmatrix} S & (\omega_C - \omega_{re}) & 0 \\ -(\omega_C - \omega_{re}) & S & 0 \\ 0 & 0 & S \end{bmatrix} * \begin{bmatrix} \lambda'_{qr} \\ \lambda'_{dr} \\ \lambda'_{0r} \end{bmatrix} + r'_r * \begin{bmatrix} i'_{qr} \\ i'_{dr} \\ i'_{0r} \end{bmatrix} \tag{8.5}$$

The stator and rotor qd0-axis flux linkage equation in arbitrary reference frame ω_C is given in Eq. 8.6.

$$\begin{bmatrix} \lambda_{qs} \\ \lambda_{ds} \\ \lambda_{0s} \\ \lambda'_{qr} \\ \lambda'_{dr} \\ \lambda'_{0r} \end{bmatrix} = \begin{bmatrix} (L_{ls}+L_m) & 0 & 0 & L_m & 0 & 0 \\ 0 & (L_{ls}+L_m) & 0 & 0 & L_m & 0 \\ 0 & 0 & L_{ls} & 0 & 0 & 0 \\ L_m & 0 & 0 & (L'_{lr}+L_m) & 0 & 0 \\ 0 & L_m & 0 & 0 & (L'_{lr}+L_m) & 0 \\ 0 & 0 & 0 & 0 & 0 & L'_{lr} \end{bmatrix} * \begin{bmatrix} i_{qs} \\ i_{ds} \\ i_{0s} \\ i'_{qr} \\ i'_{dr} \\ i'_{0r} \end{bmatrix} \tag{8.6}$$

The electromagnetic (E.M.) torque T_{em} is given in Eq. 8.7.

$$T_{em} = (3/2) * (P/2) * L_m * \left(i'_{dr} * i_{qs} - i'_{qr} * i_{ds}\right) \tag{8.7}$$

Rearranging Eqs. 8.4 and 8.5 and using Eq. 8.6, the state space representation of stator and rotor flux linkage equations can be expressed as in Eq. 8.8 [4]. In Eq. 8.8, λ_{qs_dot} is $(d\lambda_{qs}/dt)$ and so on.

Now in Eq. 8.6, representing the inductance matrix **L**, qd0-axis flux linkage matrix as **Γ** and qd0-axis stator and rotor currents as **I** and, in Eq. 8.8, angular frequency matrix containing ω_C and ω_{re} as **Ω**, stator and rotor resistance matrix as **R** and the qd0-axis stator and rotor voltage matrix as **V**, Eq. 8.8 can be expressed as in Eq. 8.9. Equation 8.9 is of the form **X_dot = A*X + B*U**.

$$\begin{bmatrix} \lambda_{qs\_dot} \\ \lambda_{ds\_dot} \\ \lambda_{0s\_dot} \\ \lambda'_{qr\_dot} \\ \lambda'_{dr\_dot} \\ \lambda'_{0r\_dot} \end{bmatrix} = \begin{bmatrix} 0 & -\omega_C & 0 & 0 & 0 & 0 \\ \omega_C & 0 & 0 & 0 & 0 & 0 \\ 0 & 0 & 0 & 0 & 0 & 0 \\ 0 & 0 & 0 & 0 & -(\omega_C - \omega_{re}) & 0 \\ 0 & 0 & 0 & (\omega_C - \omega_{re}) & 0 & 0 \\ 0 & 0 & 0 & 0 & 0 & 0 \end{bmatrix}$$

$$* \begin{bmatrix} \lambda_{qs} \\ \lambda_{ds} \\ \lambda_{0s} \\ \lambda'_{qr} \\ \lambda'_{dr} \\ \lambda'_{0r} \end{bmatrix} - \begin{bmatrix} r_s & 0 & 0 & 0 & 0 & 0 \\ 0 & r_s & 0 & 0 & 0 & 0 \\ 0 & 0 & r_s & 0 & 0 & 0 \\ 0 & 0 & 0 & r'_r & 0 & 0 \\ 0 & 0 & 0 & 0 & r'_r & 0 \\ 0 & 0 & 0 & 0 & 0 & r'_r \end{bmatrix} * \begin{bmatrix} i_{qs} \\ i_{ds} \\ i_{0s} \\ i'_{qr} \\ i'_{dr} \\ i'_{0r} \end{bmatrix} + \begin{bmatrix} v_{qs} \\ v_{ds} \\ v_{0s} \\ v'_{qr} \\ v'_{dr} \\ v'_{0r} \end{bmatrix} \tag{8.8}$$

$$\left[\overrightarrow{\mathbf{\Lambda\_dot}}\right] = \left[\vec{\mathbf{\Omega}}\right] * \left[\vec{\mathbf{\Lambda}}\right] - \left[\vec{\mathbf{R}}\right] * \left[\overrightarrow{\mathbf{L}^{-1}}\right] * \left[\vec{\mathbf{\Lambda}}\right] + \left[\vec{\mathbf{V}}\right] \tag{8.9}$$

The model of the three-phase IM using qd0-axis flux linkage equations in state space is shown in Fig. 8.3 (model file: EXAMPLE8_1). The various dialog boxes relating to this model are shown in Fig. 8.4 and the model subsystems are shown in Fig. 8.5a–c. The function of each subsystem is given below [4].

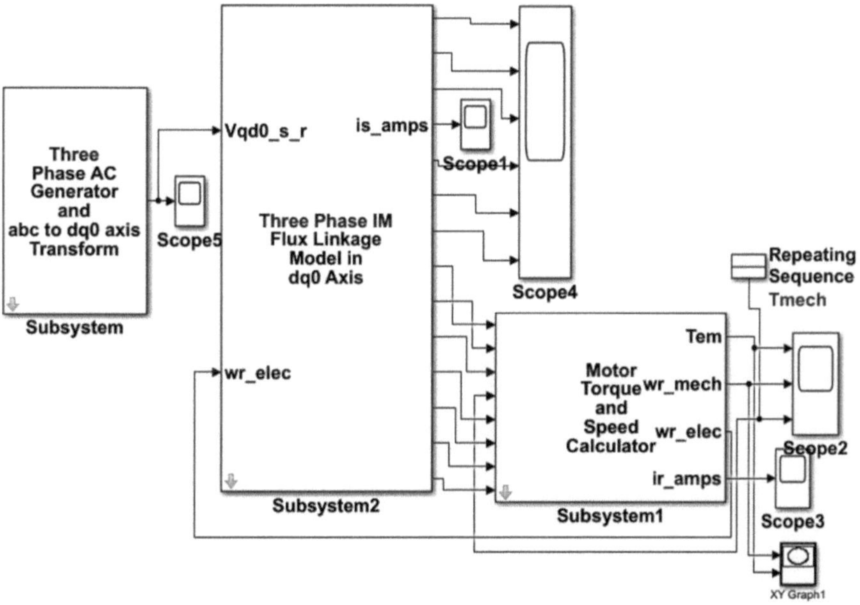

Fig. 8.3 Model of three phase IM in all reference frames using dq0-axis flux linkage equations in state space

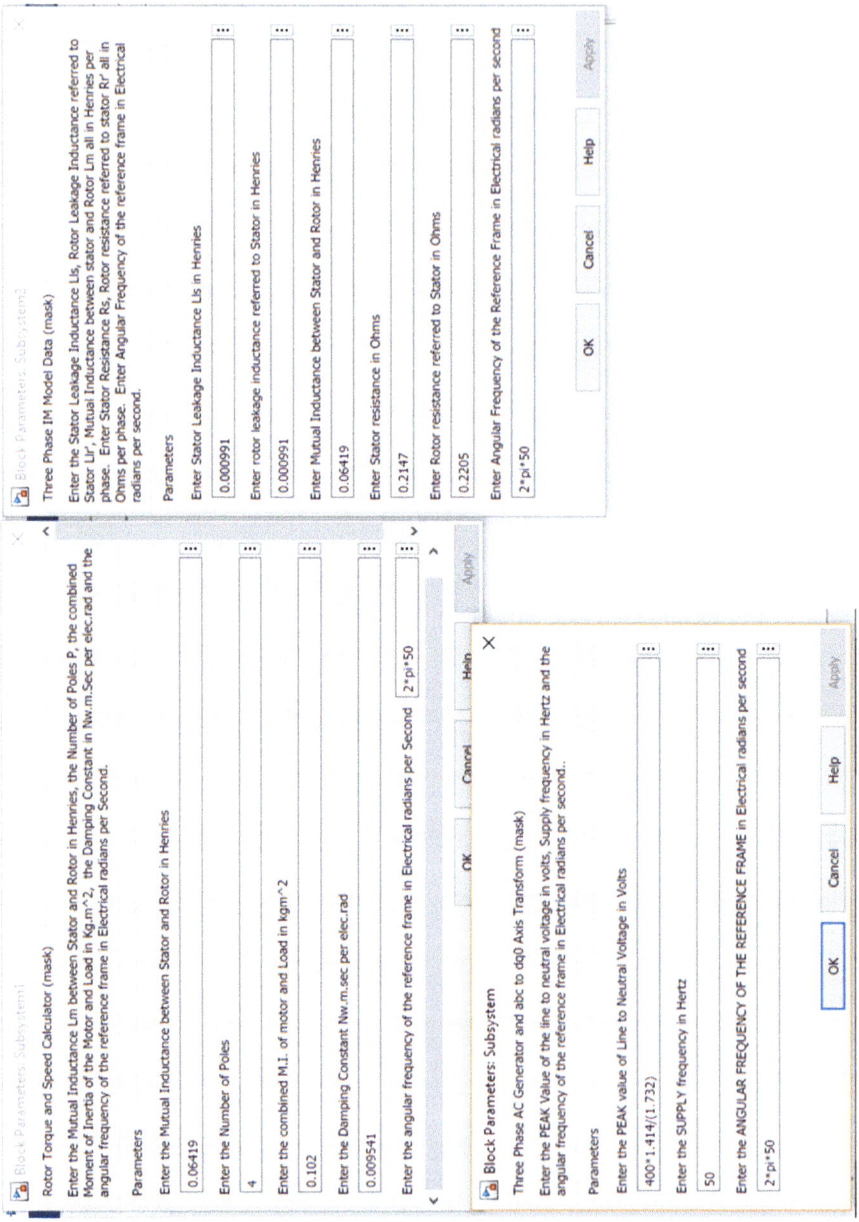

Fig. 8.4 Model of three phase IM in all reference frames using dq0-axis flux linkage equations in state space with dialog boxes

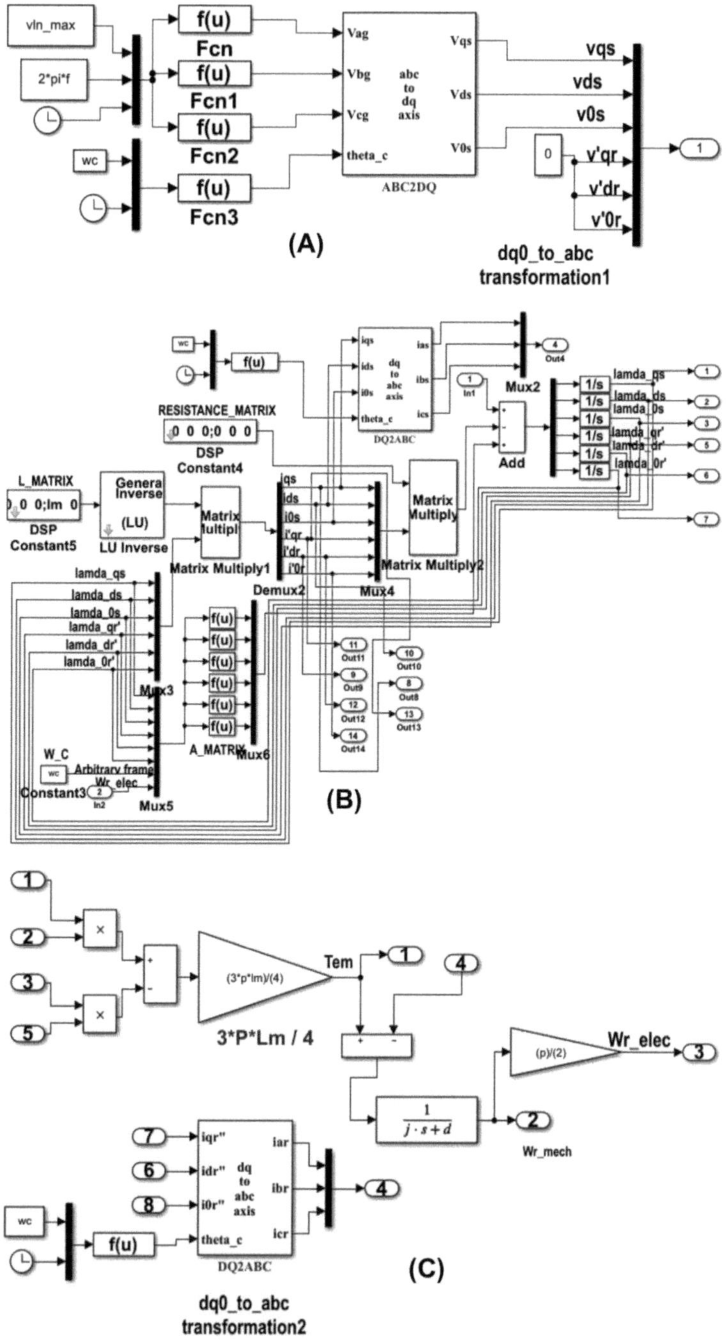

Fig. 8.5 Three phase IM flux linkage model in state space—model subsystems (**a**) Three phase AC generator and abc to dq0 axis transform. (**b**) Three phase IM model data. (**c**) Rotor torque and speed calculator

Figure 8.5a corresponds to the dialog box Three Phase AC Generator and abc- to dq0-axis transform in Fig. 8.4. The line to neutral voltage peak value, supply frequency in hertz and the angular frequency of the reference frame in electrical radians per second are entered in this box. With line to neutral peak voltage, frequency of the AC supply voltage and time as input to a three-input mux and with angular frequency of the reference frame ω_C and time as input to a two-input mux, the abc to dq0 transformation is achieved using four Fcn blocks marked Fcn, Fcn1, Fcn2 and Fcn3 as per transformation matrix in Eq. 8.10 [1, 2]. The dq0-axis rotor voltages are zero as the rotor is short-circuited. Figure 8.5b corresponds to the dialog box Three Phase IM Model data in Fig. 8.4. In this dialog box, the values of Lls, Llr′, Lm, Rs, Rr′ and ω_C relating to stator, rotor and arbitrary frame are entered in the appropriate units. Here ω_C value used is the synchronous reference frame frequency ω_e. In Fig. 8.5b, the eight-input mux marked mux5 with dq0-axis stator and rotor flux linkages Γ, reference frame frequency ω_e and rotor speed ω_{r_elec} in electrical radians per second along with the six Fcn block solves the first term on the R.H.S. of Eq. 8.9. The DSP constant block marked L_MATRIX and the general inverse (LU) block compute inverse of L matrix. This inverse of L (L^{-1}) matrix is then multiplied by Γ matrix using Matrix Multiply1 block to obtain the dq0 stator and rotor currents which form the I matrix. The DSP constant block marked RESISTANCE_MATRIX in Fig. 8.5b generates the R matrix containing diagonal elements of stator and rotor resistance referred to stator defined in Eq. 8.8. This R matrix is then multiplied by the I matrix using the Matrix Multiply2 block to obtain the second term in Eq. 8.9. The dq0-axis voltage V matrix is available from Fig. 8.5a. These three terms are then summed with appropriate signs in the Add block to obtain the derivative of the stator and rotor flux linkage vector Γ_dot matrix. This is then integrated using individual integrator block to obtain the dq0-axis stator and rotor flux linkage vector Γ matrix. Also from the I matrix output of Matrix Multiply1 block, the qd0-axis stator current and the product (ω_C*t) which forms θ_C are given as input to a four-input Mux, and the abc-axis stator currents are derived using three Fcn blocks using the transformation matrix defined in Eq. 8.11 [1, 2]. In Eqs. 8.10 and 8.11, the term f stands for either voltage, current or flux linkage, and θ is the angle θ_C between the q-axis and reference phase A of three-phase abc-axis. Figure 8.5c corresponds to the dialog box Rotor Torque and Speed Calculator in Fig. 8.4. In this dialog box, the mutual inductance Lm, number of poles *P*, moment of inertia of motor and load, the damping constant and angular frequency of reference frame ω_C are entered in the appropriate units. In Fig. 8.5c, the two multiplier blocks, the subtract block and the gain multiplier block with the multiplier constant [3*P*Lm/4] calculate the rotor electromagnetic torque Tem given by Eq. 8.7. The mechanical load torque Tmech is generated using the repeating sequence block as shown in Fig. 8.3. The value of Tem and Tmech is subtracted using the Subtract block and given to the Transfer Fcn block with the transfer function [1/((Jm + Jl)·s + D)] to obtain the rotor speed in mechanical radians per second, which is then multiplied by gain block with the multiplier [*P*/2] to obtain the rotor speed in ω_{r_elec} electrical radians per second, where Jm and Jl are the moment of inertia (M.I.) of motor and load and *P* is the number of poles. The dq0 to abc transform block in Fig. 8.5c transforms the dq0-axis rotor currents and θ_C inputs back to abc-axis rotor currents as per the transformation matrix defined in Eq. 8.11.

$$\begin{bmatrix} f_q \\ f_d \\ f_0 \end{bmatrix} = \frac{2^*}{3} \begin{bmatrix} \cos(\theta) & \cos\left(\theta - \frac{2\pi}{3}\right) & \cos\left(\theta + \frac{2\pi}{3}\right) \\ \sin(\theta) & \sin\left(\theta - \frac{2\pi}{3}\right) & \sin\left(\theta + \frac{2\pi}{3}\right) \\ \frac{1}{2} & \frac{1}{2} & \frac{1}{2} \end{bmatrix} * \begin{bmatrix} f_a \\ f_b \\ f_c \end{bmatrix} \tag{8.10}$$

$$\begin{bmatrix} f_a \\ f_b \\ f_c \end{bmatrix} = \begin{bmatrix} \cos(\theta) & \sin(\theta) & 1 \\ \cos\left(\theta - \frac{2\pi}{3}\right) & \sin\left(\theta - \frac{2\pi}{3}\right) & 1 \\ \cos\left(\theta + \frac{2\pi}{3}\right) & \sin\left(\theta + \frac{2\pi}{3}\right) & 1 \end{bmatrix} * \begin{bmatrix} f_q \\ f_d \\ f_0 \end{bmatrix} \tag{8.11}$$

8.2.2 Simulation Results

The simulation of the three-phase cage IM whose parameters are given in Table 8.1 is carried out using ode15s (stiff/NDF) variable step solver in Simulink [6]. The synchronous or excitation reference frame with angular frequency (2*pi*50) elec. rad. per second is used. The external mechanical load T_{mech} is set to zero. Simulation results for the three-phase stator currents, rotor currents, electromagnetic torque,

Table 8.1 Three-phase cage IM model parameters

Sl. no.	Three-phase IM parameters	Value	Unit
1	Rated power	40	kVA
2	Rated stator voltage Vs	400	Volts
3	Number of phase	3	–
4	Rated supply frequency f	50	Hertz
5	Rated speed Nr	1460	rpm
6	Number of poles P	4	
7	Stator resistance Rs	0.2147	Ω/ph
8	Stator leakage inductance Lls	0.000991	H/ph
9	Rotor resistance referred to stator Rr′	0.2205	Ω/ph
10	Rotor leakage inductance referred to stator Llr′	0.000991	H/ph
11	Magnetising inductance or mutual inductance between stator and rotor Lm	0.06419	H/ph
12	Moment of inertia of rotor and load Jeq(Jm+Jl)	0.102	Kg.m^2
13	Damping constant	0.009541	Nw.m-sec

rotor speed, external mechanical load applied and the electromagnetic torque versus rotor speed are shown in Fig. 8.6a–d, respectively.

8.2.3 Discussion of Results

The model of the three-phase IM is developed using flux linkage equations in state space. Model simulation results indicate that with no external mechanical load, rotor speed reaches no-load speed close to the synchronous speed, stator currents are balanced in the steady state, and the frequency is 50 Hz, and so also is the case with rotor currents. Rotor torque reaches maximum and then to zero at steady state. No-load losses such as hysteresis, eddy current, friction, windage and bearing losses are not taken into consideration in this model. Hence the no-load steady state rotor speed shown in the simulation result is only approximate.

8.3 Model of Three-Phase Sine Pulse Width Modulated Inverter-Fed Induction Motor Drive

Induction motor drives are inverter driven. To control the output voltage, PWM technique is used. These PWM techniques are presented in Chaps. 1–4. In this section the model of the three-phase IM drive fed by sine PWM inverter is presented [7, 10]. Simulation result is shown one for the under-modulation region. A method to start the three-phase IM-fed sine PWM inverter, by gradually varying the amplitude modulation (AM) index Ma, is presented. The three-phase IM data are shown in Table 8.1. For sine PWM, sine wave frequency fm is 50 Hz and amplitude 10 V. The triangle carrier has a peak value of ±10 V and frequency fc is 1200 Hz. The frequency modulation (FM) index is fc/fm which is 24. The model of the three-phase sine PWM inverter-fed IM drive is shown in Fig. 8.7 (model file: EXAMPLE8_2) [4, 7]. The three-phase IM model is the same as presented in Sect. 8.2.1. The three-phase sine PWM inverter model is already presented in Sect. 1.2.1 of Chap. 1. Here a multiport switch is used to gradually increase the amplitude of sine modulating signal with time which in turn varies the AM index for starting the three-phase IM. The first input at the top of the multiport switch is the control input and there are six data input ports from 1 to 6. A repeating sequence block is connected to control input and assume six inputs I1 to I6 are connected to data inputs from 1 to 6. In the repeating sequence block, the time intervals t1 to t6 and the corresponding port numbers 1–6 are entered in the increasing order. At any time t3, repeating sequence block output is 3, and the data I3 connected to third data port appears at the output of multiport switch.

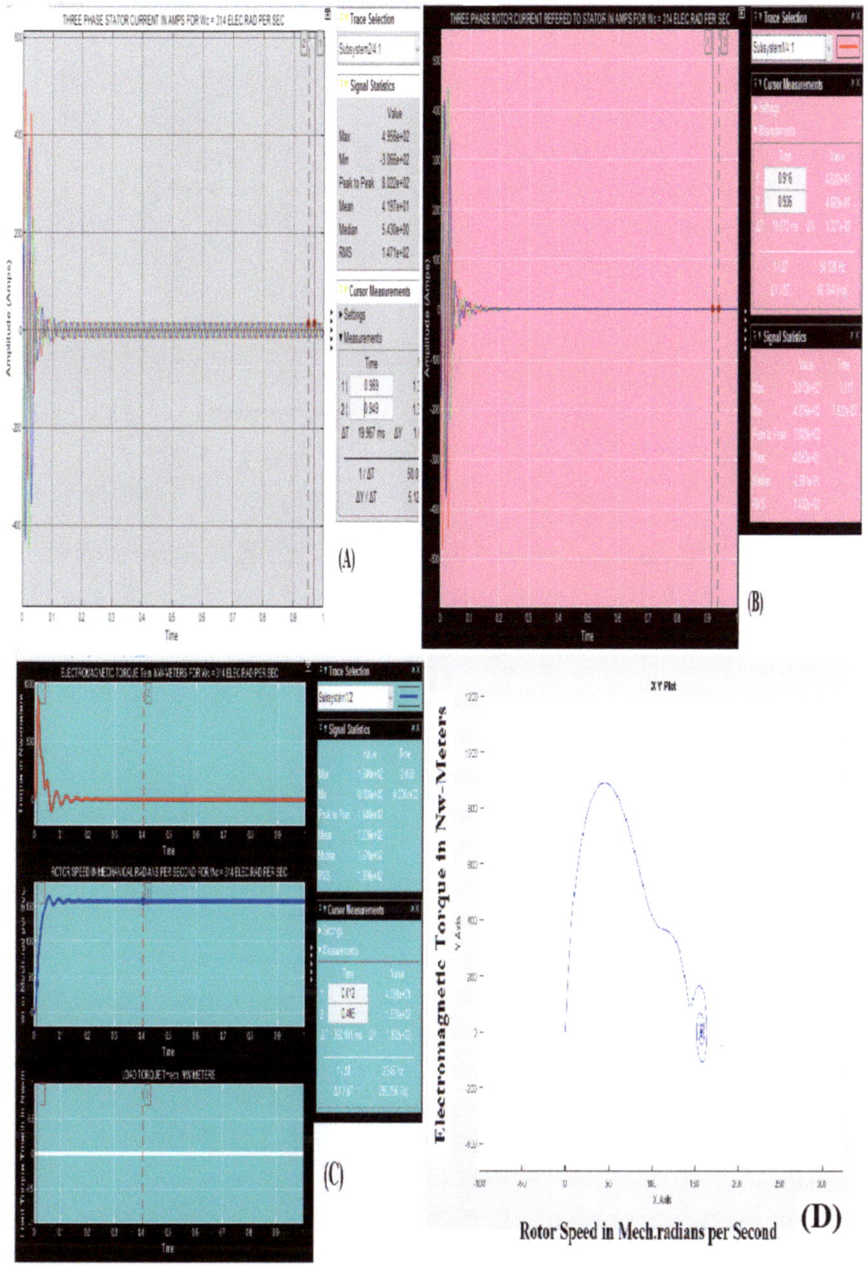

Fig. 8.6 Three-phase IM flux linkage model in state space—simulation results: (**a**) three-phase stator current in amps, (**b**) three-phase rotor current in amps, (**c**) E.M. torque in Nw-M (top), rotor speed in Mech.rad per sec. (middle) and mechanical load in Nw-M (bottom), E.M. torque in Nw-M versus rotor speed in Mech.rad per sec

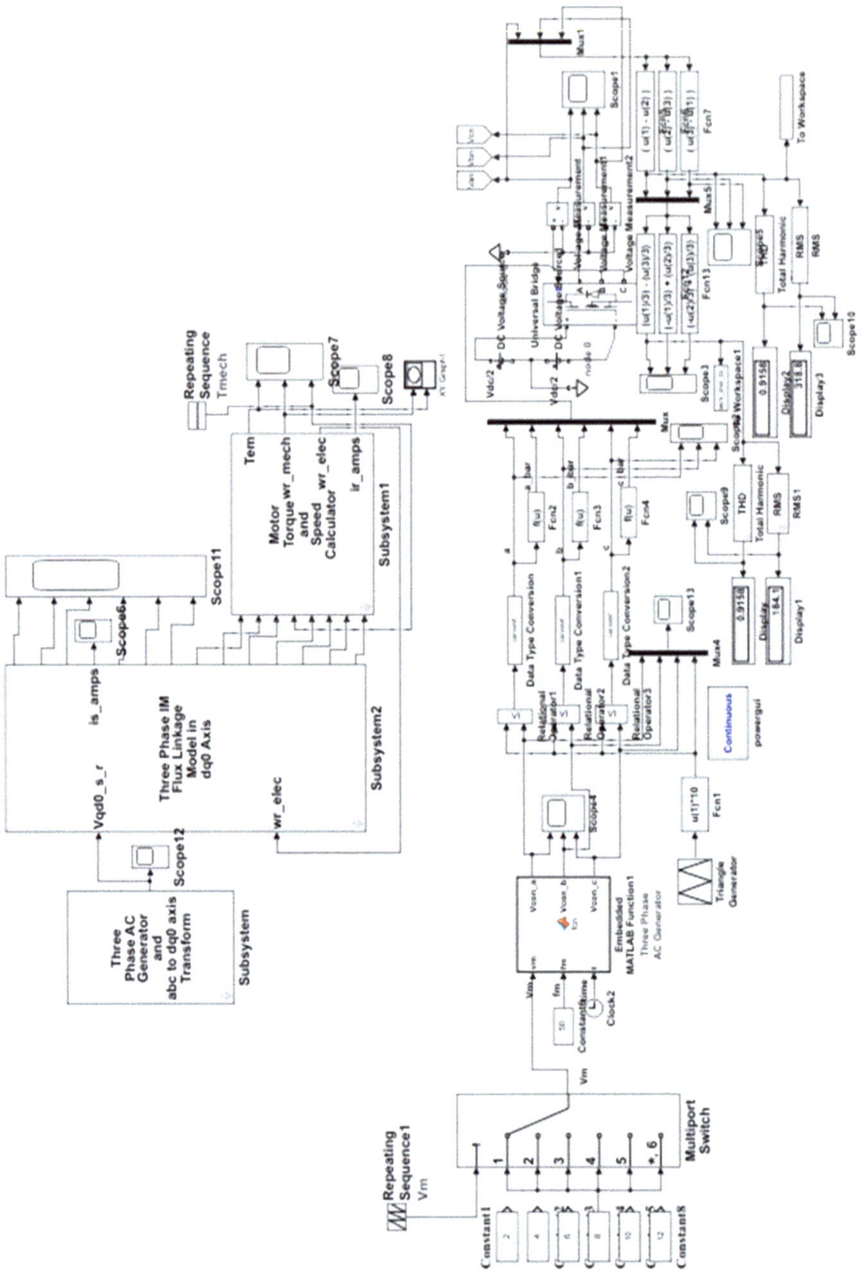

Fig. 8.7 Model of three-phase sine PWM inverter-fed induction motor drive

To study the performance of three-phase IM fed by sine PWM inverter, an AM index of 0.8 is selected. The constant block with value 8 is connected to all data input ports from 1 to 6 as shown in Fig. 8.7 and simulation is carried out for an AM index of 0.8. To study the starting of the above three-phase IM, amplitudes of modulating sine wave from 0.02 to 0.07 V in steps of 0.01 V are entered in the constant blocks and are connected to input data ports 1 to 6, respectively. In the repeating sequence block connected to control port, times 0 to 0.25 are entered at intervals of 0.05 s and the output values 1–6 are entered. Thus the AM index will vary from 0.002 to 0.007 at time intervals of 0.05 s applying a gradually increasing voltage to the stator of three-phase IM. This simulation results are also obtained.

8.3.1 Simulation Results

The simulation of the three-phase IM fed by sine PWM inverter is carried out using ode23tb (stiff/TR-BDF2) solver in Simulink [6]. Simulation results for an AM index of 0.8 are shown in Fig. 8.8a–d, respectively, and the starting phenomenon by gradually increasing the AM index is shown in Fig. 8.9a–d.

8.3.2 Discussion of Results

Here the three-phase IM is fed by square wave inverter with variable magnitude using sine PWM technique. The peak value of the stator and rotor current with AM index of 0.8 is less than that obtained by direct online starting with sine wave input voltage presented in Sect. 8.2.2. With an AM index of 0.8, the maximum electromagnetic torque is found to be 359.3 Nw-metres, and this occurs at a rotor speed of 15.63 Mech.rad per second from Fig. 8.8a, b. The starting phenomenon by gradually increasing the AM index shown in Fig. 8.9 indicates that as the AM index is gradually increased, the RMS value of the line to line voltage gradually increases from zero volts and stator currents, rotor currents, E.M. torque and rotor speed are well within the allowable range and increase gradually. Induction motor fed directly by three-phase sine wave voltage uses either autotransformers or star-delta starter for starting. Three-phase sine PWM inverter-fed IM drive can be started by varying the AM index.

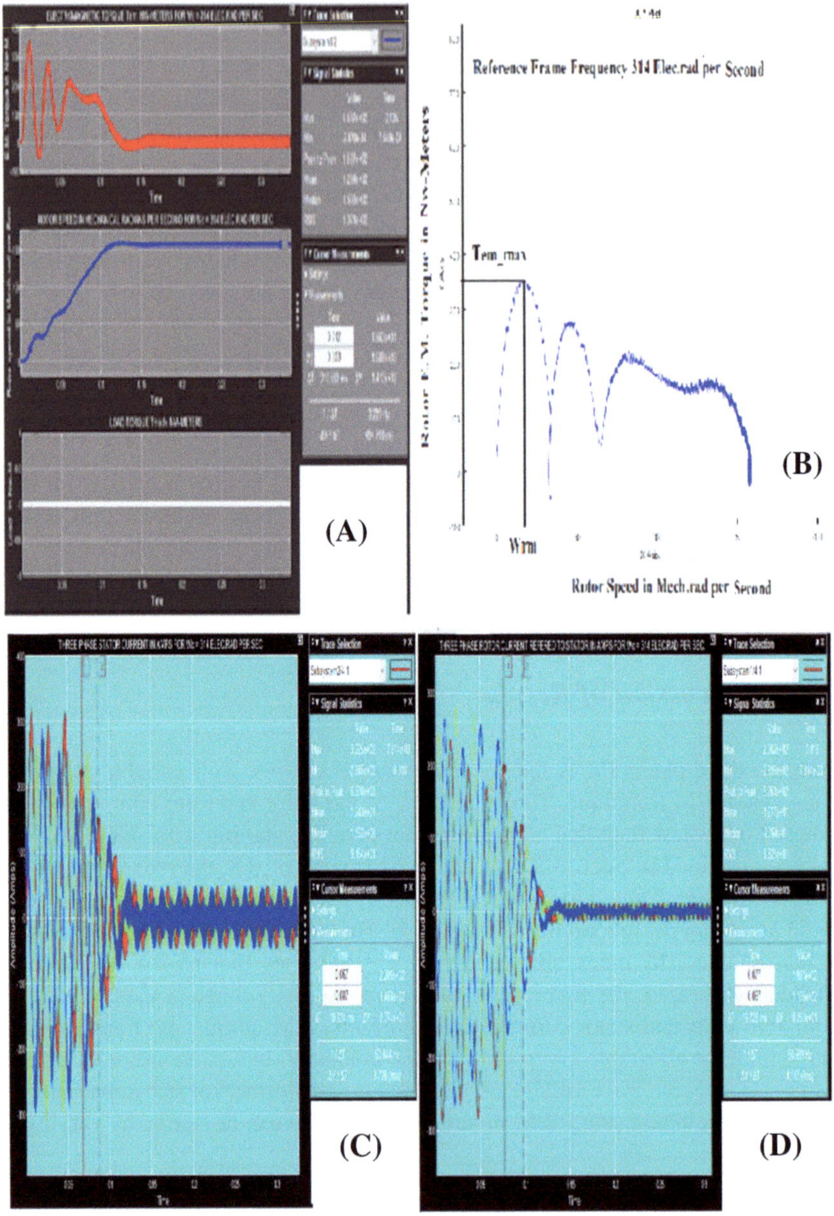

Fig. 8.8 Three-phase IM fed by sine PWM inverter—simulation results: (**a**) E.M. torque (top), rotor speed (middle) and mechanical load (bottom), (**b**) E.M. torque versus rotor speed curve, (**c**) three-phase stator currents and (**d**) three-phase rotor currents

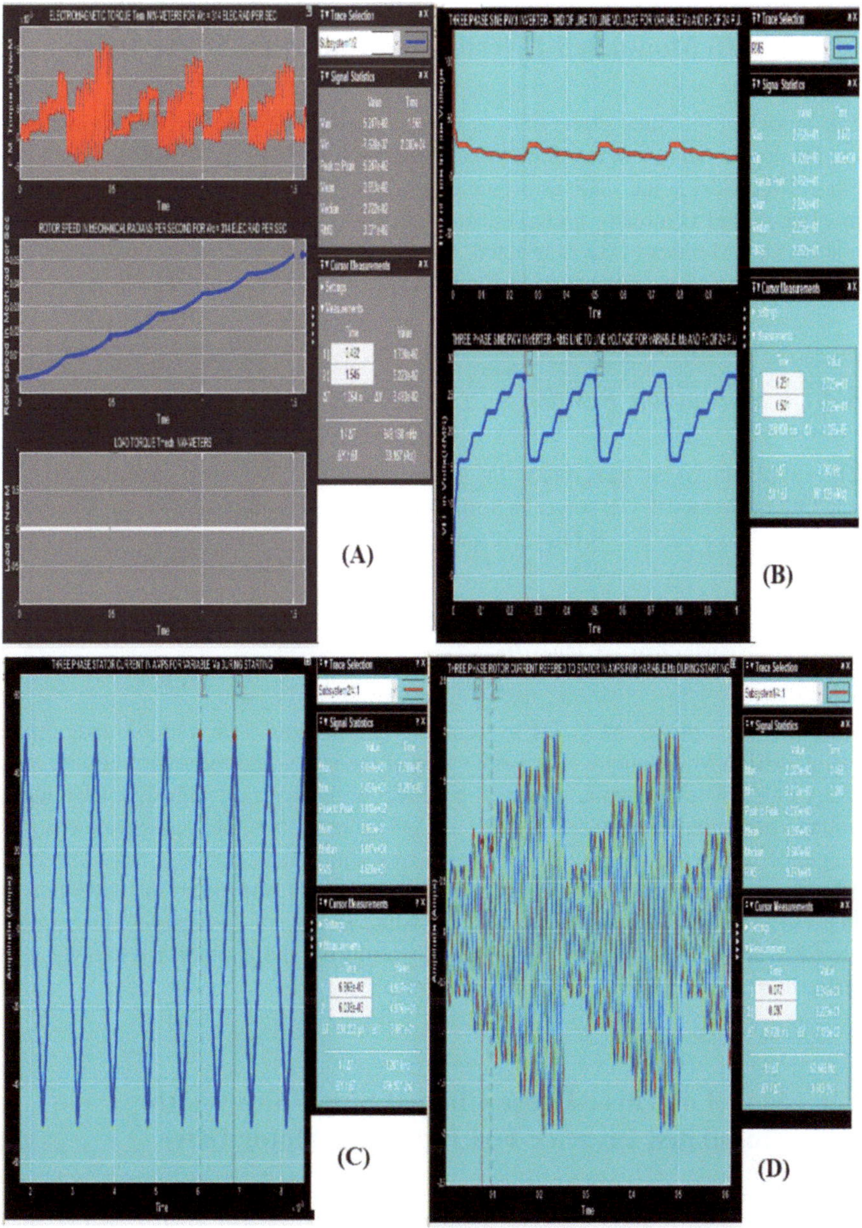

Fig. 8.9 Three-phase sine PWM inverter-fed IM drive—simulation results with variable AM index during starting: (**a**) E.M. torque (top), rotor speed (middle) and mechanical load (bottom), (**b**) THD of line to line voltage (top) and RMS value (bottom), (**c**) three-phase stator currents and (**d**) three-phase rotor currents

8.4 Model of Three-Phase Third Harmonic Injection Pulse Width Modulated Inverter-Fed Induction Motor Drive

The model of the three-phase third harmonic injection PWM (THIPWM) inverter-fed IM drive is shown in Fig. 8.10 (model file: EXAMPLE8_3) [4, 8–10]. The three-phase IM model is the same as presented in Sect. 8.2.1. The three-phase THIPWM inverter model is already presented in Sect. 2.2 of Chap. 2. To study the performance, an AM index of 0.8 is selected. A value of 8 is entered in the constant3 block corresponding to vm connected to Embedded MATLAB function in Fig. 8.10.

8.4.1 Simulation Results

The simulation of the three-phase IM fed by THIPWM inverter is carried out using ode23tb (stiff/TR-BDF2) solver in Simulink [6]. Simulation results for an AM index of 0.8 are shown in Fig. 8.11a–d, respectively.

8.4.2 Discussion of Results

Here the three-phase IM is fed by square wave inverter with variable magnitude using THIPWM technique. The peak value of the stator and rotor current with AM index of 0.8 is less than that obtained by direct online starting with sine wave input voltage presented in Sect. 8.2.2. With an AM index of 0.8, the steady state electromagnetic torque is found to be 12 Nw-metres, and this occurs at a rotor speed of 100 Mech.rad per second from Fig. 8.11a. Also for this AM index, the RMS line to line voltage across the inverter is only 70 volts. The stator peak and RMS currents are 78.9 and 30.1 amps, respectively. This confirms the applied inverter voltage, stator current, rotor torque and speed control.

8.5 Model of Three-Phase Clipped Sine Pulse Width Modulated Inverter-Fed Induction Motor Drive

The model of the three-phase clipped sine PWM (CSPWM) inverter-fed IM drive is shown in Fig. 8.12 (model file: EXAMPLE8_4) [4, 10, 11]. The three-phase IM model is the same as presented in Sect. 8.2.1. The three-phase CSPWM inverter model is already presented in Sect. 2.4 of Chap. 2. To study the performance, an AM index of 0.8 is selected. A value of 8 is entered in the Vclip constant block corresponding to v_clip input to Embedded MATLAB function in Fig. 8.12.

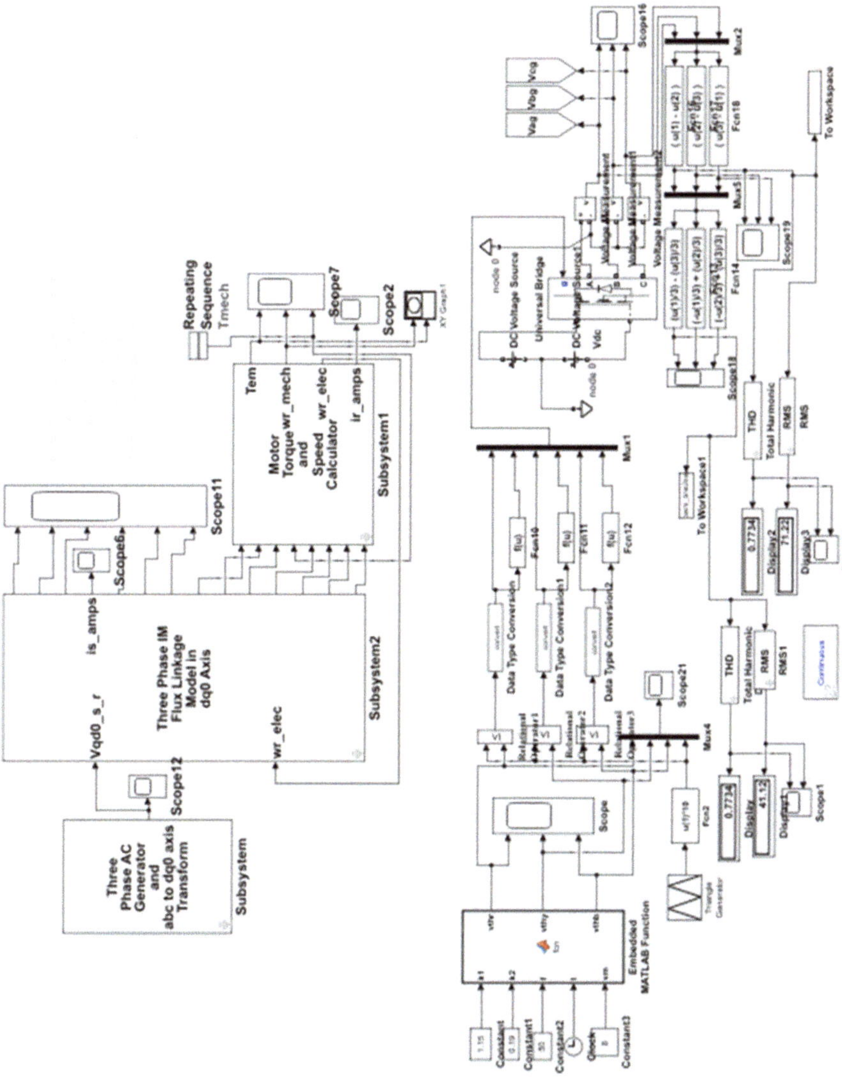

Fig. 8.10 Model of three-phase THIPWM inverter-fed induction motor drive

8.5.1 Simulation Results

The simulation of the three-phase IM fed by CSPWM inverter is carried out using ode23tb (stiff/TR-BDF2) solver in Simulink [6]. Simulation results for an AM index of 0.8 are shown in Fig. 8.13a–c, respectively.

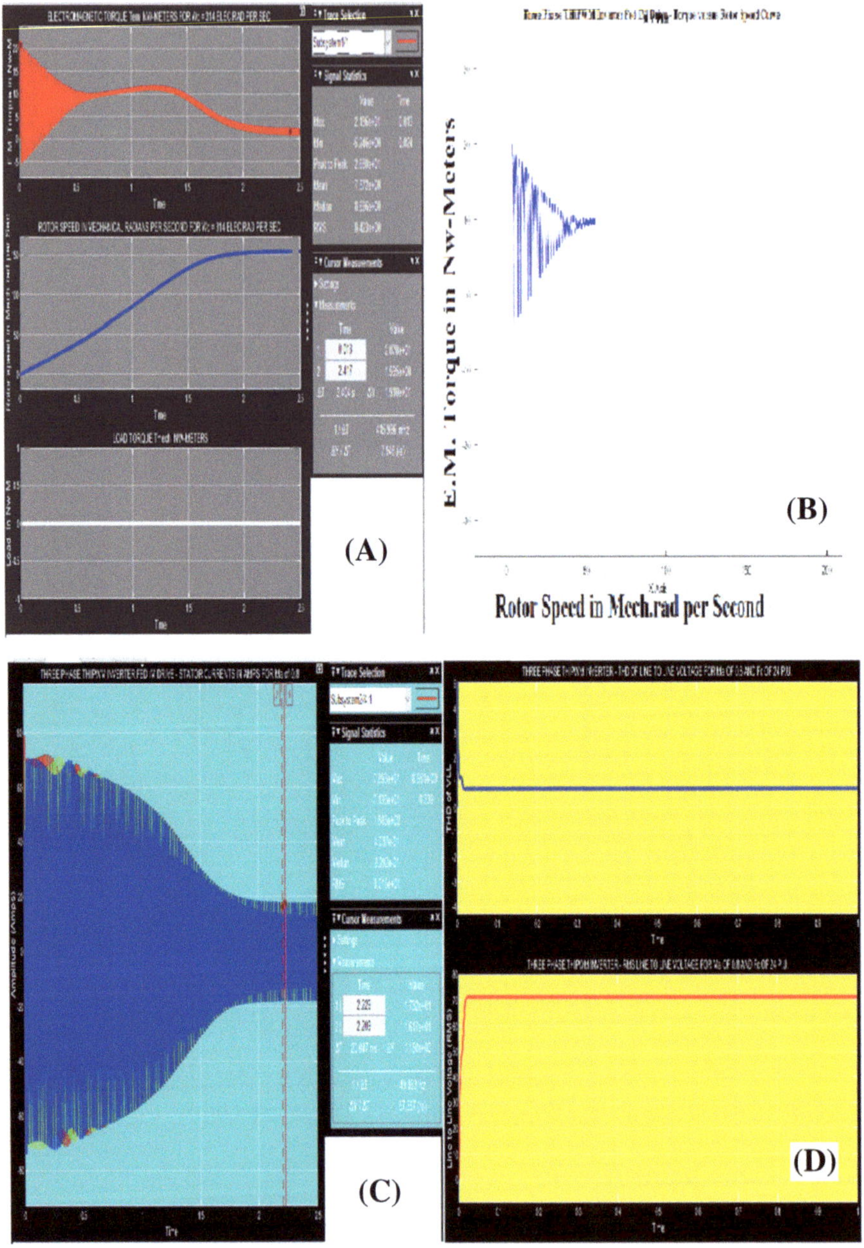

Fig. 8.11 Three-phase THIPWM inverter-fed IM drive—simulation results: (**a**) electromagnetic torque (top), rotor speed (middle) and mechanical load (bottom), (**b**) E.M. torque versus rotor speed curve. (**c**) Three-phase stator currents and (**d**) THD of line to line voltage (top) and (**d**) RMS value of line to line voltage

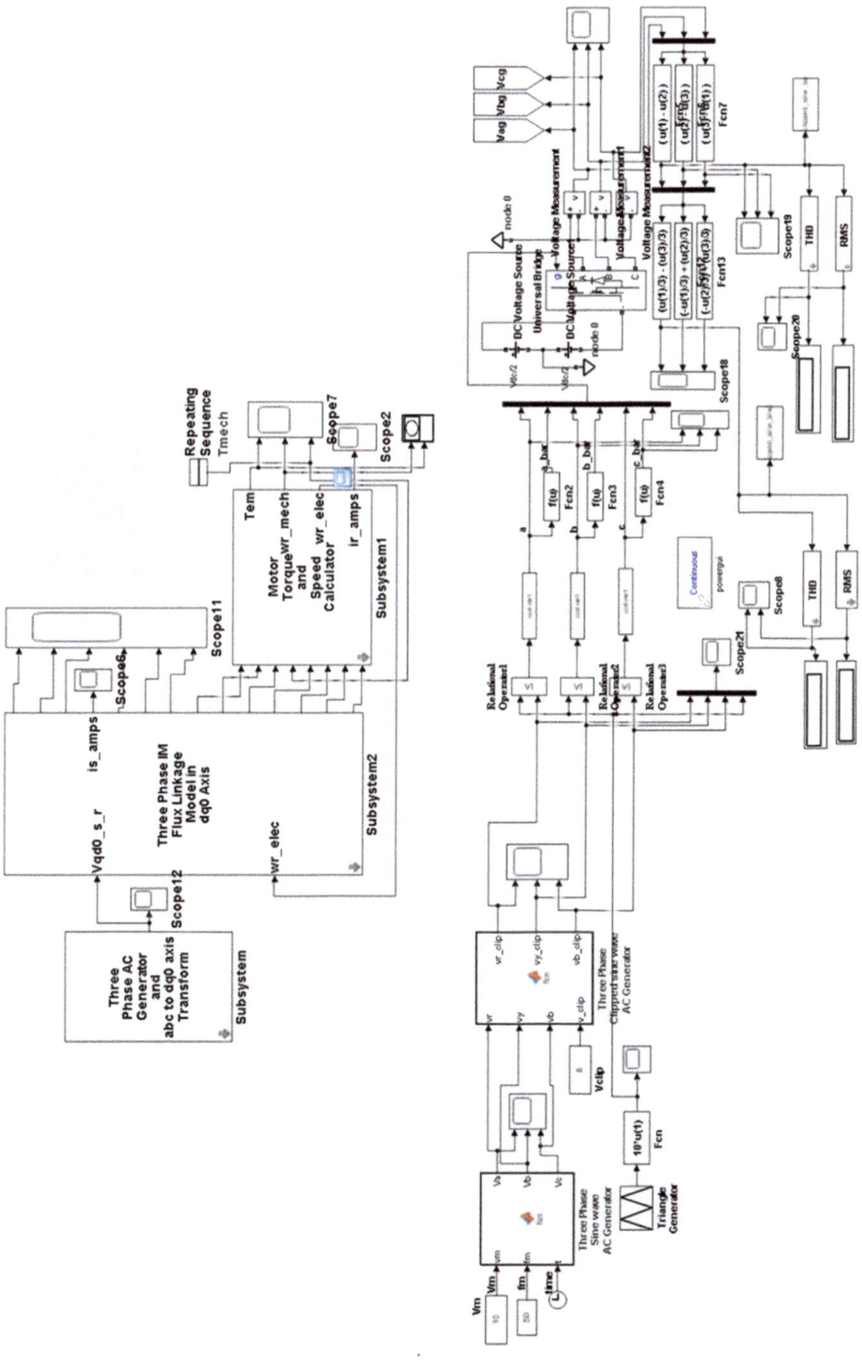

Fig. 8.12 Three-phase clipped sine PWM inverter-fed induction motor drive

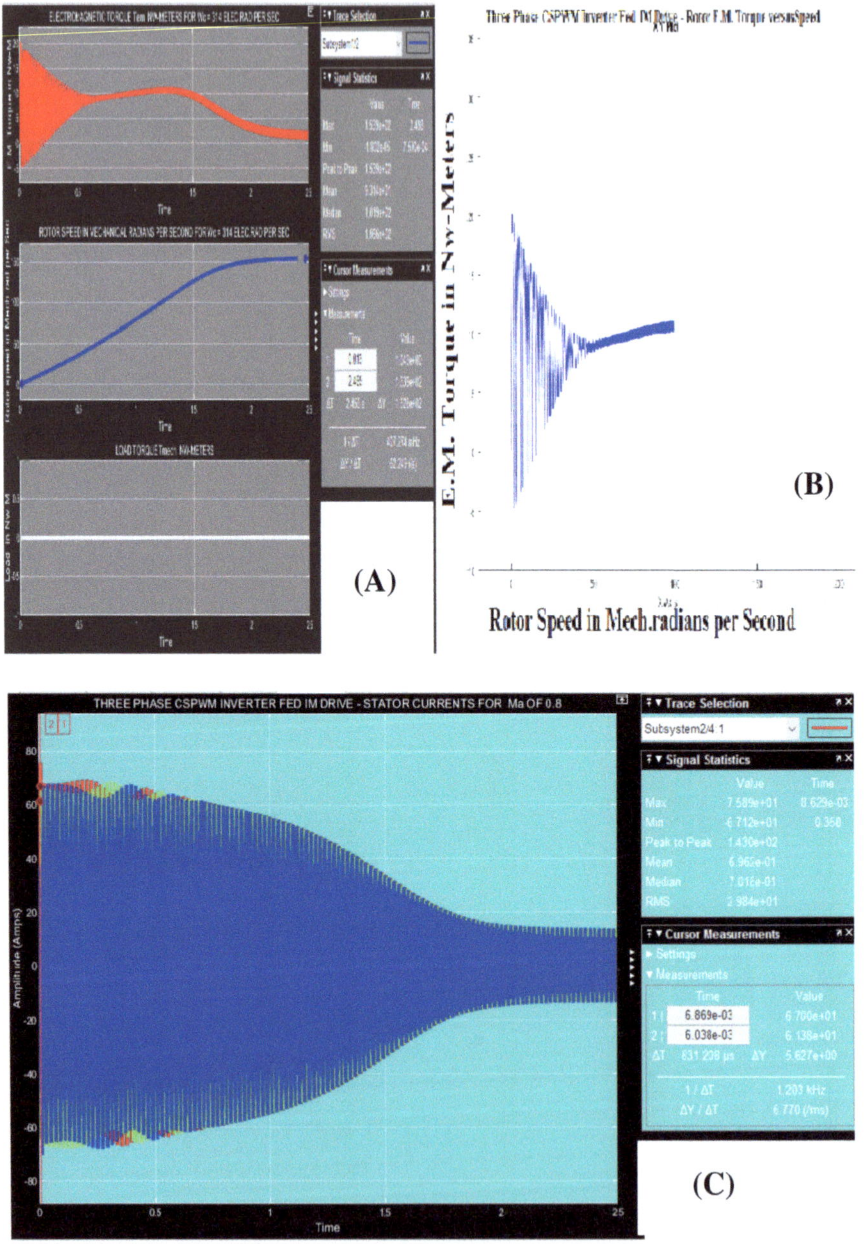

Fig. 8.13 Three-phrase CSPWM inverter-fed IM drive—simulation results: (**a**) E.M. torque (top), rotor speed (middle), mechanical load (bottom), (**b**) rotor E.M. torque versus speed curve and (**c**) three-phase stator currents

8.5.2 *Discussion of Results*

Here the three-phase IM is fed by square wave inverter with variable magnitude using CSPWM technique. The peak value of the stator and rotor current with AM index of 0.8 is less than that obtained by direct online starting with sine wave input voltage presented in Sect. 8.2.2. With an AM index of 0.8, the steady state maximum electromagnetic torque is found to be 11 Nw-metres, and this occurs at a rotor speed of 110 Mech.rad per second from Fig. 8.13a. Also for this AM index, the RMS line to line voltage across the inverter is only 70 volts. The stator peak and RMS currents are 75.89 and 29.84 amps, respectively. This confirms the applied inverter voltage, stator current, rotor torque and speed control.

8.6 Case Study: Model of Three-Phase Space Vector Pulse Width Modulated Inverter-Fed Induction Motor Drive

The space vector pulse width modulation (SVPWM) of three-phase two-level inverter is presented in Sect. 4.2 of Chap. 4 [12–14]. Here a new simplified model is developed for the space vector modulation (SVM) of three-phase two-level inverter, and the performance of the three-phase IM when fed by this inverter is studied. To develop the simplified model, referring to Table 4.3 of Chap. 4, the dwell timing for the switches in the upper and lower arm of each phase can be expressed as in Table 8.2. In Table 8.2, Sap, Sbp and Scp are the switches in the upper arms and San, Sbn and Scn are the switches in the lower arms of the three-phase inverter. Also T1, T2 and T0 are given by Eqs. 4.13–4.15, respectively, in Chap. 4. Also timings Ta2, Tb2 and Tc2 are the complementary (inverted) outputs of Ta1, Tb1 and Tc1, respectively. The model of the three-phase SVPWM inverter-fed IM drive is shown in Fig. 8.14 (model file: CASE_STUDY_EX8_1). The various subsystems are (1) sector identifier, (2) sector switch function generator and (3) gate pulse timing generator. The three-phase IM model is the same as explained in Sect. 8.2.1. The various subsystems of the three-phase SVPWM inverter are explained below:

8.6.1 *Sector Identifier*

Referring to Fig. 4.2 of Chap. 4, there are six sectors and Vref can lie in any one of the six sectors. These sectors from I to VI are displaced by $\pi/3$ radians. Embedded MATLAB function with inputs constant 1 for peak input voltage, constant 50 for frequency and time module and output sector y1 in Fig. 8.14 is used for identifying the sectors. This program is given under Program Segment 5.1 of Chap. 5 and is also shown in the model file CASE_STUDY_EX8_1. Sector output gives the sector number at any instant of time.

Table 8.2 Dwell time of switches in three-phase two-level inverter

Sl. no.	**Sector number**	**Sap**	**Sbp**	**Scp**
1	I	Ta1 = (T1 + T2 + 0.5*T0)	Tb1 = (T2 + 0.5*T0)	Tc1 = (0.5*T0)
2	II	Ta1 = (T1 + 0.5*T0)	Tb1 = (T1 + T2 + 0.5*T0)	Tc1 = (0.5*T0)
3	III	Ta1 = (0.5*T0)	Tb1 = (T1 + T2 + 0.5*T0)	Tc1 = (T2 + 0.5*T0)
4	IV	Ta1 = (05*T0)	Tb1 = (T1 + 0.5*T0)	Tc1 = (T1 + T2 + 0.5*T0)
5	V	Ta1 = (T2 + 0.5*T0)	Tb1 = (0.5*T0)	Tc1 = (T1 + T2 + 0.5*T0)
6	VI	Ta1 = (T1 + T2 + 0.5*T0)	Tb1 = (0.5*T0)	Tc1 = (T1 + 0.5*T0)
Sl. no.	**Sector number**	**San**	**Sbn**	**Scn**
7	I	Ta2 = (0.5*T0)	Tb2 = (T1 + 0.5*T0)	Tc2 = (T1 + T2 + 0.5*T0)
8	II	Ta2 = (T2 + 0.5*T0)	Tb2 = (0.5*T0)	Tc2 = (T1 + T2 + 0.5*T0)
9	III	Ta2 = (T1 + T2 + 0.5*T0)	Tb2 = (0.5*T0)	Tc2 = (T1 + 0.5*T0)
10	IV	Ta2 = (T1 + T2 + 0.5*T0)	Tb2 = (T2 + 0.5*T0)	Tc2 = (0.5*T0)
11	V	Ta2 = (T1 + 0.5*T0)	Tb2 = (T1 + T2 + 0.5*T0)	Tc2 = (0.5*T0)
12	VI	Ta2 = (0.5*T0)	Tb2 = (T1 + T2 + 0.5*T0)	Tc2 = (T2 + 0.5*T0)

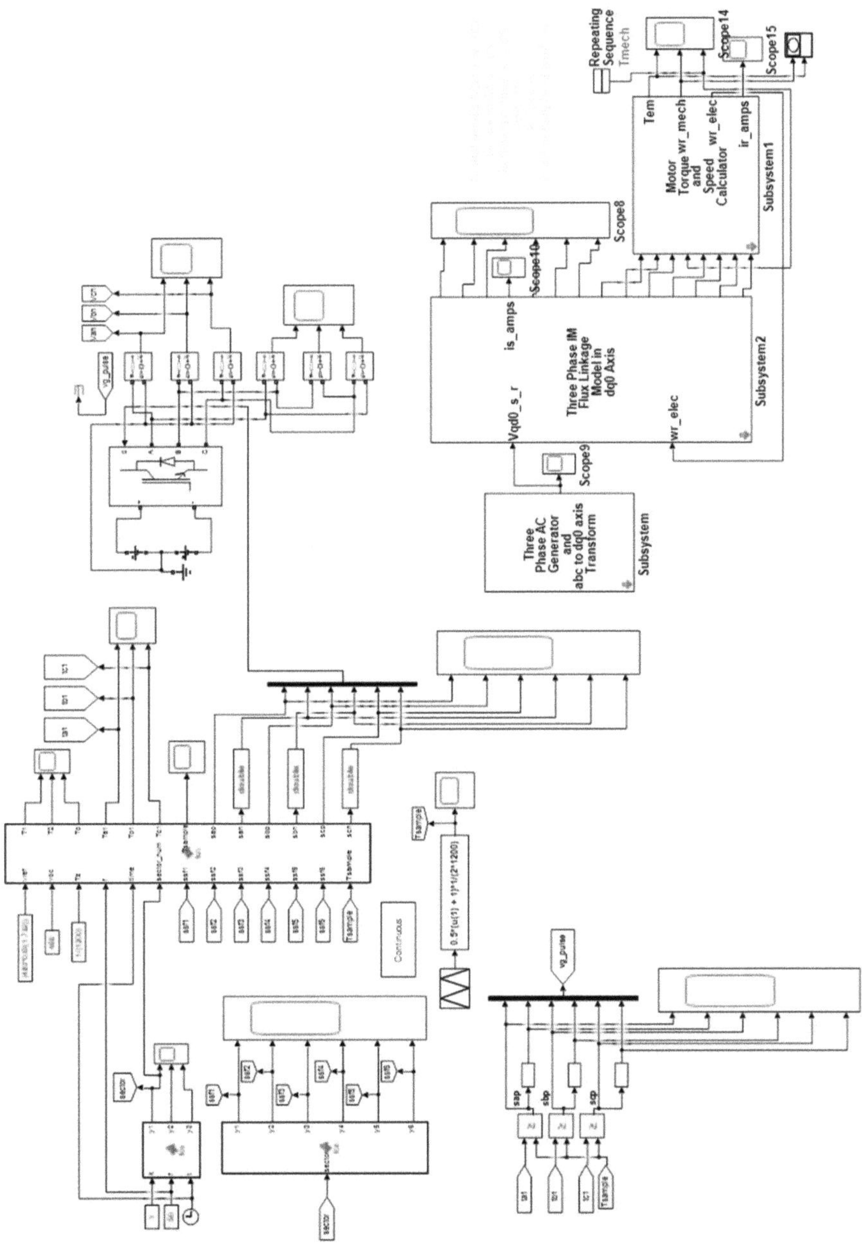

Fig. 8.14 Model of three-phase space vector PWM inverter-fed induction motor drive

8.6.2 Sector Switch Function Generator

The sector output giving sector number is given as input to another Embedded MATLAB function which gives sector switch functions ssf1 to ssf6 as outputs. For example, when Vref is in Sector I, ssf1 output is HIGH (logic 1); else its output is LOW (logic 0). The computer program to generate this sector switch functions is given under Program Segment 5.2 of Chap. 5 and is also shown in the model file CASE_STUDY_EX8_1. In Program Segment 5.2, outputs y1 to y6 correspond to sector switch functions ssf1 to ssf6.

8.6.3 Gate Pulse Timing Generator

Gate pulse timing for the three-phase inverter switches is generated using another Embedded MATLAB function This Embedded MATLAB function has the inputs Vref, Vdc, Tz, f, time t, sector_num, ssf1 to ssf6 and Tsample, and the outputs are T1, T2, T0, Ta1, Tb1, Tc1, Tsample and the six gate pulses sap, san, sbp, sbn, scp and scn. Vref is the reference voltage defined in Fig. 4.2; Vdc is the DC link voltage; Tz is the sample time Ts; f is the output frequency of the inverter, sector_num from 1 to 6, sector switch functions from 1 to 6 and Tsample which is a triangle carrier waveform with period Tz, peak value Tz/2 and minimum value zero. Tsample waveform is shown in Figs. 4.4a and 4.4b of Chap. 4. The computer program to generate gate pulse timing is shown under Program Segment 8.1 in the model file CASE_STUDY_EX8_1. In Program segment 8.1, the symbol (~) represents logical NOT operation. Timing Ta1, Tb1 and Tc1 are calculated as per Table 8.2, and the gate pulse for the upper switches sap, sbp and scp is obtained by comparison with the Tsample waveform. Gate pulses san, sbn and scn for lower switches are obtained by inverting the gate pulse for sap, sbp and scp, respectively. Alternatively timing ta1, tb1 and tc1 are compared with Tsample in three relational operator comparators, and the respective gate pulse outputs for sap, sbp and scp are obtained which are inverted using NOT gates to obtain gate pulse for san, sbn and scn as shown in Fig. 8.14. Triangle carrier Tsample is explained in Sects. 4.2.1 and 4.2.2.1.

The three-phase inverter model is already presented in Sect. 4.2.2.4 of Chap. 4.

8.6.4 Simulation Results

The three-phase IM data are the same as given in Table 8.1. A modulation index of 0.8, output frequency 50 Hz and a DC link voltage of 480 volts are selected. The Vref input in Fig. 8.14 is calculated using Eq. 4.19 of Chap. 4. The simulation of the three-phase IM fed by SVPWM inverter is carried out using ode23tb (stiff/TR-BDF2) solver in Simulink [6]. The simulation results are presented in Fig. 8.15a–d. Inverter line voltage is shown in Fig. 8.16.

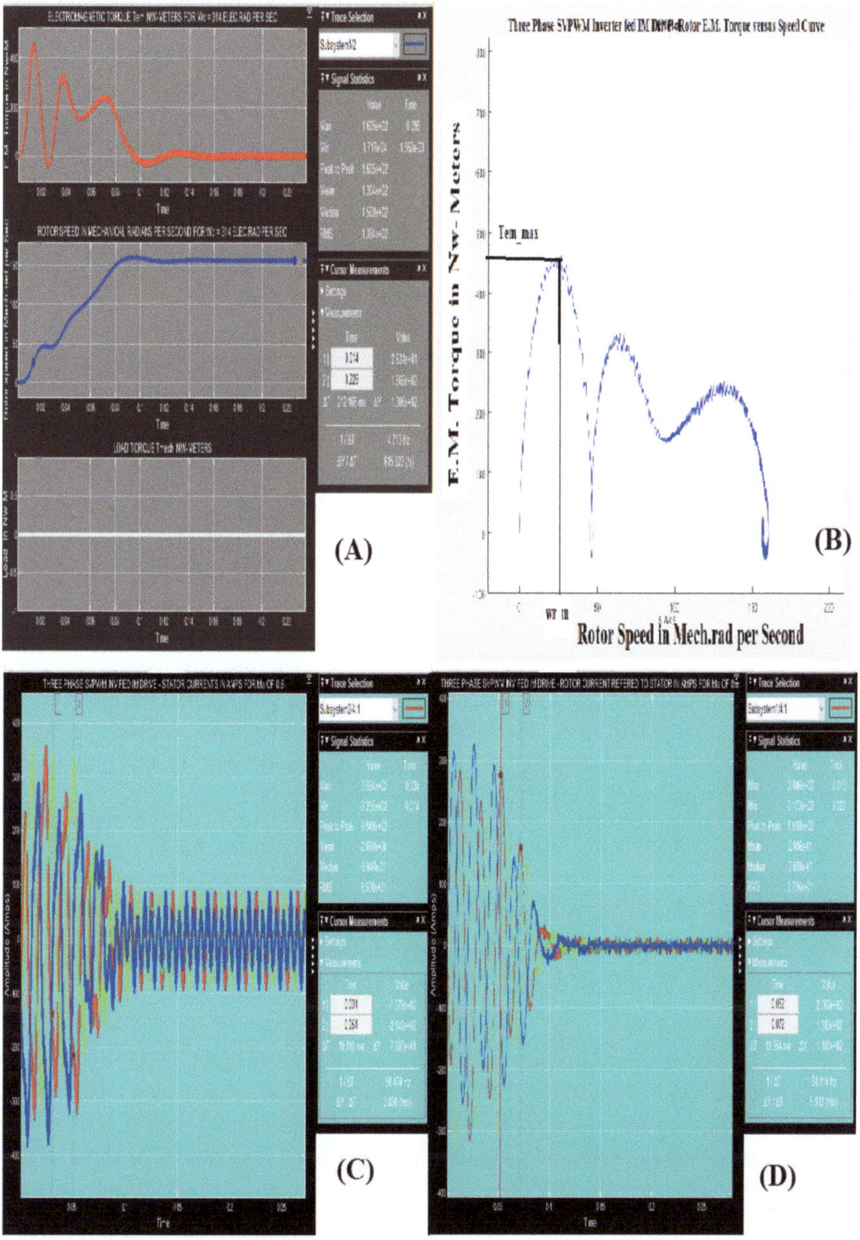

Fig. 8.15 Three-phase SVPWM inverter-fed IM drive simulation results: (**a**) rotor E.M. torque (top), rotor speed (middle), mechanical load (bottom), (**b**) rotor E.M. torque versus speed curve, (**c**) three-phase stator currents and (**d**) three-phase rotor currents

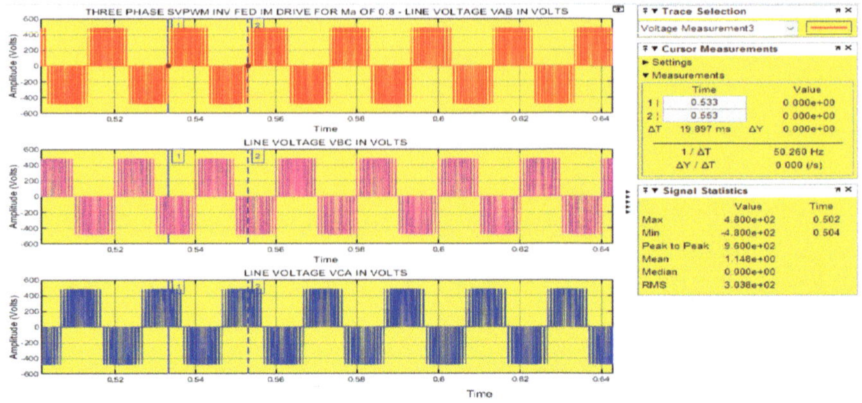

Fig. 8.16 Three-phase SVPWM inverter-fed IM drive—line to line voltage for an AM index of 0.8

8.6.5 *Discussion of Results*

Here the three-phase IM is fed by square wave inverter with variable magnitude using SVPWM technique. The peak value of the stator and rotor current with AM index of 0.8 is less than that obtained by direct online starting with sine wave input voltage presented in Sect. 8.2.2. With an AM index of 0.8, the maximum electromagnetic torque is found to be 462.5 Nw-metres, and this occurs at a rotor speed of 26.31 Mech.rad per second from Fig. 8.15a, b. The RMS line to line voltage of inverter for an AM index of 0.8 is found to be 303.8 volts.

8.7 Model of Three-Phase Multicarrier Sine Level Shift Pulse Width Modulated Diode Clamped Three-Level Inverter-Fed Induction Motor Drive

The model of three-phase multicarrier sine level shift PWM (MCSLSPWM) diode clamped three-level inverter (DCTLI) is presented in Sect. 2.8.1 of Chap. 2 [15–17]. Here in-phase disposition (IPD) of triangle carrier is used [15–17]. The three-phase IM model is the same as presented in Sect. 8.2.1. The model of three-phase MCSLSPWM inverter-fed IM drive is shown in Fig. 8.17 (model file: EXAMPLE8_5). To study the performance, an AM index of 0.8 is selected. A value of 1.6 is entered in the box for the amplitude of the three-phase 50 Hz sine wave, as per Eq. 2.5 of Chap. 2. The triangle carrier has a frequency of 1200 Hz and a peak value of ±1 Volt.

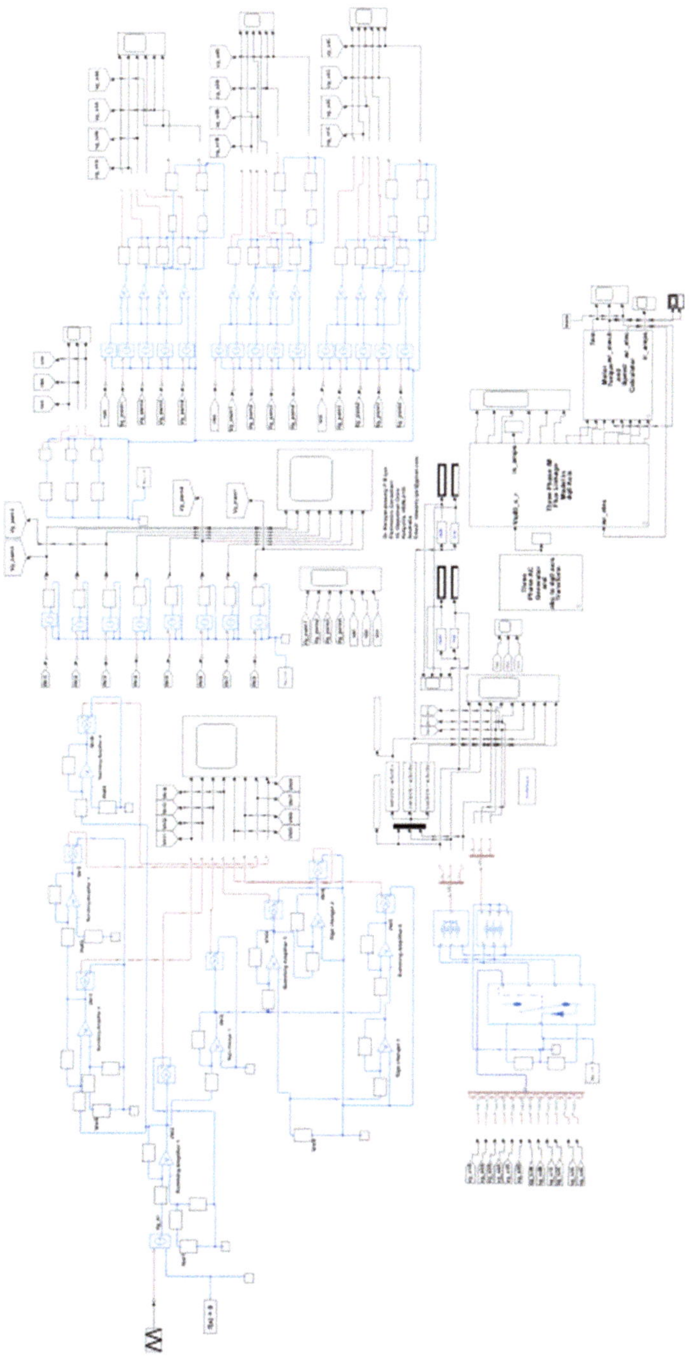

Fig. 8.17 Model of three-phase MCSLSPWM inverter-fed induction motor drive

8.7.1 Simulation Results

The simulation of the three-phase IM fed by MCSLSPWM inverter is carried out using fixed step ode 14x (Extrapolation) solver in Simulink [6]. Simulation results for an AM index of 0.8 are shown in Fig. 8.18a–d, respectively. The three-phase DCTLI output voltages are shown in Fig. 8.19.

8.7.2 Discussion of Results

Here the three-phase IM is fed by square wave DCTLI with variable magnitude using MCSLSPWM technique. The four triangle carriers used for generating PWM gate drives are having in-phase disposition (IPD). The peak value of the stator and rotor current with AM index of 0.8 is less than that obtained by direct online starting with sine wave input voltage presented in Sect. 8.2.2. With an AM index of 0.8, the maximum electromagnetic torque is found to be 691.8 Nw-metres, and this occurs at a rotor speed of 36.81 Mech.rad per second from Fig. 8.18a, b. In this case rotor electromagnetic (EM) torque goes negative as seen from Fig. 8.18b. The three-phase line to ground voltages applied to the model of IM have three levels and are shown in Fig. 8.19. The method can be extended to other carrier disposition such as Phase Opposition (PO) and Alternate Phase Opposition (APO).

8.8 Vector Control of Three-Phase Induction Motor

The dq-axis representation of three-phase IM is shown in Fig. 8.20. In this figure, θs is the angle that the stator as-axis makes with the d-axis, and θr is the angle that the rotor ar-axis makes with the d-axis. The stator and rotor currents, voltages and flux linkages in the dq-axis given the respective abc-axis values can be expressed using Eq. 8.10 where θ is replaced by θs and θr for the stator and rotor quantities, respectively. Similarly the inverse relation connecting the abc-axis quantities with that of dq-axis can be expressed using Eq. 8.11. The dq-axis stator and rotor flux linkages can be expressed as in Eq. 8.6 and the rotor electromagnetic torque is given by Eq. 8.7. The relations connecting dq-axis stator and rotor voltages, currents and flux linkages are given by Eqs. 8.4 and 8.5. In Eq. 8.5, ω_{re} represents ω_r, the rotor speed in electrical radians per second which equals $\omega_{rm}*(P/2)$, where ω_{rm} is the rotor speed in mechanical radians per second and P is the number of poles. The arbitrary reference frame ω_C is replaced by synchronous reference frame frequency ω_{syn} for vector control of three-phase IM as PI controller can be easily designed in synchronous reference frame [18, 19].

The e.m. torque in terms of the dq-axis flux linkages and currents can be expressed as follows [19]:

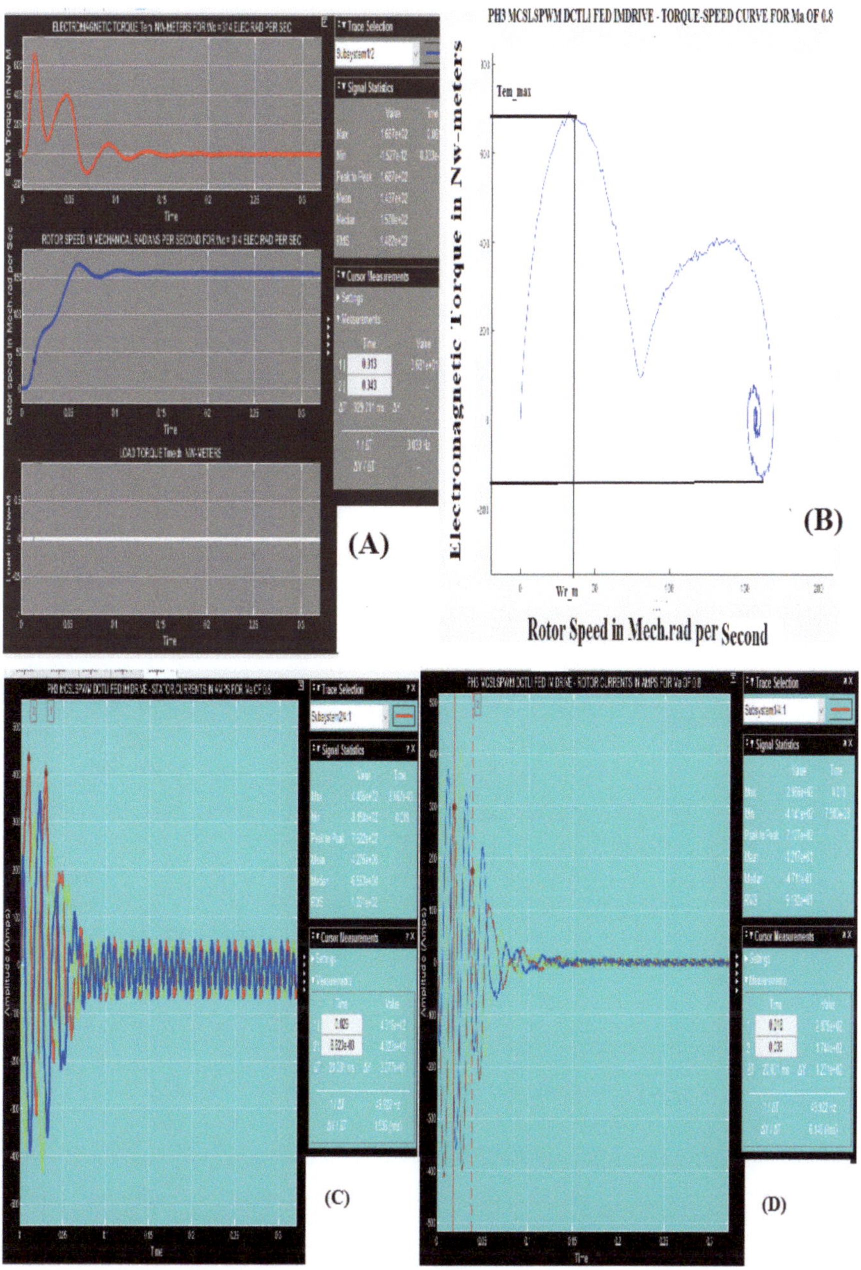

Fig. 8.18 Three-phase MCSLSPWM DCTLI-fed IM drive—simulation results: (**a**) E.M. torque (top), rotor speed (middle), mechanical load (bottom), (**b**) rotor electromagnetic torque versus speed curve, (**c**) three-phase stator currents and (**d**) three-phase rotor currents

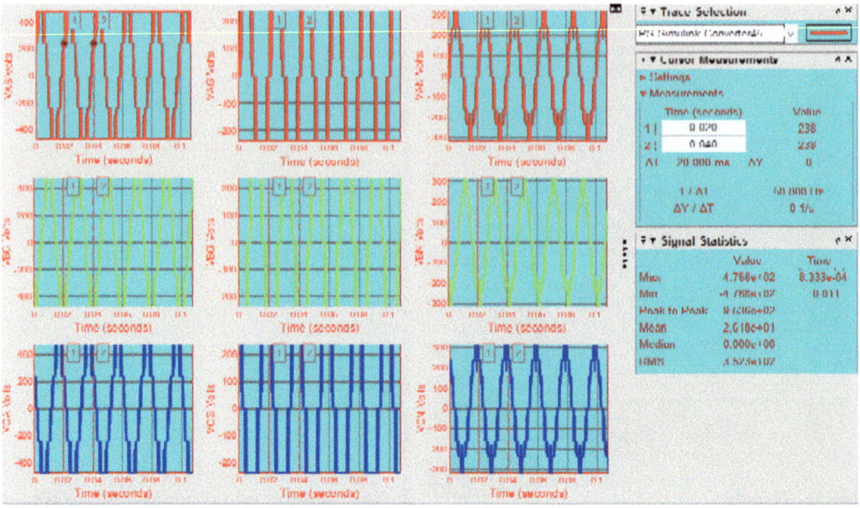

Fig. 8.19 Three-phase MCSLSPWM DCTLI-fed IM drive: three-phase line to line voltage (left column), three-phase line to ground voltage (middle column) and three-phase line to neutral voltage (right column)

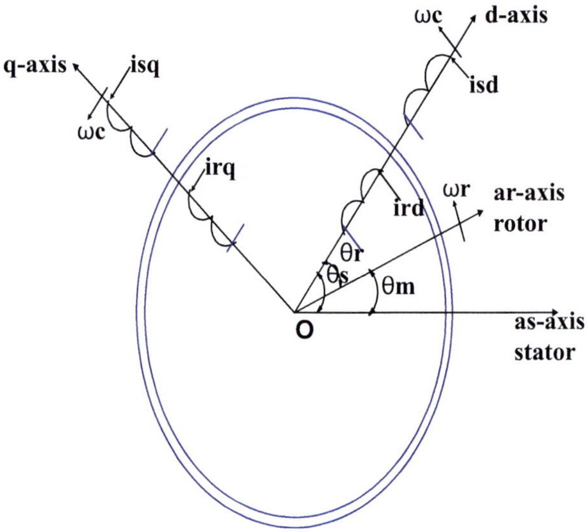

Fig. 8.20 dq-axis representation of three-phase induction motor

$$\left.\begin{aligned} T_{\text{em}} &= \frac{3}{2} * \frac{P}{2} * \left(\lambda_{qr'} * i_{dr'} - \lambda_{dr'} * i_{qr'}\right) \\ &= \frac{3}{2} * \frac{P}{2} * \left(\lambda_{ds} * i_{qs} - \lambda_{qs} * i_{ds}\right) \\ &= \frac{3}{2} * \frac{P}{2} * L_m * \left(i_{dr'} * i_{qs} - i_{qr'} * i_{ds}\right) \end{aligned}\right\} \tag{8.12}$$

The acceleration is determined by the difference of the electromagnetic torque and the load torque acting on Jeq, the combined inertia of the motor and the load. In terms of the rotor speed ωrm in mechanical radians per second, the following equation is valid [18, 19]:

$$J_{\text{eq}} * \frac{d\omega_{\text{rm}}}{dt} + B * \omega_{\text{rm}} = (T_{\text{em}} - T_{\text{L}}) \tag{8.13}$$

In Eq. 8.13, T_{L} is the load torque and B is the damping constant.

The dq winding currents can be calculated from flux linkage equations. From Eq. 8.6, this can be expressed as follows:

$$\begin{bmatrix} \lambda_{ds} \\ \lambda_{qs} \\ \lambda_{dr'} \\ \lambda_{qr'} \end{bmatrix} = \begin{bmatrix} L_s & 0 & L_m & 0 \\ 0 & L_s & 0 & L_m \\ L_m & 0 & L_r & 0 \\ 0 & L_m & 0 & L_r \end{bmatrix} * \begin{bmatrix} i_{ds} \\ i_{qs} \\ i_{dr'} \\ i_{qr'} \end{bmatrix} \tag{8.14}$$

From Eq. 8.14, the dq-axis stator and rotor currents can be expressed as follows:

$$\begin{bmatrix} i_{ds} \\ i_{qs} \\ i_{dr'} \\ i_{qr'} \end{bmatrix} = \begin{bmatrix} L_s & 0 & L_m & 0 \\ 0 & L_s & 0 & L_m \\ L_m & 0 & L_r & 0 \\ 0 & L_m & 0 & L_r \end{bmatrix}^{-1} * \begin{bmatrix} \lambda_{ds} \\ \lambda_{qs} \\ \lambda_{dr'} \\ \lambda_{qr'} \end{bmatrix} \tag{8.15}$$

In Eqs. 8.14 and 8.15, Ls is (Lls + Lm) and Lr is (Llr′ + Lm).

Accurate speed control of three-phase IM can be achieved by vector control [18, 19]. In speed and position control, the ability to produce a step change in torque on command represents total control over the drives. In vector control of three-phase IM drives, the d-axis is aligned with the rotor flux linkage space vector such that the rotor flux linkage in the q-axis is zero. Thus we have the following equations:

$$\lambda_{qr'}(t) = 0 \tag{8.16}$$

Using Eq. 8.16 in the equation for λqr′ obtained from Eq. 8.14,

$$i_{qr'} = \frac{-L_m * i_{qs}}{L_r} \tag{8.17}$$

The condition d-axis is always aligned with rotor flux linkage axis λ_r, such that $\lambda_{qr'} = 0$ also results in $\frac{d\lambda_{qr'}}{dt} = 0$. Using this value of $\lambda_{qr'}$ and $\frac{d\lambda_{qr'}}{dt}$ in the dq-axis equivalent circuit, the equivalent circuit for vector control of three-phase IM is shown in Fig. 8.21. The rotor of three-phase IM is always short-circuited and this results in $Vdr' = Vqr' = 0$. Using Eq. 8.5, the rotor angular slip frequency can be expressed as follows:

$$(\omega_C - \omega_r) = \omega_{sl} = -\frac{i_{qr'} * R_{r'}}{\lambda_{dr'}} \tag{8.18}$$

Using Eqs. 8.17 and 8.18 can be expressed as follows:

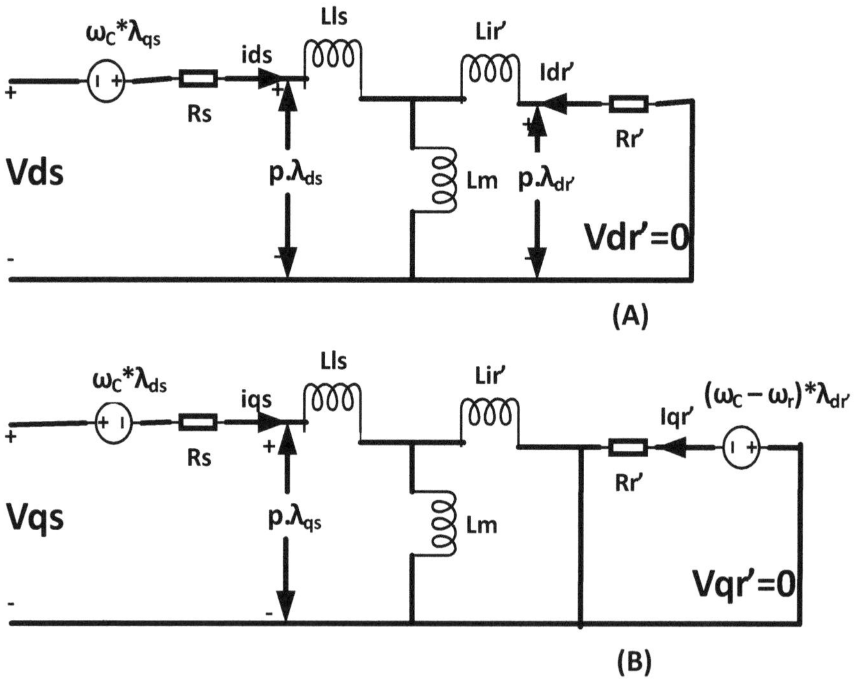

Fig. 8.21 (**a**) d-axis and (**b**) q-axis equivalent circuit of three-phase induction motor under vector control

$$(\omega_C - \omega_r) = \omega_{sl} = \frac{L_m * i_{qs} * R_{r'}}{L_r * \lambda_{dr'}} \tag{8.19}$$

Since flux linkage in the rotor q-axis is zero, the e.m. torque is produced by rotor d-axis flux linkage acting on rotor q-axis current. Using Eq. 8.12, the e.m. torque Tem is given below:

$$T_{\text{em}} = -\frac{3}{2} * \frac{P}{2} * \left(\lambda_{dr'} * i_{qr'}\right) \tag{8.20}$$

Using Eq. 8.17 in Eq. 8.20, T_{em} can be expressed as follows:

$$T_{\text{em}} = \frac{3}{2} * \frac{P}{2} * \left(\frac{\lambda_{dr'} * L_m * i_{qs}}{L_r}\right) \tag{8.21}$$

Also from Fig. 8.21a, $i_{\text{dr}'}$ can be expressed as follows:

$$i_{dr'}(s) = \frac{-i_{ds}(s) * sL_m}{(sL_m + sL_{lr'} + R_{r'})} = \frac{-i_{ds}(s) * sL_m}{(sL_r + R_{r'})} \tag{8.22}$$

From Eq. 8.14, rotor d-axis flux linkage can be expressed as follows:

$$\lambda_{dr'} = (L_m * i_{ds} + L_r * i_{dr'}) \tag{8.23}$$

Using Eq. 8.23 in Eq. 8.22 and simplifying,

$$\lambda_{dr'}(s) = \frac{L_m * R_{r'} * i_{ds}(s)}{(sL_r + R_{r'})} \tag{8.24}$$

The block diagram model of the three-phase vector controlled IM developed using Eqs. 8.13, 8.19, 8.21 and 8.24 is shown in Fig. 8.22.

8.8.1 *Indirect Vector Control of Three-Phase Induction Motor Drive*

The detailed model of the three-phase vector controlled IM using current regulated PWM inverter is shown in Fig. 8.23. The d-axis winding reference stator current i^*_{ds} controls the rotor flux linkage $\lambda dr'$, and the q-axis winding reference stator current i^*_{qs} controls the rotor e.m. torque Tem. The reference dq-axis stator currents are converted into abc-axis reference stator phase currents $i^*_a(t)$, $i^*_b(t)$ and $i^*_c(t)$ [18].

A current regulated PWM (CR-PWM) inverter supplies the current to the three-phase IM.

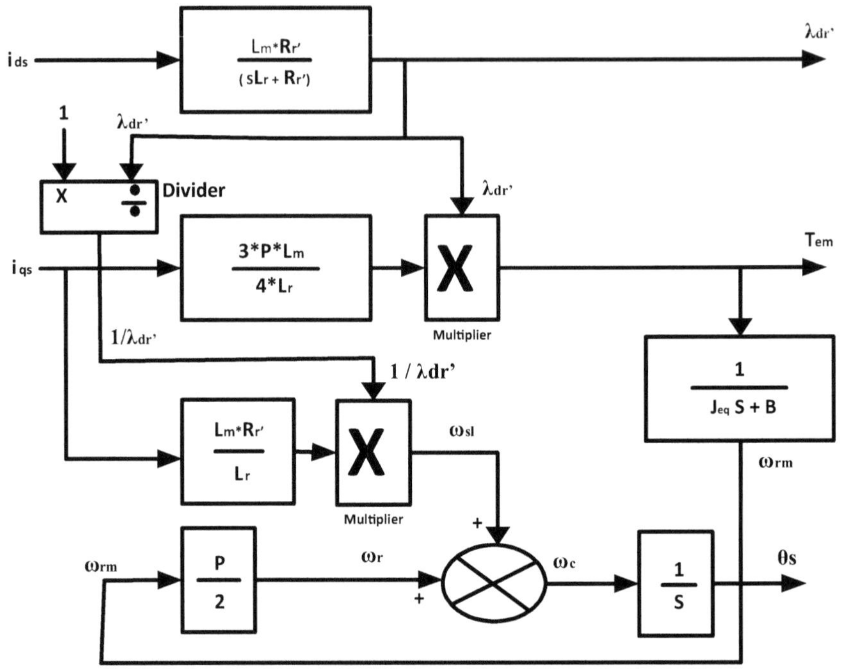

Fig. 8.22 Block diagram model of three-phase vector controlled induction motor

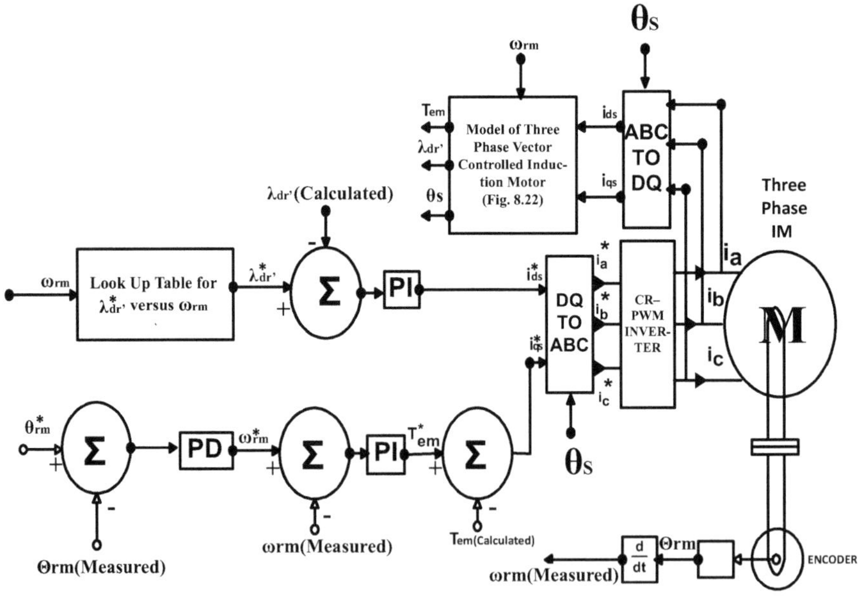

Fig. 8.23 Vector controlled three-phase induction motor drive fed by CR PWM inverter

The dq-axis current references i^*_{ds} and i^*_{qs} are generated by cascaded torque, speed and position control loops shown in the block diagram of Fig. 8.23, where θ^*_{mech} is the rotor position reference input. The actual rotor position θ_{mech} and the rotor speed ω_{mech} are measured, and the rotor flux linkage $\lambda dr'$ is calculated as shown in Fig. 8.21 and Eq. 8.24. For extended speed range operation beyond rated speed, flux weakening is implemented [18].

To design speed loop without torque loop, the torque expression is derived at the rated value of i^*_{ds}. In steady state under vector control, $i_{dr'} = 0$ from Fig. 8.21a and Eq. 8.22, by letting derivative of d-axis rotor flux linkage and stator current to zero. Using Eq. 8.14, under vector control in steady state, we have the following equation:

$$\lambda_{dr'} = L_m * i_{ds} \tag{8.25}$$

Substituting Eq. 8.25 in Eq. 8.21, we have the following:

$$T_{em} = \left[\frac{3 * P}{4} * \left(\frac{L_m^2 * i_{ds}}{L_r}\right)\right] * i_{qs} \tag{8.26}$$

The speed controller is shown in Fig. 8.24, where PI controller constants Kp and Ki are calculated on the basis of assumption of a suitable crossover frequency and phase margin for the speed controller. In Fig. 8.24, ω^*_{rm} is the reference rotor speed and ω_{rm} is the actual rotor speed. The crossover frequency or bandwidth selected for the speed loop must lie within the rotor speed range in mechanical radians per second. The stator voltages to be applied are calculated as given below.

The switching frequency of the PWM converter is maintained constant. The voltage to be applied to the PWM converter in order to get the stator current equal to the reference stator current value is derived below:

Using Eq. 8.14 and noting that $i_{dr'}$ is zero under steady state vector control, d-axis stator flux linkage can be expressed as follows:

$$\lambda_{ds} = L_S * i_{ds} \tag{8.27}$$

Similarly using Eqs. 8.14 and 8.17,

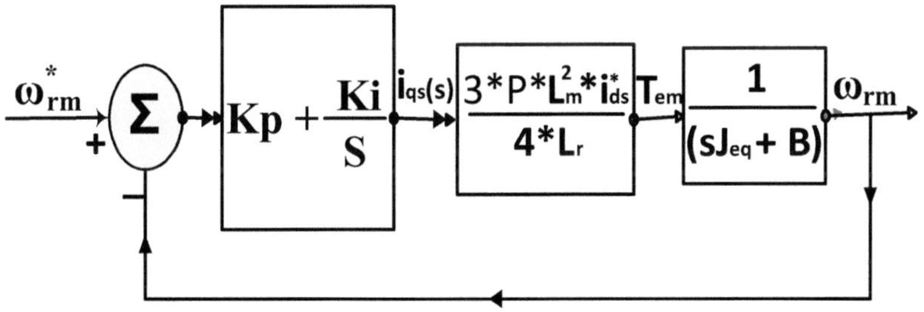

Fig. 8.24 Speed controller

$$\lambda_{qs} = L_S * i_{qs} - \frac{L_m^2 * i_{qs}}{L_r} \tag{8.28}$$

Equation 8.28 can be expressed as follows:

$$\lambda_{qs} = \left(1 - \frac{L_m^2}{L_S * L_r}\right) * L_S * i_{qs} \tag{8.29}$$

$$\text{i.e. } \lambda_{qs} = \beta * L_S * i_{qs} \tag{8.30}$$

where β is defined below:

$$\beta = \left(1 - \frac{L_m^2}{L_S * L_r}\right) \tag{8.31}$$

Multiplying Eq. 8.27 by Eq. 8.31 and adding the term ($L^2_m * i_{ds}/L_r$), we have the following:

$$\lambda_{ds} = L_S * i_{ds} * \beta + \frac{L_m^2 * i_{ds}}{L_r} \tag{8.32}$$

Using Eq. 8.25 in Eq. 8.32 and simplifying,

$$\lambda_{ds} = L_S * i_{ds} * \beta + \frac{\lambda_{dr'} * L_m}{L_r} \tag{8.33}$$

Using Eqs. 8.4, 8.30 and 8.33, dq-axis stator voltages can be expressed as follows:

$$v_{ds} = R_S * i_{ds} + L_s * \beta * S i_{ds} + \frac{L_m * S\lambda_{dr'}}{L_r} - \omega_C * \beta * L_S * i_{qs} \tag{8.34}$$

$$v_{qs} = R_S * i_{qs} + \beta * L_s * S i_{qs} + \omega_C * L_S * i_{ds} * \beta + \frac{\omega_C * \lambda_{dr'} * L_m}{L_r} \tag{8.35}$$

Equations 8.34 and 8.35 can be written as $v_{ds} = (v'_{ds} + \Delta v_{ds})$ and $v_{qs} = (v'_{qs} + \Delta v_{qs})$ where each term is defined below:

$$v'_{ds} = R_S * i_{ds} + L_s * \beta * S i_{ds} \tag{8.36}$$

$$\Delta v_{ds} = \frac{L_m * S\lambda_{dr'}}{L_r} - \omega_C * \beta * L_S * i_{qs} \tag{8.37}$$

$$v'_{qs} = R_S * i_{qs} + \beta * L_s * S i_{qs} \tag{8.38}$$

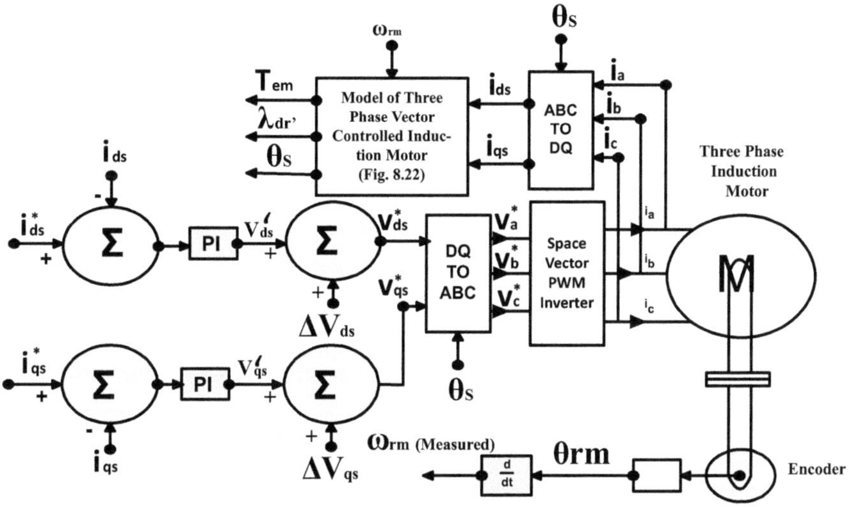

Fig. 8.25 Three-phase IM vector control with applied voltage

$$\Delta v_{qs} = \omega_C * L_S * i_{ds} * \beta + \frac{\omega_C * \lambda_{dr'} * L_m}{L_r} \tag{8.39}$$

The block diagram is shown in Fig. 8.25. As shown in Fig. 8.25, reference voltages for the stator d and q-axis are generated from stator d and q-axis reference stator currents i^*_{ds} and i^*_{qs}. Using the calculated value of θ_S, the reference phase voltages v^*_a, v^*_b and v^*_c are calculated. The actual phase voltages va, vb and vc are supplied by the three-phase PWM inverter. The voltages v'_{ds} and v'_{qs} in Fig. 8.25 are generated using PI controllers in the current loop. Assuming perfect compensation, each channel results in the block diagram shown in Fig. 8.26a, b. The transfer function ids(s)/v′ds(s) and iqs(s)/v′qs(s) are derived from Eqs. 8.36 and 8.38, respectively. The PI controller for current loop is designed assuming a suitable crossover frequency and phase margin for the current loop controller. The crossover frequency or bandwidth must be within the range used for the PWM carrier frequency which is much higher than the rotor speed range.

8.8.2 PI Controller Design for Speed Control and Current Control Loops

Referring to speed control loop in Fig. 8.24, constant k is defined below:

$$K = \frac{3 * P * L_m^2 * i_{ds}^*}{4 * L_r} \tag{8.40}$$

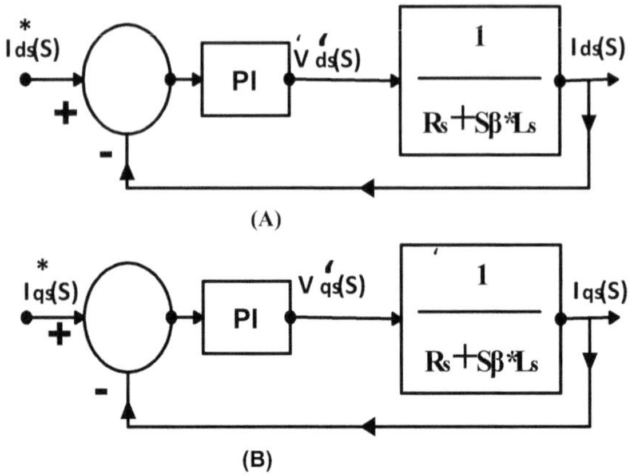

Fig. 8.26 (**a**, **b**) Three-phase IM vector control—current loop controllers

For PI controller design, the loop gain G(s)*H(s) = GH1(S) is equated to unity which is given below:

$$\left(\frac{K_pS + K_i}{S}\right) * \frac{K}{(J_{eq}S + B)} = 1 \tag{8.41}$$

Let S = $j\omega_{cr}$ be the given gain crossover frequency and PM be the desired phase margin. Then the loop gain GH1($j\omega_{cr}$) can be expressed as follows:

$$\mathrm{GH1}(j\omega_{cr}) = \frac{(j\omega_{cr} * K_p + K_i) * K}{j\omega_{cr} * (J_{eq} * j\omega_{cr} + B)} \tag{8.42}$$

Using the given phase margin PM and Eq. 8.42, the expression for phase contribution of PI controller follows:

$$\frac{K_p * \omega_{cr}}{K_i} = \tan\left[PM - \frac{\pi}{2} - \tan^{-1}\left(\frac{J_{eq} * \omega_{cr}}{B}\right)\right] \tag{8.43}$$

$$K_i = \frac{K_p * \omega_{cr}}{\tan\left[PM - \frac{\pi}{2} - \tan^{-1}\left(\frac{J_{eq} * \omega_{cr}}{B}\right)\right]} \tag{8.44}$$

In Eq. 8.44, ω_{k1} is defined below:

$$\omega_{k1} = \frac{\omega_{cr}}{\tan\left[PM - \frac{\pi}{2} - \tan^{-1}\left(\frac{J_{eq} * \omega_{cr}}{B}\right)\right]} \tag{8.45}$$

$$K_i = K_p * \omega_{k1} \tag{8.46}$$

Equating the gain of Eq. 8.42 to unity, we have the following:

$$\frac{K * \left(\sqrt{K_p^2 * \omega_{cr}^2 + K_i^2}\right)}{\omega_{cr} * \left(\sqrt{J_{eq}^2 * \omega_{cr}^2 + B^2}\right)} = 1 \tag{8.47}$$

Using Eq. 8.46 in Eq. 8.47 and simplifying,

$$K_p = \frac{\omega_{cr} * \left(\sqrt{J_{eq}^2 * \omega_{cr}^2 + B^2}\right)}{K * \left(\sqrt{\omega_{cr}^2 + \omega_{k1}^2}\right)} \tag{8.48}$$

Using Eqs. 8.46 and 8.48, the PI controller constants K_p and K_i for the speed control loop can be evaluated. Now to evaluate the PI controller constant for current loop shown in Fig. 8.26, the following procedure is adopted:

Assume that the PI controller transfer function for the current control loop is $[K_{pi} + (K_{ii}/S)]$. Further assume that the given gain crossover frequency and phase margin for the current control loops are ω_{ci} rad/s and PMi radians, respectively. Thus the loop gain G(s)*H(s) = GH2(s) = 1 results in the following equation:

$$\left(\frac{K_{pi}S + K_{ii}}{S}\right) * \frac{1}{(SL_S * \beta + R_S)} = 1 \tag{8.49}$$

$$\mathrm{GH2}(j\omega_{ci}) = \frac{(j\omega_{ci} * K_{pi} + K_{ii})}{j\omega_{ci} * (j\omega_{ci} * L_S * \beta + R_S)} \tag{8.50}$$

Using the given phase margin PMi and Eq. 8.50, the expression for phase contribution of PI controller follows:

$$\frac{K_{pi} * \omega_{ci}}{K_{ii}} = \tan\left[PMi - \frac{\pi}{2} - \tan^{-1}\left(\frac{\beta * L_S * \omega_{ci}}{R_S}\right)\right] \tag{8.51}$$

$$K_{ii} = \frac{K_{pi} * \omega_{ci}}{\tan\left[PMi - \frac{\pi}{2} - \tan^{-1}\left(\frac{\beta * L_S * \omega_{ci}}{R_S}\right)\right]} \tag{8.52}$$

In Eq. 8.52, ω_{k2} is defined below:

$$\omega_{k2} = \frac{\omega_{ci}}{\tan\left[PMi - \frac{\pi}{2} - \tan^{-1}\left(\frac{\beta * L_S * \omega_{ci}}{R_S}\right)\right]} \tag{8.53}$$

$$K_{\text{ii}} = K_{\text{pi}} * \omega_{k2} \tag{8.54}$$

Following the same procedure used for deriving Eqs. 8.47 and 8.48, K_{pi} can be expressed as follows:

$$K_{\text{pi}} = \frac{\omega_{ci} * \left(\sqrt{L_S^2 * \beta^2 * \omega_{ci}^2 + R_S^2}\right)}{\left(\sqrt{\omega_{ci}^2 + \omega_{k2}^2}\right)} \tag{8.55}$$

Using Eqs. 8.54 and 8.55, the PI controller constants K_{pi} and K_{ii} for current loop can be evaluated.

8.8.3 Model of Three-Phase Vector Controlled Induction Motor

In this section the model of the vector controlled three-phase IM is developed using Simulink [6]. The performance of the above IM is studied in detail by simulation. The three-phase IM drive used here has the parameters given in Table 8.1.

The model of vector controlled three-phase IM developed in Simulink using parameters shown in Table 8.1 is shown in Fig. 8.27 (model file: EXAMPLE8_6). The model is developed based on the equations and requirements set forth in Sects. 8.8, 8.8.1 and 8.8.2. The MATLAB program to calculate the PI controller parameters for the speed control and current control loop using the IM parameters in Table 8.1 is given in Program segment 8.2 [18]. From the simulation results, the values are tabulated in Table 8.2.

Program Segment 8.2

```
%%Calculation of Initial Conditions.
%%Vector Controlled Induction Motor Parameters.
%% Dr. Narayanaswamy P.R.Iyer
%% Reference 18
Rs=0.2147;%%Stator resistance per phase in Ohms.
Rr=0.2205;%%Rotor resistance referred to stator per phase in Ohms.
Lls=0.000991;%%Stator inductance per phase in Henries.
Llr=0.000991;%%Rotor inductance referred to stator per phase in
Henries.
Lm=0.06419;%%Mutual inductance per phase in Henries.
Jeq=0.102;%%M.I. of rotor and load.
B = 0.009541;%%Damping constant.
p=4;%%No. of poles.
%%Power output Pout = 40 kVA.
%% Steady State Operating Conditions
```

Fig. 8.27 Model of a three-phase vector controlled induction motor

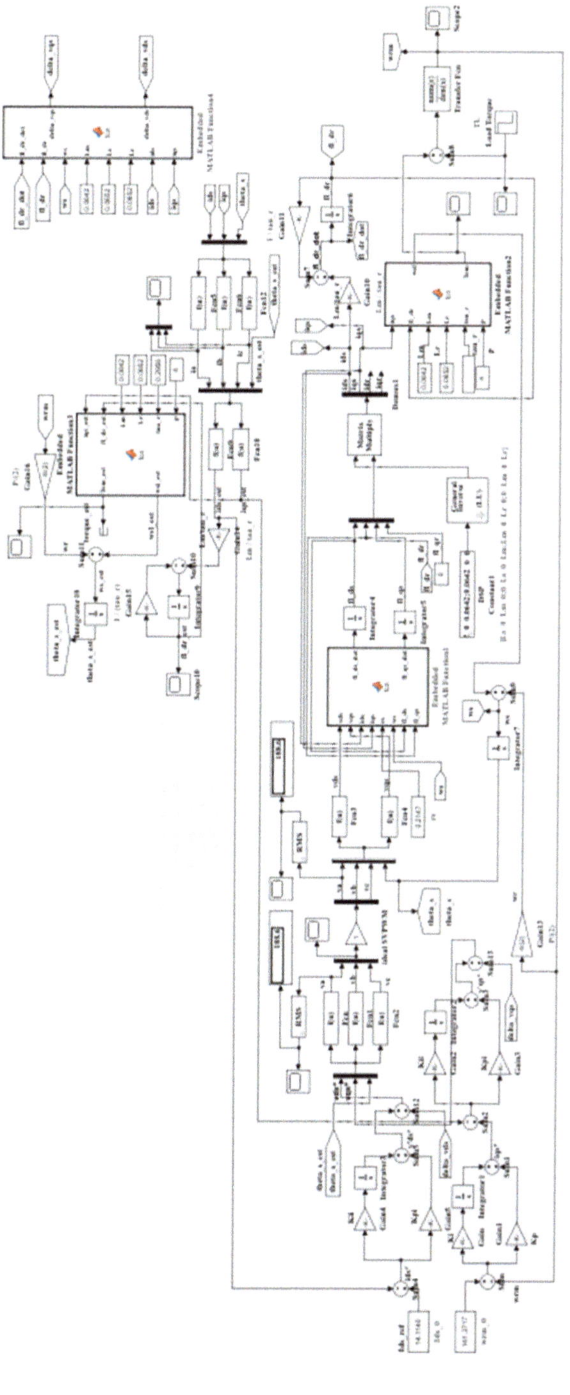

```
f=50; VLLrms= 400; s=0.1; %%Rotor slip 10%.% phase-a voltage is at its
positive peak at t=0
Wsyn=2*pi*f;                  %% synchronous speed in electrical rad/s
Wr=(1-s)*Wsyn; %%Rotor speed in electrical radians per second.
%% Phasor Calculations
Va = VLLrms * sqrt(2)/ sqrt(3);              %% Va phasor peak value.
% Space Vectors at time t=0 with stator a-axis as the reference
Vs_0 = (3/2) * Va;                     %% Vs(0) space vector
Theta_Vs_0 = angle(Vs_0);                 %% angle of Vs(0) space vector
%% We will assume that at t=0, d-axis is aligned to the stator a-axis.
Therefore, Theta_s_0=0
Theta_s_0 = 0;
Vds_0 = (2/3) * abs(Vs_0) * cos(Theta_Vs_0 - Theta_s_0);
Vqs_0 = (2/3) * abs(Vs_0) * sin(Theta_Vs_0 - Theta_s_0);
%% Calculation of machine inductances
Ls = (Lls + Lm);
Lr = (Llr + Lm);
tau_r=Lr/Rr;
%% Calculations of dq-winding currents in  steady state.
A = [Rs     -Wsyn*Ls     0      -Wsyn*Lm ;...
   Wsyn*Ls   Rs        Wsyn*Lm   0     ;...
   0      -s*Wsyn*Lm    Rr     -s*Wsyn*Lr;...   %% Matrix [A] in Eq.8.2 in
steady state
   s*Wsyn*Lm 0        s*Wsyn*Lr Rr];          %%in synchronous reference
frame.
Ainv = inv(A);                         %%s*Wsyn is (Wr + Wsl).
V_dq_0=[Vds_0; Vqs_0; 0; 0];%%Rotor is short circuited.
I_dq_0=Ainv*V_dq_0;
Ids_0=I_dq_0(1)  %% d-axis stator current initial value.
Iqs_0=I_dq_0(2)  %% q-axis stator current initial value.
Idr_0=I_dq_0(3)  %% d-axis rotor current initial value.
Iqr_0=I_dq_0(4)  %% q-axis rotor current initial value.
%% Electromagnetic Torque, which equals Load Torque in Initial Steady
State
Tem_0 = (3*p/4) * Lm * (Iqs_0 * Idr_0 - Ids_0 * Iqr_0)   %% Eq. 8.3.
TL_0 = Tem_0
%% Wrm = rotor speed in actual mechanical rad per second.
Wrm_0=(2/p)*Wr
% Inductance matrix M in Eq. 8.14.
M = [Ls 0 Lm 0 ;...
   0 Ls 0 Lm;...
   Lm 0 Lr 0 ;...
   0 Lm 0 Lr];
%% dq winding Flux Linkages with the d-axis aligned with the stator
a-axis
```

```
fl_dq_0 = M * [Ids_0; Iqs_0; Idr_0; Iqr_0]; %% dq- winding flux linkages in
vector form, Eq. %%8.14.
fl_ds_0 = fl_dq_0(1) %%d-axis stator flux initial value.
fl_qs_0 = fl_dq_0(2) %%q-axis stator flux initial value.
fl_dr_0 = fl_dq_0(3) %%d-axis rotor flux initial value.
fl_qr_0 = fl_dq_0(4) %%q-axis rotor flux initial value.
[thetar, fl_r_dq_0]=cart2pol(fl_dr_0, fl_qr_0) %%polar form of initial
dq-axis rotor flux.
[thetas, fl_s_dq_0]=cart2pol(fl_ds_0, fl_qs_0) %%polar form of initial
dq-axis stator flux.
[theta_Is_dq, Is_dq_0]=cart2pol(Ids_0, Iqs_0) %%polar form of
initial dq-axis stator current.
[theta_Vs_dq, Vs_dq_0]=cart2pol(Vds_0, Vqs_0) %%polar form of
initial dq-axis stator voltage
%% d-axis is now aligned with the rotor flux vector for vector control of
three phase IM.
fl_rd_0 = fl_r_dq_0 %%fl_rq_0 equals zero for vector control.
[fl_sd_0, fl_sq_0]=pol2cart(thetas-thetar, fl_s_dq_0) %%stator flux
linkage along d and q axis separated from resultant stator flux.
[Ids_0, Iqs_0]=pol2cart(theta_Is_dq-thetar, Is_dq_0) %%Stator
current in d and q axis separated fron resultant stator current.
[Vds_0, Vqs_0]=pol2cart(theta_Vs_dq-thetar, Vs_dq_0) %%Stator
voltages in d and q axis separated from resultant stator voltage.
%% Calculations for the controller in Fig. 8.24.
Wcr=30; %% Assumed value of crossover frequency in rad/sec
k = 3*p*Lm*Lm*Ids_0/(4*Lr); %% Eq. 8.40.
PM = 75*pi/180; %%Assumed phase margin in radians.
Wcr_kp_by_ki = tan(PM - pi/(2) - atan(Jeq*Wcr/(B))); %% kp and ki are
shown in Fig. 8.24 and %%eq. 8.43.
Wk1 = Wcr/(Wcr_kp_by_ki); %%Eq. 8.45
kp = Wcr*(sqrt(Jeq*Jeq*Wcr*Wcr + B*B))/(k*(sqrt(Wcr*Wcr + Wk1*Wk1)));
%%Eq. 8.48.
ki = kp*Wk1; %% Eq. 8.46.
%% PI controller in the current loop.
beta = 1-((Lm*Lm)/(Ls*Lr));     %% Eq. 8.31.
Wci=20*Wcr; %% Current-loop bandwidth 20 times that of the speed loop
is assumed.
PMi = PM; %%Phase margin is assumed to be the same as for speed loop.
Wci_kpi_by_kii = tan(PMi-pi/(2)-atan(Wci*Ls*beta/(Rs))); %%
Eq. 8.51.
Wk2 = Wci/(Wci_kpi_by_kii); %%Eq. 8.53.
kpi = Wci*(sqrt(Wci*Wci*Ls*Ls*beta*beta + Rs*Rs))/(sqrt(Wci*Wci +
Wk2*Wk2)); %%Eq. %%8.55.
kii = kpi*Wk2; %%Eq. 8.54.
```

Table 8.3 PI controller parameters for speed and current control loops

	Speed control loop		Current control loop		Rotor speed	Initial d-axis stator current
Sl. no.	*Kp*	*Ki*	*Kpi*	*Kii*	Mech.Rad./s	amps
1	1.1052	8.7735	1.1955	58.8381	141.3717	14.1140

The model shown in Fig. 8.27 uses the values shown in Table 8.3 for the PI controller parameters for the speed control and current control loop, initial d-axis stator current and the rotor speed set point. The initial d-axis stator current Ids_0 is subtracted from the estimated/actual d-axis stator current Ids_est from the IM model using Sum4 block, and the difference which gives the d-axis stator reference current Ids* is given to two gain blocks with constants K_{pi} and K_{ii}. The output of gain block K_{ii} is integrated and summed with the output of gain block K_{pi} to generate V′ds* as shown in Fig. 8.26a. Similarly the rotor speed set point wrm_0 is subtracted from the actual/estimated rotor speed wrm from the IM model using Sum block, and the difference speed is given to PI controller with gain blocks K_p and K_i. The output of K_i gain block after integration is added with K_p gain block output using Sum1 block to generate Iqs*, as shown in Fig. 8.24. This Iqs* is subtracted from the actual/estimated q-axis stator current from the IM model Iqs_est using Sum2 block. The output of Sum2 block is given to two gain blocks with constants Kpi and Kii. The output of Kii gain block is integrated and added with the output of Kpi gain block using Sum3 block to generate V′qs* as shown in Fig. 8.26b. With dλr′/dt, λr′, ωs, Lm, Lr, Ls, ids and iqs as inputs to Embedded MATLAB function4, outputs Δvds and Δvqs as per Eqs. 8.37 and 8.39 are generated as shown in Program segment 8.3 in the model file EXAMPLE8_6. Then v′ds* and v′qs* are, respectively, added with Δvds and Δvqs using SUM12 and SUM13 blocks to obtain vds* and vqs*, the reference dq-axis stator voltages. With vds*, vqs* and estimated θs_est from the model as input to a three-input Mux block, user defined function blocks, Fcn, Fcn1 and Fcn2, generate abc-axis voltages va, vb and vc using transformation matrix defined in Eq. 8.11. With va, vb, vc and actual value θs as inputs to a four-input Mux, user defined function blocks Fcn3 and Fcn4 generate vds and vqs using transformation matrix defined in Eq. 8.10. With vds, vqs, ids, iqs, ωs, Rs, λds and λqs as inputs to the block Embedded MATLAB Function1, outputs $d\lambda_{ds}/dt$ and $d\lambda_{qs}/dt$ defined in Eq. 8.4 are generated as per source code given in Program segment 8.4 in the model file EXAMPLE8_6. The outputs $d\lambda_{ds}/dt$ and $d\lambda_{qs}/dt$ are integrated using two integrator blocks to obtain λds and λqs. The flux linkages λds, λqs and λdr′ from model and λqr′ which is zero for vector control are given to a four-input Mux block to obtain flux linkage vector λdq_sr. The DSP Constant1 block which generates the inductance matrix [L] defined in Eq. 8.14 is inverted using General Inverse block, and this inverted [L^{-1}] matrix is multiplied with flux linkage vector λdq_sr using Matrix Multiply block to obtain dq-axis stator and rotor current vectors ids, iqs, idr′ and iqr′. This dq-axis stator and rotor current vector are available at the output of Demux1 block. Equation 8.24 can be rewritten in the form [$d\lambda_{dr'}/dt$ = (Lm*ids/τ_r) – ($\lambda_{dr'}/\tau_r$)], where τ_r = (Lr/Rr′). The ids output of Demux1 block is multiplied by gain block with constant (Lm/τ_r), and the $\lambda_{dr'}$ output is multiplied by another gain block with gain constant ($1/\tau_r$), and these two outputs are subtracted using Sum7 block to obtain $d\lambda_{dr'}/dt$ which is then integrated to get $\lambda_{dr'}$. The output iqs from Demux1, $\lambda_{dr'}$, Lm, Lr, τ_r and P are given as inputs to Embedded MATLAB function2 to get the output ωsl and Tem defined in Eqs. 8.19 and 8.21, respectively, using source code given in Program Segment 8.5 in the model file EXAMPLE8_6. The Step block is used to model mechanical load torque T_L. In this block, an external load of 200 Nw-metres is varied to 100 Nw-metres in 0.1 s. The outputs Tem and Tl are subtracted using Sum8 block, and the output is given to Transfer Fcn block with transfer function [1/(S.Jeq + B)] which gives the output ω_{rm}. This value of ωrm is multiplied by gain block with constant P/2 to get ωre. This ωre is added with ωsl using Sum9 block to get the stator reference frame frequency ωs which is integrated to get θs. With ids, iqs and θs as input to the Mux, abc-axis stator currents ia, ib and ic are generated using Fcn5, Fcn6 and Fcn12 blocks using transformation matrix defined in Eq. 8.11. This abc-axis stator currents along with estimated θs_est are given to a four-input Mux block to generate estimated ids_est and iqs_est. With ids_est as input, λdr′_est is generated in the same way for generating $\lambda_{dr'}$ explained above. With iqs_est, λdr′_est, Lm, Lr, τ_r and P as inputs, Embedded MATLAB Function3 block is used to generate Tem_est and ωsl_est defined in Eqs. 8.19 and 8.21 using source code defined in Program segment 8.5 in the model file EXAMPLE8_6, with iqs and fl_dr replaced with iqs_est and fl_dr_est. Finally ωrm multiplied by gain block P/2 is added with ωsl_est using Sum11 block to obtain ωs_est which is integrated to get estimated θs_est. The ids_est and iqs_est are fed back to the respective PI controller Sum blocks to generate vds* and vqs* voltages.

8.8.4 *Simulation Results*

The simulation of the three-phase vector controlled IM model is carried out using ode15s (stiff/NDF) solver in Simulink [6]. Simulation results for rotor E.M. torque, rotor speed, three-phase stator voltages and currents and load torque are shown in Fig. 8.28a–e, respectively.

8.8.5 *Discussion of Results*

From the simulation results, it is seen that as the load torque is varied from 200 Nw-M to 100 Nw-M in 0.6 s, the rotor E.M. torque shoots up to 1280 Nw-M and finally reaches the steady value of 201.8 Nw-M and 101.3 Nw.M before and after 0.6 s. The rotor speed shoots up to 159.8 Mech.rad per second during starting, reaches a steady speed of 140.6 Mech.rad per second in 0.6 s, shoots up to 166.2 Mech.rad per second when load is reduced at 0.6 s and then reaches a steady speed of 141.4 Mech.rad per second close to the set point value. Stator currents and voltages are well balanced, initially shoot up and finally settle to a constant value. This shows the accuracy of vector control of three-phase IM.

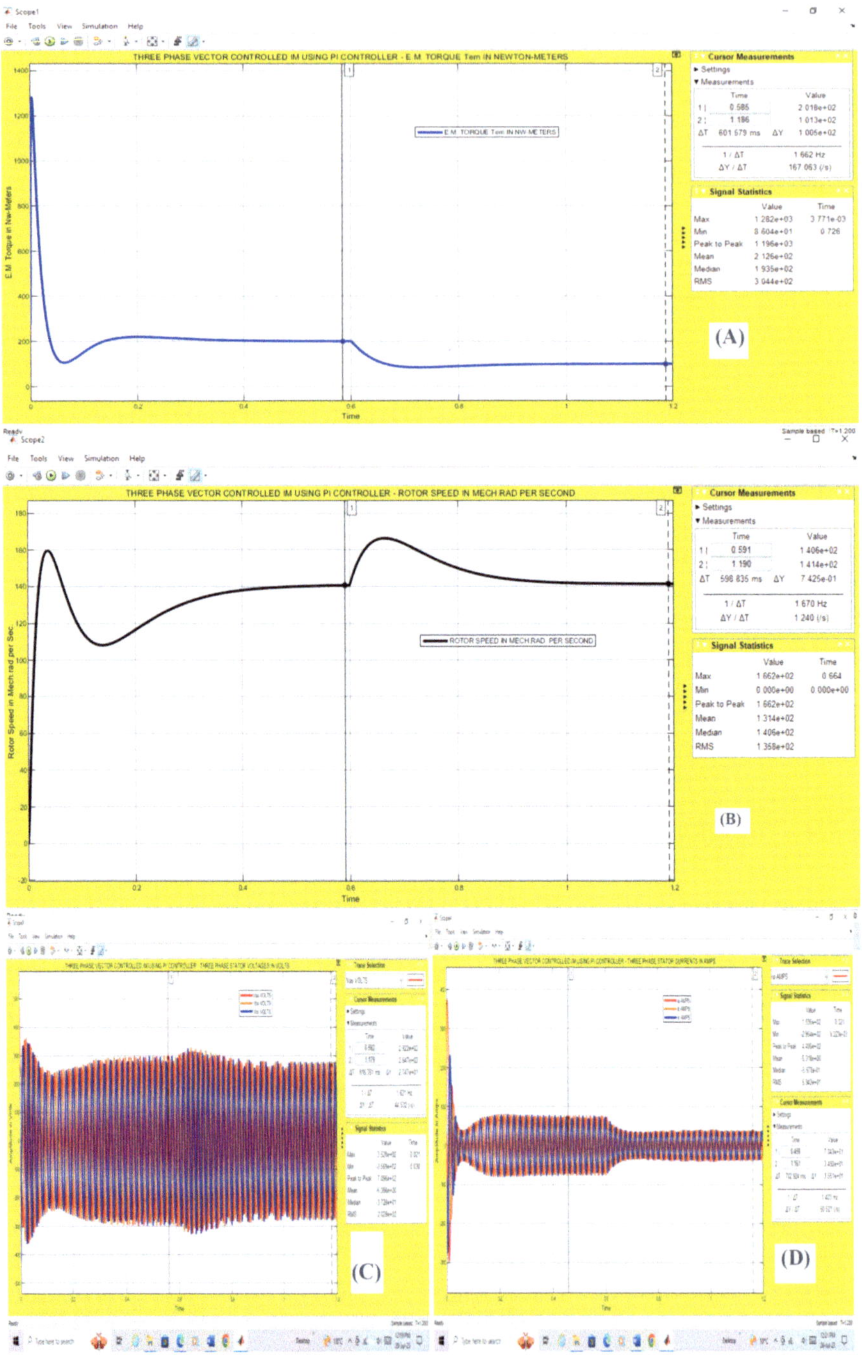

Fig. 8.28 Three-phase vector controlled IM using PI controller simulation results: (**a**) E.M. torque, (**b**) rotor speed, (**c**) stator applied voltages, (**d**) stator currents, (**e**) load torque

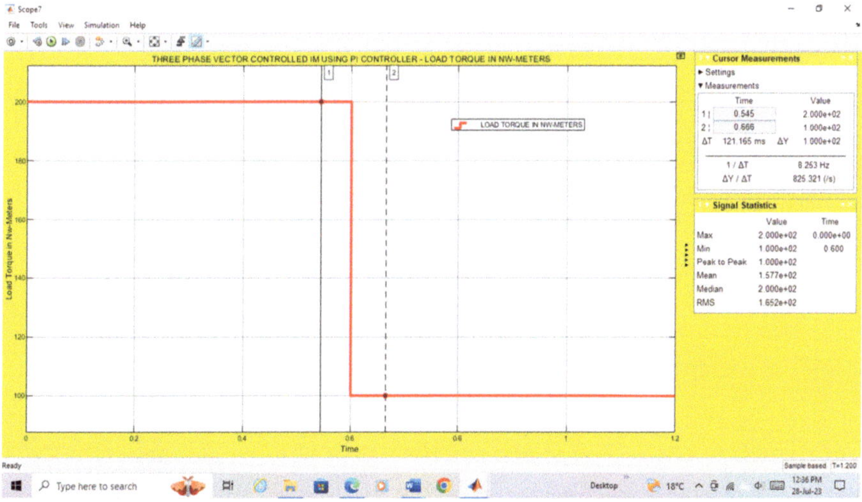

Fig. 8.28 (continued)

8.9 Case Study: Three-Phase Vector Controlled Induction Motor Drive Fed by SVPWM Inverter

The model of the vector controlled three-phase IM drive fed by SVPWM inverter is shown in Fig. 8.29 (model file: CASE_STUDY_EX8_2). Here unity value for modulation index ma is assumed. Any suitable value for ma can be assumed less than unity such that the RMS line to line voltage of the inverter is within the given value. The vector controlled three-phase IM model is the one explained in Sect. 8.8.3. This vector controlled three-phase IM model is used twice one to generate the gate pulses for the SVPWM inverter and the other to replace the hardware component of the three-phase IM. The changes in the SVPWM inverter model are explained below.

The SVPWM inverter presented in Sect. 8.6 is used here. The reference value of dq-axis stator voltages vds*, vqs* and θs_est is used to transform the reference dq-axis stator voltage to abc-axis stator voltage, and this value along with θs is used to generate dq-axis stator voltage as explained in Sect. 8.8.3. This value of vds and vqs is given to a two-input Mux and using Fcn8 block [sqrt(V^2ds + V^2qs)] which gives Vref for the SVPWM inverter. This Vref value is used to find DC link voltage Vdc with Fcn11 block using the formula [√3*Vref/ma], where ma is the AM index. Initially Vdc is calculated using the relation [VLL(RMS)*√3/√2] where RMS line to line voltage VLL(RMS) is the given value in Table 8.1. Simulation is run and Vdc value is noted using scope or display. This value of Vdc is used for the three-phase inverter and the simulation is repeated. A sampling frequency of 1200 Hz is used for the SVPWM inverter.

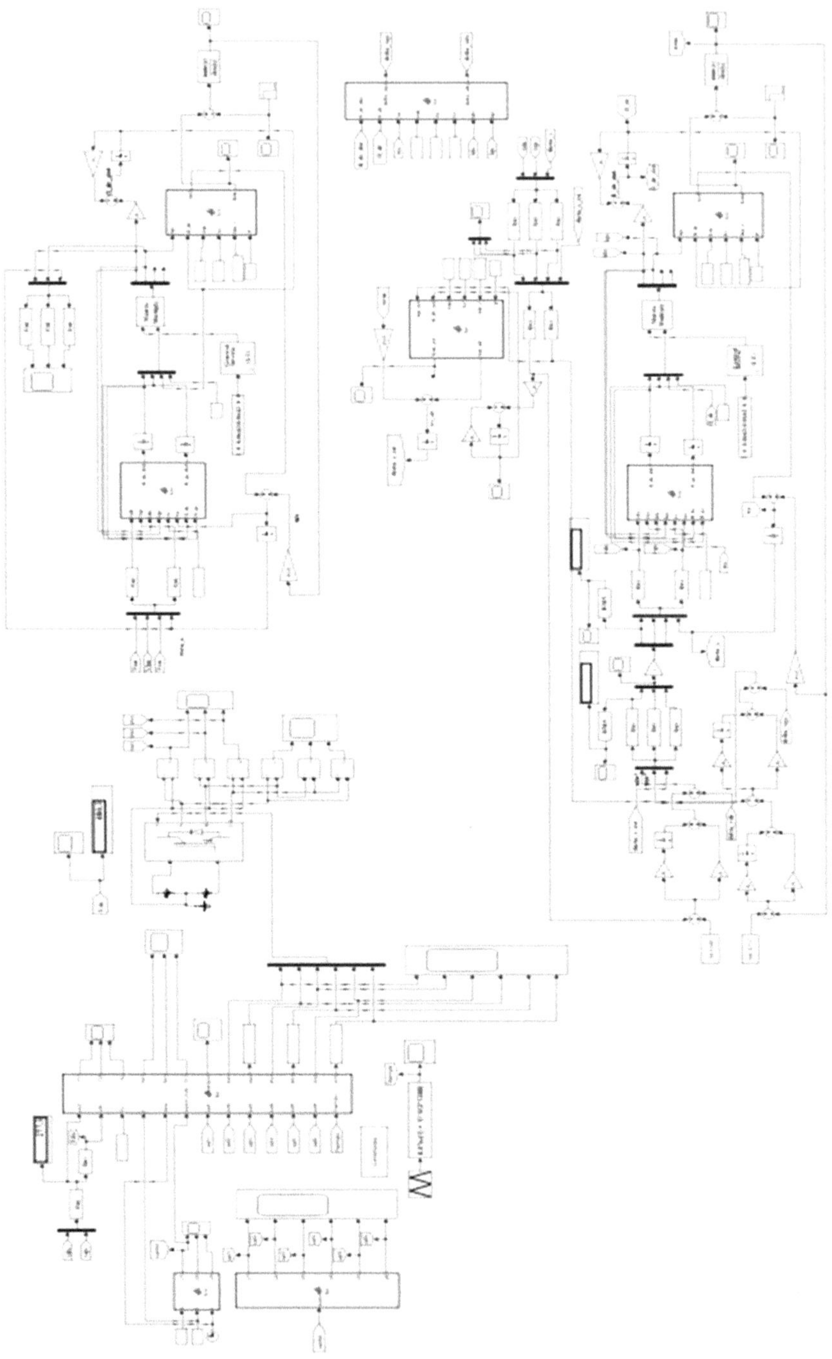

Fig. 8.29 Model of a three-phase SVPWM inverter-fed vector controlled induction motor drive

8.9.1 *Simulation Results*

The simulation of the three-phase SVPWM inverter-fed vector controlled IM drive is carried out using ode23tb (stiff/TR-BDF2) solver in Simulink [6]. Simulation results are shown in Fig. 8.30a–e.

8.9.2 *Discussion of Results*

From the simulation results, it is seen that the mean value of rotor steady state speed is 145.9 Mech.rad/s which is close to the set point speed of 141.37 mech.rad/s. Similarly the mean value rotor E.M. torque in steady state is found to be 104.6 Nw-M which is close to final steady state load torque of 100 Nw-M. By proper choice of AM index, it is possible to bring the rotor speed and electromagnetic torque very close to the set point value. This demonstrates the advantage of vector control of IM drive.

8.10 Direct Torque Control of Three-Phase Induction Motor Drive

In this direct torque control (DTC) scheme, NO dq-axis transformation of voltages and currents is needed. The estimated electromagnetic torque and stator flux are compared with their respective reference values in a hysteresis comparator, and if their error values exceed or fall below the selected hysteresis band, appropriate stator voltage space vector depending on the sector location of the stator flux is applied to the inverter to increase or decrease the stator flux and motor torque as required [20, 21]. DTC scheme is essentially voltage source inverter driven.

In the analysis of DTC, the three-phase IM model in stationary reference frame ($\omega_C = 0$) is used. This method depends on stator flux and motor torque estimate and in addition rotor speed estimate for sensorless or encoderless direct torque control. In the torque mode control, comparison of stator flux and motor torque is made with their respective reference values in a hysteresis comparator, and if the error flux and torque go above or below the specified hysteresis band depending on sector location of stator flux, appropriate switches of the six-step inverter are either turned on or off to bring them within the specified limit [20–25]. In the speed mode control, in addition reference speed is compared with estimated speed, and the speed error is given to a PI controller whose output provides the motor torque reference [18, 22]. The electromagnetic torque expression is reviewed below:

The electromagnetic (e.m.) torque expression using stator dq-axis flux linkages and currents is given in Eq. 8.12. From Eq. 8.14, the dq-axis rotor currents referred to stator can be expressed as follows:

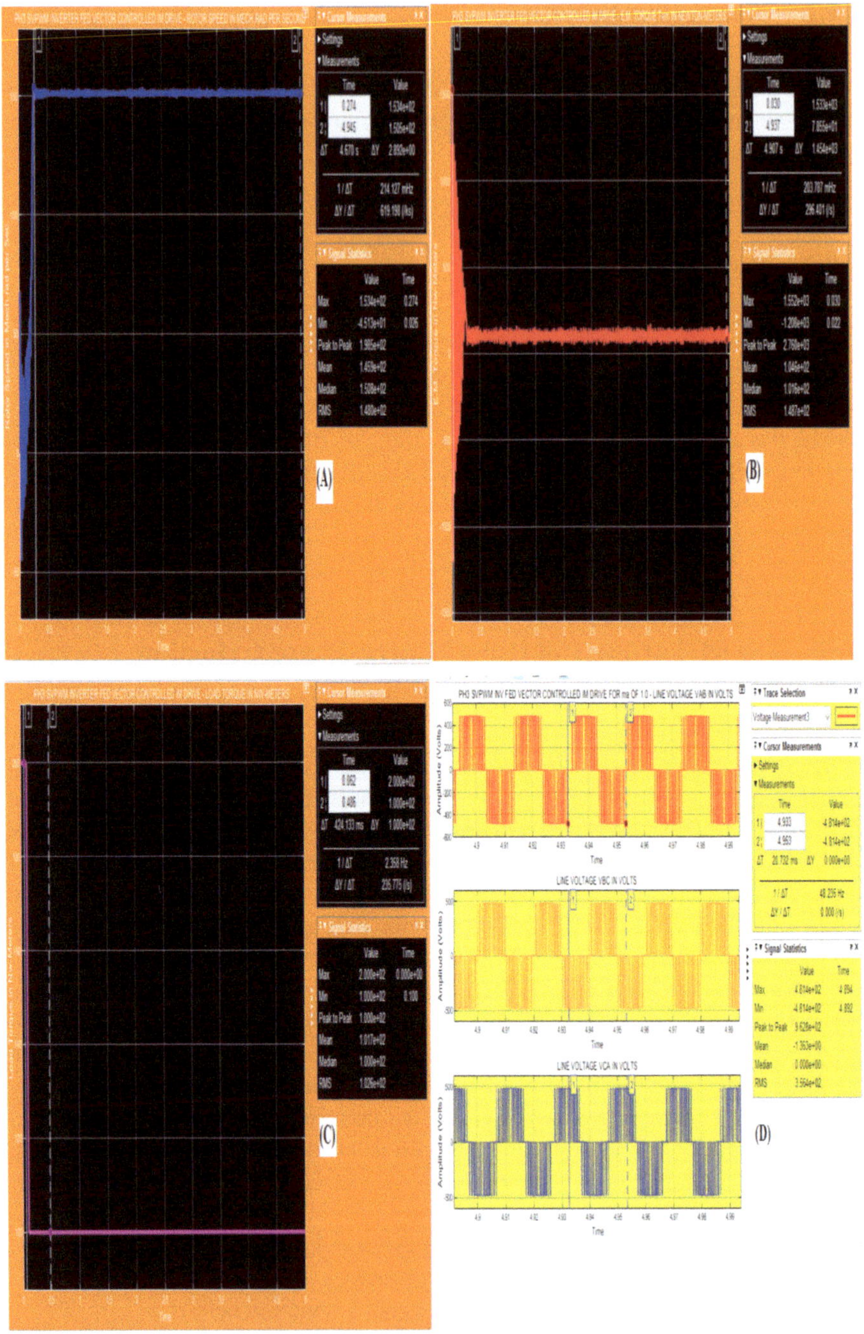

Fig. 8.30 Three-phase SVPWM inverter-fed vector controlled IM drive simulation results: (**a**) rotor speed in Mech.rad/sec. (**b**) Rotor E.M. torque in Nw-M. (**c**) Load torque in Nw-M. (**d**) Inverter line voltage in volts. (**e**) Three-phase stator current in amps

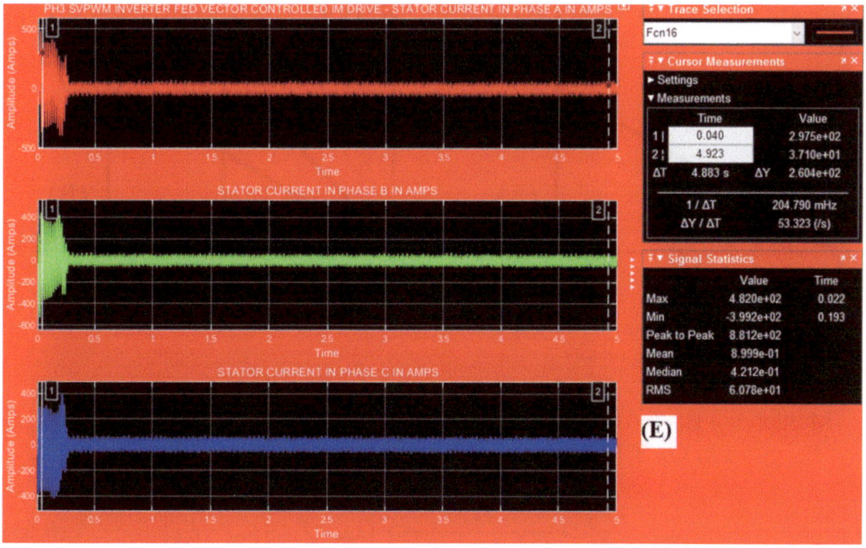

Fig. 8.30 (continued)

$$i'_{dr} = \frac{\lambda_{ds}}{L_m} - \frac{L_s \cdot i_{ds}}{L_m} \tag{8.56}$$

$$i'_{qr} = \frac{\lambda_{qs}}{L_m} - \frac{L_s \cdot i_{qs}}{L_m} \tag{8.57}$$

From the expression for λ'_{dr} and λ'_{qr} in Eq. 8.14, using Eqs. 8.56 and 8.57 and simplifying,

$$\lambda'_{dr} = \frac{i_{ds} \cdot L_\sigma}{L_m} + \frac{\lambda_{ds} \cdot L_r}{L_m} \tag{8.58}$$

$$\lambda'_{qr} = \frac{i_{qs} \cdot L_\sigma}{L_m} + \frac{\lambda_{qs} \cdot L_r}{L_m} \tag{8.59}$$

$$\text{where } L_\sigma = \left(L_m^2 - L_s \cdot L_r\right) \tag{8.60}$$

Using Eqs. 8.58 and 8.59, i_{ds} and i_{qs} can be expressed as follows:

$$i_{ds} = \frac{L_m \cdot \lambda'_{dr}}{L_\sigma} - \frac{\lambda_{ds} \cdot L_r}{L_\sigma} \tag{8.61}$$

$$i_{qs} = \frac{L_m \cdot \lambda'_{qr}}{L_\sigma} - \frac{\lambda_{qs} \cdot L_r}{L_\sigma} \tag{8.62}$$

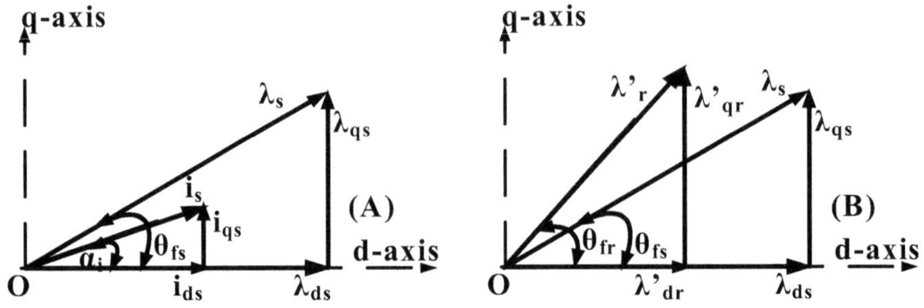

Fig. 8.31 Induction motor phasor diagram: (**a**) stator flux and stator current space vectors, (**b**) stator and rotor flux space vectors

Multiplying Eq. 8.61 by λ_{qs} and Eq. 8.62 by λ_{ds} and subtracting, using equation for T_{em} given in Eq. 8.12 and also using Eq. 8.60, we have the following:

$$T_{em} = \frac{3 * P}{4} * \frac{L_m}{\left(L_m^2 - L_s * L_r\right)} * \left[\lambda_{ds} * \lambda'_{qr} - \lambda_{qs} * \lambda'_{dr}\right] \tag{8.63}$$

Equations 8.12 and 8.63 can be represented by phasor diagrams shown in Fig. 8.31.

Referring to phasor diagrams in Fig. 8.31, Eqs. 8.12 and 8.63 can be expressed as follows:

$$T_{em} = \frac{3 * P}{4} * \lambda_s * i_s * \sin\left(\theta_{fs} - \alpha_i\right) \tag{8.64}$$

$$T_{em} = \frac{3 * P}{4} * \frac{L_m}{\left(L_m^2 - L_s * L_r\right)} * \lambda_s * \lambda'_r * \sin\left(\theta_{fs} - \theta_{fr}\right) \tag{8.65}$$

It is clear from Eq. 8.64 that stator flux angular position is varied and the angle $(\theta_{fs} - \alpha_i)$ varies, thereby increasing or decreasing the motor torque. The rotor time constant (J/B) is large and the rotor flux variation is slow and almost remains constant over a short period of time. From Eq. 8.65 it is clear that as the angular position of the stator flux is varied, angle $(\theta_{fs} - \theta_{fr})$ varies and this increases or decreases the motor torque and stator flux. This principle is used in DTC [20, 21]. The mode of implementation of DTC is given below:

The DTC has the following advantages:

- Direct control of stator flux and motor torque
- Indirect control of stator currents and voltages
- Approximately sinusoidal stator currents and stator flux
- High dynamic performance
- Absence of coordinate transformation
- Absence of voltage modulator block

- Absence of current controllers
- Inherently sensorless control

Some of the disadvantages of DTC are (1) problem during starting; (2) requirement of motor torque, stator flux and speed estimators; (3) inherent motor torque and flux ripples; and (4) sensitive to
stator resistance variations.

Referring to Eq. 8.4, the stator applied voltage v_s ($v_{ds} + jv_{qs}$) can be expressed as follows:

$$\overrightarrow{v_s} = R_s.\overrightarrow{i_s} + \frac{d}{dt}.\overrightarrow{\lambda_s} \tag{8.66}$$

$$\overrightarrow{\lambda_s} = \int\left(v_s - R_s.\overrightarrow{i_s}\right).dt = \lambda_s * e^{j\theta_{fs}} \tag{8.67}$$

Using z-transform, Eq. 8.67 can be expressed as follows:

$$\overrightarrow{\lambda_s} = \frac{z^{-1}}{(1 - z^{-1})} \cdot T_s \cdot \left(v_s - R_s \cdot \overrightarrow{i_s}\right) \tag{8.68}$$

In Eq. 8.68 T_s is the sample time. Equation 8.68 can be expressed in sample time as follows:

$$\lambda_s(k) - \lambda_s(k-1) = v_s(k-1) \cdot T_s - R_s.i_s(k-1).T_s \tag{8.69}$$

From Eqs. 8.58 and 8.59, the rotor flux $\overrightarrow{\lambda'_r}$ can be expressed as follows:

$$\overrightarrow{\lambda'_r} = \frac{L_r.\overrightarrow{\lambda_s}}{L_m} + \frac{L_\sigma.\overrightarrow{i_s}}{L_m} = \lambda'_r * e^{j\theta_{fr}} \tag{8.70}$$

where angular displacement of the rotor flux θ_{fr} from Fig. 8.31b is given below:

$$\theta_{fr} = \tan^{-1}\left(\frac{\lambda'_{qr}}{\lambda'_{dr}}\right) \tag{8.71}$$

Now rotor speed ω_{r_elec} is $d\theta_{fr}/dt$ which can be expressed in discrete time as given below:

$$\omega_{r_{elec}} = \left(\frac{1 - z^{-1}}{T_s}\right) * \theta_{fr} \tag{8.72}$$

$$\omega_{r_{elec}} = \frac{\theta_{fr}(k) - \theta_{fr}(k-1)}{T_s} \tag{8.73}$$

In Eqs. 8.72 and 8.73, T_s is the sample time. Also from Eq. 8.13, the rotor mechanical speed ω_{rm} (ω_{r_elec}*2/P) can be expressed as follows:

$$\omega_{rm} = \frac{(T_{em} - T_L)}{(J_{eq}.s + B)} \tag{8.74}$$

If stator resistance is neglected, Eq. 8.66 for stator voltage can be expressed as follows:

$$\overrightarrow{v_s} = \frac{d}{dt}.\overrightarrow{\lambda_s} \tag{8.75}$$

Thus the change in stator flux is directly related to stator voltage. For a small change in time interval Δt, stator flux change $\Delta\lambda_s = v_s \cdot \Delta t$. Thus stator flux space vector moves in the direction of the stator voltage space vector during this interval of time. Thus by appropriate choice of stator voltage space vector, it is possible to control the magnitude of stator flux as desired. Decoupled control of stator flux and motor torque is possible by varying the radial and tangential component of stator flux space vector which are proportional to the component of stator voltage acting in the same directions. By proper choice of the stator voltage space vector, the angle ($\theta_{fs} - \theta_{fr}$) in Eq. 8.65 can be increased or decreased to increase or decrease the motor torque. As rotor flux is almost constant during short interval of time due to large rotor time constant, this in effect increases or decreases the stator flux. A number of methods for switching the inverter are proposed to achieve DTC. The classical method of switching the inverter is presented below [20–24]:

The three-phase inverter configuration and its space vector diagram are shown in Figs. 4.1 and 4.2. It is shown in Sect. 4.2 how the switch states Sa, Sb and Sc and the DC link voltage Vdc are related with the six nonzero voltage space vectors V1(100), V2(110), V3(010), V4(011), V5(001) and V6(101) and the zero voltage vectors V0 (000) and V7(111). This is also summarised in Table 4.2. In classical DTC (C_DTC) method, Sectors 1 to 6 are divided as shown by dotted lines in Fig. 8.32.

Each sector is π/3 radians and is ±π/6 radians above and below the respective voltage space vector. If stator flux location corresponds to the voltage space vector V_k where k Є 1–6, then application of voltage space vectors V_{k+1} and V_{k-1} increases and that of V_{k+2} and V_{k-2} decreases the stator flux. Similarly application of voltage space vectors

V_{k+1} and V_{k+2} increases and that of V_{k-1} and V_{k-2} decreases the motor torque. This is shown in Fig. 8.33 for stator flux in Sector 1.

Similar argument applies to stator flux in other sectors. The voltage space vectors V_k and V_{k+3} are not considered because they can both increase ($-\pi/6 < \theta_{fs} <= 0$) and decrease ($0 < \theta_{fs} <= +\pi/6$) the motor torque for stator flux in the same sector 1. The general selection table for DTC is shown in Table 8.4. If zero vector V0 or V7 is applied, the torque remains within the hysteresis band limit and the stator flux magnitude and angular position don't change. The classical DTC switching scheme is shown in Table 8.5 [20]. In Table 8.5, V0 to V7 are as defined in Fig. 8.32. In

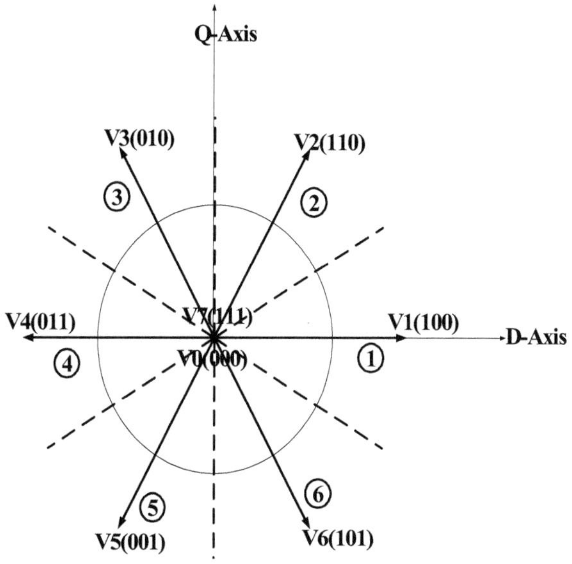

Fig. 8.32 Three-phase inverter voltage space vectors and sector allocation for classical DTC

Table 8.5, if an increase in stator flux is required, it is represented by FI and for decrease required it is represented by FD. Similarly if an increase in torque is required, it is represented by TI, and for NO change and decrease required, it is represented by TE and TD, respectively. These requirements for stator flux and torque can be expressed as follows:

$$\left.\begin{array}{c}\delta\lambda s = \text{FI True if } \lambda_s <= \lambda_{s\_ref}\text{-}H\lambda_s \\ \delta\lambda s = \text{FD True if } \lambda_s >= \lambda_{s\_ref} + H\lambda_s \\ \delta Te = \text{TI True if } T_{em} <= T_{em\_ref}\text{-}HT_{em} \\ \delta Te = \text{TE True if } T_{em} = T_{em\_ref} \\ \delta Te = \text{TD True if } T_{em} >= T_{em\_ref} + HT_{em}\end{array}\right\} \tag{8.76}$$

In Eq. 8.72, λ_{s_ref} and T_{em_ref} represent the reference or commanded values for stator flux and electromagnetic torque, and $H\lambda_s$ and HT_{em} represent their respective hysteresis band limit. The DTC schematic is presented below:

In DTC scheme, the stator flux and motor torque are estimated every sampling interval and are then compared with their respective reference values, and the error which is the difference between the respective reference value and estimated value is then fed to two separate hysteresis comparators. If the stator flux error and torque error exceed or fall below a specified hysteresis band limit, then appropriate switches of the three-phase inverter are either turned ON or OFF applying the required stator voltage to maintain the stator flux and torque within the prescribed band limit. There are two DTC schemes in use. In one method fixed torque reference and stator flux reference are used without any speed estimate and PI controller. In the second method, torque reference is calculated by finding the difference between the

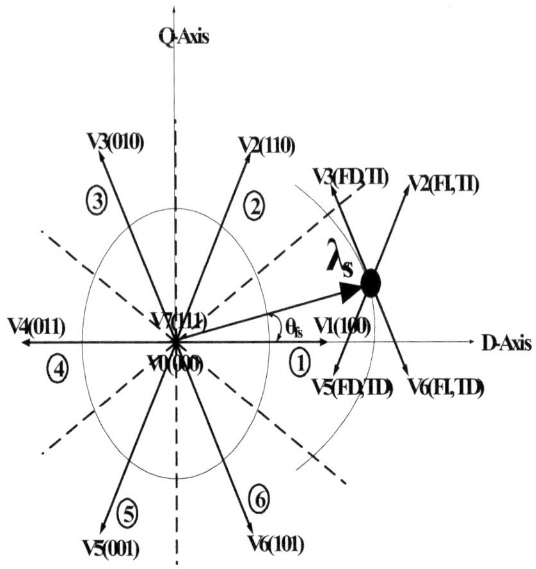

Fig. 8.33 Selection of voltage space vector for stator flux in sector 1. FI, flux increase; FD, flux decrease; TI, torque increase; TD, torque decrease

Table 8.4 General selection table for DTC for stator flux in Sector k

Sl. no.	Parameter	Voltage space vector	Action
1.	Stator flux	V_k, V_{k+1}, V_{k-1}	Increase
2.	Stator flux	V_k, V_{k+2}, V_{k-2}	Decrease
3.	E.M. torque	V_k, V_{k+1}, V_{k+2}	Increase
4.	E.M. torque	V_k, V_{k-1}, V_{k-2}	Decrease

Table 8.5 Classical DTC switching table

δλs	δTe	Sector S1	Sector S2	Sector S3	Sector S4	Sector S5	Sector S6
FI	TI	V2	V3	V4	V5	V6	V1
	TE	V7	V0	V7	V0	V7	V0
	TD	V6	V1	V2	V3	V4	V5
FD	TI	V3	V4	V5	V6	V1	V2
	TE	V0	V7	V0	V7	V0	V7
	TD	V5	V6	V1	V2	V3	V4

reference speed and actual speed, and this error speed is applied to a PI controller. The PI controller output gives the torque reference. These two schemes are shown in Figs. 8.34 and 8.35, respectively. In both these methods, stator flux amplitude during every sampling interval and its sector location are determined as follows:

Using Eqs. 8.4 and 8.68, in the stationary reference frame, the dq-axis stator flux linkage and its sector angle θ_{fs} can be expressed as follows:

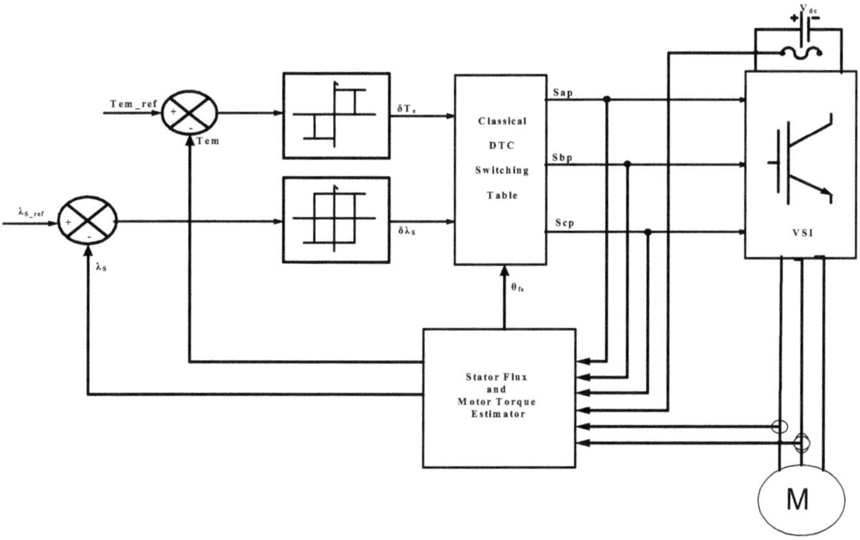

Fig. 8.34 Direct torque control scheme using torque control method

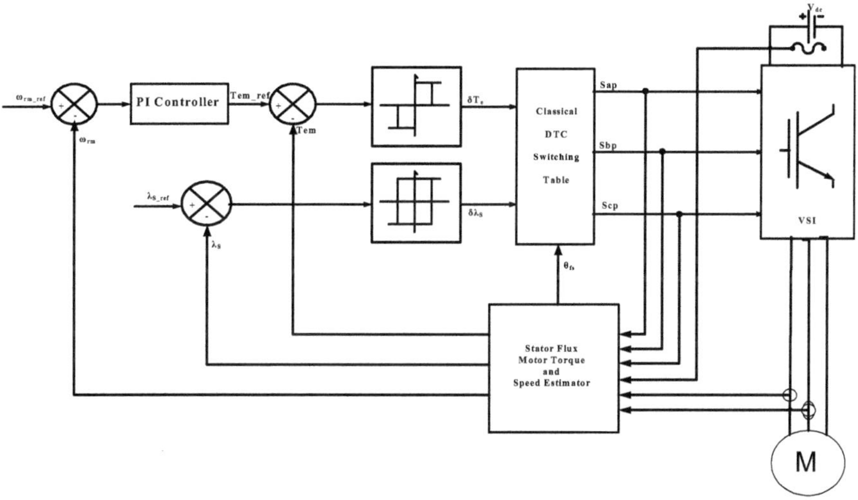

Fig. 8.35 Direct torque control scheme using speed control method

$$\overrightarrow{\lambda_{qs}} = \frac{z^{-1}}{(1 - z^{-1})} \cdot T_s \cdot \left(v_{qs} - R_s \cdot \overrightarrow{i_s} \right) \tag{8.77}$$

$$\overrightarrow{\lambda_{ds}} = \frac{z^{-1}}{(1 - z^{-1})} \cdot T_s \cdot \left(v_{ds} - R_s \cdot \overrightarrow{i_{ds}} \right) \tag{8.78}$$

$$\lambda_S = \sqrt{\left(\lambda_{ds}^2 + \lambda_{qs}^2 \right)} \tag{8.79}$$

$$\theta_{fs} = \tan^{-1}\left(\frac{\lambda_{qs}}{\lambda_{ds}} \right) \tag{8.80}$$

Torque estimate is done using values for λds, λqs, ids and iqs and number of poles P as given by Eq. 8.12. In addition, in the DTC scheme using PI controller, rotor speed estimate is done using Eq. 8.74. This method of DTC is sensorless or encoderless. Equations 8.77 to 8.80 are used to determine the stator flux amplitude λ_S and its sector location θ_{fs} during each sampling interval. The models of the C_DTC of three-phase IM drive using the torque control and speed control method are presented in the next section.

8.10.1 Model of Direct Torque Control of Three-Phase Induction Motor Drive with PI Controller Using Classical Switching Table: Speed Control Method

The model of the C_DTC of three-phase IM using PI controller in the speed control loop is presented in this section. Three-phase IM parameters are the same as in Table 8.1. Here the PI controller parameters K_p and K_i are calculated using Program Segment 8.2 and tabulated in Table 8.2. Stator flux reference is also calculated from Program Segment 8.2. The model of the C_DTC of three-phase IM with PI controller in the speed control loop in the stationary reference frame is shown in Fig. 8.36 (model file: EXAMPLE8_7). The dialog boxes are shown in Fig. 8.37. The essential model subsystem for three-phase IM model data is the one using dq0-axis flux linkage equations in state space which is shown in Fig. 8.38. The model subsystems for three-phase AC generator and dq0-axis transform are the same as shown in Fig. 8.5a. Here three-phase sine wave AC generator is removed and instead the square wave output phase voltage of three-phase inverter is used. Figure 8.5b is modified and is shown in Fig. 8.38 where two discrete integrators with sample time 10e-6 s are used to realise the dq-axis stator flux linkages. The subsystem Rotor Torque and Speed Calculator is the same as shown in Fig. 8.5c. Here angular frequency of the reference frame ω_C is entered as zero and all other data from Table 8.1 are entered in the relevant box shown in Fig. 8.37. The PI controller parameters K_p and K_i and the rotor reference speed ω_{rm_ref} from Table 8.2 are entered in the PI controller block shown in Fig. 8.36. Sample time for discrete integrator for integrating the derivative of dq-axis stator flux is entered as 10e-6 sec. The model

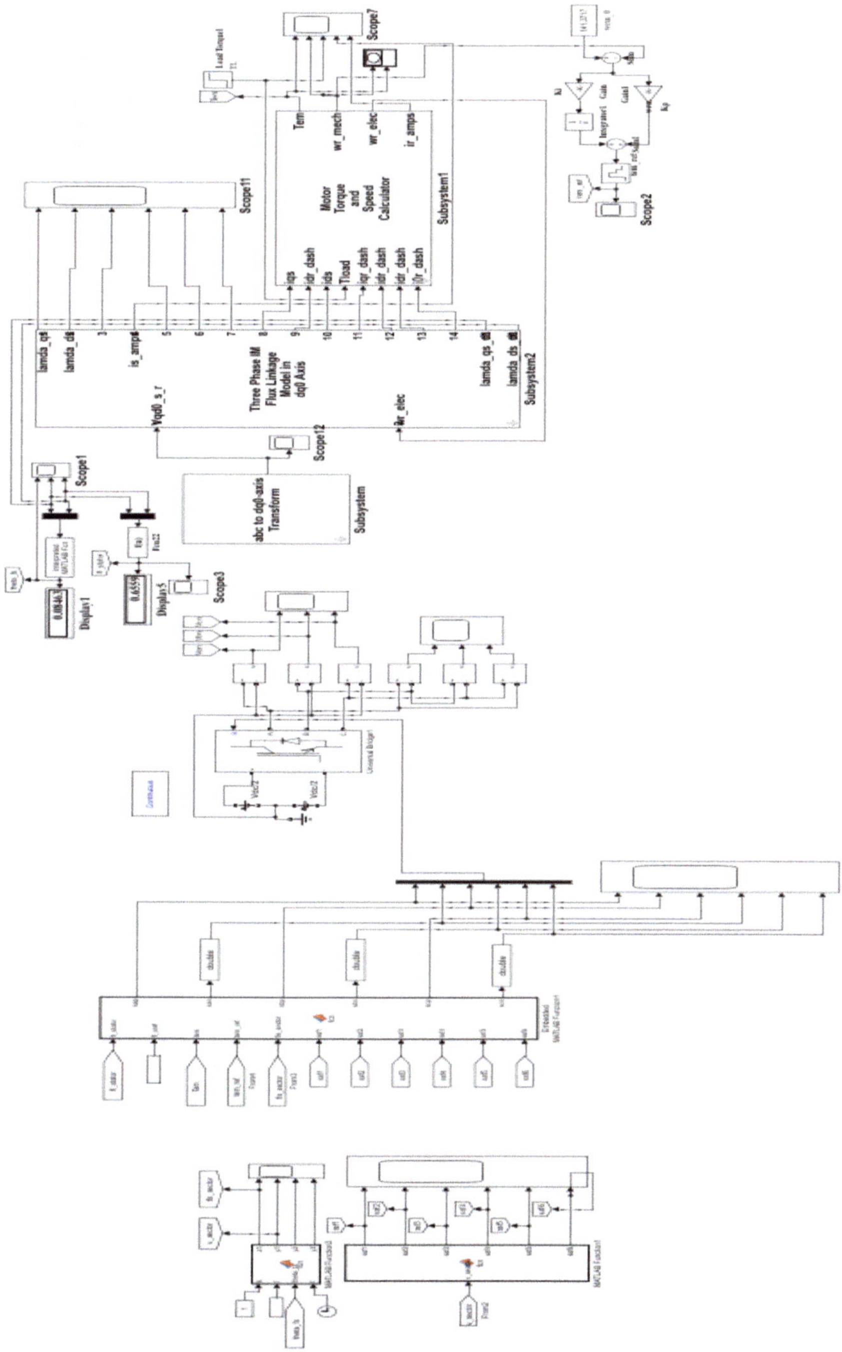

Fig. 8.36 Model of DTC of three-phase IM drive with PI controller in the speed control loop

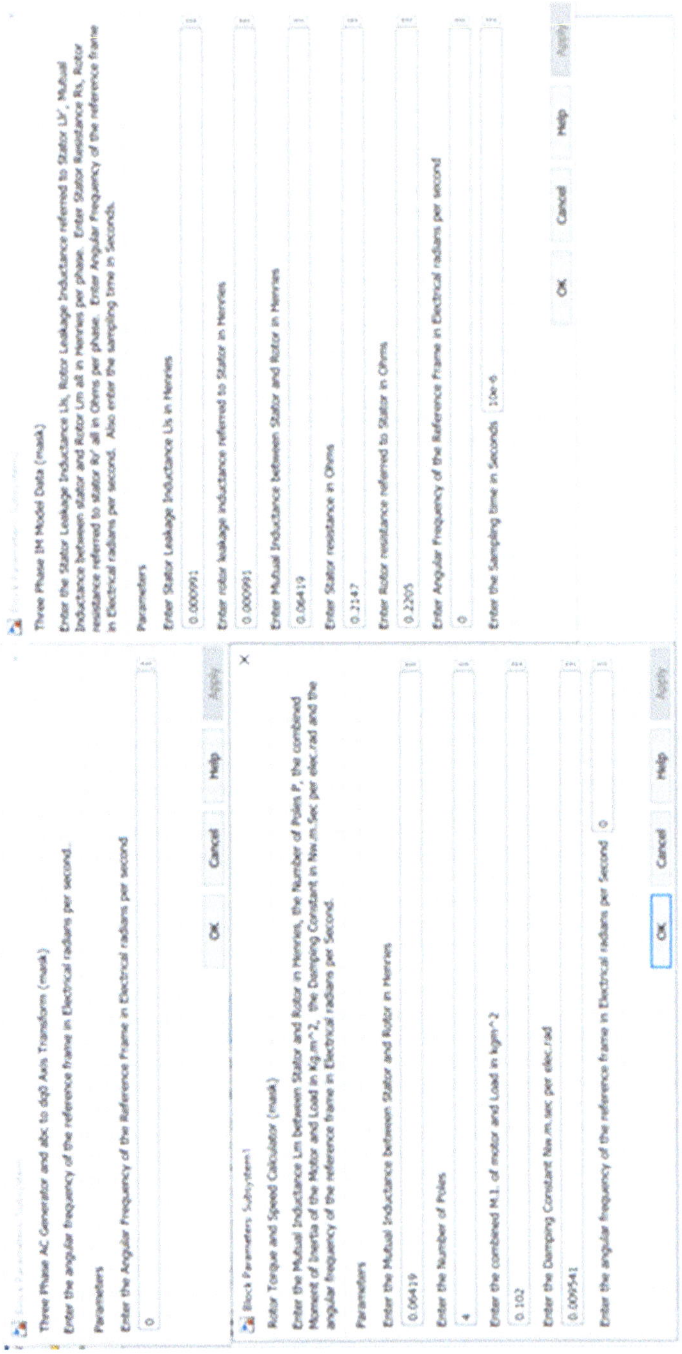

Fig. 8.37 Model of DTC of three-phase IM with PI controller in the speed control loop—dialog boxes

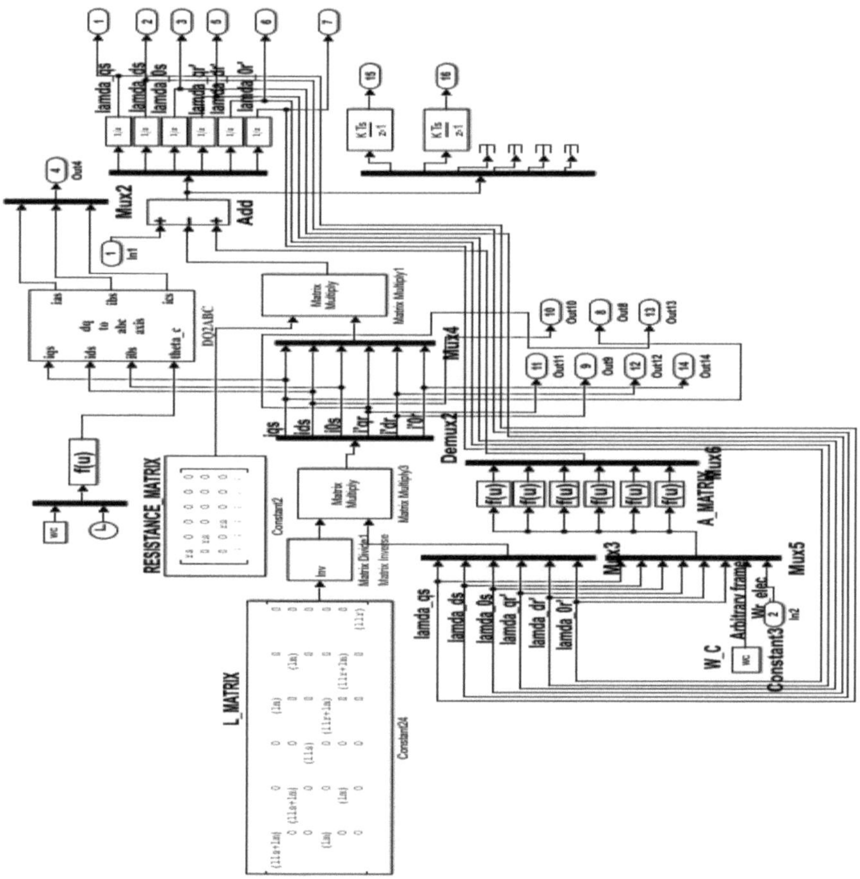

Fig. 8.38 Three-phase IM flux linkage model in state space

development is already explained under Sect. 8.2.1. Here two constant blocks from sources block set are used to enter the inductance and resistance matrix. Matrix Inverse and Matrix Multiply blocks are from Math operations block set. In the model shown in Fig. 8.36, MATLAB Function3 has the input *k* with a value of 1, frequency 50 Hz, stator flux angular position θ_{fs} and time module t as inputs and generates the outputs y1 and z1 which correspond to the stator phase voltage sector v_sector and stator flux sector fls_sector divisions as shown in Fig. 8.32. This is shown in Program Segment 8.6 in the model file EXAMPLE8_7. The y1 output v_sector is given as input to the MATLAB Function1 which generates the six sector switch functions ssf1 to ssf6 when the voltage space vector passes through the sectors 1–6 defined in Fig. 8.32. This is shown in Program segment 8.7 in the model file EXAMPLE8_7. The measured values of torque Tem, stator flux fl_stator, sector location of stator flux fls_sector from the model, reference value for torque from PI controller, reference stator flux from Program segment 8.2 and the six sector switch functions, ssf1 to ssf6, are given as input to the Embedded MATLAB Function1 which generates the six gate pulses sap, san, sbp, sbn, scp and scn for the three-phase inverter switches.

Depending on the sector location of stator flux, gate pulses for switching the inverter are generated according to Table 8.5. This is shown in Program segment 8.8 in the model file EXAMPLE8_7. The two DC voltage sources each with value +Vdc/2 volts are connected in series with midpoint grounded form the DC link voltage and the Universal Bridge1 forms the three-phase IGBT inverter. Two discrete integrators Ts/(z − 1) are used to integrate the derivative of dq-axis stator flux λds_dot and λqs_dot to determine λd and λq which are given as input to two separate two-input Mux blocks. Using user defined Fcn block and another Interpreted MATLAB Fcn block, λs and θfs are calculated as per Eqs. 8.79 and 8.80.

8.10.2 Simulation Results

The model of the DTC of three-phase IM with PI controller is carried out using ode23tb (stiff/TR-BDF2) solver in Simulink [6]. A sample time of 10e-6 s is used. A mechanical load is applied using step function. The load is initially 100 Nw-metres and changes to 200 Nw-metres at 0.5 s. A DC link voltage of 490 volts is used. The inverter switching frequency is 50 Hz. The torque hysteresis band is 1% of torque reference and stator flux hysteresis band is 10% of stator flux reference. The simulation results for the motor torque, rotor speed, mechanical load applied, three-phase stator and rotor currents are shown in Fig. 8.39. Stator flux angular position θ_{fs} and its dq-axis components λ_{ds} and λ_{qs} are shown in Fig. 8.40. The resultant stator flux λ_S is shown in Fig. 8.41. The motor reference torque is shown in Fig. 8.42. The three-phase line to line stator applied voltage is shown in Fig. 8.43. The torque speed curve is shown in Fig. 8.44.

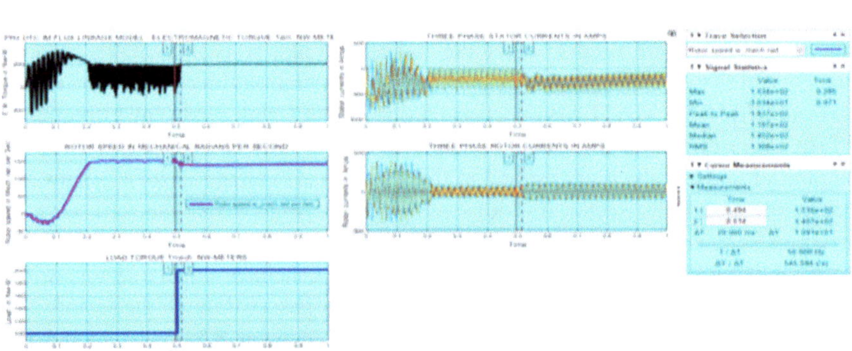

Fig. 8.39 Three-phase DTC of IM with PI controller simulation results: motor torque Tem; rotor speed wrm; load torque TL (left column, top to bottom); three-phase stator currents ias, ibs and ics; and three-phase rotor currents iar′, ibr′ and icr′ (right column top to bottom)

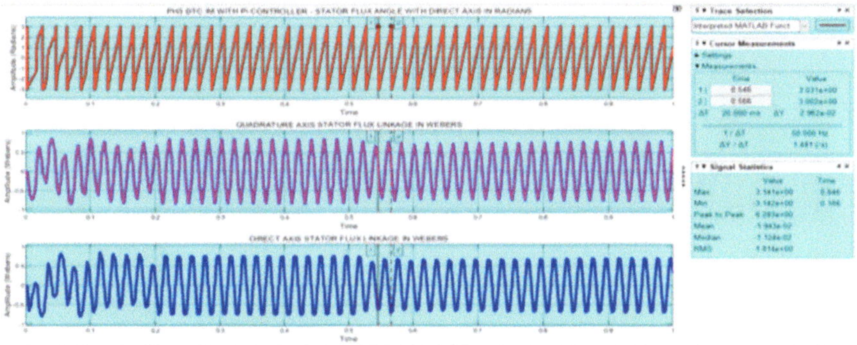

Fig. 8.40 Three-phase DTC of IM with PI controller simulation results—stator flux angular position θ_{fs} in radians, q-axis and d-axis stator flux linkage in webers (top to bottom)

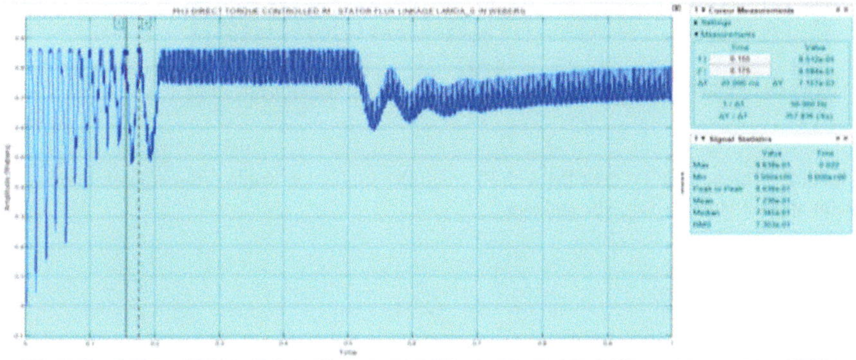

Fig. 8.41 Three-phase DTC of IM with PI controller simulation results—total stator flux ʎs in webers

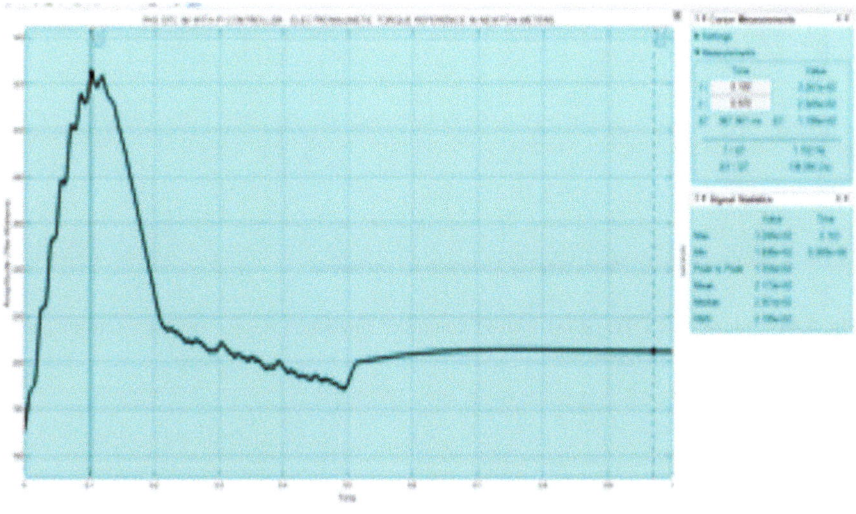

Fig. 8.42 Three-phase DTC of IM with PI controller simulation results—motor torque reference Tem_ref

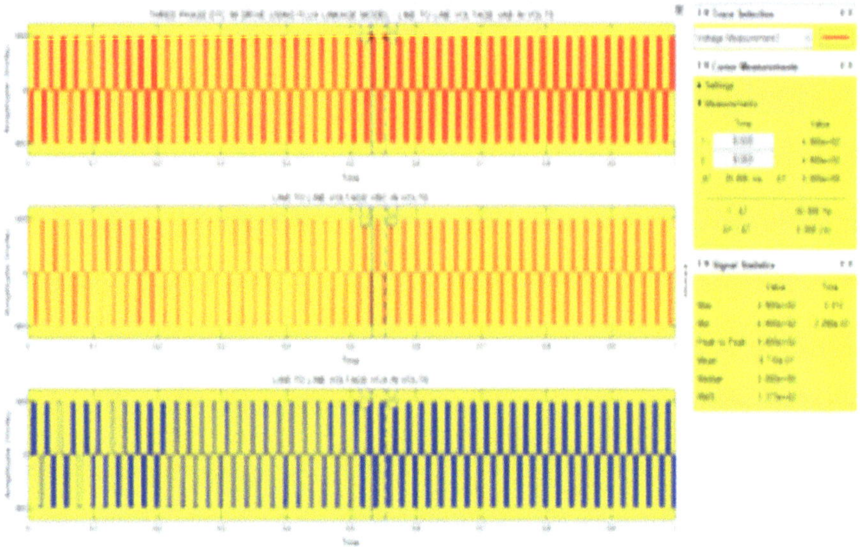

Fig. 8.43 Three-phase DTC of IM with PI controller simulation results—three-phase line to line stator applied voltages vab, vbc and vca

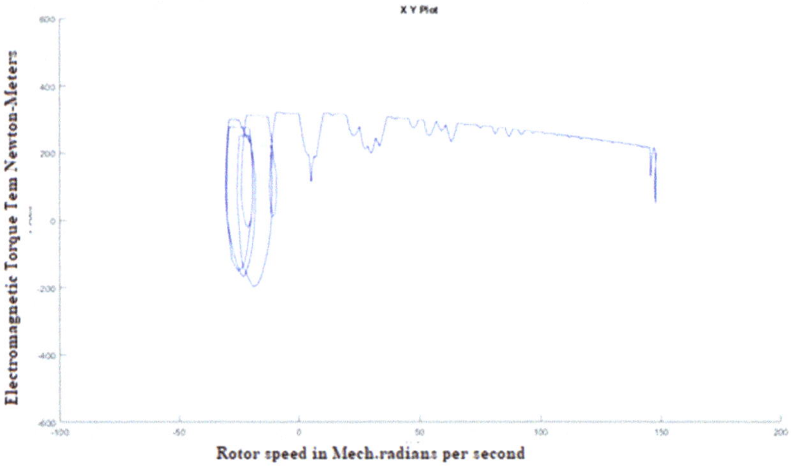

Fig. 8.44 Three-phase DTC of IM with PI controller simulation results—motor electromagnetic torque Tem versus rotor speed wrm

8.10.3 Discussion of Results

From Figs. 8.39 and 8.44 for motor torque Tem, starting torque ripples are observed. However in the steady state, torque reaches the preset value of 200 Nw-metres with NO torque ripples. From Fig. 8.40 it is seen that λqs and λds are almost sinusoidal. The total stator flux λs in the steady state has flux ripples as can be seen from Fig. 8.41. From Fig. 8.39, the rotor speed ωrm reaches the steady state value of 140.7 Mech.radians per second which is close to the set point value of 141.3717 Mech.radians per second. From Fig. 8.39 it is seen that the three-phase rotor currents are well balanced and symmetrical about the x-axis. But the three-phase stator currents although balanced are NOT symmetrical about the x-axis showing some DC offset current. But from the nature of the steady state three-phase stator currents in Fig. 8.39, it is seen that they show a tendency to go symmetrical about the x-axis with respect to time.

8.10.4 Model of Direct Torque Control of Three-Phase Induction Motor Drive Without PI Controller Using Classical Switching Table: Torque Control Method

The model of the DTC of three-phase IM without PI controller using torque control method is presented in this section. Three-phase IM parameters are the same as in Table 8.1. Stator flux reference is calculated from Program Segment 8.2. The model of the DTC of three-phase IM without PI controller in the stationary reference frame is shown in Fig. 8.45 (model file: EXAMPLE8_8). The dialog boxes are shown in Fig. 8.37. The essential model subsystem for three-phase IM model data using dq0-abc phase AC generator and dq0-axis transform is the same as shown in Fig. 8.5a. Here three-phase sine wave AC generator is removed and instead the square wave output phase voltage of three-phase inverter is used. Figure 8.5b is modified and is shown in Fig. 8.38 where two discrete integrators with sample time 10e-6 s are used to realise dq-axis stator flux linkages. The subsystem Rotor Torque and Speed Calculator is the same as shown in Fig. 8.5c. Here angular frequency of the reference frame ω_C is entered as zero and all other data from Table 8.1 are entered in the relevant box shown in Fig. 8.37. Sample time for discrete integrator for integrating the derivative of dq-axis stator flux is entered as 10e-6 s. The model development is already explained under Sect. 8.10.1.

8.10.5 Simulation Results

The model of the DTC of three-phase IM without PI controller is carried out using ode23tb (stiff/TR-BDF2) solver in Simulink [6]. A sample time of 10e-6 s is used. A

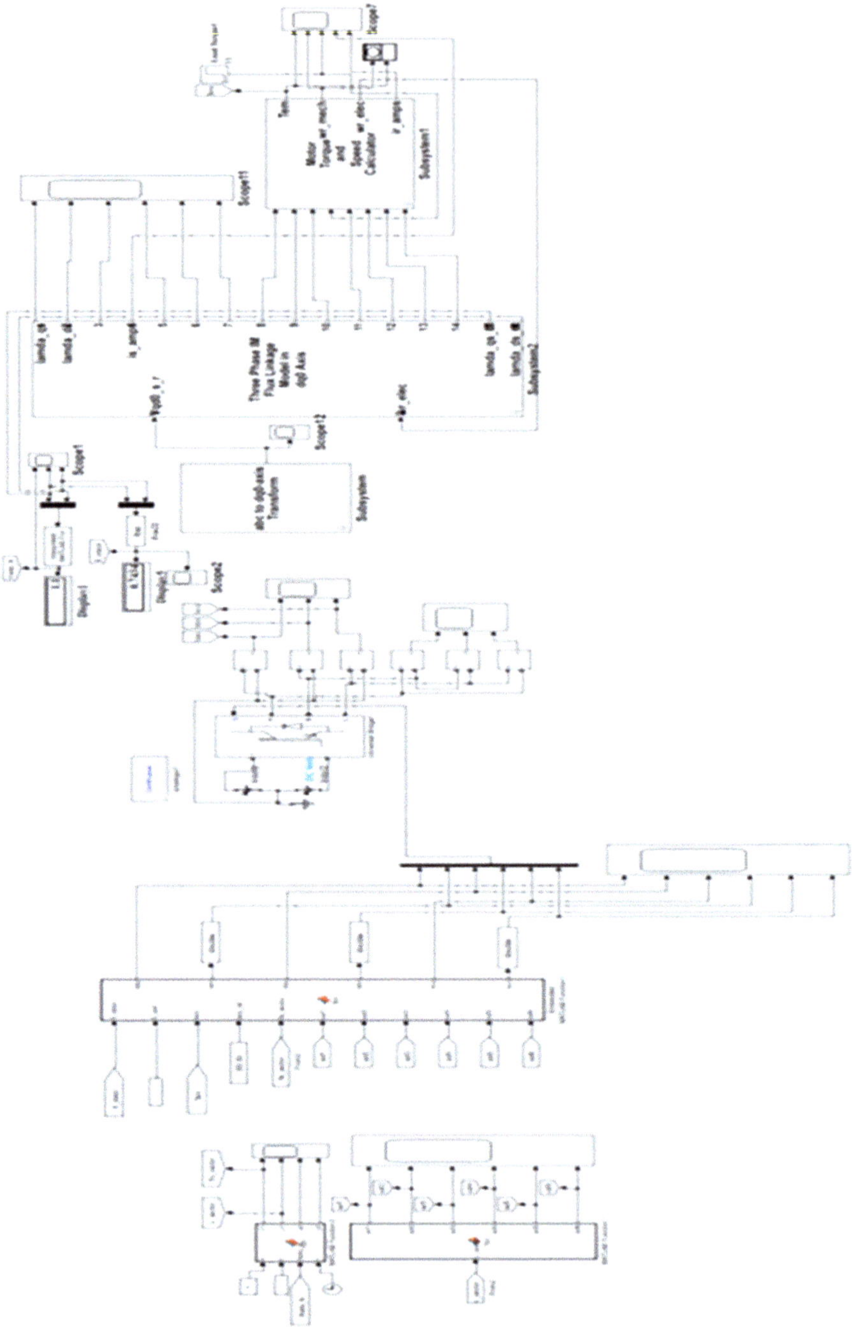

Fig. 8.45 Model of DTC of three-phase IM drive without PI controller

mechanical load is applied using step function. The load is initially 100 Nw-metres and changes to 200 Nw-metres at 0.5 s. A DC link voltage of 490 volts is used. The inverter switching frequency is 50 Hz. The torque hysteresis band is 1% of torque reference and stator flux hysteresis band is 10% of stator flux reference. The simulation results for the motor torque, rotor speed, mechanical load applied, three-phase stator and rotor currents are shown in Fig. 8.46. Stator flux angular position θ_{fs} and its dq-axis components λ_{ds} and λ_{qs} are shown in Fig. 8.47. The resultant stator flux λ_S is shown in Fig. 8.48. The motor reference torque is 350.83 Nw-metres from Program segment 8.2 and is a constant. The three-phase line to line stator applied voltage is shown in Fig. 8.49. The torque speed curve is shown in Fig. 8.50.

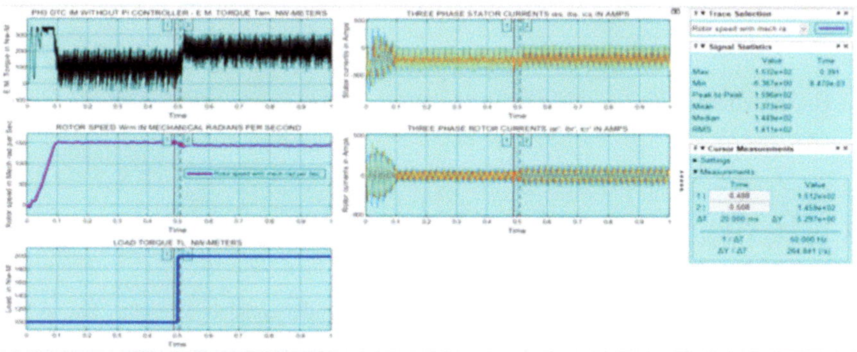

Fig. 8.46 Three-phase DTC of IM drive without PI controller simulation results—motor torque Tem; rotor speed wrm; load torque TL; stator currents ias, ibs and ics; rotor currents iar′, ibr′ and icr′

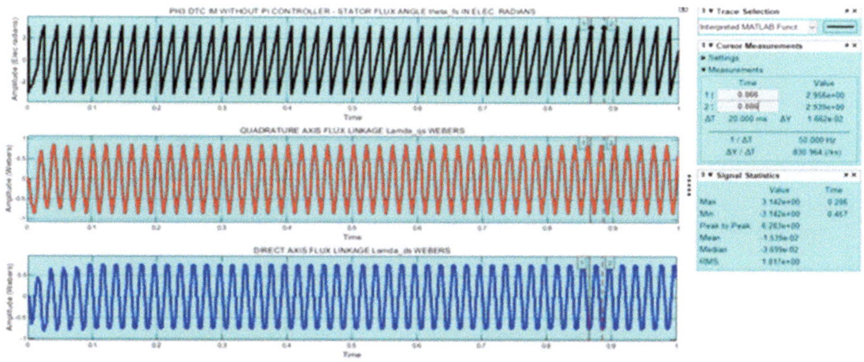

Fig. 8.47 Three-phase DTC of IM without PI controller simulation results—stator flux angle θ_{fs} in Elec.radians, quadrature axis and direct axis flux linkages λ_{qs} and λ_{ds} (top to bottom)

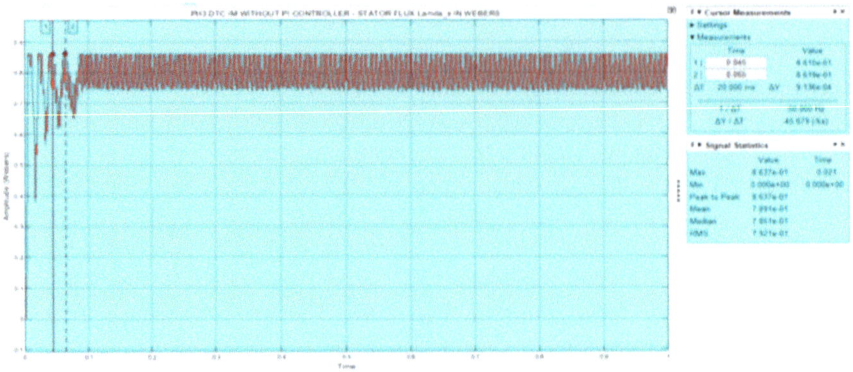

Fig. 8.48 Three-phase DTC of IM without PI controller simulation result—stator flux linkage ʎs

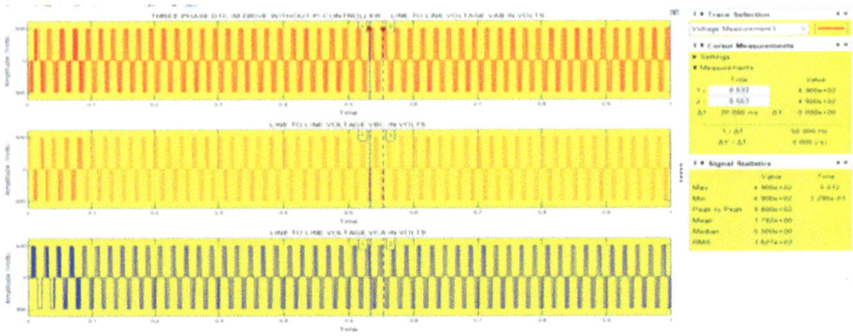

Fig. 8.49 Three-phase DTC of IM drive without PI controller simulation results—line to line output voltages Vab, Vbc and Vca of inverter

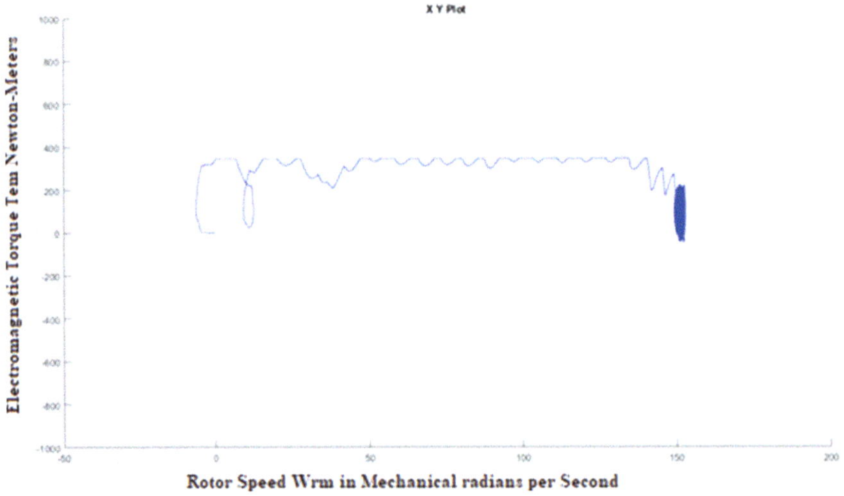

Fig. 8.50 Three-phase DTC of IM without PI controller simulation result—motor electromagnetic torque Tem versus rotor speed Wrm

8.10.6 Discussion of Results

From Figs. 8.46 and 8.50 for motor torque Tem, starting and steady state torque ripples are observed. In the steady state, torque reaches and fluctuates around 200 Nw-metres. From Fig. 8.47 it is seen that λqs and λds are almost sinusoidal. The total stator flux λs in the steady state has flux ripples as can be seen from Fig. 8.48. From Fig. 8.46, the rotor speed ωrm reaches the steady state value of 145.9 Mech.radians per second. From Fig. 8.46 it is seen that the three-phase rotor currents are well balanced and symmetrical about the x-axis. But the three-phase stator currents although balanced are NOT symmetrical about the x-axis showing some DC offset current.

8.11 Additional Switching Technique for Direct Torque Control

The classical method of switching the three-phase inverter is presented in Sect. 8.10 and the switching table is shown in Table 8.5. Yet another method of switching this inverter for implementing DTC of IM is shown in Fig. 8.51. Here Sector I is in the range 0–60 degrees, and Sectors II to VI are in the range 60 to 120, 120 to 180, 180 to 240, 240 to 300 and 300 to 360 degrees, respectively. This method of switching is known as modified DTC (M_DTC). The modified switching table is shown in Table 8.6. In this modified DTC switching, if stator flux location corresponds to the voltage space vector V_k where k Є 1 to 6, then application of voltage space vectors V_k and V_{k+1} increases and that of V_{k+3} and V_{k+4} decreases the stator flux. Similarly application of voltage space vectors V_{k+1} and V_{k+3} increases and that

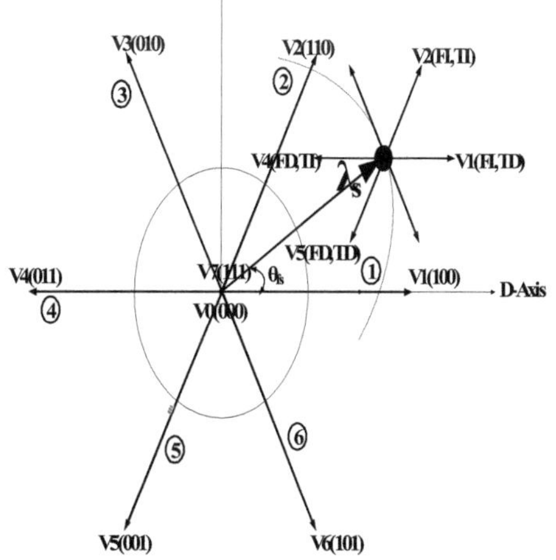

Fig. 8.51 Modified selection of voltage space vector for stator flux in sector 1. FI flux increase, FD flux decrease, TI torque increase, TD torque decrease

Table 8.6 Modified switching table

δλs	δTe	Sector S1	Sector S2	Sector S3	Sector S4	Sector S5	Sector S6
FI	TI	V2	V3	V4	V5	V6	V1
	TE	V0	V7	V0	V7	V0	V7
	TD	V1	V2	V3	V4	V5	V6
FD	TI	V4	V5	V6	V1	V2	V3
	TE	V7	V0	V7	V0	V7	V0
	TD	V5	V6	V1	V2	V3	V4

of V_k and V_{k+4} decreases the motor torque. This is shown in Fig. 8.51 for stator flux in sector 1. Similar argument applies to stator flux in other sectors. The voltage space vectors V_{k+2} and V_{k+5} are not used here. If zero-voltage vector V0 or V7 is applied, the torque remains within the hysteresis band limit, and the stator flux magnitude and angular position don't change. The modified DTC switching scheme is shown in Table 8.6. In Table 8.6, V0 to V7 are as defined in Fig. 8.51. In Table 8.6, if an increase in stator flux is required, it is represented by FI and for decrease required it is represented by FD. Similarly if an increase in torque is required, it is represented by TI, and for NO change and decrease required, it is represented by TE and TD, respectively. These requirements for stator flux and torque can be expressed as in Eq. 8.76. Equations 8.77 to 8.80 are used to determine the stator flux amplitude λ_S and its sector location θ_{fs} during each sampling interval. The models of the M_DTC of three-phase IM drive using the torque control and speed control method are presented in the next section.

8.11.1 *Case Study: Model of Direct Torque Control of Three-Phase Induction Motor Drive with PI Controller Using Modified Switching Table—Speed Control Method*

The model of the M_DTC of three-phase IM using PI controller in the speed control loop is presented in this section. Three-phase IM parameters are the same as in Table 8.1. Here the PI controller parameters Kp and Ki are calculated using Program Segment 8.2 and tabulated in Table 8.2. Stator flux reference is also calculated from Program Segment 8.2. The model of the M_DTC of three-phase IM with PI controller in the speed control loop in the stationary reference frame is shown in Fig. 8.52 (model file: CASE_STUDY_EX8_3). The dialog boxes are the same as shown in Fig. 8.37. The essential model subsystem for three-phase IM model data is the one using dq0-axis flux linkage equations in state space which is shown in Fig. 8.38. The model subsystem for three-phase AC generator and dq0-axis transform is the same as shown in Fig. 8.5a. Here three-phase sine wave AC generator is removed and instead the square wave output phase voltage of three-phase inverter is used. Figure 8.5b is modified and is shown in Fig. 8.38 where two discrete integrators with sample time 10e-6 s are used to realise the dq-axis stator flux linkages. The subsystems Rotor Torque and Speed Calculator are the same as shown in Fig. 8.5c. Here angular frequency of the reference frame ω_C is entered as zero and all other data from Tables 8.1 are entered in the relevant box, same as shown in Fig. 8.37. The PI controller parameters K_p and K_i and the rotor reference speed ωrm_ref from Table 8.2 are entered in the PI controller block shown in Fig. 8.52. Sample time for discrete integrator for integrating the derivative of dq-axis stator flux is entered as 10e-6 s. The output of the PI controller is also sampled at the same sampling rate using zero-order hold (ZOH). The model development is already explained under

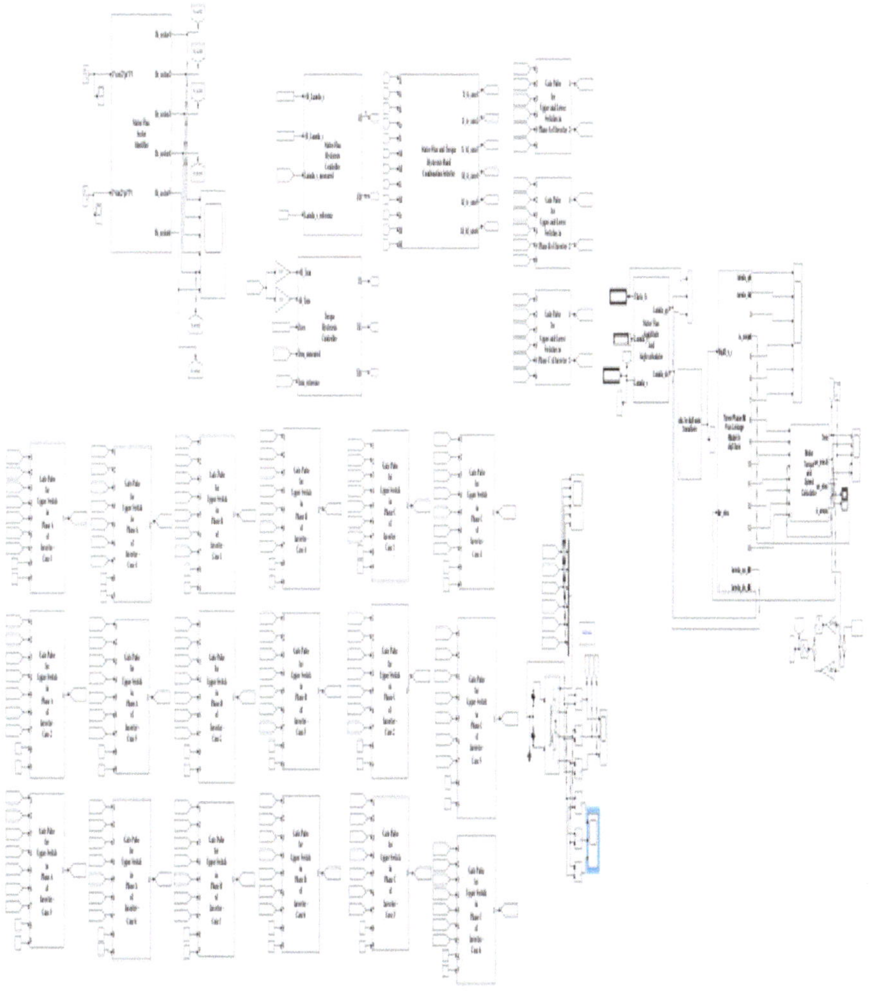

Fig. 8.52 Model of modified DTC three-phase IM using comparators and with PI controller

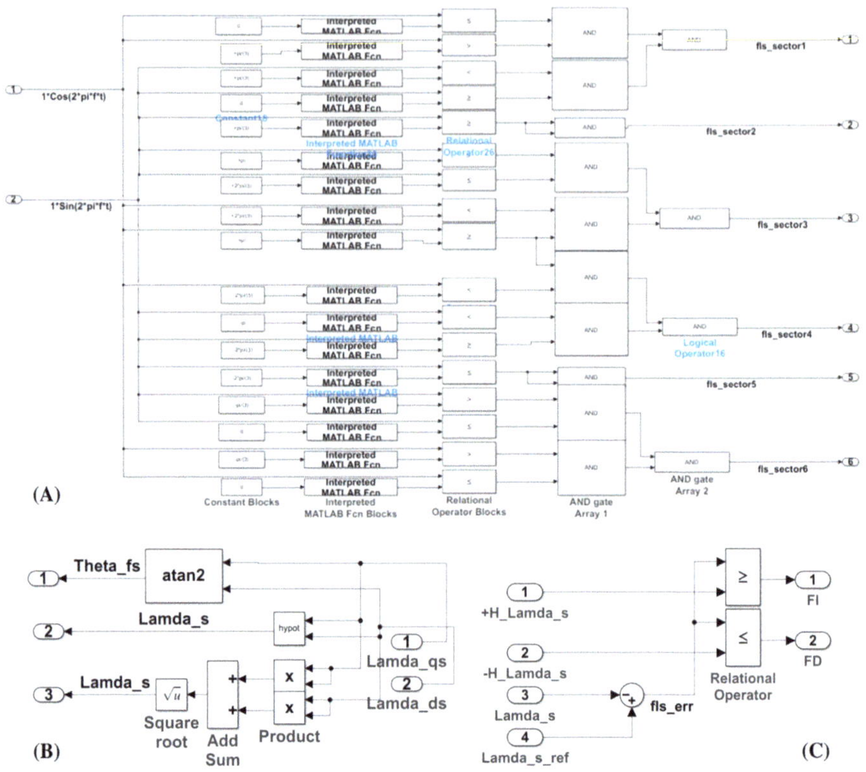

Fig. 8.53 (a–g) Modified DTC of three-phase IM with PI controller—model subsystems for inverter gate pulse generation

Sect. 8.2.1. Here two constant blocks from sources block set are used to enter the inductance and resistance matrix. Matrix Inverse and Matrix Multiply blocks are from Math operations block set.

The model subsystems for stator flux sector identification, stator flux amplitude and angle calculator, stator flux hysteresis controller, torque hysteresis controller, stator flux and torque hysteresis band combination selector, gate pulse generator for Phase A of inverter upper switches for cases 1–6 as per Table 8.6 and overall gate pulse generator for the upper and lower switches in Phase A of the three-phase inverter are shown in Fig. 8.53a–g. The three-phase two-level inverter model is the same as presented in Sect. 8.10.1.

Figure 8.53a shows the stator flux sector identifier. Here a cosine wave function generator (CFG) and a sine wave function generator (SFG) each with amplitude 1 unit and frequency 50 Hz are used along with Interpreted MATLAB Function blocks, constant blocks and logic AND gates. The sector angles from $-\pi$ to $+\pi$ at intervals of $+\pi/3$ radians are entered in the constant blocks which form the inputs to the respective MATLAB Function blocks. The CFG and SFG outputs form the first

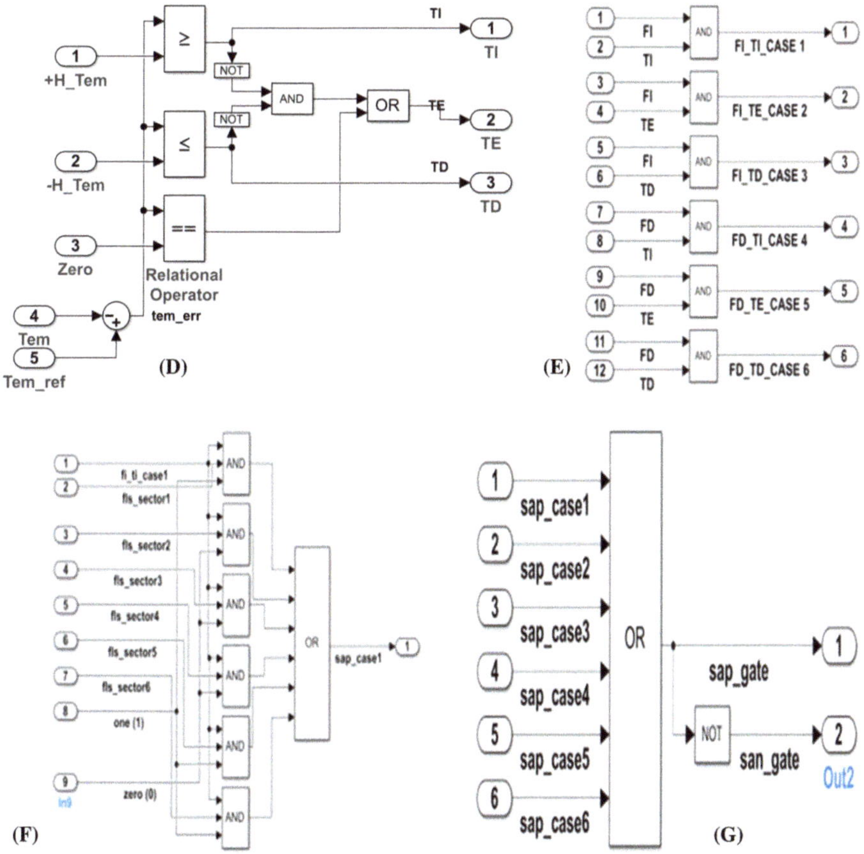

Fig. 8.53 (continued)

input, and the MATLAB Function output forms the second input to the relational operator blocks also known as comparator blocks. The relational operator blocks are marked with symbols $\leq$, $>$, $<$ and $\geq$, with meanings less than or equal to, greater than, less than and greater than or equal to, respectively. For example, if the first input to the comparator block with symbol $\leq$ is less than or equal to the second input, the comparator output is HIGH; else its output is LOW. Similar argument holds good for other comparator blocks. Thus the CFG output is compared with MATLAB Function output Cos(0) in the comparator block marked $\leq$, and if the former input value is only less than or equal to one, comparator output is HIGH. Similarly CFG output is compared with MATLAB Function output Cos(π/3) in another comparator marked $>$, and if the former output is only greater than 0.5, comparator output is HIGH. These two comparator outputs are given to the first AND gate in array 1. In the same way, SFG output is compared with MATLAB Function output Sin(0) in the comparator block marked ≥, and if the former input value is only greater than or equal to zero, comparator output is HIGH. Similarly SFG output is compared with MATLAB Function output Sin(π/3) in another comparator marked <, and if the former output is only less than 0.866, comparator output is HIGH. These two comparator outputs are given to the second AND gate in array 1. These two AND gate outputs are given to another AND gate in array 2. The output of this AND gate in array 2 gives the stator flux sector 1 marked fls_sector1. Similar procedure is used to determine stator flux sectors 2 to 6 which are marked as fls_sector2 to fls_sector6 in Fig. 8.53a.

The stator flux amplitude and angle calculator is shown in Fig. 8.53b. Here dq-axis stator flux λ_{qs} and λ_{ds} are given as input to two separate product blocks to determine their square and then their sum is determined using Add block. The output of Add block is given to Square root block which determines the amplitude λ_S of the stator flux as per Eq. 8.79. Alternatively the hypot function block can also be used to determine the amplitude of stator flux. Also λ_{qs} and λ_{ds} are given as input to atan2 function block to determine θ_{fs} as per Eq. 8.80.

Figure 8.53c shows the stator flux hysteresis controller. Here stator flux reference λ_{S_ref} and measured stator flux from model λ_S are given to a Sum block which subtracts the two values to find stator flux error fls_err. This fls_err is compared with upper hysteresis band limit for stator flux +H_ λ_S in a comparator marked ≥ with output FI which is HIGH only when the former input is greater than or equal to the latter value. Similarly fls_err is compared with lower hysteresis band limit for stator flux -H_ λ_S in a comparator marked ≤ with output FD which is HIGH only when the former input is lower than or equal to the latter value. FI and FD are the stator flux increase and decrease commands.

The torque hysteresis controller is shown in Fig. 8.53d. Here torque reference T_{em_ref} and measured electromagnetic torque from model T_{em} are given to a Sum block which finds the difference between the two values to determine the torque error tem_err. This tem_err is compared with torque upper hysteresis band limit $+H\_T_{em}$ and with torque lower hysteresis band limit $-H\_T_{em}$ in two separate comparators marked ≥ and ≤, respectively, to determine TI and TD outputs in the same way as explained for stator flux hysteresis controller. The TI and TD outputs are each inverted using NOT gate and the two NOT gate outputs are given to an AND gate. The tem_err is also compared with zero in a third comparator marked == with output HIGH only when former input tem_err is equal to the latter value zero. The output of this third comparator and that of AND gate are given as input to an OR gate. This OR gate output is TE. The TI, TD and TE are the torque increase, torque decrease and torque equal commands. The stator flux and torque hysteresis band combination selector is shown in Fig. 8.53e. These are the first two columns and six rows of Table 8.6. The six rows of Table 8.6 are the six cases from case 1 to case 6. The FI and TI commands from Fig. 8.53c, d are given to an AND gate whose output is marked FI_TI_CASE1. Similarly the stator flux and torque hysteresis band combinations FI-TE, FI-TD, FD-TI, FD-TE and FD-TD from Fig. 8.53c, d are each given as input to respective AND gates, and the outputs are marked

FI_TE_CASE2, FI_TD_CASE3, FD_TI_CASE4, FD_TE_CASE5 and FD_TD_CASE6.

The gate pulse generator for Phase A upper switch of inverter corresponding to case 1 is shown in Fig. 8.53f. Here six three-input AND gates are used. The fi_ti_case1 output from Fig. 8.53e is the first input to all the six AND gates. The second inputs to each of the six AND gates are the stator flux sectors fls_sector1 to fls_sector6 from Fig. 8.53a. The third input to each of the six AND gates is either a zero or one depending on the stator flux sector and Table 8.6. Thus referring to case 1 in Table 8.6, for the upper switch in Phase A corresponding to stator flux sectors 1–6, the third input to the six AND gates has the input values 1, 0, 0, 0, 1 and 1, respectively. All the six outputs of the AND gates are given to a six-input OR gate whose output sap_case 1 is the gate pulse for upper switch in Phase A of inverter for case 1. In this way all the gate pulses for the six cases from Table 8.6 are developed for upper switches in Phase A, B and C.

The complete gate pulse generator subsystem for the upper and lower switches in Phase A of inverter is shown in Fig. 8.53g. Here sap_case1 to sap_case6 from Fig. 8.53f are given as input to a six-input OR gate whose output is sap_gate for the upper switch in Phase A. This sap_gate is inverted using NOT gate which gives san_gate for the lower switch in Phase A. The same procedure is repeated for the upper and lower switches in Phase B and C of inverter.

8.11.2 Simulation Results

The model of the M_DTC of three-phase IM with PI controller is carried out using ode3 (Bogacki-Shampine) solver in Simulink [6]. A sample time of 10e-6 s is used. A mechanical load is applied using step function. The load is initially 100 Nw-metres and changes to 200 Nw-metres at 0.5 s. A DC link voltage of 490 volts is used. The inverter switching frequency is 50 Hz. The torque hysteresis band is 1% of torque reference and stator flux hysteresis band is 10% of stator flux reference. The simulation results for the motor torque, rotor speed, mechanical load applied, three-phase stator and rotor currents are shown in Fig. 8.54. Stator flux angular position θ_{fs} and its dq-axis components λ_{ds} and λ_{qs} are shown in Fig. 8.55. The resultant stator flux λ_S is shown in Fig. 8.56. The motor reference torque is shown in Fig. 8.57. The three-phase line to line stator applied voltage is shown in Fig. 8.58. The torque speed curve is shown in Fig. 8.59.

8.11.3 Discussion of Results

From Figs. 8.54 and 8.59 for motor torque Tem, starting torque ripples are observed. However in the steady state, torque reaches the preset value of 200 Nw-metres with NO torque ripples. From Fig. 8.55 it is seen that λqs and λds are almost sinusoidal.

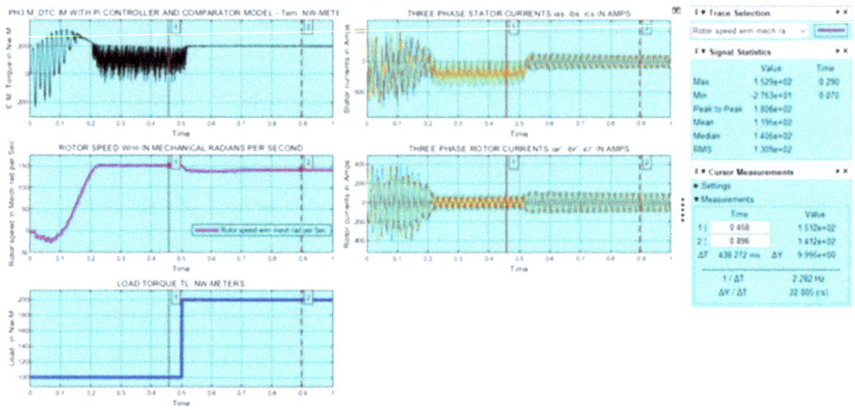

Fig. 8.54 Three-phase modified DTC of IM with PI controller simulation results: motor torque Tem; mechanical rotor speed wrm; load torque TL (left column, top to bottom); three-phase stator currents ias, ibs and ics; and three-phase rotor currents iar′, ibr′ and icr′ (right column, top to bottom)

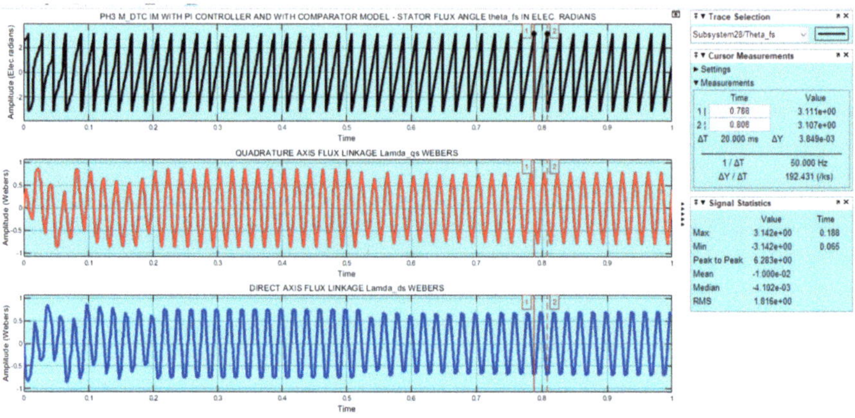

Fig. 8.55 Three-phase modified DTC of IM with PI controller simulation results: stator flux angle theta_fs, q-axis and d-axis stator flux linkage (top to bottom)

The total stator flux λs in the steady state has flux ripples as can be seen from Fig. 8.56. From Fig. 8.54, the rotor speed ωrm reaches the steady state value of 141.2 Mech.radians per second which is close to the set point value of 141.3717 Mech. radians per second. From Fig. 8.54 it is seen that the three-phase rotor currents are well balanced and symmetrical about the x-axis. But the three-phase stator currents, although balanced and symmetrical while starting, are NOT symmetrical about the x-axis during the initial period showing some DC offset current. But in steady state after applying full load torque, stator currents are found symmetrical about the x-axis. Thus three-phase stator current waveforms in Fig. 8.54 are found to be different from that of Fig. 8.39.

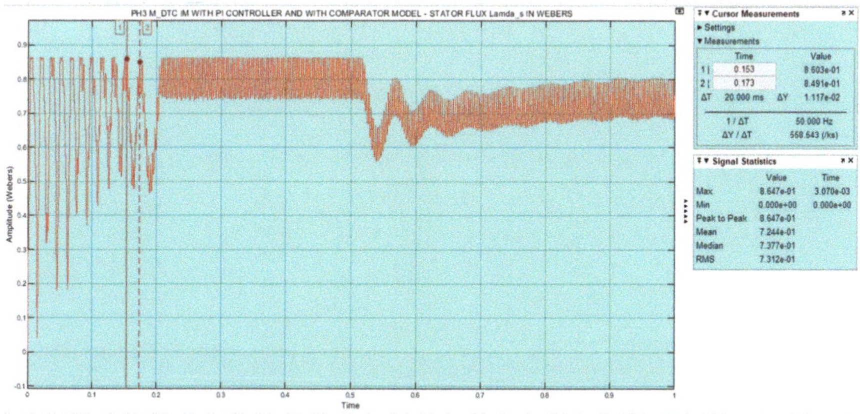

Fig. 8.56 Three-phase modified DTC of IM with PI controller simulation results: stator flux linkage Lamda_s webers

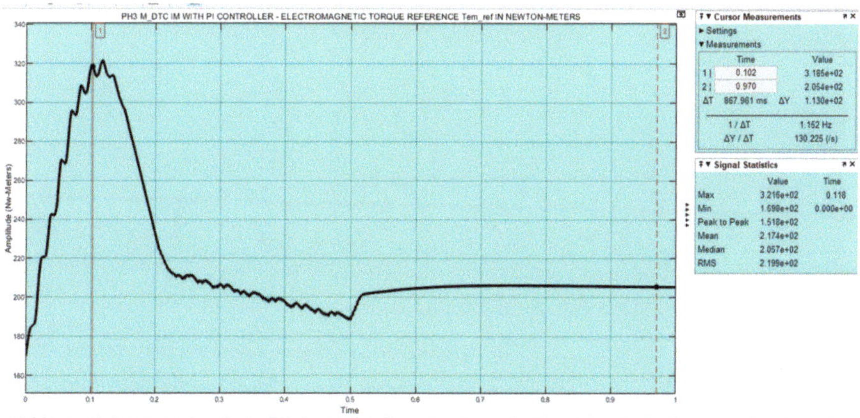

Fig. 8.57 Three-phase modified DTC IM with PI controller—motor electromagnetic torque reference Tem_ref

8.11.4 Case Study: Model of Direct Torque Control of Three-Phase Induction Motor Drive Without PI Controller Using Modified Switching Table—Torque Control Method

The model of the M_DTC of three-phase IM without PI controller using torque control method is presented in this section. Three-phase IM parameters are the same as in Table 8.1. Stator flux reference is calculated from Program Segment 8.2. The

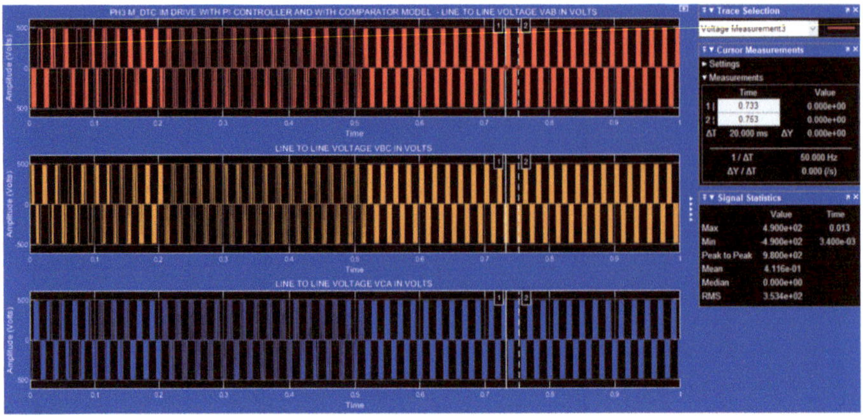

Fig. 8.58 Three-phase modified DTC of IM with PI controller simulation results: line to line voltages VAB, VBC and VCA (top to bottom)

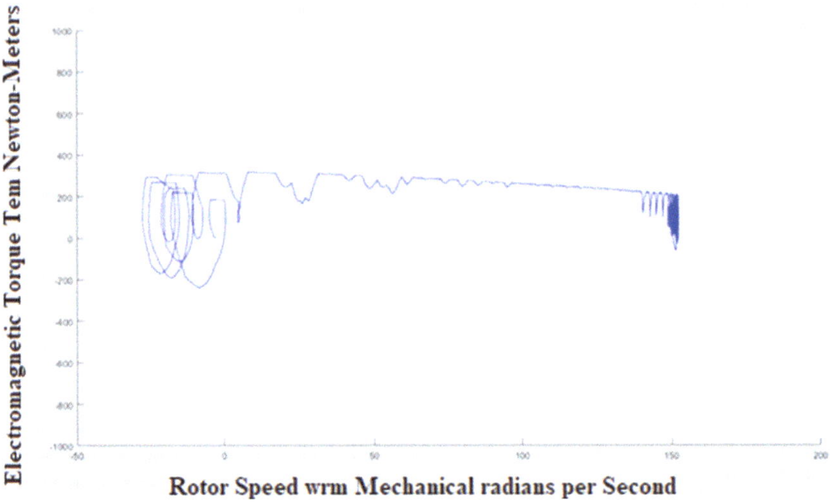

Fig. 8.59 Three-phase modified DTC of IM with PI controller simulation result: rotor electromagnetic torque Tem in Nw-metres versus rotor speed wrm in Mech.rad per second

model of the M_DTC of three-phase IM without PI controller in the stationary reference frame is shown in Fig. 8.60 (model file: CASE_STUDY_EX8_4). The dialog boxes are the same as shown in Fig. 8.37. The essential model subsystem for three-phase IM model data using dq0-axis flux linkage equations in state space is shown in Fig. 8.38. The model subsystems for three-phase AC generator and dq0-axis transform are the same as shown in Fig. 8.5a. Here three-phase sine wave AC generator is removed and instead the square wave output phase voltage of three-phase inverter is used. Figure 8.5b is modified and is shown in Fig. 8.38 where two discrete integrators with sample time 10e-6 s are used to realise dq-axis stator flux

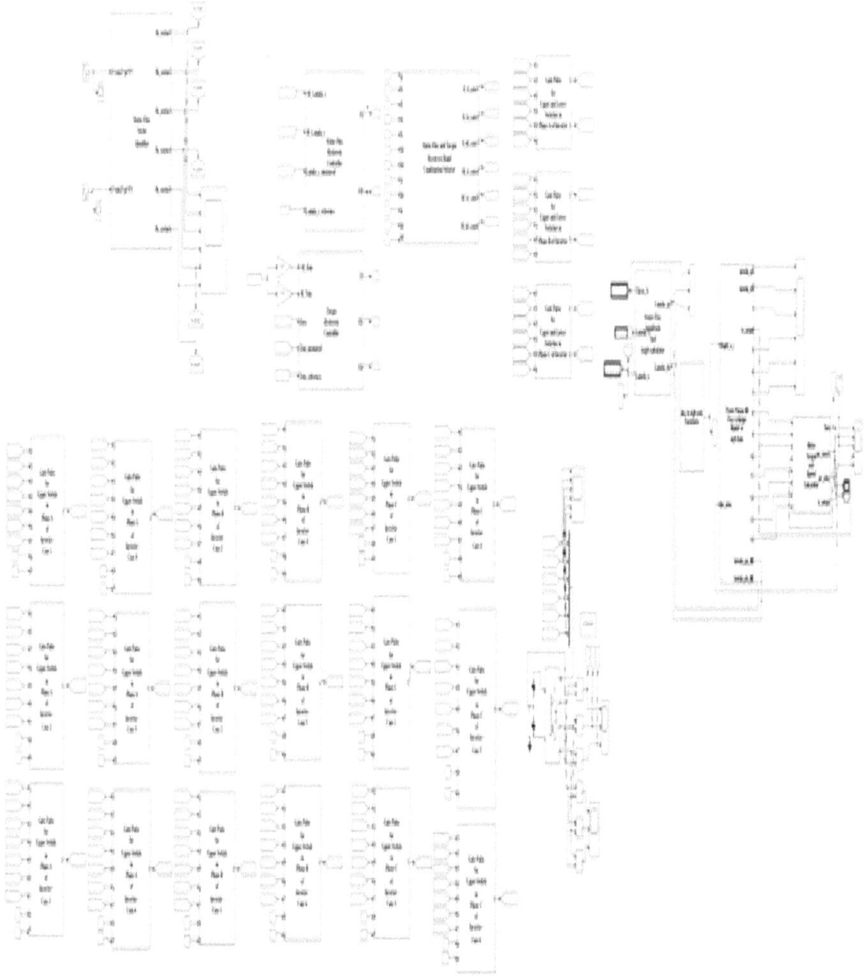

Fig. 8.60 Model of M_DTC of three-phase IM using modified switching table without PI controller—torque control method

linkages. The subsystem Rotor Torque and Speed Calculator is the same as shown in Fig. 8.5c. Here angular frequency of the reference frame ω_C is entered as zero and all other data from Table 8.1 are entered in the relevant box shown in Fig. 8.37. Sample time for discrete integrator for integrating the derivative of dq-axis stator flux is entered as 10e-6 s. The model development is already explained under Sect. 8.10.1.

The model subsystems for stator flux sector identification, stator flux amplitude and angle calculator, stator flux hysteresis controller, torque hysteresis controller, stator flux and torque hysteresis band combination selector, gate pulse generator for phase A of inverter upper switches for cases 1–6 as per Table 8.6 and overall gate pulse generator for the upper and lower switches in Phase A of the three-phase inverter are shown in Fig. 8.53a–g. The three-phase two-level inverter model is the same as presented in Sect. 8.10.1. These model subsystems have been already presented in Sect. 8.11.1.

8.11.5 Simulation Results

The model of the DTC of three-phase IM without PI controller is carried out using ode3 (Bogacki-Shampine) fixed step solver in Simulink [6]. A sample time of 10e-6 s is used. A mechanical load is applied using step function. The load is initially 100 Nw-metres and changes to 200 Nw-metres at 0.5 s. A DC link voltage of 490 volts is used. The inverter switching frequency is 50 Hz. The torque hysteresis band is 1% of torque reference and stator flux hysteresis band is 10% of stator flux reference. The simulation results for the motor torque, rotor speed, mechanical load applied and three-phase stator and rotor currents are shown in Fig. 8.61. Stator flux angular position θ_{fs} and its dq-axis components λ_{ds} and λ_{qs} are shown in Fig. 8.62. The resultant stator flux λ_S is shown in Fig. 8.63. The motor reference torque is 350.83 Nw-metres from Program segment 8.2 and is a constant. The three-phase line to line stator applied voltage is shown in Fig. 8.64. The torque speed curve is shown in Fig. 8.65.

8.11.6 Discussion of Results

From Figs. 8.61 and 8.65 for motor torque Tem, starting and steady state torque ripples are observed. In the steady state, torque reaches and fluctuates around

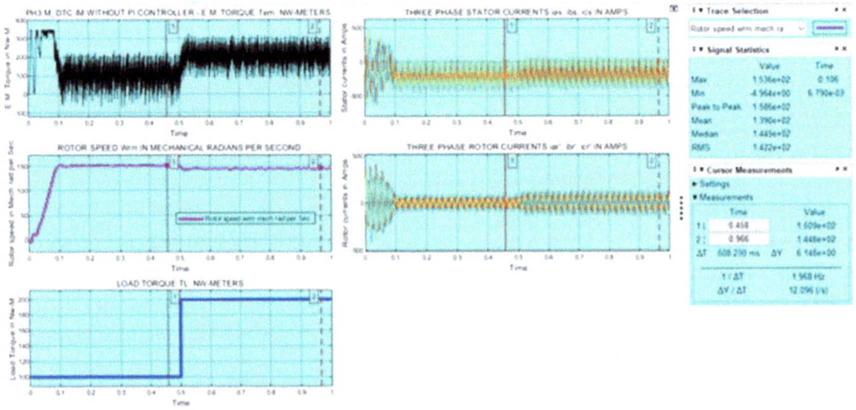

Fig. 8.61 Three-phase modified DTC of IM drive without PI controller simulation results: E.M. torque Tem; rotor speed wrm; load torque TL (left column, top to bottom); stator currents ias, ibs and ics; and rotor currents iar′, ibr′ and icr′ (right column, top to bottom)

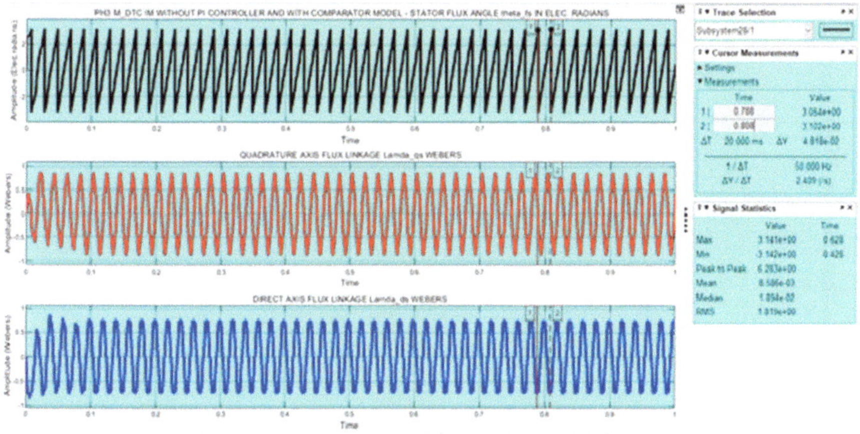

Fig. 8.62 Three-phase modified DTC of IM drive without PI controller simulation results: stator flux angle theta fs, q-axis and d-axis stator flux linkages (top to bottom)

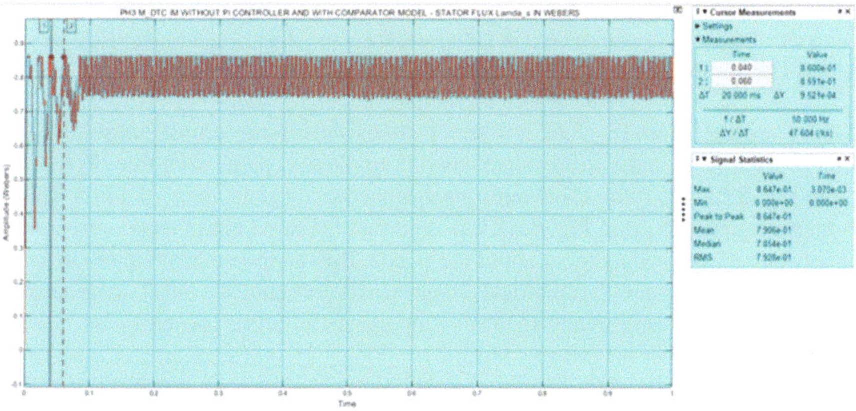

Fig. 8.63 Three-phase modified DTC of IM without PI controller simulation result: stator flux Lamda_s

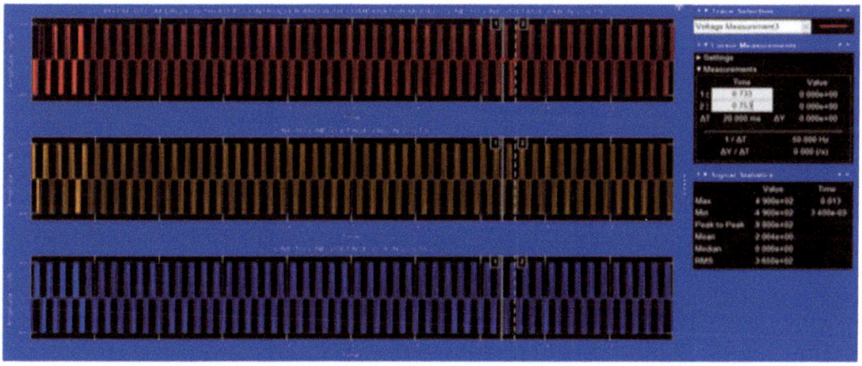

Fig. 8.64 Three-phase modified DTC of IM without PI controller simulation results: line to line voltage of inverter VAB, VBC and VCA (top to bottom)

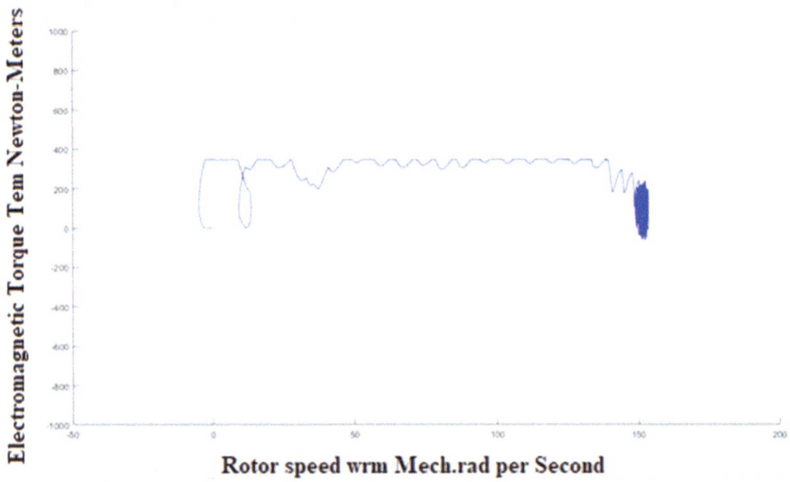

Fig. 8.65 Three-phase modified DTC of IM drive without PI controller simulation result: electromagnetic torque Tem Nw-metres versus rotor speed wrm mechanical radians per second

200 Nw-metres. From Fig. 8.62 it is seen that λqs and λds are almost sinusoidal. The total stator flux λs in the steady state has flux ripples as can be seen from Fig. 8.63. From Fig. 8.61, the rotor speed ωrm reaches the steady state value of 144.8 Mech. radians per second. From Fig. 8.61 it is seen that the three-phase rotor currents are well balanced and symmetrical about the x-axis. But the three-phase stator currents, although balanced and symmetrical about x-axis initially, are NOT symmetrical about the x-axis during steady state showing some DC offset current.

8.12 Voltage to Frequency Ratio (*V*/*f*) Control of Three-Phase Induction Motor Drive

In the above sections, vector control and DTC of IM are presented which are based on dynamic model in the dq-axis. The scalar control of speed and torque known as the voltage to frequency ratio (*V*/*f*) control is based on its steady state equivalent circuit obtainable from no-load and locked rotor tests [26, 27]. The simplified per phase equivalent circuit of the IM is shown in Fig. 8.66.

It is well known that the synchronous speed in r.p.m. Ns = (120*f/P) where f is the frequency of stator supply voltage V and P is the number of poles and that the rotor speed in r.p.m. Nr is lower than this synchronous speed. Thus for the speed control of IM, apart from stator voltage control, frequency control and pole changing methods are also used [26, 27].

A combination of stator applied voltage and frequency known as *V*/*f* control is used for three phase inverter-fed adjustable speed drives (ASDs) to control the

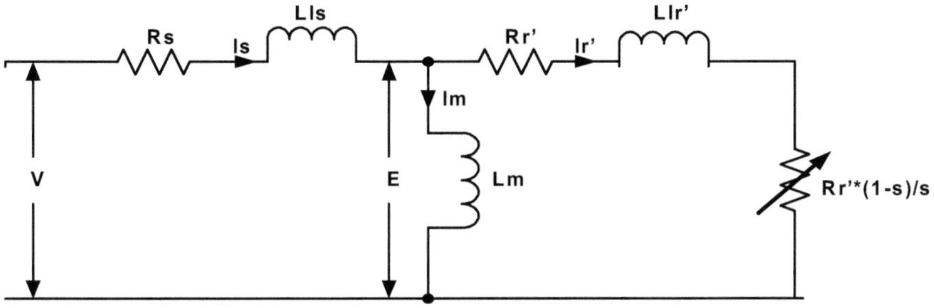

Fig. 8.66 Simplified steady state per phase equivalent circuit of three-phase induction motor

torque and speed. Referring to Fig. 8.66, the rotor current can be expressed as follows:

$$\vec{I_{r'}} = \frac{\begin{bmatrix} (R_s + j\omega.L_s) & V \\ -j\omega.L_m & 0 \end{bmatrix}}{\begin{bmatrix} (R_s + j\omega.L_s) & -j\omega.L_m \\ -j\omega.L_m & \left(\frac{R_{r'}}{S} + j\omega.L_{r'}\right) \end{bmatrix}} \tag{8.81}$$

In Eq. 8.81, $L_s = (L_{ls} + L_m)$ and $L_{r'} = (L_{lr'} + L_m)$.

$$\vec{I_{r'}} = \frac{V \cdot j\omega \cdot L_m}{\left(R_s \cdot \frac{R_{r'}}{S} - \omega^2 \cdot L_s \cdot L_{r'}\right) + j\omega \cdot \left(R_s \cdot L_{r'} + L_s \cdot \frac{R_{r'}}{S}\right)} \tag{8.82}$$

$$|I_{r'}| = I_{r'} = \frac{V \cdot \omega \cdot L_m}{\sqrt{\left(R_s \cdot \frac{R_{r'}}{S} - \omega^2 \cdot L_s \cdot L_{r'}\right)^2 + \left(\omega \cdot R_s \cdot L_{r'} + \omega \cdot L_s \cdot \frac{R_{r'}}{S}\right)^2}} \tag{8.83}$$

The total power input P_{in} and power output P_o of rotor are given below:

$$P_{in} = 3 \cdot I_{r'}^2 \cdot \frac{R_{r'}}{S} = \frac{3 \cdot V^2 \cdot \omega^2 \cdot {L_m}^2 \cdot R_{r'}/S}{\left(R_s \cdot \frac{R_{r'}}{S} - \omega^2 \cdot L_s \cdot L_{r'}\right)^2 + \left(\omega \cdot R_s \cdot L_{r'} + \omega \cdot L_s \cdot \frac{R_{r'}}{S}\right)^2} \tag{8.84}$$

$$\begin{aligned} P_o &= 3 \cdot I_{r'}^2 \cdot \frac{R_{r'} \cdot (1-S)}{S} \\ &= \frac{3 \cdot V^2 \cdot \omega^2 \cdot {L_m}^2 \cdot R_{r'} \cdot (1-S) * S}{(R_s \cdot R_{r'} - S \cdot \omega^2 \cdot L_s \cdot L_{r'})^2 + (S \cdot \omega \cdot R_s \cdot L_{r'} + \omega \cdot L_s \cdot R_{r'})^2} \end{aligned} \tag{8.85}$$

In Eqs. 8.81 to 8.85 rotor slip S is defined below:

$$S = \frac{(N_s - N_r)}{N_s} = \frac{(\omega_s - \omega_{rm})}{\omega_s} = \frac{(\omega - \omega_{re})}{\omega} \tag{8.86}$$

where $\omega_s = (2\pi \cdot N_s/60)$ and $\omega_{rm} = (2\pi \cdot N_r/60)$ both in mech.radians per second, respectively. Also $\omega = 2\pi \cdot f$ is the frequency of the stator supply voltage and $\omega_{re} = (P/2)*\omega_{rm}$ both in elec.radians per second.

The rotor developed torque $T_{em} = (Po/\omega_{rm}) = (Po.P/2\cdot\omega_{re}) = (Po\cdot P/4\cdot\pi\cdot f_{re}) = (Po\cdot P/4\cdot\pi\cdot S\cdot f)$ where $f_{re} = S\cdot f$ is the frequency of the induced currents and voltage in the rotor known as slip frequency. Thus using Eq. 8.85, T_{em} can be expressed as follows:

$$T_{em} = \frac{3 \cdot P \cdot L_m^2}{16 \cdot \pi^3 \cdot f_{re}} * \left[\frac{\left(\frac{V}{f}\right)^2}{\left(\frac{R_s \cdot R_{r'}}{S \cdot \omega^2} - L_s \cdot L_{r'}\right)^2 + \left(\frac{L_s \cdot R_{r'}}{\omega \cdot S^2} + \frac{R_s L_{r'}}{\omega}\right)^2} \right] \tag{8.87}$$

Also referring to Fig. 8.66, neglecting the stator impedance drop compared to the induced e.m.f. E in the rotor circuit, the stator voltage V and E can be expressed as follows:

$$V \triangleq E = \omega \cdot L_m \cdot I_m = 2 \cdot \pi \cdot f \cdot L_m \cdot I_m = 2 \cdot \pi \cdot f \cdot \Lambda \tag{8.88}$$

where Λ is the stator flux linkage due to air gap flux. Expression for V (Eq. 8.88) implies that by maintaining (V/f) ratio a constant value K, air gap flux can be held constant at all rotor speeds. This is achieved by varying the stator supply voltage V in proportion to the supply frequency f. For frequencies above the rated value, the applied voltage is held at the rated value which causes the stator flux to decrease. This is called field weakening. At low frequencies the applied voltage must be boosted to compensate for the voltage drop in the stator resistance.

8.12.1 Open Loop V/f Control

The open loop V/f control scheme is shown in Fig. 8.67a, b for six-step inverter mode and sine PWM inverter mode operation, respectively. Here the reference synchronous speed ω_s* in rad/s is converted to reference frequency F* in hertz by multiplying with (P/4.π) which is given to an absolute value limiter to maintain positive values for frequency F and without exceeding the rated frequency of the motor. This frequency F is multiplied with the V/f ratio K and is added to reference voltage Vo* used for compensating the stator resistance drop at low frequencies to get the stator applied voltage V. This V and F are used to generate six-step gate drive for inverter (Fig. 8.67a) or else sine PWM gate drive along with a triangle carrier (Fig. 8.67b). From the value of V, DC link voltage Vdc is obtained for six-step 180-degree mode inverter operation. This Vdc and gate pulse are given to the three-phase inverter which drives the IM.

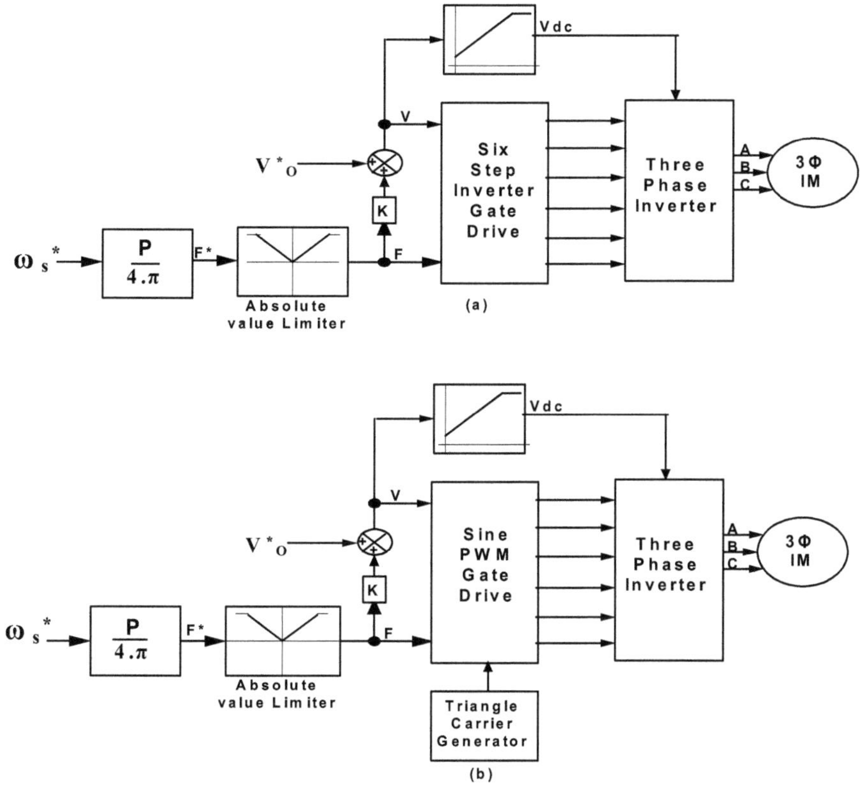

Fig. 8.67 Open loop *V/f* control of three-phase IM drive: (**a**) using six-step inverter and (**b**) using sine PWM inverter

8.12.2 Closed Loop V/f Control

The block diagram of the closed loop *V/f* control scheme is shown in Fig. 8.68a, b for the six-step inverter mode and sine PWM inverter mode, respectively. In both cases, the rotor reference speed ωrm* in mech.radians per second is compared with the actual speed of motor ωrm in an error detector, and the speed error is given to slip limiter with upper and lower limit via a PI controller. The output ωsl of slip limiter is added with rotor speed ωrm to get the synchronous speed ωs which is then multiplied by (P/4.π) to obtain the reference frequency F* in Hertz. This value of F* is given to absolute value limiter which gives positive value frequency F hertz for all rotor speeds and without exceeding the rated frequency of the motor. This frequency F is multiplied by *V/f* ratio *K* and is then added to reference voltage V_o^* which gives the voltage compensation for stator resistance drop at low frequencies, to get the line to neutral stator voltage *V*. This stator voltage is used to calculate the DC link voltage V_{dc} of inverter for six-step 180-degree mode operation using the formula ($V_{dc} = 3*V/\sqrt{2}$). This V and F are used to generate six-step gate drive for inverter (Fig. 8.68a) or

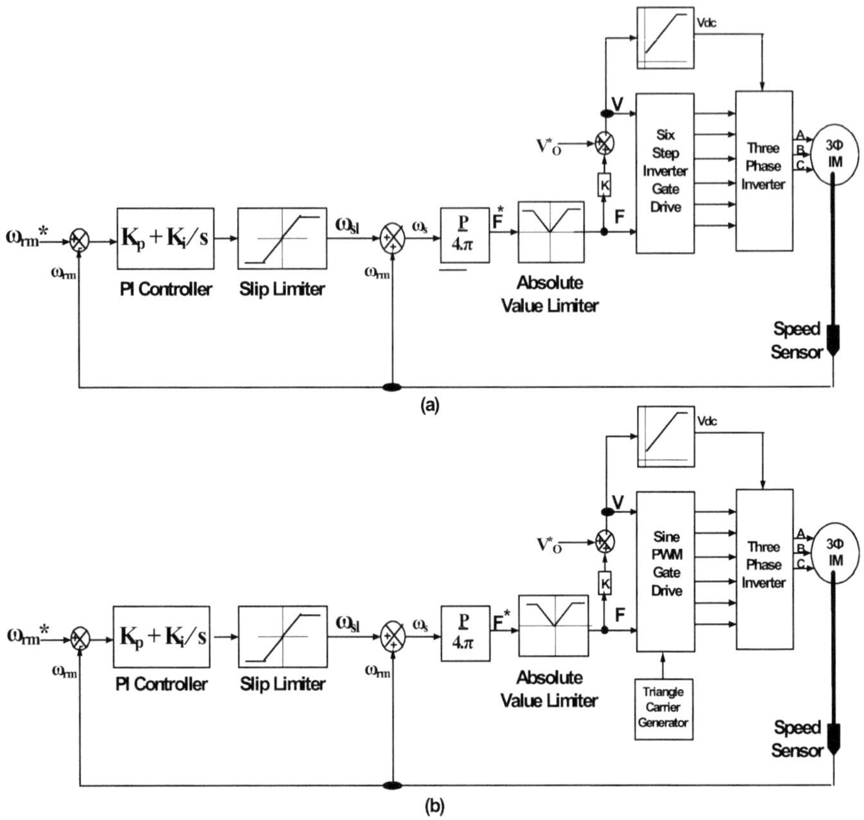

Fig. 8.68 Closed loop *V/f* control of three-phase IM drive: (**a**) using six-step inverter and (**b**) using sine PWM inverter

else sine PWM gate drive along with a triangle carrier (Fig. 8.68b). This V_{dc} and gate pulse are given to the three-phase inverter which drives the IM.

8.13 Models of Voltage to Frequency (*V/f*) Ratio Control of Three-Phase Induction Motor Drive Using PI Controller

In this section, models for the closed loop *V/f* control of three-phase IM drive using PI controller are presented (a) for six-step 180-degree mode inverter operation and (b) for sine PWM inverter operation. The data shown in Table 8.1 are used to develop the model for IM. For sine PWM inverter, a triangle carrier frequency of 1200 Hz is used.

The model of the closed loop *V/f* control of three-phase IM drive fed by six-step 180-degree mode inverter is shown in Fig. 8.69 (model file: EXAMPLE8_9). The

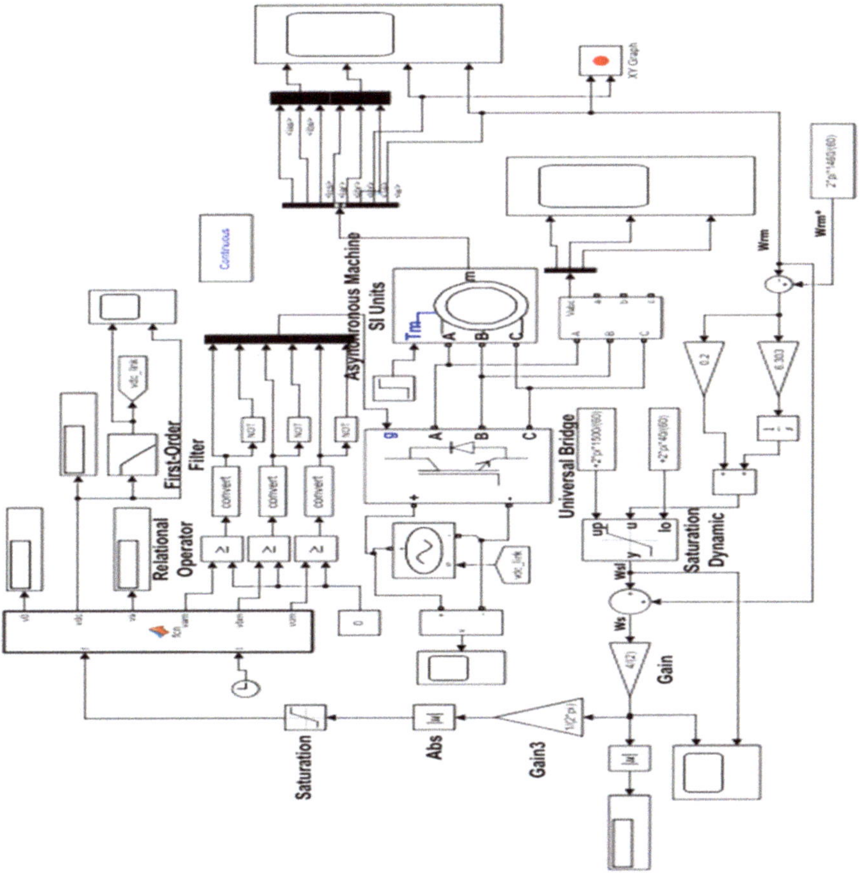

Fig. 8.69 Model of *V*/*f* control of three-phase inverter-fed induction motor drive using PI controller

model is developed as per block diagram shown in Fig. 8.68a. Here the three-phase Universal Bridge model for inverter and cage IM are from Simscape Specialized Power Systems block set [6]. Here the rated no-load speed is used as the reference and is compared with the rotor speed in an error detector block. The output of the error detector is given to the PI controller with $K_p = 0.2$ and $Ki = 6.303$ developed using two gain blocks, integrator and summing blocks. The output of the PI controller is given to the saturation dynamic block which is the slip limiter. The upper limit for the slip limiter is the synchronous speed Ns and the lower limit is the difference (Ns – Nr) in radians per second. The output of slip limiter is given to another summing block which sums the output of slip limiter with rotor speed giving the synchronous speed ω_s in radians per second. This synchronous speed ω_s is multiplied using two gain blocks with values P/2 and 1/2.π to get the frequency F* in hertz. This is given to Absolute value block and to Saturation block which limit the absolute value of frequency f within the rated value. This frequency f and time module are given as input to the Embedded MATLAB function. Program segment 8.9 shows how to generate the outputs vo (Vo* in Fig. 8.68a), vs (V in Fig. 8.68a) and vdc (Vdc) and the modulating three-phase sine wave AC voltages vam, vbm and vcm. Using the data for phase to neutral voltage vs and frequency f in Table 8.1, slope *V/f* (*K*) is calculated, and the resulting expression for vs is added with vo and limited to this maximum value. The value of vo is taken to be 2.5% of rated line to neutral voltage. A low-pass filter is used at the output of vdc to filter high-frequency oscillations. The modulating signals vam, vbm and vcm are compared with zero using three relational operator blocks which form the comparators. This output of the comparators and their respective inverted output using NOT gates are given as gate pulse to the inverter through a Mux block.

The model of the closed loop *V/f* control of three-phase IM drive fed by sine PWM inverter is shown in Fig. 8.70 (model file: EXAMPLE8_10). The model is developed as per block diagram shown in Fig. 8.68b. The model development is the same as for six-step 180-degree mode inverter-fed IM drive explained above. Program segment 8.10 is used to generate three-phase sine wave modulating voltages. Here the only difference is that the three-phase sine modulating signals vam, vbm and vcm are compared with the triangle carrier (±1 V) in three relational operator comparator blocks. The resulting output of comparator blocks and their respective inverted output using NOT gates are given to the gate input of inverter through a Mux. In the following, the design of the PI controller using Ziegler-Nichols frequency response method is presented [28].

In the model in Figs. 8.69 and 8.70, assuming only proportional controller K_p is present, the output response of the P controller is noted for various increasing value of K_p starting from above zero until sustained oscillations are observed indicating critical stability. This gain Ku and the period of oscillation Tu are observed for marginal stability. Then the value of *K* is 0.4*Ku and Ti is 0.8*Tu. The PI controller parameters are *Kp* = *K* = 0.4*Ku and K_i = *K*/Ti = (0.4*Ku/Ti) = (0.4*Ku/(0.8*Tu)).

8.13.1 Simulation Results

The simulation of the model for *V/f* controlled six-step inverter-fed IM drive using PI controller shown in Fig. 8.69 is carried out using ode23tb (stiff/TR-BDF2) solver in Simulink [6]. An initial load torque of 50 Nw-metres is applied for the first 10 s and this load torque is stepped up to 150 Nw-metres above 10 s. The simulation results for the inverter line to line voltage, stator and rotor currents, electromagnetic torque and rotor speed are shown in Fig. 8.71a, b, respectively. The expanded view of the stator currents and rotor currents is shown in Fig. 8.72a, b, respectively.

The simulation of the model for *V/f* controlled three-phase sine PWM inverter-fed IM drive using PI controller shown in Fig. 8.70 is carried out using ode23tb (stiff/TR-BDF2) solver in Simulink [6]. An initial load torque of 25 Nw-metres is applied for the first 10 s and this load torque is stepped up to 75 Nw-metres above 10 s. The

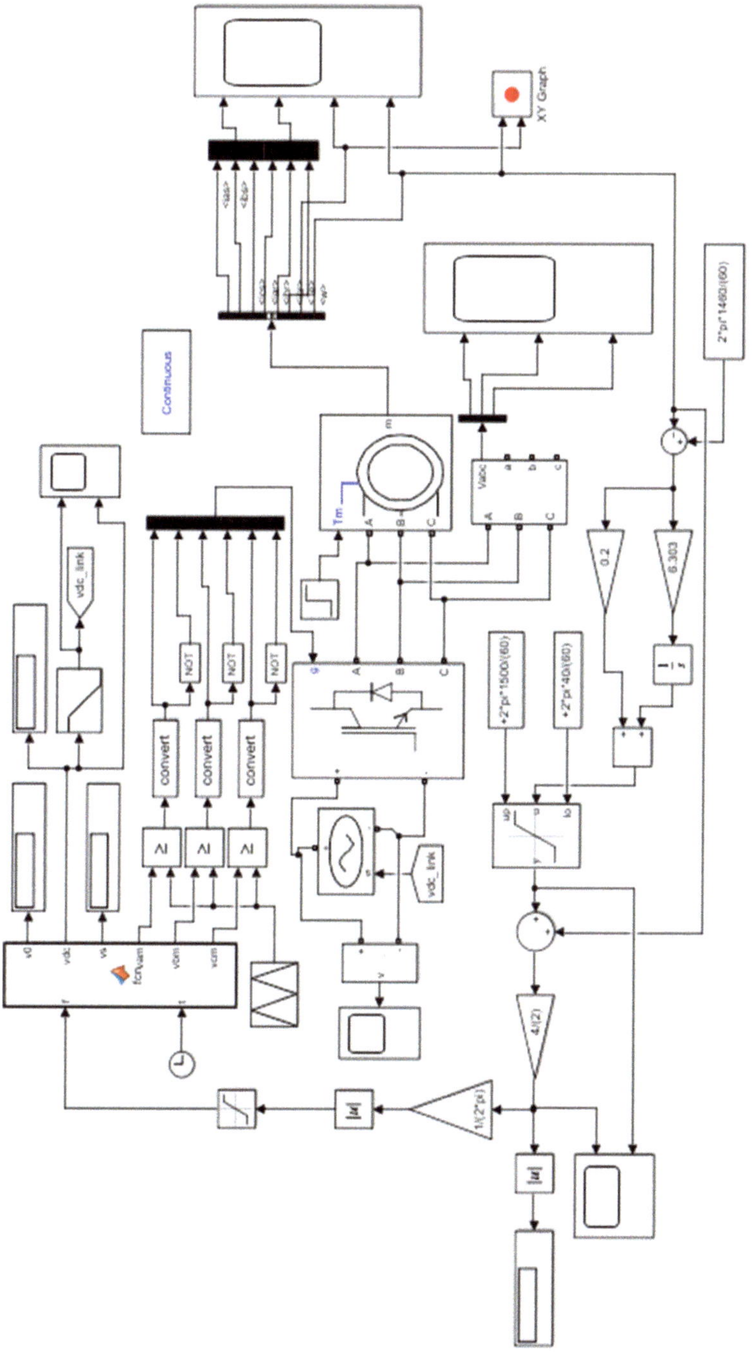

Fig. 8.70 Model of voltage to frequency ratio (*V/f*) control of three-phase sine PWM inverter-fed induction motor drive using PI controller

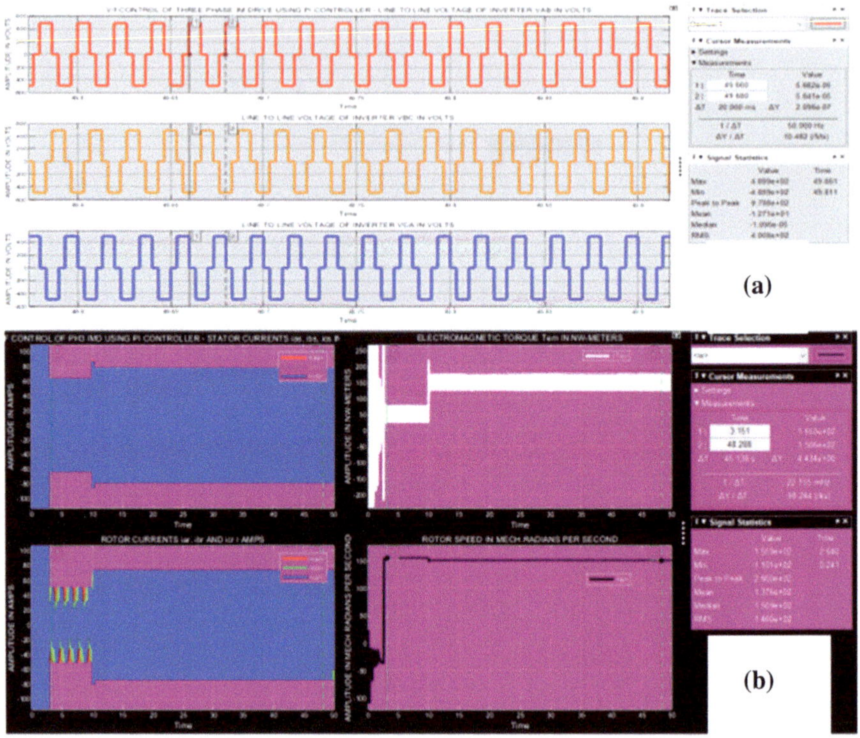

Fig. 8.71 Simulation results for closed loop *V*/*f* control of three-phase IM drive using PI controller: (**a**) Three-phase line to line voltage. (**b**) Stator and rotor currents (left column, top and bottom), E.M. torque and rotor speed (right column, top and bottom)

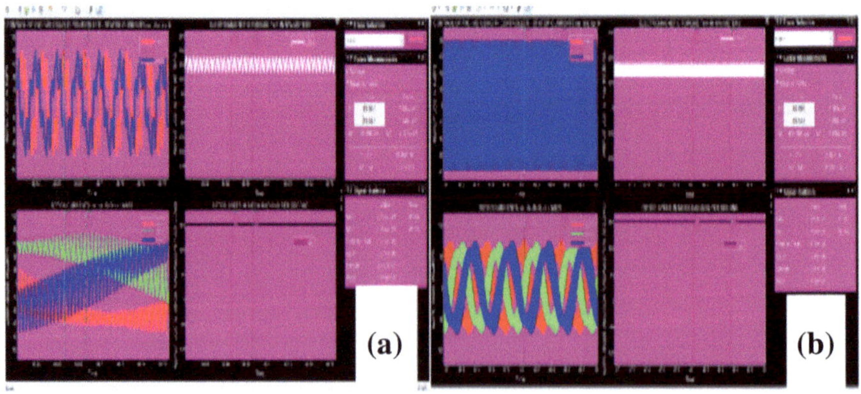

Fig. 8.72 Simulation results for closed loop *V*/*f* control of three-phase IM drive using PI controller: (**a**) stator currents (left column, top) and (**b**) rotor currents (left column, bottom)

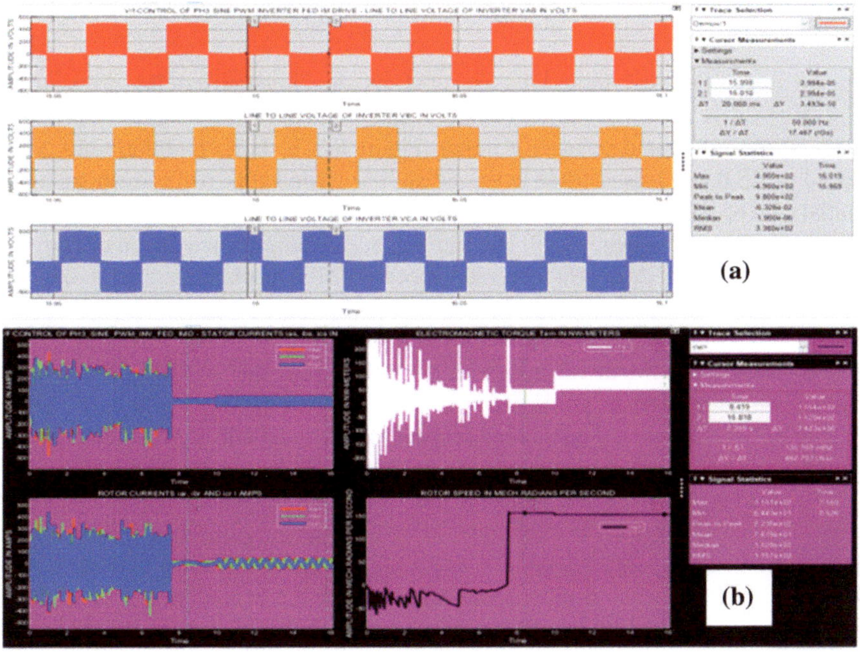

Fig. 8.73 Simulation results for closed loop *V/f* control of three-phase sine PWM inverter-fed IM drive: (**a**) line to line inverter voltage and (**b**) stator and rotor currents (left column, top and bottom), (**b**) E.M. torque and rotor speed (right column, top and bottom)

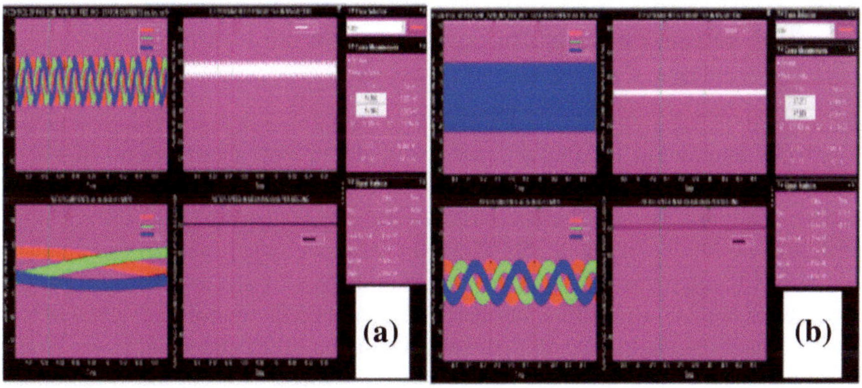

Fig. 8.74 Simulation results for closed loop *V/f* control of three-phase sine PWM inverter-fed IM drive: (**a**) stator currents (left column, top) and (**b**) rotor currents (left column, bottom)

simulation results for the inverter line to line voltage, stator and rotor currents, electromagnetic torque and rotor speed are shown in Fig. 8.73a, b, respectively. The expanded view of the stator currents and rotor currents is shown in Fig. 8.74a, b, respectively.

8.13.2 Discussion of Results

In the case of closed loop *V/f* controlled six-step inverter-fed IM drive, from Fig. 8.71a it is seen that the RMS line to line voltage of the inverter is 400.8 V very close to the rated voltage of motor. Figure 8.71b shows that the initial rotor speed is 155.0 and falls to 150.6 mech.radians per second after 10 s corresponding to initial load of 50 Nw-metres and final load of 150 Nw-metres. Figure 8.72a indicates the stator current frequency is 50 Hz, and Fig. 8.72b indicates that the rotor currents have a frequency of 2.027 Hz for a rotor speed of 150.6 mech.rad/second. Using the relation for rotor slip frequency, this value is 2.038 Hz which is close to the value obtained by simulation. In the case of closed loop *V/f* controlled three-phase sine PWM inverter-fed IM drive, from Fig. 8.73a it is seen that the RMS line to line voltage of the inverter is 338 V which is 62 V less than the rated voltage of motor. Figure 8.73b shows that the initial rotor speed is 155.4 and falls to 152 mech.radians per second after 10 s corresponding to initial load of 25 Nw-metres and final load of 75 Nw-metres. Figure 8.74a indicates the stator current frequency is 50 Hz, and Fig. 8.74b indicates that the rotor currents have a frequency of 1.618 Hz for a rotor speed of 152 mech.rad per second. Using the relation for rotor slip frequency, this value is 1.592 Hz which is close to the value obtained by simulation. Due to pulse width modulation, RMS line to line voltage input to the IM is much reduced from its rated value. This results in a lower power output of motor. Thus to maintain the same speed close to the reference speed, the load torque has to be reduced.

8.14 Case Study: PI Controller-Based Speed Control of Three-Phase Induction Motor Drive Fed by Thyristor Controller

The block diagram of the PI controller-based closed loop speed control of a star-connected three-phase IM fed by thyristor (SCR) controller is shown in Fig. 8.75a, and the configuration of the back to back or antiparallel connected SCR controller for each phase is shown in Fig. 8.75b. The rotor reference speed ωref in mech. radians per second is compared with the actual speed of motor ω in an error detector, and the speed error is given to the PI controller. The output of PI controller is given to slip limiter which maintains the slip frequency within the given upper and lower limits. The output ωsl of slip limiter is added with rotor speed ω to get the synchronous speed ωs which is then integrated and multiplied by (180/π) to obtain reference firing angle for the SCR controller. This reference firing angle is given to an absolute value limiter which limits the firing angle α within the desired limit so as to maintain the applied voltage across the stator terminals within the rated value. This value of α and the rated frequency f Hz of the motor are given as input to the gate pulse generator. The gate pulse generator output is given as trigger pulse

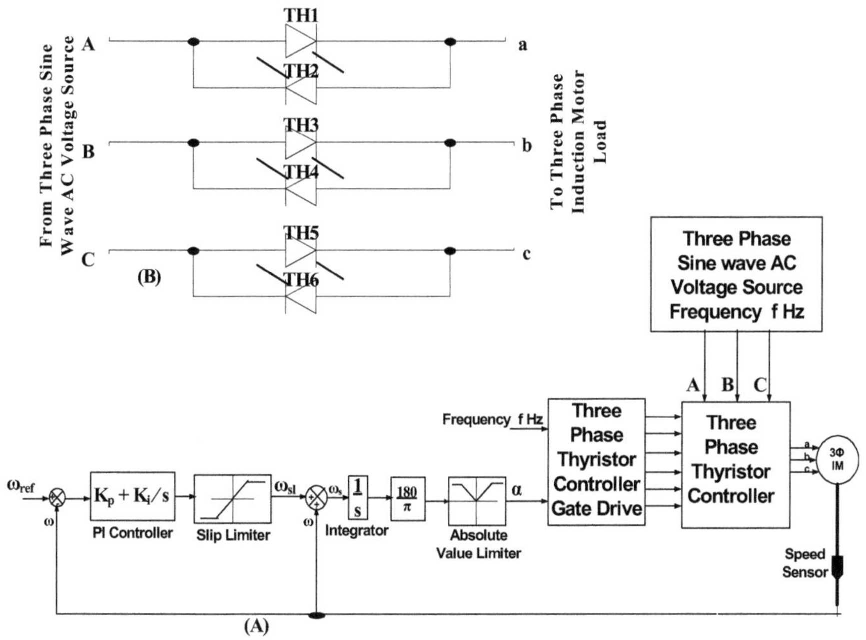

Fig. 8.75 (**a**) PI controller-based closed loop control of three-phase induction motor drive using thyristor voltage controller, (**b**) three-phase thyristor controller

to the respective gates of the SCRs in Fig. 8.75b. Here SCR controller for is connected to the three-phase star-connected IM which is an R-L load. In Fig. 8.75b, referring to SCRs TH1 and TH2 connected to input Phase A, these are fired at angles α and (π+α), respectively. Because the IM is an inductive load, TH1 will continue to conduct beyond π for an angle β called the extinction angle. If input voltages V_{AN}, V_{BN} and V_{CN} are defined as $V_{kN} = (\sqrt{2} * Vs * \mathrm{Sin}(2\pi * f * t + \gamma_k)$ where $\gamma_k = 0, -2\pi/3$ and $-4\pi/3$ for $k =$ A, B and C, respectively, then the output phase voltage V_O can be derived as follows [29]:

$$V_O = \sqrt{\frac{1}{\pi} * \int_{\alpha}^{\beta} \left[2 * V_S^2 * \sin^2(\omega \cdot t) \cdot d(\omega \cdot t)\right]} \tag{8.89}$$

$$V_O = \frac{V_S}{\pi} * \sqrt{\beta - \alpha + \frac{\sin(2 \cdot \alpha)}{2} - \frac{\sin(2 \cdot \beta)}{2}} \tag{8.90}$$

From Eq. 8.90, it is seen that varying the SCR firing angle α varies the applied voltage across the motor terminals controlling the power output and speed.

8.14.1 Model of the Speed Control of Three-Phase Induction Motor Drive Fed by Thyristor Controller Using PI Controller

The model of the PI controller-based speed control of three-phase IM fed by SCR controller is shown in Fig. 8.76. The model is developed as per block diagram shown in Fig. 8.75. The data shown in Table 8.1 are used to model the IM. The model of the closed loop speed control of three-phase IM drive fed by SCR converter using PI controller is shown in Fig. 8.76 (model file: CASE_STUDY_EX8_5). Here the thyristors and cage IM are from Simscape Specialized Power Systems block set [6]. Here the rated no-load speed is used as the reference and is compared with the rotor speed in an error detector block. The output of the error detector is given to the PI controller with $K_p = 0.2$ and $K_i = 6.303$ developed using two gain blocks, integrator and summing blocks. The output of the PI controller is given to the saturation dynamic block which is the slip limiter. The upper limit for the slip limiter is the synchronous speed Ns and the lower limit is the difference (Ns – Nr) in radians per second. The output of slip limiter is given to another summing block which sums the output of slip limiter with rotor speed giving the synchronous speed ωs in radians per second. This synchronous speed ωs is integrated and then multiplied with (180/π) using gain multiplier and given to absolute value block and Saturation block in series which form the absolute value limiter. In this Saturation block, the upper limit and lower limit of the firing angle α are entered such that the output voltage of SCR controller doesn't exceed the stator rated voltage. This value of α, frequency 50 Hz of the AC input voltage and time module are given as input to the Embedded MATLAB function block whose outputs are the modulating sine wave functions vma1, vma2, vmb1, vmb2, vmc1 and vmc2. Program segment 8.11 shows the method generating these six modulating sine functions. These six modulating functions are compared with zero in separate relational operator blocks which form the comparator. The comparator output is HIGH (logic 1) if the respective input crosses zero and goes positive and LOW (logic 0) when this input is negative. The six gate pulse outputs of comparators are given to the gates of SCRs TH1 to TH6.

8.14.2 Simulation Results

The simulation of the PI controller-based three-phase SCR controller-fed IM drive is carried out using ode23tb (stiff/TR-BDF2) solver in Simulink [6]. The initial load is 50 Nw-metres for the first 0.5 s and this steps up to 150 Nw-metres above 0.5 s. The simulation results for the three-phase stator currents, rotor currents, e.m. torque and rotor speed are shown in Fig. 8.77a. The three-phase line to line and line to neutral voltage output of the SCR controller are shown in Fig. 8.77b, c, respectively.

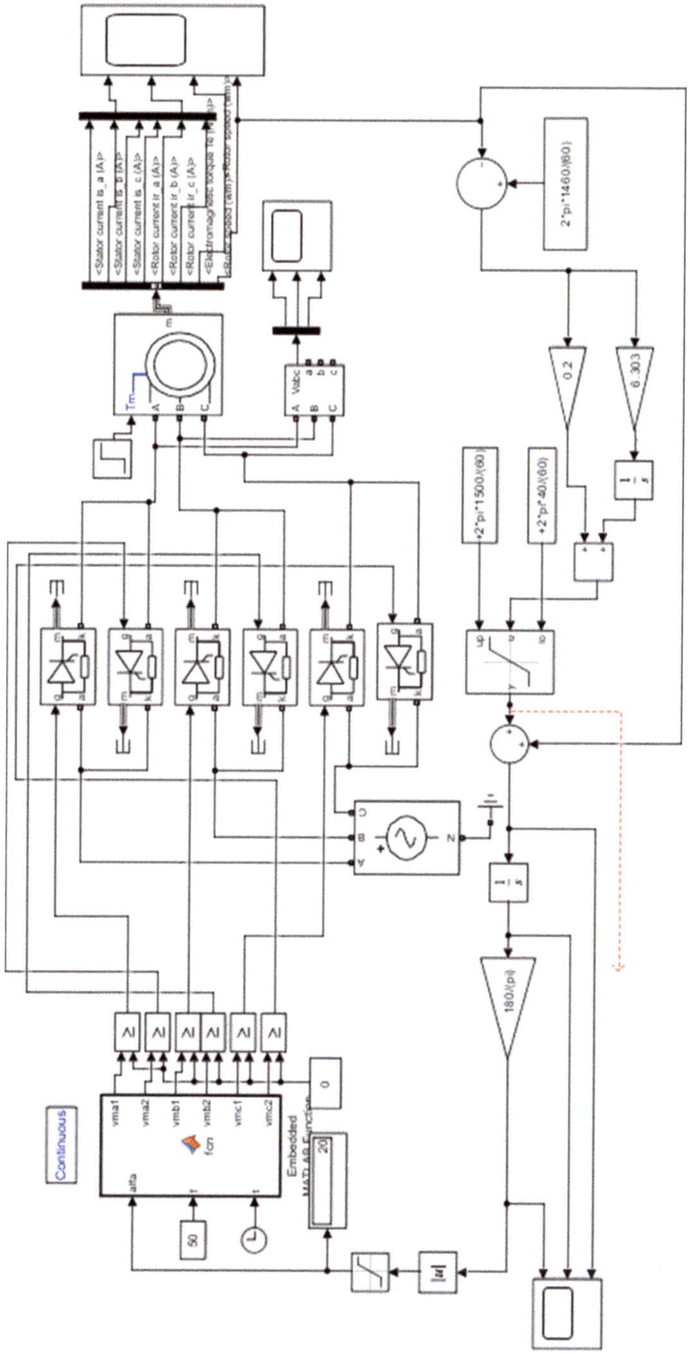

Fig. 8.76 Model of PI controller-based three-phase thyristor controller-fed induction motor drive

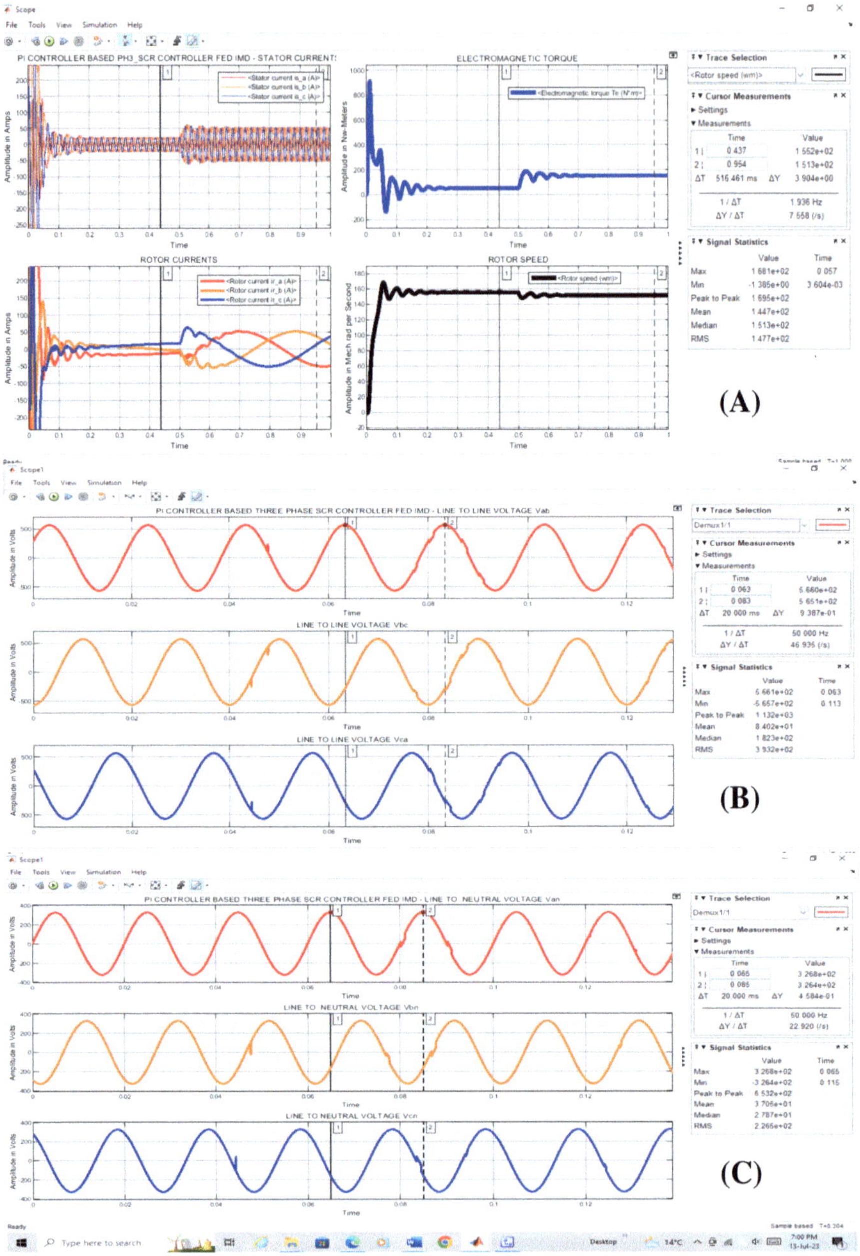

Fig. 8.77 PI controller-based three-phase SCR controller-fed IM drive simulation results—(**a**) stator currents and rotor currents (left column, top to bottom), E.M. torque and rotor speed (right column, top to bottom), (**b**) three-phase line to line output voltage and (**c**) three-phase line to neutral output voltage of SCR controller

8.14.3 Discussion of Results

The three-phase stator and rotor currents are well balanced. The steady value of e.m. torque is 50 Nw-M for the initial 0.5 s and steps up to 150 Nw-M after 0.5 s. The rotor speed is 155.2 and 151.3 mech.rad per second for the initial 0.5 s and above 0.5 s, respectively, indicating fall in speed due to load increase. The line to line voltage and the line to neutral output voltage of SCR controller are found to be 393.2 and 226.5 volts, respectively, and the switching frequency is 50 Hz. Rotor speed is close to the reference speed and takes 0.2 s to reach steady value.

8.15 Conclusions

The interactive model of the three-phase IM using flux linkage equations in state space is presented. The applied voltage, stator current, electromagnetic torque and speed control of rotor are studied by simulation for various PWM techniques for one selected value of A.M. index. The study can be easily extended to other values of AM index in the under- and over-modulation regions. The triangle carriers used for three-phase MCSLSPWM DCTLI have in-phase disposition. A case study is presented for SVPWM two-level inverter-fed IM drive. Compared to direct online starting, the PWM techniques have control over the applied voltage, stator and rotor currents, rotor torque and speed. A method of starting the three-phase inverter-fed IM using sine PWM technique by varying the AM index is also presented which can be easily extended to other PWM techniques. It is also seen that for various inverter PWM techniques, the shape of rotor torque speed curve, the maximum torque value and the rotor speed corresponding to maximum torque vary for the given three-phase IM load. Models for vector controlled three-phase IM and that for three-phase SVPWM inverter-fed vector controlled IM drive as case study are presented with simulation results, highlighting the advantage of vector control in maintaining the rotor speed and e.m. torque close to the set point value. Also models for DTC of three-phase IM with and without PI controller are presented for both classical and modified switching techniques, the latter as a case study. The DTC scheme with PI controller gives initial starting torque and flux ripples and a constant steady state torque, whereas that without PI controller gives starting and steady state torque ripples and stator flux ripples. In both cases three-phase rotor currents are found to be balanced and symmetrical about the x-axis, whereas the three-phase stator currents although balanced have DC offset currents for classical switching. With modified switching the stator currents are balanced and symmetrical with PI controller and are unsymmetrical without PI controller showing DC offset currents. The stator current and rotor current waveforms depend on the sample time and the choice of hysteresis band limit for torque and stator flux. In addition, models for scalar control technique having applied voltage to frequency ratio constant in closed loop are presented for six-step 180-degree mode inverter and that for three-phase sine PWM inverter operations maintaining rotor speed close to the reference speed with varying load condition. Also a case study is presented for PI controller-based three-phase thyristor controller-fed IM drive to illustrate the application of PI controller for closed loop speed control.

References

1. Adkins, B: "The General Theory of Electrical Machines"; Chapman & Hall Ltd., London; 1957.
2. Paul C. Krause: "Analysis of Electric Machinery"; McGraw Hill Inc.; 1986; pp. 178–197.
3. Chee-Mun-Ong: "Dynamic Simulation of Electric machinery Using MATLAB/SIMULINK"; Prentice Hall Inc.; 1998.
4. Narayanaswamy.P.R.Iyer, Venkat Ramaswamy and Jianguo Zhu: "Interactive Modeling of A Three Phase Induction Motor in All Reference Frames Using dq0-Axis Flux Linkage Equations in State Space", Australasian Universities Power Engineering Conference (AUPEC 2006), Melbourne, December 10–13, 2006, ISBN: 978 1 86272 669 7.
5. R. Krishnan: "Electric Motor Drives Modeling Analysis and Control"; Prentice Hall Inc.; 2001.
6. The MathWorks Inc, www.mathworks.com: "MATLAB Version: 9.10.0, R2021a, 2021.
7. Narayanaswamy.P.R.Iyer, Venkat Ramaswamy and Jianguo Zhu: "SIMULINK Model for Three Phase Sine PWM Inverter fed I.M. Drive", 40th International Universities Power Engineering Conference (UPEC 2005), Cork, Ireland, September 2005, pp. 1180–1184.
8. J.A. Houldsworth and D.A. Grant: "The Use of harmonic Distortion to Increase the Output Voltage of a Three Phase PWM Inverter"; IEEE Transactions on Industry Applications; Vol.IA-20; NO.5; pp. 1224–1227.
9. M.A. Boost and P.D. Ziogas: "State of the Art Carrier PWM Techniques: A Critical Evaluation"; IEEE Transactions on Industry Applications; Vol.24; No.2; pp. 271–280; March/April 1988;
10. Narayanaswamy.P.R.Iyer: "MATLAB/SIMULINK Modules for Modelling and Simulation of Power Electronic Converters and Electric Drives"; M.E.(Research) Thesis; University of Technology Sydney, NSW, 2006.
11. Narayanaswamy P.R. Iyer, Venkat Ramaswamy and Jianguo Zhu: "Three Phase Clipped Sinusoid PWM Inverter – A New technique for Pulse Width Modulation", 40th International Universities Power Engineering Conference (UPEC 2005), Cork, Ireland September 2005, pp. 1185–1189.
12. J.F. Silva and S.F. Pinto: "Advanced Control of Switching Power Converters" in "Power Electronics Handbook", Editor: M.H. Rashid, Chapter 36, pp. 1055–1058, Elsevier, 2011.
13. J.R. Espinoza: ""Inverters" in "Power Electronics Handbook", Editor: M.H. Rashid, Chapter 15, pp. 372–376, Elsevier, 2011.
14. Bin Wu: "High-Power Converters and AC Drives", IEEE Press, pp. 101–111 and pp. 143–162, 2006.
15. J.W. Dixon. "Multilevel Converters" in Power Electronic Converters and Systems Frontiers and Applications, Ed. Andrzej M. Trzynadlowski, 43–72, The Institution of Engineering and Technology, 2016.
16. Surin Khomfoi and Leon M. Tolbert.. "Multilevel Power Converters" in Power Electronics Handbook, Ed. M.H. Rashid, 455–484, Elsevier, Butterworth-Heinemann, 2011.
17. N P R Iyer: "Power Electronic Converters: Interactive Modeling using SIMULINK", CRC Press, USA, Chapter 9 and 10, pp. 243–249 and pp. 315–319., 2018.
18. Ned Mohan: "Advanced Electric Drives Analysis, Control and Modeling Using SIMULINK", MNPERE, 2001, pp. 3–1 to 3–30 and pp. 5–1 to 5–16.
19. D.W. Novotny and T.A. Lipo: "Vector Control and Dynamics of AC Drives", Oxford University Press, New York, 1996, pp. 240–243 and pp. 257–275.

20. Takahashi, I. and Noguchi, T.: "A new quick-response and high efficiency control strategy of an induction motor". IEEE Trans. Ind. Appl, IA-22(5), 1986, pp. 820–827.
21. Depenbrock, M: "Direct self-control (DSC) of inverter-fed induction motors", IEEE Trans. Power Electronics, 3(4), 1988, pp. 420–429.
22. Haitham Abu-Rub et.al.: "High Performance Control of AC Drives with Matlab/Simulink Models", John Wiley and Sons, 2012.
23. G. Buja, D. Casadei and G. Serra: "Direct Torque Control of Induction Motor Drives", IEEE-ISIE, Portugal, 1997, pp. TU-2–TU-8.
24. D. Casadei, F. Profumo, G. Serra and A. Tani: "FOC and DTC: Two Viable Schemes for Induction Motors Torque Control", IEEE Transactions on Power Electronics, Vol. 17, No. 5, September 2002, pp.779–787.
25. P. Vas and P. Tiitinen: "Sensorless Vector and Direct Torque Controlled Drives" in M.H. Rashid [Editor]: "Power Electronics Handbook", Chapter 28, pp. 743–766, Academic Press, 2001.
26. M.F. Rahman et.al.: "Motor Drives" in M.H. Rashid [Editor]: "Power Electronics Handbook", Chapter 34, pp. 923–928, Elsevier, 2011.
27. Khaled Nigim: "Induction Motor Drives" in Ali Emadi [Editor]: "Handbook of Automotive Power Electronics and Motor Drives", Chapter 18, pp. 401–402, Taylor and Francis, CRC Press, 2005.
28. K.J. Astrom and B. Wittenmark: "Computer-Controlled Systems: Theory and Design", Pearson, 1996.
29. A. K. Chattopadhyay: "AC – AC Converters" in M.H. Rashid [Editor]: "Power Electronics Handbook", Chapter 18, pp. 487–495, Elsevier, 2011.

Chapter 9
Fuzzy Logic Control of AC Drives

9.1 Introduction

Digital computers work on Boolean logic such as TRUE (logic 1) and FALSE (logic 0). But this logic using "crisp" values 1 and 0 is algorithmic and is insufficient to solve complex engineering problems. For example, consider a ceiling fan whose controller is turned off and has stopped running which is logic 0 and the controller is turned on and is running which is logic 1. But this concept is vague and imprecise. This is because with the fan controller turned on, it can be set to "VERY LOW", "LOW", "MEDIUM", "HIGH" and "VERY HIGH" speeds. So the question is "which speed corresponds to logic 1?". The answer is vague and imprecise. In 1965, L.A. Zadeh proposed "fuzzy logic" (FL) in which he argued that engineering problems and the associated human thinking are complex, vague and imprecise and Boolean logic with crisp values 1 and 0 can't address this complex thinking process [1]. FL is considered a branch of artificial intelligence (AI) and has grown to solve many vague and imprecise problems in the areas of engineering, science, medicine, stock market and so on where a Boolean logic with crisp set 1 and 0 can't find a precise solution.

In this chapter, applications of this fuzzy logic (FL) for the control of converter-fed AC motors such as induction motor (IM) and permanent magnet synchronous motor (PMSM) drives are presented. Models are developed using Fuzzy Logic Toolbox in Simulink for the control of these AC motors [2]. Model simulation results are presented.

Supplementary Information The online version contains supplementary material available at https://doi.org/10.1007/978-3-031-62784-2_9.

N. P. R. Iyer, *Inverters and AC Drives*, Power Systems,
https://doi.org/10.1007/978-3-031-62784-2_9

9.2 Fuzzy Logic Fundamentals

FL makes use of logical reasoning and intuitive thinking and decision-making of an operator into an automated system. FL works on a set of rule base to make decision. To formulate these rules, clear knowledge of the input variables and the desired output must be known. FL is a form of input-output mapping. These rules are written in English language and are known as linguistic variables. For example, consider a home air conditioning system using a reverse cycle air conditioner (AirCon). Let input fuzzy variable room temperature be assigned as VERY LOW (−10 to 0 °C), LOW (0–15 °C), MEDIUM (15–25 °C), HIGH (25–35 °C) and VERY HIGH (35–45 °C) and the AirCon temperature settings be assigned VERY COLD (16–19 °C), COLD (19–22 °C), MEDIUM (22–25 °C), HOT (25–28 °C) and VERY HOT (28–32 °C). The following fuzzy linguistic statements apply:

1. IF the weather is summer and the room temperature is 35 °C or above (VERY HIGH), THEN set AirCon to "COLD".
2. IF the weather is autumn and the room temperature is above 15 °C and below 25 ° C (MEDIUM), THEN set AirCon to "HOT".
3. IF the weather is winter and the room temperature is above 5 °C and below 15 °C (LOW), THEN set AirCon to "VERY HOT".

In general fuzzy linguistic rules R_i can be represented as follows:

$$R_i = \text{IF x is } A_i \text{ THEN y is } B_i \text{ for } i = 1, 2, 3, 4 \ldots\ldots n \tag{9.1}$$

In Eq. 9.1, x is the input variable and y is the output variable, and A_i and B_i are the fuzzy sets or fuzzy subsets in the universe of discourse X and Y that describe input x and output y, respectively. In the fuzzy statements mentioned above, x is the room temperature and y is the AirCon setting. Also in room temperature, AirCon settings are known as the linguistic variables, and the terms VERY HIGH, MEDIUM and LOW are the fuzzy subsets of input x, and VERY COLD, HOT and VERY HOT are the fuzzy subsets of output y. The fuzzy sets and subsets are also known as linguistic qualifiers [3–7].

Boolean logic belongs to classical set theory where an object or variable is member of the set having value logic 1 or it is not a member of the set having value logic 0. Thus it is a bivalued logic. On the other hand, fuzzy set theory is based on FL where an object or variable takes a value anywhere in the range 0 to 1 and hence is known as multivalued logic. In FL, logic 1 corresponds to a fuzzy variable completely in the set and logic 0 corresponds to the fuzzy variable completely not in the set. Any value a fuzzy variable takes in between 0 and 1 represents the degree of membership of the variable.

9.2.1 Fuzzyfication

Fuzzy logic makes use of membership functions (MFs) with values varying from 0 to 1. For example, consider the speed of a motor shown in Fig. 9.1 where the linguistic variable speed can be defined by fuzzy variables (fuzzy sets) VERY LOW (VL), LOW (L), MEDIUM (M), HIGH (H) and VERY HIGH (VH). These fuzzy sets are represented by trapezoidal and triangular membership functions. In Fig. 9.1, a speed of 200 rpm and below and 800 rpm and above is completely in the VERY LOW and VERY HIGH fuzzy sets, respectively, each with MF value 1. Similarly a speed of 427 rpm is in the LOW fuzzy set with MF value of 0.46 and in the MEDIUM fuzzy set with MF value of 0.16. The LOW to MEDIUM transition occurs at 450 rpm and similarly MEDIUM to HIGH transition occurs at 650 rpm. All possible values a fuzzy variable can take are known as universe of discourse and the fuzzy set represented by MF cover the whole universe of discourse. The membership value also known as degree of membership of a fuzzy variable x is represented by μ_x. All the properties of the Boolean logic are applicable to fuzzy logic as well [3–5]. These properties of fuzzy variables are given below:

Given two fuzzy sets A and B in the universe of discourse X, the UNION of these two fuzzy sets represented by A ∪ B with membership values $\mu_A(x)$ and $\mu_B(x)$ is defined below:

$$\mu_{A \cup B}(x) = \max[\mu_A(x),\ \mu_B(x)] = \mu_A(x)\ \text{v}\ \mu_B(x) \tag{9.2}$$

In Eq. 9.2, v is the max or OR operator and is equivalent to Boolean OR logic A OR B.

Given two fuzzy sets A and B in the universe of discourse X, the INTERSECTION of these two fuzzy sets represented by A ∩ B with membership values $\mu_A(x)$ and $\mu_B(x)$ is defined below:

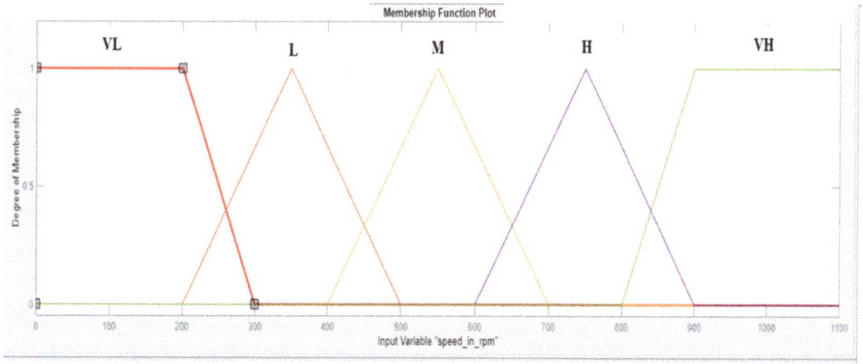

Fig. 9.1 Membership function of motor speed

$$\mu_{A\cap B}(x) = \min[\mu_A(x),\ \mu_B(x)] = \mu_A(x)^\wedge\mu_B(x) \tag{9.3}$$

In Eq. 9.3, ^ is the min or AND operator and is equivalent to Boolean AND logic A AND B.

Given a fuzzy set A in the universe of discourse X, the COMPLEMENT or NEGATION of this fuzzy set is defined below:

$$\mu_{\overline{A}}(x) = [1 - \mu_A(x)] \tag{9.4}$$

In Eq. 9.4 "–" above A is the NOT operator and is equivalent to Boolean NOT logic.

Thus if $\mu_A(x)$ is 0.8 and $\mu_B(x)$ is 0.5, then $\mu_{A\cup B}(x)$ is 0.8, $\mu_{A\cap B}(x)$ is 0.5, and $\mu_{\bar{A}}(x)$ is 0.2. All the rules of Boolean logic apply to FL as well. DeMorgan's theorems applied to FL are given below:

$$\overline{A \cup B} = \overline{A} \cap \overline{B} \tag{9.5}$$

$$\overline{A \cap B} = \overline{A} \cup \overline{B} \tag{9.6}$$

In Fig. 9.1 trapezoidal and triangular MFs are only shown. Other MFs such as Gaussian, two-sided Gaussian, bell, sigmoid-left, sigmoid-right, polynomial-Z, polynomial-Pi, polynomial-S, etc. exist [4].

9.2.2 *Fuzzy Inference System*

In the following sections, FL applications relating to control and modelling of power electronic converter-fed AC drives are presented.

A control algorithm for any given plant that is based on fuzzy logic is called fuzzy control. A fuzzy control is essentially constituted by intuitive thinking of a human operator. The conventional control such as PID control is normally based on mathematical model of a plant. If an accurate mathematical model of a plant is available in the form of transfer function, then the PID controller can be designed using step response or frequency response algorithms. With ill-defined model or models with parameter variations, multivariable, complex and non-linear such as dq0 model of three-phase AC motors, various adaptive control strategies are available. Fuzzy logic control (FLC) is found to be adaptive in nature and better than the conventional PI control from the point of view of output response overshoot and rise time to reach steady state value. FLC finds applications in the control of power electronic converter-fed electric drives.

Fuzzy control is described by a set of IF-THEN rules called implication, an example of which is given below:

IF x is A AND y is B, THEN z is C.

where x, y and z are the fuzzy variables and A, B and C are the fuzzy subsets in the universe of discourses X, Y and Z, respectively. Fuzzy inference system (FIS) is basically mapping from fuzzy input variables to the fuzzy output variable using FL rules. Thus fuzzy inference procedure consists of the following:

1. Fuzzification of input variables
2. Application of fuzzy operators such as AND, OR and NOT in the IF antecedent part of the rule
3. Implication from the antecedent part to the THEN consequent part of the rule
4. Aggregation of consequent part of all the rules
5. Defuzzification which is finding the crisp value from the fuzzy output set

Consider, for example, designing a speed controller for an electric motor using FL. The fuzzy input variables are the speed error "err" or "e" which is (ωref - ω) where ωref and ω are the reference and actual speed, respectively, and the rate of change of speed error also known as change in error "cherr" or "cie" which is (e(k) – e(k – 1)) for a constant sampling time where k and (k – 1) represent the present and previous time instants. The fuzzy rules are first formulated for the desired output response of the motor using fuzzy linguistic variables and fuzzy sets. For the purpose of understanding the design of the controller with FL, two such rules are given below:

1. IF the error e is positive medium (PM) AND change in error cie is negative small (NS), THEN control signal output cs is positive small (PS).
2. IF error e is negative small (NS) AND change in error cie is negative medium (NM), THEN control signal output cs is negative big (NB).

Rule 1 indicates that e is PM which corresponds to speed ω less than ωref and cie is NS which corresponds to e(k) less than e(k – 1). The error e in the present instant is less than that of the previous instant which implies that ω is less than ωref and is approaching towards ωref. A positive small (PS) control signal cs can increase ω closer to ωref making speed error e to zero.

Rule 2 indicates that e is NS which corresponds to speed ω greater than ωref and cie is NM which corresponds to e(k) far less than e(k – 1) in the negative direction. The error e in the present instant is far less than that of the previous instant in the negative direction which implies that ω is still greater than ωref. A negative big (NB) control signal cs can reduce ω closer to ωref making speed error e to zero.

Using the heuristic thinking as shown above for rules 1 and 2, motor speed controller can be designed using FL. Table 9.1 shows the fuzzy rules table for the motor speed controller [3–7].

The interpretation of rules 1 and 2 is shown in Fig. 9.2. If rule 1 is fired, the degree of membership for e and cie is μ1 and μ2, respectively. For the fuzzy AND operation, the minimum of the two membership values μ_1 is considered and linked to output cs, the THEN consequent part. For rule 1, output cs is positive small (PS) with membership value or height μ_1 and is shown by the shaded region under PS. If rule 2 is fired, the degree of membership for e and cie is μ_3 and μ_4, respectively. For the fuzzy

Table 9.1 Fuzzy rule base for motor speed controller

Change in error – cie ↓	Error – e →						
	PB	PM	PS	ZE	NS	NM	NB
NB	ZE	NS	NM	NB	NB	NB	NB
NM	PS	ZE	NS	NM	NB	NB	NB
NS	PM	PS	ZE	NS	NM	NB	NB
ZE	PB	PM	PS	ZE	NS	NM	NB
PS	PB	PB	PM	PS	ZE	NS	NM
PM	PB	PB	PB	PM	PS	ZE	NS
PB	PB	PB	PB	PB	PM	PS	ZE

AND operation, the minimum of the two membership values μ_4 is considered and linked to output cs, the THEN consequent part. For rule 2, output cs is negative big (NB) with membership value or height μ_4 and is shown by the shaded region under NB. Referring to Table 9.1, there are 49 rules and all the rules will have to be fired to defuzzify and to obtain a crisp output value. In general if the fuzzy input variables each have n1 and n2 fuzzy sets, then the output variable will have (n1 × n2) fuzzy sets. The individual rules are combined using UNION operator to get the overall rule R given below:

$$R = R1 \text{ U } R2 \text{ U } R3 \text{ U } R4 \ldots\ldots\ldots \text{U } Rn \tag{9.7}$$

All the rules are fired to obtain a single crisp output control action. This composition is known as MAX-MIN or SUP-MIN method [3–5] which is shown in Fig. 9.2 for the above two rules. The output membership function of each rule is given by MIN or AND operator, whereas the combined fuzzy output is given by the SUP or MAX or OR operator.

Thus considering rules 1 and 2 above, referring to Fig. 9.2, the overall rule can be stated as follows:

$$\min(\mu_1(e), \mu_2(\text{cie})) = \mu_1(e) \wedge \mu_2(\text{cie}) = \mu_1(e) \tag{9.8}$$

$$\min(\mu_3(e), \mu_4(\text{cie})) = \mu_3(e) \wedge \mu_4(\text{cie}) = \mu_4(\text{cie}) \tag{9.9}$$

$$\mu(cs) = \max(\mu_1(cs), \mu_4(cs)) = \mu_1(cs) \vee \mu_4(cs) = \mu_1(cs) \tag{9.10}$$

The values μ_1(e) and μ_4(cie) are linked to output cs membership function. Thus with n rules, the above procedures for e and cie have to be repeated to find the overall membership value for the output variable cs. The above implication method was proposed by Mamdani and hence known as Mamdani method.

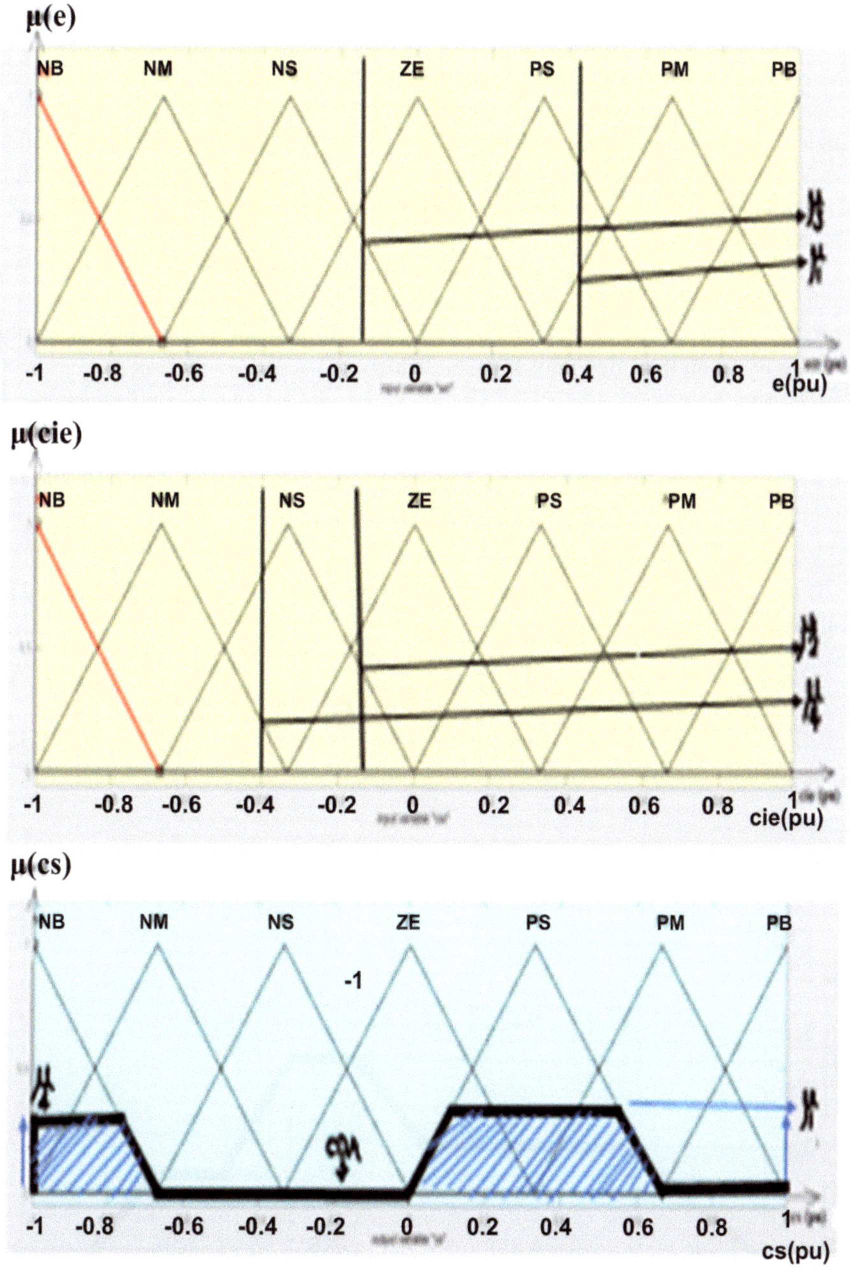

Fig. 9.2 Fuzzy membership functions for speed controller

9.2.3 Defuzzification

The result of implication and aggregation of all the fuzzy outputs using the max or union operator for all the rules fired and the conversion of this fuzzy output to a single crisp output is known as defuzzification. The various defuzzification methods are (1) centre of area (COA) also known as centre of gravity or centroid method, (2) height method and (3) mean of maximum (MOM) method [3–5]. These are presented below:

Consider the fuzzy output Z of a fuzzy system shown in Fig. 9.3. In the COA method, the crisp output Z_O of Z is the geometric centre of the area covered by μ(Z). The $\mu(Z)$ value is formed by union or maximum of the membership values contributed by firing all the rules. This is given below:

$$Z_O = \frac{\int Z * \mu(Z)}{\int \mu(Z) * dZ} \tag{9.11}$$

If the universe of discourse is taken as discrete, Eq. 9.11 can also be expressed as follows:

$$Z_O = \frac{\sum_{i=1}^{n} Z_i * \mu(Z_i)}{\sum_{i=1}^{n} \mu(Z_i)} \tag{9.12}$$

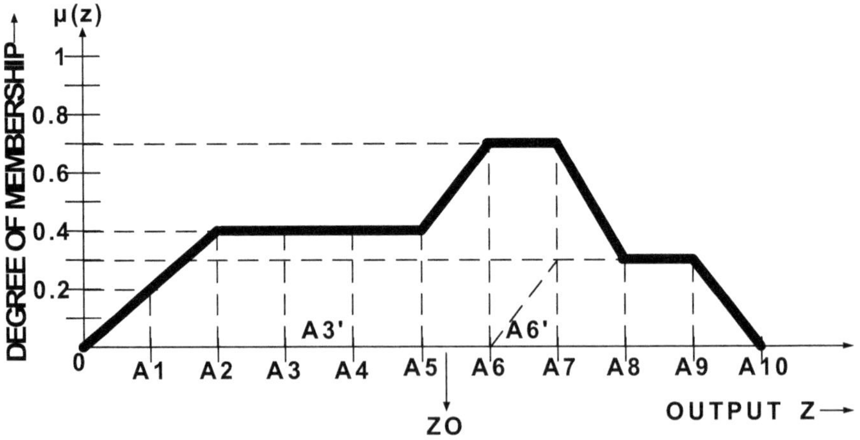

Fig. 9.3 Defuzzification of fuzzy output

Referring to Fig. 9.3, crisp output Zo using COA method can be expressed as follows:

$$Zo = \frac{[0.2 * A1 + 0.4 * (A2 + A3 + A4 + A5) + 0.7 * (A6 + A7) + 0.3 * (A8 + A9)]}{(0.2 + 0.4 + 0.4 + 0.4 + 0.4 + 0.7 + 0.7 + 0.3 + 0.3)} \tag{9.13}$$

The COA calculated using this method is shown in Figs. 9.2 and 9.3.

In the height method of defuzzification, only the height of contributing MFs at their base midpoint is taken into consideration. This is given below:

Referring to Fig. 9.3, the midpoint from A2 to A5 is A3′ and that of A6 to A7 is A6′.

$$A3' = \frac{(A3 + A4)}{2} \tag{9.14}$$

$$A6' = \frac{(A6 + A7)}{2} \tag{9.15}$$

$$Zo = \frac{(A3' * 0.4 + A6' * 0.7 + A8 * 0.3)}{(0.4 + 0.7 + 0.3)} \tag{9.16}$$

In the MOM method, only the highest of the contributing MFs are considered. If M such highest or maximum degree of membership contributing elements are present in the output $\mu(Z)$, then Z_o can be expressed as follows:

$$Zo = \frac{\sum_{i=1}^{m} Z_m}{M} \tag{9.17}$$

where Z_m is the mth element in the universe of discourse whose output degree of membership is the highest. Referring to Fig. 9.3, Z_O can be expressed as follows:

$$Zo = \frac{(A6 + A7)}{2} \tag{9.18}$$

The general structure of a fuzzy control system for motor speed control is shown in Fig. 9.4. Here the speed error e ($\omega_{ref} - \omega$) and the rate of change of speed error or change in speed error cie (de/dt) are given as input to the fuzzification block. The output of the fuzzification block is the control signal cs. Here fuzzification is carried out using the two fuzzy input variables as per IF e is X AND cie is Y, THEN cs is Z rule. This is then linked to output fuzzy variable cs in the fuzzy inference system

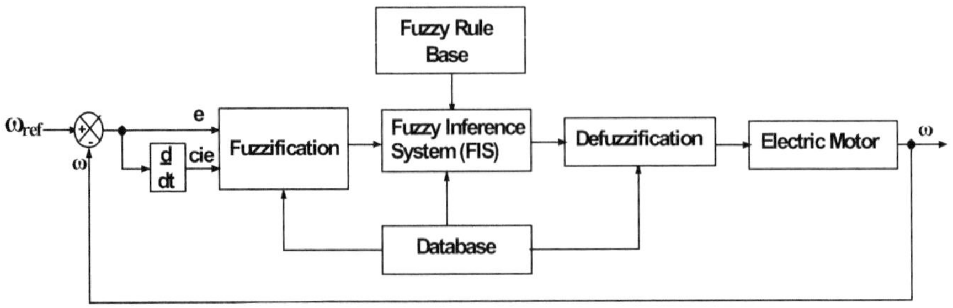

Fig. 9.4 Motor speed control system using fuzzy logic

(FIS) where firing all the rules, the union or maximum of the output MF is determined. This is given to defuzzification block which using one of the various methods such as COA, height or MOM determines the crisp output. This crisp output of defuzzifier controls the speed of motor. The e and cie values can be actual or can be in per unit, obtained by dividing the actual values by a suitable base value. The output cs of defuzzifier will also be in per unit. The blocks comprising fuzzification, FIS, defuzzification, fuzzy rule base and database all put together are known as fuzzy logic controller (FLC).

9.3 Fuzzy Logic Control of Three-Phase Inverter-Fed Induction Motor Drive

In this section fuzzy logic-based closed loop *V/f* control of three-phase IM fed by six-step 180-degree mode inverter and that by sine PWM inverter are present. The mathematical analysis relating to *V/f* control of three-phase IM is already presented in Sect. 8.12 of Chap. 8. The block diagram relating to closed loop *V/f* control of this IM using PI controller is presented in Sect. 8.12.2. The block diagram of FLC-based closed loop *V/f* control of this IM is shown in Fig. 9.5a, b. In both cases, the rotor reference speed ω_{ref} and the actual speed of motor ω are both multiplied by $(1/\omega_{base})$ to bring them to per unit values and are compared in an error detector, and the per unit values of this speed error e and rate of change of this error de/dt represented as cie are given as input to the FLC. The per unit crisp output cs of this FLC is added with the per unit actual motor speed ω, and the resulting output is multiplied by ω_{base} to get the actual value of the synchronous speed ω_s which is then multiplied by (P/4.π) to obtain reference frequency F* in hertz. The remaining parts of the block diagrams are the same as presented in Sect. 8.12.2.

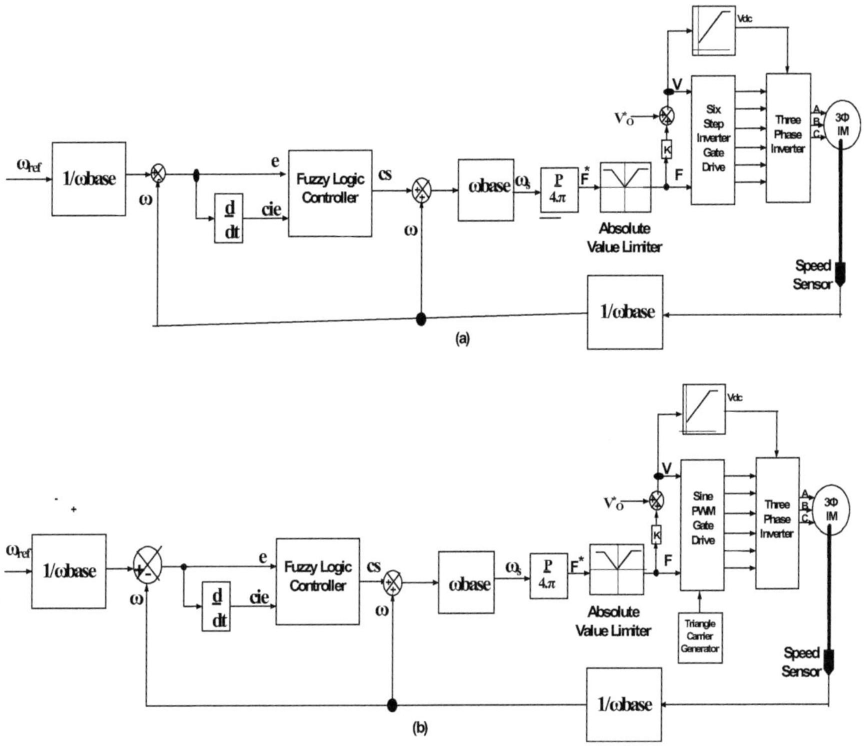

Fig. 9.5 Closed loop FLC-based *V/f* control of three-phase IM drive: (**a**) using six-step inverter and (**b**) using sine PWM inverter

9.3.1 Models of Voltage to Frequency (V/f) Ratio Control of Three-Phase Induction Motor Drive Using Fuzzy Logic Controller

In this section, models for the closed loop *V/f* control of three-phase IM drive using FLC are presented (a) for six-step 180-degree mode inverter operation and (b) for sine PWM inverter operation. The data shown in Table 8.1 are used to develop the model for IM. For sine PWM inverter, a triangle carrier frequency of 1200 Hz is used.

The model of the closed loop *V/f* control of three-phase IM drive fed by six-step 180-degree mode inverter using FLC is shown in Fig. 9.6 (model file: EXAMPLE9_1). The model is developed as per block diagram shown in Fig. 9.5a. Here the three-phase Universal Bridge model for inverter and cage IM are from Simscape Specialized Power Systems block set [2]. Here the rated no-load speed is used as the reference and is compared with the actual rotor speed in an error detector block, after converting both speeds to per unit by division with synchronous speed in radians per second. The per unit speed output of the error detector e and the rate of change of this

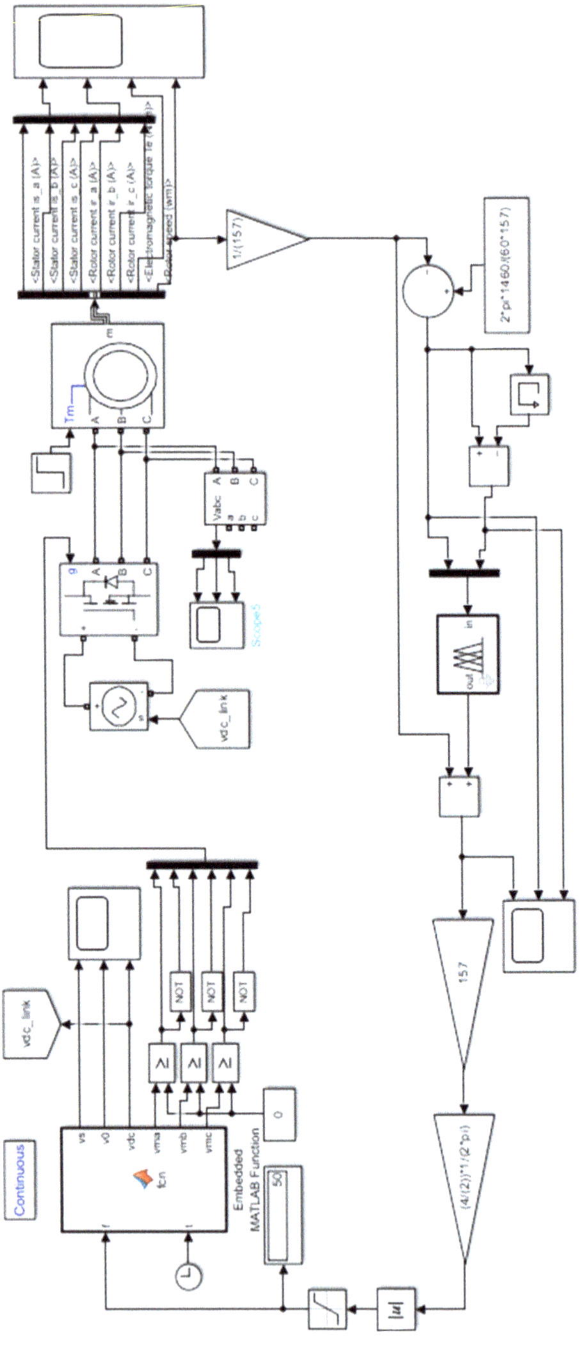

Fig. 9.6 Model of voltage to frequency ratio (*V*/*f*) control of three-phase inverter-fed induction motor drive using FLC

speed error cie are given as inputs to a two-input mux. The value of cie is obtained using the relation (e(k) – e(k − 1)) for a given sampling time where k and (k − 1) are the present and previous time instants. In this model cie value for (k − 1)th instant is obtained by using memory block. The output of mux is given to fuzzy logic controller (FLC) block. The FLC block fuzzifies the two inputs e and cie in the fuzzy inference system (FIS) using rule base and then defuzzifies to obtain a crisp value cs for the control output. The rule base is entered in the FLC block as per Table 9.1. The MFs for e, cie and cs are created using seven linguistic variables: negative big (NB), negative medium (NM), negative small (NS), zero (ZE), positive small (PS), positive medium (PM) and positive big (PB). The centroid method is used for defuzzification. The per unit output cs of FLC is added with per unit rotor speed and then multiplied by synchronous speed using gain multiplier, and this actual value is then multiplied with (P/4π) to obtain reference frequency F*. This F* is given to absolute value block to obtain only positive values and then given to a saturation block which limits the frequency within the allowable range for the motor. The frequency output of saturation block and time module are given as input to the Embedded MATLAB function. Program segment 9.1 gives the method for obtaining the line to neutral voltage vs compensation voltage vo for stator resistance drop and the DC link voltage vdc for the six-step inverter. In addition three-phase sine wave modulating signals vma, vmb and vmc are also created using a peak value (vs/vbase). The base voltage vbase is computed by adding the given rated line to neutral voltage of the motor with vo. These modulating signals vma, vmb and vmc are then compared with zero in three relational operator or comparator blocks, each giving logic 1 output when respective modulating signals are greater than or equal to zero and logic 0 output when they are less than zero. Each output of the comparator blocks is then inverted using NOT gate, and the six gate pulse outputs are given to the inverter gate terminal through a mux block. The details regarding FLC block are presented below.

Figure 9.7a shows the FIS file editor where the file name of FIS which in this case is “imcontroller”, number of samples for output discretisation which is 101 and simulation method which is “interpreted execution” are entered [2]. The seven membership functions for error e, change in error cie and output cs are shown in Fig. 9.7b–d, respectively. The required number and shape of MFs are selected by using drop-down menu. Figure 9.7e is the fuzzy rule base editor where referring to Table 9.1, the 49 rules are entered. Here Mamdani method of fuzzification is selected. For example, referring to first row and first column of Table 9.1, the rule can be entered in the rule base editor as follows: “IF e is PB and cie is NB THEN cs is ZE”. The FLC front panel is shown in Fig. 9.7f. In this front panel, AND method “min”, OR method “max”, implication method “min”, aggregation method “max” and defuzzification method “centroid” are selected using the drop-down menu [2].

The model of the closed loop *V*/*f* control of three-phase IM drive fed by sine PWM inverter using FLC is shown in Fig. 9.8 (model file: EXAMPLE9_2). The model is developed as per block diagram shown in Fig. 9.5b. The model development is the same as for six-step 180-degree mode inverter-fed IM drive using FLC explained above. Here the only difference is that the three-phase modulating signals

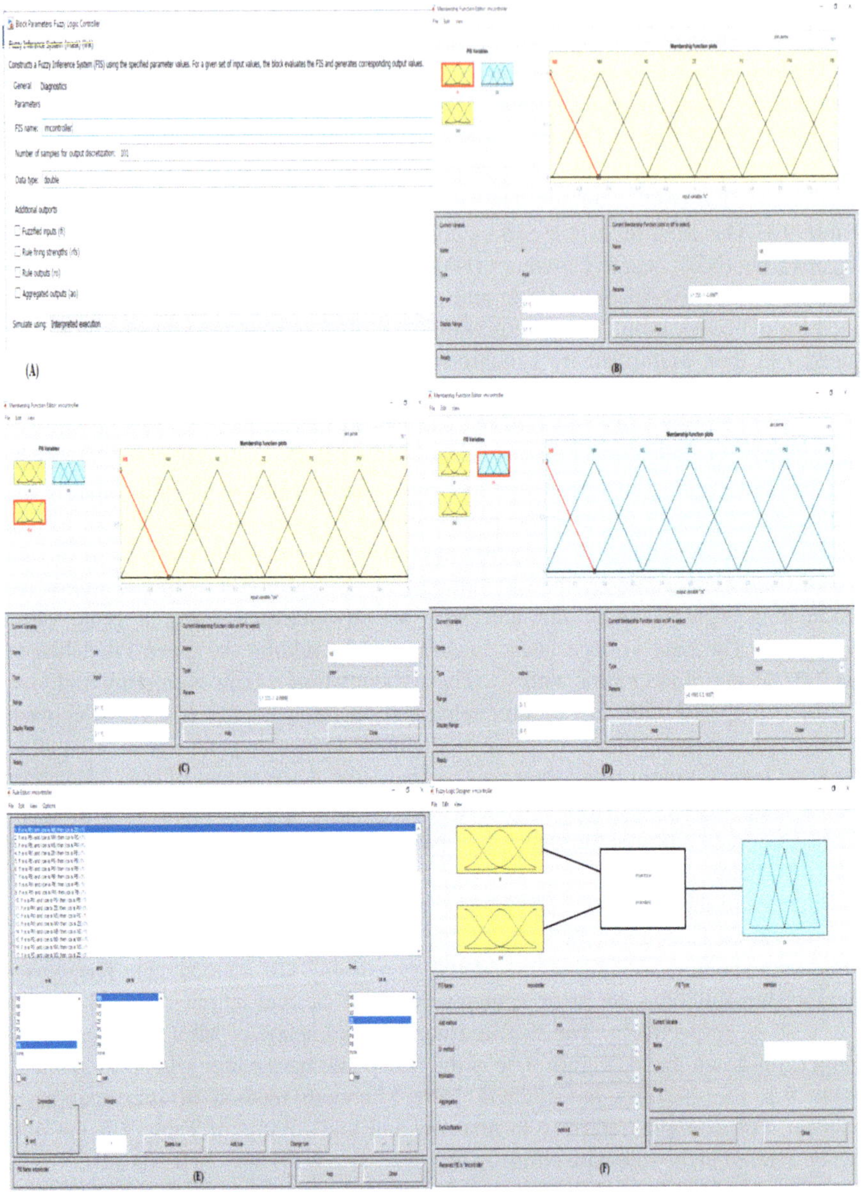

Fig. 9.7 Fuzzy logic controller units: (**a**) FIS file editor, (**b**) MFs for error e, (**c**) MFs for change in error cie, (**d**) MFs for output cs, (**e**) fuzzy rule base editor, (**f**) FLC front panel

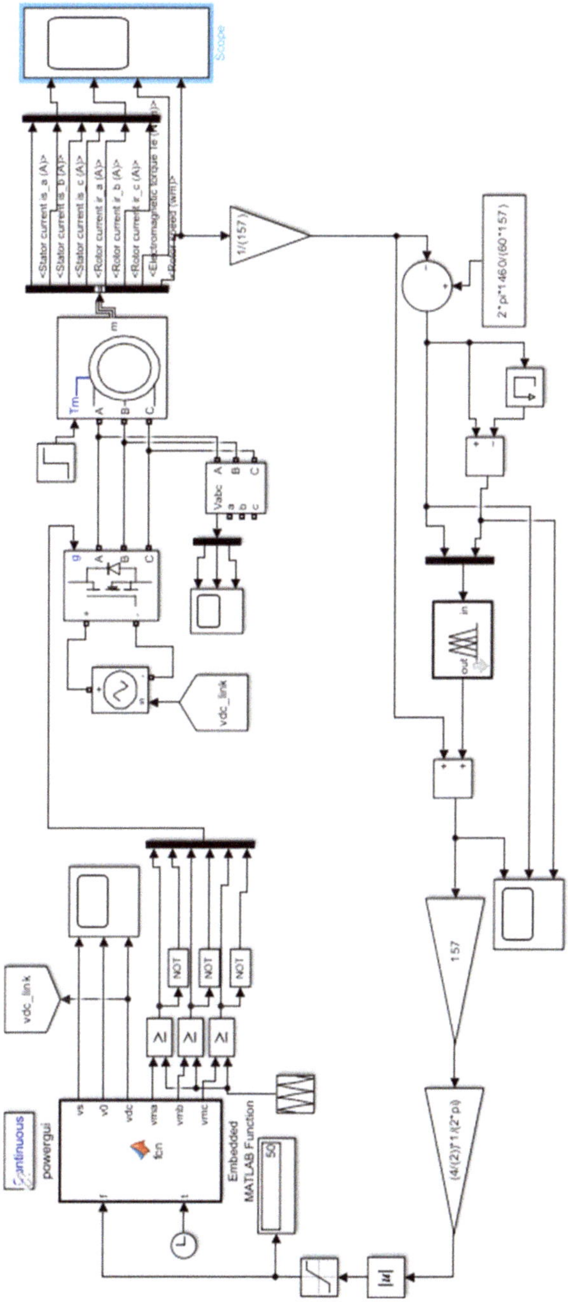

Fig. 9.8 Model of voltage to frequency ratio (*V*/*f*) control of three-phase sine PWM inverter-fed induction motor drive using FLC

vam, vbm and vcm are compared with the triangle carrier (±1 V) in three relational operator comparator blocks. The resulting output of comparator blocks and their respective inverted output using NOT gates are given to the gate input of inverter through a mux. The FIS file editor, fuzzification, rule base, defuzzification and front panel settings are the same as explained above.

9.3.2 Simulation Results

The simulation of the model for *V*/*f* controlled six-step inverter-fed IM drive using FLC shown in Fig. 9.6 is carried out using ode23tb (stiff/TR-BDF2) solver in Simulink [2]. An initial load torque of 50 Nw-metres is applied for the first 0.25 s and this load torque is stepped up to 150 Nw-metres above 0.25 s. The reference speed is set to 1460 rpm. The simulation results for the inverter line to line voltage, stator and rotor currents, electromagnetic torque and rotor speed are shown in Fig. 9.9a, b, and the expanded view of the rotor currents is shown in Fig. 9.9c, respectively.

The simulation of the model for *V*/*f* controlled three-phase sine PWM inverter-fed IM drive using FLC shown in Fig. 9.8 is carried out using ode23tb (stiff/TR-BDF2) solver in Simulink [2]. An initial load torque of 50 Nw-metres is applied for the first 0.25 s and this load torque is stepped up to 150 Nw-metres above 0.25 s. The reference speed is 1460 rpm. The simulation results for the inverter line to line voltage, stator and rotor currents, e.m. torque and rotor speed are shown in Fig. 9.10a, b, and the expanded view of rotor currents is shown in Fig. 9.10c, respectively.

9.3.3 Discussion of Results

In the case of closed loop *V*/*f* controlled six-step inverter-fed IM drive, from Fig. 9.9a it is seen that the RMS line to line voltage of the inverter is 407.7 V very slightly higher than the rated voltage of motor and frequency is 50 Hz. Figure 9.9b shows that the initial rotor speed is 155.0 and falls to 151.2 mech.radians per second after 0.25 s corresponding to initial load of 50 Nw-metres and final load of 150 Nw-metres. Figure 9.9c indicates that the rotor currents have a frequency of 1.855 Hz for a rotor speed of 151.2 mech.rad per second. Using the relation for rotor slip frequency, this value is 1.847 Hz which is close to the value obtained by simulation.

In the case of closed loop *V*/*f* controlled three-phase sine PWM inverter-fed IM drive, from Fig. 9.10a it is seen that the RMS line to line voltage of the inverter is 377.7 V which is 22.3 V less than the rated voltage of motor and frequency is 50 Hz. Figure 9.10b shows that the initial rotor speed is 153.9 and falls to 146.6 mech. radians per second after 0.25 s corresponding to initial load of 50 Nw-metres and final load of 150 Nw-metres. Figure 9.10c indicates that the rotor currents have a

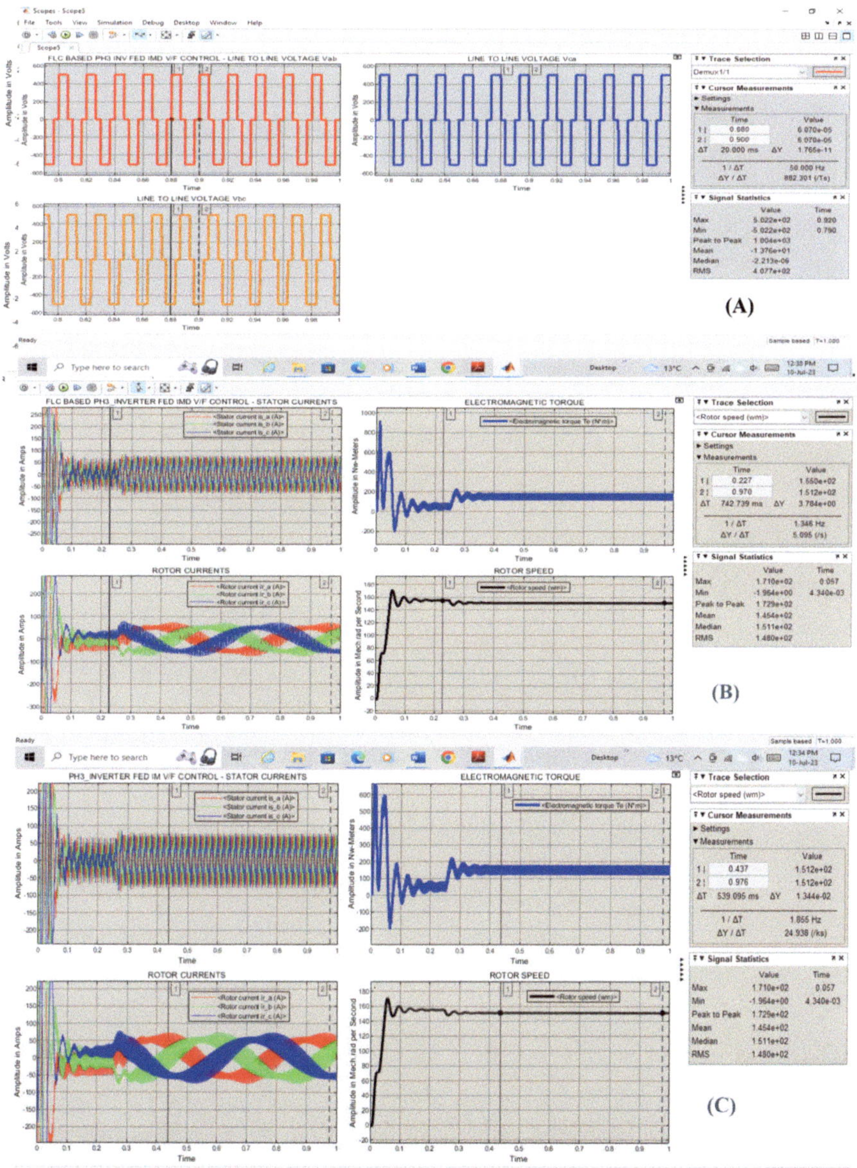

Fig. 9.9 FLC-based three-phase inverter-fed IMD *V/f* control simulation results: (**a**) line voltage Vab, Vbc (left column, top to bottom), Vca (right column), (**b**) stator currents, rotor currents (left column, top to bottom), E.M. torque and rotor speed (right column, top to bottom), (**c**) expanded view of rotor currents (left column, bottom)

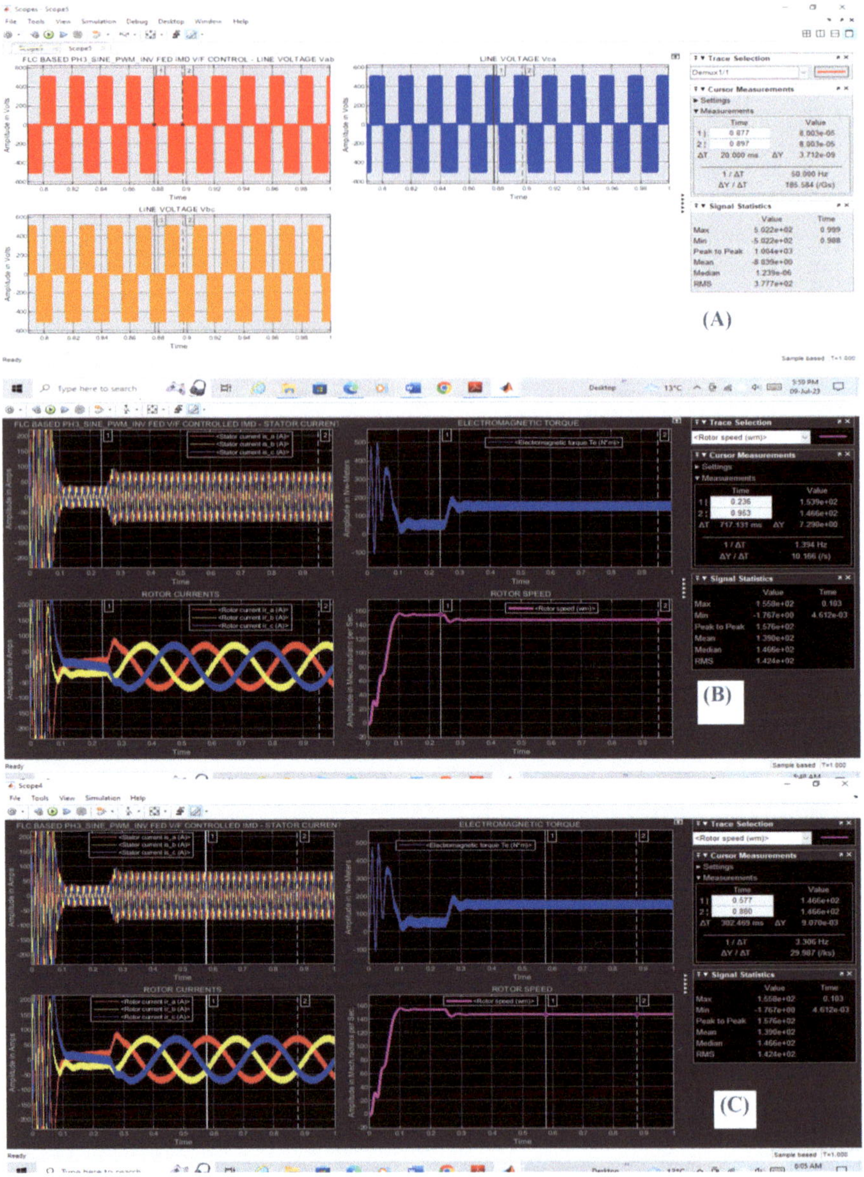

Fig. 9.10 FLC-based three-phase sine PWM inverter-fed IMD *V*/*f* control simulation results: (**a**) line voltage Vab, Vbc (left column, top to bottom), Vca (right column); (**b**) stator currents, rotor currents (left column, top to bottom), E.M. torque and rotor speed (right column, top to bottom); and (**c**) expanded view of rotor currents (left column, bottom)

frequency of 3.306 Hz for a rotor speed of 146.6 mech.rad per second. Using the relation for rotor slip frequency, this value is 3.312 Hz which is close to the value obtained by simulation. Due to pulse width modulation, RMS line to line voltage input to the IM is reduced from its rated value. This results in a lower power output of motor. Thus with the same load torque, rotor speed has reduced from that of six-step inverter-fed IM drive.

Comparing the simulation results for FLC-based six-step inverter-fed IM drive in Fig. 9.9b with that of PI controlled inverter shown in Fig. 8.71b in Chap. 8, the rotor takes only 0.15 s to reach steady speed with FLC, whereas it takes 3.15 s to reach steady speed after a series of oscillations with PI controller. Similarly comparing the simulation results for FLC-based three-phase sine PWM inverter-fed IM drive in Fig. 9.10b with that of PI controlled inverter shown in Fig. 8.73b, the rotor takes only 0.125 s to reach steady speed with FLC, whereas it takes around 7.5 s to reach steady speed after a series of oscillations with PI controller. Also the load torque is much reduced with PI controller, whereas the load torque value is the same using FLC for the case with sine PWM inverter. Also the reduction in the output voltage of inverter with Sine PWM is small with FLC compared to that with PI controller.

9.4 Case Study: Three-Phase Thyristor Controller-Fed Induction Motor Drive Using Fuzzy Logic Controller

The block diagram of FLC-based closed loop speed control of the IM drive fed by three-phase thyristor (SCR) controller is shown in Fig. 9.11. The rotor reference speed ωref and the actual speed of motor ω are both multiplied by $(1/\omega_{\text{base}})$ to bring them to per unit values and are compared in an error detector, and the per unit values of this speed error e and rate of change of this error de/dt represented as cie are given

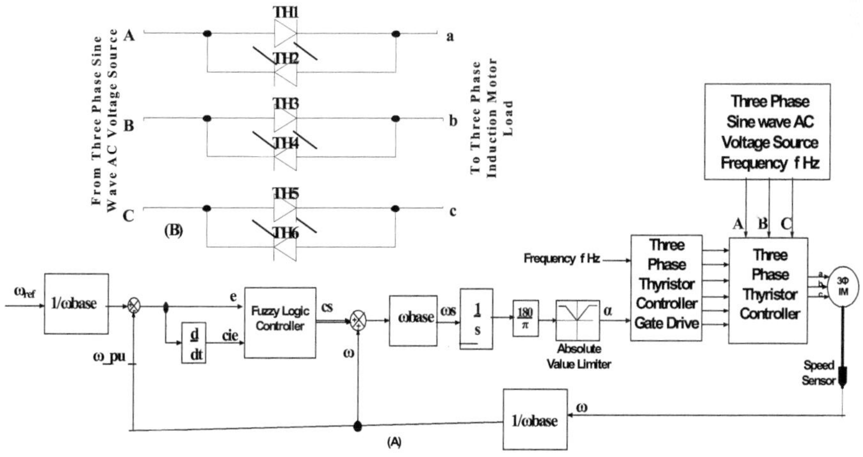

Fig. 9.11 (**a**) FLC-based closed loop control of three-phase induction motor drive using thyristor controller, (**b**) three-phase sine thyristor controller

as input to the FLC. The per unit crisp output cs of this FLC is added with the per unit actual motor speed ω, and the resulting output is multiplied by ω_{base} to get the actual value of the synchronous speed ω_s which is then integrated and multiplied by $(180/\pi)$ to obtain reference firing angle which is then given to an absolute value limiter which maintains this firing angle α so as to output rated stator voltage from SCR controller. The remaining parts of the block diagrams are the same as presented in Sect. 8.14 of Chap. 8.

9.4.1 Model of Three-Phase Thyristor Controller-Fed Induction Motor Drive Using Fuzzy Logic Controller

The model of the closed loop three-phase IM drive fed by SCR controller using FLC is shown in Fig. 9.12 (model file: CASE_STUDY_EX9_1). The model is developed as per block diagram shown in Fig. 9.11. Here the three-phase SCR controller model and cage IM are from Simscape Specialized Power Systems block set [2]. Here the rated no-load speed is used as the reference and is compared with the actual rotor speed in an error detector block, after converting both speeds to per unit by division with synchronous speed in radians per second. The per unit speed output of the error detector e and the rate of change of this speed error cie are given as inputs to a two-input mux. The value of cie is obtained using the relation (e(k) – e(k − 1)) for a given sampling time where k and (k − 1) are the present and previous time instants. In this model cie value for (k − 1)th instant is obtained by using memory block. The output of mux is given to FLC block. This FLC block fuzzifies the two inputs e and cie in the fuzzy inference system (FIS) using rule base and then defuzzifies to obtain a crisp value cs for the control output. The rule base is entered in the FLC block as per Table 9.1. The MFs for e, cie and cs are created using seven linguistic variables: negative big (NB), negative medium (NM), negative small (NS), zero (ZE), positive

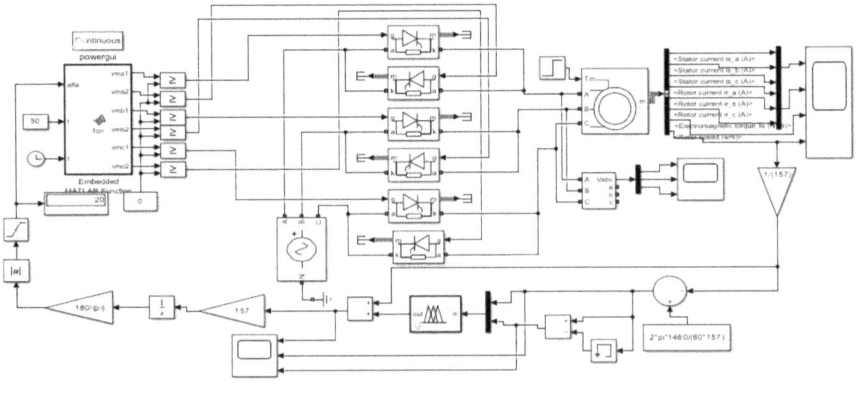

Fig. 9.12 Model of FLC-based three-phase thyristor controller-fed induction motor drive

small (PS), positive medium (PM) and positive big (PB). The centroid method is used for defuzzification. The per unit output cs of FLC is multiplied by synchronous speed using gain multiplier, and this actual value is then integrated and multiplied with (180/π) to obtain reference firing angle. This reference firing angle is given to absolute value block to obtain only positive values and then given to a saturation block which limits the firing angle α within the allowable range to output rated stator voltage from SCR controller. The firing angle output α of saturation block, rated frequency of IM and time module are given as input to the Embedded MATLAB function. Program segment 9.3 in the file "CASE_STUDY_EX9_1" gives the method for obtaining six modulating signals vma1, vma2, vmb1, vmb2, vmc1 and vmc2. These six modulating signals are then compared with zero in six separate relational operator or comparator blocks, each giving HIGH or logic 1 output when respective modulating signals are greater than or equal to zero and LOW or logic 0 output when they are less than zero. These six comparator outputs form the gate drive for the SCR controller. The details regarding FLC block are presented in Sect. 9.3.1.

9.4.2 Simulation Results

The simulation of the model for three-phase SCR controller-fed IM drive using FLC shown in Fig. 9.12 is carried out using ode23tb (stiff/TR-BDF2) solver in Simulink [2]. An initial load torque of 50 Nw-metres is applied for the first 0.5 s and this load torque is stepped up to 150 Nw-metres above 0.5 s. The reference speed is set to 1460 rpm. The simulation results for the stator and rotor currents, electromagnetic torque and rotor speed, line to line voltage and line to neutral voltage output of SCR controller are shown in Fig. 9.13a–c.

9.4.3 Discussion of Results

From the simulation results in Fig. 9.13a, it is seen that the rotor speed is initially 155.2 and this speed falls to 151.3 mech.rad per second for a load of 50 Nw-M for the first 0.5 s and 150 Nw-M above 0.5 s. The stator and rotor currents are well balanced. From Fig. 9.13b, c, the RMS value of line to line and line to neutral output voltage of SCR controller is 393.6 V and 228.8 V, respectively, and the frequency is 50 Hz. Comparing these results obtained with FLC with that obtained with PI controller shown in Fig. 8.77a of Chap. 8, it is seen that there is not much of a difference in the e.m. torque and rotor speed response and so also with the stator and rotor currents. Similarly comparing the line to line and line to neutral output voltage of SCR controller using FLC with that obtained using PI controller, the discrepancy is very small.

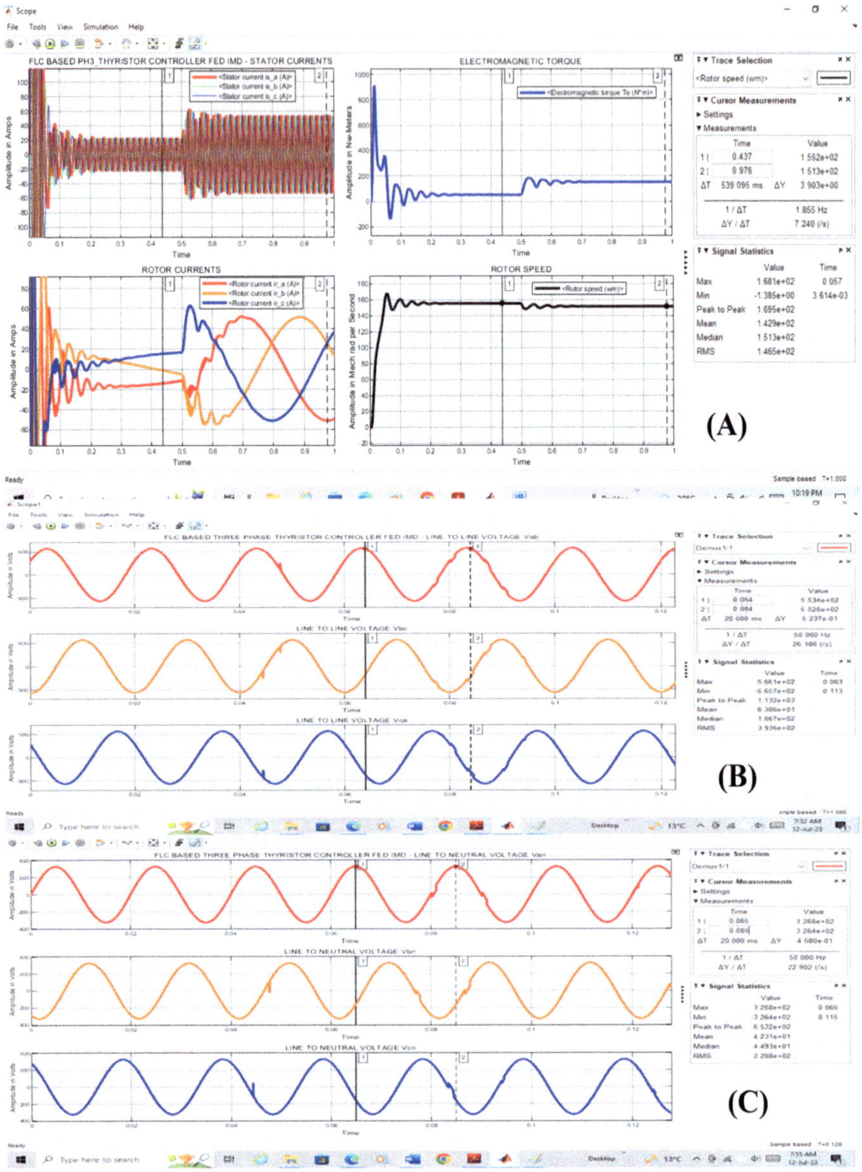

Fig. 9.13 Fuzzy logic-based speed control of three-phase IM drive fed by SCR controller simulation results: (**a**) stator and rotor currents (left column, top and bottom), E.M. torque and rotor speed (right column, top and bottom), (**b**) three-phase line to line output voltage and (**c**) three-phase line to neutral output voltage of SCR controller

9.5 Fuzzy Logic Controller-Based Direct Torque Control of Three-Phase Induction Motor Drive Using Modified Switching Table

The block diagram showing the direct torque control (DTC) of three-phase inverter-fed IM drive with FLC is shown in Fig. 9.14. Here FLC is used instead of PI controller. Here the rotor speed error e and the rate of change of this speed error cie are the fuzzy inputs to the FLC. These two inputs are fuzzified using rule base in the FIS and defuzzified to obtain a crisp output control signal cs which forms the e.m. torque reference Tem_ref. But for this difference, the model is the same as explained in Sect. 8.10 of Chap. 8.

9.5.1 Model of Fuzzy Logic Controller-Based Direct Torque Control of Three-Phase Induction Motor Drive

The model of FLC-based direct torque control of three-phase IM drive using modified switching table is shown in shown in Fig. 9.15 (model file: EXAMPLE9_3). The details regarding modified switching table and using this table to develop model for DTC of IM with PI controller are presented in Sects. 8.11 and 8.11.1 of Chap. 8. Here PI controller is replaced with FLC. The model parameters of the IM are from Table 8.1. Here the rotor reference speed and the actual speed are divided by synchronous speed to bring them to per unit and are inputs to the error detector. The rotor speed error e and rate of change of speed error cie (e(k) – e(k – 1)) are described by MFs with seven linguistic variables, negative big (NB), negative medium (NM), negative small (NS), zero (ZE), positive small (PS), positive medium (PM) and positive big (PB) and the output cs of FLC by nine linguistic variables: negative very big (NVB), negative big (NB), negative medium (NM), negative small (NS), zero (ZE), positive small (PS), positive medium (PM), positive big (PB) and positive very big (PVB). The 49 rules which form the rule base of FIS are developed

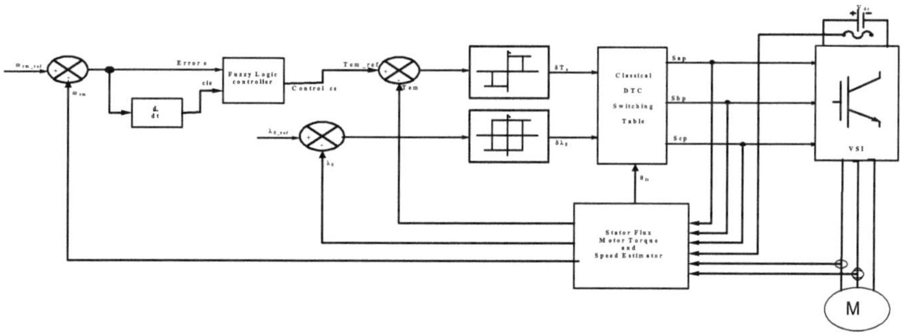

Fig. 9.14 Direct torque control scheme using fuzzy logic controller

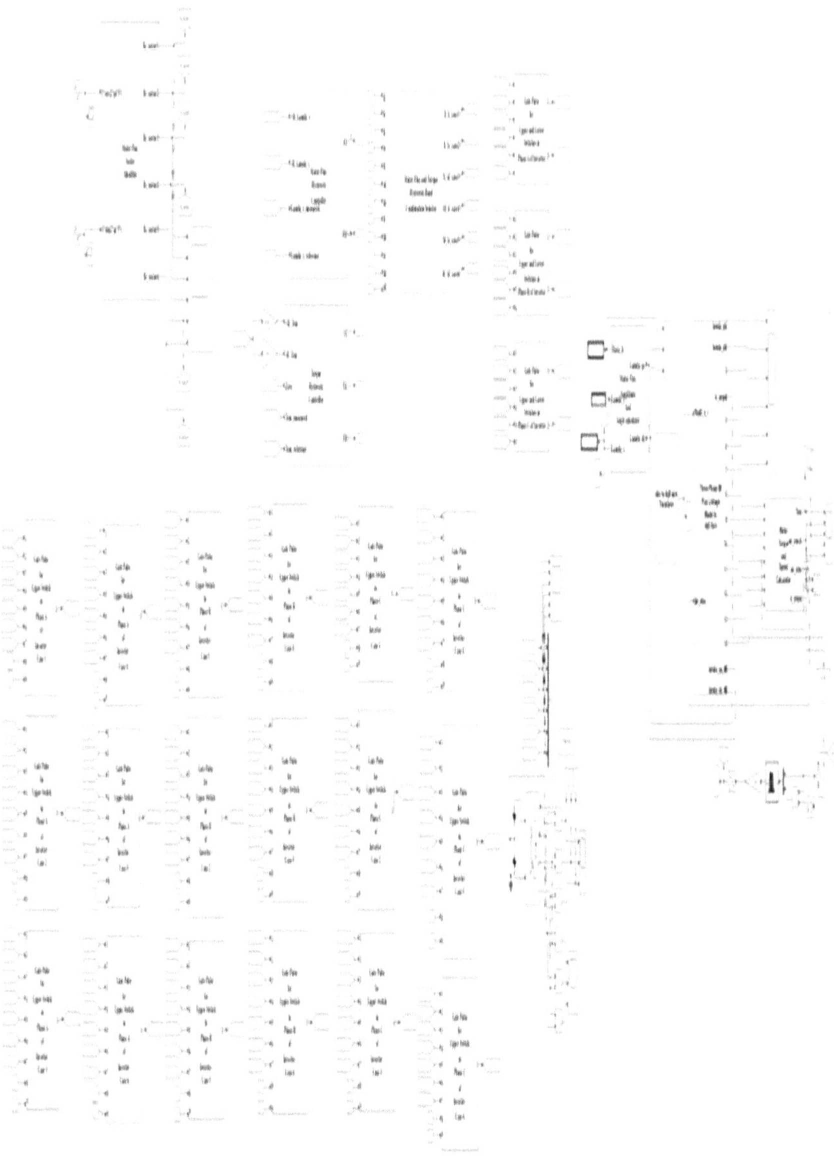

Fig. 9.15 Model of FLC-based DTC of three-phase induction motor drive using modified switching table

as per Table 9.2. The fuzzification of inputs e and cie is carried out in the FIS using Mamdani method. The defuzzification is done using COA method. The crisp per unit output cs of FLC is multiplied by synchronous speed and is given to the zero-order hold (ZOH) with sample time 10e-6 s. This output of ZOH forms the e.m. torque reference tem_ref. FLC block details are given below:

Table 9.2 Rule base for DTC of three-phase IM drive using speed control method

Change in error -	Error – e →						
cie ↓	PB	PM	PS	ZE	NS	NM	NB
NB	ZE	NS	NM	NB	NVB	NVB	NVB
NM	PS	ZE	NS	NM	NB	NVB	NVB
NS	PM	PS	ZE	NS	NM	NB	NVB
ZE	PB	PM	PS	ZE	NS	NM	NB
PS	PVB	PB	PM	PS	ZE	NS	NM
PM	PVB	PVB	PB	PM	PS	ZE	NS
PB	PVB	PVB	PVB	PB	PM	PS	ZE

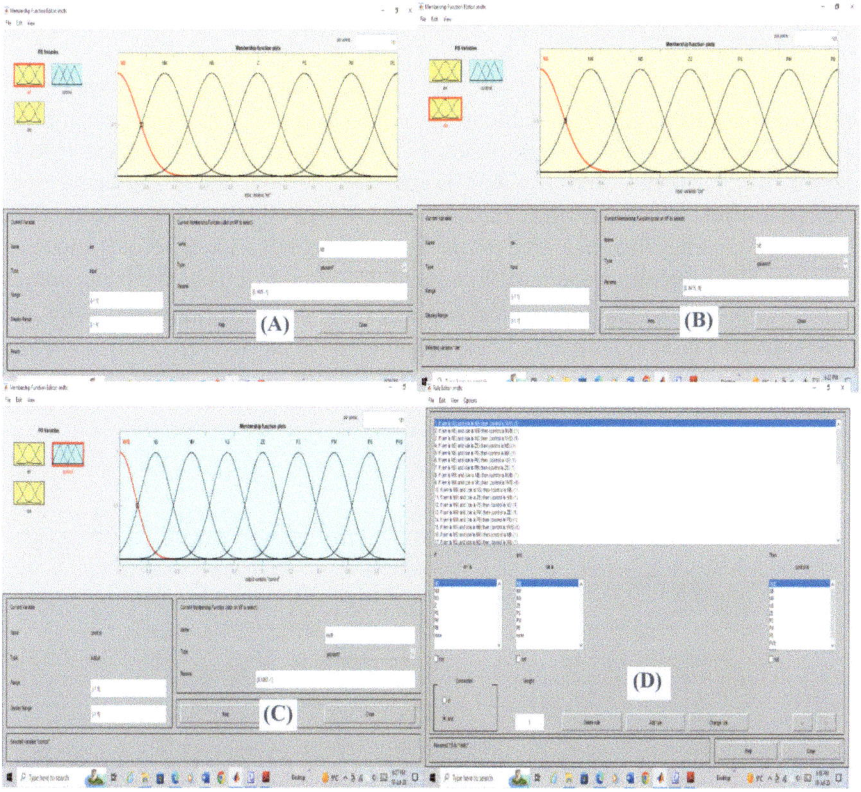

Fig. 9.16 Fuzzy logic controller units: Gaussian MFs for (**a**) error e, (**b**) change in error cie and (**c**) output cs, (**d**) fuzzy rule base editor

Figure 9.7a shows the FIS file editor with file name replaced with "imdtc", number of samples for output discretisation which is 101 and simulation method which is "interpreted execution" are entered [2]. The seven Gaussian MFs for error e, change in error cie and the nine MFs for that of output cs are shown in Fig. 9.16a–c, respectively. The required number and shape of MFs are selected by using drop-down menu. Figure 9.16d is the fuzzy rule base editor where referring to Table 9.2, the 49 rules are entered. Here Mamdani method of fuzzification is selected. For example, referring to first row and last column of Table 9.2, the rule can be entered in the rule base editor as follows: "IF e is NB and cie is NB THEN cs is NVB". The FLC front panel is the same as shown in Fig. 9.7f except here that the FIS file name is "imdtc". In this front panel, AND method "min", OR method "max", implication method "min", aggregation method "max" and defuzzification method "centroid" are selected using the drop-down menu [2].

9.5.2 Simulation Results

The simulation of the model for FLC-based DTC of three-phase inverter-fed IM drive using modified switching table is carried out using ode3 (Bogacki-Shampine) solver in Simulink [2]. A sample time of 10e-6 s is used. A mechanical load is applied using step function. The load is initially 100 Nw-metres and changes to 200 Nw-metres at 0.5 s. A DC link voltage of 490 volts is used. The inverter switching frequency is 50 Hz. The torque hysteresis band is 1% of torque reference and stator flux hysteresis band is 10% of stator flux reference. The simulation results for the motor torque, rotor speed, mechanical load applied and three-phase stator and rotor currents are shown in Fig. 9.17. The three-phase line to line stator applied voltage from output of inverter, the resultant stator flux λ_S and the motor reference torque tem_ref are shown in Fig. 9.18a–c.

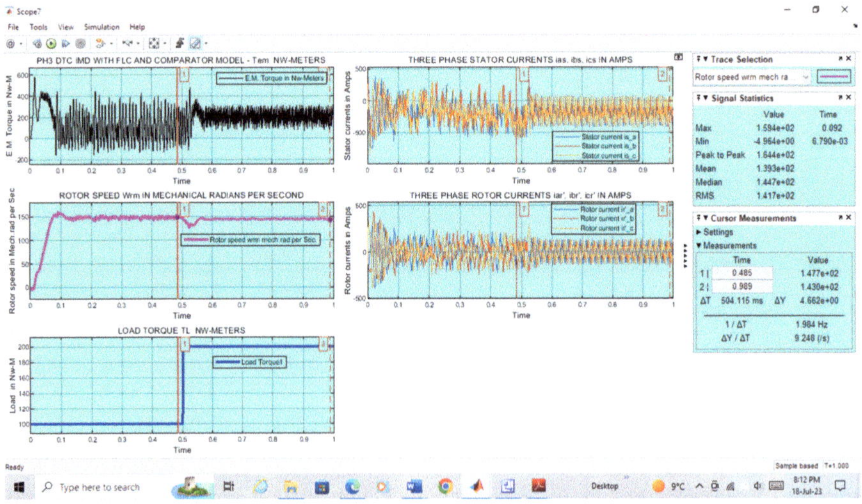

Fig. 9.17 PH3 DTC of IMD using FLC and comparator model simulation results: E.M. torque, rotor speed and load torque (left column, top to bottom), stator and rotor currents (right column, top to bottom)

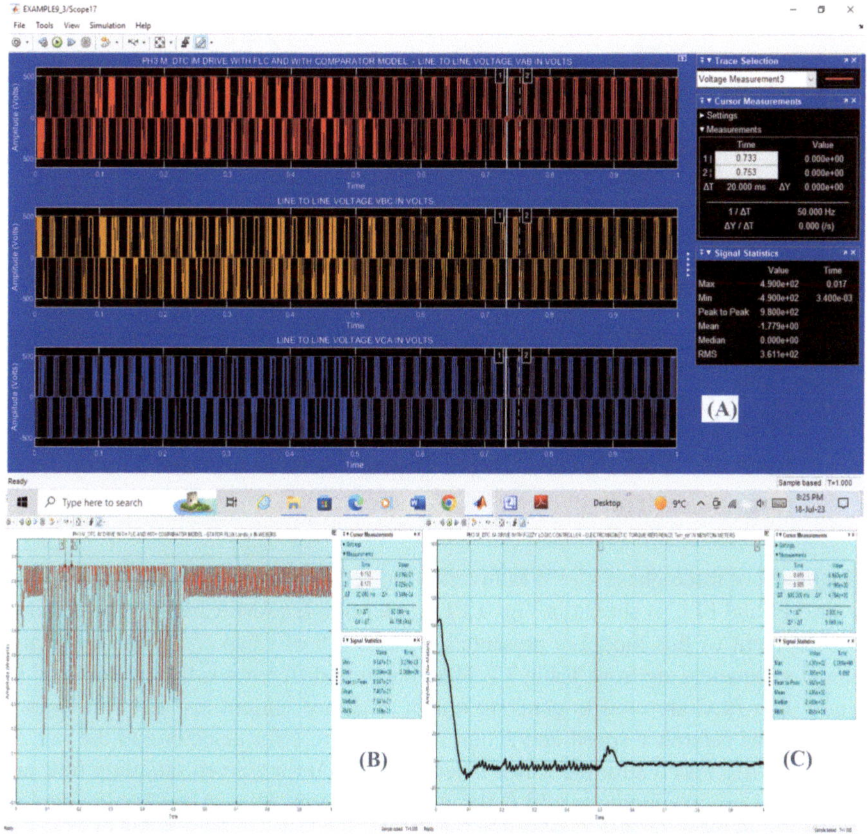

Fig. 9.18 PH3 DTC of IM drive with FLC simulation results: (**a**) three-phase line to line voltage of inverter (top to bottom), (**b**) stator flux linkage and (**c**) E.M. torque reference Tem_ref

9.5.3 Discussion of Results

Comparing the results presented in Fig. 9.17 with that of Fig. 8.54, it is seen that the starting torque ripples for the first 0.1 s are well reduced with FLC. The mean torque is 100 Nw-M within the first 0.5 s and 200 Nw-M for the time above 0.5 s. The rotor speed rise reaches steady state value within the first 0.125 s with FLC, whereas it takes 0.225 s with PI controller as seen from Fig. 8.54. The steady state rotor speed is 147.7 within the first 0.5 s and 143 mech.radians per second above 0.5 s which is closer to the set point speed. The line to line RMS output voltage of inverter is 361.1 volts with FLC. The stator flux mean value is 0.7467 webers. With FLC, the e.m. torque reference tem_ref reaches a peak value of 143.2 Nw-M and steady state values of −5.95 Nw-M before 0.5 s and −1.196 Nw-M after 0.5 s. The rotor currents are balanced, but the stator currents although balanced have DC offset, and this is due to deviation or inaccurate machine parameters.

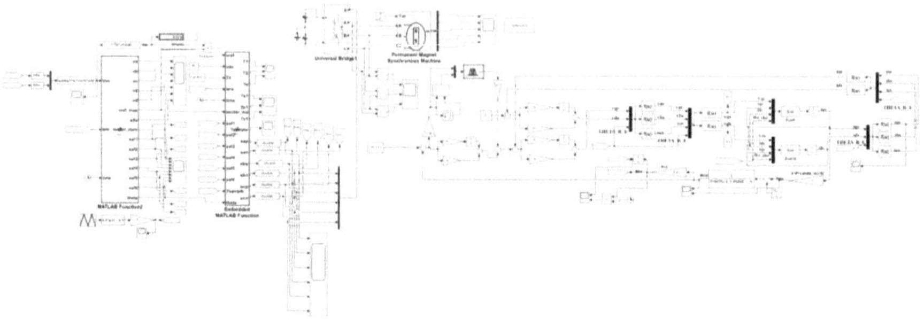

Fig. 9.19 Three-phase SVPWM inverter-fed vector controlled PMSM drive using hybrid PI and FLC

9.6 Model of Vector Control of Three-Phase SVPWM Inverter-Fed PMSM Drive Using Hybrid PI and Fuzzy Logic Controller

The model of three-phase SVPWM inverter-fed vector controlled PMSM drive using hybrid PI and FLC is shown in Fig. 9.19 (model file: EXAMPLE9_4). The detailed development of this model is presented in Sect. 7.5 of Chap. 7. The inverter and model parameters are from Table 7.2. Here the only difference is that a combined PI along with FLC are used to generate q-axis reference current iqs*. The difference between the reference speed and actual speed e from error detector and the rate of change of this speed error cie which is (e(k)-e(k − 1)) for a given sampling time are given as two fuzzy inputs to a FLC through a two-input Mux. After fuzzification in the FIS using rule base and then defuzzification, the crisp output cs of FLC along with that obtained from speed PI controller is added in a summing block. The output of this summing block forms the reference q-axis current iqs*.

Here seven linguistic variables, negative big (NB), negative medium (NM), negative small (NS), zero (ZE), positive small (PS), positive medium (PM) and positive big (PB), are used for error e and change in error cie. For the output cs, nine linguistic variables, negative very big (NVB), negative big (NB), negative medium (NM), negative small (NS), zero (ZE), positive small (PS), positive medium (PM) and positive big (PB) and positive very big (PVB), are used. The Gaussian MFs are used for e, cie and cs. The FIS file name "pmsmvectorcontrol" is used. The MFs for e, cie and cs and the rule base are the same as shown in Fig. 9.16. The rule base is developed as per Table 9.2. The centroid method is used for defuzzification.

9.6.1 Simulation Results

The simulation results of the model for three-phase SVPWM inverter-fed vector controlled PMSM drive using hybrid PI and FLC are shown in Fig. 9.20a, b. The

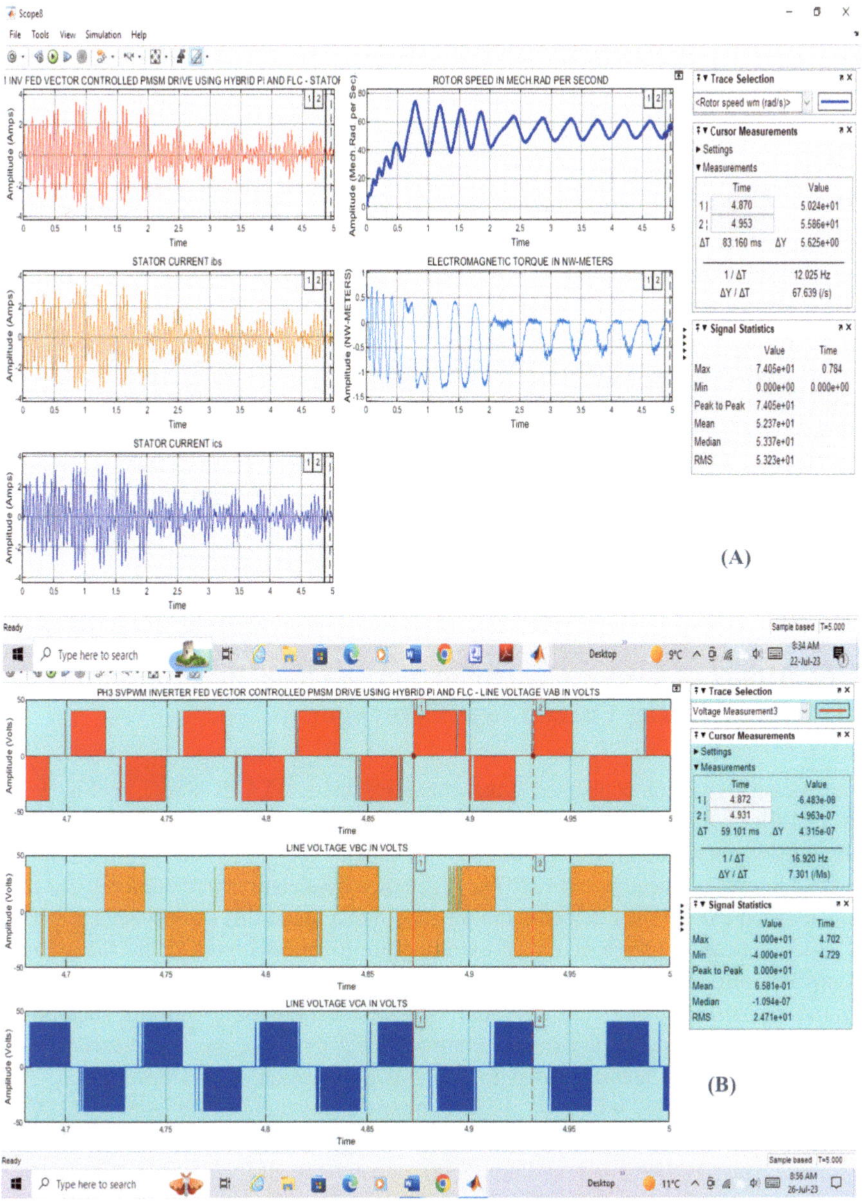

Fig. 9.20 Three-phase SVPWM inverter-fed vector controlled PMSM drive using hybrid PI and FLC simulation results: (**a**) stator currents ias, ibs and ics (left column, top to bottom), rotor speed and E.M. torque (right column, top to bottom), (**b**) three-phase line to line voltage of SVPWM inverter (top to bottom)

ode23tb (stiff/TR-BDF2) solver is used [2]. The load torque is 0.5 Nw-M for the first 2 s and is 0.25 Nw-M for time above 2 s. Triangle carrier frequency is 20 kHz for the three-phase SVPWM inverter.

9.6.2 Discussion of Results

Simulation results shown in Fig. 9.20 indicate a balanced three-phase stator current. The mean rotor speed is 52.37 mech.rad per second which is very close to the set point speed of 54.140 mech.rad per second. Also from Fig. 9.20, the line to line output voltage of three-phase SVPWM inverter is 24.71 volts (RMS) and the frequency is 16.920 Hz. Converting the rotor set point speed of 54.140 mech.rad per second to electrical frequency, this value is 17.233 Hz which closely agrees with the frequency of the output voltage of inverter. The RMS value of stator current is 0.9365 amps which is close to the value in Table 7.4. The modulation index ma is found to be 0.9258 by simulation. This model demonstrates the accuracy of vector controlled PMSM drive.

9.7 Model of Vector Control of Three-Phase Induction Motor Using Fuzzy Logic Controller

The model of vector controlled three-phase IM using FLC is shown in Fig. 9.21 (model file: EXAMPLE9_5). The model development for vector controlled three-phase IM is presented in detail in Sect. 8.8 of Chap. 8. Here PI controller in the speed control loop is replaced with FLC. The difference between rotor reference speed and actual speed e from error detector and the rate of change of this error speed cie which is $(e(k) - e(k - 1))$ for a given sample time are given as inputs to the fuzzy logic controller through a Mux. These two fuzzy inputs are fuzzified in the FIS by Mamdani method using rule base and then defuzzified using centroid method to obtain a crisp value for the output control signal cs. Here the speed error e and change in error cie are converted to per unit by division with a base speed. The output cs is multiplied with base speed to obtain the actual control signal. The MFs for e, cie and cs are formed with triangle MFs having seven linguistic variables: negative big (NB), negative medium (NM), negative small (NS), zero (ZE), positive small (PS), positive medium (PM) and positive big (PB). The FIS file name used here is "imvectorcontrol". The FIS editor; MFs for e, cie and cs; the rule base editor with 49 rules developed using Table 9.1 and FLC front panel are the same as shown in Fig. 9.7.

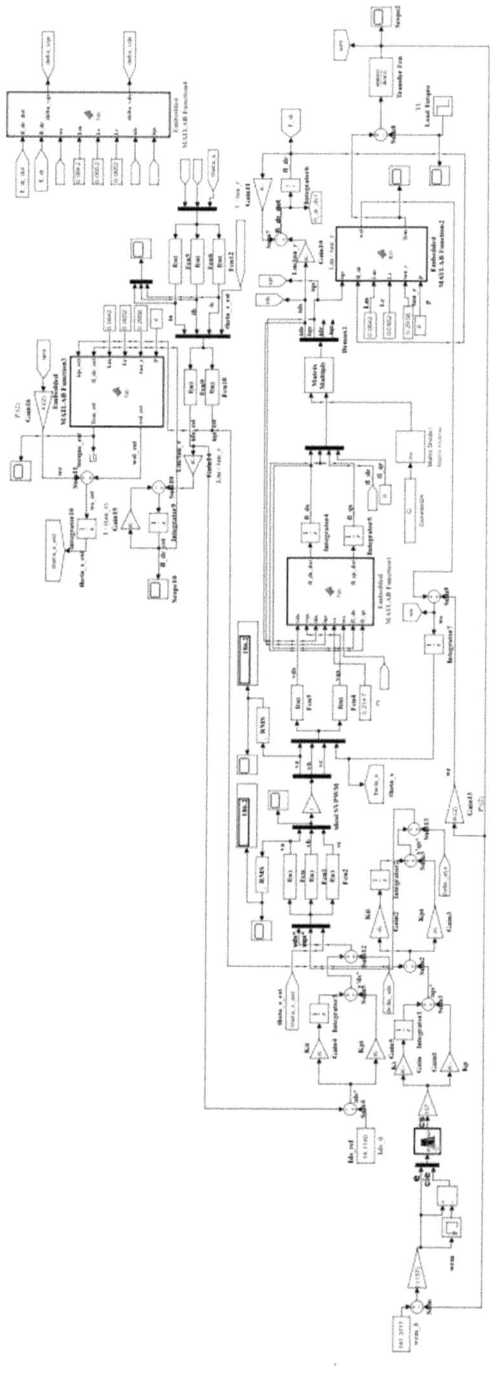

Fig. 9.21 Model of a vector controlled three-phase induction motor using fuzzy logic controller

9.7.1 Simulation Results

The simulation of the model shown in Fig. 9.21 is carried out using ode45 (Dormand-Prince) solver in Simulink [2]. The load is initially 200 Nw-M for the first 0.6 s and falls to 100 Nw-M above 0.6 s. The simulation results for the rotor e.m. torque, speed and stator currents are shown in Fig. 9.22a–c, respectively.

9.7.2 Discussion of Results

Referring to Fig. 9.22a, the peak torque is 1231 Nw-M and falls to 201.6 Nw-M within the first 0.6 s and finally reaches a value of 101.2 Nw-M above 0.6 s which is in accordance with load torque. Here the peak e.m. torque of 1231 Nw-M is much less compared to that obtained with PI controller shown in Fig. 8.28a in Chap. 8. The rotor speed in Fig. 9.22b shows an initial rise during starting when 200 Nw-M load is applied and at 0.6 s when load is reduced to 100 Nw-M shows a peak of 160 Mech. rad per second and reaches a steady state speed of 139.1 and 139.8 mech.radians per second just before and above 0.6 s, respectively. These values closely agree with the reference speed of 141.3717 mech.rad. per second. The peak shoot up of rotor speed during starting and at 0.6 s is less compared to that using PI controller shown in Fig. 8.28b. Figure 9.22c shows that the stator currents in the three-phase are well balanced.

9.8 Vector Control of Three-Phase Induction Motor Using Fuzzy Logic Controller and with Modified Membership Functions

The model of the vector controlled three-phase IM using FLC shown in Fig. 9.21 has been re-examined here using modified MFs for error e, change in error cie and output cs. These modified MFs are shown in Fig. 9.23 and also the front panel showing Mamdani method for fuzzification and centroid method for defuzzification. The FIS file name is "imvectorcontrolnew". In this pattern, triangle MFs for e, cie and cs crowd near the centre zero and are wide apart at the two ends. The fuzzy rules are formed with rule base editor using Table 9.1. These are the same as shown in Fig. 9.7.

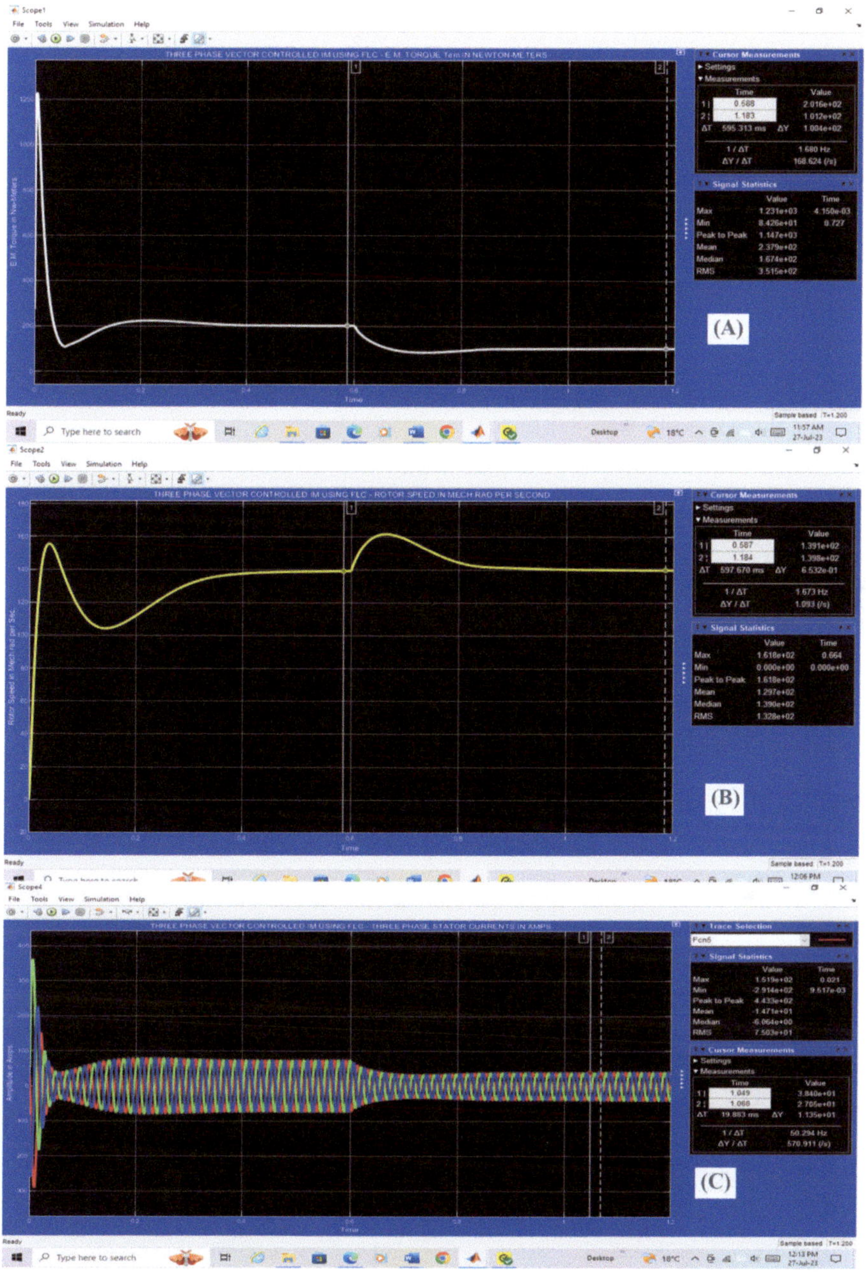

Fig. 9.22 Three phase vector controlled IM using FLC simulation results: (**a**) E.M. torque, (**b**) rotor speed and (**c**) stator currents

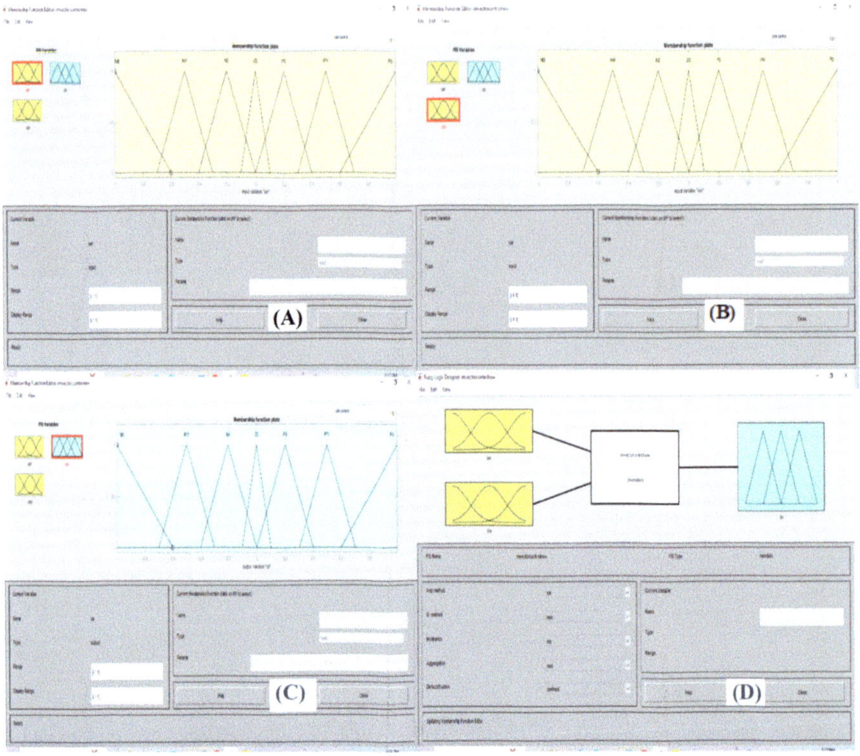

Fig. 9.23 Three-phase vector controlled IM using FLC and with modified MFs: MFs for (**a**) error e, (**b**) change in error cie, (**c**) output cs and (**d**) FLC front panel

9.8.1 Simulation Results

The simulation of the model shown in Fig. 9.21 with FIS file name for the FLC changed to "imvectorcontrolnew" is carried out using ode45 (Dormand-Prince) solver in Simulink [2]. The load is 200 Nw-M for the first 0.6 s and steps down to 100 Nw-M after 0.6 s. The simulation results for e.m. torque and rotor speed are shown in Fig. 9.24a, b.

9.8.2 Discussion of Results

Referring to Fig. 9.24a, the peak torque is 1312 Nw-M and falls to 201.5 Nw-M within the first 0.6 s and finally reaches a value of 101.3 Nw-M above 0.6 s which is in accordance with the load torque. Here the peak e.m. torque of 1312 Nw-M is much greater than that with triangle MFs for e, cie and cs presented in Sect. 9.7. The rotor speed in Fig. 9.24b shows an initial rise during starting and reaches a steady

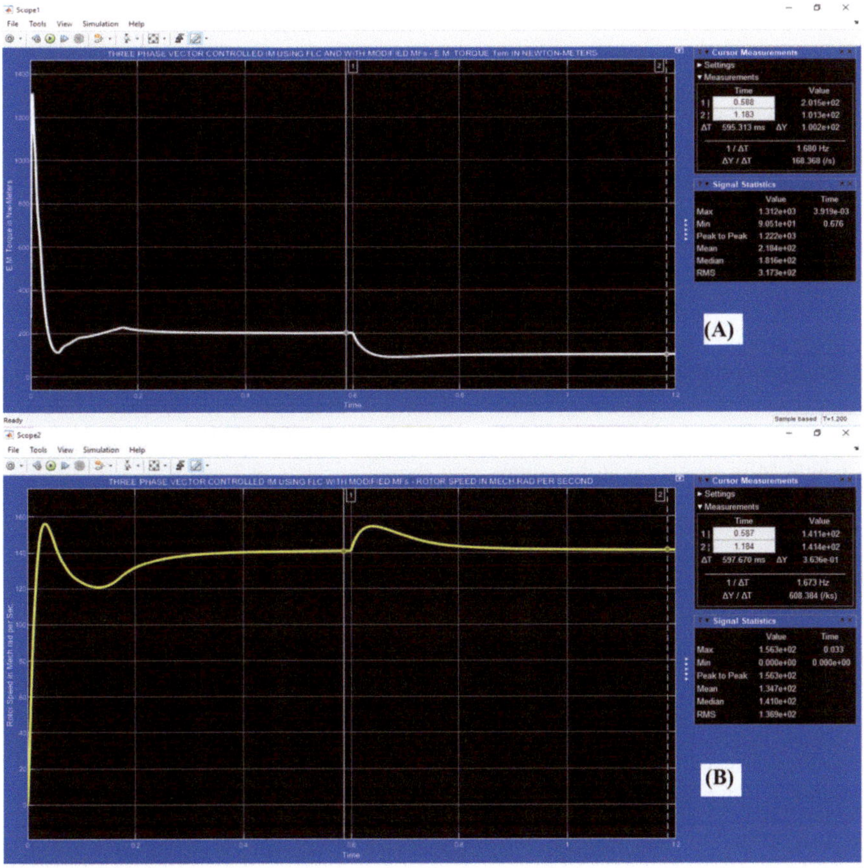

Fig. 9.24 Three-phase vector controlled IM using FLC and with modified MFs: simulation results for (**a**) E.M. torque and (**b**) rotor speed

speed of 141.1 with 200 Nw-M load within the first 0.6 s and a steady speed of 141.4 mech.radians per second with 100 Nw-M load above 0.6 s. The steady state rotor speeds are close to the reference speed. Comparing the speed response in Fig. 9.24b with that of Fig. 9.22b, it is seen that the undershoot during starting and the overshoot after 0.6 s. are much reduced with modified MFs than with triangle MFs for error e, change in error cie and output cs. Also the speed overshoot and undershoot at starting and the overshoot at 0.6 s in Fig. 9.24b are much less compared to that obtained with PI controller shown in Fig. 8.28b.

9.9 Conclusions

In this chapter models showing the application of fuzzy logic for the speed and torque control of converter-fed three-phase IM drive with applied voltage to frequency ratio held constant, direct torque controlled (DTC) and vector controlled are examined. In addition models for vector control of SVPWM inverter-fed PMSM drive using hybrid PI and FLC are presented. The superiority of FLC over PI controller is established by simulation for converter-fed IM drives. For SVPWM inverter-fed PMSM drive, the results with FLC and PI controller are almost the same. By increasing the number of MFs and rules, by changing the shape of MFs, using different MF shape combinations and also by realigning them for fuzzy inputs and output such as by crowding the MFs near centre zero, it is possible to get an improved response for motor speed and torque. This is true in the case of vector controlled three-phase IM using FLC with triangle MFs and that with modified triangle MFs crowding at the centre zero shown in Sects. 9.7 and 9.8.

References

1. Zadeh, L. A., "Fuzzy sets," Information and Control, Vol. 8, pp. 338–353, 1965.
2. The mathworks Inc., www.mathworks.com: "MATLAB Version: 9.14.0.2286388 (R2023a)"
3. Bimal K. Bose: "Expert System, Fuzzy Logic, and Neural Networks in Power Electronics and Drives" in B.K. Bose [Editor]: "Power Electronics and Variable Frequency Drives: Technology and Applications", Chapter 11, IEEE Press, 1997.
4. Bimal K. Bose: "Modern Power Electronics and AC Drives", Chapter 11, Prentice Hall PTR, 2002,
5. Bimal K. Bose: "Expert System, Fuzzy Logic, and Neural Network Applications in Power Electronics and Motion Control", Proceedings of the IEEE. Vol. 82, No. 8, August 1994, pp. 1308–1314.
6. Ahmed Rubaai et.al.: "Fuzzy Logic Applications in Electrical Drives and Power Electronics" in M.H. Rashid [Editor]: "Power Electronics Handbook: Devices, Circuits and Applications", Elsevier, 2011, Chapter 37, pp. 1115–1120 and pp. 1126–1137.
7. Ahmed Rubaai: "Fuzzy Logic in Electric Drives" in M.H. Rashid [Editor]: "Power Electronics Handbook", Academic Press, 2001, pp. 779–790.

Appendix I: Model Source Codes

Model source codes in the Embedded MATLAB function from each chapter are presented in this Appendix.

Source codes from Embedded MATLAB function blocks in the models from each chapter are given below:

```
%% Program Segment 1.1
%% Dr. Narayanaswamy P R Iyer.
function [Vcon_a, Vcon_b, Vcon_c] = fcn(vm,fm,t)
Vcon_a = vm*sin(2*pi*fm*t);
Vcon_b = vm*sin(2*pi*fm*t-2*pi/(3));
Vcon_c = vm*sin(2*pi*fm*t+2*pi/(3));
---------------------------------------------------------------------
```

```
%% Program Segment 2.1
function [vthr,vthy,vthb] = fcn(k1,k2,f,t,vm)
%% Dr.Narayanaswamy. P.R.Iyer.
vthr = vm*(k1*sin(2*pi*f*t) + k2*sin(3*2*pi*f*t));
vthy = vm*(k1*sin(2*pi*f*t-2*pi/(3)) + k2*sin(3*2*pi*f*t-2*pi));
vthb = vm*(k1*sin(2*pi*f*t+2*pi/(3)) + k2*sin(3*2*pi*f*t+2*pi));
--------------------------------------------------------------------
---------------------------------
```

```
%% Program Segment 2.2
function [vthr,vthy,vthb] = fcn(k1,k2,k3,f,t,vm)
%% Dr.Narayanaswamy. P.R.Iyer.
vthr = vm*(k1*sin(2*pi*f*t) + k2*sin(3*2*pi*f*t) - k3*sin
(9*2*pi*f*t));
```


N. P. R. Iyer, *Inverters and AC Drives*, Power Systems,
https://doi.org/10.1007/978-3-031-62784-2

```
vthy = vm* (k1*sin(2*pi*f*t-2*pi/(3)) + k2*sin(3*2*pi*f*t-2*pi)-
k3*sin(9*2*pi*f*t-6*pi));
vthb = vm* (k1*sin(2*pi*f*t+2*pi/(3))+k2*sin(3*2*pi*f*t+2*pi)-
k3*sin(9*2*pi*f*t+6*pi));
----------------------------------------------
```

```
%% Program segment 2.3
function [Va, Vb, Vc] = fcn(vm,fm,t)
%% Dr.Narayanaswamy.P.R.Iyer.
Va = vm*sin(2*pi*fm*t);
Vb = vm*sin(2*pi*fm*t-2*pi/(3));
Vc = vm*sin(2*pi*fm*t+2*pi/(3));
-----------------------------------------------
----------------------------------------
```

```
%% Program segment 2.4
function [vr_clip, vy_clip, vb_clip] = fcn(vr,vy,vb,v_clip)
%% Dr.Narayanaswamy.P.R.Iyer.
if (vr >= +v_clip)
  vr_clip = +v_clip;
else if (vr <= -v_clip)
    vr_clip = -v_clip;
  else
    vr_clip = vr;
  end
end
if (vy >= +v_clip)
  vy_clip = +v_clip;
else if (vy <= -v_clip)
    vy_clip = -v_clip;
  else
    vy_clip = vy;
  end
end
if (vb >= +v_clip)
  vb_clip = +v_clip;
else if (vb <= -v_clip)
    vb_clip = -v_clip;
  else
    vb_clip = vb;
  end
end
----------------------------------------------
```

```
%%Program segment 2.5
function [y1,y2,y3] = fcn(k,fm,t)
%% Dr. Narayanaswamy P R Iyer
y1 = 0;
y2 = k*sin(2*pi*fm*t);
y3 = k*cos(2*pi*fm*t);
x1 = y2;
x2 = y3;
if ((x1 >= 0) && (x1 < 0.866) && (x2 < 1) && (x2 >= 0.5))
  y1 = 1;
  else if ((x1 >= 0.866) && (x2 < 0.5) && (x2 >= -0.5))
    y1 = 2;
    else if ((x1 <= 0.866) && (x1 > 0) && (x2 < -0.5) && (x2 >= -1))
      y1 = 3;
      else if ((x1 <= 0) && (x1 > -0.866) && (x2 >= -1) && (x2 < -0.5))
        y1 = 4;
        else if ((x1 <= -0.866) && (x2 > -0.5) && (x2 <= 0.5))
          y1 = 5;
          else if ((x1 >= -0.866) && (x1 < 0) && (x2 > 0.5) && (x2 <= 1))
         y1 = 6;
            end
          end
        end
      end
    end
   end
----------------------------------------------------
------------------------------------
```

```
%%Program segment 2.6
function [ssf1,ssf2,ssf3,ssf4,ssf5,ssf6]=fcn(sector)
%% Dr. Narayanaswamy P R Iyer
if (sector == 1)
  ssf1 = 1;
  else
ssf1 = 0;
end
   if (sector == 2)
      ssf2 = 1;
      else
ssf2 = 0;
    end
```

```
    if (sector == 3)
ssf3 = 1;
      else

%%ProgramSegment 2.6 (Continued)
ssf3 = 0;
    end
if (sector == 4)
      ssf4 = 1;
      else
   ssf4 = 0;
end
    if (sector == 5)
      ssf5 = 1;
      else
ssf5 = 0;
    end
if (sector == 6)
      ssf6 = 1;
      else
ssf6 = 0;
    end
----------------------------------------------------------------
```

```
%%Program segment 2.7
function[f1a,f1b,f1c]= fcn(ma,fm,ssf1,ssf2,ssf3,ssf4,ssf5,ssf6,t)
%% Dr.Narayanaswamy.P.R.Iyer.
f1a =0; f1b=0; f1c=0;
if (ssf1 == 1 || ssf4 == 1)
  f1a = (ssf1||ssf4)*(ma*sin(2*pi*fm*t + pi/(3)));
  f1b = ssf4*1 - ssf1*(1);
  f1c = (ssf1||ssf4)*(ma*sin(2*pi*fm*t + 2*pi/(3)));
end
if (ssf2 == 1 || ssf5 == 1)
    f1a = ssf2*1 - ssf5*(1);
f1b = (ssf5||ssf2)*(ma*sin(2*pi*fm*t - 2*pi/(3)));
f1c = (ssf5||ssf2)*(ma*sin(2*pi*fm*t + 2*pi/(3) + pi/(3)));
end
if (ssf3 == 1 || ssf6 == 1)
f1a = (ssf3||ssf6)*(ma*sin(2*pi*fm*t));

%%Program segment 2.7(Continued)
f1b = (ssf3||ssf6)*(ma*sin(2*pi*fm*t - 2*pi/(3)
```

```
+ pi/(3)));
  f1c = ssf6*1 - ssf3*(1);
end
-----------------------------------------------------------
-----------------------------------------------
```

```
%% Program segment 2.8
function [f1a,f1b,f1c] = fcn(ma,fm,ssf1,ssf2,ssf3,ssf4,ssf5,ssf6,t)
%% Dr.Narayanaswamy.P.R.Iyer.
f1a =0; f1b=0; f1c=0;
if (ssf1 == 1 || ssf4 == 1)
  f1a = (ssf1||ssf4)*(ma*sin(2*pi*fm*t));
  f1b = ssf4*1 - ssf1*(1);
  f1c = (ssf1||ssf4)*(ma*sin(2*pi*fm*t + 2*pi/(3)));
end
if (ssf2 == 1 || ssf5 == 1)
    f1a = ssf2*1 - ssf5*(1);
    f1b = (ssf5||ssf2)*(ma*sin(2*pi*fm*t - 2*pi/(3)));
    f1c = (ssf5||ssf2)*(ma*sin(2*pi*fm*t + 2*pi/(3)));
end
if (ssf3 == 1 || ssf6 == 1)
  f1a = (ssf3||ssf6)*(ma*sin(2*pi*fm*t));
  f1b = (ssf3||ssf6)*(ma*sin(2*pi*fm*t - 2*pi/(3)));
  f1c = ssf6*1 - ssf3*(1);
end
------------------------------------------------------------
```

```
%% Program Segment 2.9
function [dvc1ua,dvc2ua,dvc3ua,dvc4ua,dvc5ua] =
MMC_UPPER_ARM(Vc1ua,Vc2ua,Vc3ua,Vc4ua,Vc5ua,vc_ref,Iua,Ila,kp)
%% Dr. Narayanaswamy P R Iyer
%%Single Phase MMC. Given 3 kW output Power P and 500 V DC link voltage %%
Vdc.
%%Neglecting switching losses idc_ref = P/Vdc = 3000/500 = 6 Amps.
idc_ref = 3000/(500);
dvc1ua = 0;dvc2ua = 0;dvc3ua = 0;dvc4ua = 0;dvc5ua = 0;
%%First cell - Upper arm.
dvc1ua = kp*(Vc1ua - vc_ref)*Iua/(vc_ref*idc_ref);
%%Second cell - Upper arm.
dvc2ua = kp*(Vc2ua - vc_ref)*Iua/(vc_ref*idc_ref);
%%Third cell - Upper arm.
dvc3ua = kp*(Vc3ua - vc_ref)*Iua/(vc_ref*idc_ref);
%%Fourth cell - Upper arm.
```

```
dvc4ua = kp* (Vc4ua - vc_ref)*Iua/(vc_ref*idc_ref);
%%Fifth cell - Upper arm.
dvc5ua = kp* (Vc5ua - vc_ref)*Iua/(vc_ref*idc_ref);
-----------------------------------------------------
-----------------------------------------------------------
%%Program segment 2.10 for the MMC_LOWER_ARM is written in the same way
as for %%Program segment 2.9.
-------------------------------------------------------
-------------------------------------------------------
%%Program segment 3.2
function [va,vb,vc,sin_theta_a,sin_theta_b,sin_theta_c] = fcn(vm,f,
t)
%%Dr.Narayanaswamy. P.R.Iyer.
va = vm*sin(2*pi*f*t);
vb = vm*sin(2*pi*f*t - 2*pi/(3));
vc = vm*sin(2*pi*f*t + 2*pi/(3));
%%theta_a, theta_b, theta_c are in radians.
sin_theta_a = sin(2*pi*f*t);
sin_theta_b = sin(2*pi*f*t - 2*pi/(3));
sin_theta_c = sin(2*pi*f*t + 2*pi/(3));


%%Program segment 3.3
function [SinAlfa1,SinAlfa2,SinAlfa3,SinAlfa4,SinAlfa5,SinAlfa6] =
fcn(alfa1,alfa2,alfa3)
%%Dr.Narayanaswamy. P.R.Iyer.
SinAlfa1 = sin(alfa1*pi/(180));
SinAlfa2 = sin(alfa2*pi/(180));
SinAlfa3 = sin(alfa3*pi/(180));
SinAlfa4 = sin(pi+alfa1*pi/(180));
SinAlfa5 = sin(pi+alfa2*pi/(180));
SinAlfa6 = sin(pi+alfa3*pi/(180));
------------------------------------------------------------


%%Program Segment 4.1
function [va,vb,vc,vq,vd,vref_max,alfa,sector_num,ssf1,ssf2,ssf3,
ssf4,ssf5,ssf6,theta] =
fcn(Vim,f,time)
%% Dr. Narayanaswamy P R Iyer.
va = Vim*sin(2*pi*f*time);
vb = Vim*sin(2*pi*f*time - 2*pi/(3));
vc = Vim*sin(2*pi*f*time + 2*pi/(3));
vd = (2/3)*(va - 0.5*vb - 0.5*vc);
```

```
vq = (2/3)*(1.732*vb/(2) - 1.732*vc/(2));
vref_max = sqrt((abs(vd))*(abs(vd))+ (abs(vq))*(abs(vq)));

%%Program Segment 4.1(Continued)
alfa = atan2(vq,vd);
sector_num = 0;
theta = 0;
ssf1=0; ssf2=0; ssf3=0; ssf4=0; ssf5=0; ssf6=0;
if (alfa >= 0 && alfa < pi/(3))
  sector_num = 1;
  theta = alfa - (sector_num - 1)*pi/(3);
  if(sector_num == 1)
    ssf1 = 1;
  else
    ssf1 = 0;
  end
end
if (alfa >= pi/(3) && alfa < 2*pi/(3))
  sector_num = 2;
  theta = alfa - (sector_num - 1)*pi/(3);
  if(sector_num == 2)
ssf2 = 1;
  else
    ssf2 = 0;
%%Program Segment 4.1(Continued)
end
end
if (alfa >= 2*pi/(3) && alfa < pi)
  sector_num = 3;
  theta = alfa - (sector_num - 1)*pi/(3);
  if(sector_num == 3)
    ssf3 = 1;
  else
ssf3 = 0;
  end
end
%%Program Segment 4.1(Continued)
if (alfa >= -pi && alfa < -2*pi/(3))
  sector_num = 4;
  theta = alfa - (sector_num - 1)*pi/(3);
  if(sector_num == 4)
    ssf4 = 1;
  else
    ssf4 = 0;
  end
```

```
end
if (alfa >= -2*pi/(3) && alfa < -pi/(3))
  sector_num = 5;
theta = alfa - (sector_num - 1)*pi/(3);
  if(sector_num == 5)
    ssf5 = 1;
else
 ssf5 = 0;
end
end
if (alfa >= -pi/(3) && alfa < 0)
  sector_num = 6;
%%Program Segment 4.1(Continued)
  theta = alfa - (sector_num - 1)*pi/(3);
  if(sector_num == 6)
    ssf6 = 1;
  else
    ssf6 = 0;
  end
end
-----------------------------------------------------------------
-----------------------------------------------
```

```
%%Program Segment 4.2
Function [s1a1,s1a2,s1a3,s1a4,s2b1,s2b2,s2b3,s2b4,s3c1, s3c2,s3c3,
s3c4,s4d1,s4d2,s4d3,s4d4,s5e1,
s5e2,s5e3,s5e4,s6f1,s6f2,s6f3,s6f4,ma] = fcn(ts,vdc,vref,theta,
sector_num,vtri)
 %% Dr. Narayanaswamy P R Iyer.
ma = sqrt(3)*vref/(vdc);
 s1a1=0; s1a2=0; s1a3=0; s1a4=0; s2b1=0; s2b2=0; s2b3=0; s2b4=0;
s3c1=0; s3c2=0; s3c3=0; s3c4=0;
 s4d1=0; s4d2=0; s4d3=0; s4d4=0; s5e1=0; s5e2=0; s5e3=0; s5e4=0;
s6f1=0; s6f2=0; s6f3=0; s6f4=0;
 if(sector_num == 1) %%odd sector 1
   ta = ts*ma*sin(pi/(3) - theta);
   tb = ts*ma*sin(theta);
   t0 = ts - ta - tb;
   if (vtri <= 0.25*t0)
     s1a1 = 1;
  else
       s1a1 = 0;
end
  if (vtri <= (0.25*t0 + 0.5*ta))
```

```
    s1a2 = 1;
  else
      s1a2 = 0;
  end
      if (vtri <= (0.25*t0 + 0.5*ta + 0.5*tb))
        s1a3 = 1;
      else
          s1a3 = 0;
      end
          if (vtri <= (0.25*t0 + 0.5*ta + 0.5*tb + 0.25*t0))
            s1a4 = 1;
          else
              s1a4 = 0;
          end
end
if(sector_num == 2) %%even sector 2
  ta = ts*ma*sin(pi/(3) - theta);
  tb = ts*ma*sin(theta);
  t0 = ts - ta - tb;
  if (vtri <= 0.25*t0)
    s2b1 = 1;
  else
      s2b1 = 0;
  end
  if (vtri <= (0.25*t0 + 0.5*tb))
    s2b2 = 1;
  else
      s2b2 = 0;
  end
      if (vtri <= (0.25*t0 + 0.5*ta + 0.5*tb))
        s2b3 = 1;
      else
          s2b3 = 0;
      end
if (vtri <= (0.25*t0 + 0.5*ta + 0.5*tb + 0.25*t0))
          s2b4 = 1;
          else
              s2b4 = 0;
          end
end

%%Program Segment 4.2(Continued)
s5e3 = 1;
      else
          s5e3 = 0;
```

```
        end
if (vtri <= (0.25*t0 + 0.5*ta + 0.5*tb + 0.25*t0))
            s5e4 = 1;
          else
              s5e4 = 0;
          end
end
if(sector_num == 6) %%even sector 6
  ta = ts*ma*sin(pi/(3) - theta);
   tb = ts*ma*sin(theta);
   t0 = ts - ta - tb;
   if (vtri <= 0.25*t0)
     s6f1 = 1;
  else
      s6f1 = 0;
  end
  if (vtri <= (0.25*t0 + 0.5*tb))
     s6f2 = 1;
else
s6f2 = 0;
  end
   if (vtri <= (0.25*t0 + 0.5*ta + 0.5*tb))
        s6f3 = 1;
      else
          s6f3 = 0;
      end
 if (vtri <= (0.25*t0 + 0.5*ta + 0.5*tb + 0.25*t0))
            s6f4 = 1;
          else
              s6f4 = 0;
          end
 end


 %%Program Segment 4.2 (Continued)
 if(sector_num == 3) %%odd sector 3
 ta = ts*ma*sin(pi/(3) - theta);
   tb = ts*ma*sin(theta);
   t0 = ts - ta - tb;
   if (vtri <= 0.25*t0)
     s3c1 = 1;
  else
      s3c1 = 0;
  end
  if (vtri <= (0.25*t0 + 0.5*ta))
```

```
s3c2 = 1;
  else
      s3c2 = 0;
  end
      if (vtri <= (0.25*t0 + 0.5*ta + 0.5*tb))
       s3c3 = 1;
else
s3c3 = 0;
end
if (vtri <= (0.25*t0 + 0.5*ta + 0.5*tb + 0.25*t0))
 s3c4 = 1;
else
s3c4 = 0;
end
end
 if(sector_num == 4) %%even sector 4
  ta = ts*ma*sin(pi/(3) - theta);
   tb = ts*ma*sin(theta);
   t0 = ts - ta - tb;
   if (vtri <= 0.25*t0)
     s4d1 = 1;
  else
      s4d1 = 0;
  end
  if (vtri <= (0.25*t0 + 0.5*tb))
     s4d2 = 1;
  else
      s4d2 = 0;
  end
      if (vtri <= (0.25*t0 + 0.5*ta + 0.5*tb))
        s4d3 = 1;
      else
          s4d3 = 0;
      end
          if (vtri <= (0.25*t0 + 0.5*ta + 0.5*tb + 0.25*t0))
            s4d4 = 1;
          else
              s4d4 = 0;
          end
 end
if(sector_num == 5) %%odd sector 5
  ta = ts*ma*sin(pi/(3) - theta);
   tb = ts*ma*sin(theta);
   t0 = ts - ta - tb;
   if (vtri <= 0.25*t0)
```

```
    s5e1 = 1;
  else
      s5e1 = 0;
  end
  if (vtri <= (0.25*t0 + 0.5*ta))
    s5e2 = 1;
  else
      s5e2 = 0;
  end
      if (vtri <= (0.25*t0 + 0.5*ta + 0.5*tb))
%%Program Segment 4.3
function [sf_s1r1,sf_s1r2,sf_s1r3,sf_s1r4,sf_s2r1,sf_s2r2,sf_s2r3,
sf_s2r4,sf_s3r1,sf_s3r2,sf_s3r3,
sf_s3r4,ma] = fcn(ts,vdc,vref,theta,sector_num)
 %%Dr. Narayanaswamy P R Iyer.
 ma = sqrt(3)*vref/(vdc);
 sf_s1r1=0; sf_s1r2=0;sf_s1r3=0;sf_s1r4=0;
 sf_s2r1=0; sf_s2r2=0; sf_s2r3=0; sf_s2r4=0;
 sf_s3r1=0; sf_s3r2=0; sf_s3r3=0; sf_s3r4=0;
x = vref*sin(theta);
 y = vref*cos(theta) - vdc/(3);
 beta = atan2(x,y);
if(sector_num == 1)  %%odd sector 1
  if(vref*sin(theta)>=0 && vref*sin(theta)<= vdc/(2*sqrt(3)) &&
beta>2*pi/(3))
sf_s1r1 = 1;%%sector 1 and region 1
  else
sf_s1r1 = 0;
  end
if(vref*sin(theta)>=0 && vref*sin(theta)<=vdc/(2*sqrt(3)) &&
beta>pi/(3) && beta<=2*pi/(3))
        sf_s1r2 = 1;%%sector 1 and region 2
      else
          sf_s1r2 = 0;
      end
if(vref*cos(theta)>vdc/(3) && beta<=pi/(3))
            sf_s1r3 = 1;%%sector 1 and region 3
          else
              sf_s1r3 = 0;
          end
if(vref*sin(theta)>vdc/(2*sqrt(3)))
                sf_s1r4 = 1;%%sector 1 and region 4
              else
                  sf_s1r4 = 0;
              end
end
```

```
%%Program Segment 4.3 (Continued)
sf_s1r1 = 0;
  end
 if(vref*sin(theta)>=0 && vref*sin(theta)<=vdc/(2*sqrt(3)) &&
beta>pi/(3) && beta<=2*pi/(3))
      sf_s1r2 = 1;%%sector 1 and region 2
      else
          sf_s1r2 = 0;
      end
if(vref*cos(theta)>vdc/(3) && beta<=pi/(3))
      sf_s1r3 = 1;%%sector 1 and region 3
          else
              sf_s1r3 = 0;
          end
if(vref*sin(theta)>vdc/(2*sqrt(3)))
      sf_s1r4 = 1;%%sector 1 and region 4
              else
                  sf_s1r4 = 0;
              end
end
if(sector_num == 2) %%even sector 2
  if(vref*sin(theta)>=0 && vref*sin(theta)<= vdc/(2*sqrt(3)) &&
beta>2*pi/(3))
    sf_s2r1 = 1;%%sector 2 and region 1
  else
      sf_s2r1 = 0;
  end
 if(vref*sin(theta)>=0 && vref*sin(theta)<=vdc/(2*sqrt(3)) &&
beta>pi/(3) && beta<=2*pi/(3))
      sf_s2r2 = 1;%%sector 2 and region 2
      else
          sf_s2r2 = 0;
      end
 if(vref*cos(theta)>vdc/(3) && beta<=pi/(3))
       sf_s2r3 = 1;%%sector 2 and region 3
          else
              sf_s2r3 = 0;
          end
%%Program Segment 4.3 (Continued)                if(vref*sin(theta)>vdc/
(2*sqrt(3)))
    sf_s2r4 = 1;%%sector 2 and region 4
else
sf_s2r4 = 0;
end
end
```

```
if(sector_num == 3)  %%odd sector 3
  if(vref*sin(theta)>=0 && vref*sin(theta)<= vdc/(2*sqrt(3)) &&
beta>2*pi/(3))
    sf_s3r1 = 1;%%sector 3 and region 1
  else
      sf_s3r1 = 0;
  end
if(vref*sin(theta)>=0 && vref*sin(theta)<=vdc/(2*sqrt(3)) &&
beta>pi/(3)&& beta<=2*pi/(3))
     sf_s3r2 = 1;%%sector 3 and region 2
      else
          sf_s3r2 = 0;
      end
  if(vref*cos(theta)>vdc/(3) && beta<=pi/(3))
sf_s3r3 = 1;%%sector 3 and region 3
          else
              sf_s3r3 = 0;
          end                 if(vref*sin(theta)>vdc/(2*sqrt(3)))
    sf_s3r4 = 1;%%sector 3 and region 4
              else
                  sf_s3r4 = 0;
              end
      end
--------------------------------------------------------------
```

```
%%Program Segment 4.4
function [sf_s4r1,sf_s4r2,sf_s4r3,sf_s4r4,sf_s5r1,sf_s5r2,sf_s5r3,
sf_s5r4,
sf_s6r1,sf_s6r2,sf_s6r3,sf_s6r4,ma] = fcn(ts,vdc,vref,theta,
sector_num)
 %%Dr. Narayanaswamy P R Iyer.
 ma = sqrt(3)*vref/(vdc);
 sf_s4r1=0; sf_s4r2=0;sf_s4r3=0;sf_s4r4=0;
 sf_s5r1=0; sf_s5r2=0; sf_s5r3=0; sf_s5r4=0;
 sf_s6r1=0; sf_s6r2=0; sf_s6r3=0; sf_s6r4=0;
 x = vref*sin(theta);
 y = vref*cos(theta) - vdc/(3);
 beta = atan2(x,y);
 if(sector_num == 4)  %%Even sector 4
 if(vref*sin(theta)>=0 && vref*sin(theta)<= vdc/(2*sqrt(3))
 &&beta>2*pi/(3))
   sf_s4r1 = 1;%%sector 4 and region 1
  else
     sf_s4r1 = 0
```

```
 end
if(vref*sin(theta)>=0 && vref*sin(theta)<=vdc/(2*sqrt(3))
&&beta>pi/(3) && beta<=2*pi/(3))
    sf_s4r2 = 1;%%sector 4 and region 2
  else
      sf_s4r2 = 0;
  end
  if(vref*cos(theta)>vdc/(3) && beta<=pi/(3))
    sf_s4r3 = 1;%%sector 4 and region 3
  else
      sf_s4r3 = 0;
  end
  if(vref*sin(theta)>vdc/(2*sqrt(3)))
        sf_s4r4 = 1;%%sector 4 and region 4
      else
          sf_s4r4 = 0;
  end
end
if(sector_num == 5) %%odd sector 5
 if(vref*sin(theta)>=0 && vref*sin(theta)<= vdc/(2*sqrt(3))
&&beta>2*pi/(3))
sf_s5r1 = 1;%%sector 5 and region 1
    else
        sf_s5r1 = 0
  end

%%Program Segment 4.4(Continued)
 if(vref*sin(theta)>=0 && vref*sin(theta)<=vdc/(2*sqrt(3))
&&beta>pi/(3) && beta<=2*pi/(3))
    sf_s5r2 = 1;%%sector 5 and region 2
  else
      sf_s5r2 = 0;
  end
  if(vref*cos(theta)>vdc/(3) && beta<=pi/(3))
sf_s5r3 = 1;%%sector 5 and region 3
else
sf_s5r3 = 0;
   end
  if(vref*sin(theta)>vdc/(2*sqrt(3)))
        sf_s5r4 = 1;%%sector 5 and region 4
      else
          sf_s5r4 = 0;
  end
end
if(sector_num == 6) %%even sector 6
```

```
if(vref*sin(theta)>=0 && vref*sin(theta)<= vdc/(2*sqrt(3)) &&
beta>2*pi/(3))
      sf_s6r1 = 1;%%sector 6 and region 1
    else
        sf_s6r1 = 0
  end
  if(vref*sin(theta)>=0 && vref*sin(theta)<=vdc/(2*sqrt(3)) &&
beta>pi/(3) && beta<=2*pi/(3))
    sf_s6r2 = 1;%%sector 6 and region 2
  else
      sf_s6r2 = 0;
  end
  if(vref*cos(theta)>vdc/(3) && beta<=pi/(3))
    sf_s6r3 = 1;%%sector 6 and region 3
  else
      sf_s6r3 = 0;
  end
  if(vref*sin(theta)>vdc/(2*sqrt(3)))
        sf_s6r4 = 1;%%sector 6 and region 4
      else
          sf_s6r4 = 0;
  end
end
%%Program segments 4.6 to 4.10 are written in the same way as for Program
segment 4.5.


%% Program Segment 4.5
function [s1r1a1,s1r1a2,s1r1a3,s1r1a4,s1r2b1,s1r2b2,s1r2b3,s1r2b4,
s1r3c1,s1r3c2,
s1r3c3,s1r3c4,s1r4d1,s1r4d2,s1r4d3,s1r4d4] = fcn(vsaw,ts,theta,ma,
sf_s1r1,sf_s1r2,sf_s1r3,sf_s1r4)
%% Dr.Narayanaswamy P R Iyer.
s1r1a1 = 0; s1r1a2 = 0; s1r1a3 = 0; s1r1a4 = 0;
s1r2b1 = 0; s1r2b2 = 0; s1r2b3 = 0; s1r2b4 = 0;
s1r3c1 = 0; s1r3c2 = 0; s1r3c3 = 0; s1r3c4 = 0;
s1r4d1 = 0; s1r4d2 = 0; s1r4d3 = 0; s1r4d4 = 0;
if(sf_s1r1 == 1)
  ta = ts*2*ma*sin(pi/(3) - theta);
  tb = ts - ts*2*ma*sin(pi/(3) + theta);
  tc = ts*2*ma*sin(theta);
  if(vsaw <= 0.25*ta)
    s1r1a1 = 1;
  else
    s1r1a1 = 0;
```

```
    end
    if(vsaw <= 0.25*ta + 0.5*tc)
      s1r1a2 = 1;
    else
      s1r1a2 = 0;
    end
    if(vsaw <= 0.25*ta + 0.5*tc + 0.5*tb)
      s1r1a3 = 1;
    else
      s1r1a3 = 0;
    end
    if(vsaw <= 0.25*ta + 0.5*tc + 0.5*tb + 0.25*ta)
      s1r1a4 = 1;
    else
%% Program Segment 4.5(Continued)
      s1r1a4 = 0;
    end
end
if(sf_s1r2 == 1)
    ta = ts - ts*2*ma*sin(theta);
    tb = -ts + ts*2*ma*sin(pi/(3) + theta);
    tc = ts - ts*2*ma*sin(pi/(3) - theta);
    if(vsaw <= 0.25*ta)
      s1r2b1 = 1;
    else
      s1r2b1 = 0;
    end
    if(vsaw <= 0.25*ta + 0.5*tc)
      s1r2b2 = 1;
    else
s1r2b2 = 0;
end
end
if(vsaw <= 0.25*ta + 0.5*tc + 0.5*tb)
      s1r2b3 = 1;
    else
      s1r2b3 = 0;
    end
    if(vsaw <= 0.25*ta + 0.5*tc + 0.5*tb + 0.25*ta)
      s1r2b4 = 1;
    else
      s1r2b4 = 0;
    end
end
if(sf_s1r3 == 1)
```

```
  ta = 2*ts - ts*2*ma*sin(pi/(3) + theta);
  tb = ts*2*ma*sin(theta);
 tc = -ts + ts*2*ma*sin(pi/(3) - theta);
  if(vsaw <= 0.25*ta)
    s1r3c1 = 1;
  else
    s1r3c1 = 0;
  end
  if(vsaw <= 0.25*ta + 0.5*tc)
    s1r3c2 = 1;
  else
s1r3c2 = 0;
  end
if(vsaw <= 0.25*ta + 0.5*tc + 0.5*tb)
s1r3c3 = 1;
else
s1r3c3 = 0;
end
if(vsaw <= 0.25*ta + 0.5*tc + 0.5*tb + 0.25*ta)
s1r3c4 = 1;
else
s1r3c4 = 0;
end
end
if(sf_s1r4 == 1)
ta = -ts + ts*2*ma*sin(theta);
tb = ts*2*ma*sin(pi/(3) - theta);
tc = 2*ts - ts*2*ma*sin(pi/(3) + theta);
if(vsaw <= 0.25*tc)
s1r4d1 = 1;
else
s1r4d1 = 0;
end
if(vsaw <= 0.25*tc + 0.5*tb)
s1r4d2 = 1;
else
s1r4d2 = 0;
end
if(vsaw <= 0.25*tc + 0.5*tb + 0.5*ta)
s1r4d3 = 1;
  else
    s1r4d3 = 0;
  end
  if(vsaw <= 0.25*tc + 0.5*tb + 0.5*ta + 0.25*tc)
    s1r4d4 = 1;
```

```
%% Program Segment 4.5(Continued)
  else
    s1r4d4 = 0;
  end
end
------------------------------------------------------------
```

```
%%Program segment 5.1
function [vtha,vthb,vthc] = fcn(M,k,f,t)
%% Dr.Narayanaswamy. P.R.Iyer.
%%Program segment 5.1(Continued)
vtha = (M*sin(2*pi*f*t) + M*k*sin(3*2*pi*f*t));
vthb = (M*sin(2*pi*f*t-2*pi/(3)) + M*k*sin(3*2*pi*f*t-2*pi));
vthc = (M*sin(2*pi*f*t+2*pi/(3)) + M*k*sin(3*2*pi*f*t+2*pi));
---------------------------------------------------------------
```

```
%%Program Segment 5.2
function [y1,y2,y3] = fcn(k,f,t)
%% Dr.Narayanaswamy P R Iyer
y1 = 0;
y2 = k*sin(2*pi*f*t);
y3 = k*cos(2*pi*f*t);
x1 = y2;
x2 = y3;
if ((x1 >= 0) && (x1 < 0.866) && (x2 < 1) && (x2 >= 0.5))
  y1 = 1;
  else if ((x1 >= 0.866) && (x2 < 0.5) && (x2 >= -0.5))
    y1 = 2;
    else if ((x1 <= 0.866) && (x1 > 0) && (x2 < -0.5) && (x2 >= -1))
      y1 = 3;
      else if ((x1 <= 0) && (x1 > -0.866) && (x2 >= -1) && (x2 < -0.5))
        y1 = 4;
        else if ((x1 <= -0.866) && (x2 > -0.5) && (x2 <= 0.5))
          y1 = 5;
          else if ((x1 >= -0.866) && (x1 < 0) && (x2 > 0.5) && (x2 <= 1))
y1 = 6;
            end
          end
        end
      end
    end
end
---------------------------------------------------------------
```

```
%%Program Segment 4.11
function [y1,ssf1a,ssf1b,ssf2a,ssf2b,ssf3a,ssf3b,ssf4a,ssf4b,
ssf5a,ssf5b,
ssf6a,ssf6b] = fcn(k,f,t)
%% Dr. Narayanaswamy P R Iyer
y1 = 0;
y2 = k*sin(2*pi*f*t);
y3 = k*cos(2*pi*f*t);
x1 = y2;
x2 = y3;
if ((x1 >= 0) && (x1 < 0.5) && (x2 < 1) && (x2 >= 0.866))
  y1 = 1;
  ssf1a = 1;
else ssf1a =0;
end
if ((x1 >= 0.5) && (x1 < 0.866)  && (x2 < 0.866) && (x2 >= 0.5))
    y1 = 1;
    ssf1b = 1;
else ssf1b =0;
end
if ((x1 > 0.866)  && (x1 < 1) && (x2 < 0.5) && (x2 >= 0))
      y1 = 2;
      ssf2a =1;
else  ssf2a = 0;
end
if ((x1 <= 1)  && (x1 > 0.866) && (x2 < 0) && (x2 >= -0.5))
%%Program Segment 4.6(Continued)
y1 = 2;
        ssf2b = 1;
else ssf2b =0;
end
if ((x1 <= 0.866) &&  (x1 > 0.5) && (x2 < -0.5) && (x2 >= -0.866))
          y1 = 3;
          ssf3a = 1;
else ssf3a = 0;
end
%%Program Segment 4.11(Continued)
if ((x1 < 0.5)  && (x1 >= 0) && (x2 < -0.866) && (x2 >= -1))
            y1 = 3;
ssf3b = 1;
else ssf3b = 0;
end
if ((x1 < 0)  && (x1 >= -0.5) && (x2 > -1) && (x2 <= -0.866))
y1 = 4;
ssf4a = 1;
```

```
else ssf4a = 0;
end
if ((x1 < -0.5) && (x1 >= -0.866) && (x2 > -0.866) && (x2 <= -0.5))
                    y1 = 4;
                    ssf4b = 1;
else ssf4b = 0;
end
if ((x1 < -0.866) && (x1 >= -1) && (x2 > -0.5) && (x2 <= 0))
                      y1 = 5;
ssf5a = 1;
else ssf5a = 0;
end
if ((x1 > -1) && (x1 <= -0.866) && (x2 > 0) && (x2 <= 0.5))
                        y1 = 5;
                        ssf5b = 1;
else ssf5b = 0;
end
if ((x1 > -0.866) && (x1 <= -0.5) && (x2 > 0.5) && (x2 <= 0.866))
                          y1 = 6;
                          ssf6a = 1;
else ssf6a = 0;
end
if ((x1 > -0.5) && (x1 <= 0) && (x2 > 0.866) && (x2 <= 1))
                            y1 = 6;
                            ssf6b = 1;
else ssf6b = 0;
end
------------------------------------------------------------
--------------------------------------
%%Program Segment 4.12
function [T1,T2,To,Ta1,Tb1,Tc1,Tsample,sap,san,sbp,sbn,scp,scn] =
fcn(vref,vdc,Tz,f,time,sector_num,ssf1a,ssf1b,ssf2a,ssf2b,ssf3a,
ssf3b,
ssf4a,ssf4b,ssf5a,ssf5b,ssf6a,ssf6b,Tsample)
%% Dr.Narayanaswamy. P.R.Iyer.
theta_dash = (2*pi*f*time - (sector_num - 1)*pi/(3));
T1 = (sqrt(3)*Tz*vref/(vdc))*(sin(pi/(3) - theta_dash)); %%eq.4.13
T2 = (sqrt(3)*Tz*vref/(vdc))*(sin(theta_dash)); %%eq.4.14
To = Tz - T1 - T2; %%eq.4.15
Ta1 = ssf1a*0.5*(T1+T2+0.5*To)+ssf1b*0.5*(T1+T2+0.5*To)+
ssf2a*0.5*(T2+0.5*To)+ssf2b*0.5*(T1+0.5*To)+ssf3a*0.5*(0.5*To)+
ssf3b*0.5*(0.5*To)+ssf4a*0.5*(0.5*To)+ssf4b*0.5*(0.5*To)+
ssf5a*0.5*(T2+0.5*To)+ssf5b*0.5*(T1+0.5*To)+
ssf6a*0.5*(T1+T2+0.5*To)+ssf6b*0.5*( T1+T2+0.5*To);
Tb1 = ssf1a*0.5*(T2+0.5*To)+ssf1b*0.5*(T1+0.5*To)+
```

```
ssf2a*0.5*(T1+T2+0.5*To)+ssf2b*0.5*(T1+T2+0.5*To)+
ssf3a*0.5*(T1+T2+0.5*To)+ssf3b*0.5*(T1+T2+0.5*To)+
ssf4a*0.5*(T2+0.5*(To)+ssf4b*0.5*(T1+0.5*To)+
ssf5a*0.5*(0.5*To)+ssf5b*0.5*(0.5*To)+ssf6a*0.5*(0.5*To)
+ ssf6b*0.5*(0.5*To);
Tc1 = ssf1a*0.5*(0.5*To)+ssf1b*0.5*(0.5*To)+ssf2a*0.5*(0.5*To)+
ssf2b*0.5*(0.5*To)+ssf3a*0.5*(T2+0.5*To)+ssf3b*0.5*(T1+0.5*To)+
ssf4a*0.5*(T1+T2+0.5*To)+ssf4b*0.5*(T1+T2+
0.5*( To)+ssf5a*0.5*(T1+T2+0.5*To)+ssf5b*0.5*(T1+T2+0.5*To)+
ssf6a*0.5*(T2+0.5*To)+ ssf6b*0.5*(T1+0.5*To);
if (Ta1 >= Tsample)
  sap = 1;
else
  sap = 0;
end
ifx (Tb1 >= Tsample)
  sbp = 1;
else
  sbp = 0;
end
if (Tc1 >= Tsample)
  scp = 1;
else
  scp = 0;
end

%%Program Segment 4.12(Continued)
  san = ~(sap);
  sbn = ~(sbp);
  scn = ~(scp);
--------------------------------------------------
-----------------------------------------------------------------
```

```
%%Program Segment 5.3
function [y1,y2,y3,y4,y5,y6]=fcn(sector)
%% Dr.Narayanaswamy P R Iyer
if (sector == 1)
  y1 = 1;
else y1 = 0;
end
    if (sector == 2)
      y2 = 1;
      else y2 = 0;
    end
```

```
if (sector == 3)
y3 = 1;
 else y3 = 0;
  end
  if (sector == 4)
    y4 = 1;
    else y4 = 0;
  end
  if (sector == 5)
    y5 = 1;
    else y5 = 0;
 end
      if (sector == 6)
        y6 = 1;
        else y6 = 0;
      end
-------------------------------------------------------------
---------------------------------------------
```

```
%%Program Segment 5.4
function [T1,T2,To,Ta,Tb,Tc,Ta1,Tb1,Tc1,Tsample,sap,sbp,scp,san,
sbn,scn] = fcn(vref,vdc,T,Ksh,f,time,sector_num,ssf1,ssf2,ssf3,
ssf4,ssf5,ssf6,Tsample)
%% Dr.Narayanaswamy. P.R.Iyer.
theta_dash = (2*pi*f*time - (sector_num - 1)*pi/(3));
T1 = (sqrt(3)*T*vref/(vdc))*(sin(pi/(3) - theta_dash));
T2 = (sqrt(3)*T*vref/(vdc))*(sin(theta_dash));
To = T - T1 - T2;
Tsho = Ksh*To;
To1 = (To - Tsho);%%zero space vector time i.e. OOO or PPP
Ta = ssf1*(T1+T2+0.5*To)+ssf2*(T1+0.5*To)+ssf3*(0.5*To)+
ssf4*(0.5*To)+ssf5*(T2+0.5*To)+ssf6*(T1+T2+0.5*To);
Tb = ssf1*(T2+0.5*To)+ssf2*(T1+T2+0.5*To)+ssf3*(T1+T2+0.5*To)+
ssf4*(T1+0.5*To)+ssf5*(0.5*To)+ssf6*(0.5*To);
Tc = ssf1*(0.5*To)+ssf2*(0.5*To)+ssf3*(T2+0.5*To)+ssf4*(T1+T2+
0.5*To)+ssf5*(T1+T2+0.5*To)+ssf6*(T1+0.5*To);
%%Program Segment 5.4 (Continued)
if (Ta >= Tsample)
  sap = 1;
else
sap = 0;
end
if (Tb >= Tsample)
  sbp = 1;
```

```
else
  sbp = 0;
end
if (Tc >= Tsample)
  scp = 1;
else
  scp = 0;
end
Ta1 = ssf1*(T1+T2+0.5*To1)+ssf2*(T1+0.5*To1)+ssf3*(0.5*To1)+
ssf4*(0.5*To1)+ssf5*(T2+0.5*To1)+ssf6*(T1+T2+0.5*To1);
Tb1 = ssf1*(T2+0.5*To1)+ssf2*(T1+T2+0.5*To1)+ssf3*(T1+T2+
0.5*To1)+ssf4*(T1+0.5*To1)+ssf5*(0.5*To1)+ssf6*(0.5*To1);

%%Program Segment 5.4(Continued)
Tc1 = ssf1*(0.5*To1)+ssf2*(0.5*To1)+ssf3*(T2+0.5*To1)+
ssf4*(T1+T2+0.5*To1)+ssf5*(T1+T2+0.5*To1)+ssf6*(T1+0.5*To1);
%%Program Segment 5.4(Continued)
if (Ta1 <= Tsample)
  san = 1;
else
  san = 0;
%%Program Segment 5.4(Continued)
end
if (Tb1 <= Tsample)
  sbn = 1;
else
  sbn = 0;
end
if (Tc1 <= Tsample)
  scn = 1;
else
  scn = 0;
end
-----------------------------------------------------------
```

```
%%Program segment 7.2
function [ea,eb,ec] = TRAPEZOID(k,f,t,phi)
%% Dr. Narayanaswamy P R Iyer
theta_e = 2*pi*f*t;
omega_e = 2*pi*f;
f1 = 0;f2 = 0;f3 = 0;ea = 0;eb = 0;ec = 0;
f1 = k*sin(theta_e + phi);
f2 = k*sin(theta_e - 2*pi/(3) + phi);
f3 = k*sin(theta_e - 4*pi/(3) + phi);
```

```
if(f1 >= 0)&& (f1 < 1)
  ea = f1;
elseif(f1 >= 1)
  ea = 1;
elseif(f1 < 1) && (f1 >= -1)
  ea = f1;
elseif(f1 <= -1)
  ea = -1;
  elseif(f1 > -1) && (f1 <= 0)
    ea = f1;
end
%%
if(f2 >= 0)&& (f2 < 1)
  eb = f2;
elseif(f2 >= 1)
  eb = 1;
elseif(f2 < 1) && (f2 >= -1)
  eb = f2;

%%Program segment 7.2(Continued)
elseif(f2 <= -1)
  eb = -1;
  elseif(f2 > -1) && (f2 <= 0)
    eb = f2;
end
%%
if(f3 >= 0)&& (f3 < 1)
  ec = f3;
elseif(f3 >= 1)
  ec = 1;
elseif(f3 < 1) && (f3 >= -1)
  ec = f3;
elseif(f3 <= -1)
  ec = -1;
  elseif(f3 > -1) && (f3 <= 0)
ec = f3;
end
--------------------------------------------------------------------------
--------------------------------------
```

```
%%Program segment 7.3
function [Tem,imax] = BLDCM_TORQUE(ea,eb,ec,ia,ib,ic,ke)
%% Dr.Narayanaswamy P R Iyer
%% ke is back EMF constant Volt/rad/Sec.
```

```
imax = 0;
if(ia >= ib) && (ia >= ic)
  imax = ia;
elseif(ib >= ia) && (ib >= ic)
  imax = ib;
elseif (ic >= ia) && (ic >= ib)
  imax = ic;
end
Tem = 2*ke*imax;
------------------------------------------------------------
```

```
%%Program Segment 7.4
function [ia_ref,ib_ref,ic_ref,sah,sal,sbh,sbl,sch,scl] =
STATOR_CURRENT_REF(imax,theta_e,ia,ib,ic,delta_i)
%%Dr. Narayanaswamy P R Iyer
ia_ref = 0;ib_ref = 0;ic_ref = 0;
x1 = sin(theta_e);
x2 = cos(theta_e);
if(x1 >= 0)&&(x1 < 0.866)
  ia_ref = +imax;
  ib_ref = -imax;
  ic_ref = 0;
end
if(x1 >= 0.866)
  ia_ref = +imax;
  ib_ref = 0;
  ic_ref = -imax;
end
if(x2 < -0.5)&&(x2 >= -1)
  ia_ref = 0;
ib_ref = +imax;
  ic_ref = -imax;
end
%%Program Segment 7.4 (Continued)
if(x1 < 0)&&(x1 >= -0.866)
  ia_ref = -imax;
  ib_ref = +imax;
  ic_ref = 0;
end
if(x1 < -0.866)
  ia_ref = -imax;
  ib_ref = 0;
  ic_ref = +imax;
end
```

```
if(x2 >= 0.5)&&(x2 < 1)
  ia_ref = 0;
  ib_ref = -imax;
  ic_max = +imax;
end
gain = 10;
sah = 0;sal = 0;sbh = 0;sbl = 0;sch = 0; scl = 0;
if(ia_ref >= 0) && (ia <= (ia_ref - delta_i))
  sah = 1*gain;
  sal = 0;
end
if(ia_ref >= 0) && (ia >= (ia_ref + delta_i))
  sah = 0;
  sal = 1*gain;
end
if(ia_ref < 0) && (ia >= (ia_ref + delta_i))
  sah = 0;
  sal = 1*gain;
end
if(ia_ref < 0) && (ia <= (ia_ref - delta_i))
  sah = 1*gain;
  sal = 0;
end
%%
if(ib_ref >= 0) && (ib <= (ib_ref - delta_i))
  sbh = 1*gain;
%%Program Segment 7.4(Continued)
  sbl = 0;
end
if(ib_ref >= 0) && (ib >= (ib_ref + delta_i))
  sbh = 0;
sbl = 1*gain;
end
if(ib_ref < 0) && (ib >= (ib_ref + delta_i))
  sbh = 0;
  sbl = 1*gain;
end
if(ib_ref < 0) && (ib <= (ib_ref - delta_i))
  sbh = 1*gain;
  sbl = 0;
end
if(ic_ref >= 0) && (ic <= (ic_ref - delta_i))
  sch = 1*gain;
  scl = 0;
end
```

```
if(ic_ref >= 0) && (ic >= (ic_ref + delta_i))
  sch = 0;
  scl = 1*gain;
end
if(ic_ref < 0) && (ic >= (ic_ref + delta_i))
  sch = 0;
  scl = 1*gain;
end
if(ic_ref < 0) && (ic <= (ic_ref - delta_i))
  sch = 1*gain;
  scl = 0;
end
----------------------------------------------------------
```

```
%%Program Segment 8.1
function [T1,T2,To,Ta1,Tb1,Tc1,Tsample,sap,san,sbp,sbn,scp,scn] =
fcn(vref,vdc,Tz,f,time,sector_num,ssf1,ssf2,ssf3,ssf4,ssf5,ssf6,
Tsample)
%% Dr.Narayanaswamy. P.R.Iyer.
theta_dash = (2*pi*f*time - (sector_num - 1)*pi/(3));
T1 = (sqrt(3)*Tz*vref/(vdc))*(sin(pi/(3) - theta_dash));
T2 = (sqrt(3)*Tz*vref/(vdc))*(sin(theta_dash));
To = Tz - T1 - T2;
Ta1 = ssf1*0.5*(T1+T2+0.5*To)+ssf2*0.5*(T1+0.5*To)+ssf3*0.5*
(0.5*To)+ssf4*0.5*(0.5*To)+ssf5*0.5*(T2+0.5*To)+ssf6*0.5*
(T1+T2+0.5*To);
Tb1 = ssf1*0.5*(T2+0.5*To)+
ssf2*0.5*(T1+T2+0.5*To)+ssf3*0.5*(T1+T2+0.5*To)+
ssf4*0.5*(T1+0.5*To)+ssf5*0.5*(0.5*To)+ssf6*0.5*(0.5*To);
Tc1 = ssf1*0.5*(0.5*To)+ssf2*0.5*(0.5*To)+
ssf3*0.5*(T2+0.5*To)+ssf4*0.5*(T1+T2+0.5*To)+
ssf5*0.5*(T1+T2+0.5*To)+ssf6*0.5*(T1+0.5*To);
if (Ta1 >= Tsample)
  sap = 1;
else
  sap = 0;
end
if (Tb1 >= Tsample)
  sbp = 1;
else
  sbp = 0;
end
if (Tc1 >= Tsample)
  scp = 1;
```

```
else
  scp = 0;
end
san = ~(sap);
sbn = ~(sbp);
scn = ~(scp);
---------------------------------------------------
----------------------------------------------------------
```

```
%% Program segment 8.3
function [delta_vqs, delta_vds] = fcn(fl_dr_dot,fl_dr,ws,Lm,Ls,Lr,ids,
iqs)
%% Dr. Narayanaswamy P R Iyer.
delta_vds = Lm*fl_dr_dot/(Lr) - ws*Ls*iqs*(1 - ((Lm*Lm)/(Ls*Lr)));
delta_vqs = Lm*fl_dr*ws/(Lr) + ws*Ls*ids*(1 - ((Lm*Lm)/(Ls*Lr)));
------------------------
%%Program segment 8.4
function [fl_ds_dot,fl_qs_dot] = fcn(vds,vqs,ids,iqs,rs,ws,fl_ds,fl_qs)
%% Dr. Narayanaswamy. P.R.Iyer.
fl_ds_dot = vds - rs*ids + ws*fl_qs;
%%Program segment 8.4(Continued)
fl_qs_dot = vqs - rs*iqs - ws*fl_ds;
--------------------------------------------------
-----------------------------------------------------------
%%Program segment 8.5
function [wsl,Tem] = fcn(iqs,fl_dr,Lm,Lr,tau_r,P)
%%Program segment 8.5(Continued)
%% Dr.Narayanaswamy. P.R.Iyer.
wsl = Lm*iqs/(tau_r*fl_dr);
Tem = 3*P*fl_dr*Lm*iqs/(4*Lr);
-------------------------------------------------------
------------------------------------------------------
```

```
%%Program Segment 8.6
function [z1,y1,y2,y3] = fcn(k,f,theta_fs,t)
%%Dr. Narayanaswamy P R Iyer
y1 = 0;
z1 = 0;
y2 = k*sin(2*pi*f*t);
y3 = k*cos(2*pi*f*t);
x1 = y2;
x2 = y3;
if ((x1 >= -0.5) && (x1 < 0.5)) && (x2 >= 0.866)
```

```
  y1 = 1;
 else if ((x1 >= 0.5) && (x1 < 1) && (x2 < 0.866) && (x2 >= 0))
 y1 = 2;
else if ((x1 < 1) && (x1 >= 0.5) && (x2 < 0) && (x2 >= -0.866))
y1 = 3;
else if ((x1 <= 0.5) && (x1 > -0.5) && (x2 <= -0.866))
y1 = 4;
----------------------------------------------------------------
%%Program Segment 8.7
function [ssf1,ssf2,ssf3,ssf4,ssf5,ssf6]=fcn(v_sector)
%% Dr. Narayanaswamy P R Iyer
if (v_sector == 1)
  ssf1 = 1;
  else ssf1 = 0;
end
    if (v_sector == 2)
      ssf2 = 1;
      else ssf2 = 0;
    end
    if (v_sector == 3)
      ssf3 = 1;
      else ssf3 = 0;
    end
    if (v_sector == 4)
      ssf4 = 1;
      else ssf4 = 0;
    end
    if (v_sector == 5)
      ssf5 = 1;
      else ssf5 = 0;
    end
 %%Program Segment 8.7(Continued)
    if (v_sector == 6)
      ssf6 = 1;
      else ssf6 = 0;
    end


 %%Program Segment 8.6(Continued)
 else if ((x1 >= -1) && (x1 < -0.5) && (x2 > -0.866) && (x2 <= 0))
 y1 = 5;
 else if ((x1 > -1) && (x1 <= -0.5) && (x2 > 0) && (x2 <= 0.866))
 y1 = 6;
 end
 end
```

```
end
end
end
end
%% For convenience add pi to theta_fs.
phi_fs = theta_fs + pi;
x3 = k*sin(phi_fs);
x4 = k*cos(phi_fs);
if ((x3 >= -0.5)  && (x3 < 0.5)) && (x4 >= 0.866)
  z1 = 4; %%As pi is added to theta_fs this corresponds to the range
-7*pi/6 to -5*pi/6 for %%theta_fs.  Sector z1=4.
  else if ((x3 >= 0.5)  && (x3 < 1) && (x4 < 0.866) && (x4 >= 0))
  z1 = 5; %%z1 = 5
  else if ((x3 < 1) && (x3 >= 0.5) && (x4 < 0)  && (x4 >= -0.866))
  z1 = 6; %%z1 = 6
 else if ((x3 <= 0.5) && (x3 > -0.5) && (x4 <= -0.866))
 z1 = 1; %%z1 = 1
 else if ((x3 >= -1) && (x3 < -0.5) && (x4 > -0.866)  && (x4 <= 0))
 z1 = 2; %%z1 = 2
 else if ((x3 > -1) && (x3 <= -0.5) && (x4 > 0)  && (x4 <= 0.866))
 z1 = 3; %%z1 = 3
 end
 end
 end
 end
 end
 end
------------------------------------------------------------
-----------------------------------------------------------
%%Program Segment 8.8
function [sap,san,sbp,sbn,scp,scn] = fcn(fl_stator,fl_sref,tem,
tem_ref,fls_sector,ssf1,ssf2,ssf3,ssf4,ssf5,ssf6)
%% Dr.Narayanaswamy. P.R.Iyer.
ton = 1; %%switch ON time
toff = 0; %%switch OFF time
%%DTC scheme of switching. Refer C_DTC switching scheme table.
%%Assume 10% hysresis band for e.m. torque and 1% hysteresis band for
stator flux.
%%tem_up = tem_ref + 0.1*tem_ref;%%tem_lo = tem_ref - 0.1*tem_ref;
%%fl_sup = fl_sref + 0.01*fl_sref;%%fl_slo = fl_sref - 0.01*fl_sref;
sap = 0; sbp = 0; scp = 0;
if((fl_sref - fl_stator) >= 0.1*fl_sref && (tem_ref - tem) >=
0.01*tem_ref)
  %%Stator flux increase and Torque increase.
  if(fls_sector == 1)
```

```
    %%S1-V2(110),S2-V3(010),S3-V4(011),S4-V5(001),S5-V6(101),S6-V1
(100)
    sap = ssf1*ton+ssf2*toff+ssf3*toff+ssf4*toff+ssf5*ton+ssf6*ton;
    sbp = ssf1*ton+ssf2*ton+ssf3*ton+ssf4*toff+ssf5*toff+ssf6*toff;
    scp = ssf1*toff+ssf2*toff+ssf3*ton+ssf4*ton+ssf5*ton+ssf6*toff;
else if(fls_sector == 2)
%%S2-V3(010),S3-V4(011),S4-V5(001),S5-V6(101),S6-V1(100),S1-V2
(110)

%%Program Segment 8.8(Continued)
        sap = ssf2*toff+ssf3*toff+ssf4*toff+ssf5*ton+ssf6*ton
+ssf1*ton;
        sbp = ssf2*ton+ssf3*ton+ssf4*toff+ssf5*toff+ssf6*toff
+ssf1*ton;
        scp = ssf2*toff+ssf3*ton+ssf4*ton+ssf5*ton+ssf6*toff
+ssf1*toff;
        else if(fls_sector == 3)
%%S3-V4(011),S4-V5(001),S5-V6(101),S6-V1(100),S1-V2(110),S2-V3
(010)

%%Program Segment 8.8(Continued)
            sap = ssf3*toff+ssf4*toff+ssf5*ton+ssf6*ton+ssf1*ton
+ssf2*toff;
            sbp = ssf3*ton+ssf4*toff+ssf5*toff+ssf6*toff+ssf1*ton
+ssf2*ton;
            scp = ssf3*ton+ssf4*ton+ssf5*ton+ssf6*toff+ssf1*toff
+ssf2*toff;
            else if(fls_sector == 4)
              %%S4-V5(001),S5-V6(10 1),S6-V1(100),S1-V2(110),S2-V3
(010),S3-V4(011)
                sap = ssf4*toff+ssf5*ton+ssf6*ton+ssf1*ton+ssf2*toff
+ssf3*toff;
               sbp = ssf4*toff+ssf5*toff+ssf6*toff+ssf1*ton+ssf2*ton
+ssf3*ton;
               scp = ssf4*ton+ssf5*ton+ssf6*toff+ssf1*toff+ssf2*toff
+ssf3*ton;
                else if(fls_sector == 5)
%%S5-V6(101),S6-V1(100),S1-V2(110),S2-V3(010),S3-V4(011),S4-V5
(001)
                sap = ssf5*ton+ssf6*ton+ssf1*ton+ssf2*toff+ssf3*toff
+ssf4*toff;
                sbp = ssf5*toff+ssf6*toff+ssf1*ton+ssf2*ton+ssf3*ton
+ssf4*toff;
               scp = ssf5*ton+ssf6*toff+ssf1*toff+ssf2*toff+ssf3*ton
+ssf4*ton;
```

```
                    else if(fls_sector == 6)
                %%S6-V1(100),S1-V2(110),S2-V3(010),S3-V4(011),S4-
V5(001),S5-V6(101)
                        sap = ssf6*ton+ssf1*ton+ssf2*toff+ssf3*toff
+ssf4*toff+ssf5*ton;
                        sbp = ssf6*toff+ssf1*ton+ssf2*ton+ssf3*ton
+ssf4*toff+ssf5*toff;
                        scp = ssf6*toff+ssf1*toff+ssf2*toff+ssf3*ton
+ssf4*ton+ssf5*ton;
                      end
                  end
              end
          end
      end
  end
end
if((fl_sref - fl_stator) >= 0.1*fl_sref && (tem_ref - tem) == 0)
  %%Stator flux increase and torque equal. S1-V7(111),S2-V0(000),S3-V7
(111),S4-V0(000),S5-V7(111),S6-V0(000)
  if(fls_sector == 1)
    sap = ssf1*ton + ssf2*toff + ssf3*ton + ssf4*toff + ssf5*ton +
ssf6*toff;
    sbp = ssf1*ton + ssf2*toff + ssf3*ton + ssf4*toff + ssf5*ton +
ssf6*toff;
    scp = ssf1*ton + ssf2*toff + ssf3*ton + ssf4*toff + ssf5*ton +
ssf6*toff;
  else if(fls_sector == 2)
      %%S2-V0(000),S3-V7(111),S4-V0(000),S5-V7(111),S6-V0(000),S1-
V7(111)
      sap = ssf2*toff+ssf3*ton+ssf4*toff+ssf5*ton+ssf6*toff
+ssf1*ton;
      sbp = ssf2*toff+ssf3*ton+ssf4*toff+ssf5*ton+ssf6*toff
+ssf1*ton;
      scp = ssf2*toff+ssf3*ton+ssf4*toff+ssf5*ton+ssf6*toff
+ssf1*ton;
    else if(fls_sector == 3)
      %%S3-V7(111),S4-V0(000),S5-V7(111),S6-V0(000),S1-V7(111),S2-
V0(000)
        sap = ssf3*ton+ssf4*toff+ssf5*ton+ssf6*toff+ssf1*ton
+ssf2*toff;
        sbp = ssf3*ton+ssf4*toff+ssf5*ton+ssf6*toff+ssf1*ton
+ssf2*toff;
        scp = ssf3*ton+ssf4*toff+ssf5*ton+ssf6*toff+ssf1*ton
+ssf2*toff;
      else if(fls_sector == 4)
```

```
        %%S4-V0(000),S5-V7(111),S6-V0(000),S1-V7(111),S2-V0(000),
S3-V7(111)
          sap = ssf4*toff+ssf5*ton+ssf6*toff+ssf1*ton+ssf2*toff
+ssf3*ton;
          sbp = ssf4*toff+ssf5*ton+ssf6*toff+ssf1*ton+ssf2*toff
+ssf3*ton;
          scp = ssf4*toff+ssf5*ton+ssf6*toff+ssf1*ton+ssf2*toff
+ssf3*ton;
        else if (fls_sector == 5)
         %%S5-V7(111),S6-V0(000),S1-V7(111),S2-V0(000),S3-V7(111),
S4-V0(000)
            sap = ssf5*ton+ssf6*toff+ssf1*ton+ssf2*toff+ssf3*ton
+ssf4*toff;
            sbp = ssf5*ton+ssf6*toff+ssf1*ton+ssf2*toff+ssf3*ton
+ssf4*toff;
            scp = ssf5*ton+ssf6*toff+ssf1*ton+ssf2*toff+ssf3*ton
+ssf4*toff;
          else if(fls_sector == 6)
              %%S6-V0(000),S1-V7(111),S2-V0(000),S3-V7(111),S4-V0
(000),S5-V7(111)
              sap = ssf6*toff+ssf1*ton+ssf2*toff+ssf3*ton+ssf4*toff
+ssf5*ton;
              sbp = ssf6*toff+ssf1*ton+ssf2*toff+ssf3*ton+ssf4*toff
+ssf5*ton;
              scp = ssf6*toff+ssf1*ton+ssf2*toff+ssf3*ton+ssf4*toff
+ssf5*ton;
            end
          end
        end
      end
    end
  end
end
if((fl_sref - fl_stator) >= 0.1*fl_sref && (tem_ref - tem) <=
-0.01*tem_ref)
  %%Stator flux increase and torque decrease.
  if(fls_sector == 1)
    %%S1-V6(101),S2-V1(100),S3-V2(110),S4-V3(010),S5-V4(011),S6-V5
(001)
    sap = ssf1*ton + ssf2*ton + ssf3*ton + ssf4*toff + ssf5*toff +
ssf6*toff;
    sbp = ssf1*toff + ssf2*toff + ssf3*ton + ssf4*ton + ssf5*ton +
ssf6*toff;
    scp = ssf1*ton + ssf2*toff + ssf3*toff + ssf4*toff + ssf5*ton +
ssf6*ton;
```

```
 %%Program Segment 8.8(Continued)
 else if(fls_sector == 2)
      %%S2-V1(100),S3-V2(110),S4-V3(010),S5-V4(011),S6-V5(001),S1-
V6(101)
      sap = ssf2*ton+ssf3*ton+ssf4*toff+ssf5*toff+ssf6*toff
+ssf1*ton;
      sbp = ssf2*toff+ssf3*ton+ssf4*ton+ssf5*ton+ssf6*toff
+ssf1*toff;
      scp = ssf2*toff+ssf3*toff+ssf4*toff+ssf5*ton+ssf6*ton
+ssf1*ton;
    else if(fls_sector == 3)
      %%S3-V2(110),S4-V3(010),S5-V4(011),S6-V5(001),S1-V6(101),S2-
V1(100)
        sap = ssf3*ton+ssf4*toff+ssf5*toff+ssf6*toff+ssf1*ton
+ssf2*ton;
        sbp = ssf3*ton+ssf4*ton+ssf5*ton+ssf6*toff+ssf1*toff
+ssf2*toff;
        scp = ssf3*toff+ssf4*toff+ssf5*ton+ssf6*ton+ssf1*ton
+ssf2*toff;
      else if(fls_sector == 4)
         %%S4-V3(010),S5-V4(011),S6-V5(001),S1-V6(101),S2-V1(100),
S3-V2(110)
          sap = ssf4*toff+ssf5*toff+ssf6*toff+ssf1*ton+ssf2*ton
+ssf3*ton;
          sbp = ssf4*ton+ssf5*ton+ssf6*toff+ssf1*toff+ssf2*toff
+ssf3*ton;
          scp = ssf4*toff+ssf5*ton+ssf6*ton+ssf1*ton+ssf2*toff
+ssf3*toff;
else if(fls_sector == 5)
%%S5-V4(011),S6-V5(001),S1-V6(101),S2-V1(100),S3-V2(110),S4-V3
(010)
             sap = ssf5*toff+ssf6*toff+ssf1*ton+ssf2*ton+ssf3*ton
+ssf4*toff;
             sbp = ssf5*ton+ssf6*toff+ssf1*toff+ssf2*toff+ssf3*ton
+ssf4*ton;
             scp = ssf5*ton+ssf6*ton+ssf1*ton+ssf2*toff+ssf3*toff
+ssf4*toff;
          else if(fls_sector == 6)
                %%S6-V5(001),S1-V6(101),S2-V1(100),S3-V2(110),S4-V3
(010),S5-V4(011)
                sap = ssf6*toff+ssf1*ton+ssf2*ton+ssf3*ton+ssf4*toff
+ssf5*toff;
                sbp = ssf6*toff+ssf1*toff+ssf2*toff+ssf3*ton+ssf4*ton
+ssf5*ton;
```

```
              scp = ssf6*ton+ssf1*ton+ssf2*toff+ssf3*toff+ssf4*toff
+ssf5*ton;
            end
          end
        end
      end
    end
  end
end
if((fl_sref - fl_stator) <= -0.1*fl_sref && (tem_ref - tem) >=
0.01*tem_ref)
  %%Stator flux decrease and torque increase.
  if(fls_sector == 1)
    %%S1-V3(010),S2-V4(011),S3-V5(001),S4-V6(101),S5-V1(100),S6-V2
(110)
    sap = ssf1*toff + ssf2*toff + ssf3*toff + ssf4*ton + ssf5*ton +
ssf6*ton;
    sbp = ssf1*ton + ssf2*ton + ssf3*toff + ssf4*toff + ssf5*toff +
ssf6*ton;
    scp = ssf1*toff + ssf2*ton + ssf3*ton + ssf4*ton + ssf5*toff +
ssf6*toff;
  else if(fls_sector == 2)
      %%S2-V4(011),S3-V5(001),S4-V6(101),S5-V1(100),S6-V2(110),S1-
V3(010)
      sap = ssf2*toff+ssf3*toff+ssf4*ton+ssf5*ton+ssf6*ton
+ssf1*toff;
      sbp = ssf2*ton+ssf3*toff+ssf4*toff+ssf5*toff+ssf6*ton
+ssf1*ton;
      scp = ssf2*ton+ssf3*ton+ssf4*ton+ssf5*toff+ssf6*toff
+ssf1*toff;
    else if(fls_sector == 3)
      %%S3-V5(001),S4-V6(101),S5-V1(100),S6-V2(110),S1-V3(010),S2-
V4(011)
        sap = ssf3*toff+ssf4*ton+ssf5*ton+ssf6*ton+ssf1*toff
+ssf2*toff;
        sbp = ssf3*toff+ssf4*toff+ssf5*toff+ssf6*ton+ssf1*ton
+ssf2*ton;
        scp = ssf3*ton+ssf4*ton+ssf5*toff+ssf6*toff+ssf1*toff
+ssf2*ton;
      else if(fls_sector == 4)
          %%S4-V6(101),S5-V1(100),S6-V2(110),S1-V3(010),S2-V4(011),
S3-V5(001)
          sap = ssf4*ton+ssf5*ton+ssf6*ton+ssf1*toff+ssf2*toff
+ssf3*toff;
          sbp = ssf4*toff+ssf5*toff+ssf6*ton+ssf1*ton+ssf2*ton
```

```
+ssf3*toff;
          scp = ssf4*ton+ssf5*toff+ssf6*toff+ssf1*toff+ssf2*ton
+ssf3*ton;
        else if(fls_sector == 5)
          %%S5-V1(100),S6-V2(110),S1-V3(010),S2-V4(011),S3-V5(001),
S4-V6(101)
            sap = ssf5*ton+ssf6*ton+ssf1*toff+ssf2*toff+ssf3*toff
+ssf4*ton;
            sbp = ssf5*toff+ssf6*ton+ssf1*ton+ssf2*ton+ssf3*toff
+ssf4*toff;
            scp = ssf5*toff+ssf6*toff+ssf1*toff+ssf2*ton+ssf3*ton
+ssf4*ton;
          else if(fls_sector == 6)
              %%S6-V2(110),S1-V3(010),S2-V4(011),S3-V5(001),S4-V6
(101),S5-V1(100)
              sap = ssf6*ton+ssf1*toff+ssf2*toff+ssf3*toff+ssf4*ton
+ssf5*ton;
              sbp = ssf6*ton+ssf1*ton+ssf2*ton+ssf3*toff+ssf4*toff
+ssf5*toff;
              scp = ssf6*toff+ssf1*toff+ssf2*ton+ssf3*ton+ssf4*ton
+ssf5*toff;
            end
          end
        end
      end
    end
  end
end
if((fl_sref - fl_stator) <= -0.1*fl_sref && (tem_ref - tem) == 0)
  %%Stator flux decrease and torque equal.
  if(fls_sector == 1)
    %%S1-V0(000),S2-V7(111),S3-V0(000),S4-V7(111),S5-V0(000),S6-V7
(111)
    sap = ssf1*toff+ssf2*ton+ssf3*toff+ssf4*ton+ssf5*toff+ssf6*ton;
    sbp = ssf1*toff+ssf2*ton+ssf3*toff+ssf4*ton+ssf5*toff+ssf6*ton;
%%Program Segment 8.8 (Continued)
    scp = ssf1*toff+ssf2*ton+ssf3*toff+ssf4*ton+ssf5*toff+ssf6*ton;
  else if(fls_sector == 2)
%%S2-V7(111),S3-V0(000),S4-V7(111),S5-V0(000),S6-V7(111),S1-V0
(000)
      sap = ssf2*ton+ssf3*toff+ssf4*ton+ssf5*toff+ssf6*ton
+ssf1*toff;
       sbp = ssf2*ton+ssf3*toff+ssf4*ton+ssf5*toff+ssf6*ton
+ssf1*toff;
      scp = ssf2*ton+ssf3*toff+ssf4*ton+ssf5*toff+ssf6*ton
```

```
+ssf1*toff;
    else if(fls_sector == 3)
%%S3-V0(000),S4-V7(111),S5-V0(000),S6-V7(111),S1-V0(000),S2-V7
(111)
        sap = ssf3*toff+ssf4*ton+ssf5*toff+ssf6*ton+ssf1*toff
+ssf2*ton;
        sbp = ssf3*toff+ssf4*ton+ssf5*toff+ssf6*ton+ssf1*toff
+ssf2*ton;
        scp = ssf3*toff+ssf4*ton+ssf5*toff+ssf6*ton+ssf1*toff
+ssf2*ton;
      else if(fls_sector == 4)
        %%S4-V7(111),S5-V0(000),S6-V7(111),S1-V0(000),S2-V7(111),
S3-V0(000)
          sap = ssf4*ton+ssf5*toff+ssf6*ton+ssf1*toff+ssf2*ton
+ssf3*toff;
          sbp = ssf4*ton+ssf5*toff+ssf6*ton+ssf1*toff+ssf2*ton
+ssf3*toff;
          scp = ssf4*ton+ssf5*toff+ssf6*ton+ssf1*toff+ssf2*ton
+ssf3*toff;
        else if(fls_sector == 5)
          %%S5-V0(000),S6-V7(111),S1-V0(000),S2-V7(111),S3-V0(000),
S4-V7(111)
            sap = ssf5*toff+ssf6*ton+ssf1*toff+ssf2*ton+ssf3*toff
+ssf4*ton;
            sbp = ssf5*toff+ssf6*ton+ssf1*toff+ssf2*ton+ssf3*toff
+ssf4*ton;
            scp = ssf5*toff+ssf6*ton+ssf1*toff+ssf2*ton+ssf3*toff
+ssf4*ton;
          else if(fls_sector == 6)
              %%S6-V7(111),S1-V0(000),S2-V7(111),S3-V0(000),S4-V7
(111),S5-V0(000)
              sap = ssf6*ton+ssf1*toff+ssf2*ton+ssf3*toff+ssf4*ton
+ssf5*toff;
              sbp = ssf6*ton+ssf1*toff+ssf2*ton+ssf3*toff+ssf4*ton
+ssf5*toff;
              scp = ssf6*ton+ssf1*toff+ssf2*ton+ssf3*toff+ssf4*ton
+ssf5*toff;
            end
          end
        end
      end
    end
  end
end
if((fl_sref - fl_stator) <= -0.1*fl_sref && (tem_ref - tem) <=
```

```
-0.01*tem_ref)
  %%Stator flux decrease and torque decrease.
  if(fls_sector == 1)
    %%S1-V5(001),S2-V6(101),S3-V1(100),S4-V2(110),S5-V3(010),S6-V4
(011)
    sap = ssf1*toff+ssf2*ton+ssf3*ton+ssf4*ton+ssf5*toff+ssf6*toff;
    sbp = ssf1*toff+ssf2*toff+ssf3*toff+ssf4*ton+ssf5*ton+ssf6*ton;
    scp = ssf1*ton+ssf2*ton+ssf3*toff+ssf4*toff+ssf5*toff+ssf6*ton;
    if(fls_sector == 2)
      %%S2-V6(101),S3-V1(100),S4-V2(110),S5-V3(010),S6-V4(011),S1-
V5(001)
      sap = ssf2*ton+ssf3*ton+ssf4*ton+ssf5*toff+ssf6*toff
+ssf1*toff;
      sbp = ssf2*toff+ssf3*toff+ssf4*ton+ssf5*ton+ssf6*ton
+ssf1*toff;
      scp = ssf2*ton+ssf3*toff+ssf4*toff+ssf5*toff+ssf6*ton
+ssf1*ton;
      if(fls_sector == 3)
      %%S3-V1(100),S4-V2(110),S5-V3(010),S6-V4(011),S1-V5(001),S2-
V6(101)
        sap = ssf3*ton+ssf4*ton+ssf5*toff+ssf6*toff+ssf1*toff
+ssf2*ton;
        sbp = ssf3*toff+ssf4*ton+ssf5*ton+ssf6*ton+ssf1*toff
+ssf2*toff;
        scp = ssf3*toff+ssf4*toff+ssf5*toff+ssf6*ton+ssf1*ton
+ssf2*ton;
        if(fls_sector == 4)
          %%S4-V2(110),S5-V3(010),S6-V4(011),S1-V5(001),S2-V6(101),
S3-V1(100)
          sap = ssf4*ton+ssf5*toff+ssf6*toff+ssf1*toff+ssf2*ton
+ssf3*ton;
          sbp = ssf4*ton+ssf5*ton+ssf6*ton+ssf1*toff+ssf2*toff
+ssf3*toff;
          scp = ssf4*toff+ssf5*toff+ssf6*ton+ssf1*ton+ssf2*ton
+ssf3*toff;
          if(fls_sector == 5)
          %%S5-V3(010),S6-V4(011),S1-V5(001),S2-V6(101),S3-V1(100),
S4-V2(110)
            sap = ssf5*toff+ssf6*toff+ssf1*toff+ssf2*ton+ssf3*ton
+ssf4*ton;
            sbp = ssf5*ton+ssf6*ton+ssf1*toff+ssf2*toff+ssf3*toff
+ssf4*ton;
            scp = ssf5*toff+ssf6*ton+ssf1*ton+ssf2*ton+ssf3*toff
+ssf4*toff;
```

```
            if(fls_sector == 6)
              %%S6-V4(011),S1-V5(001),S2-V6(101),S3-V1(100),S4-V2
(110),S5-V3(010)
              sap = ssf6*toff+ssf1*toff+ssf2*ton+ssf3*ton+ssf4*ton
+ssf5*toff;
              sbp = ssf6*ton+ssf1*toff+ssf2*toff+ssf3*toff+ssf4*ton
+ssf5*ton;
              scp = ssf6*ton+ssf1*ton+ssf2*ton+ssf3*toff+ssf4*toff
+ssf5*toff;
            end
          end
        end
      end
    end
  end
end
san = ~(sap);
sbn = ~(sbp);
scn = ~(scp);
------------------------------------------------------------
---------------------------------------------------------------------
----------------------------
```

```
%% Program Segment 8.9
function [v0,vdc,vs,vam,vbm,vcm] = fcn(f,t)
%% Dr. Narayanaswamy P R Iyer
%% Let offset voltage v0 be 2.5% of rated line to neutral voltage
%% v/f ratio is (400/1.732)*(1/50) which is 4.618
%% Program Segment 8.9(Continued)
v0 = (2.5/(100))*(400/(1.732));
vs = 4.618*f + v0;
if(vs > v0 + (400)/(1.732))
   vs = v0 + (400)/(1.732);
end
%% vs is line to neutral stator voltage which is vs = 1.414*vdc/(3)
for six step 180 degree mode conduction.
vdc = 3*(vs -v0)/(1.414);
%% Let 400/1.732 be the base voltage vbase
vbase = 400/(1.732);
vam = (vs/(vbase))*sin(2*pi*f*t);
vbm = (vs/(vbase))*sin(2*pi*f*t - 2*pi/(3));
vcm = (vs/(vbase))*sin(2*pi*f*t - 4*pi/(3));
```

```
%% Program Segment 8.10
function [v0,vdc,vs,vam,vbm,vcm] = fcn(f,t)
%% Dr. Narayanaswamy P R Iyer
%% Let offset voltage v0 be 2.5% of rated line to neutral voltage
%% v/f ratio is (400/1.732)*(1/50) which is 4.618
%% Program Segment 8.10(Continued)
v0 = (2.5/(100))*(400/(1.732));
vs = 4.619*f + v0;
if(vs > v0 + (400)/(1.732))
   vs = v0 + (400)/(1.732);
end
%% vs is line to neutral stator voltage which is vs = 1.414*vdc/(3) for six
step 180 degree mode conduction.
vdc = 3*(vs -v0)/(1.414);
%% Let 400/1.732 be the base voltage vbase
vbase = ((400)/(1.732)) + v0;
vam = (vs/(vbase))*sin(2*pi*f*t);
vbm = (vs/(vbase))*sin(2*pi*f*t - 2*pi/(3));
vcm = (vs/(vbase))*sin(2*pi*f*t - 4*pi/(3));
```

```
%% Program segment 8.11
function [vma1,vma2,vmb1,vmb2,vmc1,vmc2] = fcn(alfa,f,t)
%% Dr. Narayanaswamy P R Iyer
%% Line to Line Voltage is 400 V and frequency is 50 Hz.
alfa_rad = alfa*pi/(180);
vma1 = 10*sin(2*pi*f*t - alfa_rad);
vma2 = 10*sin(2*pi*f*t - alfa_rad - pi);
vmb1 = 10*sin(2*pi*f*t - 2*pi/(3) - alfa_rad);
vmb2 = 10*sin(2*pi*f*t - 2*pi/(3) - alfa_rad - pi);
vmc1 = 10*sin(2*pi*f*t - 4*pi/(3) - alfa_rad);
vmc2 = 10*sin(2*pi*f*t - 4*pi/(3) - alfa_rad - pi);
---------------------------------------------
---------------------------
```

```
%% Program segment 9.1
function [vs,v0,vdc,vma,vmb,vmc] = fcn(f,t)
%% Dr. Narayanaswamy P R Iyer
%% Line to Line Voltage is 400 V and frequency is 50 Hz.
%% V/f constant method.
%%V0 is assumed to be 2.5 % of line to neutral voltage.
v0 = (2.5/(100))*(400/(1.732));
vs = 4.619*f + v0;
%% vdc is DC link voltage and vs is line to neutral voltage.
%% vs = 1.414*vdc/3 for 180 degree mode inverter switching.
vdc = 3*vs/(1.414);
```

```
vbase = (400/(1.732)) +v0;
vma = (vs/(vbase))*sin(2*pi*f*t);
vmb = (vs/(vbase))*sin(2*pi*f*t - 2*pi/(3));
vmc = (vs/(vbase))*sin(2*pi*f*t - 4*pi/(3));
----------------------------------------------------
-------------------------------
 %%Program segment 9.3
%% Same as program segment 8.11.
------------------------------------------------
-------------------------------------
```

Appendix II: Model Projects

Model projects are presented in this Appendix from all the nine chapters.

Chapter 1

1. In the following problem, assume that the sine wave modulating signal frequency is 60 Hz, triangle carrier frequency is 2 kHz, and the DC link voltage of the three-phase two-level inverter is 286 volts.

Develop a model of the three-phase sine PWM inverter, and find the RMS value, THD, peak fundamental component value and third, fifth, seventh and ninth harmonic peak component values of (A) line to line voltage and (B) line to neutral voltage for modulation index 0.7 and 1.3. Also show the waveform of the three-phase line to line and line to neutral voltages.

Chapter 2

In the following problems, assume that the sine wave modulating signal frequency is 60 Hz, triangle carrier frequency is 2 kHz, and the DC link voltage of the three-phase two-level inverter is 286 volts.

1. Develop a model of the three-phase clipped sine PWM inverter, and find the RMS value, THD, peak fundamental component value and third, fifth, seventh and ninth harmonic peak component values of (A) line to line voltage and (B) line to neutral voltage for modulation indices 0.7 and 1.3. Also show the waveform of the three-phase line to line and line to neutral voltages.

N. P. R. Iyer, *Inverters and AC Drives*, Power Systems,
https://doi.org/10.1007/978-3-031-62784-2

2. Develop a model of the three-phase third harmonic injection sine PWM inverter, and find the RMS value, THD, peak fundamental component value and third, fifth, seventh and ninth harmonic peak component values of (A) line to line voltage and (B) line to neutral voltage for modulation indices 0.7 and 1.3. Also show the waveform of the three-phase line to line and line to neutral voltages.

3A. Develop a model of the three-phase DCTLI with multicarrier sine PWM having in-phase disposition (IPD) of carriers, and find the RMS value, THD, peak fundamental component value and third, fifth, seventh and ninth harmonic peak component values of (A) line to line voltage and (B) line to neutral voltage for modulation indices 0.7 and 1.3. Also show the waveform of the three-phase line to line and line to neutral voltages.

3B. Repeat problem 3(A) for a three-phase DCTLI with Phase Opposition Disposition of carriers.

4. Develop a model of the three-phase five-level cascade H-bridge inverter (FLCHBI) having two cascade H-bridges each with a DC link voltage of 143 V with multicarrier sine PWM, and find the RMS value, THD, peak fundamental component value and third, fifth, seventh and ninth harmonic peak component values of (A) line to line voltage and (B) line to neutral voltage for modulation index 0.7. Assume inverter output frequency is 60 Hz and the triangle carrier frequency is 2 kHz. Also assume (A) both level- and phase-shifted triangle carriers and (B) in-phase disposition (IPD) of the level-shifted triangle carriers. Also show the model and waveform of the three-phase line to line and line to neutral voltages.

5A. The parameters for a single-phase modular multilevel converter (MMC) using half H-bridge inverter cells are given in Table P2.1. Develop the model, and show the waveforms of the grid to ground voltage, sum of upper arm capacitor voltage, sum of lower arm capacitor voltage, voltage across upper and lower arm inductor, circulating current and upper arm and lower arm current assuming (A) NO submodule capacitor voltage compensation and (B) with submodule capacitor voltage balance.

5B. Repeat problem 5A for a three-phase MMC whose per phase parameters are given in Table P2.1.

Table P2.1 MMC Parameters

Parameter	Value
Rated power P	400 W
DC link voltage	120 V
No. of SMs per arm	3
SM capacitor	940 Mf
Arm inductance	5 m.H.
Carrier switching frequency	2500 Hz
Modulation index M	0.9

Chapter 3

1. The model parameters of a three-phase unipolar SHE-PWM inverter are given below:
DC link voltage = 100 volts; fundamental frequency = 50 Hz; modulation index $M = 0.6$; switching angles to eliminate four odd harmonics are $\alpha1 = 35.6$, $\alpha2 = 38.6$, $\alpha3 = 50.35$, $\alpha4 = 59.5$ and $\alpha5 = 64.65$ degrees, respectively. Develop a model and find the four odd harmonics eliminated from the line to ground voltage. Also find the peak fundamental value of line to ground voltage.

2. The model parameters of a three-phase bipolar SHE-PWM inverter are given below:
DC link voltage = 100 volts; fundamental frequency = 50 Hz; modulation index $M = 0.65$; switching angles to eliminate four non-triplen odd harmonics are $\alpha1 = 12.25$, $\alpha2 = 23.25$, $\alpha3 = 31.25$, $\alpha4 = 46.25$ and $\alpha5 = 52.25$ degrees, respectively. Develop a model and find the four non-triplen odd harmonics eliminated from the line to ground voltage. Also find the peak fundamental value of line to ground voltage.

3. Single-phase seven-level cascade H-bridge inverter with equal DC voltage source:
A single-phase three-cell seven-level cascade H-bridge inverter with equal DC voltage E for each cell is switched by SHE-PWM technique. The switching angles $\alpha1$, $\alpha2$ and $\alpha3$ are, respectively, 11.504, 28.717 and 57.106 degrees. Find the two odd harmonics eliminated from line to ground voltage in the range 1 to 20. Also find the peak fundamental component of the line to ground voltage, THD and modulation index M. The value of E is 100 volts and the inverter fundamental switching frequency is 50 Hz.

4. For a single-phase diode clamped five-level SHE-PWM inverter, the three switching angles are $\alpha1 = 12.462$, $\alpha2 = 34.114$ and $\alpha3 = 60.29$ degrees. Find the two odd harmonics eliminated in the range 1 to 20 and the THD of line to ground voltage. Also find the modulation index M. Assume DC link voltage is 100 volts and the inverter fundamental frequency is 50 Hz.

5. Verify the same problem given under problem 4 "single-phase diode clamped five-level SHE-PWM inverter" by developing the model for a single-phase flying capacitor five-level SHE-PWM inverter.

Chapter 4

1. A five-segment switching scheme also known as discontinuous space vector pulse width modulation switching scheme for a three-phase two-level inverter is shown in Table P4.1. For all odd sectors, the timing sequence is T0/2 -→Ta/2 -

Table P4.1 Five-segment SVPWM two-level inverter switching scheme

Sl. no.	Sector	Space vector and timing	Space vector and timing	Space vector and timing	Space vector and timing	Space vector and timing
1	I	V0 0 0 0 O O O	V1 1 0 0 P O O	V2 1 1 0 P P O	V1 1 0 0 P O O	V0 0 0 0 O O O
2	II	V0 0 0 0 O O O	V3 0 1 0 O P O	V2 1 1 0 P P O	V3 0 1 0 O P O	V0 0 0 0 O O O
3	III	V0 0 0 0 O O O	V3 0 1 0 O P O	V4 0 1 1 O P P	V3 0 1 0 O P O	V0 0 0 0 O O O
4	IV	V0 0 0 0 O O O	V5 0 0 1 O O P	V4 0 1 1 O P P	V5 0 0 1 O O P	V0 0 0 0 O O O
5	V	V0 0 0 0 O O O	V5 0 0 1 O O P	V6 1 0 1 P O P	V5 0 0 1 O O P	V0 0 0 0 O O O
6	VI	V0 0 0 0 O O O	V1 1 0 0 P O O	V6 1 0 1 P O P	V1 1 0 0 P O O	V0 0 0 0 O O O

→Tb -→Ta/2 -→T0/2, and for all even sectors, the timing sequence is T0/2 -→Tb/2 -→Ta -→Tb/2 -→T0/2, where (Ta + Tb + T0) is the sampling time Ts. The timings Ta, Tb and T0 are defined in Eqs. 4.13, 4.14 and 4.15, respectively.

Using the switching scheme for a three-phase SVPWM two-level inverter shown in Table P4.1, develop a model, and find the RMS value, peak fundamental and THD of line to line and line to neutral output voltages. Assume that the inverter DC link voltage is 100 V and fundamental frequency of inverter is 50 Hz. The sampling frequency fs is 1200 Hz and the modulation index ma is 0.8. Also find the third, fifth, seventh and ninth harmonic voltages for line to line and line to neutral output voltages.

2. Repeat the above five-segment switching scheme shown in Table P4.1 for a three-phase two-level SVPWM inverter as per data given in problem 1 above, with the triangle carrier replaced with a sawtooth carrier having frequency of 1200 Hz.
3. Consider the three-phase five-segment SVPWM two-level inverter switching scheme given in Table P4.1. Develop a model for the three-phase two-level inverter using the rigorous approach for gate pulse generation given in Sect. 4.2. Use the same parameters given in problem 1 above.

Chapter 5

1A. A Z-source three-phase inverter has the parameters shown in Table P5.1. Develop a model, and find (1) inverter DC link bridge voltage V_{PN}, (2) capacitor voltages $Vc1$ and $Vc2$, (3) inductor currents iL1 and iL2 and (4) boost factor B. Use maximum boost control technique.

1B. Repeat problem 1A when the control is using (a) simple boost, (b) third harmonic injection boost and (c) third harmonic injection maximum constant boost control techniques.

2A. A quasi Z-source three-phase inverter has the parameters shown in Table P5.2. Develop a model and find (1) inverter DC link bridge voltage V_{PN}, (2) capacitor voltages $Vc1$ and $Vc2$, (3) inductor currents iL1 and iL2 and (4) boost factor B. Use simple boost control technique. Shoot through duty ratio is D0 and triangle carrier period is T seconds.

2B. Repeat problem 2A when the control is using (a) maximum boost, (b) third harmonic injection boost and (c) third harmonic injection maximum constant boost control techniques.

3A. A single-phase Z-source seven-level cascade H-bridge inverter (SLCHBI) has the data shown in Table P5.1. Develop a model using three phase-shifted triangle carriers and simple boost control technique. Show the model and simulation results.

Table P5.1 Z-source inverter parameters

Parameter	Value	Unit
Z-source inductors L1 and L2	1.6	Millihenries
Z-source capacitors C1 and C2	1200	Microfarads
Load resistance	10	Ohms
Switching frequency	10	kHz
Inverter output frequency	60	Hz
DC source voltage	400	Volts
Shoot through duty ratio	0.2	–

Table P5.2 Quasi Z-source inverter parameters

Parameter	Value	Unit
Z-source inductors L1 and L2	1.8	Millihenries
Z-source capacitors C1 and C2	3300	Microfarads
Load resistance	50	Ohms
Filter inductor	1	Millihenries
Switching frequency	5	kHz
Inverter output frequency	60	Hz
DC source voltage	120	Volts
Shoot through duty ratio	0.2	–

3B. Repeat problem 3A for a three-phase Z-source seven-level cascade H-bridge inverter (SLCHBI) for (a) simple boost, (b) maximum boost, (c) third harmonic injection maximum boost and (d) third harmonic injection maximum constant boost control techniques. Show the model and simulation results.

4A. A single-phase quasi Z-source seven-level cascade H-bridge inverter (SLCHBI) has the parameters shown in Table P5.2. For each cascaded cell, the DC source voltage is 40 volts. Develop a model using three phase-shifted triangle carriers and simple boost control technique. Show the model and simulation results.

4B. Repeat problem 4A for a three-phase quasi Z-source seven-level cascade H-bridge inverter (SLCHBI) for (a) simple boost, (b) maximum boost, (c) third harmonic injection maximum boost and (d) third harmonic injection maximum constant boost control techniques. Show the model and simulation results.

5A. Develop a model of a single-phase Z-source NPC (ZSNPC) three-level inverter (ZSDCTLI). The parameters of the Z-source inverter are given in Table P5.1. Use simple boost control technique and two-triangle carrier Phase Opposition Disposition (POD) level shift modulation.

5B. Repeat problem 5A for a three-phase Z-source NPC three-level inverter. Use (a) simple boost and (b) maximum boost control technique and two triangle carriers in Phase Opposition Disposition (POD) level shift modulation.

6A. Develop a model for a single-phase quasi Z-source NPC three-level inverter whose parameters are given in Table P5.2. Use simple boost control technique and two triangle carriers in POD level shift modulation. Find (1) inverter DC link voltage vPN, (2) capacitor voltages Vc1 and Vc2 and (3) boost factor B.

6B. Repeat problem 6A for a three-phase quasi Z-source NPC three-level inverter. Use (a) simple boost and (b) maximum boost control technique and two triangle carriers in Phase Opposition Disposition (POD) level shift modulation. Show the model and simulation results.

Chapter 6

1A. The parameters for a three-phase switched inductor Z-source inverter (Fig. 6.1) are shown in Table P6.1. Develop a model using Third Harmonic Injection Maximum Boost (THIMB) Control algorithm, and determine (a) boost factor B; (b) capacitor voltages Vc1 and Vc2; (c) inductor currents iL1, iL2 and iL3; and (d) inverter bridge DC link voltage Vpn. Show model and the relevant simulation results.

1B. Repeat problem 1A for (a) simple boost (SB), (b) maximum boost (MB) and (c) third harmonic injection maximum constant boost (THIMCB) control techniques.

Table P6.1 Switched inductor Z-source inverter parameters

Sl. no.	Parameter	Value	Unit
1.	Z-source inductor L1, 12	1.5, 1.5	m.H., m.H.
2.	Z-source inductor L3	0.75	m.H.
3.	Z-source capacitor C1, C2	510, 510	μF., μF.
4.	Shoot through duty-ratio Do	0.3	–
5.	Modulation index M	0.846	–
6.	DC source voltage Vdc	105	Volts
7.	Carrier switching frequency fc	10	kHz
8.	Inverter output frequency fo	60	Hz
9.	Filter inductance and capacitance Lf, Cf	13, 10	m.H., μF.
10.	Load resistance and inductance R1, L1	180, 15	Ω, m.H.

Table P6.2 Switched inductor quasi Z-source SPFLCHBI parameters

Parameter	Value	Unit
Z-source inductors L1, L2 and L3	1.8	Millihenries
Z-source capacitors C1 and C2	3300	Microfarads
Load resistance	50	Ohms
Filter inductor	1	Millihenries
Switching frequency	5	kHz
Inverter output frequency	60	Hz
DC source voltage	60, 60	Volts
Shoot through duty ratio	0.2	–

2A. A switched inductor quasi Z-source single-phase cascade H-bridge five-level inverter has the parameters shown in Table P6.2. Develop a model, and find for each cascaded cell (a) boost factor B; (b) inductor currents iL1, iL2 and iL3; (c) capacitor voltages vC1 and vC2; mean AC line to neutral output voltage; and (d) mean and peak value of inverter DC link bridge voltage. Assume simple boost control (SBC) technique and two phase-shifted carriers.

2B. Repeat problem 2A for a three-phase switched inductor quasi Z-source five-level cascaded H-bridge inverter with two cascaded cells, using (a) SBC, (b) MBC, (c) THIMBC and (d) THIMCBC techniques. The parameters are the same as in Table P6.2.

3A. The three-phase switched inductor quasi Z-source boost inverter (SLQZSBI) topology is shown in Fig. P6.1. During shoot through period, switch SW1 is closed and is open during non-shoot through period. Shoot through duty ratio is Do and carrier switching period is *T*. Draw the equivalent circuit for the shoot through and non-shoot through period, and derive expression for (a) the capacitor voltage *V*c1, (b) peak value of inverter bridge DC link voltage *V*pn, (c) AC output phase to neutral voltage *V*an and (d) boost factor B.

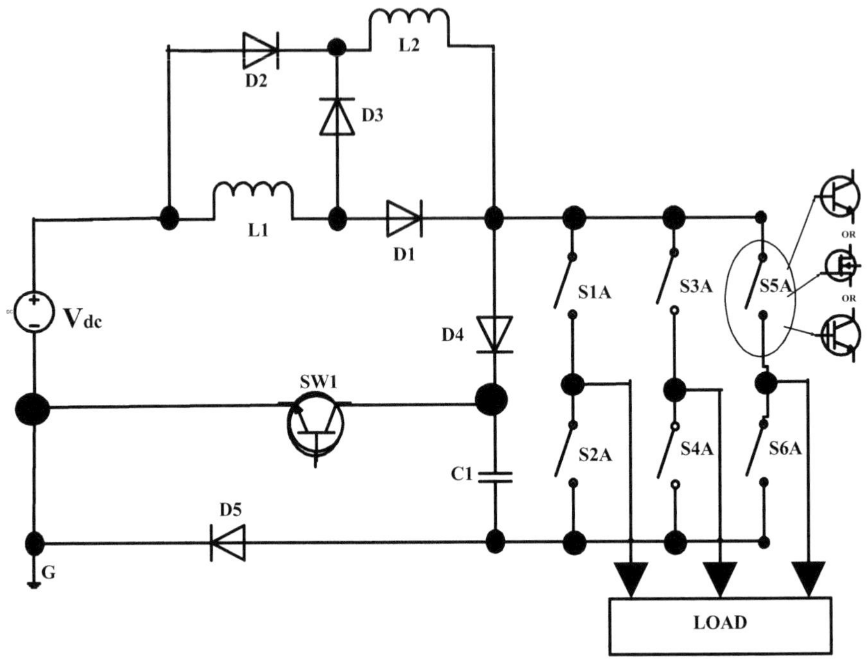

Fig. P6.1 Three-phase switched inductor quasi Z-source boost inverter

Table P6.3 SLQZSBI parameters

Parameter	Value	Unit
Quasi Z-source inductors L1, L2	1.2	Millihenries
Quasi Z-source capacitor C1	500	Microfarads
Load resistance RL	100	Ohms
Switching frequency	10	kHz
Inverter output frequency	60	Hz
DC source voltage	24	Volts
Shoot through duty ratio	0.25	–
Filter inductor	200	Micro henries
Filter capacitor	130	Microfarads

The model parameters for the above SLQZSBI are shown in Table P6.3. Develop a model and find (a) boost factor B, (b) capacitor voltage $Vc1$, (c) output line to neutral voltage Van and (d) DC link bridge voltage Vpn. Use third harmonic injection maximum boost control (THIMBC) technique.

3B. For the problem given in 3A, find the boost factor, capacitor voltage vC1, inverter bridge DC link voltage vpn and inductor currents iL1 and iL2 when the control is using (a) SB, (b) MB and (c) THIMCB techniques. Use parameters shown in Table P6.3.

Table P6.4 ASCQZSI parameters

Parameter	Value	Unit
Quasi Z-source inductors L1, L2	643	Micro henries
Quasi Z-source capacitors C1, C2, C3	100	Microfarads
Load resistance RL	40	Ohms
Load inductance LL	3.0	Millihenries
Switching frequency	5	kHz
Inverter output frequency	60	Hz
DC source voltage	30	Volts
Shoot through duty ratio	0.25	–

Table P6.5 ASCQZSI parameters

Parameter	Value	Unit
Quasi Z-source inductors L1, L2, L3	1	Millihenries
Quasi Z-source capacitors C1, C2	1000	Microfarads
Load resistance RL	100	Ohms
Switching frequency	5	kHz
Inverter output frequency	60	Hz
DC source voltage	40	Volts
Shoot through duty-ratio	0.3	–

4A. The topology of the three-phase active switched capacitor QZS inverter (ASCQZSI) is shown in Fig. P6.3. The switch SW1 is closed during the shoot through (ST) period and is open during non-shoot through (NST) period. The shoot through duty ratio is Do and the triangle carrier period is T. Derive expressions for (a) the three capacitor voltages vc1, vc2 and vc3, (b) inverter bridge DC link voltage vpn and (c) boost factor B and (d) output phase voltage van.

The model parameters for a three-phase ASCQZSI are shown in Table P6.4. Develop a model, and find by simulation (a) capacitor voltages vc1, vc2 and vc3, (b) inductor currents iL1 and iL2, (c) DC link bridge voltage of inverter vpn and (d) inverter output voltages. Use maximum boost control (MBC) technique.

4B. Repeat problem 4A by developing a model, and find by simulation (a) capacitor voltages vc1, vc2 and vc3, (b) inductor currents iL1 and iL2, (c) DC link bridge voltage of inverter vpn and (d) inverter line to line and line to neutral output voltages when the control is using (a) SB, (b) THIMB and (c) THIMCB techniques.

5. The topology of the three-phase extended active switched capacitor quasi Z-source inverter (EASCQZSI) is shown in Fig. P6.5. During the ST period, switch SW1 is closed and is open during NST period. The shoot through duty ratio is Do and the triangle carrier period is T. Derive expressions for (a) capacitor voltages *V*c1 and *V*c2, (b) inverter bridge DC link voltage *V*pn, (c) output phase to neutral voltage *V*an and (d) boost factor B. Assume inductors L2 and L3 are equal.

5A. The parameters of a three-phase ASCQZSI in Fig. 6.28 are shown in Table P6.5. Develop a model, and find by simulation (a) capacitor voltages *Vc*1 and *Vc*2, (b) inverter bridge DC link voltage *Vpn*, (c) boost factor B, (d) inductor currents iL1 and iL2 and (e) inverter output line to line and line to neutral voltages. Use third harmonic injection maximum boost control (THIMBC) technique.

5B. Repeat problem 5A when the control is using (a) SB, (b) MB and (c) THIMCB control techniques. Show relevant waveforms and tabulate the results.

6A. A single-phase diode-assisted extended boost quasi Z-source neutral point clamped three-level full H-bridge inverter (DAEB_QZS_NPCTLFHBI) topology is shown in Fig. P6.7. Draw the equivalent circuit for (a) NST period for zero output voltage, (b) NST period when delivering power to the load and (c) FST period when all the four switches of the inverter are simultaneously turned ON. Assume duty ratio Dz for the NST period when output voltage is zero, Da for the NST period when delivering power to the load and D0 for the FST period. Derive expressions for capacitor voltages vc1, vc2 and vc3, inverter bridge DC link voltage vpn, boost factor B and the line to neutral output voltage van. Use level-shifted Phase Opposition Disposition (POD) of triangle carriers with carrier period *T* seconds to generate PWM gate pulse for the inverter.

6A. The parameters of DAEB_QZS_NPCTLFHBI are shown in Table P5.2. Assume that inductor L3 and capacitor C3 have the value same as that for L1 and L2 and C1 and C2 shown in Table P5.3. Develop the model, and show waveforms for (a) capacitor voltages vc1, vc2 and vc3, (b) DC link bridge voltage vpn and (c) line to neutral output voltage van and (d) load voltage. Find the FST duty ratio D0 obtained by simulation. Tabulate the results for theoretical, calculated and model simulation results. Use two level-shifted carriers in POD and simple boost control technique to generate gate pulse.

6B. Also develop the model, tabulate results, and show all the above waveforms when NPC half H-bridge inverter (NPCHHBI) is used.

7A. Repeat problems 6A and 6B for a three-phase topology. Use the same parameters shown in Table P5.2.

Chapter 7

1. The parameters of a three-phase star-connected permanent magnet synchronous motor (PMSM) drive are shown in Table P7.1. Develop a model for the vector control of this PMSM drive when fed by a three-phase space vector PWM inverter. The load torque is varied from the initial value of 7 Nw-metres to 3.5

Table P7.1 PMSM Parameters

Sl. no.	Parameter	Value	Unit
1	Number of phases	3	–
2	DC link voltage	50	Volts
3	Stator resistance per phase Rs	1.5	Ohms
4	Stator dq-axis inductance per phase Ld, Lq	7, 5.8	Millihenries
5	Number of poles P	6	–
6	Moment of inertia J	0.0038	Kg.m^2
7	Damping constant D	0.00035	Nw.meter.Sec per Mech.rad.
8	Rotor magnet constant λm	0.1546	Volt.sec per elec.rad.
9	Torque	7	Newton-metre
10	Rotor speed	500	RPM
11	Sampling frequency Fs	20	Kilohertz

Table P7.2 PMBLDCM parameters

S. no.	Parameter	Value	Unit
1	Terminal voltage *Vdc*	100	Volts DC
2	Rated speed	2820	RPM
3	Rated torque	2.9588	Nw-M
4	Peak torque	18.1	Nw-M
5	Rated current	10.20	Amps
6	Rated power	874	Watts
7	Torque constant k_T	0.3269	Nw-metre per ampere
8	Back EMF constant k_e	34.20	Volts per KRPM
9	Stator resistance R_s	0.408	Ohms per phase
10	Stator inductance L	1.710	Millihenries per phase
11	Rotor inertia J	4940	gm-cm^2
12	Number of poles P	8	–

Nw-metres in 1 second maintaining a rotor speed of 500 RPM. Show the rotor speed response, stator abc-axis currents and the electromagnetic torque. For space vector PWM, assume a sampling frequency of 20 kHz.

2. The data sheet for a six-step 120-degree mode inverter-fed permanent magnet brushless DC motor (PMBLDCM) drive is shown in Table P7.2. The load torque is initially zero and at 50e-3 seconds the value is stepped up to 2.9588 newton-metres. Develop an open loop model, and show the waveforms for (a) three-phase stator applied voltage, (b) rotor electromagnetic torque, (c) load torque, (d) rotor speed, (e) three-phase stator currents and (f) three-phase trapezoidal back e.m.f.

3A. The data sheet for a six-step 120-degree mode inverter-fed permanent magnet brushless DC motor (PMBLDCM) drive is shown in Table P7.3. The load torque is initially zero and at 50e-3 seconds the value is stepped up to 0.7

Table P7.3 PMBLDCM parameters

S. no.	Parameter	Value	Unit
1	Terminal voltage *Vdc*	160	Volts DC
2	Rated speed	700	RPM
3	Rated torque	0.7	Nw-M
4	Number of phase	3	–
5	Type of stator connection	Star	–
6	Number of poles *P*	4	–
7	Rated power	1500	Watts
8	Torque constant k_T	0.045	Nw-metre per ampere
9	Flux linkage constant	0.105	Weber
10	Stator resistance R_s	0.7	Ohms per phase
11	Stator self-inductance *Ls*	2.72	Millihenries per phase
12	Stator mutual inductance *M*	1.5	Millihenries per phase
13	Rotor inertia *J*	0.000284	kg-m^2
14	Damping constant *B*	0.02	N-m/rad/sec.

Table P7.4 Hysteresis current controlled PMBLDCM parameters

S. no.	Parameter	Value	Unit
1	Terminal voltage *Vdc*	500	Volts DC
2	Rated torque	16	Nw-M
3	Number of phase	3	–
4	Type of stator connection	Star	–
5	Number of poles *P*	8	–
6	Torque constant k_T	0.5	Nw-metre per ampere
7	Back e.m.f. constant *Ke*	18.3260	Volts per KRPM
8	Stator resistance R_s	0.2	Ohms per phase
9	Stator inductance *L* (L_s-*M*)	8.53	Millihenries per phase
10	Load torque	16	Newton-metres
11	Rotor inertia *J*	0.0083	kg-m^2
12	PI controller constants Kp, Ki	0.1125, 9.375	–

newton-metres. Develop an open loop model, and show the waveforms for (a) three-phase stator applied voltage, (b) rotor electromagnetic torque, (c) load torque, (d) rotor speed, (e) three-phase stator currents and (f) three-phase trapezoidal back e.m.f.

3B. For the problem in 3A, compare the result with that obtained by using the three-phase BLDCM model in Simscape – Electrical – Electromechanical – Permanent Magnet block set.

4. The data for a three-phase hysteresis current controlled PMBLDCM drive is shown in Table P7.4. Design a hysteresis current controller to maintain the rotor speed of 35 Mech.rad per second. Assume the load is 16 newton-metres applied at the time of starting.

Chapter 8

1. The parameters of a three-phase induction motor (IM) are given in Table P8.1. Develop a state space model using flux linkage equations in stationary reference frame ($\omega_C = 0$), and show the waveforms of three-phase stator currents, rotor currents, electromagnetic torque, load torque and rotor speed. Assume that the load torque is initially zero and at 0.5 seconds, a step change in load torque occurs from zero to 12 newton-metres.

2. Develop the model of a space vector PWM (SVPWM) three-phase two-level inverter-fed IM drive. The parameters of the three-phase IM are given in Table P8.1. The sampling frequency fs is 2.5 kHz. The amplitude modulation index is 0.8. Use the rigorous model for three-phase SVPWM inverter shown in Sect. 4.2.2 of Chap. 4. Show the waveforms of (a) stator currents, (b) rotor currents, (c) rotor speed, (d) rotor electromagnetic torque, (d) load torque and (e) three-phase line to line voltage of the inverter. Assume that the load is initially zero and steps up to 12 Nw-metres at 0.5 seconds.

3. Develop the model of a space vector PWM (SVPWM) inverter-fed vector controlled three-phase IM drive. The parameters of the three-phase IM are given in Table P8.1. The sampling frequency fs is 2.5 kHz. The amplitude modulation index is 0.8. Use the simplified model for three-phase SVPWM inverter shown in Sect. 8.6 of Chap. 8. Show the waveforms of (a) stator currents, (b) rotor currents, (c) rotor speed, (d) rotor electromagnetic torque, (e) load torque and (e) three-phase line to line voltage of the inverter. Assume that the load is initially zero and steps up to 12 Nw-metres in 1 second.

Table P8.1 Three phase IM parameters

Sl. no.	Parameter	Value	Unit
1)	Power output	2.2	kW
2)	DC link voltage	540	Volts
3)	Phase to neutral voltage	220	Volts
4)	Frequency	50	Hz
5)	Number of poles	4	–
6)	Stator connection	Star	–
7)	Rated speed	1450	RPM
8)	Stator resistance per phase R_s	2.23	Ω
9)	Rotor resistance per phase $R_{r'}$	1.15	Ω
10)	Stator leakage inductantance Lls per phase	0.0112	H
11)	Rotor leakage inductance Llr’ per phase	0.0112	H
12)	Mutual inductance M per phase	0.1988	H
13)	Motor inertia J	0.055	Kg-m^2
14)	Damping constant D	0	Nw.m.Sec.

Table P8.2 DTC switching table

δλs	δTe	Sector	Sector	Sector	Sector	Sector	Sector
		S1	S2	S3	S4	S5	S6
FI	TI	V2	V3	V4	V5	V6	V1
	TE	V0	V7	V0	V7	V0	V7
	TD	V1	V2	V3	V4	V5	V6
FD	TI	V4	V5	V6	V1	V2	V3
	TE	V7	V0	V7	V0	V7	V0
	TD	V5	V6	V1	V2	V3	V4

4. Develop a model for the three-phase vector controlled IM whose parameters are given in Tables P8.1 and P8.2. Show the simulation results for (a) E.M. torque and (b) rotor speed.

In Table P8.2, FI = flux increase, FD = flux decrease, TI = torque increase, TE = torque equal, and TD = torque decrease. Also voltage space vectors are V1 (100), V2(110), V3(010), V4(011), V5(001), V6(101), V7(111) and V0(000).

5A. Develop the model of a space vector PWM (SVPWM) inverter-fed direct torque controlled (DTC) three-phase IM drive. Use torque control method where NO PI controller is used. The parameters of the three-phase IM are given in Table P8.1. The sampling frequency fs is 2.5 kHz. Use the modified space vector control method. Also develop the model for determining the voltage sector and stator flux sector location. Show the waveforms of (a) stator currents, (b) rotor currents, (c) rotor speed, (d) rotor electromagnetic torque, (d) load torque and (e) three-phase line to line voltage of the inverter. Assume that the load is initially 6 Nw-metres and steps up to 12 Nw-metres at 0.5 seconds. Note that in the modified space vector control strategy for voltage and stator flux, sector 1 lies between 0 and 60 degrees and sectors 2 to 6 lie in the order 60 to 120, 120 to 180, 180 to 240, 240 to 300 and 300 to 360 (0) degrees, respectively, as in Fig. 8.32 of Chap. 8. The modified DTC switching table is given in Table P8.2. Also assume stator flux reference is 0.8736 Wb and electromagnetic torque reference is 41.3718 Nw-metres. The hysteresis limit for stator flux and torque are 10% and 1%, respectively.

5B. In problem 5A for the three-phase DTC IM model without PI controller, the sampling frequency is increased to 100 kHz. Show the model response for (a) stator currents, (b) rotor currents, (c) rotor speed, (d) rotor electromagnetic torque, (e) load torque and (f) stator flux.

5C. In the problem 5A for the three-phase DTC IM without PI controller, a PI controller is now added whose data are Kp = 0.7898 and Ki = 6.3492. The rotor speed reference is 141.3717 Mech.rad per second. The stator flux reference is 0.8736 webers. The sampling frequency is 100 kHz. Develop the model using speed control method, and show the waveforms of (a) stator currents, (b) rotor currents, (c) rotor speed, (d) rotor electromagnetic torque, (e) load torque and (f) stator flux. The hysteresis limit for stator flux and torque are 10% and 1% of their respective reference values.

5D. For the DTC of three-phase IM using PI controller given in problem 5C, the sampling frequency is changed to 5 kHz. Develop a model, and show the simulation results for (a) electromagnetic torque, (b) rotor speed, (c) load torque, (d) stator currents, (e) rotor currents, (f) output voltage of inverter, (g) stator flux and e.m. torque reference Tem_ref.

Chapter 9

1. For the vector controlled PMSM drive whose data are shown in Table P7.1, develop a fuzzy logic controller for the speed control loop, and show the results for (a) three-phase stator currents, (b) rotor speed, (c) electromagnetic torque and (d) three-phase line to line voltage of inverter. Show the model and compare the results with that obtained in exercise problem 7.1 in Chap. 7. Use a suitable FIS file presented in Chap. 9.

2A. For the DTC of three-phase IM with PI controller given in exercise problem 5C in Chap. 8, the rotor speed PI controller is now replaced with a fuzzy logic controller (FLC). Develop the model, and show the simulation results for (a) E. M. torque, (b) rotor speed, (c) load torque, (d) stator currents, (e) rotor currents, (f) inverter output voltage, (g) stator flux and (h) E.M. torque reference Tem_ref. Use a suitable FIS file presented in Chap. 9.

2B. In problem 2A, the FIS file is now changed to "imvectorcontrolnew" whose MFs for e, cie and cs are shown in Fig. 9.23 in Chap. 9. Now by simulation show the model response for (a) E.M. torque, (b) rotor speed, (c) load torque, (d) stator currents, (e) rotor currents, (f) inverter output voltage, (g) stator flux and (h) E.M. torque reference Tem_ref.

3A. For the vector control of three-phase IM using PI controller shown in exercise problem 4 in Chap. 8, the rotor speed PI controller is now replaced with FLC. Develop the model and show the results for (a) E.M. torque and (b) rotor speed. Compare the results with that obtained using PI controller. Use a suitable FIS file presented in Chap. 9.

3B. For the vector control of three-phase IM using PI controller shown in exercise problem 4 in Chap. 8, the rotor speed PI controller and direct axis stator current PI controller are both replaced with FLCs. Develop the model and show the results for (a) E.M. torque and (b) rotor speed. Compare the results with that obtained using PI controller and that using FLC for speed PI controller only. Use a suitable FIS file presented in Chap. 9.

Appendix III: Answers to Selected Model Projects

Answers to selected model projects are presented in this Appendix from all the nine chapters.

Chapter 1

1. For a modulation index of 0.7, VLL(VAB): RMS = 178 V, THD = 104.7%, V1 (peak) = 173.8 V, V3(peak) = 0.898 V, V5(peak) = 0.9149 V, V7 (peak) = 0.9415 V, V9(peak) = 0.9793 V. VLN(VAN): RMS = 102.6 V, THD = 104.9%, V1(peak) = 100.1 V, V3(peak) = 0.3918 V, V5 (peak) = 0.4085 V, V7(peak) = 0.434 V, V9(peak) = 0.469 volts.
   For a modulation index of 1.3, VLL(VAB): RMS = 227.3 V, THD = 55.02%, V1(peak) = 281.6 V, V3(peak) = 1.395 V, V5(peak) = 10.2 V, V7 (peak) = 1.254 V, V9(peak) = 1.627 V. VLN(VAN): RMS = 130.9 V, THD = 55.1%, V1(peak) = 162.2 V, V3(peak) = 0.9164 V, V5(peak) = 5.432 V, V7(peak) = 1.244 V, V9(peak) = 1.086 V.

Chapter 2

1. For a modulation index of 0.7, VLL(VAB): RMS = 195.8 V, THD = 79.33%, V1(peak) = 216.9 V, V3(peak) = 1.106 V, V5(peak) = 24.73 V, V7 (peak) = 9.129 V, V9(peak) = 0.8789 V. VLN(VAN): RMS = 112.9 V, THD = 79.43 V, V1(peak) = 125 V, V3(peak) = 0.5853 V, V5(peak) = 14.67 V, V7(peak) = 5.087 V, V9(peak) = 0.5192 V.

N. P. R. Iyer, *Inverters and AC Drives*, Power Systems,
https://doi.org/10.1007/978-3-031-62784-2

For a modulation index of 1.3, VLL(VAB): RMS = 234 V, THD = 44.01%, V1 (peak) = 302.9 V, V3(peak) = 1.41 V, V5(peak) = 13.84 V, V7(peak) = 4.279 V, V9(peak) = 1.76 V. VLN(VAN): RMS = 134.8 V, THD = 44.1%, V1 (peak) = 174.5 V, V3(peak) = 0.9451 V, V5(peak) = 8.459 V, V7 (peak) = 2.86 V, V9(peak) = 1.17 V.

2. For a modulation index of 0.7, VLL(VAB): RMS = 190.8 V, THD = 90.7%, V1(peak) = 199.9 V, V3(peak) = 0.9883 V, V5(peak) = 1.012 V, V7 (peak) = 1.048 V, V9(peak) = 1.099 V. VLN(VAN): RMS = 110 V, THD = 90.81 V, V1(peak) = 115.1 V, V3(peak) = 0.5179 V, V5 (peak) = 0.5391 V, V7(peak) = 0.5715 V, V9(peak) = 0.6162 V.

   For a modulation index of 1.3, VLL(VAB): RMS = 234 V, THD = 43%, V1 (peak) = 304 V, V3(peak) = 1.426 V, V5(peak) = 17.54 V, V7(peak) = 1.282 V, V9(peak) = 1.725 V. VLN(VAN): RMS = 134.8 V, THD = 43.1%, V1 (peak) = 175.1 V, V3(peak) = 0.953 V, V5(peak) = 10.6 V, V7 (peak) = 1.229 V, V9(peak) = 1.142 V.

3A. For a modulation index of 0.7, VLL(VAB): RMS = 131.8 V, THD = 45.41%, V1(peak) = 169.7 V, V3(peak) = 0.5576 V, V5(peak) = 0.1589 V, V7 (peak) = 0.4358 V, V9(peak) = 0.6347 V. VLN(VAN): RMS = 76.16 V, THD = 45.24 V, V1(peak) = 98.13 V, V3(peak) = 0.5034 V, V5 (peak) = 0.3117 V, V7(peak) = 0.3724 V, V9(peak) = 0.5235 V.

   For a modulation index of 1.3, VLL(VAB): RMS = 203.6 V, THD = 27.84%, V1(peak) = 277.4 V, V3(peak) = 0.0866 V, V5 (peak) = 10.03 V, V7(peak) = 1.539 V, V9(peak) = 0.1454 V. VLN(VAN): RMS = 117.6 V, THD = 27.87%, V1(peak) = 160.2 V, V3(peak) = 0.0551 V, V5(peak) = 5.808 V, V7(peak) = 0.9328 V, V9(peak) = 0.0739 V.

4A. For a modulation index of 0.7, VLL(VAB): RMS = 410.6 V, THD = 17.51%, V1(peak) = 572 V, V3(peak) = 0.1926 V, V5(peak) = 15.49 V, V7 (peak) = 9.718 V, V9(peak) = 0.5626 V. VLN(VAN): RMS = 237 V, THD = 17.45%, V1(peak) = 330.2 V, V3(peak) = 0.0952 V, V5 (peak) = 8.865 V, V7(peak) = 5.601 V, V9(peak) = 0.311 volts.

4B. For a modulation index of 0.7, VLL(VAB): RMS = 408.5 V, THD = 14.11%, V1(peak) = 572 V, V3(peak) = 0.1219 V, V5(peak) = 15.67 V, V7 (peak) = 9.776 V, V9(peak) = 0.1306 V. VLN(VAN): RMS = 235.8 V, THD = 14.12%, V1(peak) = 330.2 V, V3(peak) = 0.0682 V, V5 (peak) = 9.021 V, V7(peak) = 5.697 V, V9(peak) = 0.1056 volts.

5A. There is no SM capacitor voltage compensation, i.e. Kp = 0. Simulation results are shown below:

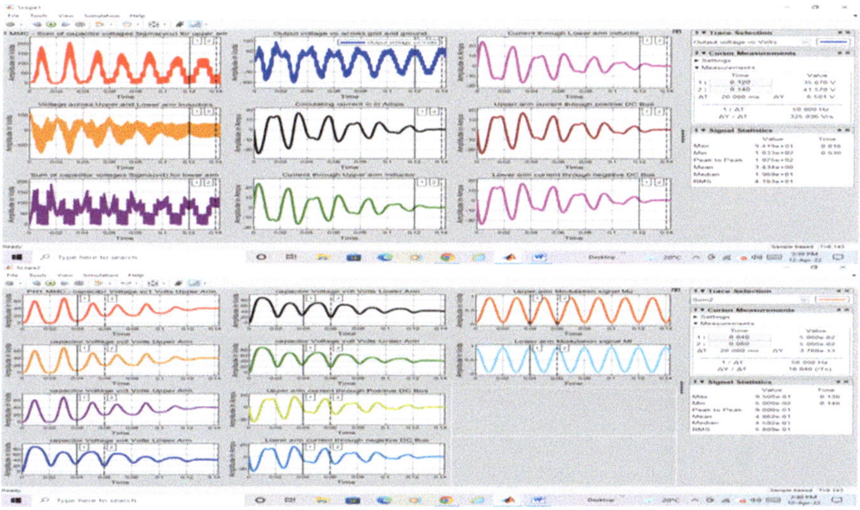

5A) (a) Simulation results for single-phase MMC with NO SM capacitor voltage compensation With SM capacitor voltage compensation by letting Kp = 0.03. The simulation results are given below:

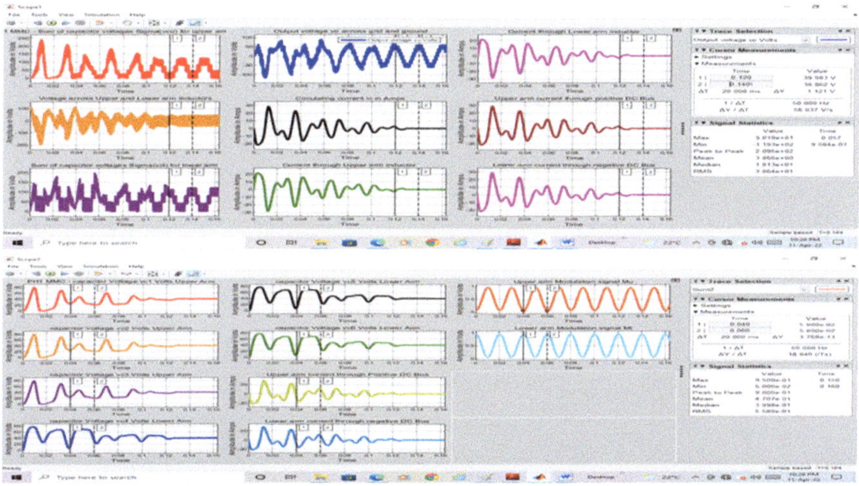

5A) (b) Simulation result for single-phase MMC with SM capacitor voltage balance

Chapter 3

1. Three-phase unipolar SHE-PWM inverter line to ground voltage with 5th, 7th, 9th and 11th harmonics eliminated is given below:

DC link voltage is 100 V, modulation index $M = 0.6$, peak fundamental voltage V1 = $(2*M*V_{dc}/\pi)$ which gives V1 = (2*0.6*100/3.14) = 38.216 V, whereas from simulation result, the value of V1 is found to be 37.58 V, and THD of VLG is 69.01%.

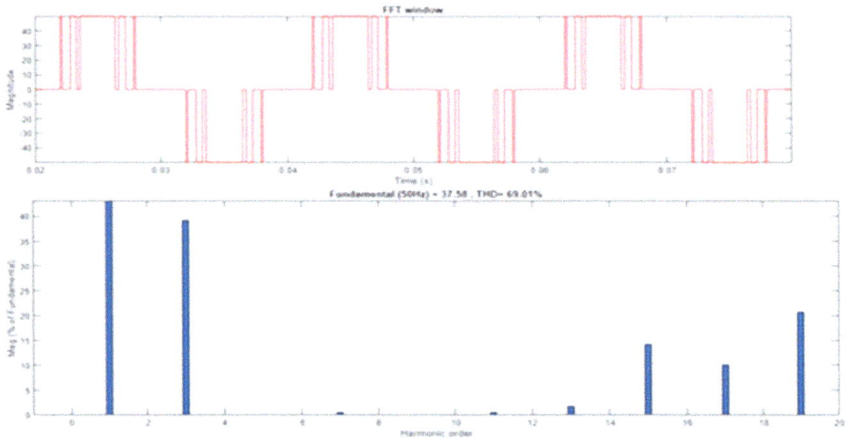

Three Phase Unipolar SHE-PWM 50 Hz Inverter Line to Ground Voltage: Modulation Index M is 0.6, Switching angles alfa1, alfa2, alfa3, alfa4 and alfa5 are [35.6 38.6 50.35 59.5 64.65] degrees. Harmonics Eliminated are 5th, 7th, 9th and 11th from line to ground voltage. Fundamental is 37.58 V (Peak).

2. Three-phase bipolar SHE-PWM inverter model and simulation results to eliminate non-triplen odd harmonics 5th, 7th, 11th and 13th from the line to ground voltage are given below:

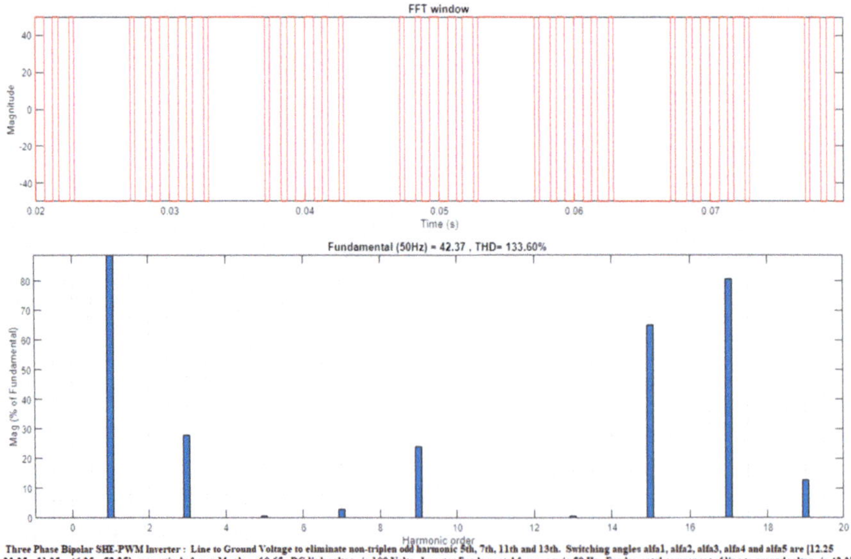

Three Phase Bipolar SHE-PWM Inverter : Line to Ground Voltage to eliminate non-triplen odd harmonic 5th, 7th, 11th and 13th. Switching angles alfa1, alfa2, alfa3, alfa4 and alfa5 are [12.25 23.25 31.25 46.25 52.25] respectively for an M value of 0.65. DC link voltage is 100 Volts. Inverter Fundamantal frequency is 50 Hz. Fundamental component of line to ground voltage is 42.37 Volts (Peak).

The DC link voltage is 100 volts. M is 0.65. For bipolar SHE-PWM switching, peak fundamental voltage V1 = (2*M*Vdc/π) which gives V1 = (2*0.65*100/3.14) = 41.40 V, whereas from simulation result, the value of V1 is found to be 42.37 V. VLG THD = 133.60%.

3. Answers: 5th and 7th harmonics eliminated. V1(peak) = 305.6 volts, THD = 12.55%, M = 0.8, 5th harmonics V5 = 0.0002435 V (peak), 7th harmonics V7 = 0.0006401 V (peak). Modulation index M = V1(peak)/(h*4*E/π) = 305.6*π/(3*4*100) = 0.8. Here h is the number of cascaded H-bridge cells and E is the DC source voltage of each cell. The RMS line to ground voltage is 217.8 volts and THD 12.55%.

4. Answers: 11th and 15th harmonics are eliminated from line to ground voltage. THD of VLG = 50%. Fundamental V1 = 41.61 V (peak), 11th harmonics = 0.904 V(peak) (2.17%) and 15th harmonics = 1.854 V(peak) (4.455) %. $M = \mathrm{V1}/(4 * V_{dc}/\pi) = 0.33$.

5. Answers: 11th and 15th harmonics are eliminated from line to ground voltage. THD of VLG = 49.9%, fundamental V1 = 41.66 V(peak), 11th harmonics = 0.9115 V(peak), 2.188%, 15th harmonics = 1.859 V(peak), 4.46%, fundamental V1 = 41.66 and $M = \mathrm{V1}/(4 * V_{dc}/\pi) = 0.33$.

Chapter 4

1. Answers: (1) VLL (VAB): RMS = 68.56 V, THD = 83.66%, V1(peak) = 74.37 V, V3(peak) = 1.08 V, V5(peak) = 1.81 V, V7(peak) = 2.89 V, V9(peak) = 1.69 V. (2) VLN (VAO): RMS = 39.21 V, THD = 86.09%, V1(peak) = 42.02 V, V3 (peak) = 0.91 V, V5(peak) = 1.35 V, V7(peak) = 1.42 V, V9(peak) = 0.72 V.

2. Answers: (1) VLL (VAB): RMS = 71.71 V, THD = 75.45%, V1 (peak) = 80.96 V, V3(peak) = 0.14 V, V5(peak) = 0.37 V, V7(peak) = 0.18 V, V9(peak) = 0.18 V. (2) VLN (VAO): RMS = 41.4 V, THD = 75.52%, V1 (peak) = 46.73 V, V3(peak) = 0.27 V, V5(peak) = 0.04 V, V7(peak) = 0.19 V, V9(peak) = 0.15 V.

3. Answers: (1) VLL (VAB): RMS = 71.32 V, THD = 77.26%, V1 (peak) = 79.82 V, V3(peak) = 6.7e-6 V, V5(peak) = 0.2188 V, V7 (peak) = 0.2571 V, V9(peak) = 1.29e-5 V. (2) VLN (VAO): RMS = 41.18 V, THD = 77.26%, V1(peak) = 46.08 V, V3(peak) = 2.4e-6 V, V5 (peak) = 0.1263 V, V7(peak) = 0.1485 V, V9(peak) = 4.57e-6 V.

Chapter 5

1A. The simulation results are tabulated in Table P5.3:

2A. The simulation results are tabulated in Table P5.4:

3A. The simulation results are tabulated in Table P5.5:

4A. The simulation results are tabulated in Table P5.6:

5A. The model and simulation results are tabulated in Table P5.7:

6A. The simulation results are tabulated in Table P5.8:

Table P5.3 Simulation result

Sl. no.	D_o	B	G B X M	V_{ac} Volts (Peak)	V_i Volts (Avg/mean)	V_i Volts (Peak)	V_c Volts (Avg/mean)	Remarks
1)	0.2 [0.216]	1.6667 [1.76]	1.6123 [1.669]	322.4627 (333.8170) [450.8]	533.3333 (552.1127) [442.4]	666.6667 (704.2254) [671.4]	533.3333 (552.1127) [533]	Theoretical (Model values) [Simulation results]

Table P5.4 Simulation result

Sl. no.	D_o	B	G B X M	V_{ac} Volts (Peak)	V_i Volts (Avg/mean)	V_i Volts (Peak)	V_{c1} Volts (Avg/mean)	V_{c2} Volts (Avg/mean)	iL1 iL2 (RMS)	Remarks
1)	0.2 [0.20017]	1.6667 [1.6676]	1.3334 [1.3338]	80 (80.028) [139.3]	160.003 (160.055) [164.6]	200.004 (200.112) [219.9]	160.003 (160.055) [165.5]	40 40.056 [45.58]	[57.71] [57.76]	Theoretical (Model values) [Simulation results]

Table P5.5 Simulation result

Sl. no.	D_o	B	G B X M	V_i Volts (Avg/mean)	V_i Volts (Peak)	Load Voltage Volts (RMS)	Remarks
1)	0.2 [0.19979]	(1.6667) [1.6655]	(1.3334) [1.3327]	(177.781) [177.699] [173.3]	(222.2266) [222.066] [256.0]	331.3	(Model values) [Simulation results] Upper H-bridge
2)	0.2 [0.20056]	1.6667 [1.66978]	1.3334 [1.3348]	(177.781) [177.985] [173.2]	(222.2266) [222.6373] [256.0]	331.3	(Model values) [Simulation results] Middle H-bridge
3)	0.2 [0.19956]	1.6667 [1.6642]	1.3334 [1.3321]	(177.781) [177.61] [173.3]	(222.2266) [221.893] [256.0]	331.3	(Model values) [Simulation results] Lower H-bridge

Table P5.6 Simulation result

Sl. no.	D_o	B	G B X M	V_i Volts (Avg/mean)	V_i Volts (Peak)	V_{c1}, V_{c2} Volts (Avg/mean)	Remarks
1)	0.2 [0.199675]	(1.6667) [1.6648]	(1.3334) [1.3324]	(53.3344) [53.295] [54.73]	(66.668) [66.59] [69.35]	(53.3344, 13.3336) [53.295, 13.297] [51.37, 12.44]	(Model values) [Simulation results] Upper H-bridge
2)	0.2 [0.19932]	1.6667 [1.6629]	1.3334 [1.3314]	(53.3344) [53.258] [54.76]	(66.668) [66.516] [69.36]	(53.3344, 13.3336) [53.258, 13.258] [51.37, 12.44]	(Model values) [Simulation results] Middle H-bridge
3)	0.2 [0.200335]	1.6667 [1.6685]	1.3334 [1.3342]	(53.3344) [53.3696] [54.78]	(66.668) [66.74] [69.36]	(53.3344, 13.3336) [53.3696, 13.370] [51.37, 12.44]	(Model values) [Simulation results] Lower H-bridge

Table P5.7 Simulation result

Sl. no.	D_o	B	G B X M	V_{ac} Volts (Peak)	V_i Volts (Avg/mean)	V_i Volts (Peak)	V_{c1}, V_{c2} Volts (Avg/mean)	Remarks
1)	0.2 [0.1993]	1.6667 [1.6628]	1.3333 [1.3314]	266.66 (266.28) [221.6]	533.344 (532.56) [564.0]	666.68 (665.12) [793.6]	533.344 (532.56) [566.7]	Theoretical (Model values) [Simulation results]

Table P5.8 Simulation result

Sl. no.	D_o	B	G B X M	V_{ac} Volts (RMS)	V_{PN} Volts (Peak)	V_{c1}, V_{c2} Volts (Avg/mean)	Remarks
1)	0.2 [0.200565]	1.6667 [1.6698]	1.3333 [1.3349]	80.0 (80.094) [69.95]	200 (200.376) [231.8]	80, 20 (80.094, 20.094) [85.49, 27.16]	Theoretical (Model values) [Simulation results]

Chapter 6

1A. The simulation results are tabulated in Table P6.6:

2A. The simulation results are tabulated in Table P6.7:

3A. For Fig. P6.1, the derived results are given below:

$$V_{pn(peak)} = V_{C1} = \frac{V_{dc} * (1 + D_0)}{(1 - 3 * D_0)}$$

$$B = \frac{(1 + D_0)}{(1 - 3 * D_0)}$$

The simulation results are tabulated in Table P6.8:

4A. For Fig. P6.3, the derived results are given below:

$$V_{pn(peak)} = V_{C1} = \frac{V_{dc}}{(1 - 3 * D)}$$

Table P6.6 Simulation result

Sl. no.	D_o	B	G B X M	V_{ac} Volts (Peak)	V_{pn} Volts (Peak)	V_{c1} Volts (Avg/mean)	V_{c2} Volts (Avg/mean)	Remarks
1)	0.3 (0.33548)	4.1935 (6.1687)	3.5477 (5.2187)	186.25 (273.98) [275.4]	440.3175 (647.71) [587.9]	237.0967 (322.2947) [300]	308.2258 (430.418) [390.8]	Theoretical (Model calculations) [Simulation results]

Table P6.7 Simulation result

Sl. no.	D_o	B	G B X M	V_{ac} Volts (Avg/mean)	V_{pn} Volts (Peak)	V_{c1} Volts (Avg/mean)	V_{c2} Volts (Avg/mean)	Remarks
1)	0.2 (0.19971)	2.143 (2.1397)	1.7144 (1.7124)	51.432 (51.372) [43.08]	128.58 (128.382) [123]	85.71 (85.639) [83.1]	42.86 (42.742) [36.9]	Theoretical (Model calculations) [Simulation results] Upper H-bridge
2)	0.2 (0.19956)	2.143 (2.138)	1.7144 (1.7113)	51.432 (51.34) [47.64]	128.58 (128.28) [120.5]	85.71 (85.6) [83.13]	42.86 (42.68) [37.32]	Theoretical (Model calculations) [Simulation results] Lower H-bridge

Table P6.8 Simulation result

S. no.	D_O	B	G B X M	V_{ac} (Volts) (RMS)	V_{PN} Volts Avg/ mean	v_{PN} Volts Peak	V_{C1} Volts Avg/ mean	iL1, iL2 Amps Avg/ mean	Remarks
1)	0.25 [0.22]	5 (3.588)	4.53 (3.3825)	54.36 (40.59) [40.19]	90.0 (67.167) [70.33]	120.0 (86.112) [107.9]	120.0 (86.112) [105.6]	3.033, 3.033	Theoretical (Model calculations) [Simulation results]

$$V_{C2} = V_{C3} = \frac{-D * V_{dc}}{(1 - 3 * D)}$$

$$B = \frac{1}{(1 - 3 * D)}$$

The simulation results are tabulated in Table P6.9:

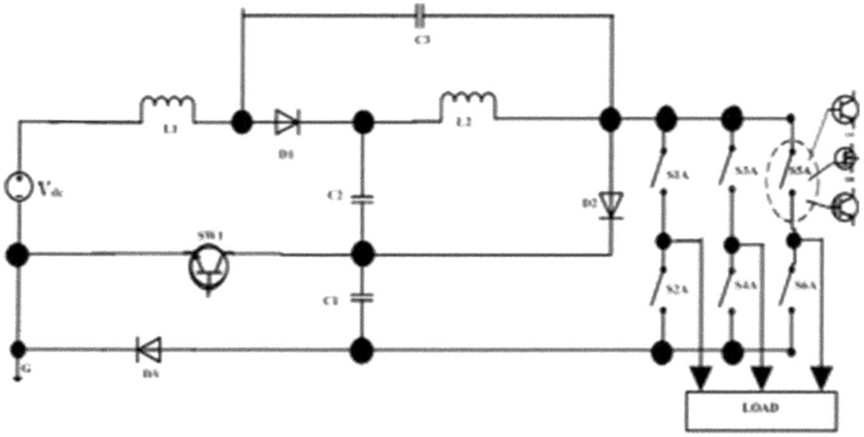

Fig. P6.3 Three-phase switched capacitor quasi Z-source boost inverter

5A. For Fig. P6.5, the derived results are given below:

$$V_{C1} = \frac{V_{dc} * (1 - D_0)}{(1 - 4 * D_0 + D_0^2)}$$

$$V_{C2} = \frac{-2 * D_0 * V_{dc}}{(1 - 4 * D_0 + D_0^2)}$$

$$V_{pn(Peak)} = \frac{V_{dc} * (1 - D_0)}{(1 - 4 * D_0 + D_0^2)}$$

$$V_{pn(Average)} = \frac{V_{dc} * (1 - D_0)^2}{(1 - 4 * D_0 + D_0^2)}$$

$$B = \frac{(1 - D_0)}{(1 - 4 * D_0 + D_0^2)}$$

The simulation results are tabulated in Table P6.10:

Table P6.9 Simulation result

S. no.	D_O	B	G B X M	V_{ac} (Volts) (RMS)	V_{PN} Volts Avg/mean	v_{PN} Volts Peak	V_{C1} Volts Avg/mean	V_{C2} Volts Avg/mean	V_{C3} Volts Avg/mean	iL1, iL2 Amps Avg/mean	Remarks
1)	0.25 [0.259]	4 (4.484)	3.624 (4.0158)	54.36 (60.237) [63.22]	90.0 (99.679) [114.7]	120.0 (134.5) [165.1]	120.0 (134.5) [158.9]	−30 (−34.84) [−44.33]	−30 (−34.84) [−43.16]	[7.019, 7.048]	Theoretical (Model calculations) [Simulation results]

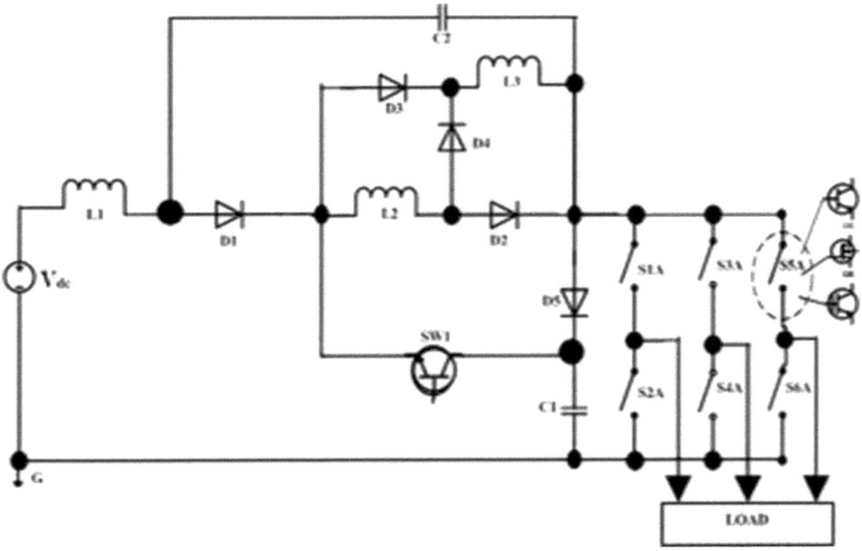

Fig. P6.5 Three-phase extended active switched capacitor quasi Z-source inverter

Table P6.10 Simulation result

S. no.	D_O	B	G B X M	V_{ac} (Volts) (RMS)	V_{PN} Volts Avg/ mean	v_{PN} Volts Peak	V_{C1} Volts Avg/ mean	V_{C2} Volts Avg/ mean	iL1, iL2 Amps Avg/ mean	Remarks
1)	0.3 [0.2803]	3.684 (3.028)	3.117 (2.634)	62.34 (52.68) [56.13]	103.15 (87.17) [100.2]	147.36 (121.12) [148.6]	147.36 (121.12) [143.5]	– 63.154 (– 47.17) [– 60.12]	[3.38, 4.995]	Theoretical (Model calculations) [Simulation results]

6A. For Fig. P6.7, the derived results are given below:

$$v_{C1} = \frac{V_{dc}}{2 * (1 - D_0)}$$

$$v_{C2} = v_{C3} = \frac{- V_{dc} * D_0}{2 * \left(1 - 3 * D_0 + 2 * D_0^2\right)}$$

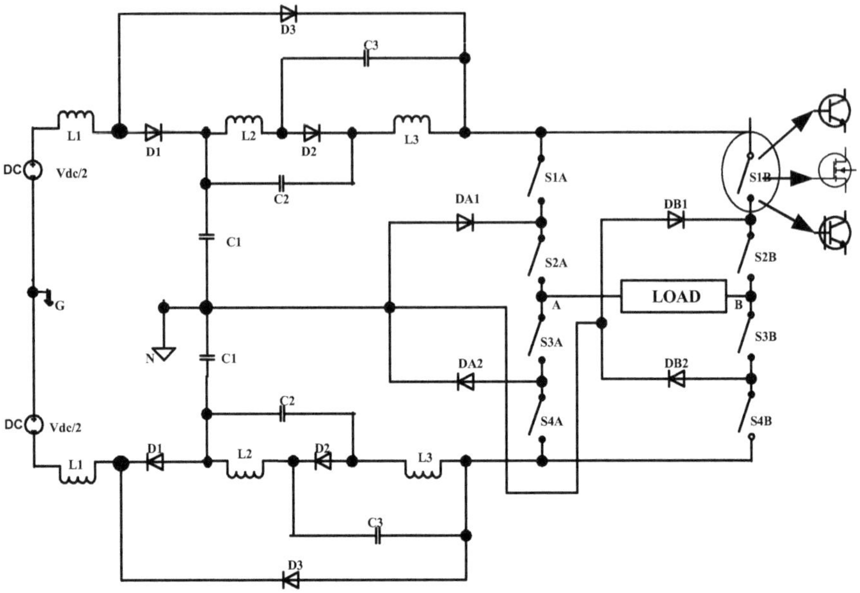

Fig. P6.7 Single-phase diode-Assisted extended boost quasi Z-source neutral point clamped three level full H-bridge inverter

Table P6.11 Simulation result

Sl. no.	D_o	B	G B X M	Van Volts (RMS)	Vpn Volts (Avg/mean)	Vpn Volts (Peak)	$Vc1$, $Vc2$, $Vc3$ Volts (Avg/mean)	Remarks
1)	0.2 [0.19929]	(2.0833) [2.0765]	(1.666) [1.6626]	(99.96) [99.756] [97.33]	(200) [199.52] [200.7]	(250) [249.18] [238.4]	(75.0, –25.0, –25.0) [74.933, –24.829, –24.829] [72.01, –21.03, –21.03]	Theoretical (Model values) [Simulation results]

$$V_{pn(peak)} = \frac{V_{dc}}{\left(1 - 3 * D_0 + 2 * D_0^2\right)}$$

$$B = \frac{1}{\left(1 - 3 * D_0 + 2 * D_0^2\right)}$$

The simulation results are tabulated in Table P6.11:

6B. The simulation results are tabulated in Table P6.12:

Table P6.12 Simulation result

Sl. no.	Do	B	G B X M	Van Volts (RMS)	Vpn Volts (Avg/mean)	Vpn Volts (Peak)	Vc1, Vc2, Vc3 Volts (Avg/mean)	Remarks
1)	0.2 [0.20019]	(2.0833) [2.0851]	(1.6667) [1.6677]	(100.00) [100.0634] [95.31]	(280) [280.25] [240.4]	(350.00) [350.4] [335.0]	(75.0, –25.0, –25.0) [75.0178, –25.0455, –25.0455] [74.36, –45.35, –45.35]	Theoretical (Model values) [Simulation results]

Chapter 7

1. The simulation results are given in Fig. P7.1:

The PI controller parameters are tabulated in Table P7.5:

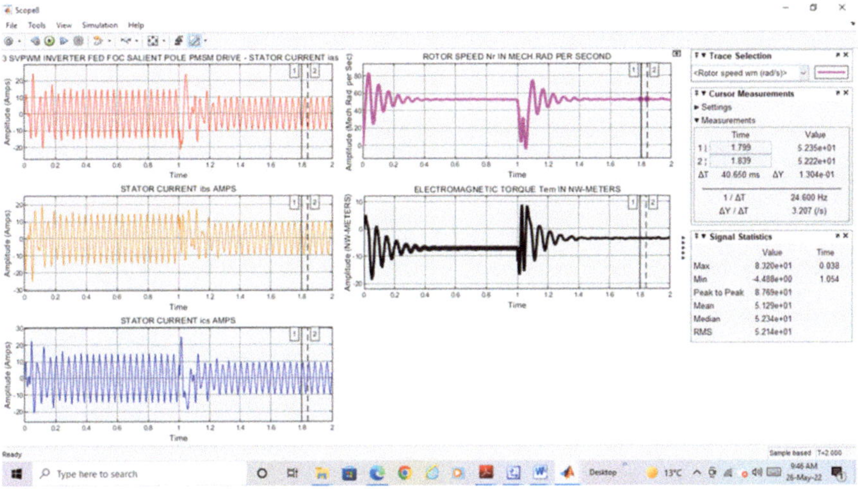

Fig. P7.1 Three-phase SVPWM inverter-fed vector controlled PMSM drive simulation results

Table P7.5 Simulation result

Sl. no.	Speed control loop Kp Ki		d-axis current control loop Kpid Kiid		q-axis current control loop Kpiq Kiiq		Rotor speed Mech. rad/sec.	Initial stator current Amps
1)	1.0553	56.4509	27.4342	2.3192E4	22.7977	1.8223E4	52.3599	10.0618

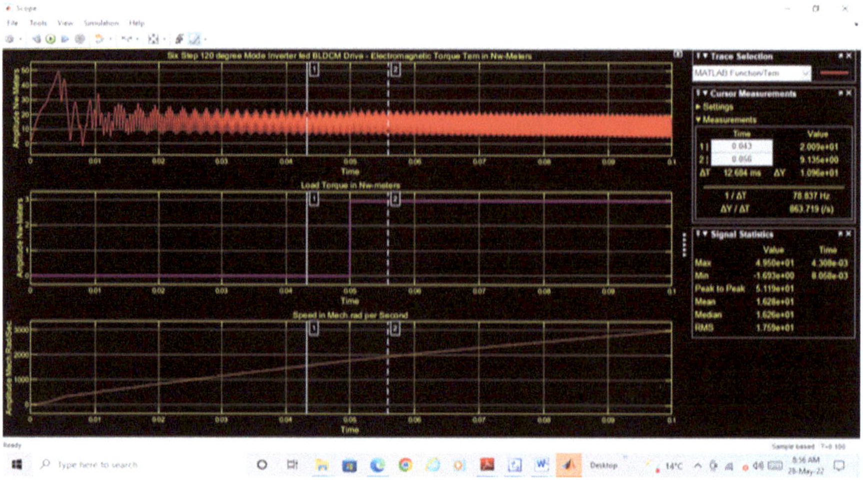

Fig. P7.2 Six-step 120-degree mode inverter-fed PMBLDCM drive—simulation result for E.M. torque, load torque and rotor speed (top to bottom). Load torque 2.9588 Nw-M at 50e-3 sec

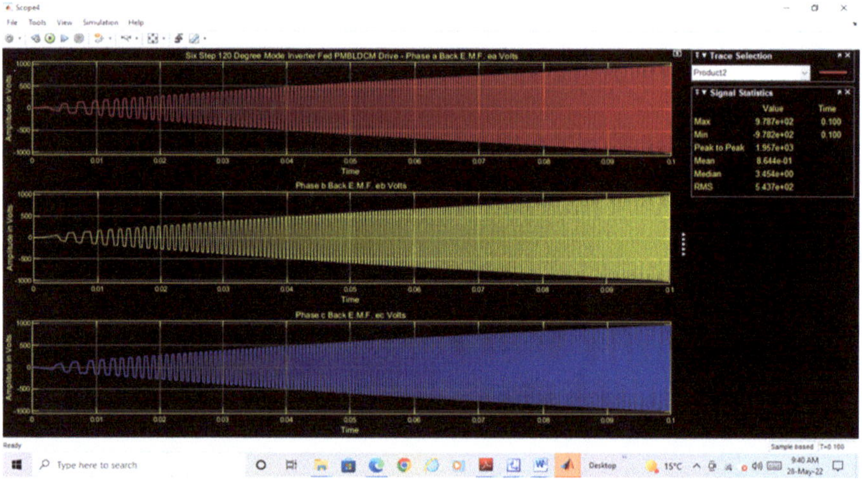

Fig. P7.3 Six-step 120-degree mode inverter-fed PMBLDCM drive—simulation result for three-phase trapezoidal back e.m.f

2. The open loop model and simulation results are given below for the load torque of 2.9588 Nw-M applied at 50e-3 seconds. The selected simulation results are given in Figs. P7.2 and P7.3.

For the load torque of 18.0774 Nw-M applied at 50e-3 seconds, selected simulation results are shown in Figs. P7.4 and P7.5:

3A. The simulation results are given in Figs. P7.6 and P7.7.

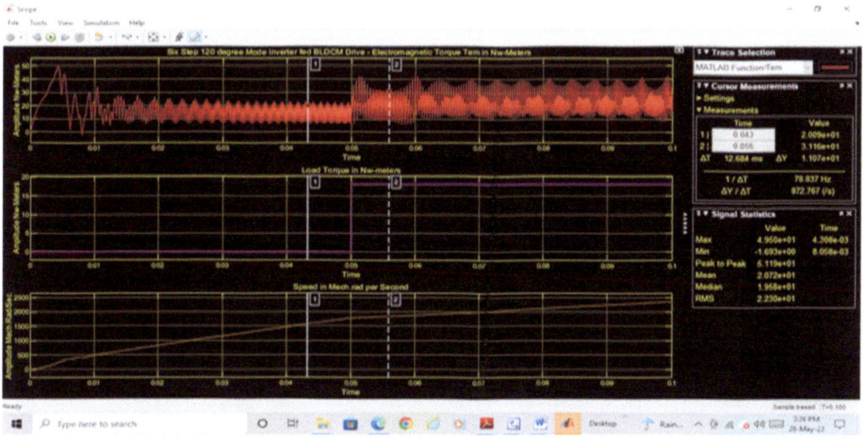

Fig. P7.4 Six-step 120-degree mode inverter-fed PMBLDCM drive—simulation result for E.M. torque, load torque and rotor speed (top to bottom). Load torque 18.0774 Nw-M at 50e-3 sec

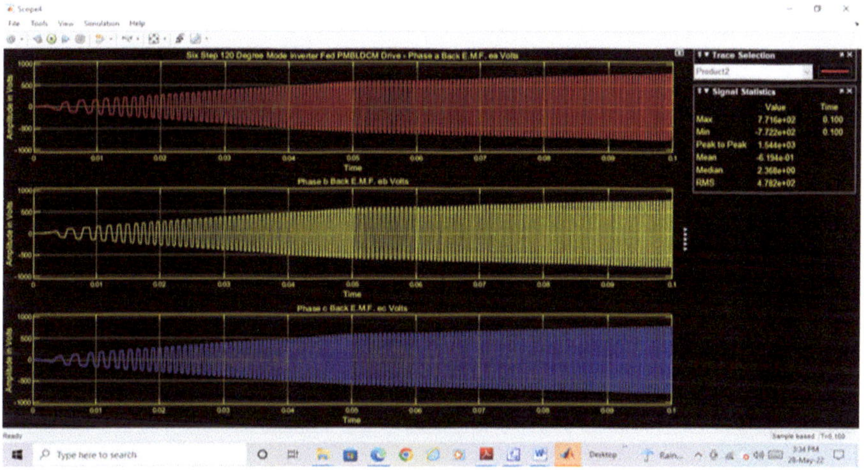

Fig. P7.5 Six-step 120-degree mode inverter-fed PMBLDCM

3B. The simulation results are given in Fig. P7.8 using built-in model of three-phase BLDCM in Simscape – Electrical – Electromechanical – Permanent Magnet block set.

Comments: In the open loop built-in model of six-step 120-degree mode inverter-fed PMBLDCM drive using Simscape – Electromechanical – PM block set, the initial speed is 302 mech.rad per second with zero load and is 313 mech.rad per second after step change in load at 50 milliseconds, whereas with the open loop model for six-step 120-degree mode inverter-fed PMBLDCM drive developed with external three-phase trapezoidal back e.m.f. generator, these values are found to be

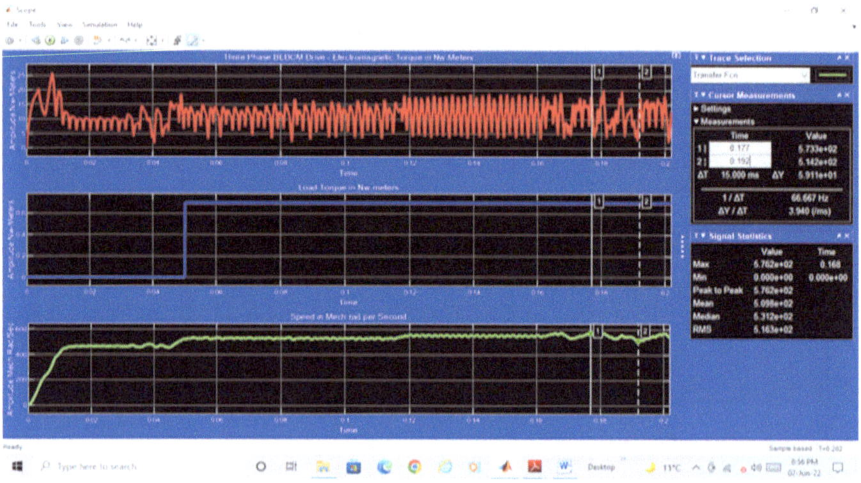

Fig. P7.6 Six-step 120-degree mode inverter-fed PMBLDCM drive—simulation result for E.M. torque, load torque and rotor speed (top to bottom). Load torque 0.7 Nw-M at 50e-3 sec

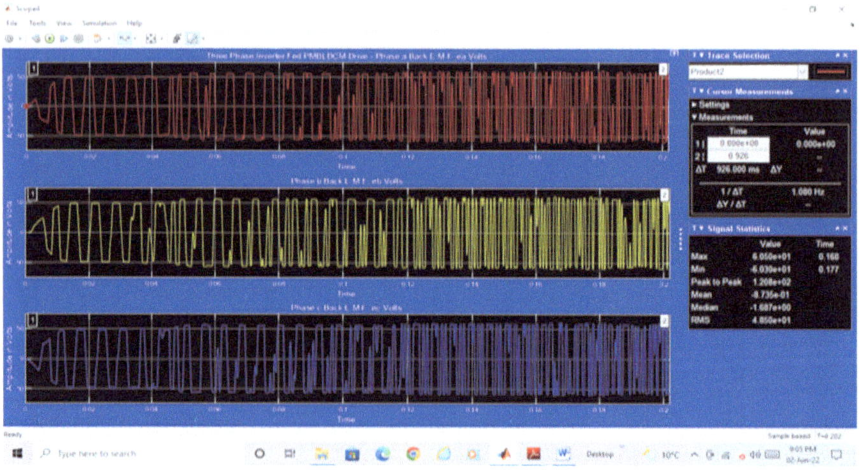

Fig. P7.7 Six-step 120-degree mode inverter-fed PMBLDCM drive—simulation result for trapezoidal back E.M.F

500 and 514 mech.rad per second, respectively. The mean value of rotor speed in the former and latter cases is 306.3 and 509.8 mech.rad per second, respectively. Similar discrepancy can be found in the peak values of three-phase stator currents and back e.m.f.s. The external three-phase trapezoidal back e.m.f. generator outputs are closely approximate compared to the actual trapezoidal back e.m.f. generator output waveforms.

4) Simulation results are given in Fig. P7.9.

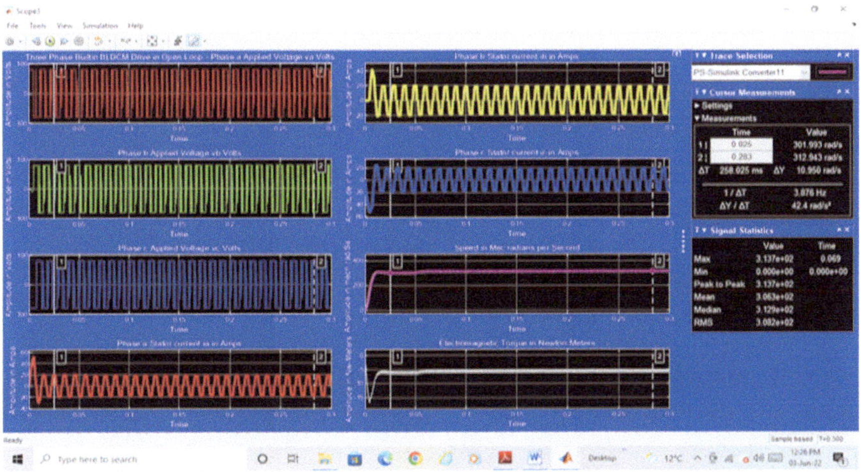

Fig. P7.8 Model of three-phase BLDCM drive using built-in Simscape PM block set—simulation results for three-phase back e.m.f. and stator phase A current (left column, top to bottom), stator Phase B, Phase C currents, rotor speed and e.m. torque (right column, top to bottom)

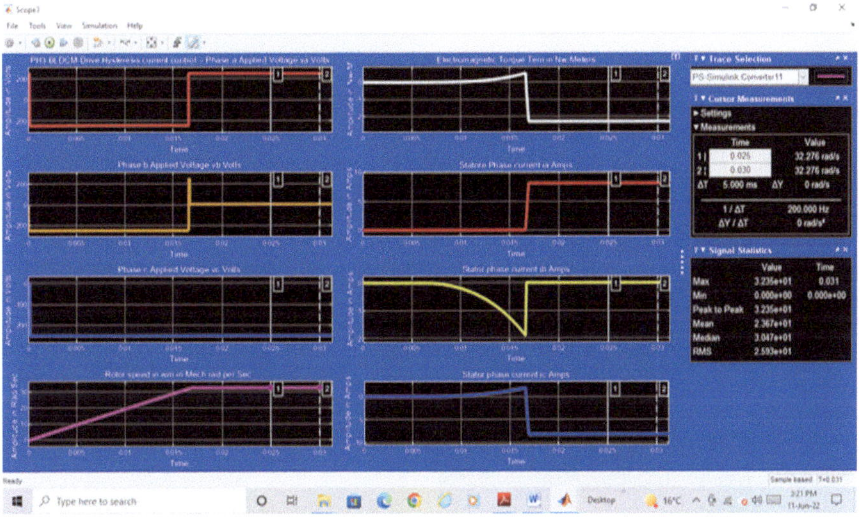

Fig. P7.9 Hysteresis current controlled three-phase BLDCM drive simulation results: three-phase stator applied voltages va, vb and vc, rotor speed (left column, top to bottom), E.M. torque, three-phase stator currents ia, ib. and ic (right column, top to bottom). Note: rotor speed displayed on extreme right in Mech.rad per second

Comments: From the plot of rotor speed with hysteresis current control, it is seen that the rotor speed remains constant at 32.276 Mech.rad per second which corresponds to a rotor speed of 308 rpm. With rotor reference speed set to 35 Mech.rad per second, the same rotor speed was observed. Thus rotor speed is held constant with hysteresis current controller.

Chapter 8

1. The selected simulation results are given in Fig. P8.10.
2. The selected simulation results are given in Fig. P8.11.

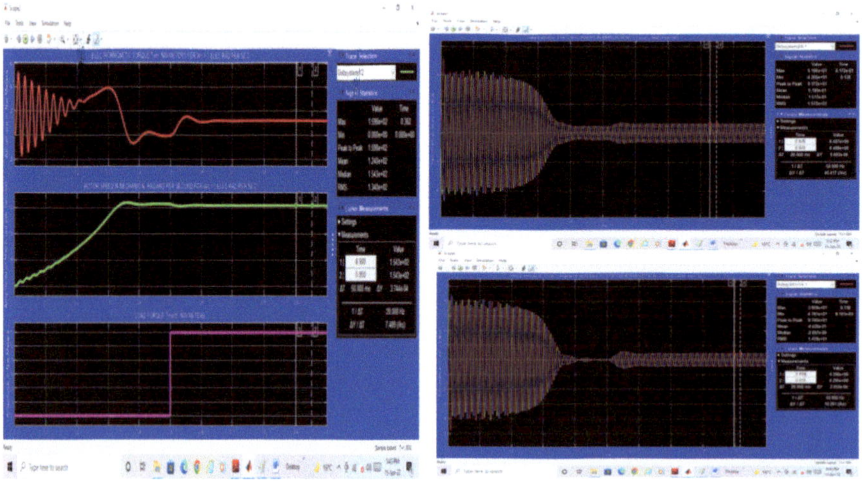

Fig. P8.10 Three-phase IM model simulation results—E.M. torque, rotor speed and load torque (left column, top to bottom), three-phase stator and rotor currents (right column, top to bottom)

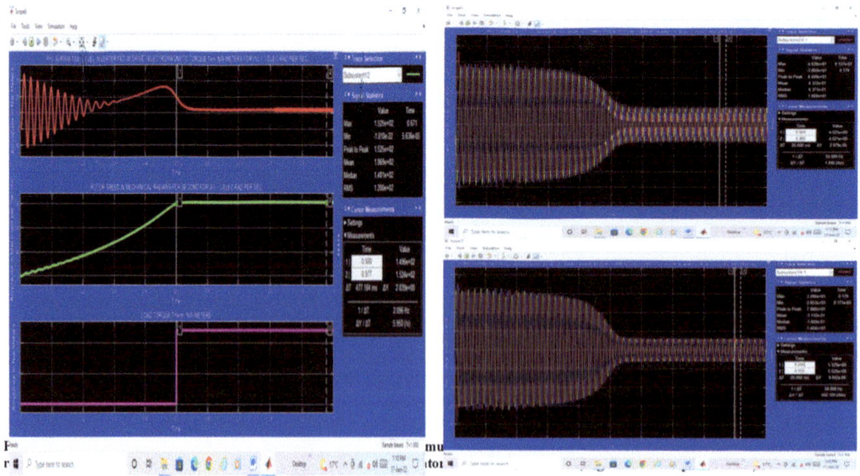

Fig. P8.11 Three-phase SVPWM inverter-fed IM drive model simulation results—E.M. torque, rotor speed and load torque (left column, top to bottom), three-phase stator and rotor currents (right column, top to bottom)

3. The selected simulation results are given in Fig. P8.12.

The PI controller parameters for the three-phase SVPWM inverter-fed vector controlled IM drive are tabulated in Table P8.3.

Comments: The rotor steady state speed from the above simulation plot is found to be 152.4 mech.rad per second, whereas the set point speed is 141.3717 mech.rad per second. The error is 7.8%. The rotor steady state e.m. torque Tem is found to be close to 12 Nw-metres.

4. Simulation results are given in Fig. P8.13.

Referring to Fig. P8.13, it is seen that after the initial undershoot for e.m. torque reaches a steady value of zero before application of load and 12 Nw-M after the application of load. The rotor speed after initial overshoot reaches a steady state speed closer to the set point speed.

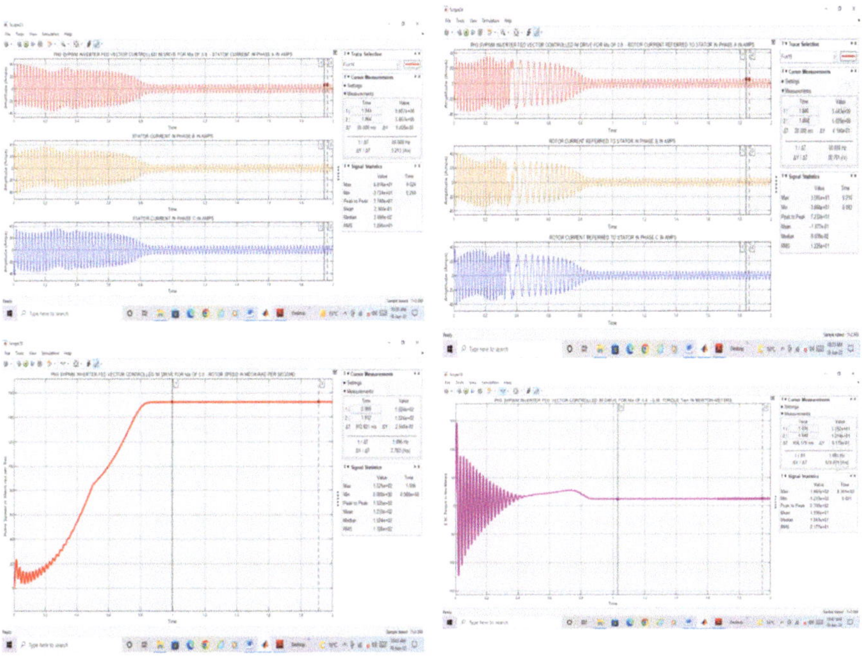

Fig. P8.12 Three-phase SVPWM inverter-fed vector controlled IM drive model simulation results—three-phase stator currents ias, ibs and ics and rotor speed (left column, top to bottom), rotor currents iar', ibr' and icr', E.M. torque (right column, top to bottom)

Table P8.3 Simulation result

Sl.no.	Speed controller Kp	Ki	Current controller Kpi	Kii	Rotor speed Mech.rad/sec.	Initial d-axis current Amps
1)	0.7898	6.3492	13.2130	739.0516	141.3717	17.0301

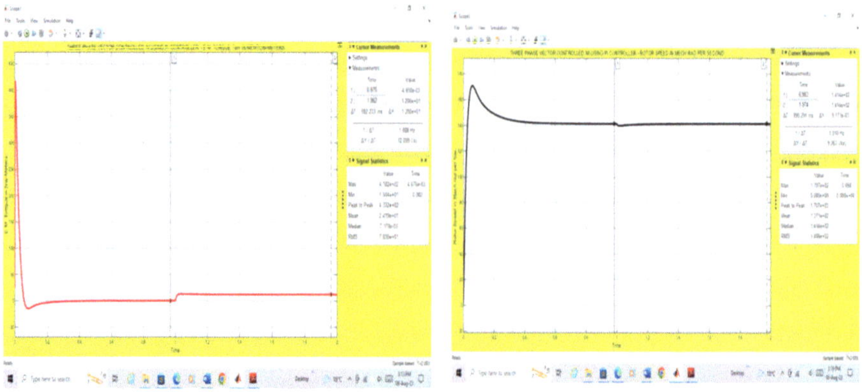

Fig. P8.13 Vector controlled three-phase IM simulation results: E.M. torque (left) and rotor speed (right)

5A. The simulation results for the DTC of SVPWM inverter-fed three-phase IM drive using torque control method are given in Fig. P8.14. Sampling frequency is 2.5 kHz.

5C. The simulation results are given in Fig. P8.15 for a sampling frequency of 100 kHz:

PI controller values are tabulated in Table P8.3.

5D. The simulation results are shown in Fig. P8.16.

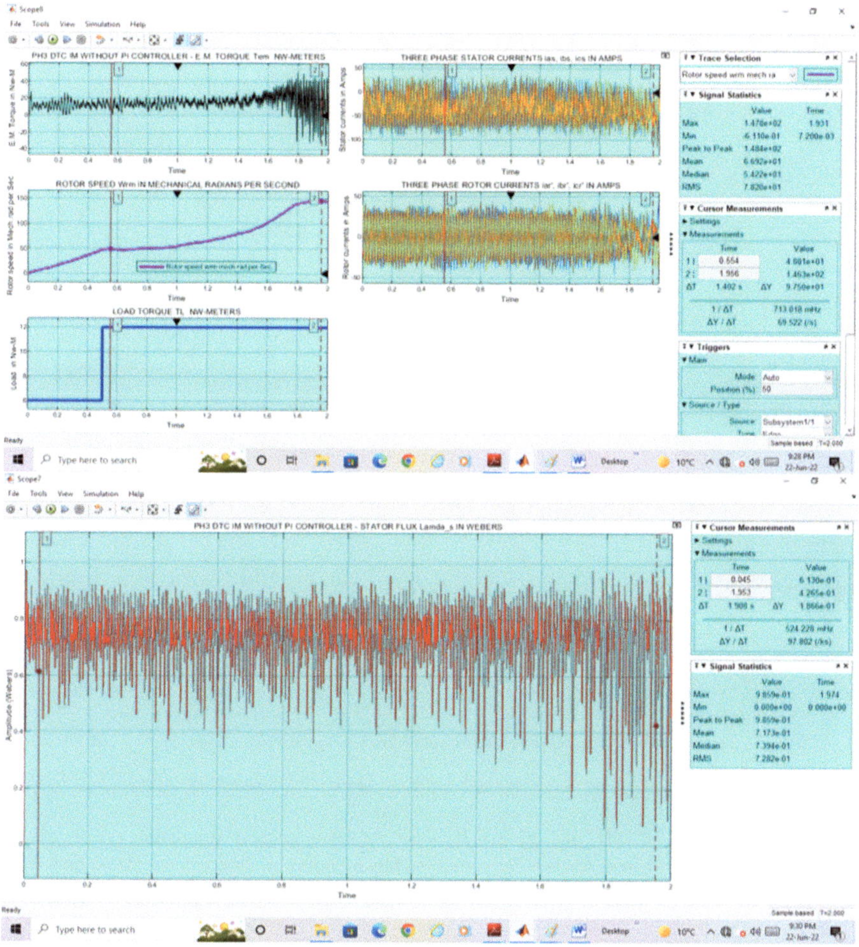

Fig. P8.14 Three-phase DTC IM drive using torque control method simulation results—E.M. torque, rotor speed, load torque (left column, top to bottom), three-phase stator and rotor currents (right column, top to bottom), stator flux linkage Lamda_s Webers (bottom row). Sampling frequency is 2.5 kHz

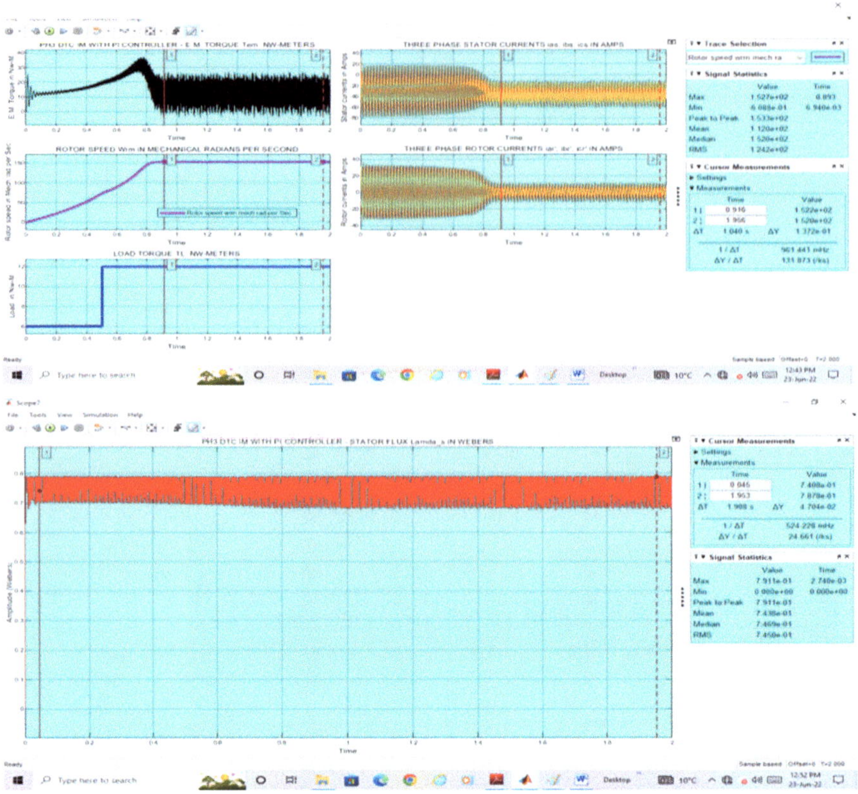

Fig. P8.15 Three-phase DTC IM drive using speed control method simulation results—E.M. torque, rotor speed and load torque (left column, top to bottom), three-phase stator and rotor currents (right column, top to bottom), stator flux linkage Lamda_s (bottom row). Sampling frequency is 100 kHz

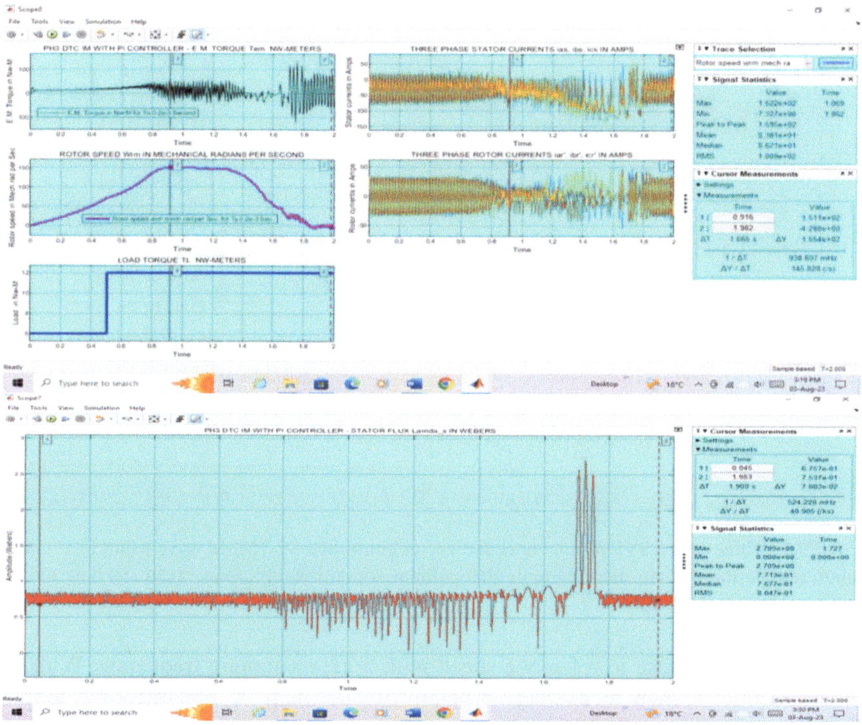

Fig. P8.16 Three-phase DTC IM drive using speed control method simulation results—E.M. torque, rotor speed and load torque (left column, top to bottom), three-phase stator and rotor currents (right column, top to bottom), stator flux linkage Lamda_s (bottom row). Sampling frequency is 5 kHz

Chapter 9

1. The simulation results are shown in Fig. P9.1.
   Here the peak amplitude and number of oscillations of rotor speed during starting are much reduced compared to that with speed PI controller shown in Fig. P7.1.

2A. The simulation results are shown in Fig. P9.2A.
   Here FLC is used. The FIS file name is "imvectorcontrol". Here per unit rotor speed error e and change in speed error cie are given to FLC through a mux. The output cs of FLC multiplied by base speed is given to the ZOH. The MFs are the same as given in Fig. 9.7 of Chap. 9. With 5 kHz sampling frequency, the final steady state rotor speed is maintained at 145.5 Mech.rad per sec. Which is close to the reference speed. Thus FLC is able to maintain

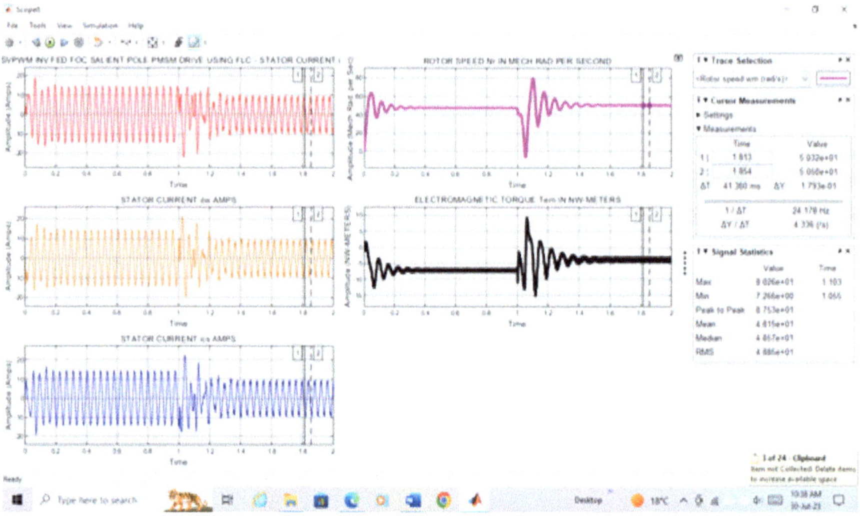

Fig. P9.1 Three-phase SVPWM inverter-fed vector controlled SP-PMSM drive using FLC simulation results: three-phase stator currents (left column, top to bottom), rotor speed and e.m. torque (right column, top to bottom)

rotor speed close to reference speed with 5 kHz sampling frequency. With PI controller shown in exercise problem 5C in Chap. 8, the reference torque Tem_ref increases enormously causing the rotor speed to fall leading to unstable operation. This is set right with FLC.

3A. The simulation results are shown in Fig. P9.3A.

Here the FIS file name “imvectorcontrolnew” is used. The MFs for speed error e, change in error cie and output cs are as shown in Fig. 9.23 of Chap. 9. The 49 fuzzy rules are based on Table 9.1. Comparing the simulation results for e.m. torque and rotor speed using FLC with that obtained using PI controller in exercise problem 4 in Chap. 8, it is seen that the e.m. torque undershoot during starting and similarly rotor speed overshoot during starting are much reduced with FLC.

The simulation results are tabulated in Table P9.1.

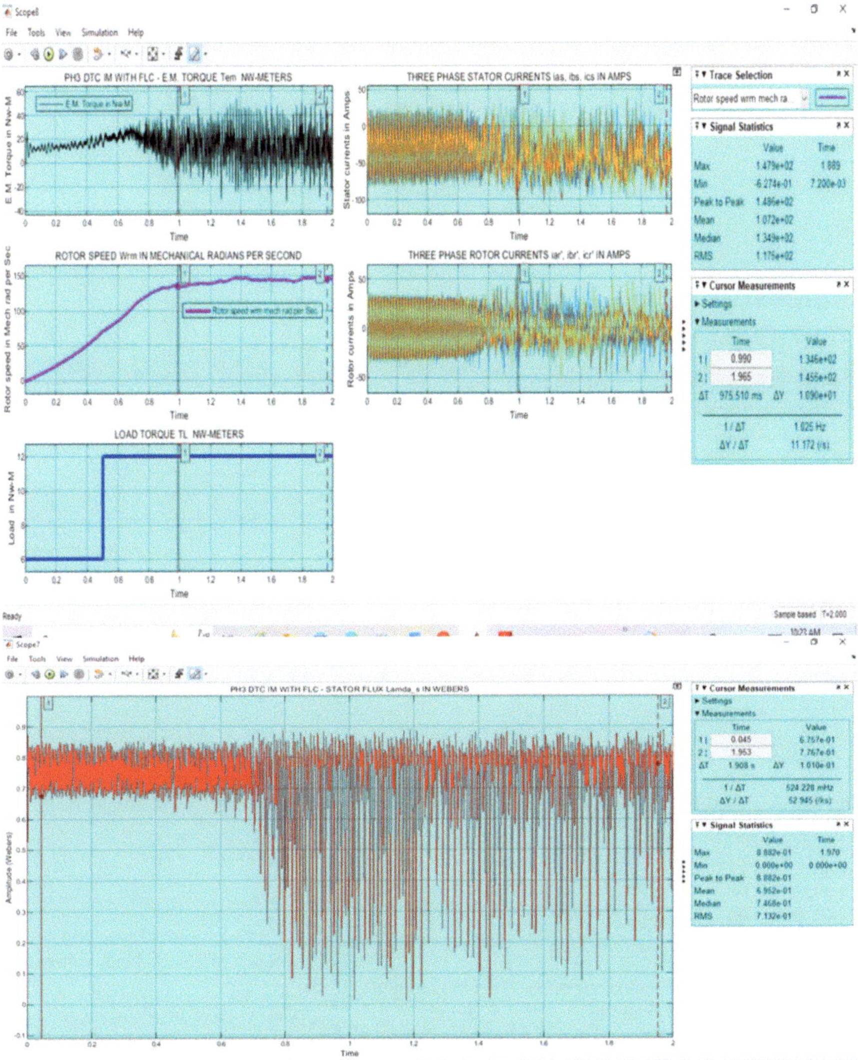

Fig. P9.2A Three-phase IM DTC using FLC simulation results: E.M. torque, rotor speed and load torque (top row, left column, top to bottom), three-phase stator and rotor currents (top row, right column, top to bottom), stator flux in webers (bottom row)

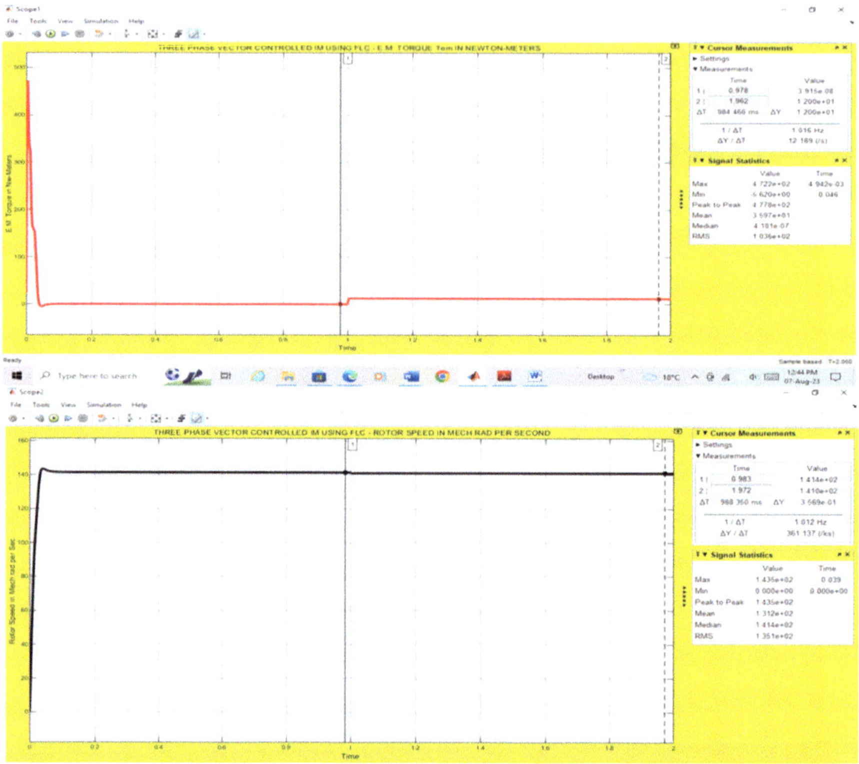

Fig. P9.3A Vector control of three-phase IM using FLC simulation results—E.M. torque (top) and rotor speed (bottom)

Table P9.1 Simulation result

Sl. no.	Rotor speed Mech.rad/ sec.	Electromagnetic torque in Nw-metres	Frequency of stator current in hertz	Frequency of stator applied voltage in hertz	Peak value of stator current in amps
1)	50.5	7 for 0 < time < =1 s 3.5 for 1 < time < =2 s	24.178	24.178	9.8

Index

N. P. R. Iyer, *Inverters and AC Drives*, Power Systems,
https://doi.org/10.1007/978-3-031-62784-2

MIX
Papier aus verantwortungsvollen Quellen
Paper from responsible sources
FSC® C105338

If you have any concerns about our products,
you can contact us on
ProductSafety@springernature.com

In case Publisher is established outside the EU,
the EU authorized representative is:
Springer Nature Customer Service Center GmbH
Europaplatz 3, 69115 Heidelberg, Germany

Printed by Libri Plureos GmbH
in Hamburg, Germany